T0214547

Lecture Notes in Computer Science 11261

Commenced Publication in 1973
Founding and Former Series Editors:
Gerhard Goos, Juris Hartmanis, and Jan van Leeuwen

More information about this series at http://www.springer.com/series/7410

Srdjan Capkun · Sherman S. M. Chow (Eds.)

Cryptology and Network Security

16th International Conference, CANS 2017
Hong Kong, China, November 30 – December 2, 2017
Revised Selected Papers

 Springer

Editors
Srdjan Capkun
ETH Zürich
Zürich, Switzerland

Sherman S. M. Chow ⓘ
Chinese University of Hong Kong
Shatin, Hong Kong

ISSN 0302-9743   ISSN 1611-3349 (electronic)
Lecture Notes in Computer Science
ISBN 978-3-030-02640-0   ISBN 978-3-030-02641-7 (eBook)
https://doi.org/10.1007/978-3-030-02641-7

Library of Congress Control Number: 2018953695

LNCS Sublibrary: SL4 – Security and Cryptology

This Springer imprint is published by the registered company Springer Nature Switzerland AG
The registered company address is: Gewerbestrasse 11, 6330 Cham, Switzerland

Preface

The 16th International Conference on Cryptology and Network Security (CANS) was held in Shatin, Hong Kong, from November 29 to December 2, 2017, organized by the Department of Information Engineering (IE), The Chinese University of Hong Kong (CUHK), in cooperation with the International Association for Cryptologic Research (IACR).

CANS is an established annual conference, focusing on all aspects of cryptology, and of data, network, and system security, attracting cutting-edge results from world-renowned scientists in the area. Earlier editions were held in Taipei (2001), San Francisco (2002), Miami (2003), Xiamen (2005), Suzhou (2006), Singapore (2007), Hong Kong (2008), Kanazawa (2009), Kuala Lumpur (2010), Sanya (2011), Darmstadt (2012), Parary (2013), Crete (2014), Marrakesh (2015), and Milan (2016).

We received 88 submissions from 28 different countries/regions (one submission was later withdrawn). The review was done via a rigorous double-blind peer review by a Program Committee (PC) consisting of 58 experts from the fields of cryptology and network security. The PC was assisted by 86 external reviewers. Each regular submission was assigned to four reviewers while a short-paper/poster submission was assigned to three reviewers. For any PC co-authored submission, an additional reviewer was assigned. In total, 351 reviews were submitted, around four on average per submission, with around 70% of the reviews prepared by the PC and the remainder by the external reviewers. The PC decided to accept 20 regular papers (an acceptance rate of 22.72%) and eight short papers. The present proceedings volume contains revised versions of all the papers presented at the conference.

The program featured two invited keynotes given by Ahmad-Reza Sadeghi (TU Darmstadt, Germany) on "Hardware-Assisted Security: From PUF to SGX" and Nico Döttling (Friedrich-Alexander-Universität Erlangen-Nürnberg, Germany) on "Identity-Based Encryption from Standard Assumptions (or the Unexpected Virtue of Garbled Circuits)." We also shared two joint invited talks with the co-located International Conference on Information Theoretic Security (ICITS), which were "Physical Assumptions for Long-Term Secure Communication" by Reihaneh Safavi-Naini (University of Calgary, Canada) and "Secret Sharing Schemes: Some New Approaches and Problems" by Huaxiong Wang (Nanyang Technological University, Singapore). Special thanks go to Stanislaw Jarecki (University of California, Irvine, USA).

We would like to thank the authors for submitting their papers to CANS, as well as the hard work of the PC and external reviewers for deciding the program. In particular, we would like to thank Amir, Aurélien, Bogdan, David, Dominique, Hong-Sheng, Jiang, Patrick, Stjepan, Tarik, and Vincenzo for volunteering to help the authors shape the revised versions of the papers. We thank EasyChair for its excellent review platform and Springer for its continuous support of CANS.

Organizing a conference is not an easy task. We would like to thank the general chair, Kehuan Zhang, publication co-chairs, publicity co-chairs, session chairs, and the supporting staff of the hosting department (IE at CUHK). This year, CANS was co-located with ICITS. We would also like to thank the ICITS 2017 PC chair, Junji Shikata, the general chair, Kenneth W. Shum, and the joint local organizing committee of CANS 2017 and ICITS 2017. Finally, we would like to thank the organizing team of CANS 2016 and the Steering Committee for helpful advice.

December 2017   Srdjan Capkun
                                                              Sherman S. M. Chow

Organization

CANS 2017 is organized by the Department of Information Engineering, The Chinese University of Hong Kong, in cooperation with the International Association for Cryptologic Research (IACR).

Program Co-chairs

Srdjan Capkun   ETH Zürich, Switzerland
Sherman S. M. Chow   Chinese University of Hong Kong, SAR China

General Chair

Kehuan Zhang   Chinese University of Hong Kong, SAR China

Publication Co-chairs

Russell W. F. Lai   Friedrich Alexander University, Erlangen Nürnberg,
                              Germany
Yongjun Zhao   Chinese University of Hong Kong, SAR China

Publicity Co-chairs

Reza Azarderakhsh   Florida Atlantic University, USA
Aniello Castiglione   University of Salerno, Italy
Aanjhan Ranganathan   ETH Zürich, Switzerland
Qian Wang   Wuhan University, China

Steering Committee

Yvo Desmedt   University of College London, UK and University
                              of Texas at Dallas, USA
Juan A. Garay   Yahoo! Labs, USA
Amir Herzberg   Bar-Ilan University, Israel
Yi Mu   University of Wollongong, Australia
David Pointcheval   CNRS and ENS Paris, France
Huaxiong Wang   Nanyang Technical University, Singapore

Program Committee

Elli Androulaki   IBM Research Zürich, Switzerland
Frederik Armknecht   University of Mannheim, Germany

Patrick Tague	Carnegie Mellon University, USA
Nils Ole Tippenhauer	Singapore University of Technology and Design
Patrick Traynor	University of Florida, USA
Mayank Varia	Boston University, USA
Ding Wang	Peking University, China
Qian Wang	Wuhan University, China
David J. Wu	Stanford University, USA
Jiang Zhang	State Key Laboratory of Cryptology, China
Yinqian Zhang	The Ohio State University, USA
Yongjun Zhao	Chinese University of Hong Kong, SAR China
Yunlei Zhao	Fudan University, China
Hong-Sheng Zhou	Virginia Commonwealth University, USA

Additional Reviewers

Muhamad Erza Aminanto	Kexin Hu	Matthew Smith
Hyeongcheol An	Michael Hölzl	Najmeh Soroush
Monir Azraoui	Peng Jiang	Martin Strohmeier
David Barrera	Jongkil Kim	Fei Tang
Carsten Baum	Panagiotis Kintis	Vincent Taylor
Mai Ben Adar - Bessos	Agnes Kiss	Sri Aravinda Thyagarajan
Alastair Beresford	Russell W. F. Lai	Orfeas Stefanos
Simon Birnbach	Iraklis Leontiadis	Thyfronitis Litos
Cecilia Boschini	Minghui Li	Alberto Trombetta
Jonas Böhler	Wenting Li	Yiannis Tselekounis
Guoxing Chen	Zengpeng Li	Binbin Tu
Haibo Cheng	Hui Ma	Cédric Van Rompay
Joseph Choi	Xinshu Ma	Dimitrios Vasilopoulos
Rakyong Choi	Giulio Malavolta	Damian Vizár
Hui Cui	Vasily Mikhalev	Boyang Wang
Joan Daemen	Martín Ochoa	Chengyu Wang
Daniel Demmler	Giorgos Panagiotakos	Jun Wang
Christoph Egger	Sikhar Patranabis	Christian Weinert
Kaoutar Elkhiyaoui	Tran Viet Xuan Phuong	Harry W. H. Wong
Katharina Fech	Ania Piotrowska	Lei Wu
Nils Fleischhacker	Sebastian Poeplau	Kang Yang
Benny Fuhry	Rakyong Rakyong	Rupeng Yang
Michael Goberman	Ruben Recabarren	Jingyue Yu
Clémentine Gritti	Manuel Reinert	Fan Zhang
Vincent Grosso	Joost Renes	Xiaokuan Zhang
Hui Guo	Katerina Samari	Lingchen Zhao
Yiming Guo	Sarah Scheffler	Minghao Zhao
Florian Hahn	Clara Schneidewind	Xinjie Zhao
Thomas Hayes	Emily Shen	Dionysis Zindros

Local Organizing Committee

Jiongyi Chen

Shuaike Dong

Yixuan Ding

Russell W. F. Lai

Fang Liu

Lilian Lun

Jack P. K. Ma

Kenneth W. Shum

Menghan Sun

Raymond K. H. Tai

Shuyang Tang

Jiafan Wang

Xiuhua Wang

Harry W. H. Wong

Huangting Wu

Fenghao Xu

Zisang Xu

Hoover H. F. Yin

Jixzin Zhang

Tao Zhang

Yongjun Zhao

Invited Talks (Abstracts)

Hardware-Assisted Security: From PUF to SGX

Ahmad-Reza Sadeghi

TU Darmstadt, Germany
ahmad.sadeghi@trust.tu-darmstadt.de

Abstract. Protecting modern software with hardware-based security is becoming increasingly more important in practice. We are witnessing this trend through recent developments in the processor technology such as Intel's SGX and AMD's SEV. Moreover, veteran hardware-based security technologies such as Physically Unclonable Functions (PUFs), Trusted Platform Modules (TPM) and ARM's TrustZone are still evolving.

However, all these solutions suffer from various shortcomings: they are afterthought and ad-hoc, require strong trust in manufacturers or their involvement, not accessible to third party developers, not scalable, or vulnerable to side-channel or runtime attacks.

On the other hand, academic research has aimed at addressing some of these problems in the recent past by providing various security architectures such as AEGIS, Bastion, Sanctum, Sancus, TrustLite, TyTAN, to name some. Unfortunately, these solutions have not found their way into practice yet.

This talk summarizes some of the recent hardware-assisted security technologies, discusses their strengths and deficiencies and future directions.

Biography: Ahmad-Reza Sadeghi is a full professor of Computer Science at the TU Darmstadt, Germany. He is the head of the Systems Security Lab at the Cybersecurity Research Center of TU Darmstadt. Since January 2012 he is also the director of the Intel Collaborative Research Institute for Secure Computing (ICRI-SC) at TU Darmstadt. He holds a Ph.D. in Computer Science from the University of Saarland, Germany. Prior to academia, he worked in R&D of Telecommunications enterprises, amongst others Ericsson Telecommunications. He has been continuously contributing to security and privacy research. For his influential research on Trusted and Trustworthy Computing he received the renowned German "Karl Heinz Beckurts" award. This award honors excellent scientific achievements with high impact on industrial innovations in Germany. He is Editor-In-Chief of IEEE Security and Privacy Magazine, and on the editorial board of ACM Books. He served 5 years on the editorial board of the ACM Transactions on Information and System Security (TISSEC).

Identity-Based Encryption from Standard Assumptions (or the Unexpected Virtue of Garbled Circuits)

Nico Döttling

Friedrich-Alexander-University Erlangen Nürnberg

Abstract. Until recently, explicit constructions of identity-based encryption (IBE) required considerably more structure or stronger assumptions than public key encryption from similar assumptions. In this talk, a framework is presented, which significantly facilitates the construction of IBE schemes and leads to new constructions from weaker assumptions. The central tool is a new primitive called compact One-Time Signatures with Encryption (OTSE), which turns out to be equivalent with IBE. This primitive can be built from weak assumptions such as the computational Diffie-Hellman problem (in groups without pairings), the Factoring problem, the Learning-with-Errors problem (with the same parameters as Regev-encryption) and the sub-exponentially hard Learning-Parity-with-Noise problem. The main technique of our framework is a novel non-black-box transformation from compact OTSE to both fully secure IBE and selectively secure hierarchical IBE. This new technique critically relies on garbled circuits and suggests new applications for this versatile primitive.

Biography: Nico Döttling studied computer science at the University of Karlsruhe (now KIT) and also obtained his PhD in Karlsruhe under the supervision of Jörn Müller-Quade. After that, he joined the crypto group of Aarhus University, working with Ivan Damgård and Jesper Buus Nielsen. He then joined UC Berkeley as a postdoc, supported by a scholarship of the DAAD, to work with Sanjam Garg. In 2016 he joined the Friedrich-Alexander-University Erlangen Nürnberg as assistant professor. He was the winner of the best-dissertation in computer-science award at KIT in 2014 and the winner of best-paper awards at ProvSec 2015 and Crypto 2017.

Contents

Web Security

Bitcoin and Blockchain

Embedded System Security

Anonymous and Virtual Private Networks

Wireless and Physical Layer Security

Short Papers

Foundation of Applied Cryptography

Forward-Security Under Continual Leakage

Mihir Bellare[1], Adam O'Neill[2], and Igors Stepanovs[1($\boxtimes$)]

[1] Department of Computer Science and Engineering,
University of California San Diego, La Jolla, USA
{mihir,istepano}@eng.ucsd.edu
[2] Department of Computer Science, Georgetown University, Washington D.C., USA
adam@cs.georgetown.edu

Abstract. Current signature and encryption schemes secure against continual leakage fail completely if the key in any time period is fully exposed. We suggest forward security as a second line of defense, so that in the event of full exposure of the current secret key, at least uses of keys prior to this remain secure, a big benefit in practice. (For example if the signer is a certificate authority, full exposure of the current secret key would not invalidate certificates signed under prior keys.) We provide definitions for signatures and encryption that are forward-secure under continual leakage. Achieving these definitions turns out to be challenging, and we make initial progress with some constructions and transforms.

1 Introduction

Classically, cryptography assumes secure endpoints and an insecure communication channel. Malware and sidechannel attacks bring the threat to the endpoints: Information about keys stored on our system can be leaked or exfiltrated to the adversary. Let us begin by reviewing two ways to address this for public-key cryptography, namely forward security and leakage resilience.

Forward Security. The threat of exposure of a secret (signing or decryption) key due to compromise of the system storing the key is not new. Forward security (FS) was developed in the late 1990s as a way to mitigate the damage. The idea of forward secure signatures was suggested by Anderson [5] and formalized by Bellare and Miner (BM) [8]. Later Canetti, Halevi and Katz (CHK) [15] formalized forward secure encryption. Subsequent work gave many schemes and extensions.

Forward-security [5,8] introduced the *key evolution* paradigm: evolve the secret key over time while keeping the public key fixed. At time period i, the secret key is sk_i, and the signing algorithm (we will discuss signatures rather than encryption as an example), applied to it and a message m, produces a signature that is a pair (i, σ), meaning the time period is explicitly included in the signature. At the start of time period $i + 1$, a (public) update function is applied to sk_i to get sk_{i+1}, and sk_i is deleted from the system. An attack

© Springer Nature Switzerland AG 2018
S. Capkun and S. S. M. Chow (Eds.): CANS 2017, LNCS 11261, pp. 3–26, 2018.
https://doi.org/10.1007/978-3-030-02641-7_1

compromising the system at some time i obtains sk_i, and automatically sk_j for any $j \geq i$ since the update function is public. Security of signatures as defined in BM [8] required that possession of sk_i does not allow forgery of signatures with time period j—meaning ones of the form (j, σ)—for $j < i$. It follows that possession of sk_i does not allow recovery of sk_j for $j < i$, meaning the update function must be one-way.

The CA Example. What does FS buy us? An illustrative example is when the signer is a certificate authority (CA). Assume the CA creates certificates using a normal (not forward secure) signature scheme, with its (single, static) secret key sk. Say that in time period 1, it creates a certificate (m, σ) for Alice, where m is Alice's certificate data (her pubic key and so on) and σ is the CA's signature on m. Suppose, in time period 168, the CA system is infiltrated by malware (a realistic possibility) and sk is exposed. Discovering this, the CA immediately revokes its public key. Now suppose in time period 169, Bob receives from Alice the (valid!) certificate (m, σ) in a TLS connection. Finding the CA public key on his revocation list, he will reject certificate (m, σ) as invalid and deny the TLS connection. Security-wise, this is the right and necessary thing to do: there is no way for Bob to know that (m, σ) is not a forgery. But the cost is enormous: all certificates the CA had issued prior to the revocation (which could be many years worth of certificates) must be discarded, and many TLS connections will be rejected, causing serious disruption to web services. Time-stamping the signature will not fix this, since, once the adversary has the secret key, it can forge the time-stamp too.

But now suppose the CA used a forward secure signature scheme instead of a normal one, so that the January 1st signature has the form $(1, \sigma)$, Alice's certificate thus being $(m, (1, \sigma))$. The infiltration in time period 168 exposes secret key sk_{168}. As before, the CA revokes its public key, and now we consider Bob receiving the certificate $(m, (1, \sigma))$ in time period 169. He sees the CA public key on his revocation list, but he also sees the revocation is marked with time period $168 > 1$. Now he can *safely accept* the certificate $(m, (1, \sigma))$ and proceed with the TLS connection, because forward-security guarantees that $(m, (1, \sigma))$ cannot be a forgery. That is, certificates created prior to the exposure are still secure and valid. This is a significant advantage in the event of compromise.

Note that FS does not prevent (or even make more difficult) exposure of a secret key. That is not its aim. Its aim is to mitigate the damage caused by an exposure, if and when the latter occurs.

Leakage-Resilience. Motivated by sidechannel attacks, leakage resilience aims to preserve security even if some information $f(sk)$ about the secret key sk is leaked. In the bounded memory leakage model of Akavia et al. [2] and extensions [21, 39], f is any function returning a number of bits enough short of the length $|sk|$ of sk. However if the adversary has some sidechannel capability, it may, over time, gather enough bits to expose the entire key, and then security is lost. To protect against this, Dodis et al. [19] and Brakerski et al. [14] propose the continual leakage (CL) model. As in FS, the secret key is updated in each time period while the public key stays fixed. In each time period i, the adversary may obtain a bounded amount of leakage $f_i(sk_i)$ on the current secret key, yet security must be maintained. The gain is that the sidechannel attack has limited time to attack a particular key before it is updated, and once that happens it must effectively start from scratch.

Security is parameterized by a leakage rate, the scheme being δ-CL if it remains secure when the number of bits leaked in any period is restricted to at most a δ fraction of the length of the secret key. Achieving δ-CL-security is not easy. One subtlety is that the update function, unlike for FS, must be randomized. Secure schemes have been provided in [14, 19] for the cases of encryption and signatures, while other works looked at extensions to basic notions and treated other primitives [4, 11, 17, 18, 27, 32, 34, 35, 38, 40].

The Problem. Security in the CL model relies on the assumption that the amount of leakage in a particular time period is bounded, in particular short of the length of the key itself. If the entire key is leaked in some time period, security is lost entirely. One could make updates more frequent to restrict the time the attacker has to expose a key before it is updated, but, while this may be reasonable for certain kinds of side-channel attacks, it may not be effective when the attack is malware on your system that can directly exfiltrate the key. Also it is not clear how to pick an update frequency or evaluate the security benefits of a choice. We introduce forward-security under continual leakage as a way to maintain the CL guarantee but add a second line of defense against full key exposure via FS.

Forward-Security Under Continual Leakage. We continue, as with both FS and CL, to work in the model where the public key is fixed but the secret key evolves with time, a public update function being applied to the period i key sk_i to produce the next key sk_{i+1}. (Indeed it is surprising that FS and CL were not explicitly connected prior to our work, given that they work in the same model.) The first definition one might consider is to ask that the scheme be both (1) CL-secure, and (2) FS-secure. We can do better. We ask that FS holds even under CL, in the following sense. In our game, in any time period, the adversary get a bounded amount of leakage $f_i(sk_i)$ on the key sk_i in that time period i, just as in CL. Additionally, it can, in some time period i of its choice, expose and obtain the *entire* secret key sk_i. The requirement is that of FS, namely security of usages of keys sk_j for $j < i$. Note a FS+CL scheme defined in this way is both CL (restrict attention to adversaries that do not make the full expose query) and FS (restrict attention to adversaries that do make this query but do not leak any information on prior keys). But it requires more than the two individually because security of keys sk_j for $j < i$ is guaranteed even when sk_i is known to the adversary and the adversary has leakage on all the keys sk_j. As with CL, security is parameterized by a leakage rate δ.

Within this template, the precise definition of security depends on the primitive. In Sect. 3, we define key-evolving signature schemes and a notion of δ-FUFCL security, for Forward Unforgeability under Continual Leakage. The definition is parameterized by the leakage rate δ. We also define key-evolving encryption schemes and a notion of δ-FINDCL security, for Forward INDistinguishability under Continual Leakage. As a tool in obtaining security against continual leakage, Dodis et al. [19] introduced the notion of relations that are one-way under CL, and in analogy, as a tool to obtain forward security under continual leakage, we define in Sect. 5 the notion of a δ-FOWCL relation, for Forward One-wanness under Continual Leakage.

The Benefits. To see the benefits provided by forward security under continual leakage over CL security alone, let us return to the CA example discussed above. Suppose that the CA is concerned about leakage and uses a CL-secure signature scheme in place of the normal secure signature scheme. Alice's certificate, produced in time period 1, has the form $(m, (1, \sigma))$. Now suppose, due to malware on the system in time period 168, the secret key sk_{168} is exposed, and the public key revoked. Bob receives the certificate $(m, (1, \sigma))$ for a TLS connection in time period 169. Seeing the CA's public key on the certificate revocation list, he cannot accept Alice's certificate, because, in a CL scheme, possession of sk_{168} could allow forgery of signatures for time period 1. Thus millions of certificates, issued over years by the CA, suddenly become obsolete. The cost in disruption to web services (gmail, amazon, ...) using TLS is huge. However if the scheme is FS+CL secure, Bob can in confidence accept Alice's certificate, because the revocation period is $168 > 1$. Thus we have provided leakage resilience with a second line of defense that significantly mitigates damage caused by full key exposures.

Challenges. We would like to give FS+CL schemes for both signatures and encryption, meaning a δ-FUFCL signature scheme and a δ-FINDCL encryption scheme. This is surprisingly challenging. The first thought one may have is that perhaps some existing CL-secure scheme is already FS+CL. This is not true, because all existing CL schemes update their secret keys by merely re-randomizing them. So a full exposure of the key sk_i in some time period i results in full recovery of the secret keys for *all* time periods, meaning that the schemes are not even FS, let alone FS+CL. The complementary question is whether any existing FS scheme happens to be FS+CL, but this seems evidently false because existing FS schemes provide no security under leakage. One reason is that the update functions are deterministic, and no scheme with a deterministic update function can be CL secure. The latter is because otherwise an adversary can repeatedly leak bits of a secret key sk_t for some future time period t, by querying $f_i(sk_i)$ for functions f_i that use sk_i to compute sk_t.

The natural next construction approach to consider is a modular one. We have CL-secure schemes, and we have FS-secure ones. Is there some way to combine a CL scheme with an FS one to get a FS+CL one? We do not know a fully general way to do this, but our first scheme is obtained by a generic transform of this ilk, as we now discuss.

Generic Transform from CL. Our constructions are summarized in Fig. 1. Our first result is a generic transform of any CL scheme into a FS+CL one in the case of signatures. FS+CL security is proven with *no extra assumptions* beyond the CL security of the base scheme. We can now use existing CL signature schemes [14, 19]. Thus we obtain the first constructions of FUFCL signatures schemes.

Our generic transform is tree-based. To get a FS+CL signature scheme, we use the binary tree FS signature construction from BM [8] with a (any) CL scheme as the base scheme. A drawback of this transform, however, is that it degrades the relative leakage parameter: If we start with a δ-CL scheme and the number of time periods is $T = 2^d$ then we get a δ'-FS+CL scheme with $\delta' = \delta/(d+1)$. In particular we do not get a FUFCL signature scheme with constant relative leakage.

Goal	$\delta' =$	Assumptions	Section
δ'-FUFCL signatures	$\delta/(d+1)$	δ-CL signatures	4
δ'-FUFCL signatures	δ	δ-FOWCL KE for T periods + WS	5
δ'-FINDCL encryption	δ	δ-FOWCL KE for T periods + WE	5

Fig. 1. Proposed constructions of δ'-FUFCL key-evolving signature schemes, and δ'-FINDCL key-evolving encryption schemes for different values of continual-leakage fraction δ' and for T time periods. We assume that $T = 2^d$ for some $d \in \mathbb{N}$.

This approach does not work in the case of encryption. For example, it is tempting to start from the CL HIBE of Lewko, Rouselakis and Waters [35] and use it to build binary-tree encryption (BTE) following the construction of FS-encryption from CHK [15], but this fails. The problem is that FS+CL security of the resulting scheme requires that multiple nodes of the BTE construction can be leaked on *jointly*, whereas the CL security of HIBE only buys us leakage on each such node *individually*. We will construct a FINDCL encryption scheme in a different way that leverages both the FUFCL signature scheme we have just built and witness encryption, as we discuss next.

Transforms Using Witness Primitives. The second set of constructions extends the paradigm of Dodis et al. [19]. They used a key evolution scheme that is one-way under continual leakage to build CL-secure signatures and encryption. We assume a key evolution scheme that is *forward* one-way under continual-leakage (FOWCL KE). Then we present a unified paradigm to get FS+CL signatures and encryption using, as an additional tool, WX, where the "W" stands for "Witness" and X = S for signatures, and X = E for encryption. In other words, we use witness signatures (WS) [7,16] to get FS+CL signatures, and witness encryption (WE) [6,26,28] to get FS+CL encryption. In this case there is no loss of relative leakage, meaning if we start with a δ-FOWCL KE scheme we get δ'-FS+CL encryption and signature schemes with $\delta' = \delta$.

To obtain FS+CL schemes from these results, we need to instantiate the components. This means we need: (1) A FOWCL key evolution scheme (2) A WS scheme to get FS+CL signatures and a WE scheme to get FS+CL encryption. For (2), witness signatures as we define them in Sect. 5 are, as explained there, easily obtained from NIZKs since they are just another name for signatures of knowledge [7,16], different from the (impossible) witness signatures of GJK [29].

In other words, these are readily available under standard assumptions. Witness encryption is more difficult since we do require (a weak form of) extractability. For (1), we can get a FOWCL KE scheme by using the tree-based FUFCL signature scheme obtained via our first (generic transform) result.

The main outcome from this is the first construction of a FINDCL encryption scheme. The assumptions are any CL signature scheme (which yields a FUFCL signature scheme via our tree-based construction, and thence a FOWCL KE scheme as noted above) plus extractable witness encryption. Due to the use of our tree-based construction, the relative leakage will again degrade compared to that of the starting CL signature scheme, so we again fail to get FINDCL encryption with constant relative leakage. Extractable witness encryption is also a suspect assumption due to the negative results of Garg et al. [25], but we view our construction as evidence that this approach has merit and as first step toward ones under more plausible assumptions.

This paradigm can be used to get FUFCL signatures too, but with the instantiation we have of the FOWCL KE scheme itself coming from our tree-based FUFCL signature scheme constructed above, we would not obtain anything over and above our generic transform result discussed above. But the transforms are still interesting both for signatures and encryption because if it were possible to find a FOWCL KE scheme withstanding constant relative leakage, we would immediately get a FUFCL signature scheme and a FINDCL encryption scheme with the same constant relative leakage, assuming the witness primitives. Thus they help to reduce the problem of FS+CL signatures and encryption to the single and hopefully simpler problem of FOWCL KE.

Related Work. Bounded leakage-resilience and its extensions were studied for various primitives including encryption and signature schemes in [2, 3, 13, 20, 21, 24, 36, 39]. Continual leakage-resilience (CL) was studied in [14, 19, 35]. In particular, [14, 19] provide CL signature schemes with leakage rate $1 - o(1)$ (i.e. arbitrarily close to 1) in bilinear groups. These schemes can be plugged into our generic transform described above. Extensions of the basic CL notions have been considered in [4, 11, 17, 18, 27, 32, 34, 35, 38, 40], which yield further CL schemes.

After the initial schemes of BM [8], various follow-up works constructed more efficient forward-secure signature schemes or gave other extensions, including [1, 10, 31, 33]. Malkin et al. [37] constructed forward-secure signatures for an unbounded number of time periods. Unfortunately, their framework does not allow to get FS+CL schemes by composing CL-schemes. Their composition methods require the secret key to contain various components (other secret keys, and random seeds) that remain unchanged for a number of time periods. If leakage on these parts of their secret key is allowed, then security is lost.

Leakage on updates, where leakage is allowed on the coins used in updating keys, is considered in [17, 22, 34]. Our model and results are a first step that do not allow leakage on updates. It is an interesting consideration for future work.

2 Preliminaries

NOTATION. We denote by $\lambda \in \mathbb{N}$ the security parameter and by 1^λ its unary representation. For $i \in \mathbb{N}$ we let $[i]$ denote the set $\{1, \dots, i\}$. We let ε denote the empty string. We denote the length of a string $x \in \{0,1\}^*$ by $|x|$. By $x \,\|\, y$ we denote the concatenation of strings x, y. Algorithms may be randomized unless otherwise indicated. Running time is worst case. "PT" stands for "polynomial-time," whether for randomized algorithms or deterministic ones. If A is an algorithm, we let $y \leftarrow A(x_1, \dots; r)$ denote running A with random coins r on inputs $x_1, \dots$ and assigning the output to y. We let $y \leftarrow\!\!{}_\$ A(x_1, \dots)$ be the result of picking r at random and letting $y \leftarrow A(x_1, \dots; r)$. We let $[A(x_1, \dots)]$ denote the set of all possible outputs of A when invoked with inputs $x_1, \dots$. We say that $f : \mathbb{N} \to \mathbb{R}$ is negligible if for every positive polynomial p, there exists $\lambda_p \in \mathbb{N}$ such that $f(\lambda) < 1/p(\lambda)$ for all $\lambda \geq \lambda_p$. We use the code based game playing framework of Bellare and Rogaway [9]. (See Fig. 2 for an example.) By $G^{\mathcal{A}}(\lambda)$ we denote the event that the execution of game G with adversary $\mathcal{A}$ and security parameter λ results in the game returning true. Booleans are assumed initialized to false and integers to 0.

NP-Relations. Relation R specifies a PT algorithm R.Vf. Witness verification algorithm R.Vf takes an instance $x \in \{0,1\}^*$ and a candidate witness $w \in \{0,1\}^*$ to return a decision in $\{\text{true}, \text{false}\}$. For any $x \in \{0,1\}^*$ we let $R(x) = \{ w \in \{0,1\}^* : \text{R.Vf}(x, w) \}$ be the witness set of x. We let $\mathcal{L}(R) = \{ x \in \{0,1\}^* : R(x) \neq \emptyset \}$ be the language defined by R. We say that R is an **NP**-relation, and $\mathcal{L}(R)$ an **NP** language, if there exists a witness-length polynomial $\text{R.wl} : \mathbb{N} \to \mathbb{N}$ such that $R(x) \subseteq \bigcup_{\ell \leq \text{R.wl}(|x|)} \{0,1\}^\ell$ for all $x \in \{0,1\}^*$.

3 Forward Security Under Continual Leakage

In this section we consider key-evolving signature and encryption schemes, for which we provide definitions of forward security under continual leakage. A key-evolving scheme has a single public key pk, while its secret key evolves with time, $sk_1 \to sk_2 \to \cdots \to sk_T$, where "$\to$" is implemented by an update algorithm and T is the number of time periods supported by the scheme.

In the continual leakage setting, in every time period t the attacker can obtain leakage on the current secret key sk_t. The security requirement from prior work [14,19] is that security under all keys be maintained. For this to be achievable, the leakage on each key must be assumed to be bounded. But this boundedness assumption is unrealistic, and in practice a key may leak entirely. In this case, prior systems require and provide no security. We propose to add forward security as a second line of defense, asking that even if a key sk_{t^*} leaks fully for any time period t^*, security under prior keys will not be compromised. This brings important gains, for example, when signing certificates, that those signed prior to the full exposure do not have to be revoked. Forward-security under continual leakage, as we define it, simultaneously implies both security under continual leakage and classical forward security.

We now define forward unforgeability under continual leakage (FUFCL) for key-evolving signatures, and forward indistinguishability under continual leakage (FINDCL) for key-evolving encryption. Security games for both are in Fig. 2.

Public and Secret Components of a Secret Key. We will parameterize security notions of key-evolving schemes by a function $\delta : \mathbb{N} \to [0,1]$ that, informally, denotes the fraction of the secret key that may leak in every time period. However, the secret key may contain some information that is necessary only for the key-evolving functionality of the scheme, but is not required to be hidden for the security of the scheme. Therefore, it is not useful to consider a leakage metric that compares different schemes based on the fraction of the *entire* secret key that may be leaked per time period: this fraction might be very small just because the secret keys of a particular scheme contain a lot of information that is not required to be kept secret. We address this by requiring that all secret keys used by key-evolving schemes can be parsed as a pair (pc, sc), where pc denotes the public component of the secret key and sc denotes the secret component of the secret key. We then define δ-leakage security of key-evolving schemes to denote the fraction of the *secret component* of the secret key that may leak in every time period. Our security games provide the public components of all secret keys to the adversary for free.

Key-Evolving Signature Schemes. A key-evolving signature scheme KES specifies PT algorithms KES.Kg, KES.Up, KES.Sig and KES.Vf, where KES.Vf is deterministic. Associated to KES are the following polynomials: public-key length KES.pkl : $\mathbb{N} \to \mathbb{N}$, secret-key length KES.skl : $\mathbb{N} \to \mathbb{N}$, public component length of the secret key KES.pcl : $\mathbb{N} \to \mathbb{N}$, secret component length of the secret key KES.scl : $\mathbb{N} \to \mathbb{N}$, message length KES.ml : $\mathbb{N} \to \mathbb{N}$, signature length KES.sigl : $\mathbb{N} \to \mathbb{N}$, and the maximum number of time periods KES.T : $\mathbb{N} \to \mathbb{N}$. For $\lambda \in \mathbb{N}$ we require that any secret key $sk \in \{0,1\}^{\text{KES.skl}(\lambda)}$ can be parsed as a pair (pc, sc) containing a public component $pc \in \{0,1\}^{\text{KES.pcl}(\lambda)}$ and a secret component $sc \in \{0,1\}^{\text{KES.scl}(\lambda)}$, such that KES.skl$(\lambda)$ = KES.pcl(λ) + KES.scl(λ). Key generation algorithm KES.Kg takes 1^λ to return a public key $pk \in \{0,1\}^{\text{KES.pkl}(1^\lambda)}$ and base (time period one) secret signing key $sk_1 \in \{0,1\}^{\text{KES.skl}(\lambda)}$. Key update algorithm KES.Up takes $1^\lambda, pk, i$ and a secret key $sk_i \in \{0,1\}^{\text{KES.skl}(\lambda)}$ for time period i to return a KES.skl(λ)-bit secret key for the next time period. Signing algorithm KES.Sig takes $1^\lambda, pk, i, sk_i$ and a message $m \in \{0,1\}^{\text{KES.ml}(\lambda)}$ to return a pair (i, σ), where $\sigma \in \{0,1\}^{\text{KES.sigl}(\lambda)}$ is a signature of m under secret key sk_i. Signature verification algorithm KES.Vf takes $1^\lambda, pk, m, (i, \sigma)$ to return a decision in {true, false} regarding whether σ is a valid signature of message m relative to public key pk and time period $i \in [\text{KES.T}(\lambda)]$. Correctness requires that KES.Vf$(1^\lambda, pk, m, (i, \sigma))$ = true for all $\lambda \in \mathbb{N}$, all $m \in \{0,1\}^{\text{KES.ml}(\lambda)}$, all $(pk, sk_1) \in [\text{KES.Kg}(1^\lambda)]$, all $i \in [\text{KES.T}(\lambda)]$, all $sk_2, \ldots, sk_i$ satisfying $sk_j \in [\text{KES.Up}(1^\lambda, pk, j - 1, sk_{j-1})]$ for $2 \leq j \leq i$, and all σ such that $(i, \sigma) \in [\text{KES.Sig}(1^\lambda, pk, i, sk_i, m)]$.

Forward Unforgeability Under Continual Leakage. Consider game FUFCL of Fig. 2 associated to a key-evolving signature scheme KES and an adversary $\mathcal{A}$, where

Game $\text{FUFCL}^{\mathcal{A}}_{\text{KES}}(\lambda)$	Game $\text{FINDCL}^{\mathcal{A}}_{\text{KEE}}(\lambda)$
$S \leftarrow \emptyset$; $t \leftarrow 1$; $t^* \leftarrow \text{KES.T}(\lambda) + 1$	$b \leftarrow\!\!{}_{\$}\, \{0,1\}$; $t \leftarrow 1$; $t^* \leftarrow \text{KEE.T}(\lambda) + 1$
$(pk, sk_1) \leftarrow\!\!{}_{\$}\, \text{KES.Kg}(1^\lambda)$	$(pk, sk_1) \leftarrow\!\!{}_{\$}\, \text{KEE.Kg}(1^\lambda)$
$(pc, sc) \leftarrow sk_1$	$(pc, sc) \leftarrow sk_1$
$(i, m, \sigma) \leftarrow\!\!{}_{\$}\, \mathcal{A}^{\text{UP,LK,EXP,SIGN}}(1^\lambda, pk, pc)$	$(i, m_0, m_1, state) \leftarrow\!\!{}_{\$}\, \mathcal{A}_1^{\text{UP,LK,EXP}}(1^\lambda, pk, pc)$
$\text{win}_1 \leftarrow (1 \le i < t^*) \wedge ((i, m, \sigma) \notin S)$	If not $(1 \le i < t^*)$ then return $\textsf{false}$
$\text{win}_2 \leftarrow \text{KES.Vf}(1^\lambda, pk, m, (i, \sigma))$	If $|m_0| \ne |m_1|$ then return $\textsf{false}$
Return $(\text{win}_1 \wedge \text{win}_2)$	$(i, c) \leftarrow\!\!{}_{\$}\, \text{KEE.Enc}(1^\lambda, pk, i, m_b)$
	$b' \leftarrow\!\!{}_{\$}\, \mathcal{A}_2(1^\lambda, state, (i, c))$
$\underline{\text{UP}()}$	Return $(b' = b)$
If $t < \text{KES.T}(\lambda)$ then	
$\quad sk_{t+1} \leftarrow\!\!{}_{\$}\, \text{KES.Up}(1^\lambda, pk, t, sk_t)$	$\underline{\text{UP}()}$
$\quad (pc, sc) \leftarrow sk_{t+1}$; $t \leftarrow t + 1$	If $t < \text{KEE.T}(\lambda)$ then
$\quad$ Return pc	$\quad sk_{t+1} \leftarrow\!\!{}_{\$}\, \text{KEE.Up}(1^\lambda, pk, t, sk_t)$
Else return $\bot$	$\quad (pc, sc) \leftarrow sk_{t+1}$; $t \leftarrow t + 1$
	$\quad$ Return pc
$\underline{\text{LK}(L)}$	Else return $\bot$
$(pc, sc) \leftarrow sk_t$; Return $L(sc)$	
	$\underline{\text{LK}(L)}$
$\underline{\text{EXP}()}$	$(pc, sc) \leftarrow sk_t$; Return $L(sc)$
$t^* \leftarrow t$; Return sk_t	
	$\underline{\text{EXP}()}$
$\underline{\text{SIGN}(m)}$	$t^* \leftarrow t$; Return sk_t
$(t, \sigma) \leftarrow\!\!{}_{\$}\, \text{KES.Sig}(1^\lambda, pk, t, sk_t, m)$	
$S \leftarrow S \cup \{(t, m, \sigma)\}$; Return (t, σ)	

Fig. 2. Games defining forward unforgeability of key-evolving signature scheme KES under continual leakage, and forward indistinguishability of key-evolving encryption scheme KEE under continual leakage.

LK takes as input a Boolean circuit $L: \{0,1\}^{\text{KES.scl}(\lambda)} \to \{0,1\}$. For $\lambda \in \mathbb{N}$ let $\text{Adv}^{\text{fufcl}}_{\text{KES},\mathcal{A}}(\lambda) = \Pr[\text{FUFCL}^{\mathcal{A}}_{\text{KES}}(\lambda)]$. We say that FUFCL adversary $\mathcal{A}$ is *valid* if it makes at most one query to its EXP oracle, and this is its last oracle query. We say that $\mathcal{A}$ is δ-bounded, where $\delta: \mathbb{N} \to [0,1]$, if $\mathcal{A}$ makes at most $\delta(\lambda) \cdot \text{KES.scl}(\lambda)$ queries to LK *per time period*. That is, leakage on the secret component of secret key in any one time period is restricted to this number of bits. We say that KES is δ-FUFCL (δ-forward unforgeable under continual leakage) if $\text{Adv}^{\text{fufcl}}_{\text{KES},\mathcal{A}}(\cdot)$ is negligible for all valid, δ-bounded PT adversaries $\mathcal{A}$.

The game begins by picking a public key pk and base secret key sk_1 for the first time period. The adversary receives pk and the public component pc of the secret key sk_1. The current time period is t and the corresponding key, sk_t, is the one under attack. The SIGN oracle allows the adversary to obtain signatures under the current key. The adversary may obtain leakage about the secret component sc of the secret key sk_t via its leakage oracle LK. The latter takes an adversary-provided boolean circuit $L: \{0,1\}^{\text{KES.scl}(\lambda)} \to \{0,1\}$ and returns $L(sc)$ as leakage. Note that the adversary is restricted to querying LK with circuits that output only one bit, but it may adaptively query the oracle multiple times to leak more bits. At any point the adversary may call UP to advance the key to the next stage, receiving as output the public component of the new secret key. Calls to SIGN, LK and UP may be adaptively interleaved. At any time the adversary also has the option of fully exposing the current secret key via its EXP oracle. The time period in which it does this is denoted t^*. At that point it is disallowed any further calls to its oracles and must terminate. To win it must output a valid message-signature pair relative to a time period prior to t^*, where valid means that the signature-verification algorithm KES.Vf accepts it, and that the message-signature pair for the particular time period was not previously received as an output of the SIGN oracle. Adversary's advantage is the probability that it wins.

Security under all keys is guaranteed as long as adversary learns at most a δ fraction of every secret key's secret component. If leakage exceeds this amount (modeled by an EXP query being made) then, rather than all being lost, forward security is provided, meaning security of prior keys is maintained. Forward-secure signatures as defined in [8] are the special case of FUFCL signatures for adversaries that make no LK queries. Signatures that are secure against continual leakage are the special case of FUFCL signatures for adversaries that make no EXP queries. Thus our model unifies the two notions under the new goal of forward unforgeability under continual leakage. Our definitions of key-evolving signatures and continual-leakage security are different from those used in the prior work [11,12,14,19,27,38], but they are equivalent up to simple transformations, as explained below. The difference is that key-evolving signatures from the prior work are defined to use signing and verification algorithms that are oblivious to the current time period. However, a key-evolving scheme as per our definition can be constructed from a standard key-evolving scheme by using the latter to sign and verify messages of the form $i \parallel m$, which is a concatenation of the current time period i and a message m. Furthermore, a standard key-evolving scheme can be constructed from a key-evolving scheme as per our definition by building the secret keys of the standard scheme as $i \parallel sk_i$, containing the current time period i and the corresponding secret key sk_i for a scheme of our type. The resulting constructions of key-evolving signature schemes inherit the continual-leakage security of the original schemes.

Key-Evolving Encryption Schemes. A key-evolving encryption scheme KEE specifies PT algorithms KEE.Kg, KEE.Up, KEE.Enc and KEE.Dec, where KEE.Dec is deterministic. Associated to KEE are the following polynomials: secret-key length KEE.skl : $\mathbb{N} \to \mathbb{N}$, public component length of the secret key KEE.pcl : $\mathbb{N} \to \mathbb{N}$, secret component length of the secret key KEE.scl : $\mathbb{N} \to \mathbb{N}$, message length KEE.ml : $\mathbb{N} \to \mathbb{N}$, and the maximum number of time periods KEE.T : $\mathbb{N} \to \mathbb{N}$. For $\lambda \in \mathbb{N}$ we require that any secret key $sk \in \{0,1\}^{\mathsf{KEE.skl}(\lambda)}$ can be parsed as a pair (pc, sc) containing a public component $pc \in \{0,1\}^{\mathsf{KEE.pcl}(\lambda)}$ and a secret component $sc \in \{0,1\}^{\mathsf{KEE.scl}(\lambda)}$, such that $\mathsf{KEE.skl}(\lambda) = \mathsf{KEE.pcl}(\lambda) + \mathsf{KEE.scl}(\lambda)$. Key generation algorithm KEE.Kg takes 1^λ to return a public key pk and base (time period one) secret signing key $sk_1 \in \{0,1\}^{\mathsf{KEE.skl}(\lambda)}$. Key update algorithm

KEE.Up takes $1^\lambda, pk, i$ and a secret key $sk_i \in \{0,1\}^{\mathsf{KEE.skl}(\lambda)}$ for time period i to return a KEE.skl(λ)-bit secret key for the next time period. Encryption algorithm KEE.Enc takes $1^\lambda, pk, i$ and a message $m \in \{0,1\}^{\mathsf{KEE.ml}(\lambda)}$ to return (i, c), where c is an encryption of m under pk for time period i. Decryption algorithm KEE.Dec takes $1^\lambda, pk, i, sk_i, (j, c)$ to return $m \in \{0,1\}^{\mathsf{KEE.ml}(\lambda)} \cup \{\bot\}$. Correctness requires that $\mathsf{KEE.Dec}(1^\lambda, pk, i, sk_i, (i, c)) = m$ for all $\lambda \in \mathbb{N}$, all $m \in \{0,1\}^{\mathsf{KEE.ml}(\lambda)}$, all $(pk, sk_1) \in [\mathsf{KEE.Kg}(1^\lambda)]$, all $i \in [\mathsf{KEE.T}(\lambda)]$, all $sk_2, \ldots, sk_i$ satisfying $sk_j \in [\mathsf{KEE.Up}(1^\lambda, pk, j-1, sk_{j-1})]$ for $2 \le j \le i$, and all c such that $(i, c) \in [\mathsf{KEE.Enc}(1^\lambda, pk, i, m)]$.

Forward Indistinguishability Under Continual Leakage. Consider game FINDCL of Fig. 2 associated to a key-evolving encryption scheme KEE and an adversary $\mathcal{A}$, where LK takes as input a Boolean circuit $L : \{0,1\}^{\mathsf{KEE.scl}(\lambda)} \to \{0,1\}$. For $\lambda \in \mathbb{N}$ let $\mathsf{Adv}^{\mathsf{findcl}}_{\mathsf{KEE},\mathcal{A}}(\lambda) = 2\Pr[\mathrm{FINDCL}^{\mathcal{A}}_{\mathsf{KEE}}(\lambda)] - 1$. We say that FINDCL adversary $\mathcal{A}$ is *valid* if it makes at most one query to its EXP oracle, and this is its last oracle query. We say that $\mathcal{A}$ is δ-bounded, where $\delta : \mathbb{N} \to [0,1]$, if $\mathcal{A}$ makes at most $\delta(\lambda) \cdot \mathsf{KEE.scl}(\lambda)$ queries to LK *per time period*. We say that KEE is δ-FINDCL (δ-forward indistinguishable under continual leakage) if $\mathsf{Adv}^{\mathsf{findcl}}_{\mathsf{KEE},\mathcal{A}}(\cdot)$ is negligible for all valid, δ-bounded PT adversaries $\mathcal{A}$.

Game FINDCL is similar to game FUFCL in terms of allowing the adversary to obtain information about the secret keys in different time periods by providing it with oracle access to UP, LK and EXP. Having finished making queries to its oracles, the adversary has to choose a time period (prior to the key exposure, if EXP was called) and a pair of challenge messages of equal length. The adversary then is given a challenge ciphertext for the specified time period, and it has to guess which of the two challenge messages was encrypted in order to win the game. Encryption secure against continual leakage as defined in [14] is the special case of FINDCL encryption for adversaries that make no EXP queries (these notions are equivalent up to simple transformations that are required due to different semantics across the definitions of key-evolving schemes, similar to the case for key-evolving signature schemes that we discussed above). Forward-secure encryption as defined in [15] is the special case of FINDCL encryption for adversaries that make no LK queries. Our model unifies the two under the new goal of forward indistinguishability under continual leakage.

Convention for Adversary Restrictions. Whenever we consider an adversary that meets certain conditions (e.g. is PT, valid, and δ-bounded), we require that it holds not just in the games defining security, but also regardless of adversary's inputs and how its oracle queries are answered. It will help us simplify the proof of Theorem 2 where an FUFCL adversary will be simulated in an environment that is different from the one it might expect from the FUFCL game.

4 FUFCL signatures from UFCL signatures

In this section we show how to construct a FUFCL signature scheme from any key-evolving signature scheme that is unforgeable under continual leakage

(UFCL). The latter is a standard continual leakage security notion that is also a special case of FUFCL with respect to adversaries that do not query EXP oracle.

Unforgeability Under Continual Leakage. Consider game FUFCL of Fig. 2 associated to a key-evolving signature scheme KES and an adversary $\mathcal{A}$, where LK takes as input a Boolean circuit $L : \{0,1\}^{\mathsf{KES.scl}(\lambda)} \rightarrow \{0,1\}$. For $\lambda \in \mathbb{N}$ let $\mathsf{Adv}^{\mathsf{ufcl}}_{\mathsf{KES},\mathcal{A}}(\lambda) = \Pr[\mathsf{FUFCL}^{\mathcal{A}}_{\mathsf{KES}}(\lambda)]$. We say that UFCL adversary $\mathcal{A}$ is valid if it makes *no queries* to its EXP oracle. We say that $\mathcal{A}$ is δ-bounded, where $\delta : \mathbb{N} \rightarrow [0,1]$, if $\mathcal{A}$ makes at most $\delta(\lambda) \cdot \mathsf{KES.scl}(\lambda)$ queries to LK *per time period*. We say that KES is δ-UFCL (δ-unforgeable under continual leakage) if $\mathsf{Adv}^{\mathsf{ufcl}}_{\mathsf{KES},\mathcal{A}}(\cdot)$ is negligible for all valid, δ-bounded PT adversaries $\mathcal{A}$.

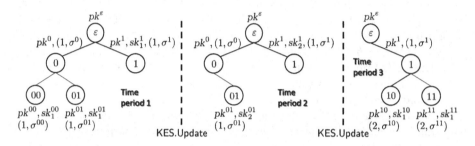

Fig. 3. The construction of a key-evolving signature scheme KES from depth-2 binary tree, showing the information stored in the tree during the first 3 out of the 4 possible time periods. Each node corresponds to an independent instance of the underlying key-evolving signature scheme SIG. Superscripts denote the positions of displayed entities in the tree, and subscripts denote the time periods of individual secret keys.

FUFCL Signatures from a Binary Tree of UFCL Signatures. We use the binary tree construction of forward secure signatures by Bellare and Miner [8], with any continual leakage secure signature scheme as the base scheme. The construction is also similar to the one by Faust et al. [23], but the goal they achieve is different.

We now describe the high-level idea of our construction. Let SIG be a key-evolving signature scheme; we compose many SIG key-pairs into a binary tree to build a new key-evolving signature scheme KES. We will then show that if SIG is UFCL then KES is FUFCL. Figure 3 shows an example of a binary tree of depth 2 for multiple subsequent time periods.

Each node of the tree containts an independently generated SIG public key. A node may also contain the corresponding secret key, but the secret key is erased as soon as both of its child nodes have been generated; this will ensure that the constructed KES scheme is forward-secure. Each non-root node contains a signature of its public key under its parent node's secret key.

The SIG public key of the root node is used as the public key of the KES scheme. The leaf nodes of the tree are used to produce KES signatures, each for a separate time period, meaning that a tree-based construction of depth h has 2^h

time periods. The secret key of the KES scheme contains all information about the current tree structure besides its root's public key; the secret component of a KES secret key contains all currently available SIG secret keys, whereas everything else is stored inside its public component. The KES signature of a message m includes a SIG signature of m for a secret key of (the current) leaf node, along with information about the path from the root node to this leaf; for each non-root node on this path, it includes this node's public key and its signature under the node's parent's secret key. This allows to verify signatures having only the public key of the root node.

The key update procedure of the KES scheme modifies the tree to generate the next leftmost leaf node (if necessary) and set it as the one that is used to generate signatures. For this to be possible, the current tree structure always contains the nodes that branch to the right of the nodes that lie on the path from the root to the current leaf node. To ensure the forward security of KES, the old leaf node is erased at the end of the update procedure. For the continual-leakage security of KES, all other SIG secret keys are updated at the end of the update procedure, meaning their individual time periods get increased (this allows to repeatedly leak on them during each separate KES time period).

For example, consider the tree-based key-evolving scheme KES from Fig. 3 in time period 1. Its public key is pk^ε. Its secret key $sk = (pc, sc)$ consists of a secret component $sc = (sk_1^{00}, sk_1^{01}, sk_1^1)$ and a public component $pc = ((pk^0, 1, \sigma^0), (pk^{00}, 1, \sigma^{00}), (pk^{01}, 1, \sigma^{01}), (pk^1, 1, \sigma^1))$. A signature of message m is $(1, \sigma)$ for $\sigma = ((pk^0, 1, \sigma^0), (pk^{00}, 1, \sigma^{00}), (1, \sigma'))$ and $\sigma' \leftarrow_\$ \mathsf{SIG.Sig}(1^\lambda, pk^{00}, 1, sk_1^{00}, m)$. The first component of $(1, \sigma)$ denotes the time period of KES, whereas the first component of $(1, \sigma')$ denotes the time period of key sk_1^{00}. The node information of the form $(pk^0, 1, \sigma^0)$ indicates that σ^0 is a signature of pk^0 under its parent node's secret key, and the latter was used when its time period was 1. A technical detail is that to distinguish parent node's left child from its right child, the signature σ^0 will be produced for a concatenation $0 \| pk^0$ where bit 0 indicates that pk^0 is the parent node's left child.

Our Construction and Its Security. We define a key-evolving signature scheme KES = KES-TREE[SIG, h] as a binary-tree construction described above, where each node of the binary tree contains a key pair of key-evolving signature scheme SIG, and the height of the binary tree is defined by a polynomial h (parameterized by a security parameter). The formal definition of KES-TREE is in the full version of the paper [30]. It is straightforward, but also very detailed.

We claim that if SIG is UFCL-secure, then KES is FUFCL-secure. Note that for any security parameter $\lambda \in \mathbb{N}$, the secret key of KES contains $h(\lambda) + 1$ secret keys of SIG. Therefore, the continual-leakage fraction supported by KES is $h(\lambda) + 1$ times worse than that of SIG.

Theorem 1. *Let $\delta : \mathbb{N} \to [0,1]$. Let SIG be a δ-UFCL key-evolving signature scheme with SIG.ml = SIG.pkl+1. Let $h : \mathbb{N} \to \mathbb{N}$ a polynomial such that $2^{h(\lambda)-1} \leq$ SIG.T(λ) for all $\lambda \in \mathbb{N}$. Let $\gamma(\lambda) = \delta(\lambda)/(h(\lambda) + 1)$ for all $\lambda \in \mathbb{N}$. Then the key-evolving signature scheme KES = KES-TREE[SIG, h] is γ-FUFCL.*

The proof is in the full version of the paper [30]. Informally, it proceeds as follows. Assume that a PT adversary $\mathcal{A}$ breaks the FUFCL-security of KES. In order to do that, it has to forge a valid message-signature pair of KES scheme for some time period i (which must be prior to the time period of full key exposure). In terms of the underlying binary tree structure, this means that $\mathcal{A}$ successfully forges a valid message-signature pair for one of the SIG verification keys that lie on the path from the root of the KES binary tree to the leaf node that is associated to time period i. We build a PT adversary $\mathcal{B}$ against the UFCL-security of SIG as follows. It attempts to guess the KES binary tree node x that will be attacked by $\mathcal{A}$ (out of the $2^{h(\lambda)+1} - 1$ possible nodes); this node will correspond to the challenge key-pair in game UFCL. It then generates SIG key pairs for all other $2^{h(\lambda)+1} - 2$ nodes, and uses its UFCL security game oracles to answer any of $\mathcal{A}$'s oracle queries that depend on the secret key of node x (which is unknown to $\mathcal{B}$). The reduction works if $\mathcal{B}$ guesses the correct challenge node.

Extensions. Note that a binary tree pre-order traversal can be used to associate *each* node of the binary tree with a separate time period of the resulting signature scheme, rather than only use the leaf nodes as we currently do. This was done in some of the previous results that used tree-based construction, such as [15, 23].

5   A Unified Paradigm for Constructing FS+CL Schemes

In this section we define *key-evolution schemes* that model the process of repeatedly evolving a secret key in the presence of a single, fixed public key, and formalize a security notion for them called *forward one-wayness under continual leakage* (FOWCL). We then show how to make a primitive FS+CL with the aid of such a key-evolution scheme and a witness version of the primitive.

Key-Evolution Schemes. A key-evolution scheme KE specifies PT algorithms KE.Kg, KE.Up and KE.Vf, where KE.Vf is deterministic. Associated to KE are the following polynomials: secret-key length $\mathsf{KE.skl} : \mathbb{N} \to \mathbb{N}$, public component length of the secret key $\mathsf{KE.pcl} : \mathbb{N} \to \mathbb{N}$, secret component length of the secret key $\mathsf{KE.scl} : \mathbb{N} \to \mathbb{N}$, and the maximum number of time periods $\mathsf{KE.T} : \mathbb{N} \to \mathbb{N}$. For $\lambda \in \mathbb{N}$ we require that any secret key $sk \in \{0,1\}^{\mathsf{KE.skl}(\lambda)}$ can be parsed as a pair (pc, sc) containing a public component $pc \in \{0,1\}^{\mathsf{KE.pcl}(\lambda)}$ and a secret component $sc \in \{0,1\}^{\mathsf{KE.scl}(\lambda)}$, such that $\mathsf{KE.skl}(\lambda) = \mathsf{KE.pcl}(\lambda) + \mathsf{KE.scl}(\lambda)$. Key generation algorithm KE.Kg takes 1^λ to return a public key pk and base (time period one) secret key $sk_1 \in \{0,1\}^{\mathsf{KE.skl}(\lambda)}$. Key update algorithm KE.Up takes $1^\lambda, pk, i$ and a secret key $sk_i \in \{0,1\}^{\mathsf{KE.skl}(\lambda)}$ for time period i to return a $\mathsf{KE.skl}(\lambda)$-bit secret key for the next time period. Key verification algorithm KE.Vf takes $1^\lambda, pk, i, sk_i$ to return a decision in {true, false} regarding whether sk_i is a valid secret key relative to public key pk and time period $i \in [\mathsf{KE.T}(\lambda)]$. Correctness requires that $\mathsf{KE.Vf}(1^\lambda, pk, i, sk_i) = \mathsf{true}$ for all $\lambda \in \mathbb{N}$, all $(pk, sk_1) \in [\mathsf{KE.Kg}(1^\lambda)]$, all $i \in [\mathsf{KE.T}(\lambda)]$ and all $sk_2, \ldots, sk_i$ satisfying $sk_j \in [\mathsf{KE.Up}(1^\lambda, pk, j-1, sk_{j-1})]$ for $2 \leq j \leq i$. That is, all secret keys that can be obtained via correct updates starting from sk_1 should pass the verification test.

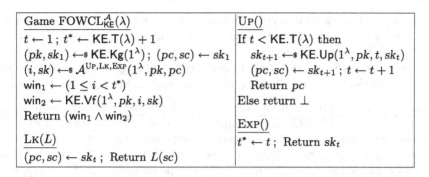

Fig. 4. Game defining forward one-wayness of key-evolution scheme KE under continual leakage.

Forward Security Under Continual Leakage. Consider game FOWCL of Fig. 4 associated to a key-evolution scheme KE and an adversary $\mathcal{A}$, where LK takes a Boolean circuit $L : \{0,1\}^{\mathsf{KE.scl}(\lambda)} \rightarrow \{0,1\}$. For $\lambda \in \mathbb{N}$ let $\mathsf{Adv}^{\mathsf{fowcl}}_{\mathsf{KE},\mathcal{A}}(\lambda) = \Pr[\mathrm{FOWCL}^{\mathcal{A}}_{\mathsf{KE}}(\lambda)]$. We say that FOWCL adversary $\mathcal{A}$ is *valid* if it makes at most one query to its EXP oracle, and this is its last oracle query. We say that $\mathcal{A}$ is δ-bounded, where $\delta : \mathbb{N} \rightarrow [0,1]$, if $\mathcal{A}$ makes at most $\delta(\lambda) \cdot \mathsf{KE.scl}(\lambda)$ queries to LK *per time period*. We say that KE is δ-FOWCL (δ-forward one-way under continual leakage) if $\mathsf{Adv}^{\mathsf{fowcl}}_{\mathsf{KE},\mathcal{A}}(\cdot)$ is negligible for all valid, δ-bounded PT adversaries $\mathcal{A}$.

Game FOWCL is similar to games FINDCL and FUFCL from Sect. 3 in terms of allowing the adversary to obtain information about the secret keys in different time periods by providing it with oracle access to UP, LK and EXP. In order to win, the adversary must output a valid secret key relative to a time period prior to t^* (when exposure happened), where valid means that the key-verification function of KE accepts it. Adversary's advantage is the probability that it wins. To recover relations that are one-way against continual leakage as defined in [19], one could consider adversaries that make no EXP queries. The two notions are equivalent up to simple transformations that are required due to different semantics between the definitions of key-evolving schemes (similar to the case of key-evolving signature schemes as discussed in Sect. 3). Considering adversaries that make no LK queries captures relations that provide forward one wayness. Our model unifies the two security notions.

The familiar requirement for security of a key is that it be indistinguishable from random. This is not achievable when the adversary is in possession of leakage on the key. The requirement we make, following [19], is very weak, namely that the adversary be unable to fully recover a valid key (one-wayness). Then the difficulty is to be able to use such a key for a cryptographic application. This will be done via witness primitives – encryption and signatures.

Witness Encryption and Witness Signatures. Witness encryption [6,26,28] for an **NP**-relation R allows anyone to encrypt messages with respect to any instance

$x \in \{0,1\}^*$. In order to decrypt a message encrypted to x, it is necessary to know a witness w such that $R.Vf(x, w) = \text{true}$. Witness signatures [7,16] for an **NP**-relation R allow to sign messages with respect to an instance-witness pair (x, w) such that $R.Vf(x, w) = \text{true}$. In order to verify a signature produced this way, it is sufficient to know the instance x that was used in the signing process.

Composing Key-Evolution Schemes with Witness Primitives. We now discuss how to use an arbitrary FOWCL key-evolution scheme in order to obtain a FUFCL signature scheme and a FINDCL encryption scheme. This is done in a generic way via a unified paradigm. Below we show that FOWCL + Witness-X yields FS+CL-secure X for X = signatures. The corresponding case of X = encryption is available in the full version of the paper [30]. Note that our security proofs will require some form of extractability from both witness primitives.

Let us explain the issues and the idea. The FOWCL key-evolution scheme provides a way to obtain keys that remain unrecoverable in the FS+CL sense. But it is not clear how to use these keys for signatures or encryption. The reason is that signature and encryption schemes usually require keys of very specific structure that varies from scheme to scheme, but here we are handed keys of a complex structure that are not obviously suitable for any particular application. But witness primitives are, in the terminology of [7], highly "key-versatile". That is, they are able to provide security of the application assuming nothing more than that secret keys are hard to recover from the public key. We will combine them with key evolution to achieve signatures and encryption. We know no direct constructions of FOWCL schemes, but any FUFCL schemes we build based on the construction in Sect. 4 are also FOWCL by definition.

The encryption and signature schemes constructed using our approach will inherit the leakage rate of the used key-evolution scheme, as opposed to the direct construction in Sect. 4 where the leakage rate detoriates logarithmically with the maximum number of time periods. Another advantage of this approach is modularity. We do not need to re-enter any details of our construction of a FOWCL key-evolution scheme, leading to conceptual simplicity. Also, should any new, more efficient or better constructions of FOWCL key-evolution schemes arise in the future, the transforms in this section can be invoked to automatically turn them into FS+CL signature and encryption schemes.

Witness Signatures. Towards detailing the transform for signatures, we define witness signatures. Let R be an **NP**-relation as defined in Sect. 2. A witness signature scheme WS for R specifies PT algorithms WS.Pg, WS.Sig, WS.Vf, WS.SimPg, WS.SimSig and WS.Ext, where WS.Vf is deterministic. Associated to WS is a message length polynomial $WS.ml : \mathbb{N} \rightarrow \mathbb{N}$. Parameter generation algorithm WS.Pg takes 1^λ to return public parameters wp. Signing algorithm WS.Sig takes 1^λ, wp, an instance $x \in \{0,1\}^*$, a witness $w \in \{0,1\}^*$ and a message $m \in \{0,1\}^{WS.ml(\lambda)}$ to return a signature σ. Signature verification algorithm WS.Vf takes 1^λ, wp, x, m, σ to return a decision in $\{\text{true}, \text{false}\}$. Correctness requires that $WS.Vf(1^\lambda, wp, x, m, \sigma) = \text{true}$ for all $\lambda \in \mathbb{N}$, all $wp \in [WS.Pg(1^\lambda)]$, all x, w such that $R.Vf(x, w) = \text{true}$, all $m \in \{0,1\}^{WS.ml(\lambda)}$ and all $\sigma \in [WS.Sig(1^\lambda, wp, x, w, m)]$.

Simulated parameter generation algorithm WS.SimPg takes 1^λ to return simulated parameters wp, a signing trapdoor std and an extraction trapdoor xtd. Simulated signing algorithm WS.SimSig takes 1^λ, wp, an instance x, signing trapdoor std and a message m (but no witness) to return a simulated signature σ. Extraction algorithm WS.Ext takes 1^λ, wp, instance x, extraction trapdoor xtd, message m and signature σ to return a candidate witness w for x.

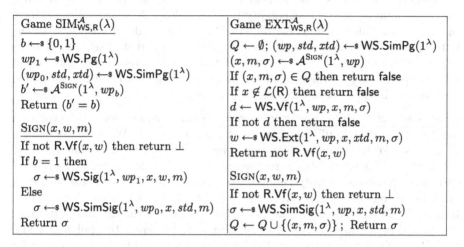

Fig. 5. Games defining signature simulatability of witness signature scheme WS for **NP**-relation R, and witness extractability of witness signature scheme WS for **NP**-relation R.

Signature Simulatability. Consider game SIM of Fig. 5 associated to an **NP**-relation R, a witness signature scheme WS for R, and an adversary $\mathcal{A}$. For $\lambda \in \mathbb{N}$ let $\mathrm{Adv}^{\mathrm{sim}}_{\mathsf{WS},\mathsf{R},\mathcal{A}}(\lambda) = 2\Pr[\mathrm{SIM}^{\mathcal{A}}_{\mathsf{WS},\mathsf{R}}(\lambda)] - 1$. We say that WS, R is signature simulatable if $\mathrm{Adv}^{\mathrm{sim}}_{\mathsf{WS},\mathsf{R},\mathcal{A}}(\cdot)$ is negligible for every PT adversary $\mathcal{A}$. This requires that the signature simulator, given simulated auxiliary parameters and a signature trapdoor, can produce a signature σ indistinguishable from the real one produced under the witness, when not just the message m, but even the instance x and witness w, are adaptively chosen by the adversary.

Witness Extractability. Consider game EXT of Fig. 5 associated to an **NP**-relation R, a witness signature scheme WS for R, and an adversary $\mathcal{A}$. For $\lambda \in \mathbb{N}$ let $\mathrm{Adv}^{\mathrm{ext}}_{\mathsf{WS},\mathsf{R},\mathcal{A}}(\lambda) = \Pr[\mathrm{EXT}^{\mathcal{A}}_{\mathsf{WS},\mathsf{R}}(\lambda)]$. We say that WS, R is witness extractable if $\mathrm{Adv}^{\mathrm{ext}}_{\mathsf{WS},\mathsf{R},\mathcal{A}}(\cdot)$ is negligible for every PT adversary $\mathcal{A}$. This requires that the witness extractor, given simulated auxiliary parameters and an extraction trapdoor, can extract from any valid forgery relative to x an underlying witness w, even when x is chosen by the adversary and the adversary can adaptively obtain simulated signatures under instances and witnesses of its choice.

Obtaining Witness Signatures. Witness signatures as we define them above are effectively another name for Signatures of Knowledge as defined by Chase and

Lysyanskaya [16] and refined by Bellare, Meiklejohn and Thomson [7]. Indeed the latter say that one might refer to this primitive as witness signatures, and we have followed that naming suggestion in order to have a unified terminology across encryption and signatures. The construction uses simulation sound extractable (SSE) NIZKs and follows [7,16,20]. Given any **NP**-relation R and polynomial p, it is possible to construct a witness signature scheme WS such that WS, R are signature simulatable and witness extractable and also WS.ml $= p$. We omit the details and assume this capability in what follows. Note that these witness signatures are different from the ones of Goyal, Jain and Khurana [29]. In the latter, public parameters are not allowed and they show that in this case witness signatures are impossible. In our case, the public parameters can simply be put into the public key of the scheme we are constructing and are not an added assumption. In this case witness signatures are easily constructed from NIZKs.

Construction of a FUFCL Signature Scheme. Assume we are given a key-evolution scheme KE that is FOWCL secure as defined in Sect. 5. We want to build a FUFCL secure key-evolving signature scheme. The difficulty is that the keys in KE may not have the structure required for any particular signature scheme and furthermore the security guarantee on them is weak, namely just that they are hard to recover in full. We achieve our ends through witness signatures. Informally, we associate to KE an **NP**-relation in which the role of the instance x is played by a triple $(1^\lambda, pk, i)$ containing security parameter, public key and time period for KE, and the role of the witness w is played by a secret key sk for KE. We then define a key-evolving signature scheme that uses this **NP**-relation to produce and verify signatures, as shown below.

NP-relation KE-REL. Let KE be a key-evolution scheme. We build an **NP**-relation R $=$ KE-REL[KE] as defined in Fig. 6, where R.wl $=$ KE.skl.

> R.Vf(x, sk)
> $(1^\lambda, pk, i) \leftarrow x$; win$_1 \leftarrow (1 \leq i \leq$ KE.T$(\lambda))$
> win$_2 \leftarrow$ (KE.Vf$(1^\lambda, pk, i, sk)$) ; Return (win$_1 \wedge$ win$_2$)

Fig. 6. NP-relation KE-REL $=$ KE-REL[KE].

Key-Evolving Signature Scheme WITNESS-KES. Let KE be a key-evolution scheme. Let WS be a witness signature scheme for the **NP**-relation R $=$ KE-REL[KE]. We build a key-evolving signature scheme KES $=$ WITNESS-KES[KE, WS] as defined in Fig. 7. The values of KES.skl, KES.pcl, KES.scl, KES.T are same as those of KE, and the values of KES.ml, KES.sigl are inherited from WS. We show that if KE is FOWCL, and if WS is signature simulatable and witness extractable for R, then KES is FUFCL-secure.

KES.Kg(1^λ)	KES.Sig(1^λ, $(pk, wp), i, sk_i, m$)
$(pk, sk_1) \leftarrow\!\!\text{\$}\ \mathsf{KE.Kg}(1^\lambda)$	$\sigma \leftarrow\!\!\text{\$}\ \mathsf{WS.Sig}(1^\lambda, wp, (1^\lambda, pk, i), sk_i, m)$
$wp \leftarrow\!\!\text{\$}\ \mathsf{WS.Pg}(1^\lambda)$	Return (i, σ)
Return $((pk, wp), sk_1)$	KES.Vf(1^λ, $(pk, wp), m, (i, \sigma)$)
KES.Up(1^λ, $(pk, wp), i, sk_i$)	$d_1 \leftarrow (1 \le i \le \mathsf{KE.T}(\lambda))$
$sk_{i+1} \leftarrow\!\!\text{\$}\ \mathsf{KE.Up}(1^\lambda, pk, i, sk_i)$	$d_2 \leftarrow \mathsf{WS.Vf}(1^\lambda, wp, (1^\lambda, pk, i), m, \sigma)$
Return sk_{i+1}	Return $(d_1 \wedge d_2)$

Fig. 7. Key-evolving signature scheme KES = WITNESS-KES[KE, WS].

Theorem 2. *Let $\delta : \mathbb{N} \to [0, 1]$. Let KE be a δ-FOWCL key-evolution scheme. Let R = KE-REL[KE] be the **NP**-relation as defined above. Let WS be a witness signature scheme for R. Assume WS, R is signature simulatable and witness extractable. Then key-evolving signature scheme KES = WITNESS-KES[KE, WS] is δ-FUFCL secure.*

Proof (Theorem 2). Let $\mathcal{A}$ be a valid, δ-bounded PT adversary attacking KES in game FUFCL. We build a valid, δ-bounded PT adversary $\mathcal{B}$ attacking KE in game FOWCL, and PT adversaries $\mathcal{A}_1, \mathcal{A}_2$ attacking signature simulatability and witness extractability of WS, R giving

$$\mathsf{Adv}^{\mathsf{fufcl}}_{\mathsf{KES}, \mathcal{A}}(\lambda) \le \mathsf{Adv}^{\mathsf{sim}}_{\mathsf{WS}, \mathsf{R}, \mathcal{A}_1}(\lambda) + \mathsf{Adv}^{\mathsf{ext}}_{\mathsf{WS}, \mathsf{R}, \mathcal{A}_2}(\lambda) + \mathsf{Adv}^{\mathsf{fowcl}}_{\mathsf{KE}, \mathcal{B}}(\lambda)$$

for all $\lambda \in \mathbb{N}$. This justifies the claim in the theorem statement.

Consider games G_0, G_1, G_2 of Fig. 8. Lines not annotated with comments are common to all games. Game G_0 is equivalent to game $\mathsf{FUFCL}^{\mathcal{A}}_{\mathsf{KES}}(\lambda)$ with the code of KES expanded according to its definition, so $\Pr[G_0] = \Pr[\mathsf{FUFCL}^{\mathcal{A}}_{\mathsf{KES}}(\lambda)]$. Game G_1 switches to using simulated parameters and signatures. Game G_2 additionally requires that the forgery produced by $\mathcal{A}$ allows to extract a valid secret key for the corresponding time period. We build PT adversaries $\mathcal{A}_1, \mathcal{A}_2, \mathcal{B}$ so that for all $\lambda \in \mathbb{N}$,

$$\Pr[G_0] - \Pr[G_1] = \mathsf{Adv}^{\mathsf{sim}}_{\mathsf{WS}, \mathsf{R}, \mathcal{A}_1}(\lambda), \tag{1}$$

$$\Pr[G_1 \text{ sets bad}] \le \mathsf{Adv}^{\mathsf{ext}}_{\mathsf{WS}, \mathsf{R}, \mathcal{A}_2}(\lambda), \tag{2}$$

$$\Pr[G_2] \le \mathsf{Adv}^{\mathsf{fowcl}}_{\mathsf{KE}, \mathcal{B}}(\lambda). \tag{3}$$

Games G_1 and G_2 are identical until bad, so by the Fundamental Lemma of Game-Playing [9] and the above, for all $\lambda \in \mathbb{N}$ we have

$$\begin{aligned}
\mathsf{Adv}^{\mathsf{fufcl}}_{\mathsf{KES}, \mathcal{A}}(\lambda) &= \Pr[\mathsf{FUFCL}^{\mathcal{A}}_{\mathsf{KES}}(\lambda)] = \Pr[G_0] \\
&= (\Pr[G_0] - \Pr[G_1]) + (\Pr[G_1] - \Pr[G_2]) + \Pr[G_2] \\
&\le \mathsf{Adv}^{\mathsf{sim}}_{\mathsf{WS}, \mathsf{R}, \mathcal{A}_1}(\lambda) + \mathsf{Adv}^{\mathsf{ext}}_{\mathsf{WS}, \mathsf{R}, \mathcal{A}_2}(\lambda) + \mathsf{Adv}^{\mathsf{fowcl}}_{\mathsf{KE}, \mathcal{B}}(\lambda).
\end{aligned}$$

```
Games G₀–G₂
S ← ∅ ; t ← 1 ; t* ← KE.T(λ) + 1
(pk, sk₁) ←$ KE.Kg(1^λ) ; (pc, sc) ← sk₁
wp ←$ WS.Pg(1^λ)                                              // G₀
(wp, std, xtd) ←$ WS.SimPg(1^λ)                              // G₁, G₂
(i, m, σ) ←$ 𝒜^{UP,LK,EXP,SIGN}(1^λ, (pk, wp), pc)
sk ←$ WS.Ext(1^λ, wp, (1^λ, pk, i), xtd, m, σ)
win₁ ← (1 ≤ i < t*) ∧ ((i, m, σ) ∉ S)
win₂ ← WS.Vf(1^λ, wp, (1^λ, pk, i), m, σ)
d ← false
If (win₁ ∧ win₂) then
    d ← true
    If not KE.Vf(1^λ, pk, i, sk) then
        bad ← true
        d ← false                                            // G₂
Return d

UP()
If t < KE.T(λ) then
    sk_{t+1} ←$ KE.Up(1^λ, pk, t, sk_t)
    (pc, sc) ← sk_{t+1} ; t ← t + 1
    Return pc
Else return ⊥

LK(L)
(pc, sc) ← sk_t ; Return L(sc)

EXP()
t* ← t ; Return sk_t

SIGN(m)
σ ←$ WS.Sig(1^λ, wp, (1^λ, pk, t), sk_t, m)                 // G₀
σ ←$ WS.SimSig(1^λ, wp, (1^λ, pk, t), std, m)               // G₁, G₂
S ← S ∪ {(t, m, σ)} ; Return (t, σ)
```

Fig. 8. Games G_0–G_2 for proof of Theorem 2.

This bounds the advantage of $\mathcal{A}$ as required for the theorem statement.

PT adversaries $\mathcal{A}_1, \mathcal{A}_2$ are defined in Fig. 9. The procedures to simulate the oracles of $\mathcal{A}$ are the same for both and thus for brevity written only once. PT adversary $\mathcal{B}$ is defined in Fig. 10. We omit the proofs of Eqs. (1)–(3) and explanations of the adversaries due to lack of space. A detailed proof is available in the full version of the paper [30]. ∎

Acknowledgments. Bellare and Stepanovs were supported in part by NSF grants CNS-1526801 and CNS-1717640, ERC Project ERCC FP7/615074 and a gift from Microsoft.

$\mathcal{A}_1^{\text{Sign}}(1^\lambda, wp)$

$S \leftarrow \emptyset$; $t \leftarrow 1$; $t^* \leftarrow \text{KE.T}(\lambda) + 1$
$(pk, sk_1) \leftarrow\!\!{}^{\$}\ \text{KE.Kg}(1^\lambda)$; $(pc, sc) \leftarrow sk_1$
$pk^* \leftarrow (pk, wp)$
$z \leftarrow\!\!{}^{\$}\ \mathcal{A}^{\text{UpSim,LkSim,ExpSim,SignSim}}(1^\lambda, pk^*, pc)$
$(i, m, \sigma) \leftarrow z$
$\text{win}_1 \leftarrow (1 \leq i < t^*) \wedge ((i, m, \sigma) \notin S)$
$\text{win}_2 \leftarrow \text{WS.Vf}(1^\lambda, wp, (1^\lambda, pk, i), m, \sigma)$
If $(\text{win}_1 \wedge \text{win}_2)$ then $b' \leftarrow 1$ else $b' \leftarrow 0$
Return b'

$\mathcal{A}_2^{\text{Sign}}(1^\lambda, wp)$

$S \leftarrow \emptyset$; $t \leftarrow 1$; $t^* \leftarrow \text{KE.T}(\lambda) + 1$
$(pk, sk_1) \leftarrow\!\!{}^{\$}\ \text{KE.Kg}(1^\lambda)$; $(pc, sc) \leftarrow sk_1$
$pk^* \leftarrow (pk, wp)$
$z \leftarrow\!\!{}^{\$}\ \mathcal{A}^{\text{UpSim,LkSim,ExpSim,SignSim}}(1^\lambda, pk^*, pc)$
$(i, m, \sigma) \leftarrow z$; $x \leftarrow (1^\lambda, pk, i)$
Return (x, m, σ)

$\underline{\text{UpSim}()}$

If $t < \text{KE.T}(\lambda)$ then
  $sk_{t+1} \leftarrow\!\!{}^{\$}\ \text{KE.Up}(1^\lambda, pk, t, sk_t)$
  $(pc, sc) \leftarrow sk_{t+1}$; $t \leftarrow t + 1$
  Return pc
Else return $\perp$

$\underline{\text{LkSim}(L)}$

$(pc, sc) \leftarrow sk_t$; Return $L(sc)$

$\underline{\text{ExpSim}()}$

$t^* \leftarrow t$; Return sk_t

$\underline{\text{SignSim}(m)}$

$\sigma \leftarrow\!\!{}^{\$}\ \text{Sign}((1^\lambda, pk, t), sk_t, m)$
$S \leftarrow S \cup \{(t, m, \sigma)\}$; Return (t, σ)

Fig. 9. Adversaries $\mathcal{A}_1$, $\mathcal{A}_2$ for proof of Theorem 2.

$\mathcal{B}^{\text{Up,Lk,Exp}}(1^\lambda, pk, pc)$

$t \leftarrow 1$; $(wp, std, xtd) \leftarrow\!\!{}^{\$}\ \text{WS.SimPg}(1^\lambda)$
$pk^* \leftarrow (pk, wp)$
$z \leftarrow\!\!{}^{\$}\ \mathcal{A}^{\text{UpSim,LkSim,ExpSim,SignSim}}(1^\lambda, pk^*, pc)$
$(i, m, \sigma) \leftarrow z$
$sk \leftarrow\!\!{}^{\$}\ \text{WS.Ext}(1^\lambda, wp, (1^\lambda, pk, i), xtd, m, \sigma)$
Return (i, sk)

$\underline{\text{SignSim}(m)}$

$\sigma \leftarrow\!\!{}^{\$}\ \text{WS.SimSig}(1^\lambda, wp, (1^\lambda, pk, t), std, m)$
Return (t, σ)

$\underline{\text{UpSim}()}$

If $t < \text{KE.T}(\lambda)$ then
  $t \leftarrow t + 1$
Return $\text{Up}()$

$\underline{\text{LkSim}(L)}$

Return $\text{Lk}(L)$

$\underline{\text{ExpSim}()}$

Return $\text{Exp}()$

Fig. 10. Adversary $\mathcal{B}$ for proof of Theorem 2.

References

1. Abdalla, M., Reyzin, L.: A new forward-secure digital signature scheme. In: Okamoto, T. (ed.) ASIACRYPT 2000. LNCS, vol. 1976, pp. 116–129. Springer, Heidelberg (2000). https://doi.org/10.1007/3-540-44448-3_10

2. Akavia, A., Goldwasser, S., Vaikuntanathan, V.: Simultaneous hardcore bits and cryptography against memory attacks. In: Reingold, O. (ed.) TCC 2009. LNCS, vol. 5444, pp. 474–495. Springer, Heidelberg (2009). https://doi.org/10.1007/978-3-642-00457-5_28

3. Alwen, J., Dodis, Y., Wichs, D.: Leakage-resilient public-key cryptography in the bounded-retrieval model. In: Halevi, S. (ed.) CRYPTO 2009. LNCS, vol. 5677, pp. 36–54. Springer, Heidelberg (2009). https://doi.org/10.1007/978-3-642-03356-8_3

4. Ananth, P., Goyal, V., Pandey, O.: Interactive proofs under continual memory leakage. In: Garay, J.A., Gennaro, R. (eds.) CRYPTO 2014. LNCS, vol. 8617, pp. 164–182. Springer, Heidelberg (2014). https://doi.org/10.1007/978-3-662-44381-1_10

5. Anderson, R.: Two remarks on public key cryptology (1997). http://www.cl.cam.ac.uk/users/rja14

6. Bellare, M., Hoang, V.T.: Adaptive witness encryption and asymmetric password-based cryptography. In: Katz, J. (ed.) PKC 2015. LNCS, vol. 9020, pp. 308–331. Springer, Heidelberg (2015). https://doi.org/10.1007/978-3-662-46447-2_14

7. Bellare, M., Meiklejohn, S., Thomson, S.: Key-versatile signatures and applications: RKA, KDM and joint Enc/Sig. In: Nguyen, P.Q., Oswald, E. (eds.) EUROCRYPT 2014. LNCS, vol. 8441, pp. 496–513. Springer, Heidelberg (2014). https://doi.org/10.1007/978-3-642-55220-5_28

8. Bellare, M., Miner, S.K.: A forward-secure digital signature scheme. In: Wiener, M. (ed.) CRYPTO 1999. LNCS, vol. 1666, pp. 431–448. Springer, Heidelberg (1999). https://doi.org/10.1007/3-540-48405-1_28

9. Bellare, M., Rogaway, P.: The security of triple encryption and a framework for code-based game-playing proofs. In: Vaudenay, S. (ed.) EUROCRYPT 2006. LNCS, vol. 4004, pp. 409–426. Springer, Heidelberg (2006). https://doi.org/10.1007/11761679_25

10. Boyen, X., Shacham, H., Shen, E., Waters, B.: Forward-secure signatures with untrusted update. In: ACM CCS 2006 (2006)

11. Boyle, E., Segev, G., Wichs, D.: Fully leakage-resilient signatures. In: Paterson, K.G. (ed.) EUROCRYPT 2011. LNCS, vol. 6632, pp. 89–108. Springer, Heidelberg (2011). https://doi.org/10.1007/978-3-642-20465-4_7

12. Boyle, E., Segev, G., Wichs, D.: Fully leakage-resilient signatures. J. Cryptol. 26(3), 513–558 (2013)

13. Brakerski, Z., Goldwasser, S.: Circular and leakage resilient public-key encryption under subgroup indistinguishability. In: Rabin, T. (ed.) CRYPTO 2010. LNCS, vol. 6223, pp. 1–20. Springer, Heidelberg (2010). https://doi.org/10.1007/978-3-642-14623-7_1

14. Brakerski, Z., Kalai, Y.T., Katz, J., Vaikuntanathan, V.: Overcoming the hole in the bucket: public-key cryptography resilient to continual memory leakage. FOCS 2010 (2010)

15. Canetti, R., Halevi, S., Katz, J.: A forward-secure public-key encryption scheme. In: Biham, E. (ed.) EUROCRYPT 2003. LNCS, vol. 2656, pp. 255–271. Springer, Heidelberg (2003). https://doi.org/10.1007/3-540-39200-9_16

16. Chase, M., Lysyanskaya, A.: On signatures of knowledge. In: Dwork, C. (ed.) CRYPTO 2006. LNCS, vol. 4117, pp. 78–96. Springer, Heidelberg (2006). https://doi.org/10.1007/11818175_5

17. Dachman-Soled, D., Dov Gordon, S., Liu, F.-H., O'Neill, A., Zhou, H.-S.: Leakage-resilient public-key encryption from obfuscation. In: Cheng, C.-M., Chung, K.-M., Persiano, G., Yang, B.-Y. (eds.) PKC 2016. LNCS, vol. 9615, pp. 101–128. Springer, Heidelberg (2016). https://doi.org/10.1007/978-3-662-49387-8_5

18. Dagdelen, Ö., Venturi, D.: A second look at Fischlin's transformation. In: Pointcheval, D., Vergnaud, D. (eds.) AFRICACRYPT 2014. LNCS, vol. 8469, pp. 356–376. Springer, Cham (2014). https://doi.org/10.1007/978-3-319-06734-6_22

19. Dodis, Y., Haralambiev, K., López-Alt, A., Wichs, D.: Cryptography against continuous memory attacks. In: FOCS (2010)

20. Dodis, Y., Haralambiev, K., López-Alt, A., Wichs, D.: Efficient public-key cryptography in the presence of key leakage. In: Abe, M. (ed.) ASIACRYPT 2010. LNCS, vol. 6477, pp. 613–631. Springer, Heidelberg (2010). https://doi.org/10.1007/978-3-642-17373-8_35

21. Dodis, Y., Kalai, Y.T., Lovett, S.: On cryptography with auxiliary input. In: STOC (2009)

22. Dodis, Y., Lewko, A., Waters, B., Wichs, D.: Storing secrets on continually leaky devices. In: FOCS (2011)

23. Faust, S., Kiltz, E., Pietrzak, K., Rothblum, G.N.: Leakage-resilient signatures. In: Micciancio, D. (ed.) TCC 2010. LNCS, vol. 5978, pp. 343–360. Springer, Heidelberg (2010). https://doi.org/10.1007/978-3-642-11799-2_21

24. Faust, S., Kohlweiss, M., Marson, G.A., Venturi, D.: On the non-malleability of the Fiat-Shamir transform. In: Galbraith, S., Nandi, M. (eds.) INDOCRYPT 2012. LNCS, vol. 7668, pp. 60–79. Springer, Heidelberg (2012). https://doi.org/10.1007/978-3-642-34931-7_5

25. Garg, S., Gentry, C., Halevi, S., Wichs, D.: On the implausibility of differing-inputs obfuscation and extractable witness encryption with auxiliary input. In: Garay, J.A., Gennaro, R. (eds.) CRYPTO 2014. LNCS, vol. 8616, pp. 518–535. Springer, Heidelberg (2014). https://doi.org/10.1007/978-3-662-44371-2_29

26. Garg, S., Gentry, C., Sahai, A., Waters, B.: Witness encryption and its applications. In: STOC (2013)

27. Garg, S., Jain, A., Sahai, A.: Leakage-resilient zero knowledge. In: Rogaway, P. (ed.) CRYPTO 2011. LNCS, vol. 6841, pp. 297–315. Springer, Heidelberg (2011). https://doi.org/10.1007/978-3-642-22792-9_17

28. Goldwasser, S., Kalai, Y.T., Popa, R.A., Vaikuntanathan, V., Zeldovich, N.: How to run turing machines on encrypted data. In: Canetti, R., Garay, J.A. (eds.) CRYPTO 2013. LNCS, vol. 8043, pp. 536–553. Springer, Heidelberg (2013). https://doi.org/10.1007/978-3-642-40084-1_30

29. Goyal, V., Jain, A., Khurana, D.: Non-malleable multi-prover interactive proofs and witness signatures. Cryptology ePrint Archive, Report 2015/1095 (2015). http://eprint.iacr.org/2015/1095

30. Bellare, M., O'Neill, A., Stepanovs, I.: Forward-security under continual leakage. Cryptology ePrint Archive, Report 2017/476 (2017). http://eprint.iacr.org/2017/476

31. Itkis, G., Reyzin, L.: Forward-secure signatures with optimal signing and verifying. In: Kilian, J. (ed.) CRYPTO 2001. LNCS, vol. 2139, pp. 332–354. Springer, Heidelberg (2001). https://doi.org/10.1007/3-540-44647-8_20

32. Katz, J., Vaikuntanathan, V.: Signature schemes with bounded leakage resilience. In: Matsui, M. (ed.) ASIACRYPT 2009. LNCS, vol. 5912, pp. 703–720. Springer, Heidelberg (2009). https://doi.org/10.1007/978-3-642-10366-7_41

33. Krawczyk, H.: Simple forward-secure signatures from any signature scheme. In: ACM CCS 2000 (2000)

34. Lewko, A., Lewko, M., Waters, B.: How to leak on key updates. In: STOC (2011)

35. Lewko, A., Rouselakis, Y., Waters, B.: Achieving leakage resilience through dual system encryption. In: Ishai, Y. (ed.) TCC 2011. LNCS, vol. 6597, pp. 70–88. Springer, Heidelberg (2011). https://doi.org/10.1007/978-3-642-19571-6_6

36. Lyubashevsky, V., Palacio, A., Segev, G.: Public-key cryptographic primitives provably as secure as subset sum. In: Micciancio, D. (ed.) TCC 2010. LNCS, vol. 5978, pp. 382–400. Springer, Heidelberg (2010). https://doi.org/10.1007/978-3-642-11799-2_23

37. Malkin, T., Micciancio, D., Miner, S.: Efficient generic forward-secure signatures with an unbounded number of time periods. In: Knudsen, L.R. (ed.) EUROCRYPT 2002. LNCS, vol. 2332, pp. 400–417. Springer, Heidelberg (2002). https://doi.org/10.1007/3-540-46035-7_27

38. Malkin, T., Teranishi, I., Vahlis, Y., Yung, M.: Signatures resilient to continual leakage on memory and computation. In: Ishai, Y. (ed.) TCC 2011. LNCS, vol. 6597, pp. 89–106. Springer, Heidelberg (2011). https://doi.org/10.1007/978-3-642-19571-6_7

39. Naor, M., Segev, G.: Public-key cryptosystems resilient to key leakage. In: Halevi, S. (ed.) CRYPTO 2009. LNCS, vol. 5677, pp. 18–35. Springer, Heidelberg (2009). https://doi.org/10.1007/978-3-642-03356-8_2

40. Nielsen, J.B., Venturi, D., Zottarel, A.: Leakage-resilient signatures with graceful degradation. In: Krawczyk, H. (ed.) PKC 2014. LNCS, vol. 8383, pp. 362–379. Springer, Heidelberg (2014). https://doi.org/10.1007/978-3-642-54631-0_21

Tightly-Secure PAK(E)

José Becerra[1], Vincenzo Iovino[1], Dimiter Ostrev[1], Petra Šala[1,2],
and Marjan Škrobot[1(✉)]

[1] University of Luxembourg, Esch-sur-Alzette, Luxembourg
{jose.becerra,vincenzo.iovino,dimiter.ostrev,petra.sala,
marjan.skrobot}@uni.lu
[2] Computer Science Department, École Normale Supérieure, Paris, France

Abstract. We present a security reduction for the PAK protocol instantiated over Gap Diffie-Hellman Groups that is tighter than previously known reductions. We discuss the implications of our results for concrete security. Our proof is the first to show that the PAK protocol can provide meaningful security guarantees for values of the parameters typical in today's world.

Keywords: Password-authenticated key exchange · PAK
Tight reductions · Random oracle

1 Introduction

1.1 PAKE Protocols

A password authenticated key exchange (PAKE) protocol allows two users who only share a password to establish a high entropy shared secret key by exchanging messages over a hostile network. PAKE protocols have only minimal requirements for the long-term secrets that users need to hold in order to succeed and therefore are interesting both theoretically and in practice. To date, there have been over twenty years of intensive research on PAKE, and PAKE protocols have recently seen more and more deployment in applications such as ad hoc networks [35] or the Internet of Things [32].

Numerous PAKE protocols have been proposed over the years. Among them, only a handful have been considered for use in real-world applications: EKE [6], SPEKE [17], SRP [36], PPK and PAK [8,25,26], KOY [19], Dragonfly [15], SPAKE2 [3] and J-PAKE [14]. The last two protocols, along with SRP and Dragonfly that have been standardized in the form of RFC2945 and RFC7664 respectively, are currently being considered by the Internet Engineering Task Force (IETF).

When evaluating different PAKE designs, two main criteria are the protocol's efficiency in terms of computation and communication, and the security guarantees that the protocol provides. Of these two criteria, the efficiency is easier to understand by just looking at the protocol description. On the other hand, it is

© Springer Nature Switzerland AG 2018
S. Capkun and S. S. M. Chow (Eds.): CANS 2017, LNCS 11261, pp. 27–48, 2018.
https://doi.org/10.1007/978-3-030-02641-7_2

difficult to judge whether a protocol is secure. A necessary condition for security is that no attacks on the protocol have been found so far, but most researchers agree that this is not sufficient.

1.2   Security Models and Reductions for PAKE

One way to rigorously discuss the security of PAKE protocols is to formally define a security challenge: an interaction between two algorithms called a challenger and an adversary. The interaction is designed to model the capabilities that a real world adversary is believed to have; the success of an adversary in the security challenge corresponds to a successful attack on the protocol. Several such security models have been introduced over the years. A few prominent ones are the indistinguishability-based models of Bellare, Pointcheval and Rogaway [5] and Abdalla, Fouque and Pointcheval [2], the simulation-based model of Boyko, MacKenzie and Patel [8][1], and the Universally Composable (UC) model of Canetti et al. [9].

In this approach, the security of a protocol is established in the following way: given an adversary $\mathcal{A}$ that runs in time t and has success probability ϵ in the security challenge, one constructs an algorithm $\mathcal{B}^{\mathcal{A}}$ known as a reduction. $\mathcal{B}^{\mathcal{A}}$ runs $\mathcal{A}$ as a subroutine and solves some known hard computational problem in time t' and with success probability ϵ'. If it is widely believed that it is impossible to solve the hard computational problem in time t' and with success probability ϵ', then one can conclude that no adversary running in time t can have a probability of ϵ to successfully attack the protocol.

1.3   Online Dictionary Attacks

Security models for PAKE must properly account for online dictionary attacks, in which an adversary guesses a password and tries to run the protocol with one of the honest users to verify the guess. Since passwords come from a small set, this attack has a non-negligible chance of success. Online dictionary attacks cannot be entirely prevented, but their effects can be mitigated to some extent for example by requiring users to choose strong passwords, limiting the number of unsuccessful login attempts, or even using machine learning to detect a pattern in login attempts that suggests an online dictionary attack might be in progress.

From the point of view of cryptographic research on PAKE, online dictionary attacks and the countermeasures listed above are taken as given; the focus is on ensuring that the adversary can do essentially no better than to run the best online dictionary attack in the circumstances. This intuitive requirement is formalized differently in indistinguishability-based and simulation-based models. In the indistinguishability-based model that we use in this paper[2], the formal

---

[1] For the relation between the indistinguishability-based and simulation-based models, see the recent work [23].

[2] A detailed description of the FtG model of Bellare, Pointcheval and Rogaway [5] can be found in Sect. 4.

requirement is that for all PPT adversaries $\mathcal{A}$ that perform at most n online dictionary attacks,

$$\text{Adv}(\mathcal{A}) \leq \mathcal{F}(\mathcal{D}, \mathcal{L}, n) + \epsilon \tag{1}$$

where $\text{Adv}(\mathcal{A})$ is the advantage[3] of the adversary in breaking the protocol, where $\mathcal{F}(\mathcal{D}, \mathcal{L}, n)$ is the maximum probability of success[4] of any password guessing strategy that uses n guesses against a password distribution $\mathcal{D}$ and login attempt policy $\mathcal{L}$, and where ϵ is a negligible term.

In the present work, we focus on the behavior of the term ϵ of equation (1). A precise theoretical or empirical characterization of the function $\mathcal{F}(\mathcal{D}, \mathcal{L}, n)$ is an important and interesting research question, but is outside the scope of this paper. Here, we merely mention that many previous works use the formulation $\mathcal{F}(\mathcal{D}, \mathcal{L}, n) = n/N$ of [2,5] which corresponds to making the simplifying assumptions that there is no login attempt policy and that passwords are independent and uniformly distributed from a dictionary of size N. On the other hand, some recent research [33] suggests that real-life passwords follow Zipf's law, and proposes [34, Sect. 3] the formulation $\mathcal{F}(\mathcal{D}, \mathcal{L}, n) = Cn^s$ where C, s are parameters that have to be estimated empirically. Our results regarding the behavior of the term ϵ of Eq. (1) hold independently of the password distribution and login attempt policy, and in particular, they hold for any of the cases mentioned above.

1.4 The PAK Protocol

One of the PAKE protocols whose security has been studied in the provable security framework is the PAK protocol [8,25,26]. It is a PAKE protocol with several desirable characteristics: low computation and communication cost, and security proofs in two different security models: the simulation-based model of Boyko, MacKenzie and Patel [8] and the so-called Find-then-Guess (FtG) model of Bellare, Pointcheval and Rogaway [5]. A modified version of PAK has been used to detect man-in-the-middle attacks against SSL/TLS without third-parties [10], and a lattice-based version of PAK has been used to provide security against quantum adversaries [11]. Moreover, variants of the PAK protocol have been included in IEEE standard [16], while the patent held by Lucent Technologies [27] is expiring soon. Therefore, the PAK protocol is a candidate for wide-scale practical deployment.

While there are security proofs for PAK in two different models, in both cases the reductions are loose, meaning either that the running time t' of the reduction $\mathcal{B}^{\mathcal{A}}$ is much larger than the running time t of the adversary $\mathcal{A}$ or that the success probability ϵ' of $\mathcal{B}^{\mathcal{A}}$ is much smaller than the success probability ϵ of $\mathcal{A}$, or some combination of the two.

A loose reduction is usually considered less than ideal. From a qualitative point of view, a reduction gives the assurance that "breaking the protocol is at most a little easier than solving the hard computational problem" [12]. However, if a reduction is loose, it leaves open the possibility that "a little easier" is in

---

[3] The advantage is twice the success probability minus one.

[4] By success we mean guessing the password of *any* user.

fact "substantially easier". From a quantitative point of view, a loose reduction means that larger security parameters must be chosen to guarantee a given level of security, which in turn increases the communication and computation cost of the protocol; therefore, a tight reduction is considered preferable [4].

We illustrate the last point by looking in detail at the best previous result for PAK [25, Theorem 6.9], which we reproduce here for convenience.

Theorem 1 (Theorem 6.9 in [25]). *Consider the PAK protocol[5] instantiated over a group $\mathbb{G} = \langle g \rangle$ of order q and with password dictionary of size N. Let $\mathcal{A}$ be an adversary that runs in time t and performs at most n_{se}, n_{ex}, n_{re}, n_{co}, n_{ro} queries of type **Send, Execute, Reveal, Corrupt, Random Oracle** and a single **Test** query. Let $\mathrm{Adv}(\mathcal{A})$ be the advantage of this adversary in the security challenge as defined in the FtG model (See footnote 2). Let t_{exp} be the time required for an exponentiation in $\mathbb{G}$. Then, for $t' = O(t + ((n_{ro})^2 + n_{se} + n_{ex})t_{exp})$*

$$\mathrm{Adv}(\mathcal{A}) = \frac{n_{se}}{N} + O\left(n_{se}\mathrm{Adv}_{\mathbb{G}}^{1\text{-}cdh}(t', (n_{ro})^2) + \frac{(n_{se} + n_{ex})(n_{ro} + n_{se} + n_{ex})}{q}\right) \tag{2}$$

where $\mathrm{Adv}_{\mathbb{G}}^{1\text{-}cdh}(t', (n_{ro})^2)$ is the maximum success probability of an algorithm that is allowed to run for time t' and to output a list of $(n_{ro})^2$ candidate solutions to the CDH problem, and succeeds if at least one solution in the list is correct.

We plug in some concrete values in the above theorem. For the order of the group q, we use the recommended $q \approx 2^{256}$ for long-term security from [12, Chapter 7]. For the number of random oracle queries, we take $n_{ro} \approx 2^{63}$, the number of SHA1 computations performed in the recent attack [31]. Next, we use the approximation that solving the discrete logarithm problem in group $\mathbb{G}$ takes about $\sqrt{q} \approx 2^{128}$ operations [22, Sect. 7]. We see that with these values of the parameters, we can estimate

$$\mathrm{Adv}_{\mathbb{G}}^{1\text{-}cdh}(t', (n_{ro})^2) \approx 1$$

and therefore the term

$$n_{se}\mathrm{Adv}_{\mathbb{G}}^{1\text{-}cdh}(t', (n_{ro})^2) \gg 1$$

makes the right hand-side of Eq. 2 meaningless in bounding $\mathrm{Adv}(\mathcal{A})$, which, by definition, is a number less than or equal to 1. This means that we cannot reasonably claim that the security proof gives the guarantee "adversary can essentially do no better than an online dictionary attack" except in the trivial case when the online dictionary attack itself succeeds with probability close to one.

1.5   Our Contribution

We provide a tight reduction for PAK instantiated over Gap Diffie-Hellman groups; these are groups in which solving the Decisional Diffie-Hellman Problem

---

[5] A detailed description of the protocol is in Sect. 3.

is easy but solving the Computational Diffie-Hellman problem is equivalent to solving the Discrete Logarithm Problem and is believed to be hard [18][6]. We employ proof techniques that have been used previously in [1,20].

The formal statement of our result can be found in Theorem refT:main1.

Theorem 2. *Consider the PAK protocol instantiated over a Gap Diffie-Hellman group $\mathbb{G}_1 = \langle g \rangle$ of order q and with password dictionary of size N. Let $\mathcal{A}$ be an adversary that runs in time t and performs at most n_{se}, n_{ex}, n_{re}, n_{co}, n_{ro} queries of type **Send, Execute, Reveal, Corrupt, Random Oracle** and a single **Test** query. Let $\mathrm{Adv}(\mathcal{A})$ be the advantage of this adversary in the security challenge as defined in the FtG model. Let t_{exp} and t_{ddh} be the time required for an exponentiation in $\mathbb{G}_1$ and deciding DDH in $\mathbb{G}_1$, respectively. Then, for $t'' = O(t + (n_{ro} + n_{se} + n_{ex})t_{exp} + (n_{se} + n_{ro})t_{ddh})$*

$$\mathrm{Adv}(\mathcal{A}) \leq \mathcal{F}(\mathcal{D}, \mathcal{L}, n_{se}) + 8\mathrm{Adv}_{\mathbb{G}_1}^{\mathrm{gap\text{-}cdh}}(t'') + O\left(\frac{(n_{se} + n_{ex})(n_{ro} + n_{se} + n_{ex})}{q}\right)$$

(3)

where $\mathrm{Adv}_{\mathbb{G}_1}^{\mathrm{gap\text{-}cdh}}(t'')$ is the maximum success probability of an algorithm that is allowed to run for time t'' in solving the Gap-Diffie-Hellman (Gap-DH) problem in group $\mathbb{G}_1$.

We perform a similar analysis of our result as in the previous section, using the same values of q, and n_{ro}. Since $t'' \ll 2^{128}$ (assuming the most powerful adversaries today have t at most $\approx 2^{80}$ to 2^{85}), we can assume that $\mathrm{Adv}_{\mathbb{G}_1}^{\mathrm{gap\text{-}cdh}}(\approx 2^{85}) \lesssim 2^{-35}$. Furthermore, the term $O\left((n_{se} + n_{ex})(n_{ro} + n_{se} + n_{ex})/q\right)$ is negligible compared to the other two terms. Thus, by using the tight reduction, we are able to obtain the guarantee: assuming that $\mathrm{Adv}_{\mathbb{G}_1}^{\mathrm{gap\text{-}cdh}}(\approx 2^{85}) \lesssim 2^{-35}$ then for all adversaries $\mathcal{A}$ with running time $t \lesssim 2^{85}$, the advantage in breaking the PAK protocol instantiated over a Gap Diffie Hellman group of order $\approx 2^{256}$ is at most $\approx 2^{-30}$ higher than the advantage of breaking the protocol using the best online dictionary attack for the given password distribution and login attempt policy.[7]

Thus, by relying on the Gap-Diffie-Hellman assumption instead of the List-Diffie-Hellman assumption as in [25] we are able to remove the degradation factors that cause the previous security proof for PAK to fail to provide meaningful guarantees for typical values of the parameters in today's world.

1.6   Organization of the Paper

The rest of the paper is organized as follows: in Sect. 2, we introduce notation and give details on the Gap Diffie-Hellman groups and hardness assumptions used in this paper. In Sect. 3, we give a detailed description of the PAK protocol. In

---

[6] More details on Gap Diffie-Hellman groups and the relevant computational problems and assumptions are given in Sect. 2.

[7] We refer to [34, Fig. 4] for an estimation of the advantage of online dictionary attacks as a function of the number of guesses for two real-world password datasets.

Sect. 4, we introduce the security model FtG of Bellare, Pointcheval and Rogaway [5]. In Sect. 5, we prove our main result. We conclude the paper in Sect. 6.

2   Preliminaries

In this section, we introduce notation, define pairings, and state the hardness assumptions upon which the security of PAK protocol rests.

2.1   Notation

We write $d \overset{\$}{\leftarrow} D$ for sampling uniformly at random from set D and $|D|$ to denote its cardinality. The output of a probabilistic algorithm A on input x is denoted by $y \leftarrow A(x)$, while $y := F(x)$ denotes a deterministic assignment of the value $F(x)$ to the variable y. Let $\{0,1\}^*$ denote the bit string of arbitrary length while $\{0,1\}^l$ stands for those of length l. Let κ be the security parameter and $negl(\kappa)$ denote a negligible function. When we sample elements from $\mathbb{Z}_q$, it is understood that they are viewed as integers in $[1 \ldots q]$, and all operations on these are performed mod q. In general, we use $\mathbb{G}$ to denote any cyclic group while $\mathbb{G}_1$ refers to a bilinear group. Let H_1 be a full-domain hash mapping $\{0,1\}^*$ to $\mathbb{G}_1$. All remaining hash functions, H_2, H_3 and H_4, map from $\{0,1\}^*$ to $\{0,1\}^\kappa$.

2.2   Cryptographic Building Blocks

Let $\mathbb{G}_1, \mathbb{G}_T$ be cyclic groups of prime order q and g a generator of $\mathbb{G}_1$.

Definition 1. *A bilinear map is a function $e : \mathbb{G}_1 \times \mathbb{G}_1 \to \mathbb{G}_T$ such that the following properties are satisfied:*

1. *Bilinear:* $\forall\ u, v \in \mathbb{G}_1,\ a, b \in \mathbb{Z}_q,\ e(u^a, v^b) = e(u,v)^{ab}$.
2. *Non-degenerate:* $e(g,g)$ *generates* $\mathbb{G}_T$.
3. *Computable:* $\forall\ u, v \in \mathbb{G}_1,\ a, b \in \mathbb{Z}_q$, *there is an efficient algorithm to compute* $e(u^a, v^b)$.

Definition 2 *(Bilinear Group).* $\mathbb{G}_1$ *is a bilinear group if there exists group* $\mathbb{G}_T$ *and a bilinear map* $e : \mathbb{G}_1 \times \mathbb{G}_1 \to \mathbb{G}_T$.

2.3   Cryptographic Hardness Assumptions

Let $\mathbb{G}$ be any multiplicative cyclic group, with generator g and $|\mathbb{G}| = q$. For $X = g^x$ and $Y = g^y$, let $DH(X,Y) = g^{xy}$, where $\{g^x, g^y, g^{xy}\} \in \mathbb{G}$.

Definition 3 *(Computational Diffie-Hellman (CDH) Problem).* *Given* (g, g^x, g^y) *compute* g^{xy}, *where* $\{g^x, g^y, g^{xy}\} \in \mathbb{G}$ *and* $(x, y) \overset{\$}{\leftarrow} \mathbb{Z}_q^2$. *Let the advantage of a PPT algorithm $\mathcal{A}$ in solving the CDH problem be:*

$$\text{Adv}_{\mathbb{G}}^{\text{cdh}}(\mathcal{A}) = \Pr\left[(x,y) \overset{\$}{\leftarrow} \mathbb{Z}_q^2, X = g^x, Y = g^y : \mathcal{A}(X,Y) = DH(X,Y)\right].$$

CDH Assumption: There exist sequences of cyclic groups $\mathbb{G}$ indexed by κ such that for all PPT adversaries $\mathcal{A}$ $\mathrm{Adv}_{\mathbb{G}}^{\mathrm{cdh}}(\mathcal{A}) \leq negl(\kappa)$, where κ is the security parameter.

Definition 4 *(List-Computational Diffie-Hellman (L-CDH) Problem).* *Given* (g, g^x, g^y) *compute* g^{xy}, *where* $\{g^x, g^y, g^{xy}\} \in \mathbb{G}$ *and* $(x, y) \xleftarrow{\$} \mathbb{Z}_q^2$. *Let* $\mathcal{A}$ *be a PPT algorithm which attempts to solve the L-CDH problem and outputs a list of n elements, its advantage is defined as follows:*

$$\mathrm{Adv}_{\mathbb{G}}^{\mathrm{l\text{-}cdh}}(\mathcal{A}, n) = \Pr\left[(x, y) \xleftarrow{\$} \mathbb{Z}_q^2, X = g^x, Y = g^y : DH(X, Y) \in \mathcal{A}(X, Y)\right].$$

L-CDH Assumption: There exist sequences of cyclic groups $\mathbb{G}$ indexed by κ such that for all PPT adversaries $\mathcal{A}$ $\mathrm{Adv}_{\mathbb{G}}^{\mathrm{l\text{-}cdh}}(\mathcal{A}, n) \leq negl(\kappa)$, where κ is the security parameter.

Definition 5 *(Decision Diffie-Hellman (DDH) Problem).* *Distinguish a tuple* (g^x, g^y, g^{xy}) *from* (g^x, g^y, g^z), *where* $\{g^x, g^y, g^z\} \in \mathbb{G}_1$ *and* $(x, y, z) \xleftarrow{\$} \mathbb{Z}_q^3$. *Let the advantage of a PPT algorithm $\mathcal{A}$ in solving DDH problem be:*

$$\mathrm{Adv}_{\mathbb{G}}^{\mathrm{ddh}}(\mathcal{A}) = |\Pr\left[(x, y) \xleftarrow{\$} \mathbb{Z}_q^2, X = g^x, Y = g^y, Z = g^{xy} : \mathcal{A}(X, Y, Z) = 1\right]$$
$$- \Pr\left[(x, y, z) \xleftarrow{\$} \mathbb{Z}_q^3, X = g^x, Y = g^y, Z = g^z : \mathcal{A}(X, Y, Z) = 1\right]|. \quad (4)$$

DDH Assumption: There exist sequences of cyclic groups $\mathbb{G}$ indexed by κ such that for all PPT adversaries $\mathcal{A}$ $\mathrm{Adv}_{\mathbb{G}}^{\mathrm{ddh}}(\mathcal{A}) \leq negl(\kappa)$, where κ is the security parameter.

Gap Diffie-Hellman (Gap-DH) groups are those where the DDH problem can be solved in polynomial time but no PPT algorithm can solve the CDH problem with advantage greater than negligible, e.g. bilinear groups from Definition 2. More formally:

Definition 6 *(Gap-Diffie-Hellman (Gap-DH) Problem).* *Given* (g, g^x, g^y) *and access to a Decision Diffie-Hellman Oracle (DDH-O) compute* g^{xy}.

$$\mathrm{Adv}_{\mathbb{G}_1}^{\mathrm{gap\text{-}cdh}}(\mathcal{A}) = \Pr\left[(x, y) \xleftarrow{\$} \mathbb{Z}_q^2, X = g^x, Y = g^y : \mathcal{A}^{ddh\text{-}o}(X, Y) = DH(X, Y)\right].$$

Gap-DH Assumption: There exists sequences of *bilinear groups* $\mathbb{G}_1$ indexed by κ, such that for all PPT $\mathcal{A}$ $\mathrm{Adv}_{\mathbb{G}_1}^{\mathrm{gap\text{-}cdh}}(\mathcal{A}) \leq negl(\kappa)$, where κ is the security parameter.

3   The PAK Protocol

In this section, we describe the PAK protocol from [25], whose mathematical description is presented in Fig. 1. A few other variants of PAK were developed in [26].

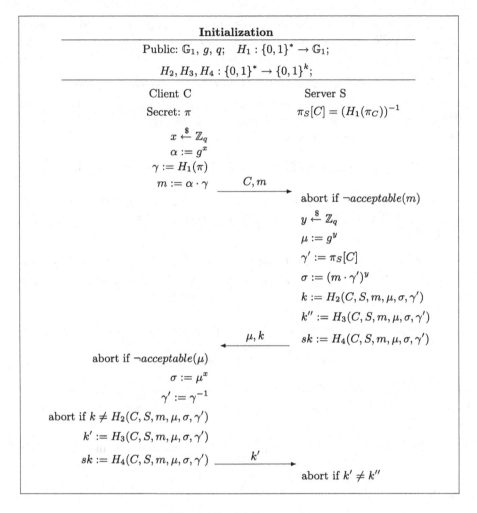

Fig. 1. The PAK protocol.

3.1 Protocol Description

Here, we make use of the same notation as in [25] (October version). Now, we describe the protocol informally.

Before any protocol execution, public parameters are fixed and passwords are shared between clients and servers during the initialization phase. More specifically, for efficiency reasons and security in case of password file compromise, servers only keep the inverse element of each password's hash value.

The PAK protocol consists of three message rounds. In the first message round, the client sends a group element m – generated by multiplying a random group element α with the mask γ (also a group element) that is derived from the shared password π – along with its ID to the server. In the second message round, upon receiving the message C, m, the server first checks with the *acceptable* function if the received value m is an element of $\mathbb{G}_1$. Then, it selects a random group element μ, removes the mask from the received m, and computes the shared secret σ, confirmation codes k, k', a session key sk and sets sid and pid values (thus accepting). Once all these values are computed, the server sends μ and k to the client. Upon receiving the second message, the client first checks if μ is valid group element. If so, it computes the shared secret and confirmation code k and checks the validity of the latter. If all checks are correct, the client computes his confirmation code k' and a session key sk, sets sid and pid values, and then it sends k' in the third message round and terminates. The server, once it receives value k' and checks its validity, also terminates.

3.2   Instantiating the Protocol over Gap Diffie-Hellman Groups

Gap-DH groups were introduced in the pioneering work of Boneh, Lynn and Shacham [7]. For instance, Gap-DH groups can be derived by the supersingular elliptic curve given by the equation $y^2 = x^3 + 2x \pm 1$ over the field $\mathbb{F}_{3^l}$. It can be seen that for some values of l the number of points in this curve divides $3^{6l} - 1$. The value 6 is called the multiplier that has to be neither too small for the CDH problem to be hard, nor too big for the Decision Diffie-Hellman Oracle (DDH-O) to be efficient. An example of DDH-O on this curve is the Weil pairing [30]. Gap-problems were also studied by Okamoto and Pointcheval [29].

In order to have efficient PAK execution, $H_1 : \{0,1\}^* \to \mathbb{G}_1$ must be an efficiently computable function. We point the reader to [26,28] for efficient implementations of H_1. Note that it is crucial for such an algorithm to run in constant time, otherwise timing attacks on a password are possible. For more details on pairings, we refer readers to [13].

4   Model

For our proof, we will use the well-known Find-then-Guess (FtG) security model from [5], which guarantees security against an adversary fully controlling the network, concurrent sessions, loss of session keys, as well as forward secrecy. Furthermore, the security model incorporates the essential requirements that PAKE protocols must satisfy: (i) an eavesdropper adversary should not learn any information about the password and (ii) an adversary can verify at most one password guess per protocol execution in an active attack.

In the FtG model, security is defined via a security experiment G_{ftg} played between a challenger $\mathcal{CH}_{ftg}$ and some adversary $\mathcal{A}$. The task of $\mathcal{CH}_{ftg}$ is to administrate the security experiment while keeping the appropriate secret information outside from $\mathcal{A}$'s view. Roughly speaking, $\mathcal{A}$ wins the security experiment if he is able to distinguish established session keys from random strings.

We will start by formally defining PAKE protocols. This will be followed by an in-depth description of the FtG security model.

PAKE Protocol. A PAKE protocol can be represented as a pair of algorithms $(genPW, P)$, where $genPW$ is a password generation algorithm and P is the description of the protocol that specifies how honest parties behave. A $genPW$ algorithm takes as input a set of possible passwords *Passwords*, together with a probability distribution $\mathcal{P}$.

Participants and Passwords. In the two-party PAKE setting, each principal U is either from a *Clients* set or a *Servers* set, both of which are finite, disjoint, nonempty sets. The set ID represents the union of *Clients* and *Servers*. Furthermore, we assume that each client $C \in Clients$ possesses a password π_C, while on the other hand each server $S \in Servers$ holds a vector of the passwords of all clients $\pi_S := \langle \pi_C \rangle_{C \in Clients}$.

Protocol Execution. P is a PPT algorithm that specifies reaction of principals to network messages. In a real scenario, each principal may run multiple executions of P with different users, thus in our model each principal is allowed an unlimited number of *instances* executing P in parallel. We denote with Π_i^U the i-th instance of a U principal. In some places, where distinction matters, we will denote client instances with Π_i^C and server instances by Π_j^S.

When assessing the security of P, we assume that the adversary $\mathcal{A}$ has complete control of the network. Practically, this means that principals solely communicate through the adversary that may consider delaying, reordering, modifying, dropping messages sent by honest principals or injecting messages of its choice in order to attack the protocol. Moreover, the adversary has access to instances of the principals through the game's interface (offered by the challenger). Thus, while playing the security game, $\mathcal{A}$ provides the inputs to the challenger $\mathcal{CH}_{ftg}$ – who parses the received messages and forwards them to corresponding instances – via the following *queries*:

- **Send**(U, i, M): $\mathcal{A}$ sends message M to instance Π_i^U. As a response, Π_i^U processes M according to the protocol description P, updates its corresponding *internal state* and outputs a reply that is given to $\mathcal{A}$. Whenever this query causes Π_i^U to *accept*, *terminate* or *abort*, it is indicated to $\mathcal{A}$. Additionally, to instruct client C to initiate a session with server S, the adversary sends a message containing the name of the server to an unused instance of C, i.e. **Send**(C, i, S).
- **Execute**(C, i, S, j): This triggers an honest run of P between instances Π_i^C and Π_j^S. The transcript of the protocol execution is given to $\mathcal{A}$. It covers passive eavesdropping on protocol flows.
- **Reveal**(U, i): As a response to this query, $\mathcal{A}$ receives the current value of the session key sk_U^i computed at Π_i^U. $\mathcal{A}$ may do this only if Π_i^U holds a session key, e.g. it is in *accept* or *terminate* state. This query captures potential session key leakage as a result of its use in higher level protocols. Also, it ensures that if some session key gets compromised, other session keys remain protected.

– **Test**(U, i): $\mathcal{CH}_{ftg}$ flips a bit b and answers this query as follows: if $b = 1$, $\mathcal{A}$ gets sk_U^i. Otherwise, it receives a random string from the session key space. This query can only be asked once by $\mathcal{A}$ at any time during the execution of G_{ftg}. This query simply measures the adversarial success and does not correspond to any real-world adversarial capability.
– **Corrupt**(U): The password π_U is given to $\mathcal{A}$ if U is a client, and the list of passwords π_U in case U is a server[8].

As can be seen above, the adversary is allowed to send multiple **Send**, **Execute**, **Reveal** and **Corrupt** queries to the challenger, and only a single **Test** query.

Accepting and Terminating. In the FtG model from [5], an instance Π_i^U accepts whenever it holds a session key sk_U^i, a session ID sid_U^i and a partner ID pid_U^i. Note that the meaning of "accept" in this context is different from the usage of "accept" in other settings such as computational complexity.

An instance Π_i^U terminates if it holds sk_U^i, sid_U^i, pid_U^i and will not send nor receive any more messages. Due to the protocol design, Π_i^U may accept once and terminate later. Note also that it is possible for a server running the PAK protocol to accept at the time it sends the second protocol flow and to later abort if it receives a wrong confirmation code in the third protocol flow.

Partnering. We say that instances Π_i^C and Π_j^S are partnered if both oracles accept holding $(sk_C^i, sid_C^i, pid_C^i)$ and $(sk_S^j, sid_S^j, pid_S^j)$ respectively and the following conditions hold:

1. $sk_C^i = sk_S^j$, $sid_C^i = sid_S^j$, $pid_C^i = S$, $pid_S^j = C$
2. no other instance accepts with the same sid.

Freshness. It captures the idea that the adversary should not trivially know the session key being tested. We incorporate forward secrecy in the definition of freshness. An instance Π_i^U is said to be fs-fresh unless i) a **Reveal** query was made to Π_i^U or its partner (if it has one) or ii) a **Corrupt**(U') query was made before the **Test** query (where U' is any participant) and **Send**(U, i, M) query was made at some point.

PAKE Security. The goal of $\mathcal{A}$ is to guess the bit b used to answer the test query. Let $\mathrm{Succ}_P^{\mathrm{FtG}}(\mathcal{A})$ be the event where $\mathcal{A}$ asks a single *test* query directed to a *fs-fresh* instance and $\mathcal{A}$ outputs his guess b', where $b' = b$. The advantage of $\mathcal{A}$ attacking P is defined as:

$$\mathrm{Adv}_P^{\mathrm{FtG}}(\mathcal{A}) = 2 \cdot \Pr\left[\mathrm{Succ}_P^{\mathrm{FtG}}(\mathcal{A})\right] - 1 \tag{5}$$

---

[8] This is the weak-corruption model of [5].

In the original formulation of the model from [5], we say that protocol P is FtG-secure if there exists a positive constant B such that for every PPT adversary $\mathcal{A}$ it holds that

$$\mathrm{Adv}_P^{\mathrm{FtG}}(\mathcal{A}) \leq \frac{B \cdot n_{se}}{N} + \varepsilon \tag{6}$$

where n_{se} is an upper bond on the number of **Send** queries $\mathcal{A}$ makes, and ε is negligible in the security parameter. Following our discussion in Sect. 1.3, we can modify the definition to allow arbitrary password distribution and login attempt policy; thus, we can define a protocol to be secure if for for every PPT adversary $\mathcal{A}$,

$$\mathrm{Adv}_P^{\mathrm{FtG}}(\mathcal{A}) \leq \mathcal{F}(\mathcal{D}, \mathcal{L}, n_{se}) + \varepsilon \tag{7}$$

where we are using n_{se} as an upper bound on the number of online password guesses the adversary can make.

The following fact can be easily verified using Eq. 5:

Fact 1

$$\Pr[\mathrm{Succ}_P^{\mathrm{FtG}}(\mathcal{A})] = \Pr[\mathrm{Succ}_{P'}^{\mathrm{FtG}}(\mathcal{A})] + \epsilon \Leftrightarrow \mathrm{Adv}_P^{\mathrm{FtG}}(\mathcal{A}) = \mathrm{Adv}_{P'}^{\mathrm{FtG}}(\mathcal{A}) + 2\epsilon. \tag{8}$$

5   Proof of Security

In this section, we prove the security of the PAK protocol instantiated over Gap Diffie-Hellman groups. Due to similarity with the proof of the original PAK protocol [25], we present an overview for those security games that remain the same as in the original protocol and focus on those that deviate from the original proof. In Fig. 2 we provide the description of the game hops and highlight those games which differ from the original security proof. The terminology regarding adversary's actions, partnering and events stays as in [25] (see Appendix A).

$\mathbf{G_0}$: Original protocol.
$\mathbf{G_1}$: Force uniqueness of instances.
$\mathbf{G_2}$: Forbid lucky guesses on hash outputs and backpatch for consistency.
$\mathbf{G_3}$: **Randomize session keys for Execute queries (L-CDH).**
$\mathbf{G_4}$: Check password guesses.
$\mathbf{G_5}$: **Randomize session keys for paired instances (L-CDH).**
$\mathbf{G_6}$: **Forbid two password guesses per online attempt on server (L-CDH).**
$\mathbf{G_7}$: Internal password oracle.

Fig. 2. Description of games for the original PAK.

The main difference between the existing proof in the FtG model and our proof is that our reduction algorithm makes use of a Decisional Diffie-Hellman

Oracle (DDH-O). Such oracle is available in gap groups, and it will output 1 on input (g, g^x, g^y, g^z) if $g^z = DH(g^x, g^y)$ and 0 otherwise. This additional information can be leveraged – in games G_3, G_5 and G_6 – to increase the success probability and reduce the running time of the reduction compared to Theorem 6.9 in [25].

Proof of Theorem 2: We will denote by P_i the protocol executed in game G_i, for i from 0 to 7. Before we start with the revised games, we will first describe in Fig. 3 how the random oracle queries to H_1 are answered by the simulator (reduction). It is important to highlight that the simulator has access to $\psi_1[\pi]$ values (see Appendix B).

Game G_0 : Original protocol. In this game, the challenger runs the original protocol P_0 for the adversary $\mathcal{A}$.

Game G_1 : Force uniqueness of instances. Let G_1 be exactly the same as G_0, except that if any of the values m and μ chosen by honest instances collide with previously generated ones, the protocol aborts and the adversary fails.

The probability of this event happening is negligible in the security parameter and limited by the birthday bound. More precisely, for all adversaries $\mathcal{A}$:

$$\mathrm{Adv}_{P_0}^{\mathrm{FtG}}(\mathcal{A}) \leq \mathrm{Adv}_{P_1}^{\mathrm{FtG}}(\mathcal{A}) + \frac{(n_{se} + n_{ex})(n_{se} + n_{ex} + n_{ro})}{q}. \tag{9}$$

Game G_2 : Forbid lucky guesses on hash outputs and backpatch for consistency. Let G_2 be the same as G_1, with the difference that now the simulator answers **Send** and **Execute** queries without making any random oracle queries, while ensuing random oracle queries are backpatched to ensure consistency in the view of the adversary.

In addition, G_2 forbids lucky guesses on hash functions. Specifically, in G_1 there are cases where an unpaired client instance Π_i^C may accept a confirmation code k, but the adversary has not asked the required random oracle queries to H_1 and H_2 in order to compute k, i.e. he proactively produced the correct one. The probability of this event happening is $\frac{\mathcal{O}(n_{ro} + n_{se})}{q}$. A similar scenario occurs when considering an unpaired server instance. Then:

$$\mathrm{Adv}_{P_1}^{\mathrm{FtG}}(\mathcal{A}) = \mathrm{Adv}_{P_2}^{\mathrm{FtG}}(\mathcal{A}) + \frac{\mathcal{O}(n_{ro} + n_{se})}{q}. \tag{10}$$

Game G_3 : Randomize session keys for Execute queries. Let G_3 be exactly the same as G_2, except that during processing of an $H_l(C, S, m, \mu, \sigma, \gamma')$ query for $l \in \{2, 3, 4\}$, there is no check for a **testexecpw**(C, i, S, j, π_c) event. As a result of this change, even if **testexecpw**(C, i, S, j, π_c) event is triggered, the simulator will answer an $H_l(C, S, m, \mu, \sigma, \gamma')$ query with a random string from $\{0, 1\}^\kappa$.

Claim 1. *For all adversaries $\mathcal{A}$ running in time t, there exists an algorithm $\mathcal{D}$ running in time $t'' = \mathcal{O}(t + (n_{ro} + n_{se} + n_{ex}) \cdot t_{exp} + n_{ro} \cdot t_{ddh})$, such that:*

$$\text{Adv}_{P_2}^{\text{FtG}}(\mathcal{A}) \leq \text{Adv}_{P_3}^{\text{FtG}}(\mathcal{A}) + 2\text{Adv}_{\mathbb{G}_1}^{\text{gap-cdh}}(t''). \tag{11}$$

Proof: Let ϵ be the probability that **testexecpw** occurs in G_2. In that case $Pr(\text{Succ}_{P_2}^{\text{FtG}}(\mathcal{A})) \leq Pr(\text{Succ}_{P_3}^{\text{FtG}}(\mathcal{A})) + \epsilon$. By Fact 1, $\text{Adv}_{P_2}^{\text{FtG}}(\mathcal{A}) \leq \text{Adv}_{P_3}^{\text{FtG}}(\mathcal{A}) + 2\epsilon$. Note that games G_2 and G_3 are indistinguishable if **testexecpw** does not occur.

Now, we will construct an algorithm $\mathcal{D}$ that attempts to win its Gap-DH game against $\mathcal{CH}_{cdh}$ by running $\mathcal{A}$ as a subroutine on a simulation of the protocol P_2. For fixed (X, Y) that are coming from $\mathcal{CH}_{cdh}$, $\mathcal{D}$ simulates G_2 to $\mathcal{A}$ with the following changes:

1. For every **Execute**(C, i, S, j) query, set $m = X \cdot g^{\rho_{iC}}$, $\mu = Y \cdot g^{\rho_{js}}$, where $(\rho_{iC}, \rho_{js}) \xleftarrow{\$} \mathbb{Z}_q^2$, while k, k' stay random strings from $\{0, 1\}^{\kappa}$.
2. Each time $\mathcal{A}$ asks a $H_l(C, S, m, \mu, \sigma, \gamma')$ query for $l \in \{2, 3, 4\}$ – where values m, μ were generated in **Execute**(C, i, S, j) query, and $H_1(\pi_c)$ query returned $(\gamma')^{-1}$ – $\mathcal{D}$ calls DDH-O with input $(m \cdot \gamma', \mu, \sigma)$. Once DDH-O returns 1, the "winning" H_l query is identified, and $\mathcal{D}$ computes Z value as follows

$$Z = \sigma \cdot X^{\rho_{js}} \cdot Y^{\rho_{iC}} \cdot g^{\rho_{iC} \cdot \rho_{js}} \cdot \mu^{\psi_1[\pi_c]}, \tag{12}$$

submits it to $\mathcal{CH}_{cdh}$ as a solution for (X, Y) challenge, and stops. The advantage of $\mathcal{D}$ in solving Gap-DH is equal to ϵ and its running time is $t'' = \mathcal{O}(t + (n_{se} + n_{ex} + n_{ro})t_{exp} + n_{ro}t_{ddh})$. □

DISCUSSION. Notice that the running time of $\mathcal{D}$ in G_3 has slightly increased (by $n_{ro}t_{ddh} - n_{ro}t_{exp}$) when comparing with MacKenzie's reduction, since $c \cdot t_{exp} = t_{ddh}$, where c is some constant. As in [25], $\epsilon' = \epsilon$. However, here only a single Z value is computed and sent, in contrast to the existing reduction where a list of size n_{ro} is submitted to $\mathcal{CH}_{cdh}$.

Game G_4 : Check password guesses. The challenger executes P_3 as in G_3, except that if **correctpw** event occurs, then the protocol execution aborts and the adversary succeeds.

As consequence, before any **Corrupt** query, whenever the simulator detects (via oracle queries) that the adversary uses the correct password to compute the confirmation code k, the protocol will be aborted and the adversary will be deemed successful, i.e., no *unpaired* client or server instance will terminate prior to **correctpw** event or **Corrupt** query.

$$\text{Adv}_{P_3}^{\text{FtG}}(\mathcal{A}) \leq \text{Adv}_{P_4}^{\text{FtG}}(\mathcal{A}). \tag{13}$$

Game G_5 : Randomize session keys for paired instances. G_5 is identical to G_4, except in case **pairedpwguess** event occurs. In that case, the game stops and adversary fails.

In this particular reduction, we will show that an adversary $\mathcal{A}$ who (i) can adaptively corrupt user (thus knowing the password π_c) and (ii) manages to compute sk for *paired* instances Π_i^C and Π_j^S, could be used as a subroutine to solve the Gap-DH problem.

Claim 2. *For any adversary $\mathcal{A}$ running in time t, an algorithm $\mathcal{D}$ running in time $t'' = \mathcal{O}(t + (n_{se} + n_{ro} + n_{exe})t_{exp} + (n_{se} + n_{ro})t_{ddh})$ can be built such that:*

$$\mathrm{Adv}_{P_4}^{\mathrm{FtG}}(\mathcal{A}) \leq \mathrm{Adv}_{P_5}^{\mathrm{FtG}}(\mathcal{A}) + 2\mathrm{Adv}_{\mathbb{G}_1}^{\mathrm{gap\text{-}cdh}}(t''). \tag{14}$$

Proof: If **pairedpwguess** does not occur, then games G_4 and G_5 are indistinguishable. Let ϵ be the probability that **pairedpwguess** event occurs, when $\mathcal{A}$ is running in G_4.

Next, we will construct an algorithm $\mathcal{D}$ that attempts to win its Gap-DH game against $\mathcal{CH}_{cdh}$ by running $\mathcal{A}$ as a subroutine on a simulation of the protocol P_4. For a given pair (X, Y), $\mathcal{D}$ simulates G_4 to $\mathcal{A}$ with the following changes:

1. In CLIENT ACTION 0 query to Π_i^C and input S, set $m = X \cdot g^{\rho_{C,i}}$ where $\rho_{C,i} \xleftarrow{\$} \mathbb{Z}_q$.
2. In SERVER ACTION 1 query to Π_j^S and input $\langle C, m \rangle$, set $\mu = Y \cdot g^{\rho_{S,j}}$, where $\rho_{S,j} \xleftarrow{\$} \mathbb{Z}_q$.
3. In CLIENT ACTION 1 to Π_i^C and input $\langle \mu, k \rangle$, if Π_i^C is unpaired, $\mathcal{D}$ first verifies k using DDH-O and the list of random oracle queries. If k is correctly constructed (DDH-O outputs 1), Π_i^C outputs k' and terminates, or rejects otherwise.
4. In SERVER ACTION 2 query to Π_j^S with input k', if Π_j^S was paired after its SERVER ACTION 1 but is now unpaired, then $\mathcal{D}$ verifies k'. If k' is correctly constructed, then Π_j^S terminates. Otherwise, it rejects.
5. After $\mathcal{A}$ terminates, the simulator selects queries of the form $H_l(C, S, m, \mu, \sigma, \pi)$, for which the following conditions are satisfied: (i) m and μ generated by some instances Π_i^C and Π_j^S respectively, (ii) Π_i^C is paired with Π_j^S and Π_j^S is paired with Π_i^C after SERVER ACTION 1, (iii) $(\gamma')^{-1} = H_1(\pi)$. For every such query, $\mathcal{D}$ calls DDH-O with input $(m \cdot \gamma', \mu, \sigma)$.

Once DDH-O returns 1, $\mathcal{D}$ computes Z value in the same way as for G_3 (Eq. 12), submits it to $\mathcal{CH}_{cdh}$ as a solution for (X, Y) challenge, and stops. The advantage of $\mathcal{D}$ in solving Gap-DH is equal to ϵ and its running time is $t'' = \mathcal{O}(t + (n_{se} + n_{ro} + n_{exe})t_{exp} + (n_{se} + n_{ro})t_{ddh})$. $\square$

DISCUSSION. To explain why the original reduction from [25] contains n_{se} degradation factor, and how we can avoid such degradation in ours, consider the following scenario:

Suppose that the adversary $\mathcal{A}$ against protocol P_4 first makes a CLIENT ACTION 0 query to Π_i^C and receives as an answer $m = X \cdot g^{\rho_{C,i}}$ value in which Diffie-Hellman challenge X is planted. Next, $\mathcal{A}$ obtains $\pi_c = \pi[C, S]$ via **Corrupt**(S) query. With this information, $\mathcal{A}$ may decide to impersonate S to C by making a CLIENT ACTION 1 query with an input $\langle \mu, k \rangle$ to Π_i^C. Since $\mathcal{A}$ knows the correct password, he could compute and send the correct confirmation code k; however, $\mathcal{A}$ could also choose to send an incorrect one. Now, the simulator faces a problem: Π_i^C has to verify k and based on the verification outcome either accept or reject. Put differently, the simulator is unable to verify whether **testpw**$(C, i, S, \pi_c, l = 2)$ is triggered; this could be done by checking if $\sigma = DH(\alpha, \mu)$, but the simulator does not know the discrete log of X.

To circumvent this obstruction, the reduction in [25] has to guess an instance that will be the target of the **Test** query: this provides guarantee that there won't be any corruption before session keys are accepted, and thus the simulator can safely plant the received Diffie-Hellman challenge (X, Y) in the **Test** session. This technique yields a factor of n_{se} in front of $\mathrm{Adv}_{\mathbb{G}}^{1\text{-}cdh}$ advantage in Theorem 1.

In contrast, by using Gap-DH groups, our simulator can query DDH-O with input (α, μ, σ) to verify if $\sigma = DH(\alpha, \mu)$ and check whether the event **testpw**$(C, i, S, \pi_c, l = 2)$ is triggered or not. Hence, we can avoid guessing of the **Test** instance, which makes our reduction tight with respect to the success probability. Compared to [25], the running time of the reduction algorithm has increased by an additive term $(n_{se} + n_{ro})t_{ddh}$, due to the invocation of DDH-O needed for the simulator to identify correct random oracle queries.

Game G_6 : Forbid two password guesses per online attempt on server. Let G_6 be identical to G_5, except that if **doublepwserver** event occurs, the protocol halts and the adversary fails. We assume that the check for **doublepwserver** occurs before the check for **pairedpwguess**.

Claim 3. *For any adversary $\mathcal{A}$ running in time t, there exists an algorithm $\mathcal{D}$ running in time $t'' = \mathcal{O}(t + (n_{se} + n_{ro} + n_{exe})t_{exp} + n_{ro}t_{ddh})$ such that*

$$\mathrm{Adv}_{P_5}^{\mathrm{FtG}}(\mathcal{A}) \leq \mathrm{Adv}_{P_6}^{\mathrm{FtG}}(\mathcal{A}) + 4\mathrm{Adv}_{\mathbb{G}_1}^{\mathrm{gap\text{-}cdh}}(t'') \tag{15}$$

Proof: We will construct an algorithm $\mathcal{D}$ that attempts to win its Gap-DH game against $\mathcal{CH}_{cdh}$ by running $\mathcal{A}$ as a subroutine on a simulation of the protocol P_5. For a given pair (X, Y), $\mathcal{D}$ simulates G_5 to $\mathcal{A}$ with the following changes:

1. In $H_1(\pi)$ query, output $X^{\psi_1[\pi]}g^{\psi_1'[\pi]}$, where $\psi_1[\pi] \xleftarrow{\$} \{0, 1\}$ and $\psi_1'[\pi] \xleftarrow{\$} \mathbb{Z}_q$.
2. In a SERVER ACTION 1 query to a server Π_j^S with input $\langle C, m \rangle$ where $acceptable(m)$ is true, set $\mu = Y \cdot g^{\rho'_{s,j}}$.
3. Tests for **correctpw** and **pairedpwguess**, from G_4 and G_5 respectively, are not made.
4. After $\mathcal{A}$ terminates, the simulator $\mathcal{D}$ using DDH-O first creates a list L_c of $H_l(C, S, m, \mu, \sigma, \gamma')$ queries, with $l \in \{2, 3, 4\}$, such that $\sigma = DH(m \cdot \gamma', \mu)$. Then $\mathcal{D}$ selects from the list L_c two different queries, say $H_l(C, S, m, \mu, \sigma, \gamma')$ and $H_{\hat{l}}(C, S, m, \mu, \hat{\sigma}, \hat{\gamma}')$, for $l, \hat{l} \in \{2, 3, 4\}$ such that there was (i) a SERVER ACTION 1 query to a server instance Π_j^S with input $\langle C, m \rangle$ and output $\langle \mu, k \rangle$, (ii) an $H_1(\pi)$ query that returned $(\gamma')^{-1}$, an $H_1(\hat{\pi})$ query that returned $(\hat{\gamma}')^{-1}$ and (iii) $\psi_1[\pi] \neq \psi_1[\hat{\pi}]$. Then $\mathcal{D}$ outputs:

$$Z = \left(\sigma \cdot \hat{\sigma}^{-1} \cdot (\gamma')^{-\rho'_{s,j}} \cdot (\hat{\gamma}')^{\rho'_{s,j}} \cdot Y^{\psi_1'[\pi] - \psi_1'[\hat{\pi}]} \right)^{\psi_1[\pi] - \psi_1[\hat{\pi}]}, \tag{16}$$

where $Z = DH(X, Y)$.

G_6 is indistinguishable from G_5 until the event **doublepwserver** occurs. Let the ϵ be the probability that **doublepwserver** occurs when $\mathcal{A}$ is running in G_5. When **doublepwserver** occurs for two passwords $\pi \neq \hat{\pi}$, the success probability of $\mathcal{D}$ is $\epsilon/2$ and its running time is $t'' = \mathcal{O}(t + (n_{se} + n_{ro} + n_{exe})t_{exp} + n_{ro}t_{ddh})$. $\square$

DISCUSSION. This game shows that $\mathcal{A}$'s probability of simultaneously guessing (discarding) more than *one* password during a single online attempt on a server executing P_6 is negligible. In most PAKE proofs (in [25] too), this reduction typically brings the highest security degradation: e.g. $1/n_{ro}^3$ appears in the case of Dragonfly [21] and SPEKE [24]. In contrast, our protocol only suffers from a constant loss (4) in the success probability.

The reason for $1/n_{ro}^2$ degradation when using L-CDH in PAK reduction is the following: $\mathcal{D}$ has to compute and output a list of *possible* DH values and he expects the solution for CDH to be contained within the list if $\mathcal{A}$ wins its game. The list is computed as follows: for particular pairs of queries $H_l(C, S, m, \mu, \sigma, \gamma')$ and $H_{\hat{l}}(C, S, m, \mu, \hat{\sigma}, \hat{\gamma}')$, for $l, \hat{l} \in \{2, 3, 4\}$, $\mathcal{D}$ computes Z as in Eq. 16 and adds it to his list of *possible* DH values. The size of the list is upper bounded by $(n_{ro})^2$, resulting in unfeasible running time for $\mathcal{D}$.

In contrast, by using Gap-DH groups, $\mathcal{D}$ can identify the right pair of H_l queries (at the cost of at most $n_{ro}t_{ddh}$ in the running time) and then compute a single, correct Z value using Eq. 16. As a result, we can remove the quadratic factor in the running time of $\mathcal{D}$.

Game G_7 : Internal password oracle. The purpose of this game is to estimate the probability of the **correctpw** event occurring, i.e. the adversary guessing the correct password π_c.

Let G_7 be as G_6, except that there is an *internal password oracle* $\mathcal{O}_{pw}$ which generates all passwords during the initialization of the users. The simulator uses it to i) handle **Corrupt** queries and ii) test whether **correctpw** occurs. More specifically, when $\mathcal{A}$ asks **Corrupt**(U), the query is simply forwarded to $\mathcal{O}_{pw}$ which returns π_U if $U \in Clients$, otherwise returns $\langle \pi_U[C] \rangle_{C \in Clients}$. To determine whether **correctpw** occurs, the simulator queries $\mathcal{O}_{pw}$ with test(π, C), which returns TRUE if $\pi = \pi_C$ and FALSE otherwise.

By definition G_6 and G_7 are perfectly indistinguishable. Then:

$$\mathrm{Adv}_{P_6}^{FtG}(\mathcal{A}) = \mathrm{Adv}_{P_7}^{FtG}. \tag{17}$$

Claim 4. *For all PPT adversaries $\mathcal{A}$:*

$$\mathrm{Adv}_{P_7}^{FtG} \leq \mathcal{F}(\mathcal{D}, \mathcal{L}, n_{se}). \tag{18}$$

Proof: Let ψ denote the **correctpw** event and ψ^c its compliment. The probability that $\mathcal{A}$ succeeds in G_7 is given by:

$$\Pr\left[\operatorname{Succ}_{P_7}^{\operatorname{FtG}}(\mathcal{A})\right] = \Pr\left[\psi\right] \cdot \Pr\left[\operatorname{Succ}_{P_7}^{\operatorname{FtG}}(\mathcal{A}) \mid \psi\right] +$$
$$\Pr\left[\psi^c\right] \cdot \Pr\left[\operatorname{Succ}_{P_7}^{\operatorname{FtG}}(\mathcal{A}) \mid \psi^c\right] \quad (19)$$

We look at the first term of Eq. 19. Since there are n_{se} **Send** queries, the probability of **correctpw** occurring is bounded by $\Pr\left[\psi\right] \leq \mathcal{F}(\mathcal{D}, \mathcal{L}, n_{se})$. Additionally, it follows from G_4 that $\Pr\left[\operatorname{Succ}_{P_7}^{\operatorname{FtG}}(\mathcal{A}) \mid \psi\right] = 1$. Now we look at the second term of Eq. 19. Given that **correctpw** does not occur, $\mathcal{A}$ succeeds by making a **Test** query to a fresh instance Π_i^U and guessing the bit b used in the **Test** query. By examining **Reveal** and H_4 queries throughout the proof, it follows that the view of $\mathcal{A}$ is independent of sk_U^i, therefore $\Pr\left[\operatorname{Succ}_{P_7}^{\operatorname{FtG}}(\mathcal{A}) \mid \psi^c\right] = 1/2$. Putting everything together, using Eqs. 5 and 19:

$$\operatorname{Adv}_{P_7}^{\operatorname{FtG}} \leq \mathcal{F}(\mathcal{D}, \mathcal{L}, n_{se}).$$

$\square$

6   Conclusion

In this paper, we proposed a new instantiation for the PAK protocol and showed that the security proof from [25] can be adapted to cover our proposal. Our reduction to the Gap Diffie-Hellman problem is significantly tighter than the previous reduction to the List Diffie-Hellman problem (see Table 1). From a theoretical point of view, this shows that the security of PAK is closely related to the security of Gap-DH assumption. In terms of concrete security, the advantage of the tighter proof is that it provides the guarantee that with typical values of the group size for today, even the most computationally powerful adversaries today cannot do significantly better than an online dictionary attack. In future work it would be interesting to see if similar techniques could lead to tighter security proofs in other existing PAKE protocols.

Table 1. Comparison of running time and success probability of PAK reduction algorithm when using different variants of CDH assumption. Variable t_{exp} represents the running time to compute exponentiation in $\mathbb{G}$, t_{ddh} the time for deciding DDH, $t_{sim} = (n_{se} + n_{ro} + n_{exe})t_{exp}$, c is a constant, n_{ro} and n_{se} are the number of random oracle and send queries respectively.

Assumption	$t' - t - t_{sim}$	ϵ'/ϵ
Standard CDH	$\mathcal{O}(c \cdot t_{exp})$	$1/n_{ro}^2$
L-CDH	$\mathcal{O}((n_{ro})^2 \cdot t_{exp})$	$1/n_{se}$
Gap-DH	$\mathcal{O}(n_{ro} \cdot t_{ddh})$	1

Acknowledgements. We would like to thank the anonymous referees for their comments. This work was supported by the Luxembourg National Research Fund (CORE project AToMS and CORE Junior grant no. 11299247).

A   Terminology from the Original Proof of PAK

First, we introduce the terminology from [25] that deals with adversary's actions and partnering.

We say "in a CLIENT ACTION κ query to Π_i^C", to refer to "in a Send query to Π_i^C that results in execution of CLIENT ACTION κ procedure" and "in a SERVER ACTION κ query to Π_j^S", to refer to "in a Send query to Π_j^S that results in execution of SERVER ACTION κ procedure". A client instance Π_i^C is paired with a server instance Π_j^S if there is a CLIENT ACTION 0 query to Π_i^C with input S and output $\langle C, m \rangle$, there is a SERVER ACTION 1 query to Π_j^S with input $\langle C, m \rangle$ and output $\langle \mu, k \rangle$ and there is a CLIENT ACTION 1 query to Π_i^C with input $\langle \mu, k \rangle$. A server instance Π_j^S *is paired* with client instance Π_j^C whenever there is a CLIENT ACTION 0 query to Π_i^C with input S and output $\langle C, m \rangle$, there is a SERVER ACTION 1 query to Π_j^S with input $\langle C, m \rangle$ and output $\langle \mu, k \rangle$, and if there is a SERVER ACTION 2 query to Π_j^S with input k', then there was previously a CLIENT ACTION 1 query to Π_i^C with input $\langle \mu, k \rangle$ and output k'.

Next we describe those events taken from [25] which are required in our proof of security.

- **testpw**(C, i, S, π, l): for some m, μ and γ', $\mathcal{A}$ makes (i) an $H_l(C, S, m, \mu, \sigma, \gamma')$ query, (ii) a CLIENT ACTION 0 query to a client instance Π_i^C with input S and output $\langle C, m \rangle$, (iii) a CLIENT ACTION 1 query to Π_i^C with input $\langle \mu, k \rangle$ and (iv) an $H_1(\pi)$ query returning $(\gamma')^{-1}$, where the last query is either the $H_l(\cdot)$ query or the CLIENT ACTION 1 query, $\sigma = DH(\alpha, \mu)$, $m = \alpha \cdot (\gamma')^{-1}$ and $l \in \{2, 3, 4\}$.
- **testpw!**(C, i, S, π): for some k, a CLIENT ACTION 1 query with input $\langle \mu, k \rangle$ causes a testpw$(C, i, S, \pi, 2)$ event to occur, with associated value k.
- **textpw**(S, j, C, π, l): for some m, μ, γ' and k, $\mathcal{A}$ makes an $H_l(C, S, m, \mu, \sigma, \gamma')$ query, and previously made (i) a SERVER ACTION 1 query to a server instance Π_j^S with input $\langle C, m \rangle$ and output $\langle \mu, k \rangle$, and (ii) an $H_1(\pi)$ query returning $(\gamma')^{-1}$, where $\sigma = DH(\alpha, \mu)$, $m = \alpha \cdot (\gamma')^{-1}$ and ACCEPTABLE(m). The associated value of this event is k, k'' or sk_s^j.
- **testpw!**(S, j, C, π): SERVER ACTION 2 query to Π_j^S is made with input k', and previously a testpw$(S, j, C, \pi, 3)$ event occurs with associated value k'.
- **testpw***(S, j, C, π): testpw(S, j, C, π, l) event occurs for some $l \in \{2, 3, 4\}$.
- **testpw**(C, i, S, j, π) : for some $l \in \{2, 3, 4\}$, both a testpw(C, i, S, π, l) and testpw(S, j, C, π, l) event occur, where Π_i^C is *paired* with Π_j^S, and Π_j^S is *paired* with Π_i^C after its SERVER ACTION 1 query.
- **testexecpw**(C, i, S, j, π): for some m, μ and γ', $\mathcal{A}$ makes an $H_l(C, S, m, \mu, \sigma, \gamma')$ query, for $l \in \{2, 3, 4\}$, and previously made (i) an Execute(C, i, S, j) query that generates m, μ, and (ii) an $H_1(\pi)$ query returning $(\gamma')^{-1}$, where $\sigma = DH(\alpha, \mu)$ and $m = \alpha \cdot (\gamma')^{-1}$.
- **correctpw**: before any Corrupt query, either a testpw!(C, i, S, π_C) event occurs for some C,i and S, or a testpw*(S, j, C, π_C) event occurs for some S, j, and C.

- **doublepwserver**: before any Corrupt query, both testpw*(S, j, C, π) event and a testpw*$(S, j, C, \hat{\pi})$, for some S, j, C and $\pi \neq \hat{\pi}$.
- **pairedpwguess**: a testpw(C, i, S, j, π_C) event occurs, for some C, i, S and j.

B   Hash Function Simulation

$\boldsymbol{H_1}$: For each hash query $H_1(\pi)$, if the same query was previously asked, the simulator retrieves the record (π, Φ, ψ_1) from the list $\mathcal{L}_{h1}$ and answers with Φ. Otherwise, the answer Φ is chosen according to the following rule:

★ **Rule H_1**

Choose $\psi_1 \xleftarrow{\$} \mathbb{Z}_q$. Compute $\Phi := g^{\psi_1}$ and write the record (π, Φ, ψ_1) to $\mathcal{L}_{h1}$.

Fig. 3. Simulation of the hash function H_1

References

1. Abdalla, M., Chevassut, O., Pointcheval, D.: One-time verifier-based encrypted key exchange. In: Vaudenay, S. (ed.) PKC 2005. LNCS, vol. 3386, pp. 47–64. Springer, Heidelberg (2005). https://doi.org/10.1007/978-3-540-30580-4_5
2. Abdalla, M., Fouque, P.-A., Pointcheval, D.: Password-based authenticated key exchange in the three-party setting. In: Vaudenay, S. (ed.) PKC 2005. LNCS, vol. 3386, pp. 65–84. Springer, Heidelberg (2005). https://doi.org/10.1007/978-3-540-30580-4_6
3. Abdalla, M., Pointcheval, D.: Simple password-based encrypted key exchange protocols. In: Menezes, A. (ed.) CT-RSA 2005. LNCS, vol. 3376, pp. 191–208. Springer, Heidelberg (2005). https://doi.org/10.1007/978-3-540-30574-3_14
4. Bellare, M.: Practice-oriented provable-security. In: Damgård, I.B. (ed.) EEF School 1998. LNCS, vol. 1561, pp. 1–15. Springer, Heidelberg (1999). https://doi.org/10.1007/3-540-48969-X_1
5. Bellare, M., Pointcheval, D., Rogaway, P.: Authenticated key exchange secure against dictionary attacks. In: Preneel, B. (ed.) EUROCRYPT 2000. LNCS, vol. 1807, pp. 139–155. Springer, Heidelberg (2000). https://doi.org/10.1007/3-540-45539-6_11
6. Bellovin, S.M., Merritt, M.: Encrypted key exchange: password-based protocols secure against dictionary attacks. In: 1992 IEEE Symposium on Research in Security and Privacy, SP 1992, pp. 72–84 (1992)
7. Boneh, D., Lynn, B., Shacham, H.: Short signatures from the weil pairing. In: Boyd, C. (ed.) ASIACRYPT 2001. LNCS, vol. 2248, pp. 514–532. Springer, Heidelberg (2001). https://doi.org/10.1007/3-540-45682-1_30
8. Boyko, V., MacKenzie, P., Patel, S.: Provably secure password-authenticated key exchange using Diffie-Hellman. In: Preneel, B. (ed.) EUROCRYPT 2000. LNCS, vol. 1807, pp. 156–171. Springer, Heidelberg (2000). https://doi.org/10.1007/3-540-45539-6_12

9. Canetti, R., Halevi, S., Katz, J., Lindell, Y., MacKenzie, P.: Universally composable password-based key exchange. In: Cramer, R. (ed.) EUROCRYPT 2005. LNCS, vol. 3494, pp. 404–421. Springer, Heidelberg (2005). https://doi.org/10.1007/11426639_24

10. Dacosta, I., Ahamad, M., Traynor, P.: Trust No One Else: detecting MITM attacks against SSL/TLS without third-parties. In: Foresti, S., Yung, M., Martinelli, F. (eds.) ESORICS 2012. LNCS, vol. 7459, pp. 199–216. Springer, Heidelberg (2012). https://doi.org/10.1007/978-3-642-33167-1_12

11. Ding, J., Alsayigh, S., Lancrenon, J., RV, S., Snook, M.: Provably secure password authenticated key exchange based on RLWE for the post-quantum world. In: Handschuh, H. (ed.) CT-RSA 2017. LNCS, vol. 10159, pp. 183–204. Springer, Cham (2017). https://doi.org/10.1007/978-3-319-52153-4_11

12. Ecrypt, I.: ECRYPT II yearly report on algorithms and keysizes. Technical report, European Network of Excellence in Cryptology II (2012)

13. Galbraith, S.D., Paterson, K.G., Smart, N.P.: Pairings for cryptographers. Discret. Appl. Math. 156(16), 3113–3121 (2008)

14. Hao, F., Ryan, P.: J-PAKE: authenticated key exchange without PKI. Trans. Comput. Sci. 11, 192–206 (2010)

15. Harkins, D.: Simultaneous authentication of equals: a secure, password-based key exchange for mesh networks. In: Proceedings of the 2008 Second International Conference on Sensor Technologies and Applications, SENSORCOMM 2008, pp. 839–844. IEEE Computer Society (2008)

16. Standard Specifications for Password-Based Public Key Cryptographic Techniques: Standard. IEEE Standards Association, Piscataway, NJ, USA (2002)

17. Jablon, D.P.: Strong password-only authenticated key exchange. ACM SIGCOMM Comput. Commun. Rev. 26(5), 5–26 (1996)

18. Joux, A., Nguyen, K.: Deparating decision Diffie-Hellman from computational Diffie-Hellman in cryptographic groups. J. Cryptol. 16(4), 239–247 (2003)

19. Katz, J., Ostrovsky, R., Yung, M.: Efficient password-authenticated key exchange using human-memorable passwords. In: Pfitzmann, B. (ed.) EUROCRYPT 2001. LNCS, vol. 2045, pp. 475–494. Springer, Heidelberg (2001). https://doi.org/10.1007/3-540-44987-6_29

20. Krawczyk, H.: HMQV: a high-performance secure Diffie-Hellman protocol. In: Shoup, V. (ed.) CRYPTO 2005. LNCS, vol. 3621, pp. 546–566. Springer, Heidelberg (2005). https://doi.org/10.1007/11535218_33

21. Lancrenon, J., Škrobot, M.: On the provable security of the dragonfly protocol. In: Lopez, J., Mitchell, C.J. (eds.) ISC 2015. LNCS, vol. 9290, pp. 244–261. Springer, Cham (2015). https://doi.org/10.1007/978-3-319-23318-5_14

22. Lenstra, A.K.: Key lengths. Technical report, Wiley (2006)

23. Lopez Becerra, J.M., Iovino, V., Ostrev, D., Škrobot, M.: On the relation between SIM and IND-RoR security models for PAKEs. In: SECRYPT 2017. SCITEPRESS (2017)

24. MacKenzie, P.: On the security of the speke password authenticated key exchange protocol. Cryptology ePrint Archive, Report 2001/057 (2001). http://eprint.iacr.org/2001/057

25. MacKenzie, P.: The PAK suite: protocols for password-authenticated key exchange. DIMACS Technical report 2002-46 (2002)

26. MacKenzie, P.: More efficient password-authenticated key exchange. In: Naccache, D. (ed.) CT-RSA 2001. LNCS, vol. 2020, pp. 361–377. Springer, Heidelberg (2001). https://doi.org/10.1007/3-540-45353-9_27

27. MacKenzie, P.: Methods and apparatus for providing efficient password authenticated key exchange (2002). Publication number US20020194478 A1. https://www.google.com/patents/US20020194478
28. Mrabet, N.E., Joye, M.: Guide to Pairing-Based Cryptography. Chapman & Hall/CRC, Boca Raton (2016)
29. Okamoto, T., Pointcheval, D.: The gap-problems: a new class of problems for the security of cryptographic schemes. In: Kim, K. (ed.) PKC 2001. LNCS, vol. 1992, pp. 104–118. Springer, Heidelberg (2001). https://doi.org/10.1007/3-540-44586-2_8
30. Silverman, J.H.: The Arithmetic of Elliptic Curves. GTM, vol. 106. Springer, New York (2009). https://doi.org/10.1007/978-0-387-09494-6
31. Stevens, M., Bursztein, E., Karpman, P., Albertini, A., Markov, Y.: The first collision for full SHA-1. IACR Cryptology ePrint Archive 2017, 190 (2017). http://eprint.iacr.org/2017/190
32. Thread-Group: Thread Protocol (2015). http://threadgroup.org/
33. Wang, D., Cheng, H., Wang, P., Huang, X., Jian, G.: Zipf's law in passwords. IEEE Trans. Inf. Forensics Secur. **12**, 2776–2791 (2017)
34. Wang, D., Wang, P.: On the implications of Zipf's law in passwords. In: Askoxylakis, I., Ioannidis, S., Katsikas, S., Meadows, C. (eds.) ESORICS 2016. LNCS, vol. 9878, pp. 111–131. Springer, Cham (2016). https://doi.org/10.1007/978-3-319-45744-4_6
35. Warner, B.: Magic Wormhole (2016). https://github.com/warner/magic-wormhole
36. Wu, T.D.: The secure remote password protocol. In: Proceedings of the Network and Distributed System Security Symposium, NDSS 1998. The Internet Society (1998)

Processing Encrypted Data

On the Security of Frequency-Hiding Order-Preserving Encryption

Matteo Maffei[1], Manuel Reinert[2]([⊠]), and Dominique Schröder[3]

[1] TU Wien, Wien, Austria
matteo.maffei@tuwien.ac.at
[2] CISPA, Saarland University, Saarbrücken, Germany
reinert@cs.uni-saarland.de
[3] Friedrich-Alexander Universität Erlangen-Nürnberg, Nürnberg, Germany
dominique.schroeder@fau.de

Abstract. Order-preserving encryption (OPE) is an encryption scheme with the property that the ordering of the plaintexts carry over to the ciphertexts. This primitive is particularly useful in the setting of encrypted databases because it enables efficient range queries over encrypted data. Given its practicality and usefulness in the design of databases on encrypted data, OPE's popularity is growing. Unfortunately, nearly all computationally efficient OPE constructions are vulnerable against ciphertext frequency-leakage, which allows for inferring the underlying plaintext frequency. To overcome this weakness, Kerschbaum recently proposed a security model, designed a frequency-hiding OPE scheme, and analyzed its security in the programmable random oracle model (CCS 2015).

In this work, we demonstrate that Kerschbaum's definition is imprecise and using its natural interpretation, we describe an attack against his scheme. We generalize our attack and show that his definition is, in fact, not satisfiable. The basic idea of our impossibility result is to show that any scheme satisfying his security notion is also IND-CPA-secure, which contradicts the very nature of OPE. As a consequence, no such scheme can exist. To complete the picture, we rule out the imprecision in the security definition and show that a slight adaption of Kerschbaum's tree-based scheme fulfills it.

1 Introduction

Outsourcing databases is common practice in today's businesses. The reasons for that are manifold, varying from the sharing of data among different offices of the same company to saving on know-how and costs that would be necessary to maintain such systems locally. Outsourcing information, however, raises privacy concerns with respect to the service provider hosting the data. A first step towards a privacy-preserving solution is to outsource encrypted data and to let the database application operate on ciphertexts. However, simply encrypting all entries does in general not work because several standard queries on the database

© Springer Nature Switzerland AG 2018
S. Capkun and S. S. M. Chow (Eds.): CANS 2017, LNCS 11261, pp. 51–70, 2018.
https://doi.org/10.1007/978-3-030-02641-7_3

do no longer work. To maintain as much functionality of the database as possible while adding confidentiality properties, researchers weakened the security properties of encryption schemes to find a useful middle ground. Examples include encryption schemes that support plaintext equality checks, or order-preserving encryption. In this work, we re-visit the recent work on frequency-hiding order preserving encryption by Kerschbaum [11] (CCS 2015).

Background and Related Work. Order-preserving encryption (OPE) [3,20] is arguably the most popular building block for databases on encrypted data, since it allows for inferring the order of plaintexts by just looking at the respective ciphertexts. More precisely, for any two plaintexts p_1 and p_2, whenever $p_1 < p_2$, we have that $E(p_1) < E(p_2)$. Hence, OPE allows for efficient range queries and keyword search on the encrypted data. The popularity of this scheme is vouched for by plenty of industrial products (e.g., Ciphercloud[1], Perspecsys[2], and Skyhigh Networks[3]) and research that investigates OPE usage in different scenarios [1,2,10,13,17,18]. Despite the growing popularity and usage in practice, OPE security is debatable. The ideal security notion for OPE is called *indistinguishability against ordered chosen plaintext attacks* (IND-OCPA), which intuitively says that two equally ordered plaintext sequences should be indistinguishable under encryption. Boldyreva *et al.* [3] show that stateless OPE cannot achieve IND-OCPA, unless the ciphertext size is exponential in the plaintext size. Consequently, either one has to relax the security notion or to keep a state.

The former approach has been explored in the context of classical OPE [3, 4,21] as well as a slightly different notion called *order-revealing encryption* (ORE) [5,6,14,19]. ORE is more general than OPE in the sense that comparison on the ciphertexts can happen by computing a comparison function different from "<". Either way, those schemes do not achieve IND-OCPA but target different, weaker security notions, which allow them to quantify the leakage incurred by a scheme or to restrict the attacker's capabilities. For instance, the scheme by Boldyreva *et al.* [3] is known to leak about the first half of the plaintexts and the scheme by Chenette *et al.* [6] leaks the first bit where two encrypted target plaintexts differ. To date, there exist several works that exploit this extra leakage in order to break OPE applied to different data sets such as medical data and census data [7–9,15]. For instance, using a technique based on bipartite graphs, Grubbs *et al.* [9] have recently shown how to break the schemes of Boldyreva *et al.* [3,4], thereby achieving recovery rates of up to 98%. As opposed to earlier work, this technique works even for large plaintext domains such as first names, last names, and even zip codes.

With regards to the latter approach based on stateful OPE schemes, Popa *et al.* [16] introduced a client-server architecture, where the client encrypts plaintexts using a deterministic encryption scheme and maintains a search tree on the server into which it inserts the ciphertexts. The server exploits the search tree when computing queries on encrypted data. This approach requires a significant

---

[1] http://www.ciphercloud.com/.
[2] http://perspecsys.com/.
[3] https://www.skyhighnetworks.com/.

amount of communication between the client and the server both for encryption and queries. Similarly, but rather reversed, Kerschbaum and Schroepfer [12] present an OPE scheme where the client stores a search tree that maps plaintexts to ciphertexts. The ciphertexts are chosen such that ordering is preserved and then inserted along with the plaintexts in the search tree. The server only learns the ciphertexts. This approach has less communication between client and server but requires the client to keep a state that is linear in the number of encrypted plaintexts. Both of these schemes are provably IND-OCPA-secure.

Even though these schemes achieve the ideal IND-OCPA security notion, Kerschbaum [11] raises general doubts about the security definition of OPE. Aside the leakage that is introduced by many schemes on top of the order information (e.g., leaking half of the plaintext [3] or the first bit where two plaintexts differ [6]), one central problem of OPE is the leakage of the plaintext frequency. It is easy to distinguish the encryption of data collections in which elements occur with different frequencies. For instance, the encryption of the sequences $1, 2, 3, 4$ and $1, 1, 2, 2$ are not necessarily indistinguishable according to the IND-OCPA security definition.

In order to solve the frequency-leakage problem, Kerschbaum has recently strengthened the IND-OCPA definition of OPE so as to further hide the frequency of plaintexts under the encryption, thus making the encryptions of the above two sequences indistinguishable [11] (CCS 2015). To this end, Kerschbaum introduces the notion of *randomized order*, which is a permutation of the sequence $1, \ldots, n$ where n is the length of the challenge plaintext sequence. Such a permutation is called randomized order if, when applied to a plaintext sequence, the resulting plaintext sequence is ordered with respect to "$\leq$". The original IND-OCPA security definition requires that the two challenge plaintext sequences agree on all such common randomized orders, which implies that every pair of corresponding plaintexts in the two sequences occurs with the same frequency. For instance, this does not hold for the above two sequences $1, 2, 3, 4$ and $1, 1, 2, 2$, since the former can only be ordered using the permutation $(1, 2, 3, 4)$ while the latter can be ordered by any of $(1, 2, 3, 4)$, $(1, 2, 4, 3)$, $(2, 1, 3, 4)$, or $(2, 1, 4, 3)$. Kerschbaum's insight to make the definition frequency-hiding is that the existence of one common randomized order should be sufficient in order not to be able to distinguish them. For instance, the above sequences both share the randomized order $(1, 2, 3, 4)$ and should thus be indistinguishable when encrypted. This intuition is captured by the security notion of *indistinguishability against frequency-analyzing ordered chosen plaintext attacks* (IND-FA-OCPA). Besides devising a novel definition, Kerschbaum also presents a cryptographic instantiation of an OPE scheme and analyzes its security with respect to the new definition in the programmable random oracle model.

Despite the seeming improvement added by Kerschbaum's scheme, Grubbs *et al.* [9] show that using auxiliary information, such as the plaintext distribution that is likely to underlie a certain ciphertext collection, this scheme can be broken with significant recovery rates. In contrast to the practical attacks in [9], our work targets the purely theoretic side of frequency-hiding OPE and we do not consider having auxiliary information at disposal.

Our Contributions. In this work, we present both negative and positive results for frequency-hiding order-preserving encryption. On the negative side, we observe that the original definition of IND-FA-OCPA is imprecise [11], which leaves room for interpretation. In particular, the security proof for the scheme presented in [11] seems to suggest that the game challenger chooses a randomized order according to which one of the challenge sequences is encrypted. This fact, however, is not reflected in the definition. Hence, according to a natural interpretation of the definition, we show that it is, in fact, not achievable. We develop this impossibility result for the natural interpretation of IND-FA-OCPA step by step. Investigating on Kerschbaum's frequency-hiding scheme [11], we show that it can actually be attacked–without using auxiliary information as done by Grubbs *et al.* [9]–allowing an adversary to win the IND-FA-OCPA game with very high probability. We further observe that this concrete attack can be generalized into a result that allows us to precisely quantify an attacker's advantage in winning the security game for two arbitrary plaintext sequences that adhere to the security game restrictions. Since Kerschbaum provides formal security claims for his construction [11], we identify where the security proof is incorrect. All these considerations on the concrete scheme finally lead to our main negative result: IND-FA-OCPA security is impossible to achieve or, more precisely, any IND-FA-OCPA secure OPE scheme is also secure with respect to IND-CPA, which clearly contradicts the very functionality of OPE. Hence, such an OPE scheme cannot exist.

As mentioned above, the impossibility of IND-FA-OCPA is bound to an imprecision in the definition in [11], which is only presented informally and lacks necessary information to make it achievable. We hence clarify those imprecisions. The underlying problem of the original definition lies in the capability of the game challenger, which, when reading the definition naturally, is very restricted. The challenger has, for instance, no means to ensure that the encryption algorithm chooses a common randomized order of the two challenge plaintext sequences. To remedy those shortcomings, we devise a more formal definition that removes the consisting imprecisions and makes it possible to devise a frequency-hiding OPE scheme. In particular, we first augment the OPE model, allowing for specifying a concrete ordering when encrypting plaintexts, e.g., to concretely say that the sequence $1, 1, 2, 2$ should be encrypted sticking to the randomized order $(1, 2, 4, 3)$. Secondly, we show that an extension of Kerschbaum's scheme [11], adapted to the new model, is provably secure with respect to the correct definition.

To summarize, our contributions are as follows.

– We show that the original definition of IND-FA-OCPA is imprecise. We then demonstrate that the frequency-hiding OPE scheme of [11] is insecure under a natural interpretation of IND-FA-OCPA. We further generalize the attack, which allows us to rigorously quantify the success probability of an attacker for two arbitrary plaintext sequences. To conclude on the concrete scheme, we identify and explain the problem in the security proof.

- Going one step beyond the concrete scheme, we prove a general impossibility result showing that IND-FA-OCPA cannot be achieved by any OPE scheme.
- We clarify the imprecise points in the original security definition and provide a corrected version called IND-FA-OCPA*.
- To define IND-FA-OCPA* in the first place, we have to augment the OPE model, adding a concrete random order as input to the encryption function.
- Finally, we prove that an extension of [11] fulfills our new definition.

Overall, we believe this work yields a solid foundation for order-preserving encryption, showing that a state-of-the-art security definition is impossible to realize along with an attack on a previously published scheme, and presenting an achievable definition and a concrete realization.

Outline. The rest of the paper is structured as follows. We recall the OPE model and its security definitions in Sect. 2. In Sect. 3 we describe the relevant parts of Kerschbaum's scheme [11]. We present our attack, its generalization, and the problem in the security proof in Sect. 4. Section 5 proves the impossibility result. In Sect. 6 we present the augmented OPE model and the definition of IND-FA-OCPA*. We show that an adaption of [11] to the new model achieves IND-FA-OCPA* in Sect. 7. Finally, we conclude this work in Sect. 8.

2   Order-Preserving Encryption

In this section, we briefly review the formal definitions of order-preserving encryption, originally proposed in [20], following the definition adopted in [11].

Definition 1 ((Order-Preserving) Encryption). *An* (order-preserving) *encryption scheme* $\mathcal{OPE} = (\mathsf{K}, \mathsf{E}, \mathsf{D})$ *is a tuple of* PPT *algorithms where* $S \leftarrow \mathsf{K}(\kappa)$. *The* key generation *algorithm takes as input a security parameter* κ *and outputs a secret key (or state)* S.
$(S', y) \leftarrow \mathsf{E}(S, x)$. *The* encryption *algorithm takes as input a secret key* S *and a message* x. *It outputs a new key* S' *and a ciphertext* y;
$x \leftarrow \mathsf{D}(S, y)$. *The* decryption *algorithm takes as input a secret key* S *and a ciphertext* y *and outputs a message* x.

An OPE scheme is *complete* if for all S, S', x, and y we have that if $(S', y) \leftarrow \mathsf{E}(S, x)$, then $x \leftarrow \mathsf{D}(S', y)$.

The next definition formalizes the property of order preservation for an encryption scheme. Roughly speaking, this property says that the ordering on the plaintext space carries over to the ciphertext space.

Definition 2 (Order-Preserving). *An encryption scheme* $\mathcal{OPE} = (\mathsf{K}, \mathsf{E}, \mathsf{D})$ *is* order-preserving *if for any two ciphertexts* y_1 *and* y_2 *with corresponding messages* x_1 *and* x_2 *we have that whenever* $y_1 < y_2$ *then also* $x_1 < x_2$.

This general definition allows for modeling both stateful as well as stateless versions of OPE. We focus on the stateful variant in this paper, hence, the *key* S defined above is actually the state. The definition, moreover, does not specify where the state has to reside, allowing us to model client-server architectures.

2.1 Security Definitions

Indistinguishability Against Ordered Chosen Plaintext Attacks. The standard security definition for order-preserving encryption is indistinguishability against ordered chosen plaintext attacks (IND-OCPA) [3]. Intuitively, an OPE scheme is secure with respect to this definition if for any two equally ordered plaintext sequences, no adversary can tell apart their corresponding ciphertext sequences. IND-OCPA is fulfilled by several schemes (e.g., [12,16]). We recall the selective version of the definition in the following.

Definition 3 (IND-OCPA). *An order-preserving encryption scheme* $\mathcal{OPE} =$ (K, E, D) *has* indistinguishable *ciphertexts under ordered chosen plaintext attacks (IND-OCPA) if for any* PPT *adversary* $\mathcal{A}$, *the following probability is negligible in the security parameter* κ:

$$\left| \Pr[\mathsf{Exp}^{\mathcal{A}}_{\mathsf{OCPA}}(\kappa, 1) = 1] - \Pr[\mathsf{Exp}^{\mathcal{A}}_{\mathsf{OCPA}}(\kappa, 0) = 1] \right|$$

where $\mathsf{Exp}^{\mathcal{A}}_{\mathsf{OCPA}}(\kappa, b)$ *is the following experiment:*

Experiment $\mathsf{Exp}^{\mathcal{A}}_{\mathsf{OCPA}}(\kappa, b)$
  $(X_0, X_1) \leftarrow \mathcal{A}$ *where* $|X_0| = |X_1| = n$ *and*
    $\forall 1 \leq i, j \leq n.\ x_{0,i} < x_{0,j} \iff x_{1,i} < x_{1,j}$
  $S_0 \leftarrow \mathsf{K}(\kappa)$
  For all $1 \leq i \leq n$ *run* $(S_i, y_{b,i}) \leftarrow \mathsf{E}(S_{i-1}, x_{b,i})$
  $b' \leftarrow \mathcal{A}(y_{b,1}, \ldots, y_{b,n})$
  Output 1 if and only if $b = b'$.

Definition 3 requires that the challenge plaintext sequences are ordered exactly the same, which in particular implies that the plaintext frequency must be the same.

Indistinguishability Under Frequency-Analyzing Ordered Chosen Plaintext Attacks. A drawback of the previous definition is that it can be achieved by schemes that leak the plaintext frequency, although any two sequences in which plaintexts occur with different frequencies, e.g., $1, 2, 3, 4$ and $1, 1, 1, 1$, are trivially distinguishable by the attacker. In order to target even such sequences, Kerschbaum [11] proposes a different security definition: instead of requiring the sequences to have exactly the same order, it is sufficient for them to have a common *randomized* order. For a plaintext list X of length n, a randomized order is a permutation of the plaintext indices $1, \ldots, n$ which are ordered according to a sorted version of X. This is best explained by an example: consider the plaintext sequence $X = 1, 5, 3, 8, 3, 8$. A randomized order thereof can be any of $\Gamma_1 = (1,4,2,5,3,6)$, $\Gamma_2 = (1,4,3,5,2,6)$, $\Gamma_3 = (1,4,2,6,3,5)$, or $\Gamma_4 = (1,4,3,6,2,5)$, because the order of 3 and 3 as well as the order of 8 and 8 does not matter in a sorted version of X. Formally, a randomized order is defined as follows.

Definition 4 (Randomized order). *Let n be the number of not necessarily distinct plaintexts in sequence $X = x_1, x_2, \ldots, x_n$ where $x_i \in \mathbb{N}$ for all i. For a randomized order $\Gamma = \gamma_1, \gamma_2, \ldots, \gamma_n$, where $1 \leq \gamma_i \leq n$ and $i \neq j \implies \gamma_i \neq \gamma_j$ for all i, j, of sequence X it holds that*

$$\forall i, j. \ (x_i > x_j \implies \gamma_i > \gamma_j) \ \wedge \ (\gamma_i > \gamma_j \implies x_i \geq x_j)$$

Using this definition, Kerschbaum [11] defines security of OPE against *frequency-analyzing ordered chosen plaintext attacks*. Since the definition is informal in [11], we report the natural way to read the definition.

Definition 5 (IND-FA-OCPA). *An order-preserving encryption scheme $\mathcal{OPE} = (\mathsf{K}, \mathsf{E}, \mathsf{D})$ has* indistinguishable ciphertexts under frequency-analyzing ordered chosen plaintext attacks *(IND-FA-OCPA) if for any* PPT *adversary $\mathcal{A}$, the following probability is negligible in the security parameter κ:*

$$\left| \Pr[\mathsf{Exp}^{\mathcal{A}}_{\mathsf{FA-OCPA}}(\kappa, 1) = 1] - \Pr[\mathsf{Exp}^{\mathcal{A}}_{\mathsf{FA-OCPA}}(\kappa, 0) = 1] \right|$$

where $\mathsf{Exp}^{\mathcal{A}}_{\mathsf{FA-OCPA}}(\kappa, b)$ is the following experiment:

Experiment $\mathsf{Exp}^{\mathcal{A}}_{\mathsf{FA-OCPA}}(\kappa, b)$
   $(X_0, X_1) \leftarrow \mathcal{A}$ *where* $|X_0| = |X_1| = n$ *and* X_0 *and* X_1
      have at least one common randomized order Γ
   $S_0 \leftarrow \mathsf{K}(\kappa)$
   For all $1 \leq i \leq n$ *run* $(S_i, y_{b,i}) \leftarrow \mathsf{E}(S_{i-1}, x_{b,i})$
   $b' \leftarrow \mathcal{A}(y_{b,1}, \ldots, y_{b,n})$
   Output 1 if and only if $b = b'$

It is clear that while the standard IND-OCPA definition could be achieved, in principle, by a deterministic encryption scheme, the frequency-hiding variant can only be achieved by using randomized ciphertexts since otherwise frequencies are trivially leaked.

Discussion. Comparing the two definitions, we observe that IND-FA-OCPA is a generalization of IND-OCPA since the constraint on the sequences X_0 and X_1 allows for a greater class of instances. In order to see that, we have to consider the constraint, which is

$$\forall 1 \leq i, j \leq n. \ x_{0,i} < x_{0,j} \iff x_{1,i} < x_{1,j}.$$

This constraint is an alternative way of saying that X_0 and X_1 should agree on all randomized orders. Hence, duplicate plaintexts may occur in any of the sequences, but they should occur symmetrically in the other sequence as well.

3   Kerschbaum's Construction

We review the OPE scheme of [11]. At a high level, encryption works by inserting plaintexts into a binary search tree that stores duplicates as often as they occur. When an element arrives at its designated node, a ciphertext is selected according to this position.

More formally, let T be a binary tree. We denote by ρ the root of T. For a node $t \in T$ we write $t.m$ to denote the message stored at t and $t.c$ to denote the respective ciphertext. We further use $t.left$ and $t.right$ to denote the left and right child of t, respectively. There are several other parameters: N is the number of distinct plaintexts, n is the number of plaintexts in the sequence that is to be encrypted, $k = \log(N)$ is the required number of bits necessary to represent a plaintext in a node, $\ell = k\kappa$ is this number expanded by a factor of κ and refers to the size of the ciphertexts. Finally, the construction requires a source of randomness, which is called in terms of a function RandomCoin() (hereafter called RC() for brevity). According to Kerschbaum, this function can be implemented as a PRF that samples uniformly random bits.

We refer in the following to the client as the one storing the binary tree. This is well motivated in the cloud setting where the client outsources encrypted data to the cloud server who may not have access to the actual message-ciphertext mapping. One may wonder why a client that anyway has to store a mapping of plaintexts to ciphertexts cannot simply store the data itself: Kerschbaum also presents an efficient compression technique for the tree which in some cases can lead to compression ratios of 15.

Implementation of $S \leftarrow \mathsf{K}(\kappa)$. The client sets up an empty tree T. The state S consists of the tree T as well as all parameters k, ℓ, n, and N. Furthermore, S contains the minimum ciphertext $min = -1$ and the maximum ciphertext $max = 2^{\kappa \log(n)}$. These minimum and maximum numbers are only necessary to write the encryption procedure in a recursive way. Usually, n is not known upfront, so it has to be estimated. If the estimation is too far from reality, the tree has to be freshly setup with new parameters.

Algorithm 1. $\mathsf{E}(S, x)$ where $S = t, min, max$

1: **if** $t = \bot$ **then**	10: **if** $x = t.m$ **then**
2: $t.m = x$	11: $b \leftarrow \mathsf{RC}()$
3: $t.c = min + \lfloor \frac{max - min}{2} \rfloor$	12: **end if**
4: **if** $t.c = 0$ **then**	13: **if** $b = 1 \vee x > t.m$ **then**
5: rebalance the tree	14: $\mathsf{E}(t.right, t.c + 1, max, x)$
6: **end if**	15: **else**
7: **return** $t.c$	16: **if** $b = 0 \vee x < t.m$ **then**
8: **end if**	17: $\mathsf{E}(t.left, min, t.c - 1, x)$
9: $b \leftarrow -1$	18: **end if**
	19: **end if**

Implementation of $(S', y) \leftarrow \mathsf{E}(S, x)$. To encrypt a plaintext x, the client proceeds as follows. Whenever the current position in the tree is empty (especially in the beginning when the tree is empty), the client creates a new tree node and inserts x as the plaintext (lines 1.1–1.8). The ciphertext is computed as the mean value of the interval from min to max (line 1.3). In particular, the first ciphertext will be $2^{\kappa \log(n)-1}$. Whenever there is no ciphertext available (line 1.4), the estimation of n has shown to be wrong and the tree has to be rebalanced. We do not detail this step here since it is not important for our attack; instead we refer the interested reader to [11] for a detailed explanation. If instead, the current position in the tree is already occupied and the message is different from x, then we either recurse left (line 1.17) or right (line 1.14) depending on the relation between the occupying plaintext and the one to be inserted. The same happens in case x is equal to the stored message, but then we use the RC procedure to decide where to recurse (lines 1.9–1.12).

Implementation of $x \leftarrow \mathsf{D}(S, y)$. To decrypt a given ciphertext y, we treat the tree as a binary search tree where the key is $t.c$ and search for y. We return $t.m$ as soon as we reach a node t where $t.c = y$.

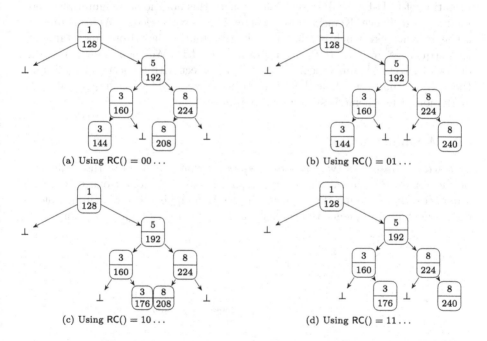

Fig. 1. An example for different binary search trees after inserting the sequence $X = 1, 5, 3, 8, 3, 8$, depending on the output of RC.

Example 1. To simplify the access to the construction, we review a detailed example. Figure 1 shows the four possible resulting binary search trees after inserting $X = 1, 5, 3, 8, 3, 8$, depending on the output of RC. We use a ciphertext space of $\{1, \ldots, 256\}$. Each different output of RC corresponds to one of the four possible randomized orders Γ_i for $1 \leq i \leq 4$.

Security. The scheme is proven secure against frequency-analyzing ordered chosen plaintext attack. To this end, [11] constructs a simulator which, given the two challenge plaintext sequences, produces identical views independent of which of two sequences is chosen. We investigate on the proof in the next section.

4   An Attack on Kerschbaum's FH-OPE Scheme

In this section, we investigate on the security achieved by Kerschbaum's construction [11]. In order to start, we observe that Kerschbaum proves his construction secure. However, as we show later in this section, the security proof makes an extra assumption on the game challenger's capabilities, namely that the challenger can dictate the randomized order used by the encryption algorithm to encrypt either challenge plaintext sequence. Using the natural interpretation of IND-FA-OCPA (see Definition 5), this additional assumption is not justified and, hence, Kerschbaum's scheme is no longer secure. We thus present a concrete attack, which is related to the distribution based on which randomized sequences are chosen for encryption (Sect. 4.1). We then explain more in detail why Kerschbaum's security result is incorrect with respect to Definition 5 (Sect. 4.2). Finally, we show that even if randomized orders are chosen uniformly at random, the scheme is still vulnerable (Sect. 4.3).

4.1   A Simple Attack

Our attack consists of two plaintext sequences that are given to the challenger of the FA-OCPA game, who encrypts step-by-step randomly one of the two sequences. By observing the sequence of resulting ciphertexts, we will be able to determine which sequence the challenger chose with very high probability.

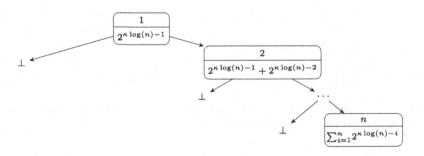

Fig. 2. The resulting binary search tree when encrypting sequence X_0.

Consider the two plaintext sequences $X_0 = 1, 2, 3, \ldots, n$ and $X_1 = 1, \ldots, 1$ such that $|X_0| = |X_1| = n$. Clearly, both X_0 and X_1 have a common randomized order, namely $\Gamma = 1, 2, 3, \ldots, n$, i.e., the identity function applied to X_0. Moreover, consider the binary search tree produced by the scheme when encrypting X_0 in Fig. 2. This tree is generated fully deterministically since the elements in X_0 are pairwise distinct, or equivalently, X_0 has only a single randomized order. Now let us investigate how X_1 would be encrypted by Algorithm 1. In every step, the RC procedure has to be called in order to decide where to insert the incoming element. If coins are drawn uniformly at random then only with a probability of $1/2^{n(n-1)/2}$ RC will produce the bit sequence $1 \ldots 1$ of length $n(n-1)/2$, which is required in order to produce the exact same tree as in Fig. 2. Notice that the RC sequence must be of length $n(n-1)/2$ since for every plaintext that is inserted on the right, the function has to be called once more. Hence, the Gaussian sum $\sum_{i=1}^{n-1} i$ describes the number of required bits. Consequently, an adversary challenging the FH-OCPA challenger with X_0 and X_1 will win the game with probability $(1/2)(1 - 1/2^{n(n-1)/2})$ where the factor $1/2$ accounts for the probability that the challenger chooses X_1. Hence, if X_1 is chosen, our attacker wins with overwhelming probability, otherwise he has to guess. Notice that the combined probability is nevertheless non-negligible.

In conclusion, the central observation is that the number of calls to RC strongly depends on how many equal elements are met on the way to the final destination when encrypting an element. Therefore, not all ciphertext trees are equally likely.

4.2   Understanding the Problem

In this section, we analyze the core of the problem.

Artifacts of the Construction. The analysis in the previous section shows that randomized orders are not drawn uniformly random. Otherwise, the adversary's success probability would be $(1/2)(1-1/n!)$ since X_1 has $n!$ many randomized orders and the probability that a specific one is chosen uniformly is $1/n!$. Instead, we analyzed the probability that the *encryption algorithm* chooses that specific randomized order, which depends on the number of calls to RC and its results, which should all be 1.

In order to exemplify this artifact, we consider the sequence $1, 1, 1$. We depict the different trees when encrypting the sequence in Fig. 3. As we can see, different trees require a different number of calls to RC, and have thus a different probability of being the encryption tree. On the one hand, the trees in Fig. 3a and Fig. 3c–e all have a probability of $1/8$ to be the result of the encryption since each of them requires RC to be called three times. On the other hand, the two trees in Fig. 3b have a probability of $1/4$ of being chosen each since RC has to be called only twice.

To formally capture the probability range of different randomized orders, we want to understand which randomized orders are most probable and which ones are least probable. Before we start the analysis, we observe that it does not matter whether we consider the probability or the number of calls to RC, since every call to RC adds a factor of 1/2 to the probability that a certain randomized order is chosen. So as we have seen in the concrete counter-example in the previous section and the example above, a tree with linear depth represents the least likely randomized order since it requires the most calls to RC, which increases by one every time a new element is encrypted. Conversely, randomized orders represented by a perfectly balanced binary tree are more likely since they require the minimum number of calls to RC. Let H be the histogram of plaintext occurrences in a sequence. Then, as before, the number of calls to RC can be computed as the sum over every node's depth in the subtree in which all duplicate elements reside, which is at least

$$\sum_{p \in X} \frac{\sum_{i=1}^{H(p)} \log(i)}{H(p)} = \sum_{p \in X} \frac{\log(H(p)!)}{H(p)} \geq \sum_{p \in X} \frac{\frac{H(p)}{2} \log\left(\frac{H(p)}{2}\right)}{H(p)}$$

$$= -\frac{|X|}{2} + \frac{1}{2} \sum_{p \in X} \log(H(p))$$

where we make use of the fact that $\left(\frac{n}{2}\right)^{\frac{n}{2}} \leq n! \leq n^n$ in the first inequality. All other randomized orders lie probability-wise somewhere in between.

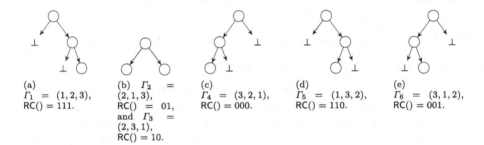

(a) $\Gamma_1 = (1, 2, 3)$, RC() = 111.

(b) $\Gamma_2 = (2, 1, 3)$, RC() = 01, and $\Gamma_3 = (2, 3, 1)$, RC() = 10.

(c) $\Gamma_4 = (3, 2, 1)$, RC() = 000.

(d) $\Gamma_5 = (1, 3, 2)$, RC() = 110.

(e) $\Gamma_6 = (3, 1, 2)$, RC() = 001.

Fig. 3. The trees displaying different randomized orders for the sequence $1, 1, 1$.

Proof Technique. Despite our counter-example, the scheme is proven secure in the programmable random oracle model [11]; this obviously constitutes a contradiction. To understand the problem in depth, we have to have a closer look at the security proof. The idea behind the proof is as follows: the challenger selects uniformly at random a common randomized order with respect to the two challenge sequences. This common randomized order is then given as source of randomness to the random oracle which answers questions accordingly. More precisely, let $\Gamma = (\gamma_1, \ldots, \gamma_n)$ be the selected order. Whenever the algorithm cannot decide where to place a plaintext x_j in the search tree, i.e., $x_i = x_j$ for $i < j$, meaning that x_i is an entry that is already encrypted in the tree, then the challenger asks the random oracle for a decision on i and j. The oracle answers with 1 if $\gamma_i < \gamma_j$ and with 0 if $\gamma_i > \gamma_j$ (notice that $\gamma_i \neq \gamma_j$ by Definition 4). In this way, the challenger produces a search tree and corresponding ciphertexts that are valid for both challenge sequences and which are in fact independent of the bit. Hence, the adversary has no chance of determining which sequence has been chosen other than guessing.

The Simulation is Incorrect. The proof strategy clearly excludes the attack described in the previous section. The reason is that the RC function is supposed to output *uniformly random* coins. As we have seen, even if RC outputs truly random coins then not every possible randomized order of the chosen sequence is equally likely. Hence, the choice of the random sequence is in two aspects unfaithful: first, the challenger restricts the number of randomized orders to those that both sequences have in common while RC does not know the two sequences and can, hence, not choose according to this requirement. Second, the fact that the choice is uniform does not reflect the reality. As we have seen, one artifact of the construction is that not every randomized order is equally likely. Consequently, forcing RC to generate output based on a common randomized order changes the distribution from which coins are drawn and, hence, neither all randomized orders are possible nor are their probabilities of being chosen correctly distributed. In the extreme case described in Sect. 4.1, it even would disallow all but one randomized order to be the result of the encryption routine. As a consequence, the proof technique changes the behavior of RC to an extent that makes the simulation incorrect.

4.3   Generalizing the Attack in an Ideal Setting

Since the scheme is vulnerable to an attack and it chooses the randomized order under which a sequence is encrypted in a non-uniform way, we find it interesting to also investigate whether the scheme is still vulnerable in an ideal setting where the choice of the randomized order happens uniformly.

The answer to this question is unfortunately positive, as the following result shows. Concretely, only if two sequences agree on essentially all randomized orders, the adversary has a negligible advantage of distinguishing them.

Theorem 1. *Let X_0 and X_1 be two plaintext sequences of length n. Further assume that X_0 has m_0 and X_1 has m_1 randomized orders, respectively, and that they have m randomized orders in common. Then, for the idealized construction of [11] which encrypts plaintexts under a uniformly chosen randomized order, there exists an adversary whose success probability in winning the IND-FA-OCPA game is at least $1 - m\frac{m_0+m_1}{2m_0m_1}$.*

Proof. We construct an adversary, which submits both X_0 and X_1 to the FA-OCPA challenger. Since X_0 has m_0 randomized orders, the probability that one of those in common with X_1 is chosen by the encryption procedure is $\frac{m}{m_0}$ due to the uniformly random behavior. Likewise, for X_1, the probability that a common randomized order is chosen by the encryption procedure is $\frac{m}{m_1}$. Hence, depending on the challenger's bit b, the adversary sees a non-common randomized order with probability $1 - \frac{m}{m_b}$, which also reflects its success probability for winning the game when the challenger picks b. Consequently,

$$\Pr[\mathcal{A} \text{ wins}] = \left| \Pr[\mathsf{Exp}^{\mathcal{A}}_{\mathsf{FA-OCPA}}(\kappa, 1) = 1] - \Pr[\mathsf{Exp}^{\mathcal{A}}_{\mathsf{FA-OCPA}}(\kappa, 0) = 1] \right|$$

$$= \frac{1}{2}\left(1 - \frac{m}{m_0}\right) + \frac{1}{2}\left(1 - \frac{m}{m_1}\right) = 1 - m\frac{m_0 + m_1}{2m_0 m_1}$$

In the example from the previous section, we have parameters $m_0 = 1$, $m_1 = n!$, and $m = 1$. Substituting those into Theorem 1, we get the aforementioned non-negligible success probability of

$$1 - m\frac{m_0 + m_1}{2m_0 m_1} = 1 - \frac{1 + n!}{2n!} = \frac{1}{2}\left(1 - \frac{1}{n!}\right).$$

5   Impossibility of IND-FA-OCPA

The previously presented results raise the question if IND-FA-OCPA, as presented in [11] can be achieved at all. It turns out that this is not the case: in this section, we prove an impossibility result for frequency-hiding order-preserving encryption as defined in Definition 5. Formally we prove the following theorem.

Theorem 2. *Let X_0 and X_1 be two arbitrary plaintext sequences of the same length that do not share any randomized order. Let furthermore $\mathcal{OPE}$ be an order-preserving encryption scheme secure against IND-FA-OCPA. Then, the probability of distinguishing whether X_0 is encrypted or whether X_1 is encrypted with $\mathcal{OPE}$ is negligibly close to 1/2.*

Before we prove the theorem using Definition 5, we argue why it implies the impossibility of frequency-hiding OPE. According to the theorem, no adversary can distinguish the encryptions of two arbitrary sequences of his choice that are ordered in a completely different manner. This constitutes a formulation of the IND-CPA property for multiple messages, restricted to sequences that do not share a randomized order. The restriction, however, is not necessary since sequences that share a randomized order are trivially indistinguishable by IND-FA-OCPA. To exemplify, let the two sequences be $X_0 = 1, 2, \ldots, n$ and $X_1 = n, n - 1, \ldots, 1$, so X_1 is the reverse of X_0. According to Theorem 2, no adversary can distinguish which one of the two is encrypted. However, due to the correctness of OPE it must be the case that the encryption Y^* fulfills for all i and j and $b \in \{0, 1\}$

$$y_i^* \geq y_j^* \implies x_i^b \geq x_j^b.$$

Consequently, if X_0 is encrypted we have $y_i^* < y_j^*$ for $i < j$ and vice versa for X_1. Hence, an adversary could trivially distinguish which of the two sequences is encrypted. Hence, by contraposition of Theorem 2, an IND-FA-OCPA-secure OPE scheme cannot exist.

Proof (Theorem 2). In the security game $G(b)$, on input X_0 and X_1 by $\mathcal{A}$, the challenger encrypts sequence X_b, gives the ciphertexts Y^* to $\mathcal{A}$, who replies with a guess b'. $\mathcal{A}$ wins if $b = b'$.

We define three games. $G_1 = G(0)$. For G_2, we select a sequence X^* which has a randomized order in common with both X_0 and X_1. Notice that such a sequence always exists, e.g., take the series $a, a, \ldots, a$ (n times) for arbitrary a in an appropriate domain. Instead of encrypting X_0 as in G_1, we now encrypt X^*. Finally, $G_3 = G(1)$.

In order to show that $G_1 \approx G_2$, assume that there exists a distinguisher $\mathcal{A}$ that can distinguish between G_1 and G_2 with non-negligible probability. Then we construct a reduction $\mathcal{B}$ that breaks IND-FA-OCPA. On $\mathcal{A}$'s input, $\mathcal{B}$ forwards X_0 and a sequence X^* to the IND-FA-OCPA challenger. The challenger answers with Y^*, which $\mathcal{B}$ again forwards to $\mathcal{A}$. $\mathcal{A}$ outputs a bit b', which $\mathcal{B}$ again forwards to the challenger. The simulation is obviously efficient. If the internal bit of the IND-FA-OCPA challenger is 0, we perfectly simulate G_1, while we simulate G_2 when the bit is 1. Hence, the success probability of $\mathcal{A}$ carries over to $\mathcal{B}$ since $\mathcal{B}$ only forwards messages. Since we assumed that $\mathcal{A}$ can successfully distinguish G_1 and G_2 with non-negligible probability, it must be the case that $\mathcal{B}$ wins the IND-FA-OCPA game with non-negligible probability. This is a contradiction.

The proof of $G_2 \approx G_3$ is symmetric to the one above. In conclusion, we have that $G_1 \approx G_2 \approx G_3$, and hence, $G(0) \approx G(1)$, meaning that every adversary can distinguish between encryptions of X_0 and X_1 that do not share a randomized order only with negligible probability.

6 An Achievable Definition: IND-FA-OCPA*

Since the notion of indistinguishable ciphertexts under frequency-analyzing ordered chosen plaintext attacks is not achievable, it is desirable to understand the problem of the original definition and try to come up with a suitable one that still captures the idea of frequency-hiding but is achievable.

Interestingly enough, the solution to our problem can be found by investigating again Kerschbaum's security proof. The proof builds a random oracle that overcomes the issues of the definition. Even though this construction of the oracle is incorrect, as we have shown previously, it helps us identify the problem with the definition. In the definition, the challenger has no means to tell the encryption algorithm which randomized order to choose when encrypting the chosen challenge sequence. Hence, it could be the case that the algorithm chooses an order that is not common to both challenge sequences. Had the challenger a way to decide which order to encrypt with, the problem were gone.

Consequently, we tackle the problem from two angles: (1) we augment the OPE model by one more input to the encryption function, namely, the randomized order that is supposed to be used and (2) we strengthen the challenger's capabilities during the security game: it may now, additionally to selecting which sequence to encrypt, also choose a common randomized order as input to the encryption algorithm. This new definition still captures the notion of frequency-hiding in the same way, it just excludes the attacks presented in this work and makes the definition, thus, achievable.

6.1   Augmented OPE Model

We present the augmented model in the following definition. Notice that the only difference to Definition 1 is the additional input Γ to the encryption function. This additional input serves the purpose of deciding from outside of the function, which randomized order should be used to encrypt the plaintexts. In contrast, standard OPE decides about the ordering randomly inside of the function. We stress that augmented OPE is more general than OPE since the input Γ can be replaced by the result of a call to a random function.

Definition 6 (Augmented OPE). *An augmented order-preserving encryption scheme* $\mathcal{OPE}^* = (\mathsf{K}, \mathsf{E}, \mathsf{D})$ *is a tuple of* PPT *algorithms where* $S \leftarrow \mathsf{K}(\kappa)$. *The* key generation *algorithm takes as input a security parameter* κ *and outputs a secret key (or state)* S; $(S', y) \leftarrow \mathsf{E}(S, x, \Gamma)$. *The* encryption *algorithm takes as input a secret key* S, *a message* x, *and an order* Γ *and outputs a new key* S' *and a ciphertext* y; $x \leftarrow \mathsf{D}(S, y)$. *The* decryption *algorithm* $x \leftarrow \mathsf{D}(S, y)$ *takes as input a secret key* S *and a ciphertext* y *and outputs a message* x.

6.2   The New Definition IND-FA-OCPA*

The new security game is close in spirit to Definition 5. The difference is that (1) it is defined over an augmented OPE scheme which makes the randomized order used for encryption explicit and (2) the challenger chooses that order uniformly at random from the orders that both challenge sequences have in common. Since we define the notion adaptively, we introduce some new notation with respect to randomized orders.

In the following definition, we let $\Gamma = \gamma_1, \ldots, \gamma_n$ and we use the notation $\Gamma \downarrow_i$ to denote the order of the sequence $\gamma_1, \ldots, \gamma_i$. Notice that this order is unique since Γ is already an order. For instance, take the randomized sequence $\Gamma = 1, 6, 4, 3, 2, 5$. Then, $\Gamma \downarrow_3 = 1, 3, 2$, which is the order of $1, 6, 4$.

Definition 7 (IND-FA-OCPA*). *An augmented order-preserving encryption scheme* $\mathcal{OPE}^* = (\mathsf{K}, \mathsf{E}, \mathsf{D})$ *has* indistinguishable ciphertexts under frequency-analyzing ordered chosen plaintext attacks *if for any* PPT *adversary* $\mathcal{A}$, *the following probability is negligible in the security parameter* κ:

$$\left| \Pr[\mathsf{Exp}_{\mathsf{FA-OCPA}^*}^{\mathcal{A}}(\kappa, 1) = 1] - \Pr[\mathsf{Exp}_{\mathsf{FA-OCPA}^*}^{\mathcal{A}}(\kappa, 0) = 1] \right|$$

where $\mathsf{Exp}_{\mathsf{FA-OCPA}^*}^{\mathcal{A}}(\kappa, b)$ *is the following experiment:*

Experiment $\mathsf{Exp}^{\mathcal{A}}_{\mathsf{FA-OCPA^*}}(\kappa, b)$

$(X_0, X_1) \leftarrow \mathcal{A}$ *where* $|X_0| = |X_1| = n$ *and* X_0 *and* X_1
    have at least one common randomized order
Select Γ *uniformly at random from the common randomized orders of* X_0, X_1
$S_0 \leftarrow \mathsf{K}(\kappa)$
For all $1 \le i \le n$ *run* $(S_i, y_{b,i}) \leftarrow \mathsf{E}(S_{i-1}, x_{b,i}, \Gamma{\downarrow}i)$
$b' \leftarrow \mathcal{A}(y_{b,1}, \ldots, y_{b,n})$
Output 1 if and only if $b = b'$.

7 Constructing Augmented OPE

We show how to construct an augmented OPE scheme. Interestingly enough, the key observation to $\mathcal{OPE}^*$ is that the scheme of [11], which we presented in Sect. 3 can be modified so as to fit the new model.

As we introduce a third input to the encryption function, namely an order that is as long as the currently encrypted sequence plus one, we have to show how to cope with this new input in the construction. The key idea is quite simple: usually, the encryption scheme draws randomness from a PRF when the plaintext to be encrypted is already encrypted, in order to decide whether to move left or right further in the tree. The additional input solves this decision problem upfront, so there is no need for using randomness during the encryption.

While the setup and re-balancing algorithms are as described in [11], we describe the new encryption algorithm in Algorithm 2. Furthermore, we require that every node in the tree stores its index in the plaintext sequence, i.e., we add an attribute *index* to each node t. We further assume that the index of the message that is currently to be encrypted is the length of the order Γ. As we can see, the only difference between Algorithms 1 and 2 is the behavior when the message to be inserted is equal to the message currently stored at t. Then, the order Γ is considered so as to decide whether to traverse the tree further to the left or right.

Algorithm 2. $\mathsf{E}(S, x, \Gamma)$ where $S = t, min, max$ and $\Gamma = \gamma_1, \ldots, \gamma_k$

1: **if** $t = \perp$ **then**	11: **if** $x = t.m$ **then**
2: $t.m = x$	12: $b \leftarrow \gamma_k > \gamma_{t.index}$
3: $t.index = k$	13: **end if**
4: $t.c = min + \lfloor \frac{max - min}{2} \rfloor$	14: **if** $b = 1 \vee x > t.m$ **then**
5: **if** $t.c = 0$ **then**	15: $\mathsf{E}(t.right, t.c + 1, max, x, \Gamma)$
6: rebalance the tree	16: **else**
7: **end if**	17: **if** $b = 0 \vee x < t.m$ **then**
8: **return** $t.c$	18: $\mathsf{E}(t.left, min, t.c - 1, x, \Gamma)$
9: **end if**	19: **end if**
10: $b \leftarrow -1$	20: **end if**

The encryption algorithm also nicely demonstrates that it does not matter from which domain the ordering draws its elements. The only important property of such an ordering is the relation of the single elements to each other, i.e., that (1) all elements of the order are distinct and (2) whether $\gamma_i < \gamma_j$ or the other way around. We do, hence, not require the shrinking function $\Gamma\!\downarrow_i$ for this construction: when encrypting an overall sequence of plaintexts with a predetermined randomized order Γ, it is sufficient to just cut the Γ to size i when encrypting the i-th element. The reason is that the relative ordering of the first i elements is not changed after shrinking, which suffices to let the algorithm decide about where to branch.

7.1 Formal Guarantees

We argue in this section that the tree-based scheme presented in the previous section is IND-FA-OCPA* secure. The reason for that is as follows: first, the challenger can select a randomized order that is in common to both sequences and give that chosen order to the encryption algorithm. Second, no matter which of the two sequences is encrypted according to the order, the resulting ciphertexts are equivalent in both cases. Hence, the adversary cannot do better than guessing the bit, since the ciphertexts are independent of the underlying plaintexts.

Theorem 3. *The $\mathcal{OPE}^*$ scheme presented in Sect. 7 is IND-FA-OCPA* secure.*

Proof. Let $\mathcal{A}$ be an arbitrary adversary for the game in Definition 7. Let furthermore X_0 and X_1 be the two plaintext sequences chosen by $\mathcal{A}$. By definition, those sequences share at least one common randomized order. Let Γ be one of those common orders, selected uniformly at random from the universe of common randomized orders. When encrypting either X_0 or X_1, Algorithm 2 uses Γ to decide where to branch. Hence, Algorithm 2's decisions are independent of the input plaintext sequence, and thus independent of the chosen bit b. Consequently, all information that $\mathcal{A}$ receives from the challenger are independent of b and he can thus, only guess what b is. This concludes the proof. $\quad\blacksquare$

8 Conclusion

Order-preserving encryption (OPE) is an enabling technology to implement database applications on encrypted data: the idea is that the ordering of ciphertexts matches the one of plaintexts so that inequalities on encrypted data are efficiently computable. Unfortunately, recent works showed that various attacks can be mounted by exploiting the inherent leakage of plaintext frequency. Frequency-hiding OPE [11] (CCS 2015) is a stronger primitive that aims at solving this problem by hiding the frequency of plaintexts.

We contribute to this line of work with the following results. First, we present an attack against the construction presented in [11], identifying the corresponding problem in the security proof. Second, we formulate a more general impossibility result, proving that the security definition introduced in [11] cannot be achieved by any OPE scheme. Third, to complete the picture and assess which theoretical security is achievable at all, we make the definition in [11] more precise by giving the challenger more capabilities and augmenting the OPE model so as to receive randomized orders as inputs which are used to break ties. We finally show that the more precise version of the definition can be achieved by a variant of the construction introduced in [11].

Despite this seemingly positive results, in the presence of the plethora of empirical attacks against (FH-)OPE and its variants (e.g., ORE), we suggest to not use any of those schemes for actual deployment since the security guarantees achieved do not reflect practical requirements. We recommend to move away from OPE in general, more towards other alternatives, even if there are none that solve the problem so conveniently; at the price of low to no security.

Acknowledgements. This research is based upon work supported by the German research foundation (DFG) through the collaborative research center 1223, by the German Federal Ministry of Education and Research (BMBF) through the Center for IT-Security, Privacy and Accountability (CISPA), and by the state of Bavaria at the Nuremberg Campus of Technology (NCT). NCT is a research cooperation between the Friedrich-Alexander-Universität Erlangen-Nürnberg (FAU) and the Technische Hochschule Nürnberg Georg Simon Ohm (THN). Dominique Schröder is supported by the German Federal Ministry of Education and Research (BMBF) through funding for the project PROMISE. Finally, we thank the reviewers for their helpful comments and our shepherd for the excellent and valuable feedback, which improved the paper.

References

1. Arasu, A., Blanas, S., Eguro, K., Kaushik, R., Kossmann, D., Ramamurthy, R., Venkatesan, R.: Orthogonal security with cipherbase. In: Proceedings of the Biennial Conference on Innovative Data Systems Research (CIDR 2013) (2013)
2. Bellare, M., Keelveedhi, S., Ristenpart, T.: DupLESS: server-aided encryption for deduplicated storage. In: Proceedings of the USENIX Security Symposium (USENIX 2013), pp. 179–194. USENIX Association (2013)
3. Boldyreva, A., Chenette, N., Lee, Y., O'Neill, A.: Order-preserving symmetric encryption. In: Joux, A. (ed.) EUROCRYPT 2009. LNCS, vol. 5479, pp. 224–241. Springer, Heidelberg (2009). https://doi.org/10.1007/978-3-642-01001-9_13
4. Boldyreva, A., Chenette, N., O'Neill, A.: Order-preserving encryption revisited: improved security analysis and alternative solutions. In: Rogaway, P. (ed.) CRYPTO 2011. LNCS, vol. 6841, pp. 578–595. Springer, Heidelberg (2011). https://doi.org/10.1007/978-3-642-22792-9_33
5. Boneh, D., Lewi, K., Raykova, M., Sahai, A., Zhandry, M., Zimmerman, J.: Semantically secure order-revealing encryption: multi-input functional encryption without obfuscation. In: Oswald, E., Fischlin, M. (eds.) EUROCRYPT 2015. LNCS, vol. 9057, pp. 563–594. Springer, Heidelberg (2015). https://doi.org/10.1007/978-3-662-46803-6_19
6. Chenette, N., Lewi, K., Weis, S.A., Wu, D.J.: Practical order-revealing encryption with limited leakage. In: Peyrin, T. (ed.) FSE 2016. LNCS, vol. 9783, pp. 474–493. Springer, Heidelberg (2016). https://doi.org/10.1007/978-3-662-52993-5_24

7. Durak, F.B., DuBuisson, T.M., Cash, D.: What else is revealed by order-revealing encryption? In: Proceedings of the Conference on Computer and Communications Security (CCS 2016), pp. 1155–1166. ACM Press (2016)
8. Grubbs, P., McPherson, R., Naveed, M., Ristenpart, T., Shmatikov, V.: Breaking web applications built on top of encrypted data. In: Proceedings of the Conference on Computer and Communications Security (CCS 2016), pp. 1353–1364. ACM Press (2016)
9. Grubbs, P., Sekniqi, K., Bindschaedler, V., Naveed, M., Ristenpart, T.: Leakage-abuse attacks against order-revealing encryption. In: Proceedings of the IEEE Symposium on Security and Privacy (S&P 2017). IEEE Computer Society Press (2017)
10. He, W., Akhawe, D., Jain, S., Shi, E., Song, D.: ShadowCrypt: encrypted web applications for everyone. In: Proceedings of the Conference on Computer and Communications Security (CCS 2014), pp. 1028–1039. ACM Press (2014)
11. Kerschbaum, F.: Frequency-hiding order-preserving encryption. In: Proceedings of the Conference on Computer and Communications Security (CCS 2015), pp. 656–667. ACM Press (2015)
12. Kerschbaum, F., Schroepfer, A.: Optimal average-complexity ideal-security order-preserving encryption. In: Proceedings of the Conference on Computer and Communications Security (CCS 2014), pp. 275–286. ACM Press (2014)
13. Lau, B., Chung, S., Song, C., Jang, Y., Lee, W., Boldyreva, A.: Mimesis aegis: a mimicry privacy shield–a system's approach to data privacy on public cloud. In: Proceedings of the USENIX Security Symposium (USENIX 2014), pp. 33–48. USENIX Association (2014)
14. Lewi, K., Wu, D.J.: Order-revealing encryption: new constructions, applications, and lower bounds. In: Proceedings of the Conference on Computer and Communications Security (CCS 2016), pp. 1167–1178. ACM Press (2016)
15. Naveed, M., Kamara, S., Wright, C.V.: Inference attacks on property-preserving encrypted databases. In: Proceedings of the Conference on Computer and Communications Security (CCS 2015), pp. 644–655. ACM Press (2015)
16. Popa, R.A., Li, F.H., Zeldovich, N.: An ideal-security protocol for order-preserving encoding. In: Proceedings of the IEEE Symposium on Security and Privacy (S&P 2013), pp. 463–477. IEEE Computer Society Press (2013)
17. Popa, R.A., Redfield, C.M.S., Zeldovich, N., Balakrishnan, H.: CryptDB: protecting confidentiality with encrypted query processing. In: Proceedings of the ACM Symposium on Operating Systems Principles (SOSP 2011), pp. 85–100. ACM Press (2011)
18. Popa, R.A., Stark, E., Valdez, S., Helfer, J., Zeldovich, N., Balakrishnan, H.: Building web applications on top of encrypted data using Mylar. In: Proceedings of the USENIX Symposium on Networked Systems Design and Implementation (NSDI 2014), pp. 157–172. USENIX Association (2014)
19. Roche, D.S., Apon, D., Choi, S.G., Yerukhimovich, A.: POPE: partial order preserving encoding. In: Proceedings of the Conference on Computer and Communications Security (CCS 2016), pp. 1131–1142. ACM Press (2016)
20. Song, D.X., Wagner, D., Perrig, A.: Practical techniques for searches on encrypted data. In: Proceedings of the IEEE Symposium on Security and Privacy (S&P 2000), pp. 44–55. IEEE Computer Society Press (2000)
21. Teranishi, I., Yung, M., Malkin, T.: Order-preserving encryption secure beyond one-wayness. In: Sarkar, P., Iwata, T. (eds.) ASIACRYPT 2014. LNCS, vol. 8874, pp. 42–61. Springer, Heidelberg (2014). https://doi.org/10.1007/978-3-662-45608-8_3

Privacy-Preserving Whole-Genome Variant Queries

Daniel Demmler[✉], Kay Hamacher, Thomas Schneider,
and Sebastian Stammler[✉]

Technische Universität Darmstadt, Darmstadt, Germany
{daniel.demmler,kay.hamacher,thomas.schneider,
sebastian.stammler}@cysec.de

Abstract. Medical research and treatments rely increasingly on genomic data. Queries on so-called variants are of high importance in, e.g., biomarker identification and general disease association studies. However, the human genome is a very sensitive piece of information that is worth protecting. By observing queries and responses to classical genomic databases, medical conditions can be inferred. The *Beacon project* is an example of a public genomic querying service, which undermines the privacy of the querier as well as individuals in the database.

By secure outsourcing via secure multi-party computation (SMPC), we enable privacy-preserving genomic database queries that protect sensitive data contained in the queries and their respective responses. At the same time, we allow for multiple genomic databases to combine their datasets to achieve a much larger search space, without revealing the actual databases' contents to third parties. SMPC is generic and allows to apply further processing like aggregation to query results.

We measure the performance of our approach for realistic parameters and achieve convincingly fast runtimes that render our protocol applicable to real-world medical data integration settings. Our prototype implementation can process a private query with 5 genetic variant conditions against a person's exome with 100,000 genomic variants in less than 180 ms online runtime, including additional range and equality checks for auxiliary data.

1 Introduction

Genomic data holds the key to the understanding of many diseases and medical conditions. Some genomic variations in individuals in particular might be crucial in the diagnosis of a disease or a treatment regime. As a first step, a doctor or researcher might want to query the world's pool of sequenced genomes for such a variation in order to identify if it has been encountered and studied before. For this purpose, the *Beacon project* was established by the Global Alliance for Genomics & Health[1] to evaluate the eagerness of institutions around the globe

---

[1] http://genomicsandhealth.org/.

ⓒ Springer Nature Switzerland AG 2018
S. Capkun and S. S. M. Chow (Eds.): CANS 2017, LNCS 11261, pp. 71–92, 2018.
https://doi.org/10.1007/978-3-030-02641-7_4

to engage in a distributed variant query service. Participating institutions can be queried for a variation and confirm or deny its existence in their database.

However, this open form of collaboration quickly raises privacy concerns about, e.g., re-identification risk [34]. Thus, it would be highly desirable to secure participants of such a variant querying service, as well as individuals in their genome databases. Furthermore, it can be assumed that many small institutions will not be comfortable in joining the Beacon scheme in its current form, since re-identification risk impacts small databases even more severely.

We address these (genomic) privacy demands by developing a secured form of a federated variant query service – eventually an extension of the Beacon project – in which the count of matches is learned, while hiding which institution contributed to what extent. Our solution can optionally apply a threshold t on the count of matches so as to only release data if there are more than t matches. This can mitigates re-identification risk when querying for rare mutations.

Here, secure multi-party computation (SMPC) gives us the powerful ability to run arbitrary computations on sensitive data, while protecting the privacy of this data. In this work, we use two parties, called proxies, to achieve high efficiency. We rely on SMPC for *secure outsourcing* of genomic data from arbitrarily many sources to two proxies that enable clients to run *private queries* on the data. The proxies are assumed to not collude and therefore learn nothing about the outsourced data or the client's query and its response. We focus on the use-case of privately querying a large, aggregated genome database.

1.1 Our Contributions

In this paper, we provide the following contributions:

- We allow private queries to multi-center genome databases. We hide the query, which elements it accesses, and what elements match the query.
- Our protocol allows to privately aggregate data from multiple data sources. Usually this is prohibited by patient privacy laws, which aim at protecting sensitive medical data. We retain privacy, while at the same time providing a larger search space that leads to more expressive query results.
- Due to the generic nature of our protocol we can perform additional multi-property queries that add only negligible overhead to the database lookup.
- We develop a custom format (Variant Query Format – VQF) for the lossy storage of genomic variants that also allows *similar* variants to match. The widely used Variant Call Format (VCF) can easily be compressed to VQF, providing a bridge to existing genomic variant databases.
- We present a prototypical implementation of our protocol in C++ using the ABY framework [9] and achieve practical runtimes for real-world inputs.

1.2 Deployment Setting

We depict our setting in Fig. 1. An arbitrary number of genomic database providers DB_i privately outsource their data to two non-colluding proxies D'

and D'', who simply aggregate the received data as one large dataset. Privacy is achieved by using XOR-based secret sharing, as described in more detail in Sect. 4. DB_i can extend the database by simply sending new entries D'_i and D''_i to the proxies at any time. Updates of existing entries require sending only the difference as bitwise XOR to one of the proxies.

A client C who wants to query entries from the aggregated databases sends an XOR-secret-shared query to both proxies. Privacy-preserving computation is made possible by the protocol of Goldreich, Micali and Wigderson [15], which enables efficient computation on secret-shared data. Optionally, we allow results to be t-threshold released, i.e., the client receives a query response only if more than t database entries overall match the query criteria.

As a special case, our setting can also be used by a client that runs private queries on a single genomic database held by a server without involving additional parties. For this, the client runs C and D', and the server runs DB_1 and D''.

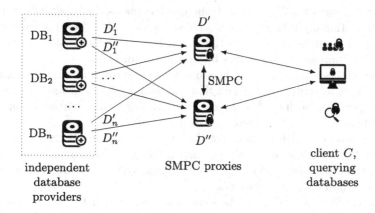

Fig. 1. Deployment setting overview. The plus sign denotes the federation of genomic databases and the lock symbol marks secret-shared data. The client only interacts with the SMPC proxies, which hold XOR secret-shared copies of the genetic variant data.

1.3   Related Work in Genomic Privacy

There exists a long line of work in genomic privacy, starting with [19]. In the subsequent section we provide an overview of recent work related to our solution. We compare the key aspects of related work with our proposal in Table 1.

In [20], count queries on fixed-indexed single-nucleotide polymorphisms (SNPs) databases are performed. They convert the database into an index tree structure and perform an encrypted tree traversal using Paillier's additively homomorphic encryption scheme [32] and Yao's garbled circuits [38] for comparison. The queried SNP indices are leaked and the index tree needs to be built by a single and trusted certified institution that must know all the data in plaintext (coming from possibly multiple data sources as in our setup).

In the field of privacy-preserving genomic testing many approaches securely calculate the edit distance (ED) with protected genetic databases. ED is a way of measuring the closeness of two strings by calculating the minimum number of operations to transform between the two. Most of the time, the Levensthein distance is meant by ED and it is defined as the minimum amount of substitutions, insertions and deletions to transmute between the strings. It is an important measure in genomics to estimate the closeness of two genomes, and finds applications in similar patient querying (SPQ).

The authors of [1] implemented a form of secure count- and ranked similar patient queries, running in seconds. However, their setup is quite different from ours since the query is known in plaintext to each data center, and the output is learned by a central server (CS). The aggregation/sorting is done on the CS and only individual contributions are hidden from the CS and querier.

In [23], the authors developed a privacy-preserving ED algorithm leveraging Yao's garbled circuits. While their technique calculates the exact edit distance, it is computationally infeasible on a whole-genome scale. Their implementation takes several minutes to calculate the ED of just a few hundred-character long strings. [21] improved those results by up to a factor of 29.

In the whole-genome context it is often more sensible to only approximate the ED, taking advantage of the fact that the human genome is mostly preserved (>99% of genomic positions) and most variations are simple substitutions. Two important works leveraging these factors are [2,37].

The authors of [37] test their distributed query system GenSets over, theoretically, 250 distributed hospitals, each holding 4,000 genome sequences of around 75 million nucleotides each. It took their system 200 minutes to search through one million cancer patients.

In [2], the authors partition the sequences into smaller blocks and then precompute the ED within the blocks. Since the human genome shows high preservation, this greatly reduces the number of distinct blocks. E.g., for a realistic data set of 10,000 genome snippets of length about 3,500 from a region of high divergence, after partitioning them into 15-letter blocks, for each group of blocks they observed at most 40 unique blocks, instead of the theoretic maximum of 10,000. Still, they could determine the t best matching sequences against a query sequence with high success.

In [25], an attack by Goodrich [16] on genome matching queries was investigated. They developed a detection technique against such inference attacks employing zero-knowledge proofs to ensure querier honesty.

The authors of [8] developed a novel method for *secure genomic testing with size- and position-hiding private substring matching* (SPH-PSM). In their setup, a testing facility holds a DNA substring (e.g., a marker) and a patient possesses their full genome sequence. The patient sends their homomorphically encrypted genome and public key to the facility, which computes an accumulator applying their substring. The still encrypted accumulator is sent back to the patient, who decrypts it to learn the binary answer—whether the substring is included in her genome. In the process, the facility doesn't learn anything and the patient doesn't learn the substring or its position in their genome.

The work [40] presents an innovative and efficient approach to outsourced pattern matching employing a new outsourced discrete Fourier transform protocol called OFFT. It solves a similar problem to the aforementioned method [8] and scales logarithmically in the string to be matched.

In [35], a genomic cloud storage and query solution is presented. VCF data is symmetrically encrypted and sent to cloud storage. Using a custom method based on private information retrieval (PIR, see also Sect. 2.2), the data owner, or an authorized entity, can query the cloud storage for a specific variant (utilizing a homomorphically encrypted 0–1-array mask). This solution cannot be generalized to multiple patients as the querier would need access to all VCFs' symmetric keys. A generalization of this work [12] allows computations that are similar to ours and offers strong security guarantees. However, in this case, the runtime depends on the size of the response and thus reveals meta information about the query, which we specifically intend to hide in our work. Padding could solve this issue, but would increase the runtime drastically.

The protocol [4, Sect. 4.2] uses Authorized Private Set Intersection [6] to allow for authorized queries of a list of specific SNPs (a SNP profile, e.g., of SNPs relevant for drug selection and dosage): The querier first submits the SNP profile to a certified institution, which sends back an authorization. The querier can then use this authorization to query for his SNP profile in a patient's genome, learning the matching SNPs. The query is hidden from the patient/database. The protocol [33] extends this idea and uses additively homomorphic Elliptic Curve based El-Gamal [10] to calculate a weighted sum over a set of SNPs, where the weights are authorized by a certified institution.

In conclusion, several solutions have addressed the problem of secure genome queries (see Table 1), but most work directly on the sequence instead of called variants and none grant the easy extensibility that our solution provides and at the same time deliver whole-genome scale protection for all included parties.

2 Preliminaries

2.1 Secure Multi-Party Computation (SMPC)

Computation on multiple parties' secret data was first proposed in the 1980's with Yao's seminal garbled circuits protocol [38] and the work of Goldreich, Micali, and Wigderson (GMW) [15]. More formally, for the case of two parties, an SMPC protocol can be described as the evaluation of a function f on the two parties' respective private inputs x and y. The parties learn nothing but what can be inferred from the function's output $z = f(x, y)$. Due to limited computation power, SMPC protocols were at first seen as merely theoretical results. However, a long line of research has lead to significant algorithmic improvements that, combined with an increase in computing power in the last decades, has shown that SMPC can indeed be used for practical purposes. Most SMPC protocols

Table 1. Comparison of features and limitations of related work to our solution.

	[20]	[1]	[21,23]	[2,37]	[8,40]/[4]/[33]	[35]	[12]	Our work
Variants (V)/ Sequences(S)/ Both(*)	S	S	S	S	S	V	*	V
Single trusted party eliminated	✗	✗	✓	✓	✓	✓	✓	✓
Query protected	✗	✗	✓	✓	✓	✓	✓	✓
Top similar patient query	✗	✓	n.a.	✓	n.a.	✗	✗	✗
Whole-genome scale	✗	✓	✗	✓	✓	✓	✓	✓
Easy extensibility	✗	(✓)	✗	✗	✗	✗	✗	✓
Output	Count	Count & similar patients	Exact ED	Similar patients	Substr. match? (Y/N)/ matching SNPs/ weighted average over SNPs	Variant present? (Y/N)	Count & sum[a]	(Thresh.) count/ matches[b]

[a] Optionally differentially private
[b] Arbitrary (aggregation) function f

rely heavily on oblivious transfer (OT) as a building block. OT extension [3,22] is one of the key results that enables practical performance for SMPC.

In this work we consider the special case of secure two-party computation with security against semi-honest (passive) adversaries in the offline/online model. This means that we divide the protocol into two phases: First, an offline phase that is independent of the private inputs and requires just some upper bound of their size. It can be pre-computed and stored before the actual computation takes place. Second, a very efficient online phase that uses the pre-computed data to compute the function on the parties' private inputs.

While Yao's garbled circuits protocol offers a constant number of communication rounds, it requires the evaluation of symmetric cryptographic operations in both offline and online phase. The GMW protocol allows the preprocessing of so called Beaver multiplication triples [5] that can be pre-computed in the offline phase. The online phase of GMW requires one communication round per layer of dependent AND gates (circuit depth), but involves only very efficient bit operations (XOR) and no symmetric cryptographic operations.

Since our circuits can be optimized for low depth, we decided to implement our protocol using the GMW protocol. There also exist extensions for security against stronger malicious adversaries that can be applied to our protocol [31].

2.2 Related Privacy-Preserving Technologies

Aside from SMPC there are many other privacy-preserving technologies that could potentially achieve similar results as our solution. In this section we briefly provide an overview and motivate why we chose SMPC.

Private information retrieval (PIR) is a technique that allows private queries to a *public* database. This is not applicable in our scenario where the database consists of sensitive genomic data by multiple providers. Furthermore, PIR does not allow for more than the private query, i.e., additional operations like comparisons on the data and threshold checking are not possible.

Homomorphic encryption (HE) is a powerful tool that allows for operations on encrypted values. While somewhat homomorphic encryption is reasonably fast, it only supports a limited set of operations. Fully homomorphic encryption overcomes this limitation. This is an active field of research but implementations still lack the efficiency to be of practical use [30].

Oblivious RAM (ORAM) is another technique that could be used for our use case and was indeed shown to be applicable for, e.g., Bayesian statistics [24] frequently used in bio-informatics [39]. ORAM allows multiple private write accesses to a database, which we do not require, as we need to securely store the genome only once. Thus, our approach is simpler and we expect it to perform better than ORAM-based solutions.

Intel's Software Guard Extensions (SGX) is a recent instruction set extension that allows programs to run in a protected private environment and can be used as an alternative to SMPC [18]. While this is also specifically designed for cloud computing, additional care must be taken to not reveal memory access patterns and timing patterns, which especially affects our use case of privately querying a data set. Observing such access or timing patterns could allow the cloud provider to learn information about the data or the query itself. Learning the query, however, might also reveal critical information about patients, e.g., if a physician queries the database(s) to confirm a particular diagnosis.

Recently the survey [13] categorized the entire field of encrypted or protected search, which offers several additional alternatives to our work. We can imagine a combination of the presented techniques with our ideas as possible future work.

3   Genetic Variant Queries on Distributed Databases

We consider a federation of databases storing genomic variants in our custom Variant Query Format (VQF, see Sect. 3.2). They jointly offer a privacy-preserving *Beacon Service*: individual privacy is guaranteed in the sense that it is not learned which dataset from which data-center contributed to which extent to the final count. Optionally, privacy can be strengthened by enforcing a threshold criterion on the query: the actual count will only be returned if it is larger than a predefined threshold parameter t.

3.1   Beacon Network and Potential Extensions

The term *Beacon* stems from the Beacon Network Project, which is "a global search engine for genetic mutations". It was instantiated by the Global Alliance for Genomics & Health to "test the willingness of international sites to share genetic data". The joint service answers queries of the form "Do you have any genomes with an 'A' at position 100,735 on chromosome 3?" and participants each give a simple *Yes/No* answer.

Privacy Concerns. It has been shown in [34] that in its original form, Beacon queries are susceptible to re-identification attacks. The authors showed that with 5,000 queries, a person can be re-identified from a Beacon holding 1,000 genomes. Our proposed framework lowers this risk twofold: Firstly, the querier doesn't learn immediately which database contributes to the count. Only in a follow-up consultation can the databases and querier reveal a match should they mutually agree to do so. And secondly, because of an optional t-threshold check in the final step, it is harder for the querier to craft individual-identifying queries. hile recent changes to the project's architecture address some privacy concerns by access-control checks[2], those solutions cannot withstand a malicious man-in-the-middle or potential exploits of access-control vulnerabilities. That is, the project doesn't offer privacy by design since queries are sent in clear to the beacons and central web-interface, which also learns all beacons' answers during aggregation of the results.

3.2   Genomic Variant Representation Format

Variant Call Format (VCF). The most popular format to store genomic variants is in the Variant Call Format (VCF) [7], as used by the 1,000 Genomes Project and the Broad institute. It stores the results of a process called *variant calling* on aligned reads (usually stored in *SAM/BAM* files [28] or more recently in the *CRAM* format [11]). It describes in precise detail genetic mutations versus a reference genome.

However, for the application of genetic querying, the VCF is too precise in the sense that a lot of auxiliary data is stored per variation and exact matches are unlikely, especially for complex structural variations. We therefore developed a machine friendly and simplistic format for storing genetic variants in the context of variant querying that also makes it possible for *similar* variations to match.

Variant Query Format (VQF). We store a genome's set of variation against a common reference genome in the following custom format which we will call Variant Query Format (VQF). Variations are stored in a fixed-size dictionary (possibly with multiple entries for the same key), where the key is a variant's position in the reference genome and the value encodes the variation. We set the key size κ to 32 bit, which suffices to address every locus in any human reference genome, which has about 3.2 billion (haploid) loci[3].

For the dictionary's value we choose a size ψ of 16 bit. Mapping any genomic variant information into 16 bit is not straightforward and must incur some information loss that we can bound from above (see below). Table 2 lists the proposed

---

[2] Beacon FAQs, 2).
[3] $\log_2(3.2 \cdot 10^9) \approx 31.6 < 32$.

mappings from genetic variations to a 16 bit value in our dictionary. The total number of attainable values is $4 + 4s + 2 \cdot 4^{s_{ins}} + 1 = 32,837$, which fit into the 16 bit value space.

Table 2. Details of the variant compression in VQF.

Variant	Stored information	#Values
SNP/SNV	We store the alternative nucleotide	4
Deletion/CNV/Inversion	Store two entries: the up and down rounded logarithm to the base $b = 2$ of deletion/CNV[a]/inversion length, up to log-length s. Also store one frameshift bit for deletions	$s = 16$
Insertion	Save up to $s_{ins} = 7$ inserted nucleotides and a frameshift bit	$2 \cdot 4^{s_{ins}} = 7$
Other	Only flag as *other* variation. This flag captures all other more complex variations, like rearrangements with breakends etc.	1

[a] For CNVs like tandem repeats, we record the absolute number of repeats, not the relative change

Note that the chosen logarithm base $b = 2$ for deletion/CNV/inversion events is just exemplary and could be adjusted differently, even per event, without changing the number of possible values. The maximum length logarithm s to store could also be different per event. Storing the up *and* down rounded logarithms for those events increases the chances of a match for closely related variations. E.g., if, at a given locus, one genome has a deletion of length 20 and another of length 60, those would be mapped to two entries each, $\{16, 32\}$ and $\{32, 64\}$, so a query for 32 would match in both genomes. For insertions and deletions (InDels) we also store a frameshift bit, i.e., whether its length is divisible by 3. Substitutions are just stored as combined InDels, if longer than a SNP/SNV.

Also note that for CNVs, we do not store the motif, no matter if a short CNV ($\sim$ tandem repeat) or long CNV was called, possibly resulting in false positives. For diploid variations, we create two entries with the same key (same locus) and each value encoding one of the two variations coming from either parent.

The compression of variant information might seem strong, but the information loss mainly concerns complex structural variations, where an exact query match is very unlikely in the overwhelming majority of application scenarios. We deem it sensible to rather have a small number of "false positives", but still closely related matches. When querying for similar patients, it is actually desirable to match less specific conditions, as is the case for copy number variations. E.g., it doesn't matter if a short tandem repeat gained 20 or 30 motifs in length. Only the magnitude by which it increased or decreased is important when searching for similar genomes.

Note that a mapping (which could also be described as a compression) from the very detailed VCF to our VQF encoding scheme is straightforward.

Reference Genome. Since variations are indexed by a reference genome, it is important that all genomes are called against the same reference when querying several individuals' variations. This is assumed for all databases in our setup.

Alternative Variant Format. Note that a different variant compression scheme akin to the hashing method found in [35] could be used. For a single VCF row, one could hash the ID, REF and ALT entry and some important fields from the INFO column, and project it to the 16 bit (or larger) value space. Ignoring collisions, this scheme could also serve as a variant compression method with reasonable low probability of false positives. However, since this method doesn't incorporate domain specific knowledge like our VQF, *similar* variations wouldn't have a chance to match.

Typical Sizes. There are typically two ways to analyze a person's genetic variation. The first option is to store the full genetic variation, which leads to around $N = 3$ million entries. The second, and more viable, option is to only store variation from coding regions, i.e., a person's exome. While a person's exome makes up only about 1% of their genome, this region is the most important in the context of research and disease testing. As such, exome sequencing is a common quicker and cheaper alternative to full-genome sequencing. A person's exome has about $N = 100,000$ variations. In our database model, a person's full genetic variation is mapped to N entries of size $\kappa + \psi = 48$ bit.

Auxiliary Data. Besides information on genetic variation, the databases can also hold auxiliary data like sex, age, weight and health data like blood pressure or disease indicators. For those fields, our protocol allows for range or threshold conditions in the queries, where desired.

3.3   Queries

The queries that we support are of the form

```
SELECT f(*) FROM Variants
WHERE ((locus₁, var₁), ..., (locusₘ, varₘ)) IN Genome
    AND cancertype = X AND ... AND ageₘᵢₙ ≤ age ≤ ageₘₐₓ ...
```

when expressed in SQL for illustration purposes. Here, m is the number of variant equalities to check and the remaining auxiliary queries can be more versatile and include ranges, besides equalities. For most of our benchmarks, we choose $m = 5$.

The querier specifies the values for the highlighted parts of the query. These values are secret-shared and remain private, such that the proxies do not learn them. The *structure* of f cannot be chosen by the querier and is fixed a priori. However, since we use the ABY SMPC framework [9], the function f can generally be an arbitrary (aggregation) function on the bit string of matching genomes, like the identity (simple output of matches) or the count (Hamming weight).

Technically, any query prompts the two proxies to generate a secret-shared bit string representing the matching genomes. As can later be seen in Sect. 6.1, this task causes the bulk computation and communication cost. Next, the proxies apply function f to this bit string and output it to the querier (or any other predefined party).

Query Scenario. For our experiments, we choose a scenario in which the querier receives the count of matches, together with a random query ID. Each database with at least one match receives the corresponding sub-bit mask of the genomes only in their own database together with the same query ID. This way, the database doesn't learn the query which matched their genomes but can contact the querier for a follow-up discussion of the matched patients.

Optionally, we can substitute the simple count by a t-threshold count, which returns the count only if it is larger than t. This would mitigate the re-identification risk as presented in [34]. However, since Beacon queries are often used to query for rare mutations, this extension brings its own problems. Our solution can easily realize both options and due to the generic nature of SMPC it can also be adapted for additional requirements.

4 Our Protocol for Private Genome Variant Queries

Our protocol ensures the privacy of both, the query to a genomic database and its response. At the same time, the entire database is hidden from the two proxies. This matches closely the cloud computing paradigm where computation and data storage is outsourced to a powerful set of machines that are maintained by an external party. In Fig. 1 on page 3, we depict our setting where multiple databases $DB_1, \ldots, DB_n$ outsource their data to two proxies D' and D'' that run the SMPC protocol and are assumed to not collude. A client C queries the combined data from all databases for the counts of entries that (a) match the query criteria and optionally (b) fulfill a pre-defined t-threshold level.

We achieve privacy by using XOR-based secret-sharing between the SMPC parties. More precisely, for every plaintext bit p, we choose a random masking bit r. We send r to D' and $s = p \oplus r$ to D''. The values r and s are called *shares* of the plaintext value p. For bit strings of length ℓ we apply this technique ℓ times in parallel. To further improve communication, we could send to D' a single seed for a pseudo-random generator instead of the values r.

We use the protocol of Goldreich, Micali and Wigderson (GMW) [15] to *privately* evaluate a Boolean circuit that corresponds to our functionality, i.e., querying a genome database. GMW operates directly on XOR-secret-shared values. We use the GMW protocol in the offline/online computation model, i.e., an *offline phase* that is preprocessed at any time before the actual private inputs are known. The data from the offline phase is then used in the efficient *online phase* to compute the function on the private data.

4.1 Protocol Description

In this section, we describe the phases of our overall protocol and depict it graphically in Fig. 2. Optionally, before the protocol, the database providers and proxies agree on a privacy threshold t that defines the minimum amount of matching records that a query response must contain. If a query matches $\leq t$

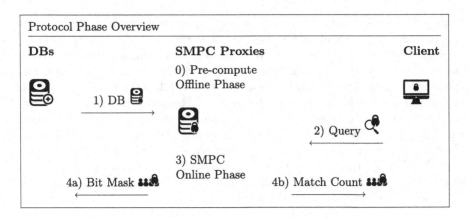

Fig. 2. Protocol phases. Note that all communication with the SMPC proxies happens secret-shared, such that the proxies never gain access to any plaintext.

records, the query response will be the empty set $\emptyset$, i.e., the same as if no record matched the query.

Phase (0) SMPC Offline Phase. At any point before phase 3, D' and D'' pre-compute the SMPC offline phase, which is independent of the actual inputs from other parties.

Phase (1) Database Outsourcing. Each database provider DB_i is assumed to hold their data in the VQF format (see Sect. 3.2). The provider then generates a random mask of the size of its database and sends the random mask to proxy D' and the XOR of the mask and its database to D''. The proxies concatenate the shares they receive from all database providers and keep track of the mapping of shares to DB providers. Note that this phase needs to be performed only once. The secret-shared database can be queried multiple times. Databases need only send new entries or updates to existing entries if they have changed.

Phase (2) Client Query. Client C secret-shares its query between D' and D'' and sends in plain text the type of auxiliary queries it wants to run. Note that we only reveal the *operation* of the auxiliary queries and not the values that are compared with the datasets.

Phase (3) SMPC Online Phase. The proxies D' and D'' run the SMPC protocol on the databases and the query they received. Due to memory constraints, the client query is run on a single patient dataset, consisting of up to 3 million variants at a time. Multiple patients' datasets can be evaluated in separate SMPC instances, which can be run sequentially or in parallel. Given enough hardware, this step can be ideally parallelized. The individual outcomes (match/no match) are stored in a bit mask, still secret-shared and thus unknown to the proxies.

In our scenario, after the query was run against all patients' datasets, the resulting bit mask is processed by a final circuit that counts the number of matches, i.e., the Hamming weight, optionally compared to a threshold.

Phase (4) Output reconstruction. Both proxies D' and D'' hold the output of the computation in secret-shared form and send their output shares to C, who computes the XOR and thereby receives the plaintext output. If the optional threshold t privacy is enforced, C will only receive the count if there are more than t matches. Each database also receives the two shares of its sub-bit mask of matches, together with a random query ID for potential follow-up. Reconstruction will reveal the matching genomes, if any, or an all-zero bit string otherwise.

4.2   Security Considerations

We discuss the security of our scheme and the attacker model in this section.

The goal of our protocol is to ensure the privacy of the queries clients send to the service, as well as the corresponding responses they receive. At the same time, our protocol ensures the privacy of the genomic databases that outsource their data to our service. We achieve these properties by directly relying on the proven security of the GMW protocol [15], which allows to privately evaluate any computable function that is represented as a Boolean circuit. GMW uses XOR-based secret-sharing as underlying primitive, which protects private values. In its original form secret sharing offers information theoretic security since plaintexts are masked with randomness of the same length. We use a PRG to expand a short seed to the length of the plaintext, thus reducing information theoretic security to computational security of the PRG. In our implementation we rely on AES as PRG.

We make the assumption that adversaries behave semi-honestly and corrupt at most one of the two proxies at the same time. The latter corresponds to a non-collusion assumption between the two proxies. Given the semi-honest non-collusion assumptions we can proof that our protocol is secure, since the transcript of every party can be simulated given their respective inputs and outputs. We consider the following cases of semi-honest corrupt parties, malicious clients or external adversaries:

Corrupt Semi-honest Client/Corrupt Database: A corrupt client or corrupt database with input query Q and response R can be simulated by a simulator playing the role of the two proxies, running the GMW protocol. This implies that a corrupt client or database learns no additional information from the protocol execution.

Corrupt Semi-honest Proxy: For a set of databases and a single client query each separate proxy's view consists of a share of the client's query and a share of each database. In all cases XOR-secret-sharing is used, which makes the corresponding strings appear uniformly distributed and leaks no information about the content. Each proxy's view also contains the other proxy's inputs for the GMW protocol. These are proven not to leak information about private inputs in [14].

Malicious Clients: Since the proxies and the client interactively agree on the query structure beforehand, a malicious client can only influence values within the boundaries of the query, i.e., the client can only send an input bit string of the pre-defined length. Malicious clients can send bogus queries with the correct length, which will be processed by the proxies. This leads to corrupt outputs that leak no more information than valid queries. The number of queries a client can send can be controlled by using rate-limiting.

External Adversaries: Data in transit is protected from external adversaries by using state-of-the-art secure channels, e.g., TLS, to ensure confidentiality, integrity and authenticity between all communicating parties.

5   Implementation

In this section we describe our implementation decisions and the software design of our application, as well as its limitations.

5.1   The ABY Framework

We implemented our protocols within the ABY framework for secure two-party computation [9]. It provides efficient C++ implementations of recent secure two-party computation protocols and includes many recent optimizations. ABY offers abstractions of the underlying protocols and building blocks and is thus a viable option for implementing SMPC applications. More specifically, we rely on the included implementation of the GMW protocol, which is based on XOR secret-sharing and thus is well-suited for our outsourcing scenario.

5.2   Boolean Circuit Design

The GMW protocol operates on Boolean circuits. Our protocol contains two circuit designs that we optimized for a low multiplicative depth in order to minimize the number of communication rounds between the proxies. The biggest circuit is the query circuit that checks if a user query matches the genome of a patient. It consists of an equality gate that compares every patient variant with a query variant. On the circuit level, this is done in parallel on a person's entire variants and all query variants. From this we get 1 bit per patient variant, which is fed into an OR tree that returns 1 if at any position the query matched with a patient's variant. For all query variants the results from the OR trees are fed into an AND tree that returns 1 if all query variants are in the patient's variants. These gate trees are the reason for the logarithmic circuit depth. The circuit also checks for auxiliary patient properties, which are implemented as single equality or comparison gates.

The circuit's final result is a single bit that indicates if all query variants are in the person's variant dataset and if all auxiliary queries matched. The circuit output for each patient record is stored (still secret-shared). As soon as all patient queries have been run, the previously stored result shares are fed into a smaller (threshold-)counting circuit (that optionally checks that the count is larger than the threshold t). It consists of a Hamming weight circuit that counts the number of 1-bits, and a comparison gate that controls if a multiplexer gate outputs the string of matches or an all-zero bitstring. Its output is a bit string with one bit per patient. Bits are set to one at the indices where the query matched. If it contains more than t ones, it is output, otherwise an all-zero bit string is output, in case threshold t counting is applied.

5.3 Limitations of Our Approach

While we only ran queries, where *all* conditions must be met, i.e., all conditions are connected with AND ($\land$) expressions, the ABY framework would easily allow for more complex or nested queries, such as $A \lor ((B \land C) \lor D)$. The performance impact would be minimal and would only depend on the size of the formula. Universal circuits [17,26,27,29,36] could be used to also hide the structure of the formula.

The translation of variants into 16 bit strings (see Sect. 3.2) certainly is a limitation and cannot reflect the full spectrum of possible variations. However, as elaborated before, we used this compression as a way to match similar variations while only using strict equality queries.

6 SMPC Benchmarks

In this section we provide benchmark results of our implementation. We performed runtime benchmarks of our SMPC implementation on two identical desktop computers with 16 GiB RAM and a 3.6 GHz Intel Core i7-4790 CPU, connected via local Gbps network. We ran a 64-bit Debian Jessie with Linux kernel 3.16 and used gcc v4.9.2 to compile our code. For all measurements we instantiate the parameters to achieve a symmetric security level of 128 bits. All results are averaged over 25 iterations. The provided communication numbers are the sum of sent and received data of one SMPC proxy. In the following section we use the term offline phase to refer to step (0) from Figure 2, while online phase refers to step (3). We did not measure the time for steps (1), (2), and (4), i.e., the conversion and transmission of databases and query/response, as these are simple and efficient plaintext operations and data transmissions that scale linearly with size and available bandwidth.

6.1   Variant Query Performance

In this section we measure the performance of running a query with a certain length against a person's dataset with a given number of variants. As default parameters we use $N = 100,000$ variants, query length $= 5$, and $\kappa + \psi = 48$ bit, which corresponds to a query on a person's exome. In Figure 3 we show the runtimes of the offline and the online phase for *varying number of patient variants* with queries of length 5. In Figure 4 we fix the number of a patient's variants to $N = 100,000$ entries and show how the runtimes scale for *varying query length*. Figure 5 shows how a *varying entry bitlen* $(\kappa + \psi)$ influences the protocol runtimes. We provide the corresponding detailed numbers in Tables 3, 4, and 5.

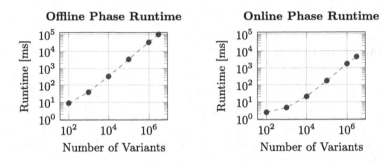

Fig. 3. Offline and online runtime in ms for *varying number of variants per patient* and fixed key length $\kappa = 32$ bit, value length $\psi = 16$ bit, and a query with 5 components.

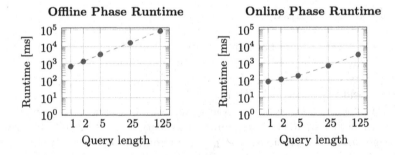

Fig. 4. Offline and online runtime in ms for *varying query length* and fixed key length $\kappa = 32$ bit, value length $\psi = 16$ bit, and variant count of $N = 100,000$ entries.

In all cases the offline and online runtime and circuit size (number of AND gates) increase linearly with the database size, query length, and entry bitlength. The circuit depth, i.e., the number of communication rounds between the two proxies, scales only with the logarithm of the input sizes.

Offline Phase Runtime   **Online Phase Runtime**

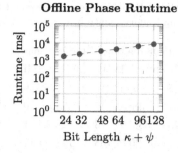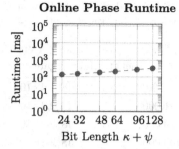

Fig. 5. Offline and online runtime in ms for *varying total element size* $\kappa + \psi$ at a fixed variant count of $N = 100,000$ entries and query length of 5.

Regarding the auxiliary queries, we fixed five different equality and range queries, which didn't have any noticeable performance impact. They are thus omitted in the following discussions.

For querying a patient's exome variants ($N = 100,000$), assuming a query of length 5 and our proposed entry format with key length $\kappa = 32$ bit, and value length $\psi = 16$ bit, we achieve an offline runtime of 3.4 s and an online runtime of 178 ms. In this case we need to transfer 733 MiB in the offline phase and the online phase requires 28 communication rounds with a total transmission of 11 MiB. The circuit for these parameters consists of 24 million AND gates. Using the same parameters to query a patient's full genome ($N = 3,000,000$) requires an offline and online runtime of 99.5 s and 4.6 s, respectively.

Our performance is comparable to the single-variant query in Sousa et al. [35], which takes 2.4–4.3 s online runtime for 5 million variants. However, their query is not extensible and can only answer whether a single variant is present, without the option for further privacy-preserving aggregation.

6.2 Count Performance

The circuit (f in Sect. 3.3) that determines the total number of matches and compares this to the privacy threshold t is very small. For processing the results of 100,000 patient records, the t-threshold count circuit consists of 100,040 AND gates and has a depth of 22. It requires 75 ms (55 ms) runtime and 439 KiB (2,124 KiB) communication in the online (offline) phase. Communication, runtime, and circuit size scale linearly with the input size, while circuit depth grows logarithmically. Since these numbers are negligible for the total runtime, we omit a more detailed analysis at this point.

6.3 Conclusions from the Benchmarks

We consider our solution practical for typical private genome queries. While the computation of responses for databases with thousands of patients still do not answer instantaneously, we can run these private queries over night or increase

Table 3. Benchmark results and circuit properties for *varying variant count* at fixed key length $\kappa = 32$ bit, value length $\psi = 16$ bit, and query length 5.

Variants N	Query length [bit]	$\kappa + \psi$	#AND gates	Circuit depth	Offline phase [ms]	[MiB]	Online phase [ms]	[MiB]
100	5	48	$2.4 \cdot 10^4$	18	9	1	2.4	0
1,000	5	48	$2.4 \cdot 10^5$	21	38	7	4.5	0
10,000	5	48	$2.4 \cdot 10^6$	25	335	73	20.2	1
100,000	5	48	$2.4 \cdot 10^7$	28	3,420	733	177.9	11
1,000,000	5	48	$2.4 \cdot 10^8$	31	34,297	7,325	1,756.2	114
3,000,000	5	48	$7.2 \cdot 10^8$	33	99,507	21,975	4,567.7	343

Table 4. Benchmark results and circuit properties for *varying query length* at fixed key length $\kappa = 32$ bit, value length $\psi = 16$ bit, and variant count $N = 100,000$.

Variants N	Query length [bit]	$\kappa + \psi$	#AND gates	Circuit depth	Offline phase [ms]	[MiB]	Online phase [ms]	[MiB]
100,000	1	48	$4.8 \cdot 10^6$	27	669	146	83.2	2
100,000	2	48	$9.6 \cdot 10^6$	27	1,327	293	111.5	5
100,000	5	48	$2.4 \cdot 10^7$	28	3,420	733	177.9	11
100,000	25	48	$1.2 \cdot 10^8$	29	16,629	3,662	687.5	57
100,000	125	48	$6.0 \cdot 10^8$	32	83,141	18,312	3,096.9	286

Table 5. Benchmark results and circuit properties for *varying total element size* $\kappa + \psi$ at fixed query length of 5 and variant count $N = 100,000$.

Variants N	Query length [bit]	$\kappa + \psi$	#AND gates	Circuit depth	Offline phase [ms]	[MiB]	Online phase [ms]	[MiB]
100,000	5	24	$1.2 \cdot 10^7$	27	1,662	366	133.2	6
100,000	5	32	$1.6 \cdot 10^7$	27	2,241	488	147.9	8
100,000	5	48	$2.4 \cdot 10^7$	28	3,420	733	177.9	11
100,000	5	64	$3.2 \cdot 10^7$	28	4,409	977	205.6	15
100,000	5	96	$4.8 \cdot 10^7$	29	6,640	1,465	269.6	23
100,000	5	128	$6.4 \cdot 10^7$	29	8,860	1,953	321.3	31

throughput by running them in parallel on dedicated hardware and faster networks. As we can see from our performance evaluation, both runtime and communication complexity scale linearly with the input size. Our circuit constructions are optimized for use with the GMW protocol and their depth grows only logarithmically with increasing input size. Network traffic and memory requirements are well within the limits of modern hardware.

The generation of the bit string representing the matching genomes takes the bulk computation and communication cost, while the cost for evaluating the auxiliary conditions, aggregation, and threshold comparison is negligible. Therefore, possibly complex and versatile post-processing functions can be applied to the matches thanks to the use of generic SMPC techniques. This ability sets our system apart from related works in this field.

7 Summary

In this work we have presented a new, privacy-preserving protocol to allow multi-center variant queries on genomic databases. The achieved performance renders this approach applicable in real-world scenarios with some dozens of centers. Full genome studies are supported based on the state-of-the-art VCF format. Our approach leverages the custom Variant Query Format, which can be built from existing VCF data. Variants have to be called against a pre-defined reference genome to be agreed upon before setting up the federated data analysis platform. An interesting and demanding research question immediately arises: how would one use our (and many previously developed) genomic privacy techniques on genomic data while facing the problem of different reference genomes? The simple answer is to regenerate the genomic data against the new reference genome via the same pipeline. But this approach might not always be feasible or possible, if the pipeline is not fully automated or recorded. Note that this applies to almost all previous work where genomic data from different patients is compared, as is the case in the Beacon project. This "transcription" to other reference genomes is beyond the scope of the present work but will be addressed in the future.

Another open problem is how to effectively mitigate the re-identification risk as presented in [34], while still handling queries of rare mutations in a sensible way. We described two query types which our framework supports: a regular count, which is susceptible to the aforementioned re-identification risk, even though to a lesser extent, since the querier doesn't learn in which databases the matches occurred, and a threshold-t-count, which only outputs the count if it is larger than t. While the latter method provides more privacy, it may render the system unusable for very rare mutations.

Acknowledgments. We thank the anonymous reviewers and our shepherd for their valuable feedback on our paper. This work has been supported by the German Federal Ministry of Education and Research (BMBF) and by the Hessian State Ministry for Higher Education, Research and the Arts (HMWK) within CRISP (www.crisp-da.de), by the DFG as part of project E4 within the CRC 1119 CROSSING, as well as by collaborations within the BMBF-funded HiGHmed consortium.

References

1. Aziz, A., Momin, M., Hasan, M.Z., Mohammed, N., Alhadidi, D.: Secure and efficient multiparty computation on genomic data. In: 20th International Database Engineering and Applications Symposium (IDEAS 2016), pp. 278–283. ACM (2016)

2. Asharov, G., Halevi, S., Lindell, Y., Rabin, T.: Privacy-preserving search of similar patients in genomic data. Cryptology ePrint Archive, Report 2017/144 (2017). https://eprint.iacr.org/2017/144

3. Asharov, G., Lindell, Y., Schneider, T., Zohner, M.: More efficient oblivious transfer and extensions for faster secure computation. In: 20th ACM Conference on Computer and Communications Security (CCS 2013), pp. 535–548. ACM (2013)

4. Baldi, P., Baronio, R., De Cristofaro, E., Gasti, P., Tsudik, G.: Countering GATTACA: efficient and secure testing of fully-sequenced human genomes. In: 18th ACM Conference on Computer and Communications Security (CCS 2011), pp. 691–702. ACM (2011)

5. Beaver, D.: Efficient multiparty protocols using circuit randomization. In: Feigenbaum, J. (ed.) CRYPTO 1991. LNCS, vol. 576, pp. 420–432. Springer, Heidelberg (1992). https://doi.org/10.1007/3-540-46766-1_34

6. De Cristofaro, E., Tsudik, G.: Practical private set intersection protocols with linear complexity. In: Sion, R. (ed.) FC 2010. LNCS, vol. 6052, pp. 143–159. Springer, Heidelberg (2010). https://doi.org/10.1007/978-3-642-14577-3_13

7. Danecek, P., et al.: The variant call format and VCFtools. Bioinformatics 27(15), 2156–2158 (2011)

8. De Cristofaro, E., Faber, S., Tsudik, G.: Secure genomic testing with size- and position-hiding private substring matching. In: 12th ACM Workshop on Privacy in the Electronic Society (WPES 2013), pp. 107–118. ACM (2013)

9. Demmler, D., Schneider, T., Zohner, M.: ABY - a framework for efficient mixed-protocol secure two-party computation. In: 22th Network and Distributed System Security Symposium (NDSS 2015). The Internet Society (2015)

10. El Gamal, T.: A public key cryptosystem and a signature scheme based on discrete logarithms. IEEE Trans. Inf. Theory 31(4), 469–472 (1985)

11. Fritz, M.H.Y., Leinonen, R., Cochrane, G., Birney, E.: Efficient storage of high throughput DNA sequencing data using reference-based compression. Genome Res. 21(5), 734–740 (2011)

12. Froelicher, D., et al.: UnLynx: a decentralized system for privacy-conscious data sharing. Proc. Priv. Enhancing Technol. 4, 152–170 (2017)

13. Fuller, B., et al.: SoK: cryptographically protected database search. In: 38th IEEE Symposium on Security and Privacy (S&P 2017), pp. 172–191 (2017)

14. Goldreich, O.: The Foundations of Cryptography - Volume 2, Basic Applications. Cambridge University Press, Cambridge (2004)

15. Goldreich, O., Micali, S., Wigderson, A.: How to play any mental game or a completeness theorem for protocols with honest majority. In: 19th ACM Conference on Theory of Computing (STOC 1987), pp. 218–229. ACM (1987)

16. Goodrich, M.T.: The mastermind attack on genomic data. In: 30th IEEE Symposium on Security and Privacy (S&P 2009), pp. 204–218. IEEE (2009)

17. Günther, D., Kiss, Á., Schneider, T.: More efficient universal circuit constructions. In: Takagi, T., Peyrin, T. (eds.) ASIACRYPT 2017. LNCS, vol. 10625, pp. 443–470. Springer, Cham (2017). https://doi.org/10.1007/978-3-319-70697-9_16. http://thomaschneider.de/papers/GKS17.pdf. Full version: http://ia.cr/2017/798

18. Gupta, D., Mood, B., Feigenbaum, J., Butler, K., Traynor, P.: Using intel software guard extensions for efficient two-party secure function evaluation. In: Clark, J., Meiklejohn, S., Ryan, P.Y.A., Wallach, D., Brenner, M., Rohloff, K. (eds.) FC 2016. LNCS, vol. 9604, pp. 302–318. Springer, Heidelberg (2016). https://doi.org/ 10.1007/978-3-662-53357-4_20
19. Hamacher, K., Hubaux, J.P., Tsudik, G.: Genomic Privacy (Dagstuhl Seminar 13412). Dagstuhl Rep. **3**(10), 25–35 (2014). http://drops.dagstuhl.de/opus/ volltexte/2014/4426
20. Hasan, M.Z., Mahdi, M.S.R., Mohammed, N.: Secure count query on encrypted genomic data. In: 3rd International Workshop on Genome Privacy and Security (GenoPri 2016) (2017). https://arxiv.org/abs/1703.01534
21. Huang, Y., Evans, D., Katz, J., Malka, L.: Faster secure two-party computation using garbled circuits. In: 20th USENIX Security Symposium (USENIX Security 2011). USENIX (2011)
22. Ishai, Y., Kilian, J., Nissim, K., Petrank, E.: Extending oblivious transfers efficiently. In: Boneh, D. (ed.) CRYPTO 2003. LNCS, vol. 2729, pp. 145–161. Springer, Heidelberg (2003). https://doi.org/10.1007/978-3-540-45146-4_9
23. Jha, S., Kruger, L., Shmatikov, V.: Towards practical privacy for genomic computation. In: 29th IEEE Symposium on Security and Privacy (S&P 2008), pp. 216–230. IEEE (2008)
24. Karvelas, N., Peter, A., Katzenbeisser, S., Tews, E., Hamacher, K.: Privacy-preserving whole genome sequence processing through proxy-aided ORAM. In: 13rd Workshop on Privacy in the Electronic Society (WPES 2014), pp. 1–10. ACM (2014)
25. Kerschbaum, F., Beck, M., Schönfeld, D.: Inference control for privacy-preserving genome matching. CoRR abs/1405.0205 (2014). https://arxiv.org/abs/1405.0205
26. Kiss, Á., Schneider, T.: Valiant's universal circuit is practical. In: Fischlin, M., Coron, J.-S. (eds.) EUROCRYPT 2016. LNCS, vol. 9665, pp. 699–728. Springer, Heidelberg (2016). https://doi.org/10.1007/978-3-662-49890-3_27
27. Kolesnikov, V., Schneider, T.: A practical universal circuit construction and secure evaluation of private functions. In: Tsudik, G. (ed.) FC 2008. LNCS, vol. 5143, pp. 83–97. Springer, Heidelberg (2008). https://doi.org/10. 1007/978-3-540-85230-8_7. http://thomaschneider.de/papers/KS08UC.pdf. Code: http://encrypto.de/code/FairplayPF
28. Li, H., et al.: The Sequence Alignment/Map format and SAMtools. Bioinformatics **25**(16), 2078–2079 (2009)
29. Lipmaa, H., Mohassel, P., Sadeghian, S.S.: Valiant's universal circuit: improvements, implementation, and applications. IACR Cryptology ePrint Archive 2016(17) (2016). http://ia.cr/2016/017
30. Naehrig, M., Lauter, K.E., Vaikuntanathan, V.: Can homomorphic encryption be practical? In: 3rd ACM Cloud Computing Security Workshop (CCSW 2011), pp. 113–124. ACM (2011)
31. Nielsen, J.B., Nordholt, P.S., Orlandi, C., Burra, S.S.: A new approach to practical active-secure two-party computation. In: Safavi-Naini, R., Canetti, R. (eds.) CRYPTO 2012. LNCS, vol. 7417, pp. 681–700. Springer, Heidelberg (2012). https://doi.org/10.1007/978-3-642-32009-5_40
32. Paillier, P.: Public-key cryptosystems based on composite degree residuosity classes. In: Stern, J. (ed.) EUROCRYPT 1999. LNCS, vol. 1592, pp. 223–238. Springer, Heidelberg (1999). https://doi.org/10.1007/3-540-48910-X_16
33. Perillo, A.M., Cristofaro, E.D.: PAPEETE: private, authorized, and fast personal genomic testing. Technical report 770 (2017). https://ia.cr/2017/770

34. Shringarpure, S., Bustamante, C.: Privacy risks from genomic data-sharing beacons. Am. J. Hum. Genet. **97**(5), 631–646 (2015)

35. Sousa, J.S., et al.: Efficient and secure outsourcing of genomic data storage. BMC Med. Genomics **10**(2), 46 (2017)

36. Valiant, L.G.: Universal circuits (preliminary report). In: 8th ACM Symposium on Theory of Computing (STOC 1976), pp. 196–203. ACM (1976)

37. Wang, X.S., Huang, Y., Zhao, Y., Tang, H., Wang, X., Bu, D.: Efficient genome-wide, privacy-preserving similar patient query based on private edit distance. In: 22nd ACM Conference on Computer and Communications (CCS 2015), pp. 492–503. ACM (2015)

38. Yao, A.C.C.: How to generate and exchange secrets. In: 27th Symposium on Foundations of Computer Science (FOCS 1986), pp. 162–167. IEEE (1986)

39. You, N., et al.: SNP calling using genotype model selection on high-throughput sequencing data. Bioinformatics **28**(5), 643 (2012)

40. Zhou, J., Cao, Z., Dong, X.: PPOPM: more efficient privacy preserving outsourced pattern matching. In: Askoxylakis, I., Ioannidis, S., Katsikas, S., Meadows, C. (eds.) ESORICS 2016. LNCS, vol. 9878, pp. 135–153. Springer, Cham (2016). https://doi.org/10.1007/978-3-319-45744-4_7

A New Secure Matrix Multiplication from Ring-LWE

Lihua Wang$^{(\boxtimes)}$, Yoshinori Aono, and Le Trieu Phong

National Institute of Information and Communications Technology, Tokyo, Japan
{wlh,aono,phong}@nict.go.jp

Abstract. Matrix multiplication is one of the most basic and useful operations in statistical calculations and machine learning. When the matrices contain sensitive information and the computation has to be carried out in an insecure environment, such as a cloud server, secure matrix multiplication computation (MMC) is required, so that the computation can be outsourced without information leakage. Dung et al. apply the Ring-LWE-based somewhat public key homomorphic encryption scheme to secure MMC [TMMP2016], whose packing method is an extension of Yasuda et al.'s methods [SCN2015 and ACISP2015] for secure inner product. In this study, we propose a new packing method for secure MMC from Ring-LWE-based secure inner product and show that ours is efficient and flexible.

Keywords: Secure matrix multiplication · Ring-LWE
Public key homomorphic encryption (PHE) · Secure inner product

1 Introduction

Motivation. Matrix multiplication is one of the most basic and useful operations in statistical calculations and machine learning. When the matrices A and B contain sensitive information and the computation has to be carried out in an insecure environment, such as a cloud server, secure matrix multiplication computation (secure MMC) is required, as the data and the computation can be outsourced without information leakage.

Several secure MMC schemes are proposed, in which there are two types according to being based on secure inner product (e.g., [2]) or not (e.g., [3,4]). Since matrix multiplication AB can be calculated from inner product of row vectors of the left matrix A and column vectors of the right matrix B (Fig. 1), secure matrix multiplication can be naturally constructed from secure inner product. In detail, the (i,j)-element of the multiplication AB is $\langle A_i^{(r)}, B_j^{(c)} \rangle$, the inner product of the i-th row vector of A and the j-th column vector of B.

In this study, we focus on constructing secure MMC based on secure inner product from "ring learning with errors" (Ring-LWE)-based somewhat public-key homomorphic encryption (PHE) (see, for instance, [5]) that have the potential to be safe against quantum computers.

© Springer Nature Switzerland AG 2018
S. Capkun and S. S. M. Chow (Eds.): CANS 2017, LNCS 11261, pp. 93–111, 2018.
https://doi.org/10.1007/978-3-030-02641-7_5

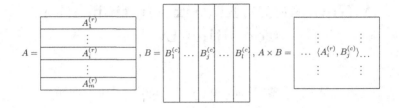

Fig. 1. Matrix multiplication from vector inner products.

Related Works for Secure Inner Product from Ring-LWE-Based PHE.
Several secure inner product protocols have been presented, such as [6,8–10].

- Two-Pack-encoding Method: Yasuda et al. proposed special packing methods for secure inner product [8–10] that can be used for somewhat PHE schemes based on the polynomial ring $\mathcal{R} = \mathbb{Z}[x]/(x^n+1)$. Under their packing method, $\mathsf{pm}^{(1)}$ and $\mathsf{pm}^{(2)}$ denote two different mappings. The basic idea is packing a plaintext $m \in \mathcal{R}_p = \mathbb{Z}_p[x]/(x^n+1)$, whose coefficient vector is $(m_0, m_1, ..., m_{n-1}) \in \mathbb{Z}_p^n$, into the following two different forms:

$$\mathsf{pm}^{(1)}(m) := \sum_{i=0}^{n-1} m_i x^i, \quad \mathsf{pm}^{(2)}(m) := -\sum_{i=0}^{n-1} m_i x^{n-i}. \tag{1}$$

Since $x^n = -1$, for plaintexts $u = \sum_{i=0}^{n-1} u_i x^i$ and $v = \sum_{i=0}^{n-1} v_i x^i$, the product of $\mathsf{pm}^{(1)}(u)$ and $\mathsf{pm}^{(2)}(v)$ is

$$\mathsf{pm}^{(1)}(u) \cdot \mathsf{pm}^{(2)}(v) = \sum_{i=0}^{n-1} u_i v_i + (\text{inconstant terms}). \tag{2}$$

Therefore, by mod x, one can compute the inner product of the coefficient vectors $(u_0, ..., u_{n-1})$ and $(v_0, ..., v_{n-1})$ by $\mathcal{R}_p$-multiplication.
- One-Pack-encoding Method: Recently, Wang et al. proposed a generic yet efficient method for secure inner product [6] that can be also applied to Ring-LWE-based somewhat PHE scheme constructed on the polynomial ring $\mathcal{R} = \mathbb{Z}[x]/(x^n+1)$ (named *WHAP secure inner product* in short). For each plaintext m, this method only requires the first type packing $\mathsf{pm}^{(1)}(m)$ showed in Eq. (1), whereas $\mathsf{pm}^{(2)}(m)$ is not needed for plaintext (See Sect. 2.2 for detail). Correspondingly, using this method, computation cost on client, the storage size on cloud server, and communication cost are reduced. Therefore, it is more efficient than the method introduced in [8] whose key idea is the same with that in [7] and [10].

Previous Secure MMC from Ring-LWE-Based Somewhat PHE Scheme. Recently, Dung et al. in [2] constructed the first secure MMC based on the Ring-LWE-based somewhat PHE scheme [5] which is a variant of [1]. Their packing method (*DMY approach*) for secure matrix multiplication is a generic extension of Yasuda et al.'s methods for secure inner product in [7] and [10]. They proposed two packing methods: for two $m \times m$ square matrices Packing-1 encodes all row vectors of left matrix into one ring element and each column vector of ring matrix into a difference ring element, so $m + 1$ times encryption and m times decryption operations are required in Packing-1. Packing-2 requires twice encryption operations, since it pack all row vectors of a matrix into an element, and all column vectors into another element of ring $\mathcal{R}$ (See Sect. 2.3 for detail). Then, according to Eq. (2), each entry of AB can be obtained from the multiplication of the two elements. Summarily, a matrix needs two types encoding: one is for left matrix and another is for right matrix of multiplication computation. Therefore, two encryption operations for a matrix and one time decryption operation for a MMC are needed. Packing-2 is more efficient on computation/communication cost, and applies to secure MMC for two square matrices as described in [2].

Our Contribution. In this study, by combining and extending packing approaches introduced in [2] and [6], we propose a new secure MMC from Ring-LWE-based secure inner product. Our packing method, named one-pack-encoding method (all row and column vectors of a matrix are embedded into one element of ring), is more flexible and efficient in some cases (Sect. 4) than the existing method introduced by Dung et al. [2]:

(1) *Flexible for size.* We describe a concrete packing method under our scheme to show how to compute the multiplication of rectangular matrices $A = (a_{ij})_{m \times k}$ and $B = (b_{ij})_{k \times l}$;
(2) *Flexible for left/right matrix.* One-pack-encoding for each matrix implies that it can act as either left matrix or right one of secure MMC. For matrix $A = (a_{ij})_{m \times k}$, $B = (b_{ij})_{k \times l}$ and $C = (c_{ij})_{h \times m}$, though there only exits one packing for A, it can be used to compute both AB and CA. Especially, if $A = (a_{ij})_{m \times l}$, $B = (b_{ij})_{l \times m}$, then both AB and BA can be computed without extra encoding.
(3) *Efficiency.* For the special case of both A and B are square matrices of size $\beta \times \beta$ with $\beta^3 \leq n/4$, we compare our method and the existing one to show the efficiency.

Since our packing method can be applied to the same Ring-LWE-based somewhat PHE scheme with Dung et al.'s, ours provides a trade-off to realize more flexible and efficient secure MMC in the case of $\beta^3 \leq n/4$.

Paper Outline. In Sect. 2, we recall WHAP secure inner product scheme and DMY approach to construct secure MMC on a Ring-LWE-based PHE scheme (the LNV-PHE scheme). Then, in Sect. 3, we propose a new secure MMC based on WHAP secure inner product from the above Ring-LWE-based scheme. After that, we show the flexibility and efficiency of our secure MMC in Sect. 4, and make brief concluding remarks in Sect. 5.

2 Related Works

Lauter, Naehrig, and Vaikuntanathan proposed a PHE scheme from Ring-LWE [5], which is a variant of [1], we call the LNV-PHE scheme. The related works of this study - WHAP secure inner product [6] and DMY secure MMC [2] - can be constructed based on the LNV-PHE scheme.

In this section, we describe the Ring-LWE definition and the LNV-PHE scheme as the preliminary (Sect. 2.1), then recall the WHAP secure inner product (Sect. 2.2) and the DMY secure MMC (Sect. 2.3).

PKE part:			
ParamGen(1^λ):	KeyGen(1^λ, pp):	Enc(pk, $m \in \mathcal{R}_p$):	Dec(S, (c_1, c_2)):
Fix $q = q(\lambda) \in \mathbb{Z}^+$	Take $s = s(\lambda, pp) \in \mathbb{R}^+$	$e_1, e_2, e_3 \xleftarrow{g} \mathcal{R}_{(0,s^2)}$	$\tilde{m} = c_1 S + c_2$
$n = n(\lambda) \in \mathbb{Z}^+$	$r, S \xleftarrow{g} \mathcal{R}_{(0,s^2)}$, $a \xleftarrow{\$} \mathcal{R}_q$	$c_1 = e_1 a + p e_2 \in \mathcal{R}_q$	$\in \mathcal{R}_q$
prime $p < q$,	$P = pr - aS \in \mathcal{R}_q$	$c_2 = e_1 P + p e_3 + m \in \mathcal{R}_q$	$m = \tilde{m} \mod p$
Return	Return	Return	Return
$pp = (q, n, p)$	$pk = (a, P)$, $sk = S$	$c = (c_1, c_2) \in \mathcal{R}_q^2$	$m \in \mathcal{R}_p$
Homomorphic part: Given Enc(m) = (c_1, c_2) = c and Enc(m') = (d_1, d_2) = c'			
Add(c, c'):		DecA(S, c_{add}) = Dec(S, c_{add}):	
Return $c_{add} = c + c' \in \mathcal{R}_q$		Return $m + m' \in \mathcal{R}_p$	
Mul(c, c'):		DecM(S, c_{mul}):	
$c_{mul} = (\text{Mul}_1, \text{Mul}_2, \text{Mul}_3)$		mm'	
$= (c_1 d_1, c_1 d_2 + c_2 d_1, c_2 d_2) \mod q$		$\text{Mul}_1 S^2 + \text{Mul}_2 S + \text{Mul}_3 \mod q \mod p$	
Return $c_{mul} \in \mathcal{R}_q$		Return $mm' \in \mathcal{R}_p$	

Here, $\mathcal{R} = \mathbb{Z}[x]/(x^n + 1)$, $\mathcal{R}_q = \mathcal{R}/q$ and $\mathcal{R}_p = \mathcal{R}/p$. Moreover, $\mathcal{R}_{(0,s^2)}$ stands for polynomials in $\mathcal{R}$ with small Gaussian coefficients $\xleftarrow{g} \mathbb{Z}_{(0,s^2)}$.

Fig. 2. The Ring-LWE-based LNV scheme [5].

2.1 Preliminary

Security of the LNV-PHE scheme is based on the Ring-LWE assumption.

The Ring-LWE Assumption. Let $\mathcal{R} := \mathbb{Z}[x]/f(x)$ and $\mathcal{R}_q := \mathbb{Z}_q[x]/f(x)$ for some degree n irreducible integer polynomial $f(x) \in \mathbb{Z}[x]$ and a prime integer $q \in \mathbb{Z}$. Addition in these rings is done component-wise in their coefficients. Multiplication is polynomial multiplication modulo $f(x)$ (and also q, in the case of the ring $\mathcal{R}_q$).

Given parameters (n, q, s), the Ring-LWE assumption asserts that it is infeasible to distinguish the following two distributions:

- one samples $(a_i, b_i) \xleftarrow{\$} (\mathcal{R}_q)^2$, i.e., uniformly from $(\mathcal{R}_q)^2$;
- one samples $(a_i, b_i = a_i s + e_i)$, where $a_i \xleftarrow{\$} \mathcal{R}_q$, and $s, e_i \xleftarrow{g} \mathcal{R}_{(0,s^2)}$, i.e., coefficients sampled from Gaussian distribution $\mathbb{Z}_{(0,s^2)}^n$.

The LNV-PHE Scheme ([5]). We recall Lauter et al.'s Ring-LWE based PHE scheme in Fig. 2, in which the Ring-LWE problem for specific choices of the polynomial $f(x) = x^n + 1$ for n a power of two.

The scheme consists of four algorithms: ParamGen (parameter generation), KeyGen (public/secret key pair generation), Enc (encryption), and Dec (decryption). In addition, algorithms Add, DecA and Mul, DecM show that the scheme allows addition and multiplication homomorphic computations.

Theorem 1. *The ciphertexts of the scheme in Fig. 2 are indistinguishable from random under the Ring-LWE assumption.*

Proof. The proof is folklore and is given for completeness. First, we have $p^{-1}P = r + (-p^{-1}a)S \in \mathcal{R}_q$ so that $p^{-1}P$ is uniformly random under the Ring-LWE assumption with secret S. Therefore P is also uniformly random under the Ring-LWE assumption. Second, $p^{-1}c_1 = e_1(p^{-1}a) + e_2 \in \mathcal{R}_q$ and $p^{-1}c_2 = e_1(p^{-1}P) + e_3 + p^{-1}m \in \mathcal{R}_q$, so that $p^{-1}c_1$ and the mask $e_1(p^{-1}P) + e_3$ are uniformly random under the Ring-LWE assumption with secret e_1. Therefore, $p^{-1}c_1$ and $p^{-1}c_2$ are random, so is the ciphertext (c_1, c_2) as claimed. □

2.2 WHAP Secure Inner Product Based on Ring-LWE-Based PHE

Wang, Hayashi, Aono, and Phong proposed a generic yet efficient method for secure inner product [6], which can be applied to a secure inner product from the LNV scheme - named *WHAP secure inner product scheme* in short. The following list of notation should be useful.

Notation List-1.

Coefficient vector: $\mathsf{Vec}(\cdot)$ denotes the coefficient vector of an element of $\mathcal{R}$, e.g., $\mathsf{Vec}(u) = (u_0, u_1, ..., u_{n-1})$ for $u = u_0 + u_1 x + ... + u_{n-1}x^{n-1}$.

Transpose: $u^{(t)} := -\sum_{i=0}^{n-1} u_i x^{n-i} \in \mathcal{R}$, named u's transpose, can be obained easily for any given $u \in \mathcal{R}$.

The WHAP Secure Inner Product scheme consists of algorithms $\mathsf{ParamGen}(1^\lambda) \to pp$, $\mathsf{KeyGen}(1^\lambda) \to (pk, sk)$, $\mathsf{Enc}(pk, m) \to c$, $\mathsf{Dec}(sk, c) \to m$ (as in Fig. 2), and $\mathsf{InnerP}(c, c') \to ip$ and $\mathsf{DecIP}(sk, ip) \to \langle \mathsf{Vec}(m), \mathsf{Vec}(m') \rangle$ defined as follows:

- $\mathsf{InnerP}(c, c')$: For $c = \mathsf{Enc}(pk, m) = (c_1, c_2)$ and $c' = \mathsf{Enc}(pk, m') = (d_1, d_2) \in \mathcal{R}_q^2$, define
$$ip := ((W_1, W_2), \xi),$$
where $W_1 = c_1^{(t)}d_1 \in \mathcal{R}_q$, $W_2 = d_1^{(t)}c_2 + c_1^{(t)}d_2 \in \mathcal{R}_q$, $\xi = c_2^{(t)}d_2 \mod x \in \mathbb{Z}_q$.
- $\mathsf{DecIP}(S, ip)$:
$$IP := (W_1 S^* + W_2 S^{(t)} \mod x) + \xi \mod q,$$
where $S^* = SS^{(t)} \in \mathcal{R}_q$ is precomputed. Then return the inner product of the coefficient vectors of plaintexts m and m'
$$\langle \mathsf{Vec}(m), \mathsf{Vec}(m') \rangle := IP \mod p.$$

2.3   DMY Approach for Secure MMC from Ring-LWE-Based PHE

Dung, Mishra and Yasuda proposed the first secure MMC from Ring-LWE-based PHE [2]. We recall how to use their approaches (packing-1 and packing-2) - *the DMY approach* - to construct secure MMC from the LNV scheme. Their secure MMC constructed by packing-2 applies to multiplication of two square matrices of size $m \times m$. In this subsection, for any matrix M of size $m \times m$, the following list of notation should be useful.

Notation List-2.

> Row vector: $M_i^{(r)} \in \mathbb{Z}^{1 \times m}$ denotes the i-th row vector of M, for $i = 1, ..., m$.
>
> Column vector: $M_j^{(c)} \in \mathbb{Z}^{1 \times m}$ denotes transpose of the j-th column vector
> $(M_j^{(c)})^T \in \mathbb{Z}^{m \times 1}$ of M, for $j = 1, ..., m$.
>
> Variant vectors: $X = (1, x, ..., x^{m-1})^T$ and $Y = (-x^n, -x^{n-1}, ..., -x^{n-(m-1)})^T$.

Packing-1. For matrix $M \in \mathbb{Z}_p^{m \times m}$, packing all row vectors of M into one element $\widetilde{M}[rows](x) \in \mathcal{R}_p$ and m column vectors of M into m elements $\widetilde{M}_j[column](x) \in \mathcal{R}_p$ $(j = 1, ..., m)$ as follows:

$$\widetilde{M}[rows](x) = M_1^{(r)} X + ... + x^{(i-1)m} M_i^{(r)} X + ... + x^{(m-1)m} M_m^{(r)} X, \quad (3)$$

$$\widetilde{M}_j[column](x) = M_j^{(c)} Y \quad (j = 1, ..., m).$$

For given two matrices A and $B \in \mathbb{Z}_p^{m \times m}$ to compute the multiplication AB, according to Eq. (2), it can be checked that

$$\widetilde{A}[rows](x)\widetilde{B}_j[column](x)$$
$$= \langle A_1^{(r)}, B_j^{(c)} \rangle + ... + \langle A_i^{(r)}, B_j^{(c)} \rangle x^{(i-1)m} + ... + \langle A_m^{(r)}, B_j^{(c)} \rangle x^{(m-1)m} + ...$$

holds for $j = 1, ..., m$.

Therefore, the corresponding secure MMC for two matrices with size of $m \times m$ under the above Packing-1 method requires: $m+1$ encryption operations for each matrix and m decryption operations for each MMC.

As an enhanced DMY approach, the following Packing-2 method is also proposed to reduce Encryption/Decryption operations in [2].

Packing-2. For matrix $M \in \mathbb{Z}_p^{m \times m}$, packing all row vectors of M into one element $\widetilde{M}[rows](x) \in \mathcal{R}_p$ via Eq. (3) similar to that in Packing-1, and all column vectors of M into another element $\widetilde{M}[columns](x) \in \mathcal{R}_p$ as follows:

$$\widetilde{M}[columns](x) = M_1^{(c)} Y + ... + x^{(j-1)m^2} M_j^{(c)} Y + ... + x^{(m-1)m^2} M_m^{(c)} Y. (4)$$

For given two matrices A and $B \in \mathbb{Z}_p^{m \times m}$ to compute the multiplication AB, according to Eq. (2), it can be checked that

$$\widetilde{A}[rows](x)\widetilde{B}[columns](x) = \langle A_1^{(r)}, B_1^{(c)} \rangle + ... + \langle A_i^{(r)}, B_j^{(c)} \rangle x^{(i-1)m+(j-1)m^2}$$
$$+ ... + \langle A_m^{(r)}, B_m^{(c)} \rangle x^{(m-1)m+(m-1)m^2} + ... \in \mathcal{R}_p$$

holds.

Now we describe the *DMY secure MMC* under the Packing-2 method based on the LNV scheme (Fig. 2) step by step below:

Step 1 (Encode-then-Encrypt). Input matrices A and B, output ciphertexts (c_1, c_2) and (d_1, d_2) of their corresponding encoded elements $\widetilde{A}[rows](x)$ from Eq. (3) and $\widetilde{B}[columns](x)$ from Eq. (4), respectively.

$$\mathsf{Enc}(pk, \widetilde{A}[rows](x)) = (c_1, c_2), \mathsf{Enc}(pk, \widetilde{B}[columns](x)) = (d_1, d_2) \in \mathcal{R}_q \times \mathcal{R}_q,$$

where Enc is encryption algorithm defined in Fig. 2.

Step 2 (Multiplication Homomorphic Computation). Input (c_1, c_2) and (d_1, d_2), output

$$\mathsf{Mul} := (\mathsf{Mul}_1, \mathsf{Mul}_2, \mathsf{Mul}_3) = (c_1 d_1, c_2 d_1 + c_1 d_2, c_2 d_2) \in \mathcal{R}_q \times \mathcal{R}_q \times \mathcal{R}_q.$$

Step 3 (Decrypt-then-Decode). Input $\mathsf{Mul} = (\mathsf{Mul}_1, \mathsf{Mul}_2, \mathsf{Mul}_3)$, output $AB = \left(\langle A_i^{(r)}, B_j^{(c)} \rangle \right)_{m \times m}$ as follows:

$$\mathsf{DecM}(sk, \mathsf{Mul}) := \mathsf{Mul}_1 S^2 + \mathsf{Mul}_2 S + \mathsf{Mul}_3 \mod q \mod p$$
$$= w_0 + w_1 x + \dots + w_{n-1} x^{n-1} \in \mathcal{R}_p,$$

then set $\langle A_i^{(r)}, B_j^{(c)} \rangle = w_{(i-1)m + (j-1)m^2} \in \mathbb{Z}_p$, for $i, j = 1, 2, \dots, m$.

Therefore, the DMY secure MMC under Packing-2 method requires two encryption operations for each matrix and one decryption operation for each MMC. In [2], Dung et al. only showed how to apply the method to compute the multiplication of two square matrices.

3   Our Secure MMC Based on the WHAP Secure Inner Product from a Ring-LWE-Based PHE Scheme

In this section, by combining and extending the WHAP secure inner product [6] recalled in Sect. 2.2 and the DMY packing-2 approach [2] recalled in Sect. 2.3, we propose a secure MMC from Ring-LWE-based secure inner product.

Here, we point out that our encoding method is not a trivial combination of existing techniques. The shift and transpose properties of encrypted data that we introduce and prove in Lemmas 1 and 2 (Sect. 3.1) are novel, which act as the key to realize secure MMC under one decryption. Indeed, thanks to these properties, our one-pack-encoding method works as described in Theorem 2 and Corollary 1-2 (Sect. 3.2). After that, we present our secure MMC step by step and prove the correctness (Sect. 3.3).

3.1   Shifting Under Encrypted Form and Transpose Properties on $\mathcal{R}$

Our MMC is constructed on the LNV scheme from polynomial ring $\mathcal{R} = \mathbb{Z}[x]/(x^n + 1)$ (recalled in Fig. 2), which supposes the shift and transpose properties under encrypted form. We summarize and prove the properties below.

Lemma 1 (Shifting). *Given* $\mathsf{Enc}(m) = (c_1, c_2) \in \mathcal{R}_q$, *a ciphertext of message* $m \in \mathcal{R}_p$ *under LNV scheme (Fig. 2). Then*

$$x^\alpha \mathsf{Enc}(m) = Enc(x^\alpha m),$$

for integer α, $1 \leq \alpha \leq n - 1$.

Proof. Since $(c_1, c_2) = (e_1 a + pe_2, e_1 P + pe_3 + m)$, we have

$$
\begin{aligned}
x^\alpha (c_1, c_2) &= (x^\alpha (e_1 a + pe_2), x^\alpha (e_1 P + pe_3 + m)) \\
&= ((x^\alpha e_1)a + p(x^\alpha e_2), (x^\alpha e_1)P + p(x^\alpha e_3) + (x^\alpha m)) \\
&= (e'_1 a + pe'_2, e'_1 P + pe'_3 + (x^\alpha m)),
\end{aligned}
$$

where e'_i are elements in $\mathcal{R}_q$ with Gaussian coefficients, because they are with coefficients shifted from that of $e_i \overset{g}{\leftarrow} \mathbb{Z}_{(0,s^2)}$ $(i = 1, 2, 3)$. In details, assume $e_i = e_{io} + e_{i1}x + \ldots + e_{i,n-1}x^{n-1}$, then for $i = 1, 2, 3$, we have

$$
\begin{aligned}
e'_i = x^\alpha e_i &= e_{io}x^\alpha + e_{i1}x^{\alpha+1} + \ldots + e_{i,n-1}x^{\alpha+n-1} \\
&= -e_{i,n-\alpha} - e_{i,n-\alpha+1}x - \ldots - e_{i,n-1}x^{\alpha-1} + e_{io}x^\alpha + \ldots + e_{i,n-\alpha-1}x^{n-1} \\
&= e'_{io} \qquad\quad + e'_{i1}x + \ldots + e'_{i,\alpha-1}x^{\alpha-1} + e'_{i,\alpha}x^\alpha + \ldots + e'_{i,n-1}x^{n-1},
\end{aligned}
$$

then $e'_{ij} = -e_{i,j+(n-\alpha)}$ when $j = 0, \ldots, \alpha - 1$, and $e'_{ij} = e_{i,j-\alpha}$ when $j = \alpha, \ldots, n - 1$, i.e., the coefficients of e'_i are $\pm$ the coefficients of e_i that are selected from $\overset{g}{\leftarrow} \mathbb{Z}_{(0,s^2)}$. So, the coefficients of e'_i are also from $\overset{g}{\leftarrow} \mathbb{Z}_{(0,s^2)}$. Therefore,

$$x^\alpha \mathsf{Enc}(m) = Enc(x^\alpha m),$$

ending the proof. □

Lemma 2 (Properties of Element Transpose). *For any element* $u, v \in \mathcal{R} = \mathbb{Z}[x]/(x^n + 1)$, *we have*

(P-1) $(x^\alpha u)^{(t)} = -x^{n-\alpha}u^{(t)}$, *where* $1 \leq \alpha \leq n - 1$.
(P-2) $(uv)^{(t)} = u^{(t)}v^{(t)}$.

Proof. Since $x^n = -1$, we have

$$x^\alpha u = -u_{n-\alpha} - u_{n-\alpha+1}x - \ldots - u_{n-1}x^{\alpha-1} + u_0 x^\alpha + u_1 x^{\alpha+1} + \ldots + u_{n-\alpha-1}x^{n-1}$$

over $\mathcal{R}$. Then, the transpose element is

$$
\begin{aligned}
(x^\alpha u)^{(t)} &= u_{n-\alpha}x^n + u_{n-\alpha+1}x^{n-1} + \ldots + u_{n-1}x^{n-(\alpha-1)} - u_0 x^{n-\alpha} - u_1 x^{n-(\alpha+1)} \\
&\quad - \ldots - u_{n-\alpha-1}x \\
&= -x^{n-\alpha}x^\alpha(-u_0 x^{n-\alpha} - u_1 x^{n-(\alpha+1)} - \ldots - u_{n-\alpha-1}x + u_{n-\alpha}x^n \\
&\quad + u_{n-\alpha+1}x^{n-1} + \ldots + u_{n-1}x^{n-(\alpha-1)}) \\
&= -x^{n-\alpha}(-u_0 x^n - u_1 x^{n-1} - \ldots - u_{n-\alpha-1}x^{\alpha+1} - u_{n-\alpha}x^\alpha \\
&\quad - u_{n-\alpha+1}x^{\alpha-1} - \ldots - u_{n-1}x) \\
&= -x^{n-\alpha}u^{(t)}.
\end{aligned}
$$

Therefore, (P-1) holds. Furthermore, since

$$uv = \left(\sum_{i=0}^{n-1} u_i x^i\right)\left(\sum_{j=0}^{n-1} v_j x^j\right)$$

$$= \sum_{i,j=0}^{n-1} u_i v_j x^{i+j} = \sum_{i+j=0}^{n-1} u_i v_j x^{i+j} - \sum_{i+j=n}^{2n-2} u_i v_j x^{i+j-n},$$

we have

$$(uv)^{(t)} = \left(\sum_{i+j=0}^{n-1} u_i v_j x^{i+j}\right)^{(t)} - \left(\sum_{i+j=n}^{2n-2} u_i v_j x^{i+j-n}\right)^{(t)}$$

$$= -\sum_{i+j=0}^{n-1} u_i v_j x^{n-(i+j)} + \sum_{i+j=n}^{2n-2} u_i v_j x^{2n-(i+j)} = \sum_{i+j=0}^{2n-2} u_i v_j x^{n-i} x^{n-j}$$

$$= \left(\sum_{i=0}^{n-1} u_i x^{n-i}\right)\left(\sum_{j=0}^{n-1} v_j x^{n-j}\right) = u^{(t)} v^{(t)},$$

so that (P-2) holds and the proof is complete. $\qquad\square$

3.2 Our New Packing Method for Secure MMC

In this subsection, we present our one-pack-encoding method for secure MMC, and show that it is useful not only for square matrices but also for rectangular ones. For any $m \times k$ matrix A and $k \times l$ matrix B, AB can be computed. Let $\beta = \max\{m, k, l\}$, and assume $\beta^3 \leq n/4$ holds in the following context.

Notation List-3.

Variant vector: $X = (1, x, ..., x^{\beta-1})^T$.
Coefficient vector: $\mathsf{Vec}(\cdot)$ denotes the coefficient vector of an element of $\mathcal{R}$, e.g.,
$\mathsf{Vec}(u) = (u_0, u_1, ..., u_{n-1})$ for $u = u_0 + u_1 x + ... + u_{n-1} x^{n-1}$.

For a square matrix M with dimension size of $\beta \times \beta$, different from the DMY approach, column vectors are also encoded by using variant vector X but not Y as follows:

$$\widetilde{M}[rows](x) = (M_1^{(r)} + ... + x^{(i-1)\beta} M_i^{(r)} + ... + x^{(\beta-1)\beta} M_\beta^{(r)})X,$$
$$\widetilde{M}[columns](x) = (x^{2\times\beta^2-\beta} M_1^{(c)} + ... + x^{2j\beta^2-\beta} M_j^{(c)} + ... + x^{2\beta\times\beta^2-\beta} M_\beta^{(c)})X. \qquad (5)$$

Encode matrix M by embedding all row vectors and column vectors of matrix M into one element that denoted by

$$\begin{aligned}\widetilde{M}(x) &:= \widetilde{M}[rows](x) + \widetilde{M}[columns](x) \\ &= (M_1^{(r)} + x^\beta M_2^{(r)} + ... + x^{(\beta-1)\beta} M_\beta^{(r)})X \\ &\quad + (x^{2\times\beta^2-\beta} M_1^{(c)} + x^{4\times\beta^2-\beta} M_2^{(c)} + ... + x^{2\beta\times\beta^2-\beta} M_\beta^{(c)})X.\end{aligned} \qquad (6)$$

Theorem 2. *For given two square matrices A and B with the same size of $\beta \times \beta$, they are encoded into elements of polynomial ring $\mathcal{R} = \mathbb{Z}[x]/(x^n + 1)$ using the method shown in Eq. (6), then both $AB = \left(\langle A_i^{(r)}, B_j^{(c)} \rangle \right)_{\beta \times \beta}$ and $BA = \left(\langle B_i^{(r)}, A_j^{(c)} \rangle \right)_{\beta \times \beta}$ can be obtained by computing coefficient vector inner products of shifted $\widehat{A}(x)$ and $\widehat{B}(x)$ when $4\beta^3 \leq n$. In detail,*

$$\langle \mathsf{Vec}(x^{2j\beta^2 - \beta}\widehat{A}(x)), \mathsf{Vec}(x^{(i-1)\beta}\widehat{B}(x)) \rangle = \langle A_i^{(r)}, B_j^{(c)} \rangle,$$
$$\langle \mathsf{Vec}(x^{2j\beta^2 - \beta}\widehat{B}(x)), \mathsf{Vec}(x^{(i-1)\beta}\widehat{A}(x)) \rangle = \langle B_i^{(r)}, A_j^{(c)} \rangle,$$

for $i, j = 1, ..., \beta$.

The proof of correctness is described in Appendix A. Furthermore, we show that this encoding method can be extended to calculate multiplication for two rectangular matrices[1] in the following Corollary 1 and 2.

Corollary 1. *For given matrix A with size of $m \times k$ and matrix B with size of $k \times l$, let $\beta = \max\{m, k, l\}$. When $4\beta^3 \leq n$, they are encoded into elements of polynomial ring $\mathcal{R} = \mathbb{Z}[x]/(x^n + 1)$ using the method shown in Eq. (6), then $AB = \left(\langle A_i^{(r)}, B_j^{(c)} \rangle \right)_{m \times l}$ can be obtained by computing coefficient vector inner products of shifted $\widehat{A}(x)$ and $\widehat{B}(x)$. In detail,*

$$\langle \mathsf{Vec}(x^{2j\beta^2 - \beta}\widehat{A}(x)), \mathsf{Vec}(x^{(i-1)\beta}\widehat{B}(x)) \rangle = \langle A_i^{(r)}, B_j^{(c)} \rangle,$$

for $i = 1, ..., m$ and $j = 1, ..., l$.

Example. Given 2×4 matrix A and 4×3 matrix B, to compute multiplication matrix $M = AB$ as follows:

$$A = \begin{bmatrix} a_{11} & a_{12} & a_{13} & a_{14} \\ a_{21} & a_{22} & a_{23} & a_{24} \end{bmatrix}, \quad B = \begin{bmatrix} b_{11} & b_{12} & b_{13} \\ b_{21} & b_{22} & b_{23} \\ b_{31} & b_{32} & b_{33} \\ b_{41} & b_{42} & b_{43} \end{bmatrix} \rightarrow AB = M := \begin{bmatrix} m_{11} & m_{12} & m_{13} \\ m_{21} & m_{22} & m_{23} \end{bmatrix}.$$

According to Eq. (6), where $\beta = 4$, we have

$$\widehat{A}(x) = a_{11} + a_{12}x + a_{13}x^2 + a_{14}x^3 + a_{21}x^4 + a_{22}x^5 + a_{23}x^6 + a_{24}x^7 + a_{11}x^{28}$$
$$+ a_{21}x^{29} + a_{12}x^{60} + a_{22}x^{61} + a_{13}x^{92} + a_{23}x^{93} + a_{14}x^{124} + a_{24}x^{125},$$
$$\widehat{B}(x) = b_{11} + b_{12}x + b_{13}x^2 + b_{21}x^4 + b_{22}x^5 + b_{23}x^6 + b_{31}x^8 + b_{32}x^9 + b_{33}x^{10}$$
$$+ b_{41}x^{12} + b_{42}x^{13} + b_{43}x^{14} + b_{11}x^{28} + b_{21}x^{29} + b_{31}x^{30} + b_{41}x^{31} + b_{12}x^{60}$$
$$+ b_{22}x^{61} + b_{32}x^{62} + b_{42}x^{63} + b_{13}x^{92} + b_{23}x^{93} + b_{33}x^{94} + b_{43}x^{95}.$$

---

[1] When encode a rectangular matrix, add zero terms to last rows (or/and columns) if the row (or/and column) number is smaller than β, e.g., $m < k = \beta$,

$$\widetilde{A}[rows](x) = A_1^{(r)}X + ... + x^{(i-1)\beta}A_i^{(r)}X + ... + x^{(m-1)\beta}A_m^{(r)}X$$
$$\widetilde{A}[columns](x) = A_1^{(c)}X' + ... + x^{(j-1)\beta^2}A_j^{(c)}X' + ... + x^{(\beta-1)\beta^2}A_\beta^{(c)}X',$$

where $X' = (1, x, ..., x^{m-1})^T$.

It can be easily checked that

$$m_{ij} = \langle \mathsf{Vec}(x^{2j\beta^2-\beta}\widehat{A}(x)), \mathsf{Vec}(x^{(i-1)\beta}\widehat{B}(x)) \rangle \quad (i=1,2; j=1,2,3).$$

For example, when $i = 2, j = 1$, we have

$$x^{28}\widehat{A}(x)$$
$$= a_{11}x^{28} + a_{12}x^{29} + a_{13}x^{30} + a_{14}x^{31} + a_{21}x^{32} + a_{22}x^{33} + a_{23}x^{34} + a_{24}x^{35}$$
$$\quad + a_{11}x^{56} + a_{21}x^{57} + a_{12}x^{88} + a_{22}x^{89} + a_{13}x^{120} + a_{23}x^{121} + a_{14}x^{152} + a_{24}x^{153},$$
$$x^4\widehat{B}(x)$$
$$= b_{11}x^4 + b_{12}x^5 + b_{13}x^6 + b_{21}x^8 + b_{22}x^8 + b_{23}x^{10} + b_{31}x^{12} + b_{32}x^{13}$$
$$\quad + b_{33}x^{14} + b_{41}x^{16} + b_{42}x^{17} + b_{43}x^{18} + b_{11}x^{32} + b_{21}x^{33} + b_{31}x^{34} + b_{41}x^{35}$$
$$\quad + b_{12}x^{64} + b_{22}x^{65} + b_{32}x^{66} + b_{42}x^{67} + b_{13}x^{96} + b_{23}x^{97} + b_{33}x^{98} + b_{43}x^{99}.$$

Therefore,

$$\langle \mathsf{Vec}(x^{28}\widehat{A}(x)), \mathsf{Vec}(x^4\widehat{B}(x)) \rangle = \langle (a_{21}, a_{22}, a_{23}, a_{24}), (b_{11}, b_{21}, b_{31}, b_{41}) \rangle = m_{21}.$$

Corollary 2. *For given matrix A with size of $m \times k$ and matrix B with size of $k \times m$, let $\beta = \max\{m, k\}$. When $4\beta^3 \le n$, they are encoded into elements of polynomial ring $\mathcal{R} = \mathbb{Z}[x]/(x^n + 1)$ using the method shown in Eq. (6), then both $AB = \left(\langle A_i^{(r)}, B_j^{(c)} \rangle \right)_{m \times m}$ and $BA = \left(\langle B_i^{(r)}, A_j^{(c)} \rangle \right)_{k \times k}$ can be obtained by computing coefficient vector inner products of shifted $\widehat{A}(x)$ and $\widehat{B}(x)$ when $4\beta^3 \le n$. In detail,*

$$\langle \mathsf{Vec}(x^{2j\beta^2-\beta}\widehat{A}(x)), \mathsf{Vec}(x^{(i-1)\beta}\widehat{B}(x)) \rangle = \langle A_i^{(r)}, B_j^{(c)} \rangle,$$

for $i, j = 1, ..., m$; and

$$\langle \mathsf{Vec}(x^{2j\beta^2-\beta}\widehat{B}(x)), \mathsf{Vec}(x^{(i-1)\beta}\widehat{A}(x)) \rangle = \langle B_i^{(r)}, A_j^{(c)} \rangle,$$

for $i, j = 1, ..., k$.

3.3   Our Secure MMC

Using the one-pack-encoding method proposed in Sect. 3.2 and the WHAP secure inner product recalled in Sect. 2.2, we construct a new secure MMC scheme based on the LNV Ring-LWE based PHE scheme (see Fig. 2) in this subsection. We describe *our secure MMC* on two rectangular matrices[2] step by step below:

Step 1 (Encode-then-Encrypt). Input matrices $A \in \mathbb{Z}_p^{m \times k}$ and $B \in \mathbb{Z}_p^{k \times l}$, let $\beta = \max\{m, k, l\}$ that satisfies $\beta^3 \le n/4$, output ciphertexts (c_1, c_2) and

---

[2]   We show the MMC AB of $A \in \mathbb{Z}_p^{m \times k}$ and $B \in \mathbb{Z}_p^{k \times l}$ for flexible m, k, l under Corollary 1 here, for example. The MMC AB and BA under the cases of $m = k = l$ (Theorem 2) and $m = l \neq k$ (Corollary 2) can be similarly done.

(d_1, d_2) of their corresponding encoded elements $\widehat{A}(x)$ and $\widehat{B}(x)$ from Eq. (6), respectively.

$$\mathsf{Enc}(pk, \widehat{A}(x)) = (c_1, c_2), \mathsf{Enc}(pk, \widehat{B}(x)) = (d_1, d_2) \in \mathcal{R}_q \times \mathcal{R}_q,$$

where Enc is encryption algorithm defined in Fig. 2.

Step 2 (WHAP Secure Inner Product Computation). Input (c_1, c_2) and (d_1, d_2), run InnerP and output $ip := (W_1, W_{21}, W_{22}, \xi)$, where $W_1, W_{21}, W_{22} \in \mathcal{R}_q$ and $\xi = (\xi_{ij})_{m \times l} \in \mathbb{Z}_q^{m \times l}$ as follows:

$$W_1 = c_1 d_1^{(t)}, \quad W_{21} = c_1^{(t)} d_2, \quad W_{22} = d_1^{(t)} c_2 \mod q \in \mathcal{R}_q,$$
$$\xi_{i,j} = x^{2j\beta^2 - i\beta} c_2 d_2^{(t)} \mod q \mod x \in \mathbb{Z}_q \quad (i = 1, ..., m; j = 1, ..., l).$$

Step 3 (Decrypt-then-Decode). Input ip, output $AB = \left(\langle A_i^{(r)}, B_j^{(c)} \rangle \right)_{m \times l}$ as follows:

$\mathsf{DecIP}^*(sk, ip)$
$$= \left(W_1 S^*, W_{21} S^{(t)}, W_{22} S^{(t)} \right) \mod q = \left(\sum_{k=0}^{n-1} w_k x^k, \sum_{k=0}^{n-1} u_k x^k, \sum_{k=0}^{n-1} v_k x^k \right),$$

where $S^{(t)}$ and S^* are the same with that defined in Sect. 2.2. Then,

$$\langle A_i^{(r)}, B_j^{(c)} \rangle = -w_{n-(2j\beta^2 - i\beta)} + u_{2j\beta^2 - i\beta} - v_{n-(2j\beta^2 - i\beta)} + \xi_{ij} \mod q \mod p,$$

for $i = 1, 2, ..., m; j = 1, 2, ..., l$.

Correctness. We prove the correctness of our secure MMC below

Proof. First, for given $\mathsf{Enc}(pk, \widehat{A}(x)) = (c_1, c_2)$, $\mathsf{Enc}(pk, \widehat{B}(x)) = (d_1, d_2)$, according to Lemma 1, we have

$$c_{i,j} := (x^{2j\beta^2 - \beta} c_1, x^{2j\beta^2 - \beta} c_2) = \mathsf{Enc}(pk, x^{2j\beta^2 - \beta} \widehat{A}(x)),$$
$$c'_{i,j} := (x^{(i-1)\beta} d_1, x^{(i-1)\beta} d_2) = \mathsf{Enc}(pk, x^{(i-1)\beta} \widehat{B}(x)).$$

Second, for $i = 1, 2, ..., m; j = 1, 2, ..., l$, according to Corollary 1,

$$\langle A_i^{(r)}, B_j^{(c)} \rangle = \langle \mathsf{Vec}(x^{2j\beta^2 - \beta} \widehat{A}(x)), \mathsf{Vec}(x^{(i-1)\beta} \widehat{B}(x)) \rangle.$$

Next, let $ip^{(i,j)} := (W_1^{(i,j)}, W_2^{(i,j)}, \xi^{(i,j)}) = \mathsf{InnerP}(c_{i,j}, c'_{i,j})$, where InnerP is the inner product homomorphic computation algorithm defined in the WHAP secure inner product in Sect. 2.2. According to Lemma 2, we have

$$W_1^{(i,j)} = (x^{2j\beta^2-\beta}c_1)(x^{(i-1)\beta}d_1)^{(t)} \mod q$$
$$= (x^{2j\beta^2-\beta}c_1)(-x^{n-(i-1)\beta}d_1^{(t)}) \mod q$$
$$= x^{2j\beta^2-i\beta}W_1 \mod q \in \mathcal{R}_q,$$
$$W_2^{(i,j)} = (x^{2j\beta^2-\beta}c_1)^{(t)}(x^{(i-1)\beta}d_2) + (x^{(i-1)\beta}d_1)^{(t)}(x^{2j\beta^2-\beta}c_2) \mod q$$
$$= -x^{n-(2j\beta^2-\beta)}c_1^{(t)}x^{(i-1)\beta}d_2 + (-x^{n-(i-1)\beta})d_1^{(t)}x^{2j\beta^2-\beta}c_2 \mod q$$
$$= -x^{n-(2j\beta^2-i\beta)}W_{21} + x^{2j\beta^2-i\beta}W_{22} \mod q \in \mathcal{R}_q,$$
$$\xi^{(i,j)} = (x^{(i-1)\beta}d_2)^{(t)}(x^{2j\beta^2-\beta}c_2) \mod q \mod x$$
$$= (-x^{n-(i-1)\beta}d_2^{(t)}x^{2j\beta^2-\beta}c_2 \mod q \mod x$$
$$= x^{2j\beta^2-i\beta}c_2 d_2^{(t)} \mod q \mod x$$
$$= \xi_{i,j} \in \mathbb{Z}_q,$$

where $W_1 = c_1 d_1^{(t)}, W_{21} = c_1^{(t)}d_2, W_{22} = d_1^{(t)}c_2$. Therefore,

$$\langle A_i^{(r)}, B_j^{(c)} \rangle = \mathsf{DecIP}(sk, ip^{(i,j)})$$
$$= (W_1^{(i,j)}S^* + W_2^{(i,j)}S^{(t)} \mod q \mod x) + \xi^{(i,j)} \mod q \mod p$$
$$= (x^{2j\beta^2-i\beta}W_1 S^* - x^{n-(2j\beta^2-i\beta)}W_{21}S^{(t)} + x^{2j\beta^2-i\beta}W_{22}S^{(t)} \mod q \mod x)$$
$$+ \xi_{i,j} \mod q \mod p$$
$$= (x^{2j\beta^2-i\beta}\sum_{k=0}^{n-1} w_k x^k - x^{n-(2j\beta^2-i\beta)}\sum_{k=0}^{n-1} u_k x^k + x^{2j\beta^2-i\beta}\sum_{k=0}^{n-1} v_k x^k \mod x)$$
$$+ \xi_{i,j} \mod q \mod p$$
$$= - w_{n-(2j\beta^2-i\beta)} + u_{2j\beta^2-i\beta} - v_{n-(2j\beta^2-i\beta)} + \xi_{i,j} \mod q \mod p,$$

ending the proof. $\qquad\qquad\square$

Security Analysis. For a PHE scheme, indistinguishability under chosen plaintext attack (IND-CPA) is the basic security requirement. The ciphertexts of the LNV PHE scheme recalled in Fig. 2 is proved indistinguishable from random (see Theorem 1), which implies the IND-CPA security under the Ring-LWE assumption. In our proposed secure MMC, as the input of the homomorphic computation consists of only ciphertexts, the IND-CPA security will not be weakened.

4   Evaluation and Application

How to use our packing method to compute the multiplication of rectangular matrices $A = (a_{ij})_{m\times k}$ and $B = (b_{ij})_{k\times l}$ under the condition $\beta^3 \le n/4$ ($\beta = \max\{m, k, l\}$) is described in Sect. 3.2. Thanks to the one-pack-encoding method, our scheme is more flexible than the existing packing method introduced by Dung et al. [2] in the following cases (refer to Theorem 2, Corollary 1, 2 for detail):

- One type encoding for each matrix implies that the encoded ring element can be used to compute secure MMC whenever the matrix acts as left or right matrix of multiplication. I.e., for matrix $A = (a_{ij})_{m\times k}$, $B = (b_{ij})_{k\times l}$ and $C = (c_{ij})_{h\times m}$, one packing element $\widehat{A}(x)$ can be used to compute both AB and CA.

– Especially, when $m = l$, i.e., $A = (a_{ij})_{m \times k}$, $B = (b_{ij})_{k \times m}$, then using packing elements $\widehat{A}(x)$ and $\widehat{B}(x)$, both AB and BA can be computed without extra encoding.

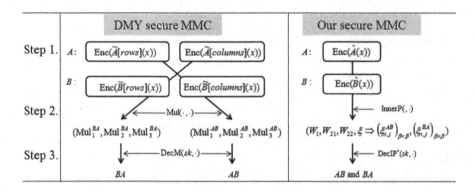

Fig. 3. Outline of the difference between two packing methods.

Efficiency. Figure 3 shows the difference between the DMY packing-2 approach and ours. If only computing multiplication AB of two $\beta \times \beta$ square matrices A and B, our scheme is more efficient than the DMY approach on encryption operations and saving storage space, but has no superior in other computation operations. However, when both AB and BA need to be computed, our secure MMC is more efficient than the DMY scheme not only for communication cost but also for computation cost (refer to Table 1 for detail).

Accordingly, our work provides a trade-off to realize efficient secure MMC as follows. For a fix parameter n,

- when $\beta^3 \leq n/4$ ($\beta = \max\{m, k, l\}$), it is better to select our secure MMC to save storage space and encryption operations. Especially, when both AB and BA need to be computed.

In Table 1, we summarize the flexibilty[3] and efficiency via comparing our packing approach with the DMY approach (*Packing-2*) on the LNV scheme [5].

- when $n/4 < m^3 \leq n$, where m is dimension parameter of square matrix, DMY secure MMC works well.

Though directly using our method is impractical when the dimension is large, block matrices and parallel computation method can be used to improve the

---

[3] The notation "−" in Table 1 means that no concrete packing method for rectangular matrix multiplication was explicitly considered in [2]. However, the approach similar to ours considering rectangular matrices may work for [2].

Table 1. Comparison on the LNV Scheme [5].

	DMY Secure MMC [2]	Ours
Encoding method	Two-pack method	One-pack method
Appliable MMC (flexiblity): $A = (a_{ij})_{m \times k}$, $B = (b_{ij})_{k \times l}$		
when $m = k = l$	AB and BA	AB and BA
when $m = l \neq k$	-	AB and BA
when $m \neq l$	-	AB
Computation cost (operations): $\beta = \max\{m, k, l\}$		
Encryption for each matrix (Step 1)	4 $\mathcal{R}_q$-Mul	2 $\mathcal{R}_q$-Mul
Outsourced computation (Step 2)		
for AB	4 $\mathcal{R}_q$-Mul	4 $\mathcal{R}_q$-Mul
for AB and BA	8 $\mathcal{R}_q$-Mul	4 $\mathcal{R}_q$-Mul
Decryption for MMC (Step 3)		
for AB	2 $\mathcal{R}_q$-Mul	3 $\mathcal{R}_q$-Mul
for AB and BA	4 $\mathcal{R}_q$-Mul	3 $\mathcal{R}_q$-Mul
Communication cost for Secure MMC (bits):		
Ciphertexts for input	$2 \cdot 4n\lceil \log_2 q \rceil$	$4n\lceil \log_2 q \rceil$
Output ciphertext		
for AB	$3n\lceil \log_2 q \rceil$	$(3n + \beta^2)\lceil \log_2 q \rceil$
for AB and BA	$6n\lceil \log_2 q \rceil$	$(3n + 2\beta^2)\lceil \log_2 q \rceil$

n and q are system parameters; $\beta = \max\{m, k, l\}$ with the condition $\beta^3 \leq n/4$ holds; In order to easily compare computation/communication costs, we assume both A and B are $\beta \times \beta$ square matrices. $\mathcal{R}_q$-Mul denotes multiplication operations over $\mathcal{R}_q$.

efficiency of computation. For example, matrix U of size 10×30, V of size 30×20, then U and V can be blocked into several 10×10 submatrices as follows:

$$U = \begin{bmatrix} U_{11} \ U_{12} \ U_{13} \end{bmatrix}, \ V = \begin{bmatrix} V_{11} \ V_{12} \\ V_{21} \ V_{22} \\ V_{31} \ V_{32} \end{bmatrix} \rightarrow UV = \begin{bmatrix} \sum_{j=1}^{3} U_{1j}V_{j1}, \ \sum_{j=1}^{3} U_{1j}V_{j2} \end{bmatrix}$$

Using the parallel computational technique, *six* secure MMC $U_{1j}V_{jk}$ for $j = 1, 2, 3; k = 1, 2$ can be operated efficiently with the computation cost similar to one MMC of two 10×10 matrices. More details and the corresponding implementation will be included in the full version paper.

Application. Multiplication of rectangular matrices is useful on data compression. So, the secure MMC for rectangular matrices can be applied to secure data compression as follows: Server has $\mathsf{Enc}(U)$, where matrix U is of size $k \times m$ ($k < m$). Clients send $\mathsf{Enc}(x_i)$ for $1 \leq i \leq N$, where x_i is of size $m \times 1$. A data analyst wants to securely outsource the computation $y_i = Ux_i$ to the server. Therefore, the data analyst instructs the server to compute:

$$CT_i = \mathsf{Enc}(U)\mathsf{Enc}(x_i)$$

for all i and then decrypts CT_i to get y_i of size $k \times 1$. For example, let $m = 10, k = 2$, then 10-dimension vectors can be compressed into 2-dimension vectors. In other words, data size can be compressed from $10N \to 2N$ (for instance, $10\,\text{GB} \to 2\,\text{GB}$) securely.

5   Concluding Remarks

In this study, we proposed a new secure matrix multiplication computation scheme by combining and extending Dung et al.'s packing approaches and Wang et al.'s secure inner product method. Similar to the existing scheme, ours is also constructed on the Lauter et al.'s somewhat homomorphic encryption, which is over polynomial ring $\mathcal{R} = \mathbb{Z}[x]/(x^n + 1)$. Our scheme is efficient and flexible in the case that the biggest dimension of row and column vectors of two rectangular matrices is smaller than $0.63n^{1/3}$. For larger dimentions, block matrices and parallel computation method can be used to improve the efficiency of computation.

Acknowledgement. This work was partially supported by JSPS KAKENHI Grant Number JP15K00028 and JST CREST Number JPMJCR168A. We thank Takuya Hayashi and Mishra Pradeep Kumar for the useful discussion. We also greatly appreciate the anonymous reviewers for their thoughtful comments that helped improving the manuscript.

A   Correctness of Theorem 2

Apart from Notation list-1.~3., the following list of notation should be useful:

Notation List-4. For a vector $V = (v_1, ..., v_\gamma)$, $\mathsf{pol}(V) = v_1 x^{\delta_1} + ... + v_\gamma x^{\delta_\gamma}$, let

$$V[k] := v_k,$$
$$\mathsf{pol}(V[k]) := v_k x^{\delta_k},$$
$$\deg(\mathsf{pol}(V[k])) := \delta_k, \text{ for } k = 1, ..., \gamma. \text{ Moreover, let}$$
$$\deg(\mathsf{pol}(V)) := \{\delta_1, ..., \delta_k, ..., \delta_\gamma\} (\delta_1 < ... < \delta_\gamma),$$
$$\mathsf{left} - \deg(\mathsf{pol}(V)) := \delta_1, \text{ and}$$
$$\mathsf{right} - \deg(\mathsf{pol}(V)) := \delta_\gamma.$$

When V is a row or column vector of a matrix of size $\beta \times \beta$, $\gamma = \beta$.

Proof. According to Eqs. (6) and (5), we have

$$x^{2j\beta^2 - \beta} \widehat{A}(x) = x^{2j\beta^2 - \beta} \widetilde{A}(x)[rows] + x^{2j\beta^2 - \beta} \widetilde{A}(x)[columns]$$

$$x^{2j\beta^2-\beta}\widetilde{A}(x)[rows]$$

$$= (a_{11}x^{2j\beta^2-\beta} + ... + a_{1\beta}x^{2j\beta^2-1}) + ... + (a_{i-1,1}x^{2j\beta^2-(i-3)\beta}$$

$$+... + a_{i-1,\beta}x^{2j\beta^2+(i-2)\beta-1}) + (a_{i,1}x^{2j\beta^2-(i-2)\beta} + ... + a_{i,\beta}x^{2j\beta^2+(i-1)\beta-1})$$

$$+(a_{i+1,1}x^{2j\beta^2-(i-1)\beta} + ... + a_{i+1,\beta}x^{2j\beta^2+i\beta-1}) + ... + (a_{\beta,1}x^{2j\beta^2-(\beta-2)\beta}$$

$$+... + a_{\beta,\beta}x^{2j\beta^2+\beta^2-\beta-1})$$

$$= \mathsf{pol}(A_1^{(r)}) + ... + \mathsf{pol}(A_{i-1}^{(r)}) + \mathsf{pol}(A_i^{(r)}) + \mathsf{pol}(A_{i+1}^{(r)}) + ... + \mathsf{pol}(A_\beta^{(r)}),$$

$$x^{2j\beta^2-\beta}\widetilde{A}(x)[columns]$$

$$= x^{2j\beta^2-\beta}(x^{2\times\beta^2-\beta}A_1^{(c)} + ... + x^{2i\times\beta^2-\beta}A_i^{(c)} + ... + x^{2\beta\times\beta^2-\beta}A_\beta^{(c)})X$$

$$= x^{2\beta^2-\beta}(a_{11}x^{2j\beta^2-\beta} + ... + a_{\beta,1}x^{2j\beta^2-1}) + ... + x^{2i\beta^2-\beta}(a_{1,i}x^{2j\beta^2-\beta}$$

$$+... + a_{\beta,i}x^{2j\beta^2-1}) + ... + x^{2\beta\beta^2-\beta}(a_{1,\beta}x^{2j\beta^2-\beta} + ... + a_{\beta,\beta}x^{2j\beta^2-1})$$

$$= \mathsf{pol}(A_1^{(c)}) + ... + \mathsf{pol}(A_i^{(c)}) + ... + \mathsf{pol}(A_\beta^{(c)});$$

and

$$x^{(i-1)\beta}\widehat{B}(x) = x^{(i-1)\beta}\widetilde{B}(x)[rows] + x^{(i-1)\beta}\widetilde{B}(x)[columns]$$

$$x^{(i-1)\beta}\widetilde{B}(x)[rows]$$

$$= (b_{11}x^{(i-1)\beta} + ... + b_{1\beta}x^{i\beta-1}) + (b_{21}x^{i\beta} + ... + b_{2\beta}x^{(i+1)\beta-1})$$

$$+... + (b_{\beta,1}x^{(i-1)\beta+(\beta-1)\beta} + ... + b_{\beta,\beta}x^{(i-1)\beta+\beta^2-1})$$

$$= \mathsf{pol}(B_1^{(r)}) + \mathsf{pol}(B_2^{(r)}) + ... + \mathsf{pol}(B_\beta^{(r)}),$$

$$x^{(i-1)\beta}\widetilde{B}(x)[columns]$$

$$= x^{(i-1)\beta}(x^{2\times\beta^2-\beta}B_1^{(c)} + ... + x^{2j\times\beta^2-\beta}B_i^{(c)} + ... + x^{2\beta\times\beta^2-\beta}B_\beta^{(c)})X$$

$$= x^{2\beta^2-\beta}(b_{11}x^{(i-1)\beta} + ... + b_{\beta 1}x^{i\beta-1}) + x^{4\beta^2-\beta}(b_{12}x^{(i-1)\beta} + ... + b_{\beta 2}x^{i\beta-1})$$

$$+... + x^{2j\beta^2-\beta}(b_{1j}x^{(i-1)\beta} + ... + b_{\beta j}x^{i\beta-1}) + x^{2(j+1)\beta^2-\beta}(b_{1,j+1}x^{(i-1)\beta}$$

$$+... + b_{\beta,j+1}x^{i\beta-1}) + ... + x^{2\beta\beta^2-\beta}(b_{1\beta}x^{(i-1)\beta} + ... + b_{\beta\beta}x^{i\beta-1})$$

$$= \mathsf{pol}(B_1^{(c)}) + \mathsf{pol}(B_2^{(c)}) + ... + \mathsf{pol}(B_j^{(c)}) + \mathsf{pol}(B_{j+1}^{(c)}) + ... + \mathsf{pol}(B_\beta^{(c)}).$$

We should prove for any $i, j = 1, ..., \beta$, $\mathsf{pol}(A_i^{(r)})$ in $x^{2j\beta^2-\beta}\widehat{A}(x)$ and $\mathsf{pol}(B_j^{(c)})$ in $x^{(i-1)\beta}\widehat{B}(x)$ satisfy exactly

$$\deg(\mathsf{pol}(A_i^{(r)}[k])) = \deg(\mathsf{pol}(B_j^{(c)}[k])) \quad (k = 1, ..., \beta).$$

Case 1: when $j = 1$, i.e., $\langle \mathsf{Vec}(x^{2\beta^2-\beta}\widehat{A}(x)), \mathsf{Vec}(x^{(i-1)\beta}\widehat{B}(x)) \rangle = \langle A_i, B_1 \rangle$. It can be easily check that

$$\mathsf{left} - \deg(\mathsf{pol}(A_1^{(r)})) < \cdots < \mathsf{right} - \deg(\mathsf{pol}(A_{i-1}^{(r)})) = 2\beta^2 + (i-2)\beta - 1 < $$
$$< 2\beta^2 + (i-2)\beta \leq$$

$$\deg(\mathsf{pol}(A_i^{(r)}[k])) = \deg(\mathsf{pol}(B_1^{(c)}[k])) = 2\beta^2 + (i-2)\beta + (k-1) \text{ for } k = 1, ..., \beta$$

$$\leq 2\beta^2 + (i-1)\beta - 1 < 2\beta^2 + (i-1)\beta = \mathsf{left} - \deg(\mathsf{pol}(A_{i+1}^{(r)}))$$
$$< \cdots < \mathsf{left} - \deg(\mathsf{pol}(A_1^{(c)})) < \mathsf{right} - \deg(\mathsf{pol}(A_1^{(c)})) < \mathsf{left} - \deg(\mathsf{pol}(B_2^{(c)}))$$
$$< \cdots < \deg(\mathsf{pol}(B_i^{(c)})) < \deg(\mathsf{pol}(A_i^{(c)})) < \cdots < \deg(\mathsf{pol}(B_\beta^{(c)})) < \deg(\mathsf{pol}(A_\beta^{(c)})).$$

Case 2: when $j \geq 2$, i.e., $\langle \mathsf{Vec}(x^{2j\beta^2 - \beta}\widehat{A}(x)), \mathsf{Vec}(x^{(i-1)\beta}\widehat{B}(x)) \rangle = \langle A_i, B_j \rangle$. It can be easily check that

$$\text{right} - \deg(\mathsf{pol}(B_{j-1}^{(c)})) < \text{left} - \deg(\mathsf{pol}(A_1^{(r)})) < \cdots < \text{right} - \deg(\mathsf{pol}(A_{i-1}^{(r)}))$$
$$= 2j\beta^2 + (i-2)\beta - 1 < 2j\beta^2 + (i-2)\beta \leq$$

$$\deg(\mathsf{pol}(A_i^{(r)}[k])) = \deg(\mathsf{pol}(B_j^{(c)}[k])) = 2j\beta^2 + (i-2)\beta + (k-1), \text{ for } k = 1, ..., \beta$$

$$\leq 2j\beta^2 + (i-1)\beta - 1 < 2j\beta^2 + (i-1)\beta = \text{left} - \deg(\mathsf{pol}(A_{i+1}^{(r)}))$$
$$< \cdots < \text{left} - \deg(\mathsf{pol}(A_1^{(c)})) < \text{right} - \deg(\mathsf{pol}(A_1^{(c)})) < \text{left} - \deg(\mathsf{pol}(B_{j+1}^{(c)}))$$
$$< \cdots < \deg(\mathsf{pol}(B_{j+l-1}^{(c)})) < \deg(\mathsf{pol}(A_l^{(c)})) < \cdots < \deg(\mathsf{pol}(B_\beta^{(c)}))$$
$$< \deg(\mathsf{pol}(A_{\beta-j+1}^{(c)})) < ... < \deg(\mathsf{pol}(A_\beta^{(c)})).$$

Note. Since
$$\text{right} - \deg(x^{2j\beta^2 - \beta}\widehat{A}(x)) = 4\beta^3 - \beta - 1,$$
$$\text{right} - \deg(x^{(i-1)\beta}\widehat{B}(x)) = 2\beta^3 + \beta^2 - \beta - 1,$$

we have
$$\max_{i,j}\{\deg(x^{2j\beta^2 - \beta}\widehat{A}(x)), \deg(x^{(i-1)\beta}\widehat{B}(x))\} < 4\beta^3.$$

Therefore, our packing method works if

$$4\beta^3 \leq n.$$

The proof for correctness of the MMC AB is complete. Correctness of the MMC BA can be proved similarly.   □

References

1. Brakerski, Z., Vaikuntanathan, V.: Fully homomorphic encryption from ring-LWE and security for key dependent messages. In: Rogaway, P. (ed.) CRYPTO 2011. LNCS, vol. 6841, pp. 505–524. Springer, Heidelberg (2011). https://doi.org/10.1007/978-3-642-22792-9_29
2. Dung, D.H., Mishra, P.K., Yasuda, M.: Efficient secure matrix multiplication over LWE-based homomorphic encryption. Tatra Mt. Math. Publ. **67**, 69–83 (2016)
3. Fu, S., Yu, Y., Xu, M.: A secure algorithm for outsourcing matrix multiplication computation in the cloud. In: SCC 2017, pp. 27–33. ACM (2017)
4. Lei, X., Liao, X., Huang, T., Heriniaina, F.: Achieving security, robust cheating resistance, and high-efficiency for outsourcing large matrix multiplication computation to a malicious cloud. Inf. Sci. **280**, 205–217 (2014)
5. Lauter, K.E., Naehrig, M., Vaikuntanathan, V.: Can homomorphic encryption be practical? In: CCSW 2011, pp. 113–124. ACM (2011)
6. Wang, L., Hayashi, T., Aono, Y., Phong, L.T.: A generic yet efficient method for secure inner product. In: Yan, Z., Molva, R., Mazurczyk, W., Kantola, R. (eds.) NSS 2017. LNCS, vol. 10394, pp. 217–232. Springer, Cham (2017). https://doi.org/10.1007/978-3-319-64701-2_16

7. Yasuda, M., Shimoyama, T., Kogure, J., Yokoyama, K., Koshiba, T.: New packing method in somewhat homomorphic encryption and its applications. Secur. Commun. Netw. **8**(13), 2194–2213 (2015)
8. Yasuda, M., Shimoyama, T., Kogure, J., Yokoyama, K., Koshiba, T.: Practical packing method in somewhat homomorphic encryption. In: Garcia-Alfaro, J., Lioudakis, G., Cuppens-Boulahia, N., Foley, S., Fitzgerald, W.M. (eds.) DPM/SETOP -2013. LNCS, vol. 8247, pp. 34–50. Springer, Heidelberg (2014). https://doi.org/10.1007/978-3-642-54568-9_3
9. Yasuda, M., Shimoyama, T., Kogure, J., Yokoyama, K., Koshiba, T.: Secure pattern matching using somewhat homomorphic encryption. In: CCSW 2013, pp. 65–76. ACM (2013)
10. Yasuda, M., Shimoyama, T., Kogure, J., Yokoyama, K., Koshiba, T.: Secure statistical analysis using RLWE-based homomorphic encryption. In: Foo, E., Stebila, D. (eds.) ACISP 2015. LNCS, vol. 9144, pp. 471–487. Springer, Cham (2015). https://doi.org/10.1007/978-3-319-19962-7_27

Predicate Encryption

Subset Predicate Encryption and Its Applications

Jonathan Katz[1], Matteo Maffei[2], Giulio Malavolta[3(✉)],
and Dominique Schröder[3]

[1] University of Maryland, College Park, USA
[2] TU Vienna, Vienna, Austria
[3] Friedrich-Alexander-University Erlangen-Nürnberg, Erlangen, Germany
malavolta@cs.fau.de

Abstract. In this work we introduce the notion of Subset Predicate Encryption, a form of attribute-based encryption (ABE) in which a message is encrypted with respect to a set s' and the resulting ciphertext can be decrypted by a key that is associated with a set s if and only if $s \subseteq s'$. We formally define our primitive and identify several applications. We also propose two new constructions based on standard assumptions in bilinear groups; the constructions have very efficient decryption algorithms (consisting of one and two pairing computations, respectively) and small keys: in both our schemes, private keys contain only two group elements. We prove selective security of our constructions without random oracles. We demonstrate the usefulness of Subset Predicate Encryption by describing several black-box transformations to more complex primitives, such as identity-based encryption with wildcards and ciphertext-policy ABE for DNF formulas over a small universe of attributes. All of the resulting schemes are as efficient as the base Subset Predicate Encryption scheme in terms of decryption and key generation.

1 Introduction

Attribute-Based Encryption (ABE), and more generically functional encryption, introduces a new communication paradigm where the sender is allowed to specify a certain policy that the receiver must satisfy in order to read the data. Since its introduction in [15], ABE has had a tremendous impact in the research community and a plethora of different construction have been proposed, from different assumption and with different security notions and functionalities. However, more effort is needed towards the adoption of ABE schemes on a large scale as we only know a bunch of schemes that are efficient enough to be deployed in practice. In this work we contribute to the understanding of efficiency trade-offs in ABE (and weaker instances of functional encryption) by proposing a new prospective for the construction of efficient schemes. With this aim in mind, we introduce the notion of Subset Predicate Encryption (SPE). In a SPE scheme, sets are defined over some finite universe of elements. A user with a secret key for the set s can decrypt a ciphertext encrypted with the public key s' if and

S. Capkun and S. S. M. Chow (Eds.): CANS 2017, LNCS 11261, pp. 115–134, 2018.
https://doi.org/10.1007/978-3-030-02641-7_6

only if s defines a *subset* of s'. A SPE scheme must enforce that an adversary knowing the key for some set s cannot derive a valid key for any set different from s (e.g., by stripping off part of s from its original key). In particular, users must not be allowed to combine different keys in a meaningful manner (e.g., to decrypt any ciphertext that no user could have decrypted individually). A perhaps more natural way to look at SPE is as a generalization of broadcast encryption (BE): In this perspective BE can be seen as a special case of SPE where secret keys are associated with singleton subsets, i.e., $|s| = 1$. SPE opens the possibility to efficiently enforce expressive access control policies in several interesting scenarios, as described below.

CONCISE ACCESS CONTROL. An important aspect of SPE is that it enables access control over data in a very concise fashion. For instance, let us consider a corporate setting, where all users of the system encrypt all messages under the sets corresponding to the attributes of the fields "{sender, receiver, department, current-date}". Deriving keys in a hierarchical fashion is straightforward, however our system allows us to assign keys for more complex policies in a concise way. As an example, we can generate a key for decrypting all messages exchanged on a certain day across multiple departments by simply deriving a key for current-date. Furthermore, we can generate a key to read all the messages sent *from and to* Alice with a single key for the set corresponding to the element Alice.

PATTERN MATCHING. Imagine a scenario where each email is encrypted under the set corresponding to the words of the subject (assuming a subject of a fixed length). We could disclose the content of all emails containing a certain word (buy, as an example) in the subject by simply creating a key for the set corresponding to the element buy. It is important to note that the position of the word must not be necessarily known in advance, since the decryption is successful if the set encoded in the key matches any subset of the set encoded in the ciphertext.

A blackbox instantiation of SPE from Identity Based Encryption seems not be easily achievable: One could express the same functionality by encrypting the same message for the powerset of a given identity, but it is easy to see that the size of the ciphertext would grow exponentially in the length of the identity. While we conjecture that SPE is strictly more expressive than IBE, it is not hard to show that SPE is implied by any generic ABE system. However, due to the simplicity of our primitive, there is hope to create a SPE scheme in a more efficient manner, without resorting to generic ABE solutions. In particular, we are interested in maximizing the efficiency of the system for the end-users, both in terms of computation and in terms of storage. An efficient decryption algorithm is an important feature of any encryption scheme as it allows computationally-constrained devices to be integrated in the system: Since decryption is arguably the most recurrent operation (a user typically encrypts once for multiple receivers), its running time is fundamental for the scalability of the system. Additionally, small private keys are convenient as they are often stored in tamper-resistant memory, which in general is very costly. This can be especially critical in small devices, such as sensors, for which low cost solutions are often required.

In this work, we focus on the improvement over these two aspects and we present two cryptographic constructions for SPE with a very efficient decryption algorithm and constant-size private keys. Perhaps surprisingly, our abstraction turns out to subsume more complex primitives, such as ABE for DNF formulas over a small universe of attributes, and we show how to generically instantiate them from a SPE scheme. All of the resulting schemes inherit the efficiency of our constructions.

1.1 Our Contributions

We formalize the notion of Subset Predicate Encryption and its security guarantees using standard game-based definitions. We provide two instantiations for a SPE scheme from bilinear maps. Both of the schemes are proven secure in the selective security model without random oracles. Our first construction offers an extremely efficient decryption operation consisting of only a *single* pairing. Moreover, the secret keys are very compact as each key is composed of a group element and an integer value. The security of this scheme relies on the hardness of the Decisional q-Bilinear Diffie-Hellman Inversion assumption over bilinear groups. Our second scheme has a slightly less efficient decryption procedure (where two pairings are computed) but is based on the Decisional Bilinear Diffie-Hellman assumption. In this scheme, each private key is as large as two group elements.

We describe several generic black-box transformations that turn SPE into more expressive primitives. Our first transformation turns any SPE into an Identity-Based Encryption scheme with wildcards (WIBE), whereas our second transformation yields an ABE scheme for formulas in their DNF over for a small universe of attributes. A nice feature of these transformations is that the resulting schemes maintain the same decryption algorithm and key sizes of the base construction. Beyond being an interesting primitive on its own right, we believe that the conceptual simplicity of SPE might help in the future design of efficient WIBE and ABE schemes.

We summarize a comparison of our instantiations against the most efficient known WIBE schemes in Table 1: Our transformation yields the first scheme with constant-size keys and the decryption of our first construction is roughly 50% faster than the best instantiation of [1]. The performance of the ABE schemes derived generically from our instantiations of SPE are shown in Table 2. With respect to the best known instance of ABE in terms of key-size [11], both of our instantiations cut the size of the keys down to 50%. Furthermore the decryption algorithm of our first construction computes only one pairing and one modular ' exponentiation (while the second computes two pairings). This is unprecedented in the context of ABE, where in the fastest known scheme [16] the amount of modular exponentiations is linear in the size of the universe of attributes. This means that our schemes have an arbitrarily more efficient decryption, depending on the size of the universe of attributes. For a fair comparison we shall mention that the aforementioned schemes are more expressive the ours and satisfy stronger security notions.

Table 1. Comparison amongst the most efficient wildcard IBE schemes in the literature in terms of size of the public parameters ($|\mathsf{pk}|$), size of the decryption keys ($|\mathsf{sk}|$), size of the ciphertexts ($|\mathsf{c}|$), number of operations required for decrypting (Decrypt), and complexity assumptions. Here ω denotes the depth of the hierarchy, P denotes the number of pairing operations and E the number of modular exponentiations.

WIBE scheme	$|\mathsf{pk}|$	$|\mathsf{sk}|$	$|\mathsf{c}|$	Decrypt	Assumption
BBG-WIBE [1]	$(\omega+4)\mathbb{G}$	$(\omega+2)\mathbb{G}$	$(\omega+2)\mathbb{G}+\mathbb{G}_T$	2P	ω-BDHI
Waters-WIBE [1]	$(n+1)(\omega+3)\mathbb{G}$	$(\omega+1)\mathbb{G}$	$(n+1)\omega\mathbb{G}+\mathbb{G}_T$	$(\omega+1)$P	DBDH
Construction 1	$(2\omega+2)\mathbb{G}_1+\mathbb{G}_T$	$\mathbb{G}_2+\mathbb{Z}_p$	$(2\omega+1)\mathbb{G}_1+\mathbb{G}_T$	1P+1E	q-BDHI
Construction 2	$(2\omega+1)\mathbb{G}_1+2\mathbb{G}_2$	$\mathbb{G}_1+\mathbb{G}_2$	$2\omega\mathbb{G}_1+\mathbb{G}_2+\mathbb{G}_T$	2P	DBDH

Table 2. Comparison amongst the most efficient ABE schemes in the literature. Here we additionally compare the schemes by the family of predicates supported by the scheme (f), which can either be arbitrary Boolean formulas (Bool), zero inner-product predicates (InnerProd), or formulas in their DNF. We denote the number of disjunctive clauses in a DNF formula by γ.

ABE scheme	$|\mathsf{pk}|$	$|\mathsf{sk}|$	$|\mathsf{c}|$	Decrypt	Assumption	f
CP-ABE [11]	$(2\mathrm{U}+3)\mathbb{G}_1+\mathbb{G}_T$	$(2\mathrm{U}+4)\mathbb{G}_2$	$(2\mathrm{U}+2)\mathbb{G}_2+\mathbb{G}_T$	4P+4UE	SXDH	Bool
ZIPE [11]	$(2\mathrm{U}+4)\mathbb{G}_1+\mathbb{G}_T$	$4\mathbb{G}_2$	$(2\mathrm{U}+2)\mathbb{G}_2+\mathbb{G}_T$	4P+2UE	SXDH	InnerProd
KP-ABE [16]	$(\mathrm{U}+1)\mathbb{G}+\mathbb{G}_T$	$2\mathrm{U}\mathbb{G}+\mathrm{U}^2\mathbb{G}$	$(\mathrm{U}+1)\mathbb{G}+\mathbb{G}_T$	2P+2UE	U-BDHE	Bool
Construction 1	$(\mathrm{U}+2)\mathbb{G}_1+\mathbb{G}_T$	$\mathbb{G}_2+\mathbb{Z}_p$	$\gamma((\mathrm{U}+1)\mathbb{G}_1+\mathbb{G}_T)$	1P+1E	q-BDHI	DNF
Construction 2	$(\mathrm{U}+1)\mathbb{G}_1+2\mathbb{G}_2$	$\mathbb{G}_1+\mathbb{G}_2$	$\gamma(2\mathrm{U}\mathbb{G}_1+\mathbb{G}_2+\mathbb{G}_T)$	2P	DBDH	DNF

2   Related Work

Identity-Based Encryption was first proposed by Shamir [22], and the first efficient realization was presented in the seminal work of Boneh and Franklin [8], where they suggested the usage of bilinear maps for cryptographic purposes. Canetti *et al.* [10] introduced the first construction that was provably secure without random oracles: the authors defined a slightly weaker security model (selective security) where the attacker is required to commit to the challenge identity prior to the beginning of the experiment. In the same settings, Boneh and Boyen [5] showed two efficient and practical schemes in the standard model. The first scheme with full security was presented by Boneh and Boyen [6] and later Waters [24] constructed a more efficient variant with an elegant security proof. Several other schemes have followed, such as [9]. It is worth mentioning that the notion of IBE has also been extended to support a hierarchical key-derivation structure [7,13].

Attribute-Based Encryption was envisioned by Sahai and Waters [21] as a generalization of IBE, where keys and ciphertext are generated under sets of attributes and it is possible to encode arbitrary access formulas. The concept of ABE was refined by Goyal et al. in [15], where the authors proposed two complementary notions: (i) Key-Policy ABE (KP-ABE) allows one to encode sets of attributes in ciphertexts and embed access formulas in users' secret keys, whereas in (ii) Ciphertext Policy ABE (CP-ABE) formulas are attached to the ciphertexts. Goyal et al. [15] described a selectively-secure construction of KP-ABE that allows polices to be expressed by any monotonic formula. The first efficient CP-ABE system was proposed by Bethencourt et al. [4] with a security proof in the generic group model, while the first CP-ABE scheme in the standard model is due to Waters [23]. In [14] Goyal et al. showed how to generically transform a KP-ABE into a CP-ABE. Until recently, all of the known attribute-based systems were proven secure only in the selective sense: a fully secure ABE was first proposed by Lewko et al. [18], leveraging the dual system encryption technique. In light of this, several efficient and adaptively-secure ABE schemes were recently proposed by Chen et al. [11] in the prime-order settings, constructed on a novel framework based on clever predicate encodings. ABE was further generalized as Predicate Encryption (PE) [17], where the ciphertext is required to hide the set of attributes associated to it, in addition to the message.

An ABE scheme with an efficient decryption algorithm was introduced by Attrapadung et al. [3], where the authors presented an ABE system with constant-size ciphertexts. As a result, the decryption algorithm requires a constant number of pairings. In this perspective, Hohenberger and Waters [16] improved this result with a scheme that computes only two pairings in the decryption algorithm. However, this comes at the cost of an increase in the size of the secret keys. A revocation system with small keys was proposed by Lewko et al. [19], along with an ABE system where the size of the secret keys grows linearly in the number of attributes. Finally, it is worth mentioning that Okamoto and Takashima provided an inner-product encryption scheme with constant-size keys [20] (later improved in [11]). However, the generic transformation from inner-products to arbitrary Boolean formulas introduces an overhead in the encoding of attributes exponential in the number of variables (see [17]), making this primitive less appealing for practical purposes.

On a different line of research, Abdalla et al. [1] proposed the notion of IBE with wildcards (WIBE): in this primitive, one is allowed to specify certain positions of the identity associated to a ciphertext that are not required to match with the secret key. A related notion was formalized and instantiated by Abdalla et al. in [2], where one can include wildcards in the key generation phase. Both of these works build on top of various Hierarchical IBE schemes and therefore inherit the long size of the keys, typically linear in the depth of the hierarchy.

Hence, this work improves the state-of-the-art by presenting the most efficient constructions in terms of key size *and* decryption operations supporting complex functionalities (beyond the simple IBE). We stress that in this work we consider only the notion of selective security.

3   Preliminaries

We denote by $\lambda \in \mathbb{N}$ the security parameter and by $\mathsf{poly}(\lambda)$ any function that is bounded by a polynomial in λ. We address any function that is *negligible* in the security parameter with $\mathsf{negl}(\lambda)$. We say that an algorithm is PPT if it is modelled as a probabilistic Turing machine whose running time is bounded by some function $\mathsf{poly}(\lambda)$. Given a set S, we denote by $x \leftarrow S$ the sampling of and element uniformly at random in S. For an arbitrary pair of binary strings (a, b) of the same length ℓ, we write $a \subseteq b$ if for all $i \in \{1, \ldots, \ell\}$ such that $a_i = 1$ then $b_i = 1$. Given a binary string a, we say that an index $i \in a$ if $a_i = 1$.

3.1   Bilinear Maps

Let $\mathbb{G}_1$ and $\mathbb{G}_2$ be two cyclic groups of large prime order p. Let $g_1 \in \mathbb{G}_1$ and $g_2 \in \mathbb{G}_2$ be respective generators of $\mathbb{G}_1$ and $\mathbb{G}_2$. Let $e : \mathbb{G}_1 \times \mathbb{G}_2$ be a function that maps pairs of elements $\in (\mathbb{G}_1, \mathbb{G}_2)$ to elements of some cyclic group $\mathbb{G}_T$ of order p. Throughout the following sections we write all of the group operations mutiplicatively, with identity elements denoted by 1. We further require that:

- The map e and all the group operations in $\mathbb{G}_2$, $\mathbb{G}_2$, and $\mathbb{G}_T$ are efficiently computable.
- The map e is non degenerate, i.e., $e(g_1, g_2) \neq 1$.
- The map e is bilinear, i.e., $\forall u \in \mathbb{G}_1, \forall v \in \mathbb{G}_2, \forall (a, b) \in \mathbb{Z}_p^2, e(u^a, v^b) = e(u, v)^{ab}$.

3.2   Complexity Assumptions

In the following we formally define the Decisional q-Bilinear Diffie-Hellman Inversion assumption and the Decisional Bilinear Diffie-Hellman assumption. Both of the conjectures are widely used in pairing-based cryptographic constructions, among the others we mention the work of Boneh and Boyen [5]. We must point out that a sub-exponential attack is known for the former assumption [12], and therefore the security parameter of any scheme based on such a conjecture must be increased correspondingly. This, however, does not have a severe impact on the efficiency of the constructions, as discussed in [5].

Definition 1 (q-Decision-BDHI   Assumption). *The q-Decision-BDHI assumption holds in $(\mathbb{G}_1, \mathbb{G}_2)$ if, for all PPT algorithms $\mathcal{A}$, there exists a negligible function negl such that*

$$\left| \Pr\left[1 \leftarrow \mathcal{A}\left(g_1, g_1^x, g_2, g_2^x, \ldots, g_2^{x^q}, e(g_1, g_2)^{1/x}\right)\right] - \Pr\left[1 \leftarrow \mathcal{A}\left(g_1, g_1^x, g_2, g_2^x, \ldots, g_2^{x^q}, T\right)\right] \right| \leq \mathsf{negl}(\lambda)$$

where the probability is taken over the random choice of the generators $g_1 \in \mathbb{G}_1$ and $g_2 \in \mathbb{G}_2$, the random choice of $x \in \mathbb{Z}_p^*$, the random choice of $T \in \mathbb{G}_T$, and the random coins of $\mathcal{A}$.

Definition 2 (DBDH Assumption). *The DBDH assumption holds in $(\mathbb{G}_1, \mathbb{G}_2)$ if, for all* PPT *algorithms $\mathcal{A}$, there exists a negligible function* negl *such that*

$$\left| \begin{array}{c} \Pr\left[1 \leftarrow \mathcal{A}\left(g_1, g_1^a, g_1^b, g_1^c, g_2, g_2^a, g_2^b, g_2^c, e(g_1, g_2)^{abc}\right)\right] - \\ \Pr\left[1 \leftarrow \mathcal{A}\left(g_1, g_1^a, g_1^b, g_1^c, g_2, g_2^a, g_2^b, g_2^c, e(g_1, g_2)^z\right)\right] \end{array} \right| \leq \mathsf{negl}(\lambda)$$

where the probability is taken over the random choice of the generators $g_1 \in \mathbb{G}_1$ and $g_2 \in \mathbb{G}_2$, the random choice of $(a, b, c, z) \in (\mathbb{Z}_p^*)^4$, and the random coins of $\mathcal{A}$.

4   Subset Predicate Encryption

In this section, we formally introduce the concept of Subset Predicate Encryption. Our definition is very close to the standard Identity-Based Encryption, except that we do not necessarily require the string associated with the secret key to match the string embedded in the ciphertext. In fact, we allow anybody who owns a key for a string that matches any *subset* of the string of the ciphertext, to decrypt the latter.

Definition 3 (Subset Predicate Encryption). *A Subset Predicate Encryption (SPE) scheme consists of four* PPT *algorithms* Setup, KeyGen, Encrypt, *and* Decrypt *such that:*

$(\mathsf{pk}, \mathsf{msk}) \leftarrow \mathsf{Setup}(1^\lambda, 1^n)$ *The setup algorithm takes as input the security parameter 1^λ and a length parameter n. It outputs public parameters* pk *and the master secret key* msk.

$\mathsf{sk}_s \leftarrow \mathsf{KeyGen}(\mathsf{msk}, \mathsf{pk}, s)$ *The key-derivation algorithm takes as input the master secret key* msk, *public parameters* pk, *and a string $s \in \{0,1\}^n$. It outputs a private key* sk_s. *We assume that s can be recovered from* sk_s.

$\mathsf{c} \leftarrow \mathsf{Encrypt}(\mathsf{pk}, \mathsf{m}, s)$ *The encryption algorithm takes as input public parameters* pk, *a message* m, *and a string $s \in \{0,1\}^n$. It outputs a ciphertext* c. *We assume that s can be recovered from* c.

$\mathsf{m} \leftarrow \mathsf{Decrypt}(\mathsf{sk}_s, \mathsf{pk}, \mathsf{c})$ *The decryption algorithm takes as input the private key* sk_s, *the public parameters* pk, *and a ciphertext* c. *It outputs a message* m *or a designated failure symbol $\perp$.*

Our notion of correctness for SPE is defined as follows:

Definition 4 (Correctness). *Correctness requires that for all security parameters λ, all n, all $(\mathsf{pk}, \mathsf{msk})$ output by $\mathsf{Setup}(1^\lambda, 1^n)$, all $s' \in \{0,1\}^n$, all s such that $s \subseteq s'$, all sk_s output by $\mathsf{KeyGen}(\mathsf{msk}, \mathsf{pk}, s)$, all* m, *and all* c *output by $\mathsf{Encrypt}(\mathsf{pk}, \mathsf{m}, s')$, we have $\mathsf{Decrypt}(\mathsf{sk}_s, \mathsf{pk}, \mathsf{c}) = \mathsf{m}$.*

SECURITY. In the following, we define the security model for SPE schemes. Informally, the adversary should be unable to learn anything about the content of a ciphertext associated with some set s^* even if it has obtained secret keys corresponding to arbitrary sets $s_1, \ldots, s_q$, so long as none of those satisfies $s_i \subseteq s^*$. Our definition corresponds to "selective" security, whereby the attacker is required to commit to the s^* that he wants to be challenged on before seeing the public parameters of the scheme. Alternatively one could consider the stronger "adaptive" notion, where the challenge set is revealed by the adversary only in the challenge phase.

Consider the following experiment parameterized by λ:

1. The attacker specifies a universe of elements $\{0,1\}^n$ (i.e., a bound n on the size of the universe) and a challenge set $s^* \in \{0,1\}^n$.
2. Setup($1^\lambda, 1^n$) is run to obtain (pk, msk), and the adversary is given pk.
3. The adversary is allowed to query for private keys for arbitrary sets $s_1, \ldots, s_q$ such that for all $i \in \{1, \ldots q\}$ it holds that $s_i \not\subseteq s^*$.
4. The adversary outputs a message pair (m_0, m_1) with $|m_0| = |m_1|$. A uniform bit $b \in \{0,1\}$ is chosen, and the ciphertext $c \leftarrow$ Encrypt(pk, m_b, s^*) is computed and given to the adversary.
5. The adversary may continue to request private keys for arbitrary sets, subject to the same restriction as before.
6. Finally, the adversary outputs a guess b' for b.

The advantage of the adversary in this experiment is defined as $|\Pr[b' = b] - \frac{1}{2}|$.

Definition 5 (Selective Security). *A SPE scheme is selectively secure if the advantage of any* PPT *adversary in the above experiment is negligible.*

4.1 Generic Instantiations

Before presenting our schemes we first describe some potential approaches to instantiate Subset Predicate Encryption and we show their drawbacks.

SPE FROM PE. One can instantiate SPE from any predicate encryption for inner products as follows: Given a universe of n attributes, keys for a set s are associated with the binary vector $(s_1, \ldots, s_n)$. A ciphertext for a set s' is encrypted under the vector $(s'_1 \oplus 1, \ldots, s'_n \oplus 1)$. The inner product of the two vectors is 0 if and only if $s \subseteq s'$, therefore correctness and security follow. This instantiation however, inherently generates ciphertexts and secret keys that grow linearly with the size of the universe of elements.

SPE FROM WIBE. We observe that if we encode a set s in a ciphertext as a string where the 1s are substituted with the wildcard symbol, then an IBE with wildcards supports the same functionality as a SPE. Given the current state-of-the-art for WIBE schemes, this approach suffers from the same drawbacks as described above.

SPE FROM FUZZY IBE. It is an easy exercise to instantiate SPE from the Fuzzy Identity-Based Encryption of Sahai and Waters [21]: Setting the degree d

of the polynomial associated with the secret key to be equal to the number of components of the key itself, one can ensure that the decryptor needs to use all of the components of the secret key in order to decrypt a ciphertext. It follows that a key associated with a string s can decrypt any ciphertext encrypted under any s' such that $s \subseteq s'$. However the decryption algorithm requires to interpolate the polynomial in the exponent, which incurs in one pairing operation per element associated with the secret key. Additionally, the size of the key grows linearly with the number of elements associated with it, which is, in the average case, linear in the security parameter. Our observation is that our primitive does not require the flexibility of a Fuzzy IBE scheme, and therefore we can hope to achieve better performance at the cost of sacrificing the malleability in the manipulation of the secret keys.

5   Our Constructions

In this section we present our two instantiations from bilinear maps.

5.1   First Scheme

In the following we describe our first construction, inspired by second scheme presented in [5]. The key difference is that our ciphertexts is composed by disjoint components, each corresponding to an element of the public parameters. This additional flexibility allows one to choose an arbitrary subset of elements in the decryption phase.

Construction 1. Our first construction consists of the following algorithms.

Setup($1^\lambda, 1^n$): To generate the SPE system given a bilinear group pair $(\mathbb{G}_1, \mathbb{G}_2)$, with respective generators (g_1, g_2), the setup algorithm selects a random generator $h \in \mathbb{G}_2$ and it computes $v = e(g_1, h)$. Then it samples a random $x_0 \in \mathbb{Z}_p^*$ and a random vector $(x_1, \dots, x_n) \in (\mathbb{Z}_p^*)^n$ and sets $X_0 = g_1^{x_0}$ and for all $i \in \{1, \dots, n\} : X_i = g_1^{x_i}$. The public parameters pk and the master secret key are given by

$$\mathsf{pk} = (g_1, X_0, X_1, \dots, X_n, v) \in \mathbb{G}_1^{n+2} \times \mathbb{G}_T$$

$$\mathsf{msk} = (x_0, x_1, \dots, x_n, h) \in (\mathbb{Z}_p^*)^{n+1} \times \mathbb{G}_2$$

KeyGen($\mathsf{msk}, \mathsf{pk}, s$): To generate the private key associated with the set s, the key generation algorithm picks a random $\kappa \in \mathbb{Z}_p$ such that $\sum_{i \in s} x_i + \kappa x_0 \neq 0$ mod p and computes $K = h^{\frac{1}{\sum_{i \in s} x_i + \kappa x_0}}$. The private key is defined as

$$\mathsf{sk}_s = (\kappa, K) \in \mathbb{Z}_p \times \mathbb{G}_2$$

$\mathsf{c} \leftarrow$ Encrypt(pk, m, s): The encryption of a message $m \in \mathbb{G}_T$ for a given set s is done by picking a random $\rho \in \mathbb{Z}_p^*$ and returning the following ciphertext

$$\mathsf{c} = (m \cdot v^\rho, X_0^\rho, \forall_{i \in s} : X_i^\rho) \in \mathbb{G}_T \times \mathbb{G}_1^{|s|+1}$$

$m \leftarrow \mathsf{Decrypt}(\mathsf{sk}_s, \mathsf{pk}, \mathsf{c})$: To decrypt a ciphertext $\mathsf{c} = (A, B, C_1, \ldots, C_\ell)$, for some positive integer $\ell \leq n$, using the private key $\mathsf{sk}_s = (\kappa, K)$, return

$$\frac{A}{e\left(B^\kappa \prod_{i \in s} C_i, K\right)}$$

To check that the system is consistent it is enough to observe that, for a valid private key sk_s and a valid ciphertext encoded under a string s' such that $s \subseteq s'$, there always exists an element C_i for all $i \in s$, thus we have

$$\frac{A}{e\left(B^\kappa \prod_{i \in s} C_i, K\right)} = \frac{A}{e\left((X_0^\rho)^\kappa g_1^{\rho \cdot \sum_{i \in s} x_i}, h^{\frac{1}{\sum_{i \in s} x_i + \kappa x_0}}\right)}$$

$$= \frac{m \cdot v^\rho}{e\left(g_1^{\rho \cdot (\kappa x_0 + \sum_{i \in s} x_i)}, h^{\frac{1}{\sum_{i \in s} x_i + \kappa x_0}}\right)}$$

$$= \frac{m \cdot e(g_1, h)^\rho}{e(g_1, h)^\rho}$$

$$= m$$

Here we elaborate the formal guarantees of our construction. The security proof is non-trivial as our reduction is required to include in the challenge ciphertext each group element separately (as opposed to their product), this arises subtle issues in the generation of the secret key that we address in the following.

Theorem 1. *Assume that the q-Decision-BDHI assumption holds in groups $(\mathbb{G}_1, \mathbb{G}_2)$ of size p. Then Construction 1 is a selectively-secure SPE scheme.*

Proof. Assume towards contradiction that there exists an adversary $\mathcal{A}$ that has advantage $\epsilon(\lambda)$ in attacking the SPE system, for some non negligible function $\epsilon(\lambda)$. Then we can construct the following reduction $\mathcal{R}$ against the q-Decision-BDHI assumption in $(\mathbb{G}_1, \mathbb{G}_2)$.

The reduction $\mathcal{R}$ takes as input a tuple $\left(g_1, g_1^\alpha, g_2, g_2^\alpha, \ldots, g_2^{\alpha^q}, T\right)$, where T is either $e(g_1, g_2)^{1/\alpha}$ or a random element of $\mathbb{G}_T$. The algorithm $\mathcal{R}$ interacts with $\mathcal{A}$ in the selective-security game as follows:

Preparation: The reduction $\mathcal{R}$ samples a vector $(w_0, \ldots, w_{q-1}) \in (\mathbb{Z}_p^*)^q$, let

$$f(\alpha) = w_0 \prod_{j=1}^{q-1} (\alpha + w_j) = \sum_{j=0}^{q-1} c_j \alpha^j$$

for some coefficients c_j where $c_0 \neq 0$. The algorithm sets $h = \prod_{j=1}^{q-1} \left(g_2^{\alpha^j}\right)^{c_j} = g_2^{f(\alpha)}$. The variable w_0 ensures that h is a uniformly distributed generator of $\mathbb{G}_2$. Note that we can assume that $h \neq 1$ otherwise it must have been the case that there exists a $j \in \{1 \ldots q - 1\}$ such that $w_j = -\alpha$ and thus the algorithm can efficiently output a solution to the decisional problem. We observe that for all $j \in \{1 \dots q-1\}$ it is easy for $\mathcal{R}$ to compute the tuple $\left(w_j, h^{\frac{1}{\alpha+w_j}}\right)$ by considering

$$\frac{f(\alpha)}{(\alpha+w_j)} = \sum_{j=0}^{q-2} d_j \alpha^j$$

and setting $h^{\frac{1}{\alpha+w_j}} = g_2^{\frac{f(\alpha)}{(\alpha+w_j)}} = \prod_{j=0}^{q-2} \left(g_2^{\alpha^j}\right)^{d_j}$. Additionally, the reduction $\mathcal{R}$ computes

$$T_h = T^{c_0} \cdot \prod_{j=1}^{q-1} e\left(g_1, g_2^{c_j \alpha^{j-1}}\right)$$

It is easy to see that whenever T is uniformly distributed in $\mathbb{G}_T$ then so is T_h, whereas whenever $T = e(g_1, g_2)^{1/\alpha}$ then $T_h = e(g_1, h)^{1/\alpha}$.

Initialization: The experiment begins with $\mathcal{A}$ outputting bound n on the universe of elements and a challenge set $s^* \in \{0,1\}^n$.

Setup: To generate the public parameters, the algorithm $\mathcal{R}$ proceeds by uniformly sampling for all $i \in s^*$ an element $a_i \in \mathbb{Z}_p^*$ and setting $X_i = g_1^{x_i} = g_1^{-a_i \cdot \alpha}$. For all $i \in \overline{s^*}$ the reduction picks a pair $(a_i, b_i) \in (\mathbb{Z}_p^*)^2$ and sets $X_i = g_1^{x_i} = g_1^{-a_i \cdot (\alpha+b_i)}$. The public parameters provided to the adversary are

$$(g_1, X_0 = g_1^\alpha, X_1, \dots, X_n, v = e(g,h))$$

where h is defined as specified above. We remark that h is a uniformly distributed element in $\mathbb{G}_T$. Since all of the other elements of the public parameters are uniformly distributed over $\mathbb{G}_1$ to the view of the adversary, we can conclude that the public parameters are correctly distributed according to our construction.

Phase 1: The adversary can issue up to $q-1$ private key queries for some sets s^j under the constraint that for all $j \in \{1, \dots, q-1\}$ it holds that $s^j \not\subseteq s^*$. The algorithm $\mathcal{R}$ responds to each query j as follows: let $\left(w_j, h^{\frac{1}{\alpha+w_j}}\right)$ the j-th pair constructed in the preparation phase, the reduction computes an $r \in \mathbb{Z}_p$ that satisfies

$$\left(r - \sum_{i \in s^j} a_i\right)(\alpha + w_j) = -\alpha \sum_{i \in s^j} a_i - \sum_{i \in s^j \cap \overline{s^*}} a_i b_i + \alpha r$$

Expanding the equation we obtain

$$r = \sum_{i \in s^j} a_i - \frac{\sum_{i \in s^j \cap \overline{s^*}} a_i b_i}{w_j}$$

Note that the unknown α cancels out of the equation and the algorithm can evaluate the expression. The secret key for the set s^j is set to be

$$\mathsf{sk}_{s^j} = \left(r, h^{\frac{1}{(\alpha+w_j)(r - \sum_{i \in s^j} a_i)}}\right).$$

We note that the key is functional, as

$$h^{\frac{1}{(\alpha+w_j)(r-\sum_{i\in s^j} a_i)}} = h^{\frac{1}{-\alpha\sum_{i\in s^j} a_i - \sum_{i\in s^j \cap \overline{s^*}} a_i b_i + \alpha r}} = h^{\frac{1}{\sum_{i\in s^j} x_i + \alpha r}}$$

it allows the adversary to decrypt the ciphertexts that he is intended to. To argue about the correct distribution of the key it is enough to observe that the value w_j is sampled uniformly at random from $\mathbb{Z}_p^*$, therefore whenever $\sum_{i\in s^j \cap \overline{s^*}} a_i b_i \neq 0$ then r is a uniformly distributed element of $\mathbb{Z}_p$. First we point out that the set $s^j \cap \overline{s^*}$ is never empty due to the non-triviality of the game, i.e., $s^j \not\subseteq s^*$, secondly we observe that the expression $\sum_{i\in s^j \cap \overline{s^*}} a_i b_i$ can return at most 2^n different results, due to the total number of elements' combinations. Therefore by choosing a large enough size of p, e.g. $2^{2 \cdot n}$, we ensure that the probability of $\sum_{i\in s^j \cap \overline{s^*}} a_i b_i$ returning 0 is negligible in the security parameter (recall that for all $i \in \{1, \ldots, n\}$ it holds that a_i and b_i are elements uniformly distributed in $\mathbb{Z}_p^*$). For completeness, we note that this procedure will fail to produce a private key for an $s \subseteq s^*$ since in that case we obtain $r = \sum_{i\in s^j} a_i$ and therefore $h^{\frac{1}{(\alpha+w_j)(r-\sum_{i\in s^j} a_i)}} = h^{\frac{1}{(\alpha+w_j)\cdot 0}}$.

Challenge: The adversary outputs two messages $(m_0, m_1) \in \mathbb{G}_T^2$. The reduction $\mathcal{R}$ samples a random $b \in \{0,1\}$ and a random $z \in \mathbb{Z}_p^*$ and hands over to the attacker the challenge ciphertext

$$c^* = \left(m_b \cdot T_h^z, g_1^z, \forall_{i\in s^*} : g_1^{-a_i z}\right)$$

Consider $\rho = z/\alpha$. We shall note that whenever $T_h = e(g_1, h)^{1/\alpha}$ then c^* is a valid ciphertext as

$$m_b \cdot T_h^z = m_b \cdot e(g_1, h)^{z/\alpha} = m_b \cdot v^\rho$$
$$g_1^z = \left(X_0^{1/\alpha}\right)^z = X_0^\rho$$
$$\forall_{i\in s^*} : g_1^{-a_i z} = \left(X_i^{1/\alpha}\right)^z = X_i^\rho$$

On the other hand, whenever T_h is uniform in $\mathbb{G}_T$, then the message m_b is hidden from the view of the adversary in an information theoretic sense.

Phase 2: The adversary can issue additional private key queries for a total of at most $q - 1$. The reduction answer as specified in Phase 1.

Guess: The adversary outputs a guess b' and the reduction returns $b = b'$ to the challenger.

As argued above, when the input tuple contains a $T = (g_1, g_2)^{1/\alpha}$, then the view of the adversary perfectly resembles the inputs that he is expecting in the standard experiment for SPE security. It follows that the advantage of the adversary is, as assumed, greater than some non negligible $\epsilon(\lambda)$. On the other hand, when the input tuple contains a T uniformly distributed in $\mathbb{G}_T$, then the view of the adversary contains no information about the secret bit b. Thus in this case $\mathcal{A}$ cannot do better than guessing. It follows that

$$\left| \Pr\left[1 \leftarrow \mathcal{R}\left(g_1, g_1^x, g_2, g_2^x, \ldots, g_2^{x^q}, e(g_1, g_2)^{1/x}\right)\right] - \right.$$
$$\left. \Pr\left[1 \leftarrow \mathcal{R}\left(g_1, g_1^x, g_2, g_2^x, \ldots, g_2^{x^q}, T\right)\right] \right| \geq$$
$$|1/2 + \epsilon(\lambda) - 1/2| = \epsilon(\lambda)$$

This represents a contradiction to the q-Decision-BDHI assumption and it concludes our proof. □

5.2   Second Scheme

Our second scheme can be seen as a descendant of the celebrated IBE of Waters [24]. On a very high-level, our main observation is that the scheme satisfies our notion of security if the group elements of the ciphertext are not multiplied together. This change, together with our different notion of security, forces us to develop a different proof strategy.

Construction 2. Our second construction consists of the following algorithms.

Setup($1^\lambda, 1^n$):   To generate the SPE system given a bilinear group pair $(\mathbb{G}_1, \mathbb{G}_2)$ with respective generators (g_1, g_2), the setup algorithm selects a random $\alpha \in \mathbb{Z}_p^*$ and sets $h = g_2^\alpha$. Then it samples a random vector $(x_1, \ldots, x_n) \in (\mathbb{Z}_p^*)^n$ and sets for all $i \in \{1, \ldots, n\} : X_i = g_1^{x_i}$. The public parameters pk and the master secret key msk are given by

$$\mathsf{pk} = (g_1, X_1, \ldots, X_n, g_2, h) \in \mathbb{G}_1^{n+1} \times \mathbb{G}_2^2$$

$$\mathsf{msk} = g_1^\alpha \in \mathbb{G}_1$$

KeyGen(msk, pk, s):   To generate the private key associated with the set s, the key generation algorithm picks a random $r \in \mathbb{Z}_p$ and defines the private key as

$$\mathsf{sk}_s = \left(g_1^\alpha \left(\prod_{i \in s} X_i\right)^r, g_2^r\right) \in \mathbb{G}_1 \times \mathbb{G}_2$$

$\mathsf{c} \leftarrow$ Encrypt(pk, m, s):   The encryption of a message $m \in \mathbb{G}_T$ for a given set s is done by picking a random $\rho \in \mathbb{Z}_p^*$ and returning the following ciphertext

$$\mathsf{c} = (m \cdot e(g_1, h)^\rho, g_2^\rho, \forall_{i \in s} : X_i^\rho) \in \mathbb{G}_T \times \mathbb{G}_2 \times \mathbb{G}_1^{|s|}$$

$m \leftarrow$ Decrypt(sk_s, pk, c):   To decrypt a ciphertext $\mathsf{c} = (A, B, C_1, \ldots, C_\ell)$, for some positive integer $\ell \leq n$, using the private key $\mathsf{sk}_s = (K, R)$, return

$$A \cdot \frac{e\left(\prod_{i \in s} C_i, R\right)}{e(K, B)}$$

To check that the system is correct we observe, as before, that

$$A \cdot \frac{e\left(\prod_{i \in s} C_i, R\right)}{e(K, B)} = m \cdot e(g_1, h)^\rho \cdot \frac{e\left(\prod_{i \in s} X_i^\rho, g_2^r\right)}{e\left(g_1^\alpha \left(\prod_{i \in s} g_1^{x_i}\right)^r, g_2^\rho\right)}$$

$$= m \cdot e(g_1, g_2)^{\alpha\rho} \cdot \frac{e\left(\prod_{i \in s} X_i^\rho, g_2^r\right)}{e(g_1^\alpha, g_2^\rho) e\left(\left(\prod_{i \in s} g_1^{x_i}\right)^r, g_2^\rho\right)}$$

$$= m \cdot e(g_1, g_2)^{\alpha\rho} \cdot \frac{e\left(\prod_{i \in s} X_i, g_2\right)^{r\rho}}{e(g_1, g_2)^{\alpha\rho} e\left(\prod_{i \in s} X_i, g_2\right)^{r\rho}}$$

$$= m$$

The construction above is a secure SPE scheme if the DBDH assumption holds.

Theorem 2. *Assume that the DBDH assumption holds in groups $(\mathbb{G}_1, \mathbb{G}_2)$ of size p. Then Construction 2 is a selectively-secure SPE scheme.*

Proof. Assume towards contradiction that there exists an adversary $\mathcal{A}$ that has advantage $\epsilon(\lambda)$ in attacking the SPE system, for some non negligible function $\epsilon(\lambda)$. Then we can construct the following reduction $\mathcal{R}$ against the DBDH assumption in $(\mathbb{G}_1, \mathbb{G}_2)$.

The reduction $\mathcal{R}$ takes as input a tuple $(g_1, A_1, B_1, C_1, g_2, A_2, B_2, C_2, Z)$, where Z is either $e(g_1, g_2)^{abc}$ or a random element of $\mathbb{G}_T$. The algorithm $\mathcal{R}$ interacts with $\mathcal{A}$ in the selective-security game as follows:

Initialization: The experiment begins with $\mathcal{A}$ outputting bound n on the universe of elements and a challenge set $s^* \in \{0, 1\}^n$.

Setup: To generate the public parameters, the algorithm $\mathcal{R}$ proceeds by uniformly sampling for all $i \in s^*$ an pair of elements $y_i \in \mathbb{Z}_p^*$ and setting $X_i = g_1^{y_i}$. For all $i \in \overline{s^*}$ the reduction picks a pair $(y_i, w_i) \in (\mathbb{Z}_p^*)^2$ and sets $X_i = A_1^{w_i} g_1^{y_i} = g_1^{a \cdot w_i + y_i}$. The public parameters provided to the adversary are

$$(A_1, X_1, \ldots, X_n, g_2, B_2)$$

Since all of the elements of the public parameters are uniformly distributed over the corresponding group to the view of the adversary, we can conclude that the public parameters are correctly distributed according to our construction.

Phase 1: The adversary can issue up to $q - 1$, for some polynomial bound q, private key queries for some sets s^j under the constraint that for all $j \in \{1, \ldots, q-1\}$ it holds that $s^j \not\subseteq s^*$. The algorithm $\mathcal{R}$ responds to each query j as follows: the reduction samples an $r \in \mathbb{Z}_p$ and sets

$$sk_{s^j} = \left(B_1^{\frac{-\sum_{i \in s^j} y_i}{\sum_{i \in s^j} w_i}} \cdot \left(\prod_{i \in s^j} X_i'\right)^r, B_2^{\frac{-1}{\sum_{i \in s^j} w_i}} \cdot g_2^r \right)$$

We observe that

$$
\begin{aligned}
sk_{s^j} &= \left(g_1^{\, b \cdot \frac{-\sum_{i \in s^j} y_i}{\sum_{i \in s^j} w_i} + r\left(\sum_{i \in s^j} aw_i + y_i\right)} \, , \, g_2^{\, r - \frac{b}{\sum_{i \in s^j} w_i}} \right) \\[2mm]
&= \left(g_1^{\, ab + b \cdot \left(\frac{-\sum_{i \in s^j} y_i}{\sum_{i \in s^j} w_i} - a \right) + r\left(\sum_{i \in s^j} aw_i + y_i\right)} \, , \, g_2^{\, r - \frac{b}{\sum_{i \in s^j} w_i}} \right) \\[2mm]
&= \left(g_1^{\, ab - \frac{b}{\sum_{i \in s^j} w_i} \cdot \left(\sum_{i \in s^j} y_i + a \cdot \sum_{i \in s^j} w_i \right) + r\left(\sum_{i \in s^j} aw_i + y_i\right)} \, , \, g_2^{\, r - \frac{b}{\sum_{i \in s^j} w_i}} \right) \\[2mm]
&= \left(g_1^{\, ab + \left(\sum_{i \in s^j} aw_i + y_i\right)\left(r - \frac{b}{\sum_{i \in s^j} w_i} \right)} \, , \, g_2^{\, r - \frac{b}{\sum_{i \in s^j} w_i}} \right) \\[2mm]
&= \left(g_1^{\, ab} \cdot \left(\prod_{i \in s^j} X_i \right)^{\left(r - \frac{b}{\sum_{i \in s^j} w_i} \right)} \, , \, g_2^{\, r - \frac{b}{\sum_{i \in s^j} w_i}} \right)
\end{aligned}
$$

which gives us a functional and correctly distributed key, as r is uniformly distributed over $\mathbb{Z}_p^*$. For completeness we note that the procedure fails whenever $\sum_{i \in s^j} w_i = 0$, which happens only in case the queried set s^j is a subset of the challenge set s^*, except with negligible probability (for a large enough p).

Challenge: The adversary outputs two messages $(m_0, m_1) \in \mathbb{G}_T^2$. The reduction $\mathcal{R}$ samples a random $\mathsf{b} \in \{0,1\}$ and hands over to the attacker the challenge ciphertext

$$
\mathsf{c}^* = (m_{\mathsf{b}} \cdot Z, C_2, \forall_{i \in s^*} : C_1^{y_i})
$$

We shall note that whenever $Z = e(g_1, h)^{abc}$ then c^* is a valid ciphertext as

$$
\begin{aligned}
m_{\mathsf{b}} \cdot Z &= m_{\mathsf{b}} \cdot e(g_1, g_2)^{abc} = m_{\mathsf{b}} \cdot e(A_1, B_2)^c \\
C_2 &= g_2^c \\
\forall_{i \in s^*} : C_1^{y_i} &= (g_1^{y_i})^c = X_i^c
\end{aligned}
$$

On the other hand, whenever Z is uniform in $\mathbb{G}_T$, then the message m_{b} is hidden from the view of the adversary in an information theoretic sense.

Phase 2: The adversary can issue additional private key queries for a total of at most $q - 1$. The reduction answer as specified in Phase 1.

Guess: The adversary outputs a guess b' and the reduction returns $\mathsf{b} = \mathsf{b}'$ to the challenger.

As argued above, when the input tuple contains a $Z = (g_1, g_2)^{abc}$, then the view of the adversary perfectly resembles the inputs that he is expecting in the standard experiment for SPE security. It follows that the advantage of the adversary is, as assumed, greater than some non negligible $\epsilon(\lambda)$. On the other hand, when the input tuple contains a Z uniformly distributed in $\mathbb{G}_T$, then the view of the adversary contains no information about the secret bit b. Thus in this case $\mathcal{A}$ cannot do better than guessing. It follows that

$$\left| \begin{array}{l} \Pr\left[1 \leftarrow \mathcal{A}\left(g_1, g_1^a, g_1^b, g_1^c, g_2, g_2^a, g_2^b, g_2^c, e(g_1, g_2)^{abc}\right)\right] - \\ \Pr\left[1 \leftarrow \mathcal{A}\left(g_1, g_1^a, g_1^b, g_1^c, g_2, g_2^a, g_2^b, g_2^c, e(g_1, g_2)^z\right)\right] \end{array} \right| \geq \\ |1/2 + \epsilon(\lambda) - 1/2| = \epsilon(\lambda)$$

This represents a contradiction to the DBDH assumption and it concludes our proof. □

LARGE UNIVERSE CONSTRUCTION. We note that we can extend our second construction to support elements that were not considered in the setup phase, assuming the existence of a random oracle. Assume that all parties have access to the function $H : \{0,1\}^* \to \mathbb{G}_1$, we can remove the group elements from the public parameters and substitute them with the description of H. This extension yields a scheme with *constant-size* public parameters for an *exponentially-large* universe of elements.

6 Generic Transformations

In the following we describe some black-box transformations from SPE to well known cryptographic primitives.

IDENTITY BASED ENCRYPTION. As an easy warm up we show how to deploy SPE in order to achieve standard Identity Based Encryption (IBE). Although not surprising, this will guide us through the subsequent transformations. We first initialize the system by running the Setup algorithm with a length parameter of $2 \cdot n$ and the corresponding security parameter λ. The KeyGen algorithm, on input $\mathsf{ID} \in \{0,1\}^n$, generates $s \in \{0,1\}^{2 \cdot n}$ by setting, for all $i \in \{1, \ldots, 2 \cdot n\}$:

$$s_i = \begin{cases} 1 - \mathsf{ID}_{i/2} & \text{if } i = 0 \\ \mathsf{ID}_{(i+1)/2} & \text{if } i = 1 \end{cases} \pmod 2$$

Then the standard KeyGen algorithm is executed on s and the corresponding output is returned. The same modification is applied to the Encrypt algorithm.

To better visualize this transformation one can imagine the Setup algorithm to return two arrays of elements $(x_1^0, \ldots, x_n^0)$ and $(x_1^1, \ldots, x_n^1)$. The identities ID in the set $\{0,1\}^n$ index the binary choice of each element $x_i^{\mathsf{ID}_i}$ between the two arrays. It is important to note that all of the valid sets contain the same amount of elements, i.e. n many, and that any two sets differ in at least one position. This implies that no valid identity is a subset of any other and the security of the IBE scheme follows from the security of the underlying SPE.

IDENTITY BASED ENCRYPTION WITH WILDCARDS. Here show how to modify our primitive in a black-box fashion to handle *wildcards* in both the ciphertexts and the keys: this allows us to specify certain positions of the identity encoded in the ciphertext (in the key, respectively) that are not required to match the key (the ciphertext, respectively) for the decryption to be successful. IBE schemes that allow for wildcards in the ciphertexts are known in the literature as WIBE [1], whereas schemes that support wildcards in the keys are called Wicked IBE [2]. We stress that, as opposed to the original proposals, our generic transformation does not support a hierarchical structure of identities, since it is not clear how to delegate keys in the general settings. In the following we describe how to modify the Encrypt and the KeyGen algorithms to handle wildcards. We denote the wildcard with the distinguished symbol $*$.

We first initialize the system by running the Setup algorithm with a parameter of $2 \cdot n$ and the corresponding security parameter λ. The Encrypt algorithm is modified to take as input $\mathsf{ID} \in \{0, 1, *\}^n$ and generate $s \in \{0, 1\}^{2 \cdot n}$ as follows:

$$
s_i = \begin{cases}
1 - \mathsf{ID}_{i/2} & \text{if } \mathsf{ID}_{i/2} \in \{0, 1\} \wedge i = 0 \\
\mathsf{ID}_{(i+1)/2} & \text{if } \mathsf{ID}_{(i+1)/2} \in \{0, 1\} \wedge i = 1 \\
1 & \text{if } \mathsf{ID}_{i/2} = * \wedge i = 0 \\
1 & \text{if } \mathsf{ID}_{(i+1)/2} = * \wedge i = 1
\end{cases} \pmod 2
$$

for all $i \in \{1, \ldots, 2 \cdot n\}$. As before, if we consider Setup to output two vectors $(x_1^0, \ldots, x_n^0)$ and $(x_1^1, \ldots, x_n^1)$, then the identities $\mathsf{ID} \in \{0, 1, *\}^n$ represent the binary choice over the elements of the two vectors except when $s_i = *$, in which case both x_i^0 and x_i^1 are included in the set. We observe that the decryption is successful whenever one owns a key that encodes a subset of s, which matches the policy enforced by the WIBE scheme. Therefore the security of the SPE carries over.

We can encode wildcards in the decryption keys applying a similar modification to the KeyGen algorithm, that differs in assigning $s_i = 0$, as opposed to 1, whenever the corresponding bit of ID is $*$. We note that the two modifications are not mutually exclusive and can coexist for a hybrid of the two approaches.

CIPHERTEXT POLICY ATTRIBUTE BASED ENCRYPTION. Perhaps the most interesting feature of our primitive is that it can be used to obtain a Ciphertext-Policy Attribute Based Encryption (CP-ABE) scheme for a small universe of attributes. The transformation is as follows. Fix a universe of attributes $\mathbb{U}$ of size n, we uniquely assign to each attribute $a \in \mathbb{U}$ an index $i \in \{1, \ldots, n\}$. The private key associated with a set of attributes A will be the key associated with the set $\mathbb{U} \setminus A$. Specifically, we construct the private key for A by executing KeyGen on input s^A, where

$$
s_i^A = \begin{cases}
0 & \text{if } a_i \in A \\
1 & \text{if } a_i \notin A
\end{cases}
$$

To encrypt a message m using a DNF formula $C_1 \vee \cdots \vee C_t$, where each C_j represents a conjunction over some subset of the attributes, the sender processes each of the t clauses independently. For the ith clause C_j, the sender encrypts the message running Encrypt on input s^{C_j}, where

$$s_i^{C_j} = \begin{cases} 0 & \text{if } a_i \in C_j \\ 1 & \text{if } a_i \notin C_j \end{cases}$$

The algorithm returns the concatenation of the ciphertexts corresponding to each clause. To decrypt, the receiver finds some clause C_j that is satisfied by his attributes A. Note that this means $C_j \subseteq \mathsf{A}$, or equivalently $\mathbb{U} \setminus \mathsf{A} \subseteq \mathbb{U} \setminus C_j$. Thus, the receiver will be able to decrypt the ciphertext corresponding to that clause if and only if its key is associated with a set s such that $s \subseteq s^{C_j}$.

Acknowledgements. This research is based upon work supported by the German research foundation (DFG) through the collaborative research center 1223, by the German Federal Ministry of Education and Research (BMBF) through the project PROMISE (16KIS0763), and by the state of Bavaria at the Nuremberg Campus of Technology (NCT). NCT is a research cooperation between the Friedrich-Alexander-Universität Erlangen-Nürnberg (FAU) and the Technische Hochschule Nürnberg Georg Simon Ohm (THN). We thank the anonymous reviewers for their valuable comments that helped to improve our paper. We thank Vincenzo Iovino for the insightful discussions on this work.

References

1. Abdalla, M., Catalano, D., Dent, A.W., Malone-Lee, J., Neven, G., Smart, N.P.: Identity-based encryption gone wild. In: Bugliesi, M., Preneel, B., Sassone, V., Wegener, I. (eds.) ICALP 2006. LNCS, vol. 4052, pp. 300–311. Springer, Heidelberg (2006). https://doi.org/10.1007/11787006_26

2. Abdalla, M., Kiltz, E., Neven, G.: Generalized key delegation for hierarchical identity-based encryption. In: Biskup, J., López, J. (eds.) ESORICS 2007. LNCS, vol. 4734, pp. 139–154. Springer, Heidelberg (2007). https://doi.org/10.1007/978-3-540-74835-9_10

3. Attrapadung, N., Libert, B., de Panafieu, E.: Expressive key-policy attribute-based encryption with constant-size ciphertexts. In: Catalano, D., Fazio, N., Gennaro, R., Nicolosi, A. (eds.) PKC 2011. LNCS, vol. 6571, pp. 90–108. Springer, Heidelberg (2011). https://doi.org/10.1007/978-3-642-19379-8_6

4. Bethencourt, J., Sahai, A., Waters, B.: Ciphertext-policy attribute-based encryption. In: 2007 IEEE Symposium on Security and Privacy, Oakland, CA, USA, pp. 321–334. IEEE Computer Society Press, 20–23 May 2007

5. Boneh, D., Boyen, X.: Efficient selective-ID secure identity-based encryption without random oracles. In: Cachin, C., Camenisch, J.L. (eds.) EUROCRYPT 2004. LNCS, vol. 3027, pp. 223–238. Springer, Heidelberg (2004). https://doi.org/10.1007/978-3-540-24676-3_14

6. Boneh, D., Boyen, X.: Secure identity based encryption without random oracles. In: Franklin, M. (ed.) CRYPTO 2004. LNCS, vol. 3152, pp. 443–459. Springer, Heidelberg (2004). https://doi.org/10.1007/978-3-540-28628-8_27

7. Boneh, D., Boyen, X., Goh, E.-J.: Hierarchical identity based encryption with constant size ciphertext. In: Cramer, R. (ed.) EUROCRYPT 2005. LNCS, vol. 3494, pp. 440–456. Springer, Heidelberg (2005). https://doi.org/10.1007/11426639_26

8. Boneh, D., Franklin, M.: Identity-based encryption from the Weil pairing. In: Kilian, J. (ed.) CRYPTO 2001. LNCS, vol. 2139, pp. 213–229. Springer, Heidelberg (2001). https://doi.org/10.1007/3-540-44647-8_13

9. Boneh, D., Gentry, C., Hamburg, M.: Space-efficient identity based encryption without pairings. In: 48th Annual Symposium on Foundations of Computer Science, 20–23 October 2007, Providence, RI, USA, pp. 647–657. IEEE Computer Society Press (2007)

10. Canetti, R., Halevi, S., Katz, J.: A forward-secure public-key encryption scheme. In: Biham, E. (ed.) EUROCRYPT 2003. LNCS, vol. 2656, pp. 255–271. Springer, Heidelberg (2003). https://doi.org/10.1007/3-540-39200-9_16

11. Chen, J., Gay, R., Wee, H.: Improved dual system ABE in prime-order groups via predicate encodings. In: Oswald, E., Fischlin, M. (eds.) EUROCRYPT 2015. LNCS, vol. 9057, pp. 595–624. Springer, Heidelberg (2015). https://doi.org/10.1007/978-3-662-46803-6_20

12. Cheon, J.H.: Security analysis of the strong Diffie-Hellman problem. In: Vaudenay, S. (ed.) EUROCRYPT 2006. LNCS, vol. 4004, pp. 1–11. Springer, Heidelberg (2006). https://doi.org/10.1007/11761679_1

13. Gentry, C., Silverberg, A.: Hierarchical ID-based cryptography. In: Zheng, Y. (ed.) ASIACRYPT 2002. LNCS, vol. 2501, pp. 548–566. Springer, Heidelberg (2002). https://doi.org/10.1007/3-540-36178-2_34

14. Goyal, V., Jain, A., Pandey, O., Sahai, A.: Bounded ciphertext policy attribute based encryption. In: Aceto, L., Damgård, I., Goldberg, L.A., Halldórsson, M.M., Ingólfsdóttir, A., Walukiewicz, I. (eds.) ICALP 2008. LNCS, vol. 5126, pp. 579–591. Springer, Heidelberg (2008). https://doi.org/10.1007/978-3-540-70583-3_47

15. Goyal, V., Pandey, O., Sahai, A., Waters, B.: Attribute-based encryption for fine-grained access control of encrypted data. In: Juels, A., Wright, R.N., De Capitani di Vimercati, S. (eds.) 13th Conference on Computer and Communications Security, ACM CCS 2006, 30 October– 3 November 2006, Alexandria, Virginia, USA, pp. 89–98. ACM Press (2006). Cryptology ePrint Archive Report 2006/309

16. Hohenberger, S., Waters, B.: Attribute-based encryption with fast decryption. In: Kurosawa, K., Hanaoka, G. (eds.) PKC 2013. LNCS, vol. 7778, pp. 162–179. Springer, Heidelberg (2013). https://doi.org/10.1007/978-3-642-36362-7_11

17. Katz, J., Sahai, A., Waters, B.: Predicate encryption supporting disjunctions, polynomial equations, and inner products. In: Smart, N. (ed.) EUROCRYPT 2008. LNCS, vol. 4965, pp. 146–162. Springer, Heidelberg (2008). https://doi.org/10.1007/978-3-540-78967-3_9

18. Lewko, A., Okamoto, T., Sahai, A., Takashima, K., Waters, B.: Fully secure functional encryption: attribute-based encryption and (hierarchical) inner product encryption. In: Gilbert, H. (ed.) EUROCRYPT 2010. LNCS, vol. 6110, pp. 62–91. Springer, Heidelberg (2010). https://doi.org/10.1007/978-3-642-13190-5_4

19. Lewko, A.B., Sahai, A., Waters, B.: Revocation systems with very small private keys. In: 2010 IEEE Symposium on Security and Privacy, 16–19 May 2010, Berkeley/Oakland, CA, USA, pp. 273–285. IEEE Computer Society Press (2010)

20. Okamoto, T., Takashima, K.: Achieving short ciphertexts or short secret-keys for adaptively secure general inner-product encryption. In: Lin, D., Tsudik, G., Wang, X. (eds.) CANS 2011. LNCS, vol. 7092, pp. 138–159. Springer, Heidelberg (2011). https://doi.org/10.1007/978-3-642-25513-7_11

21. Sahai, A., Waters, B.: Fuzzy identity-based encryption. In: Cramer, R. (ed.) EUROCRYPT 2005. LNCS, vol. 3494, pp. 457–473. Springer, Heidelberg (2005). https://doi.org/10.1007/11426639_27

22. Shamir, A.: Identity-based cryptosystems and signature schemes. In: Blakley, G.R., Chaum, D. (eds.) CRYPTO 1984. LNCS, vol. 196, pp. 47–53. Springer, Heidelberg (1985). https://doi.org/10.1007/3-540-39568-7_5

23. Waters, B.: Ciphertext-policy attribute-based encryption: an expressive, efficient, and provably secure realization. In: Catalano, D., Fazio, N., Gennaro, R., Nicolosi, A. (eds.) PKC 2011. LNCS, vol. 6571, pp. 53–70. Springer, Heidelberg (2011). https://doi.org/10.1007/978-3-642-19379-8_4

24. Waters, B.: Efficient identity-based encryption without random oracles. In: Cramer, R. (ed.) EUROCRYPT 2005. LNCS, vol. 3494, pp. 114–127. Springer, Heidelberg (2005). https://doi.org/10.1007/11426639_7

Multi-client Predicate-Only Encryption for Conjunctive Equality Tests

Tim van de Kamp[1](✉), Andreas Peter[1], Maarten H. Everts[1,2],
and Willem Jonker[1]

[1] University of Twente, Enschede, The Netherlands
{t.r.vandekamp,a.peter,maarten.everts,w.jonker}@utwente.nl
[2] TNO, Groningen, The Netherlands

Abstract. We propose the first multi-client predicate-only encryption scheme capable of efficiently testing the equality of two encrypted vectors. Our construction can be used for the privacy-preserving monitoring of relations among multiple clients. Since both the clients' data and the predicates are encrypted, our system is suitable for situations in which this information is considered sensitive. We prove our construction plaintext and predicate private in the generic bilinear group model using random oracles, and secure under chosen-plaintext attack with unbounded corruptions under the symmetric external Diffie–Hellman assumption. Additionally, we provide a proof-of-concept implementation that is capable of evaluating one thousand predicates defined over the inputs of ten clients in less than a minute on commodity hardware.

Keywords: Multi-client functional encryption
Predicate-only encryption · Privacy-preserving multi-client monitoring

1 Introduction

Predicate encryption (PE) [17] is a special type of encryption that supports the evaluation of functions on encrypted data. On a conceptual level, in predicate encryption a ciphertext of a message m is associated with a descriptive value x and a decryption key SK_f with a predicate f. The decryption of a ciphertext using a key SK_f only succeeds if the predicate $f(x)$ evaluates to TRUE. Special-purpose variants of this notion include identity-based encryption (IBE) [3], attributebased encryption (ABE) [28], and hidden vector encryption (HVE) [6]. Another variant of PE is *predicate-only encryption* [17,30]. In predicate-only encryption, ciphertexts do not contain a message m, but merely consist of an encryption of the descriptive value x. In this case, the decryption algorithm returns the outcome of the predicate f evaluated on the predicate subject x, that is, $f(x)$.

The concept of PE can be generalized to functional encryption (FE) [5,25], in which the decryption of a ciphertext using a key SK_f for a (not necessarily predicate) function f does not return the original plaintext m, but the

S. Capkun and S. S. M. Chow (Eds.): CANS 2017, LNCS 11261, pp. 135–157, 2018.
https://doi.org/10.1007/978-3-030-02641-7_7

value $f(m)$ instead. More recently, Goldwasser et al. [15] formally defined multiclient functional encryption (MC-FE). MC-FE is a type of secret key encryption in which n distinct clients can individually encrypt a message m_i using their secret encryption key usk_i. Using a decryption key for an n-ary function f, the decryption algorithm takes as input the n ciphertexts of the clients and returns $f(m_1, \ldots, m_n)$. Although FE for generalized functionalities [14,15] is an active field of research and of great theoretical interest, FE constructions for a restricted family of functions (such as predicates) are often far more efficient than FE schemes for arbitrary polynomially sized circuits. For example, most works in the area of MC-FE for generalized functionalities rely on inefficient primitives such as indistinguishability obfuscation or multilinear maps.

In this work, we propose the first *multi-client* predicate-only encryption scheme. Our construction can evaluate an n-ary predicate f on the descriptive values x_i coming from n distinct clients. The type of predicates that we can evaluate using our construction is restricted to conjunctive equality tests. To put it simply, our multi-client predicate-only encryption (MC-PoE) scheme is capable of testing the equality of two encrypted vectors. One of these vectors is determined by the decryption key, while the other vector is composed of ciphertexts from several distinct clients. We also provide an extension to our construction in which the decryption keys may contain wildcard components. A wildcard component in the decryption key indicates that it does not matter what the client corresponding to that vector component encrypts: any value matches the wildcard. An attentive reader familiar with the concept of HVE [6] will recognize the functional similarity between the two concepts. However, a crucial difference in our construction is that the ciphertext vector is composed of the ciphertexts from multiple clients, instead of being generated by a single party. A further comparison of related work is discussed in Sect. 1.2.

Our multi-client predicate-only encryption construction uses pairing-based cryptography and satisfies two distinct security notions. The first notion encompasses both the *attribute-hiding* [17] (also termed *plaintext-privacy* [30]) and *predicate-privacy* [30] properties of predicate encryption. Informally, these properties guarantee that an adversary can neither learn the value x of a ciphertext, nor learn the predicate from a given decryption key. Since we construct a multiclient scheme, we choose to adapt the established MC-FE security requirement [15] for our *full security* notion of multi-client predicate-only encryption. This full security notion protects against an attacker that has oracle access to both the key generation algorithm and the encryption algorithm. In the associated security game, the adversary is additionally allowed to statically corrupt clients. We prove our construction secure in the generic bilinear group model using random oracles. We also propose the (intuitively weaker) *chosen-plaintext security* notion, in which an attacker has only oracle access to the encryption algorithm, but can instead corrupt an *unbounded* number of clients. We prove our construction secure under this second notion in the standard model using the symmetric external Diffie–Hellman (SXDH) assumption.

Our construction is designed to be simple and fast. We have implemented and analyzed our construction to evaluate whether it is efficient enough to run in practice. In our proof-of-concept implementation, clients can encrypt their values in about 2.6 ms, while decryption keys, depending on the number of vector components, can be created in less than a second. The Test algorithm, used to evaluate the predicate on the multiple inputs, scales linearly in the number of inputs and requires only 0.10 s for the comparison of vectors of length 20.

1.1 Motivating Use Cases

Privacy-preserving monitoring over encrypted data is one of the main applications for multi-client predicate-only encryption. For example, consider the monitoring of a system comprised of various independent subsystems. We want to raise an alarm when a dangerous combination of events at the various subsystems occurs. By centrally collecting status messages of the individual systems, we can check for such situations. Such a central collection of status messages additionally avoids the need for costly interactions between the various systems. However, if these status messages are considered sensitive, the monitoring cannot be done on the cleartext messages. Multi-client predicate-only encryption overcomes this problem by allowing a monitor to evaluate an n-ary predicate over multiple ciphertexts and raise an alarm when the predicate returns TRUE.

A careful reader might realize that encryption of the status messages is not a sufficient requirement. If the monitor can check arbitrary predicates, it can as well recover the individual plaintext status messages[1], making its encryption useless. Therefore, we have to require that another party issues the decryption keys to the monitor. Since we can consider the monitor to be a third party, it is unlikely that it is allowed to learn the predicates, making a strong case for the requirement of both plaintext privacy and predicate privacy.

The functionality of our construction is developed with the applications in the critical infrastructure (CI) domain in mind. The benefits of information sharing are widely acknowledged [27], but stakeholders still very reluctant in sharing their information with other parties [12,23,33]. We give two concrete use cases.

- *Detection of coordinated attacks.* While a single failure of a system in CI may occur occasionally, a sudden failure of multiple systems from distinct CI operators, could be an indication of a large scale cyberattack. By centrally monitoring the "failure"/"running" status messages of the CI operators, a warning can be given to the national computer emergency response team whenever a combination of systems fails, allowing further investigation of the failures. Additionally, instead of sharing just binary messages to indicate whether a system has failed, it is also helpful to share and monitor *cyberalert levels*. These cyberalert levels from different clients are used to get an improved situational overview [20].

---

[1] For example, the monitor could create a decryption key for a predicate evaluation of a single message, e.g., $f(x_1, \ldots, x_n) = \text{TRUE}$ if and only if $x_1 = 0$.

- *Monitoring of dependencies among* CI *operators.* There exist many dependencies among various CIs [21], making it possible for disruptions to easily propagate from one infrastructure to another [11]. By timely reporting status messages on supply, a central authority can determine whether supply will meet demand and otherwise instruct parties to prepare their backup resources. Similarly, the sharing of compliance status (e.g., whether they can be met or not) can be used to take the right security measures at another party [20].

1.2   Related Work

A multi-*input* functional encryption (MI-FE) [15] scheme is FE scheme that supports the computation of functions over multiple encrypted inputs. Examples of special-purpose MI-FE include property-preserving encryption [26], such as for ordering [4,10] or equality [35], and multi-input inner product encryption (MI-IPE) [2]. The MI-IPE scheme by Abdalla et al. [2] is capable of computing the inner product of two vectors, i.e.,the decryption algorithm returns a scalar. This should not be confused with an inner-product *predicate* encryption scheme where *predicates* (with a TRUE/FALSE result) can be evaluated by an inner product. A private-key, multi-*client* FE (MC-FE) scheme [15,16] is a variant of MI-FE. There are two key differences between the two notions. Firstly, MC-FE requires that the ciphertexts for the function inputs are generated by individual *distinct* parties, while in MI-FE it is allowed to have only a single encryptor for all the inputs. Secondly, in MC-FE the ciphertexts are associated with a *time-step* [15] or *identifier*. Such an identifier is used to prevent mix-and-match attacks: decryption only works when all ciphertexts are associated with the same identifier.

Although not recognized as such, several special-purpose MC-FE schemes have already been proposed in literature. Shi et al. [31] propose a construction for the privacy-preserving aggregation of time-series data. Their construction allows a central party to compute and learn the sum over encrypted numbers, without learning the individual numbers themselves. Decentralized multi-authority attribute-based encryption (MA-ABE) [19] can also be considered a form of MC-FE. In MA-ABE, several decryption keys, issued by different authorities and associated with an identifier, need to be combined to decrypt a single ciphertext. The similarity becomes apparent once we swap the roles of the ciphertext and decryption keys.

Wildcards have been used in PE before by Abdalla et al. [1] in IBE and by Boneh and Waters [6] in HVE. These works differ from our work in several aspects. Most importantly, our construction is a multi-client variant instead of single-client. If we would apply a single-client construction in a multi-client setting, we would leak the individual predicate results for each party. Secondly, we achieve both plaintext privacy and predicate privacy, which is known to be impossible to accomplish in the public-key setting [30] ([1,6] are in the public-key setting). Finally, we look at predicate-only encryption, not at regular PE in which the ciphertexts may also contain an encrypted *payload* message.

Numerous PE schemes are used for searchable encryption (SE) [7]. However, we see no great benefit in applying MC-PoE as SE scheme. MC-PoE enables us to compute a predicate over multiple inputs from several explicitly chosen clients. In SE, this would correspond to a search over documents where the query specifies which keywords have to be set by which parties. This is also the reason why existing multi-*writer* [7] schemes, do not consider searching over documents using queries which, for example, specify that party p_1 should have added keyword w_1, while party p_2 should have added keyword w_2.

2 Preliminaries

Throughout this paper, we use $x \xleftarrow{R} S$ to denote that x is chosen uniformly at random from the finite set S. We denote the ith component of a vector $\boldsymbol{v}$ as v_i. For a set of indices I, we write $\boldsymbol{v}_I$ for the subvector of $\boldsymbol{v}$. Instead of consistently using the vector notation, we use set notation when this is more convenient.

2.1 Primitives and Assumptions

Our construction uses *asymmetric bilinear maps*.

Definition 1 (Bilinear Map). *Let $\mathbb{G}_1$, $\mathbb{G}_2$, and $\mathbb{G}_T$ be cyclic multiplicative groups of prime order p. The map $e \colon \mathbb{G}_1 \times \mathbb{G}_2 \to \mathbb{G}_T$ is an asymmetric bilinear map if the following two conditions hold.*

- *The map is bilinear; $\forall g_1 \in \mathbb{G}_1, g_2 \in \mathbb{G}_2, a, b \in \mathbb{Z}_p \colon e(g_1^a, g_2^b) = e(g_1, g_2)^{ab}$.*
- *The map is non-degenerate; generators g_1 and g_2 are chosen such that the order of the element $e(g_1, g_2) \in \mathbb{G}_T$ equals p, the order of group $\mathbb{G}_T$.*

More specifically, we use a Type 3 pairing [13], where no efficiently computable homomorphisms between the groups $\mathbb{G}_1$ and $\mathbb{G}_2$ can be found.

We use the function $\mathcal{G}(1^\kappa)$ to generate the parameters for a Type 3 bilinear group for the security parameter κ.

Additionally, we use a *pseudorandom permutation* (PRP) over $\mathcal{M} \subseteq \mathbb{Z}_p$.

Definition 2 (Pseudorandom Function). *For key space $\mathcal{K}$ and message space $\mathcal{M}$ define the function $\pi \colon \mathcal{K} \times \mathcal{M} \to \mathcal{M}$. The function π is a pseudorandom permutation (PRP) if the output of π is indistinguishable from the output of a permutation chosen uniformly at random from the set of all possible permutations over $\mathcal{M}$.*

The security of our construction is based on the *decisional Diffie–Hellman* (DDH) problem and the *symmetric external Diffie–Hellman* (SXDH) problem.

Assumption 1. *The decisional Diffie–Hellman (DDH) assumption states that, given $(\mathbb{G}, g \in \mathbb{G}, g^a, g^b, Z)$ for uniformly at random chosen a and b, it is hard to distinguish $Z = g^{ab}$ from $Z \xleftarrow{R} \mathbb{G}$.*

Assumption 2. *Given the bilinear groups $\mathbb{G}_1$ and $\mathbb{G}_2$, the symmetric external Diffie–Hellman (SXDH) assumption states that the DDH problem in both group $\mathbb{G}_1$ and group $\mathbb{G}_2$ is hard.*

3   Multi-client Predicate-Only Encryption

A multi-client predicate-only encryption scheme is a collection of the following four polynomial-time algorithms.

Setup($1^\kappa, n$). This algorithm defines the public parameters pp, a master secret key msk, and the encryption keys usk_i for every client $1 \leq i \leq n$. The algorithm also defines the finite message space $\mathcal{M}^n$ and the predicate family $\mathcal{F}$, which predicates are efficiently computable on $\mathcal{M}^n$.

Encrypt(usk_i, id, x_i). A client i can encrypt a value $x_i \in \mathcal{M}$ using its encryption key usk_i and an identifier id. Different clients can use the same identifier, however, each client can only use an identifier at most once. The algorithm returns a ciphertext $ct_{id,i}$. We usually omit the index id when there is no ambiguity. Furthermore, we introduce the following simplification of notation for a set of ciphertexts associated with the same id: For an ordered set $S \subseteq \{1, \ldots, n\}$ of indices, we write the set of ciphertexts $\{ \text{Encrypt}(usk_j, id, x_j) \mid j \in S \}$ as $\text{Encrypt}(usk_S, id, \boldsymbol{x}_S)$. If $S = \{1, \ldots, n\}$, we simply write $\text{Encrypt}(usk, id, \boldsymbol{x})$ or $ct_{\boldsymbol{x}}$.

GenToken(msk, f). The key generator can create a decryption key, termed *token*, for predicate $f \in \mathcal{F}$ using the msk. The algorithm returns the token tk_f.

Test($tk_f, ct_{\boldsymbol{x}}$). The Test algorithm requires a vector of ciphertexts $ct_{\boldsymbol{x}}$ and a token tk_f as input. The algorithm outputs a Boolean value.

Definition 3 (Correctness). *A multi-client predicate-only encryption scheme is correct if $\text{Test}(tk_f, ct_{\boldsymbol{x}}) = f(\boldsymbol{x})$. Formally, we require for all $n \in \mathbb{N}$, $\boldsymbol{x} \in \mathcal{M}^n$, and $f \in \mathcal{F}$,*

$$\Pr \left[\text{Test}(ct_{\boldsymbol{x}}, tk_f) \neq f(\boldsymbol{x}) : \begin{array}{c} (pp, msk, \{usk_i\}) \leftarrow \text{Setup}(1^\kappa, n) \\ ct_{\boldsymbol{x}} \leftarrow \text{Encrypt}(usk, id, \boldsymbol{x}) \\ tk_f \leftarrow \text{GenToken}(msk, f) \end{array} \right]$$

is negligible in the security parameter κ, where the probability is taken over the coins of Setup, Encrypt, and GenToken.

Note that we do not impose any restriction on the output of Test if it operates on messages encrypted under different identifiers.

3.1   Security

A commonly considered security game for private-key functional encryption is an indistinguishability-based notion under which the adversary may query both the Encrypt and the GenToken oracles [15,17,30]. Since our MC-PoE is a special case of MC-FE, we start from the security notion from Goldwasser et al. [15]. However, they only consider the indistinguishability of plaintexts (*plaintext privacy* [17, 30]) and not of functions (*function* or *predicate privacy* [8,30]) in their security definition. In the following *full security* notion, we combine the plaintext-privacy and predicate-privacy notions, similarly to Shen et al. [30].

Because an evaluation of a predicate on a set of messages reveals some information about the messages in relation to the predicate (and vice versa), we cannot allow the adversary to query for all combinations of messages and predicates. For example, an adversary can distinguish an encryption of message x_0 from an encryption of x_1 if it has a token for a predicate f such that $f(x_0) \neq f(x_1)$. Even if we require $f(x_0) = f(x_1)$ for all predicates f that the adversary queried, a similar situation can still appear. To see this, consider an adversary corrupting client i so that it can encrypt any message m_i as ith input. This means that the adversary can also trivially distinguish the two messages if there exists a value m_i, such that if it replaces the ith input of x_0 and x_1 by m_i (resulting in inputs x_0' and x_1' respectively), the predicate has different outputs, i.e., $f(x_0') \neq f(x_1')$. Likewise, we also have to require that the predicates f_0 and f_1 yield the same result on a queried input x, even if the adversary replaces some of the corrupted clients' inputs by another value.

In our security definition, we use the term *static corruptions* to indicate that the adversary announces the corrupted clients at the beginning of the game and cannot corrupt additional clients during the rest of the game. We let I be the set of indices of the uncorrupted clients and, similarly, indicate the indices of the corrupted clients by the set $\bar{I}$. Recall that we use the notation x_I to denote the subvector of x containing only the components from the set I. We denote with $f(x_I, \cdot)$ a predicate f with the pre-filled inputs x_I.

Definition 4 (Full Security). *A multi-client predicate-only encryption scheme is adaptive full secure under static corruptions if every probabilistic polynomial time adversary $\mathcal{A}$ has at most a negligible advantage in winning the following game.*

Initialization. *The adversary $\mathcal{A}$ submits a set of indices $\bar{I}$ to the challenger. We define the complement set $I = \{1, \ldots, n\} \setminus \bar{I}$.*

Setup. *The challenger runs Setup($1^\kappa, n$) to get the pp, msk, and $\{usk_i\}_{1 \leq i \leq n}$. It gives the public parameters pp and corrupted clients' keys $\{usk_i \mid i \in \bar{I}\}$ to the adversary.*

Query 1. *The adversary $\mathcal{A}$ may query the challenger for ciphertexts or tokens.*

- *In case of a ciphertext query for (i, id, x_i), the challenger returns $ct_{id,i} \leftarrow$ Encrypt(usk_i, id, x_i).*
- *In case of a token query for f, the challenger returns $tk_f \leftarrow$ GenToken(msk, f).*

Challenge. *The challenger picks a random bit b. The adversary can either request a ciphertext challenge or a token challenge.*

- *In case of a ciphertext challenge, the adversary sends $(id^*, x_{0,I}^*, x_{1,I}^*)$ to the challenger. The challenger returns the challenge $Ch_I \leftarrow$ Encrypt($usk_I, id^*, x_{b,I}^*$).*

– *In case of a token challenge, the adversary sends* (f_0^*, f_1^*) *to the challenger. The challenger returns the challenge* Ch ← GenToken(msk, f_b^*).

Query 2. *The adversary may query the challenger again, similar to* **Query 1**.

Guess. *The adversary outputs its guess* $b' \in \{0, 1\}$ *for the bit* b.
We say that adversary $\mathcal{A}$ *wins the game, if* $b' = b$ *and*

– *in case of a ciphertext challenge,* $\mathcal{A}$ *did not query for a ciphertext using identifier* id* *in any of the two query phases, nor query for a predicate* f, *such that* $f(x_{0,I}^*, \cdot) \neq f(x_{1,I}^*, \cdot)$;
– *in case of a token challenge,* $\mathcal{A}$ *did not query for* (i, id, x_i), *for uncorrupted clients* $i \in I$, *such that it can combine these inputs* x_i *for the same* id, *into a vector* x_I, *where* $f_0^*(x_I, \cdot) \neq f_1^*(x_I, \cdot)$.

Note that in the above defined game, in case of a ciphertext challenge, the challenger only returns challenge ciphertexts for the uncorrupted clients. The adversary can still evaluate predicates on the received challenge by generating the ciphertext values for the corrupted clients using their encryption keys.

It is important to realize that the challenger can decide whether the adversary wins the game or not in polynomial time. This is possible because the adversary $\mathcal{A}$ can only query for a polynomial number of ciphertexts and tokens. Moreover, the challenger is able to efficiently check if $f(x_I, \cdot) = f'(x_I', \cdot)$ as both n and $\mathcal{M}^n$ are finite and fixed by Setup($1^\kappa, n$).

Definition 5 (Selective Full Security). *The definition of a selective full secure under static corruptions multi-client predicate-only encryption scheme is similar to the adaptive full security notion of Definition 4. The difference between the two, is that in selective security game, the challenge request (i.e., either* (id*, $x_{0,I}^*, x_{1,I}^*$) *or* (f_0^*, f_1^*)*) is announced during* Initialization.

As explained before, the full security definition actually defines two security notions. We say that MC-POE scheme is *adaptive (selective) plaintext private* if no adversary can win the adaptive (selective, respectively) full security game with a ciphertext challenge. Similarly, POE scheme is *adaptive (selective) predicate private* if no adversary can win the adaptive (selective, respectively) full security game with a token challenge.

Chosen-Plaintext Security. The definition of full security is very strong as it allows an adversary to query for both ciphertexts and tokens. This is similar to the chosen-ciphertext attack (CCA) security notion used in public-key cryptography, where the adversary can query both the encryption and decryption[2] oracle. To accommodate for a different attacker model, we define a *chosen-plaintext* security notion, where the adversary only has access to the encryption oracle and is asked to distinguish between two ciphertexts. Such a notion is similar to

---

[2] In MC-POE, an adversary can use a token and the public Test algorithm to learn more about the encrypted plaintext.

chosenplaintext attack (CPA) security as defined in public-key cryptography and is also related to the *offline security* notion of Lewi and Wu [18], in which an attacker has only access to ciphertexts and not to decryption keys. To make our notion stronger, we give the adversary access to all clients' encryption keys (but not to the internal randomness of the clients).

Definition 6 (Chosen-Plaintext Security). *A multi-client predicate-only encryption scheme is chosen-plaintext secure under unbounded corruptions if any probabilistic polynomial time algorithm $\mathcal{A}$ has at most a negligible advantage in winning the following game.*

Setup. *The challenger runs $Setup(1^\kappa, n)$ to get the pp, msk, and $\{usk_i\}_{1 \leq i \leq n}$. It gives the public parameters pp and all clients' keys $\{usk_i\}_{1 \leq i \leq n}$ to the adversary. Note that the adversary $\mathcal{A}$ can encrypt any message x_i for identifier id using the key usk_i by computing $Encrypt(usk_i, id, x_i)$.*

Challenge. *The adversary sends the challenge request (id^*, x_0^*, x_1^*) to the challenger. The challenger picks a random bit b and returns $Encrypt(usk, id^*, x_b^*)$ to the adversary.*

Guess. *The adversary outputs its guess $b' \in \{0, 1\}$ for the bit b.*
We say that adversary $\mathcal{A}$ wins the game if $b' = b$.

Observe that in this game the adversary is given *every* client's private key. This security requirement is quite strong and corresponds to a following situation: Even if an attacker compromises a client and steals its encryption keys, it *remains* hard for the attacker to determine the plaintexts of the ciphertexts created *before and after* the compromise.

4   Our Construction

We construct a multi-client predicate-only encryption scheme for the functionality of a conjunctive equality test. To test if n messages $x_1, \ldots, x_n$, encrypted by distinct clients, equal the values $y_1, \ldots, y_n$, we evaluate the predicate

$$\mathsf{Match}(x, y) = \begin{cases} \text{TRUE} & \text{if } \bigwedge_{i=1}^n (x_i = y_i), \\ \text{FALSE} & \text{otherwise.} \end{cases}$$

As discussed in Sect. 1.1, this functionality turns out to be surprisingly useful in the domain of critical infrastructure protection. In this setting, a monitor combines the ciphertexts associated with the same identifier and evaluates all its tokens (corresponding to various predicates) on the ciphertext vector to see if there is a match. If a match is found, the monitor may raise an alarm or take other appropriate actions. A schematic overview of relations among all parties of such a multi-client monitoring system is shown in Fig. 1.

We now describe our multi-client predicate-only encryption construction for conjunctive equality tests over multiple clients.

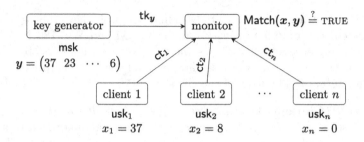

Fig. 1. In this example of a multi-client monitoring system, there are n distinct clients (with keys $\mathsf{usk}_1, \ldots, \mathsf{usk}_n$) that determine the values $x_1, \ldots, x_n$. The monitor computes the functionality $\mathsf{Match}(\boldsymbol{x}, \boldsymbol{y})$ using the encrypted values $\mathsf{ct}_1, \ldots, \mathsf{ct}_n$ and a token $\mathsf{tk}_{\boldsymbol{y}}$. The monitor is only able to compute the functionality if all clients encrypted their value x_i using the same identifier id (not shown in the figure).

Setup$(1^\kappa, n)$. Let $(p, \mathbb{G}_1, \mathbb{G}_2, \mathbb{G}_T, e, g_1, g_2) \leftarrow \mathcal{G}(1^\kappa)$ be the parameters for a bilinear group. Choose a pseudorandom permutation $\pi \colon \mathcal{K} \times \mathcal{M} \to \mathcal{M}$ for message space $\mathcal{M} \subseteq \mathbb{Z}_p$ and a cryptographic hash function $H \colon \{0,1\}^* \to \mathbb{G}_1$. The bilinear group parameters together with both functions form the public parameters. To generate the keys, select $\alpha_i, \gamma_i \xleftarrow{R} \mathbb{Z}_p^*$ and $\beta_i \xleftarrow{R} \mathcal{K}$ for $1 \le i \le n$. The master secret key is

$$\mathsf{msk} = \left\{ (g_2^{\alpha_i}, \beta_i, g_2^{\gamma_i}) \right\}_{i=1}^n.$$

The secret encryption key for client i is

$$\mathsf{usk}_i = (g_1^{\alpha_i}, \beta_i, \gamma_i).$$

Encrypt$(\mathsf{usk}_i, \mathsf{id}, x_i)$. Client i can encrypt its message $x_i \in \mathcal{M}$ for identifier id using usk_i and $r_i \xleftarrow{R} \mathbb{Z}_p^*$,

$$\mathsf{ct}_i = \left(H(\mathsf{id}), g_1^{r_i}, g_1^{\alpha_i \pi(\beta_i, x_i) r_i} H(\mathsf{id})^{\gamma_i} \right).$$

GenToken$(\mathsf{msk}, \boldsymbol{y})$. The token generator can encrypt a vector $\boldsymbol{y} \in \mathcal{M}^n$ using its key msk. Choose $u_i \xleftarrow{R} \mathbb{Z}_p^*$ for $1 \le i \le n$ and output

$$\mathsf{tk}_{\boldsymbol{y}} = \left(\left\{ g_2^{u_i}, g_2^{\alpha_i \pi(\beta_i, y_i) u_i} \mid 1 \le i \le n \right\}, \prod_{1 \le i \le n} (g_2^{\gamma_i})^{u_i} \right).$$

Test$(\mathsf{tk}_{\boldsymbol{y}}, \{\mathsf{ct}_i\}_{1 \le i \le n})$. Output the result of the test

$$\prod_{1 \le i \le n} e\left(g_1^{\alpha_i \pi(\beta_i, x_i) r_i} H(\mathsf{id})^{\gamma_i}, g_2^{u_i} \right) \stackrel{?}{=}$$

$$\prod_{1 \le i \le n} e\left(g_1^{r_i}, g_2^{\alpha_i \pi(\beta_i, y_i) u_i} \right) e\left(H(\mathsf{id}), \prod_{1 \le i \le n} (g_2^{\gamma_i})^{u_i} \right).$$

4.1   Correctness

Correctness follows from the definition of Test. We remark that the output of Test is completely determined by $\sum_{1 \leq i \leq n} (\pi(\beta_i, x_i) - \pi(\beta_i, y_i)) \stackrel{?}{=} 0$. Since the function π is a PRP, the probability of $\text{Test}(\text{tk}_y, \text{ct}_x) \neq \text{Match}(x, y)$ is negligible.

4.2   Security

To get an intuition for the security of our construction, observe that the clients' messages itself are first encrypted using the PRP π. By using the output of the PRP as an exponent and randomizing it with the value r, we create a probabilistic encryption of the message. The PRP's randomized output also prevents malleability attacks. Similarly, the vector components of the vector y are individually encrypted in a similar way. Because part of the clients' keys (i.e., $g_1^{\alpha_i}$) and the master secret key (i.e., $g_2^{\alpha_i}$) reside in different groups, it is hard for a client to create a token and hard for the token generator to create a ciphertext.

The formal security analysis can be found in Appendix A. We prove our construction selective plaintext private and adaptive predicate private. Additionally, we prove the chosen-plaintext security property of the construction. Plaintext and predicate privacy are proven in the generic group model using random oracles. This combination of models has been successfully applied in other works before [9,34]. Chosen-plaintext security can be proven in the standard model and under the DDH assumption in group $\mathbb{G}_1$. We formulate the following two theorems.

Theorem 1. *Let $\mathcal{A}$ be an arbitrary probabilistic polynomial time adversary having oracle access to the group operations and the encryption and token generation algorithms, while it is bounded in receiving at most q distinct group elements. The adversary $\mathcal{A}$ has at most an advantage of $O(q^2/p)$ in winning either the selective plaintext-privacy (see Definition 5) or the adaptive predicate-privacy game (see Definition 4) in the random oracle model.*

Theorem 2. *The construction presented above is chosen-plaintext secure with an unbounded number of corruptions (Definition 6) under the DDH assumption in group $\mathbb{G}_1$.*

Both plaintext privacy and predicate privacy are proven secure through a series of hybrid games. In every game hop, a component of the challenge vector (either the ciphertext or token challenge vector) is replaced by a random one. In the final game, once all components are replaced by random elements, no adversary can gain an advantage since it is impossible to distinguish a random vector from another random one.

However, in the selective plaintext-privacy game, not *every* component of the challenge vector can be replaced by a random component. If a component $x_{b,i}^*$ of the challenge vector x_b^* is deterministic, i.e.,the challenge inputs were the same for that component, $x_{0,i}^* = x_{1,i}^* = m$, the adversary may query for a token to match this single component for the value $y_i = m$. Note that if this component is replaced by a random element, Match will, with overwhelming probability, return FALSE, while it should have returned TRUE. Hence, the deterministic components of the challenge vector have to remain untouched in every game hop. This implies that the number of game hops depends on the challenge inputs, requiring the challenger to know the challenge inputs a priori. This limitation does not appear for predicate privacy, making it possible to prove adaptive security instead.

4.3   Extension Allowing Wildcards

Although a construction for the described conjunctive equality matching functionality would suffice, it may be very inefficient when a predicate is defined over a subset of the clients' inputs. For example, suppose the token generator has a predicate for which it actually does not care what client i sends. Now, if we have only conjunctive equality matching, we would need to create a token for every possible message that client i can send. Besides that this will be very inefficient if client i could send many different messages, it would also reveal whenever client i has sent the same values multiple times: whenever a client sends the same value multiple times, the same token will match multiple times as well!

We can extend our construction with the ability to test for the equality of vectors with the additional feature that the predicate vector $\boldsymbol{y}$ can now contain wildcard components. Such a wildcard component matches against any value of the corresponding ciphertext component. This makes the testing functionality similar to the one used in HVE [6], however our system combines the ciphertexts from multiple clients. Formally, the clients encrypt their messages from the message space $\mathcal{M} \subseteq \mathbb{Z}_p$, where the token generator uses the space $\mathcal{M}^* = \mathcal{M} \cup \{\star\}$. The multi-client predicate-only encryption construction now evaluates the function

$$\mathsf{Match}^\star(\boldsymbol{x}, \boldsymbol{y}) = \begin{cases} \text{TRUE} & \text{if } \forall i: (x_i = y_i) \vee (y_i = \star), \\ \text{FALSE} & \text{otherwise.} \end{cases}$$

To achieve this additional functionality, we have to change the GenToken and Test algorithms, the other algorithms remain unchanged.

GenToken*(msk, $\boldsymbol{y}$). The token generator can encrypt a predicate vector $\boldsymbol{y} \in (\mathcal{M}^*)^n$ using the master secret key msk. Let $S_{\boldsymbol{y}}$ be the set of indices of the non-wildcard components of the vector $\boldsymbol{y}$. Choose $u_i \xleftarrow{R} \mathbb{Z}_p^*$ for $i \in S_{\boldsymbol{y}}$ and output

$$\mathsf{tk}_{\boldsymbol{y}} = \left(\left\{ g_2^{u_i}, g_2^{\alpha_i \pi(\beta_i, y_i) u_i} \mid i \in S_{\boldsymbol{y}} \right\}, \prod_{i \in S_{\boldsymbol{y}}} (g_2^{\gamma_i})^{u_i} \right).$$

Test*(tk$_{\boldsymbol{y}}$, $\{$ct$_i\}_i \in S_{\boldsymbol{y}}$). Output the result of the test

$$\prod_{i \in S_y} e\left(g_1^{\alpha_i \pi(\beta_i, x_i) r_i} H(\mathsf{id})^{\gamma_i}, g_2^{u_i}\right) \stackrel{?}{=}$$

$$\prod_{i \in S_y} e\left(g_1^{r_i}, g_2^{\alpha_i \pi(\beta_i, y_i) u_i}\right) e\left(H(\mathsf{id}), \prod_{i \in S_y} (g_2^{\gamma_i})^{u_i}\right).$$

In this adapted construction, the wildcards are made possible by allowing the token generator to specify which clients need to contribute a ciphertext before one can evaluate the predicate over the subset of clients. This idea is encoded in the token by the value $\prod_{i \in S_y} (g_2^{\gamma_i})^{u_i}$ and in the ciphertext by the value $H(\mathsf{id})^{\gamma_i}$. The latter also prevents the monitor to combine ciphertext for different identifiers.

The addition of wildcards to the scheme should be mainly considered an efficiency improvement, rather than a security improvement, although the ciphertext security actually slightly improves when one uses wildcards – the wildcard components do not leak any information about the matched ciphertext, as discussed above. However, we point out that this adapted construction is not predicate private. In fact, if wildcards are used in the proposed construction, the token would leak their positions: by looking at a token, it is possible to tell which components encode a wildcard. But, if we accept this fact, yet still want to assure that no other information is leaked, we can define a *restricted* predicate-privacy game. In this restricted game, we restrict the adversary to only provide challenge inputs with wildcards in the same position, i.e., we require for challenge inputs $f_0^* = \boldsymbol{y}_0^*$, $f_1^* = \boldsymbol{y}_1^*$ that for all $1 \le i \le n$, $y_{0,i} = \star \iff y_{1,i} = \star$.

It is trivial to see that changing the GenToken or Test algorithm does not influence the chosen-ciphertext security. In Appendix A we give the security proofs for the construction with wildcards.

4.4   Efficiency

Since the Encrypt and GenToken algorithms do not use any expensive pairing operations, they can efficiently run on less powerful hardware. For the Encrypt algorithm it is only needed to compute the PRP π and three modular exponentiations. The computational complexity of GenToken* depends on the number of non-wildcard components in the predicate. For every non-wildcard component one evaluation of the PRP π and three modular exponentiations are needed.

The Test algorithm is the only algorithm that requires pairings. To evaluate a token with n non-wildcard components, $2n + 1$ pairing evaluations are needed.

In the next section we discuss a concrete implementation of the construction and evaluate its performance.

5   Implementation and Evaluation

We have implemented a prototype of our construction with wildcards to get a better understanding of its performance. The implementation[3] uses the Pairing-Based Cryptography Library[4] that allows one to easily change the underlying curve and its parameters.

Instantiating the Pseudorandom Permutation. Our construction uses a PRP π to permute an element in $\mathbb{Z}_p$. However, since we use the outcome of the permutation to exponentiate a generator in $\mathbb{G}_1$ and $\mathbb{G}_2$, we can instead directly map values in $\mathbb{Z}_p$ to one of these groups respectively. The pseudorandom function (PRF) proposed by Naor and Reingold [24] exactly achieves this. Their PRF maps a message $x \in \mathcal{M} \subseteq \{0, \ldots, 2^m - 1\} \subseteq \mathbb{Z}_p$ using a key $\boldsymbol{b} = \left\{ b_i \stackrel{R}{\leftarrow} \mathbb{Z}_p^* \mid 0 \leq i \leq m \right\}$ to an element in a group $\langle g \rangle$ of prime order p. The PRF F is defined as

$$F(\boldsymbol{b}, x) = g^{b_0 \prod_{i=1}^m b_i^{x[i]}},$$

where $x[i] \in \{0, 1\}$ denotes the ith bit of message x. The advantage of using this PRF over PRP is that it is relatively simple to compute while it is provably secure under the DDH assumption.

We apply the PRF to both the Encrypt and the GenToken* algorithms to obtain ciphertexts of the form

$$\mathsf{ct}_i = \left(H(\mathsf{id}), g_1^{r_i}, g_1^{\alpha_i \prod_{j=1}^m \beta_{i,j}^{x_i[j]} r_i} H(\mathsf{id})^{\gamma_i} \right),$$

and tokens of the form

$$\mathsf{tk}_y = \left(\left\{ g_2^{u_i}, g_2^{\alpha_i \prod_{j=1}^m \beta_{i,j}^{y_i[j]} u_i} \mid i \in S_y \right\}, \prod_{i \in S_y} (g_2^{\gamma_i})^{u_i} \right).$$

Notice that we use $b_0 = \alpha_i$ and $b_j = \beta_{i,j}$. In addition, observe that it is not necessary to know the value α_i to compute a ciphertext or token, as long the value $g_1^{\alpha_i}$, or $g_2^{\alpha_i}$ respectively, is known.

Performance Measurements. We ran several performance evaluations on a notebook containing an Intel Core i5 CPU, running on Debian GNU/Linux. We chose to evaluate the system using an MNT curve [22] over a 159 bit base field size with embedding degree 6.

As expected from the theoretical performance analysis in Sect. 4.4, both the GenToken* and Test* algorithms scale linearly in the number of non-wildcard components used. The GenToken* algorithm spends, on average, 19 ms to encrypt a non-wildcard component. To evaluate a token that contains no wildcards using

---

[3] https://github.com/CRIPTIM/multi-client-monitoring.
[4] https://crypto.stanford.edu/pbc/.

n ciphertexts, takes $4.5n + 10$ ms on average. The Setup algorithm scales linearly as well, spending on average 18 ms per client to create their public and private keys. The Encrypt algorithm is the fastest, taking only 2.6 ms for an individual client to encrypt a message $x_i \in \{0, \ldots, 15\}$.

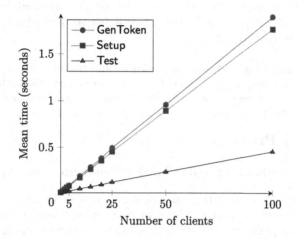

Fig. 2. Performance measurements of the implementation using an MNT-159 curve.

In Fig. 2 the average computational time is plotted against the number of clients involved in the computation. No wildcards were used in the GenToken* and Test* algorithms to obtain these timing results, meaning that the algorithms are identical to GenToken and Test, respectively.

Considering an example of the monitoring of several CIs, we remark that a typical information-sharing community (e.g., ISAC) consists of about 10 parties. So, if every party sends 5 distinct messages for each identifier (e.g., every party has five subsystems to be monitored), we would require a system of about 50 clients. We see that in such a realistically sized system we can evaluate about 250 predicates per minute. Optimizations such as the preprocessing of pairings can increase the number of predicate evaluations per minute.

6   Conclusion

By designing a special-purpose multi-client functional encryption scheme, it is possible to create a practical privacy-preserving monitoring system. To achieve this, we defined multi-client predicate-only encryption (MC-PoE) and corresponding security definitions for the protection of both the messages of the individual clients and the predicates. Our proposed construction for such PoE scheme is capable of conjunctive equality testing over vector components which can include wildcards. The performance evaluation of our implementation shows that the evaluation time of a predicate scales linearly in the number of clients, where a predicate defined over 20 clients can be evaluated in a tenth of a second. Additionally, we see that the encryption algorithm is very lightweight, making it suitable to run on resource-constrained devices.

Future work will include the construction of MC-PoE scheme which will allow for more expressive functionality, while remaining efficient enough to run in practice and keeping the confidentiality of both the messages and the predicates. Additionally, further research is needed to construct MC-PoE scheme that is fully secure in the standard model.

Acknowledgment. This work was supported by the Netherlands Organisation for Scientific Research (NWO) in the context of the CRIPTIM project. The authors additionally thank the reviewers and shepherd for their suggested improvements.

A   Security Proofs

A.1   Selective Plaintext and Adaptive Predicate Security

We prove Theorem 1, stating that the construction without wildcards is secure, by using the following lemma and by proving that the construction with wildcards is selective plaintext private and restricted adaptive predicate private. Recall that the restricted predicate-private game is almost identical to our predicate-private game. However, in the restricted game, we additionally require $y_{0,i}^* = \star \iff y_{1,i}^* = \star$ for the challenge inputs $\boldsymbol{y}_0^*, \boldsymbol{y}_1^*$.

Lemma 1. *If the construction* with *wildcards is selective plaintext private and* restricted *adaptive predicate private, then the construction* without *wildcards is selective plaintext private and adaptive predicate private.*

Proof. First, let us look at the selective plaintext privacy. Assume $\mathcal{A}$ is a probabilistic polynomial time adversary, having a non-negligible advantage in winning the selective plaintext-privacy game without wildcards. It is clear that $\mathcal{A}$ is also an adversary that has an identical, non-negligible, advantage in winning the selective plaintext-privacy game with wildcards (however, it chooses not to use any). This contradicts with the given statement that no such adversary exists.

For the other part, assume that $\mathcal{A}$ is a probabilistic polynomial time adversary, making no wildcard queries, and having a non-negligible advantage in winning the predicate-privacy game. Note that $\mathcal{A}$ is also an adversary that has an identical, non-negligible, advantage in winning the predicate-privacy game with wildcards (however, it chooses not to use any). Specifically, since $\mathcal{A}$ chooses its challenge inputs without wildcards, $\mathcal{A}$ also satisfied the extra requirement in the restricted predicate-privacy game.

We now give a proof for both selective plaintext privacy as well as restricted predicate privacy for the construction with wildcards.

Proof (sketch). We first define the generic group model setting and all oracle interactions, including the oracles for encryption and token generation.

Generic Group Model. Let ϕ_1, ϕ_2, ϕ_T be distinct random injective mappings from the domain $\mathbb{Z}_p$ to $\{0,1\}^\kappa$, where $\kappa > 3\log p$. We write $\mathbb{G}_1$ for $\{\phi_1(x) \mid x \in \mathbb{Z}_p\}$, $\mathbb{G}_2$ for $\{\phi_2(x) \mid x \in \mathbb{Z}_p\}$, and $\mathbb{G}_T$ for $\{\phi_T(x) \mid x \in \mathbb{Z}_p\}$. The adversary is given access to an oracle to compute the group actions on $\mathbb{G}_1$, $\mathbb{G}_2$, and $\mathbb{G}_T$. Additionally, it is given access to an oracle capable of computing a non-degenerate bilinear map $e \colon \mathbb{G}_1 \times \mathbb{G}_2 \to \mathbb{G}_T$. Lastly, we also define a random oracle to model the hash function $H \colon \{0,1\} \to \mathbb{G}_1$.

Instead of writing $\phi_1(x)$, we write g_1^x. Similarly, we write g_2^x for $\phi_2(x)$ and $e(g_1, g_2)^x$ for $\phi_T(x)$.

Hash Oracle H. The challenger keeps track of oracle queries it received before by maintaining a table. If it has not received an oracle query for the value id before, it chooses a random value $t_{id} \in \mathbb{Z}_p$ and stores this value in its table. It returns the value $g_1^{t_{id}}$ to the querier.

Game Interactions. The adversary's first interaction with the challenger is to receive the group parameters and the secret keys of the corrupted clients.

Setup. The challenger chooses $\alpha_i, \gamma_i \overset{R}{\leftarrow} \mathbb{Z}_p^*$ and $\beta_i \overset{R}{\leftarrow} \mathcal{K}$ for $1 \le i \le n$, just like in the actual scheme. It also defines the secret keys usk_i and master secret key msk according to the scheme.

Corruptions. The adversary submits its choices for the corrupted clients $\bar{I}$ to the challenger. In the selective plaintext-privacy game, the adversary additionally submits its challenge inputs $(\mathsf{id}^*, \boldsymbol{x}_{0,I}^*, \boldsymbol{x}_{1,I}^*)$. The challenger gives the secret keys $\mathsf{usk}_{\bar{I}}$ of the corrupted clients to the adversary.

Queries. The adversary interacts with the challenger by asking the challenger to encrypt a messages or to generate a token for some predicate. To be able to refer to a specific query later on in the proof, we label every query with a query number. Let j represent this query number.

Encrypt The challenger answers valid Encrypt queries for a message $x_i^{(j)}$ for client i and identifier $\mathsf{id}^{(j)}$ similar as in the scheme. It chooses $r_i^{(j)} \overset{R}{\leftarrow} \mathbb{Z}_p^*$ and returns the ciphertext $\mathsf{ct}_{i,\mathsf{id}}^{(j)}$,

$$\left(g_1^{t_{\mathsf{id}(j)}}, g_1^{r_i^{(j)}}, g_1^{\alpha_i \pi(\beta_i, x_i^{(j)}) r_i^{(j)}} g_1^{t_{\mathsf{id}(j)} \gamma_i} \right).$$

GenToken* Similarly, token queries for $\boldsymbol{y}^{(j)}$ are answered according to the scheme as well. The challenger chooses $u_i^{(j)} \overset{R}{\leftarrow} \mathbb{Z}_p^*$ for $i \in S_{\boldsymbol{y}^{(j)}}$ and returns the token $\mathsf{tk}_{\boldsymbol{y}}^{(j)}$,

$$\left(\left\{ g_2^{u_i^{(j)}}, g_2^{\alpha_i \pi(\beta_i, y_i^{(j)}) u_i^{(j)}} \mid i \in S_{\boldsymbol{y}} \right\}, \prod_{i \in S_{\boldsymbol{y}}} g_2^{u_i^{(j)} \gamma_i} \right),$$

to the adversary.

Proof Structure. We prove both selective plaintext privacy and restricted adaptive predicate privacy through a series of hybrid games.

For selective plaintext privacy the number of games depends on the number of differentiating components of the challenge inputs – hence the *selective* game type. Let $\overline{X}$ denote the set of indices where the components of $\boldsymbol{x}_0^*$ differ from $\boldsymbol{x}_1^*$, $\overline{X} = \{ i \mid x_{0,i}^* \neq x_{1,i}^* \}$. Let game k be identical to the original game, except that in the challenge phase now the first $k - 1$ components of $\overline{X}$ in the returned challenge vector are chosen at random. Note that game $k = 1$ is identical to the original game and that in game $k = |\overline{X}|$ not even an unbounded adversary is able to gain an advantage in winning the game.

For restricted adaptive predicate privacy, we assume w.l.o.g. that $\boldsymbol{y}_{0,I}^* \neq \boldsymbol{y}_{1,I}^*$, because if $\boldsymbol{y}_{0,I}^* = \boldsymbol{y}_{1,I}^*$, the adversary would not be able to gain an advantage in the game since this implies $\boldsymbol{y}_0^* = \boldsymbol{y}_1^*$. Note that this means that the result of Match* with any allowed ciphertext vector will be FALSE. We define game k identical to the original game, except that in the challenge phase now the first $k - 1$ components of the returned challenge vector are chosen at random. Note that game $k = 1$ is identical to the original game and that in game $k = n$ not even an unbounded adversary is able to gain an advantage in winning the game.

For both the selective plaintext-privacy as well as the restricted adaptive predicate-privacy game, we show that an adversary has at most an advantage of $O(q^2/p)$ in distinguishing between game k and game $k + 1$. Furthermore, we use another hybrid game to change to a real-or-random based challenge instead of a left-or-right based challenge. It is not difficult to see that an adversary gaining an advantage ϵ in the left-or-right based game, gains an advantage of at least $\frac{\epsilon}{2}$ in the real-or-random based game.

Challenges. Since we changed the game to a real-or-random based game, the challenge phase changes slightly. The challenger now chooses a bit $b \xleftarrow{R} \{0,1\}$ that is used to determine whether to return the encryption of the submitted value or a random one. In case of the selective plaintext-privacy game, the adversary submits a vector $\boldsymbol{x}_I^{(c)}$ together with an identifier $\text{id}^{(c)}$ to the challenger. In case of the restricted predicate-privacy game, the adversary submits a vector $\boldsymbol{y}^{(c)}$ to the challenger. The challenger chooses values $\nu_i, \nu_i' \xleftarrow{R} \mathbb{Z}_p^*$ for $1 \leq i \leq n$. For a ciphertext challenge it returns the challenge

$$\mathsf{ct_{Ch}} = \left\{ \left(g_1^{t_{\text{id}(c)}}, g_1^{\nu_i}, \mathsf{ct}'_{\mathsf{Ch},i} \right) \mid i \in I \right\},$$

where

$$\mathsf{ct}'_{\mathsf{Ch},k} = \begin{cases} g_1^{\nu_k \alpha_k \pi(\beta_k, \nu_k') + t_{\text{id}(c)} \gamma_k} & \text{if } b = 0 \\ g_1^{\nu_k \alpha_k \pi(\beta_k, x_k^{(c)}) + t_{\text{id}(c)} \gamma_k} & \text{if } b = 1. \end{cases}$$

For a token challenge, it returns the challenge

$$\mathsf{tk_{Ch}} = \left(\left\{ (g_2^{\nu_i}, \mathsf{tk'_{Ch},i}) \mid i \in S_y \right\}, \prod_{i \in S_y} g_2^{\nu_i \gamma_i} \right),$$

where, if $k \in S_y$,

$$\mathsf{tk'_{Ch},k} = \begin{cases} g_2^{\nu_k \alpha_k \pi(\beta_k, \nu'_k)} & \text{if } b = 0 \\ g_2^{\nu_k \alpha_k \pi(\beta_k, y_k^{(c)})} & \text{if } b = 1. \end{cases}$$

Indistinguishability. We now show that an adversary has at most a negligible advantage of $O(q^2/p)$ in distinguishing between game k and game $k+1$, i.e., it is unable to distinguish $g_1^{\nu_k \alpha_k \pi(\beta_k, x_k^{(c)}) + t_{\mathsf{id}(c)} \gamma_k}$ from $g_1^{\nu'_k}$ for ciphertext challenges and $g_2^{\nu_k \alpha_k \pi(\beta_k, y_k^{(c)})}$ from $g_2^{\nu'_k}$ for token challenges.

As is common in the generic bilinear group model [32], we consider the challenger keeping record of all group elements the adversary has. It does so by keeping lists $P_{\mathbb{G},l}$ of linear polynomials in $\mathbb{Z}_p$ for each of the groups $\mathbb{G}_1$, $\mathbb{G}_2$, and $\mathbb{G}_T$. These polynomials use indeterminates for γ_i, $\alpha_i \pi(\beta_i, c_i)$, the $t_{\mathsf{id}(j)}$'s, $\alpha_i \pi(\beta_i, x_i^{(j)})$'s, $\alpha_i \pi(\beta_i, y^{(j)})$'s, $r_i^{(j)}$'s, and the $u_i^{(j)}$'s.

To simplify our reasoning, we will only look at polynomials $P_{\mathbb{G}_T,l}$ in $\mathbb{G}_T$. This is justified as we can transform any polynomial in $\mathbb{G}_1$ or $\mathbb{G}_2$ to a polynomial $P_{\mathbb{G}_T,l}$ in $\mathbb{G}_T$ through an additional query to the pairing oracle.

We can now say that the adversary wins the game if for a random assignment to all the indeterminates, any $P_{\mathbb{G}_T,i} \neq P_{\mathbb{G}_T,j}$ evaluates to the same value. We will show that the adversary is not able to query for distinct polynomials $P_{\mathbb{G}_T,i}, P_{\mathbb{G}_T,j}$ such that, if the challenger plays the 'real' experiment and if the indeterminates get assigned with random values, they will evaluate to the same value, except for negligible probability. Then, by the Schwartz lemma [29] and the extended result of Shoup [32], we can bound this probability of $P_{\mathbb{G}_T,i} \neq P_{\mathbb{G}_T,j}$ evaluating to the same value by $O(q^2/p)$ if at most q group elements are given to the adversary.

In the case of a ciphertext challenge, we first have to bring the challenge response, which is an element of $\mathbb{G}_1$, to the target group $\mathbb{G}_T$. Since the adversary only has (linear combinations of) the elements g_2, $g_2^{u_i^{(j)}}$, $g_2^{\alpha_i \pi(\beta_i, y_i^{(j)}) u_i^{(j)}}$, and $\prod_{i \in S_y} g_2^{u_i^{(j)} \gamma_i}$ in $\mathbb{G}_2$, it can only bring the challenge to $\mathbb{G}_T$ by pairing with one of these. Similarly, for token challenges, the adversary can only pair with the elements g_1, $g_1^{t_{\mathsf{id}(j)}}$, $g_1^{r_i^{(j)}}$, or $g_1^{\alpha_i \pi(\beta_i, x_i^{(j)}) r_i^{(j)} + t_{\mathsf{id}(j)} \gamma_i}$ in $\mathbb{G}_1$.

The resulting polynomials for these challenge responses are summarized in Table 1. Since the group elements are represented by uniformly independent values, the adversary can only distinguish between game k and game $k+1$ with more than a negligible advantage if it can construct at least one of the polynomials in this table.

Table 1. Target polynomials in both indistinguishability games.

Ciphertext challenge

$\nu_k \alpha_k \pi(\beta_k, x_k^{(c)}) + t_{\mathrm{id}(c)} \gamma_k$	$u_i^{(j)}(\nu_k \alpha_k \pi(\beta_k, x_k^{(c)}) + t_{\mathrm{id}(c)} \gamma_k)$
$u_i^{(j)} \alpha_i \pi(\beta_i, y_i^{(j)})(\nu_k \alpha_k \pi(\beta_k, x_k^{(c)}) + t_{\mathrm{id}(c)} \gamma_k)$	$(\nu_k \alpha_k \pi(\beta_k, x_k^{(c)}) + t_{\mathrm{id}(c)} \gamma_k) \sum_{i \in S_{\boldsymbol{y}(j)}} u_i^{(j)} \gamma_i$

Token challenge

$\nu_k \alpha_k \pi(\beta_k, y_k^{(c)})$	$t_{\mathrm{id}} \nu_k \alpha_k \pi(\beta_k, y_k^{(c)})$
$r_i^{(j)} \nu_k \alpha_k \pi(\beta_k, y_k^{(c)})$	$\left(r_i^{(j)} \alpha_i \pi(\beta_i, x_i^{(j)}) + t_{\mathrm{id}} \gamma_i\right) \nu_k \alpha_k \pi(\beta_k, y_k^{(c)})$

Linear Combinations. We now argue that the adversary cannot construct any of these challenges by looking at the components it has. We summarize the polynomials the adversary has access to, again by only looking at the elements in the target group $\mathbb{G}_T$, in Table 2.

Table 2. Elements the adversary can query for in an indistinguishability game (up to linear combinations).

1	$t_{\mathrm{id}(j)}$
$u_{i'}^{(j')}$	$u_{i'}^{(j')} t_{\mathrm{id}(j)}$
$u_{i'}^{(j')} \alpha_{i'} \pi(\beta_{i'}, y_{i'}^{(j')})$	$u_{i'}^{(j')} \alpha_{i'} \pi(\beta_{i'}, y_{i'}^{(j')}) t_{\mathrm{id}(j)}$
$\sum_{i' \in S_{\boldsymbol{y}(j')}} u_{i'}^{(j')} \gamma_{i'}$	$t_{\mathrm{id}(j)} \sum_{i' \in S_{\boldsymbol{y}(j')}} u_{i'}^{(j')} \gamma_{i'}$

$\cdots$

$r_i^{(j)}$	$r_i^{(j)} \alpha_i \pi(\beta_i, x_i^{(j)}) + t_{\mathrm{id}(j)} \gamma_i$
$u_{i'}^{(j')} r_i^{(j)}$	$u_{i'}^{(j')}\left(r_i^{(j)} \alpha_i \pi(\beta_i, x_i^{(j)}) + t_{\mathrm{id}(j)} \gamma_i\right)$
$u_{i'}^{(j')} \alpha_{i'} \pi(\beta_{i'}, y_{i'}^{(j')}) r_i^{(j)}$	$u_{i'}^{(j')} \alpha_{i'} \pi(\beta_{i'}, y_{i'}^{(j')})\left(r_i^{(j)} \alpha_i \pi(\beta_i, x_i^{(j)}) + t_{\mathrm{id}(j)} \gamma_i\right)$
$r_i^{(j)} \sum_{i' \in S_{\boldsymbol{y}(j')}} u_{i'}^{(j')} \gamma_{i'}$	$\left(r_i^{(j)} \alpha_i \pi(\beta_i, x_i^{(j)}) + t_{\mathrm{id}(j)} \gamma_i\right) \sum_{i' \in S_{\boldsymbol{y}(j')}} u_{i'}^{(j')} \gamma_{i'}$

$\cdots$

We show in the full version of the paper that no linear combination of the polynomials in Table 2 equals any of the polynomials in Table 1.

A.2   Chosen-Plaintext Security

The proposed construction is also chosen-plaintext secure as stated in Theorem 2. We remark that the proof does not rely on the use of random oracles.

Proof. We construct a challenger $\mathcal{B}$ capable of breaking the DDH assumption in $\mathbb{G}_1$ by using an adversary $\mathcal{A}$ that is able to win the chosen-plaintext with corruptions game with more than a negligible advantage.

We proof this though a series of hybrid games. Let game j be the game as defined in Definition 6, but where the first $j - 1$ components of the challenge query are replaced by random elements. Note that game 1 is identical to the original game and that it is not possible for any adversary to gain an advantage in game $n + 1$. We are left to show that an adversary has at most a negligible advantage in distinguishing game j from game $j + 1$.

Setup. The challenger $\mathcal{B}$ receives the bilinear group parameters and the DDH instance $(A = g_1^a, B = g_1^b, Z) \in (\mathbb{G}_1)^3$. It chooses the hash function H and the encryption keys usk_i. It sets encryption key $\mathsf{usk}_j = (A, \beta_j \xleftarrow{R} \mathcal{K}, \gamma_j \xleftarrow{R} \mathbb{Z}_p^*)$ and chooses the rest of the encryption keys according to the scheme. The public parameters and the encryption keys usk_i are given to the adversary.

Challenge. The adversary $\mathcal{A}$ submits an identifier id^* and two vectors $\boldsymbol{x}_0^*, \boldsymbol{x}_1^*$ to the challenger. The challenger chooses $b \xleftarrow{R} \{0, 1\}$ and sets $g_1^{r_j} = B$. Additionally, it picks values $r_i \xleftarrow{R} \mathbb{Z}_p^*$ for $1 \leq i \neq j \leq n$. It gives the challenge

$$
\mathsf{ct}_i = \begin{cases} \left(H(\mathsf{id}^*), g_1^{r_i}, R \xleftarrow{R} \mathbb{G}_1\right) & \text{if } i < j \\ \left(H(\mathsf{id}^*), B, Z^{\pi(\beta_i, x_{b,i}^*)} H(\mathsf{id}^*)^{\gamma_j}\right) & \text{if } i = j \\ \left(H(\mathsf{id}^*), g_1^{r_i}, g_1^{\alpha_i r_i \pi(\beta_i, x_{b,i}^*)} H(\mathsf{id}^*)^{\gamma_i}\right) & \text{if } i > j \end{cases}
$$

for $1 \leq i \leq n$ to the adversary.

If the challenger is given $Z = g_1^{ab}$, then challenge ciphertext is identically distributed as the challenge ciphertext in game j and component j is a real encryption. If the challenger is given $Z \xleftarrow{R} \mathbb{G}_1$, then challenge ciphertext is identically distributed as the challenge ciphertext in game $j + 1$ and component j is a random encryption.

Guess. The challenger outputs its guess that $Z = g_1^{ab}$ if the adversary guesses that it is playing game j, and outputs its guess that $Z \xleftarrow{R} \mathbb{G}_1$ if the adversary guesses that it is playing game $j + 1$.

If the adversary has a non-negligible advantage in distinguishing between game j and game $j + 1$, the challenger obtains a non-negligible advantage in solving the DDH problem in group $\mathbb{G}_1$.

References

1. Abdalla, M., Catalano, D., Dent, A.W., Malone-Lee, J., Neven, G., Smart, N.P.: Identity-based encryption gone wild. In: Bugliesi, M., Preneel, B., Sassone, V., Wegener, I. (eds.) ICALP 2006. LNCS, vol. 4052, pp. 300–311. Springer, Heidelberg (2006). https://doi.org/10.1007/11787006_26
2. Abdalla, M., Gay, R., Raykova, M., Wee, H.: Multi-input inner-product functional encryption from pairings. In: Coron, J.-S., Nielsen, J.B. (eds.) EUROCRYPT 2017. LNCS, vol. 10210, pp. 601–626. Springer, Cham (2017). https://doi.org/10.1007/978-3-319-56620-7_21
3. Boneh, D., Franklin, M.: Identity-based encryption from the weil pairing. In: Kilian, J. (ed.) CRYPTO 2001. LNCS, vol. 2139, pp. 213–229. Springer, Heidelberg (2001). https://doi.org/10.1007/3-540-44647-8_13

4. Boneh, D., Lewi, K., Raykova, M., Sahai, A., Zhandry, M., Zimmerman, J.: Semantically secure order-revealing encryption: multi-input functional encryption without obfuscation. In: Oswald, E., Fischlin, M. (eds.) EUROCRYPT 2015. LNCS, vol. 9057, pp. 563–594. Springer, Heidelberg (2015). https://doi.org/10.1007/978-3-662-46803-6_19

5. Boneh, D., Sahai, A., Waters, B.: Functional encryption: definitions and challenges. In: Ishai, Y. (ed.) TCC 2011. LNCS, vol. 6597, pp. 253–273. Springer, Heidelberg (2011). https://doi.org/10.1007/978-3-642-19571-6_16

6. Boneh, D., Waters, B.: Conjunctive, subset, and range queries on encrypted data. In: Vadhan, S.P. (ed.) TCC 2007. LNCS, vol. 4392, pp. 535–554. Springer, Heidelberg (2007). https://doi.org/10.1007/978-3-540-70936-7_29

7. Bösch, C., Hartel, P., Jonker, W., Peter, A.: A survey of provably secure searchable encryption. CSUR 47(2), 18:1–18:51 (2014)

8. Brakerski, Z., Segev, G.: Function-private functional encryption in the private-key setting. In: Dodis, Y., Nielsen, J.B. (eds.) TCC 2015. LNCS, vol. 9015, pp. 306–324. Springer, Heidelberg (2015). https://doi.org/10.1007/978-3-662-46497-7_12

9. Chase, M., Meiklejohn, S., Zaverucha, G.: Algebraic MACs and keyed-verification anonymous credentials. In: CCS, pp. 1205–1216. ACM (2014)

10. Chenette, N., Lewi, K., Weis, S.A., Wu, D.J.: Practical order-revealing encryption with limited leakage. In: Peyrin, T. (ed.) FSE 2016. LNCS, vol. 9783, pp. 474–493. Springer, Heidelberg (2016). https://doi.org/10.1007/978-3-662-52993-5_24

11. Conrad, S.H., LeClaire, R.J., O'Reilly, G.P., Uzunalioglu, H.: Critical national infrastructure reliability modeling and analysis. Bell Labs Tech. J. 11(3), 57–71 (2006)

12. Dunn-Cavelty, M., Suter, M.: Public-private partnerships are no silver bullet: an expanded governance model for critical infrastructure protection. Int. J. Crit. Infrast. Prot. 2(4), 179–187 (2009)

13. Galbraith, S.D., Paterson, K.G., Smart, N.P.: Pairings for cryptographers. Discrete Appl. Math. 156(16), 3113–3121 (2008). Applications of Algebra to Cryptography

14. Garg, S., Gentry, C., Halevi, S., Raykova, M., Sahai, A., Waters, B.: Candidate indistinguishability obfuscation and functional encryption for all circuits. In: FOCS, pp. 40–49 (2013)

15. Goldwasser, S., et al.: Multi-input functional encryption. In: Nguyen, P.Q., Oswald, E. (eds.) EUROCRYPT 2014. LNCS, vol. 8441, pp. 578–602. Springer, Heidelberg (2014). https://doi.org/10.1007/978-3-642-55220-5_32

16. Gordon, S.D., Katz, J., Liu, F.H., Shi, E., Zhou, H.S.: Multi-input functional encryption. Cryptology ePrint Archive, Report 2013/774 (2013)

17. Katz, J., Sahai, A., Waters, B.: Predicate encryption supporting disjunctions, polynomial equations, and inner products. In: Smart, N. (ed.) EUROCRYPT 2008. LNCS, vol. 4965, pp. 146–162. Springer, Heidelberg (2008). https://doi.org/10.1007/978-3-540-78967-3_9

18. Lewi, K., Wu, D.J.: Order-revealing encryption: new constructions, applications, and lower bounds. In: CCS. ACM (2016)

19. Lewko, A., Waters, B.: Decentralizing attribute-based encryption. In: Paterson, K.G. (ed.) EUROCRYPT 2011. LNCS, vol. 6632, pp. 568–588. Springer, Heidelberg (2011). https://doi.org/10.1007/978-3-642-20465-4_31

20. Luiijf, E., Klaver, M.: On the sharing of cyber security information. In: Rice, M., Shenoi, S. (eds.) ICCIP 2015. IAICT, vol. 466, pp. 29–46. Springer, Cham (2015). https://doi.org/10.1007/978-3-319-26567-4_3

21. Luiijf, E., Nieuwenhuijs, A., Klaver, M., van Eeten, M., Cruz, E.: Empirical findings on critical infrastructure dependencies in Europe. In: Setola, R., Geretshuber, S. (eds.) CRITIS 2008. LNCS, vol. 5508, pp. 302–310. Springer, Heidelberg (2009). https://doi.org/10.1007/978-3-642-03552-4_28

22. Miyaji, A., Nakabayashi, M., Takano, S.: Characterization of elliptic curve traces under FR-reduction. In: Won, D. (ed.) ICISC 2000. LNCS, vol. 2015, pp. 90–108. Springer, Heidelberg (2001). https://doi.org/10.1007/3-540-45247-8_8

23. Moteff, J.D., Stevens, G.M.: Critical infrastructure information disclosure and homeland security (2002). http://www.dtic.mil/docs/citations/ADA467310

24. Naor, M., Reingold, O.: Number-theoretic constructions of efficient pseudo-random functions. ACM **51**(2), 231–262 (2004)

25. O'Neill, A.: Definitional issues in functional encryption. Cryptology ePrint Archive, Report 2010/556 (2010)

26. Pandey, O., Rouselakis, Y.: Property preserving symmetric encryption. In: Pointcheval, D., Johansson, T. (eds.) EUROCRYPT 2012. LNCS, vol. 7237, pp. 375–391. Springer, Heidelberg (2012). https://doi.org/10.1007/978-3-642-29011-4_23

27. President's Commission on Critical Infrastructure Protection: Critical foundations: Protecting America's infrastructures (1997). https://www.fas.org/sgp/library/pccip.pdf

28. Sahai, A., Waters, B.: Fuzzy Identity-based encryption. In: Cramer, R. (ed.) EUROCRYPT 2005. LNCS, vol. 3494, pp. 457–473. Springer, Heidelberg (2005). https://doi.org/10.1007/11426639_27

29. Schwartz, J.T.: Fast probabilistic algorithms for verification of polynomial identities. ACM **27**(4), 701–717 (1980)

30. Shen, E., Shi, E., Waters, B.: Predicate privacy in encryption systems. In: Reingold, O. (ed.) TCC 2009. LNCS, vol. 5444, pp. 457–473. Springer, Heidelberg (2009). https://doi.org/10.1007/978-3-642-00457-5_27

31. Shi, E., Chan, T.H., Rieffel, E.G., Chow, R., Song, D.: Privacypreserving aggregation of time-series data. In: NDSS. The Internet Society (2011). https://www.ndss-symposium.org/ndss2011/privacy-preserving-aggregation-of-time-series-data/

32. Shoup, V.: Lower bounds for discrete logarithms and related problems. In: Fumy, W. (ed.) EUROCRYPT 1997. LNCS, vol. 1233, pp. 256–266. Springer, Heidelberg (1997). https://doi.org/10.1007/3-540-69053-0_18

33. Skopik, F., Settanni, G., Fiedler, R.: A problem shared is a problem halved: a survey on the dimensions of collective cyber defense through security information sharing. Comput. Secur. **60**, 154–176 (2016)

34. Smart, N.P.: The exact security of ECIES in the generic group model. In: Honary, B. (ed.) Cryptography and Coding 2001. LNCS, vol. 2260, pp. 73–84. Springer, Heidelberg (2001). https://doi.org/10.1007/3-540-45325-3_8

35. Yang, G., Tan, C.H., Huang, Q., Wong, D.S.: Probabilistic public key encryption with equality test. In: Pieprzyk, J. (ed.) CT-RSA 2010. LNCS, vol. 5985, pp. 119–131. Springer, Heidelberg (2010). https://doi.org/10.1007/978-3-642-11925-5_9

Credentials and Authentication

Revisiting Yasuda et al.'s Biometric Authentication Protocol: Are You Private Enough?

Elena Pagnin$^{(\boxtimes)}$, Jing Liu, and Aikaterini Mitrokotsa

Chalmers University of Technology, Gothenburg, Sweden
elenap@chalmers.se

Abstract. Biometric Authentication Protocols (BAPs) have increasingly been employed to guarantee reliable access control to places and services. However, it is well-known that biometric traits contain sensitive information of individuals and if compromised could lead to serious security and privacy breaches. Yasuda et al. [23] proposed a distributed privacy-preserving BAP which Abidin et al. [1] have shown to be vulnerable to biometric template recovery attacks under the presence of a malicious computational server. In this paper, we fix the weaknesses of Yasuda et al.'s BAP and present a detailed instantiation of a distributed privacy-preserving BAP which is resilient against the attack presented in [1]. Our solution employs Backes et al.'s [4] verifiable computation scheme to limit the possible misbehaviours of a malicious computational server.

Keywords: Biometric authentication · Verifiable delegation
Privacy-preserving authentication

1 Introduction

Biometric authentication has become increasingly popular as a fast and convenient method of authentication that does not require to remember and manage long and cumbersome passwords. However, the main advantage of biometrics, i.e., their direct and inherent link with the identity of individuals, also rises serious security and privacy concerns. Since biometric characteristics can not be changed or revoked, unauthorised leakage of this information leads to irreparable security and privacy breaches such as identity fraud and individual profiling or tracking [18]. Thus, there is an urgent need for efficient and reliable privacy-preserving biometric authentication protocols (BAPs).

The design of privacy-preserving BAPs is by itself a very delicate procedure. It becomes even more challenging when one considers the distributed setting in which a resource-constrained client outsources the computationally heavy authentication process to more powerful external entities. In this paper, we focus on Yasuda et al.'s protocol for privacy-preserving BAPs in the distributed setting [23] and show how to mitigate the privacy attacks presented by Abidin et al. [1] by employing Backes et al.'s verifiable computation scheme [4].

© Springer Nature Switzerland AG 2018
S. Capkun and S. S. M. Chow (Eds.): CANS 2017, LNCS 11261, pp. 161–178, 2018.
https://doi.org/10.1007/978-3-030-02641-7_8

1.1   Background and Related Work

Distributed privacy-preserving BAPs usually involve the following entities: *(i)* a client/user $\mathcal{C}$, *(ii)* a database $\mathcal{DB}$, *(iii)* a computational server $\mathcal{CS}$, and *(iv)* an authentication server $\mathcal{AS}$. The granularity of roles and entities in the biometric authentication process facilitates the privacy-preservation of the sensitive information. This distributed setting, indeed guarantees that no single entity has access to both the biometric templates (fresh and stored ones) and the identity of the querying user.

Several existing proposals of privacy-preserving BAPs use the distributed setting, e.g., [5,21–23], and make leverage on advanced cryptographic techniques such as homomorphic encryption [7,23], oblivious transfer [8] and garbled circuits [14]. In particular, Yasuda et al.'s protocol [23] was claimed to be privacy-preserving since it is based on the distributed setting and relies on a novel somewhat homomorphic encryption scheme based on ideal lattices. Abildin et al. [1] showed that Yasuda et al.'s BAP is privacy-preserving only in the honest-but-curious model and described an algorithm that enables a malicious $\mathcal{CS}$ to recover a user's biometric template. Intuitively, Abidin et al.'s attack succeeds because $\mathcal{AS}$ does not detect that the malicious $\mathcal{CS}$ returns a value different from the one corresponding to the output of the (honest) outsourced computation, leaving space for *hill-climbing strategies* [20] that may lead to the disclosure of the stored reference biometric template.

Verifiable delegation of computation (VC) is a cryptographic primitive that enables a client to securely and efficiently offload computations to an untrusted server [11]. Verification of arbitrary complex computations was initially achieved via interactive proofs [2,13] and then moved towards more flexible and efficient schemes such as [3,9,10,19]. The setting of VC schemes is by nature distributed and thus perfectly fits the basic requirement of privacy-preserving BAPs. For this reason, Bringer et al. [6] suggested to use VC techniques to detect malicious behaviours in BAP.

In this paper, we provide the first explicit instantiation of a distributed privacy-preserving BAP which achieves security against malicious $\mathcal{CS}$ thanks to the verifiability of the delegated computation.

1.2   Our Contributions

In this paper, we mitigate Abidin et al.'s attack [1] against Yasuda et al.'s privacy-preserving biometric authentication protocol [23] by the means of the verifiable computation scheme by Backes et al. [4]. We combine the two schemes in an efficient and secure way, and obtain a modification of Yasuda et al.'s protocol with strong privacy guarantees. As a result, we obtain a new BAP which builds on top of Yasuda et al.'s and is truly privacy-preserving in the distributed setting.

From a general point of view, this paper offers a strategy to transform privacy-preserving BAPs that are secure in the honest-but-curious model into schemes that can tolerate a malicious CS by addressing the most significant challenges in privacy-preserving BAPs: to guarantee integrity and privacy of both the data and the computation. Despite the idea of combining VC and BAP is quite natural and intuitive [6], the actual combination needs to be done carefully in order to avoid flawed approaches.

Organisation. The paper is organized as follows. Section 2 describes the background notions used in the rest of the paper. Section 3 contains our modification of Backes et al.'s VC scheme to combine it with the somewhat homomorphic encryption scheme used in [23]. Section 4 presents an improved version of Yasuda et al.'s BAP together with a security and efficiency analysis. The proposed privacy-preserving BAP incorporates the new construction of VC on encrypted data of Sect. 3. Section 5 is an important side-note to our contributions, as it demonstrates how naïve and straight-forward compositions of VC and homomorphic encryption may lead to leakage of private information. Section 6 concludes the paper.

2 Preliminaries

Notations. We denote by $\mathbb{Z}$ and $\mathbb{Z}_p = \mathbb{Z}/p\mathbb{Z}$ the ring of integers and the integers modulo p, respectively. For two integers $x, d \in \mathbb{Z}$, $[x]_d$ denotes the reduction of x modulo d in the range of $[-d/2, d/2]$. We write vectors with capital letters, e.g., A, and refer to the i-th component of A as A_i. The symbol $x \xleftarrow{\$} \mathcal{X}$ denotes selecting x uniformly at random from the set $\mathcal{X}$.

We denote the Hadamard product for binary vectors as $\diamond : \mathbb{Z}_2^n \diamond \mathbb{Z}_2^n \rightarrow \mathbb{Z}_2^n$, with $A \diamond B = C$, $C_i = A_i \cdot B_i \in \mathbb{Z}_2$ for $i = 1, 2, \ldots, n$. The Hadamard product is similar to the inner product of vectors except that the output is a vector rather than an integer.

Bilinear Maps. A symmetric bilinear group is a tuple $(p, \mathbb{G}, \mathbb{G}_T, g, g_T, e)$, where $\mathbb{G}$ and $\mathbb{G}_T$ are groups of prime order p. The elements $g \in \mathbb{G}$ and $g_T \in \mathbb{G}_T$ are generators of the group they belong to, and $e : \mathbb{G} \times \mathbb{G} \longrightarrow \mathbb{G}_T$ is a bilinear map, i.e., $\forall A, B \in \mathbb{G}$ and $x, y \in \mathbb{Z}_p$ it holds that $e(xA, yB) = e(A, B)^{xy}$ and $e(g, g) \neq 1_{\mathbb{G}_T}$. In the setting of VC, the map e is cryptographically secure, i.e., it should be defined over groups where the discrete logarithm problem is assumed to be hard or it should be hard to find inverses. In bilinear groups there exists a natural isomorphism between $\mathbb{G}$ and $(\mathbb{Z}_p, +)$ given by $\phi_g(x) = g^x$; similarly for $\mathbb{G}_T$. Since ϕ_g and ϕ_{g_T} are isomorphisms, there exist inverses $\phi_g^{-1} : \mathbb{G} \rightarrow \mathbb{Z}_p$ and $\phi_{g_T}^{-1} : \mathbb{G}_T \rightarrow \mathbb{Z}_p$, that can be used to homomorphically evaluate any arithmetic circuit $f : \mathbb{Z}_p^n \rightarrow \mathbb{Z}_p$, from $\mathbb{G}$ to $\mathbb{G}_T$. More precisely, there exists a map **GroupEval** (as defined in [4]):

$$\mathbf{GroupEval}(f, X_1, \ldots, X_n) = \phi_{g_T}(f(\phi_g^{-1}(X_1), \ldots, \phi_g^{-1}(X_n))).$$

For security, we assume ϕ_g and ϕ_{g_T} are not efficiently computable.

Homomorphic MAC Authenticators. In this paper, we make use of Backes, Fiore and Reischuk's verifiable computation scheme based on homomorphic MAC authenticators [4], which we refer to as BFR. The BFR scheme targets functions f that are quadratic polynomials over a large number of variables. Figure 1 contains a succinct description of the BFR scheme.

KeyGen$(\lambda) \rightarrow (ek, vk)$: Given the security parameter λ, the key generation algorithm outputs a secret verification key vk and a public evaluation key ek.

Auth$(vk, L, m) \rightarrow \sigma$: Given the secret verification key vk, a multi-label $L = (\Delta, \tau)$ and a valid message m, this algorithm outputs an authentication tag σ.

Ver$(vk, \mathcal{P}_\Delta, m, \sigma) \rightarrow \{0, 1\}$: Given the secret key vk, a multi-label program $\mathcal{P} = ((f, \tau_1, \ldots, \tau_n), \Delta)$, a valid message m and a tag σ, the verification algorithm returns an acceptance bit *acc*: "0" for rejection and "1" for acceptance.

Eval$(ek, f, \boldsymbol{\sigma}) \rightarrow \sigma$: Given the public evaluation key ek, a circuit of a quadratic polynomial f and a vector of tags $\boldsymbol{\sigma} = (\sigma_1, \ldots, \sigma_n)$, the evaluation algorithm produces a new tag $\sigma \leftarrow$ **GroupEval**$(f, \sigma_1, \ldots, \sigma_n)$. The evaluation of **GroupEval** proceeds gate-by-gate through the arithmetic circuit of f with the following rules:

Fan-in-2 addition gate:

   i. $X_1, X_2 \in \mathbb{G}$, output $X = X_1 \cdot X_2 = g^{x_1} \cdot g^{x_2} = g^{x_1 + x_2} \in \mathbb{G}$.

   ii. $\hat{X}_1, \hat{X}_2 \in \mathbb{G}_T$, output $\hat{X} = \hat{X}_1 \cdot \hat{X}_2 = e(g, g)^{x_1} \cdot e(g, g)^{x_2} = e(g, g)^{x_1 + x_2} \in \mathbb{G}_T$.

   iii. $\hat{X}_1 \in \mathbb{G}_T, X_2 \in \mathbb{G}$, output $\hat{X} = \hat{X}_1 \cdot e(X_2, g) = e(g, g)^{x_1 + x_2} \in \mathbb{G}_T$.

   iv. $X_1 \in \mathbb{G}, \hat{X}_2 \in \mathbb{G}_T$, output $\hat{X} = e(X_1, g) \cdot \hat{X}_2 = e(g, g)^{x_1 + x_2} \in \mathbb{G}_T$.

Fan-in-2 multiplication gate:

   i. $X_1, X_2 \in \mathbb{G}$, output $\hat{X} = e(X_1, X_2) = e(g, g)^{x_1 x_2} \in \mathbb{G}_T$.

   ii. $X_1 \in \mathbb{G} \cup \mathbb{G}_T, c \in \mathbb{Z}_p$ constant, output $X = (X_1)^c = e(g, g)^{x_1 c} \in \mathbb{G}_T$.

The output of **GroupEval** is the output of the last gate of the arithmetic circuit.

Fig. 1. The BFR verifiable delegation of computation scheme.

For further details we refer the reader to the main paper [4].

Homomorphic Encryption. Let $\mathcal{M}$ denote the space of plaintexts that support an operation $\boxdot$, and $\mathcal{C}$ be the space of ciphertexts with $\odot$ as operation. An encryption scheme is said to be homomorphic if for any key, the encryption function **Enc** satisfies: **Enc**$(m_1 \boxdot m_2) \leftarrow$ **Enc**$(m_1) \odot$ **Enc**(m_2), for all $m_1, m_2 \in \mathcal{M}$, where $\leftarrow$ means computed without decryption. In this paper, we only use Somewhat Homomorphic Encryption schemes (SHE). As the name suggests these schemes only support a limited number of homomorphic operations, e.g., indefinite number of homomorphic additions and finite number of multiplications. The choice to use SHE instead of Fully Homomorphic Encryption [12] is due to efficiency: SHE, if used appropriately, can be much faster and more compact [15].

The Yasuda et al. Protocol. Yasuda et al. [23] proposed a privacy-preserving biometric authentication protocol that targets one-to-one authentication and relies on somewhat homomorphic encryption based on ideal lattices. Two packing methods facilitate efficient calculations of the secure Hamming distance, which is a common metric used for comparing biometric templates. The protocol uses a distributed setting with three parties: a client C, a computation server CS (which contains the database DB) and an authentication server AS. The protocol is divided into three phrases.

Setup Phase: AS generates the public key pk and the secret key sk of the SHE scheme in [23]. AS gives pk to C and CS and keeps sk.

Enrollment phase: C provides a feature vector A from the client's biometric data (e.g., fingerprints), runs the type-1 packing method and outputs the encrypted feature vector $\mathsf{vEnc}_1(A)$. The computation server stores $(\mathsf{ID}, \mathsf{vEnc}_1(A))$ in DB as the reference template for the client ID.

Authentication phase: upon an authentication request, C provides a fresh biometric feature vector B encrypted with the type-2 packing method and sends $(\mathsf{ID}, \mathsf{vEnc}_2(B))$ to the computational server. CS extracts from the database the tuple $(\mathsf{ID}, \mathsf{vEnc}_1(A))$ using ID as the search key. CS calculates the encrypted Hamming distance ct_{HD} and sends it to the authentication server. CS decrypts ct_{HD} and retrieves the actual Hamming distance $\mathsf{HD}(A,B) = \mathsf{Dec}(sk, ct_{\mathsf{HD}})$. AS returns yes if $\mathsf{HD}(A,B) \leq \kappa$ or no if $HD(A,B) > \kappa$, where κ is the predefined accuracy threshold of the authentication system.

Figure 2 depicts the authentication phase of Yasuda et al.'s BAP.

For additional details on biometric authentication protocols and systems we refer the reader to [16].

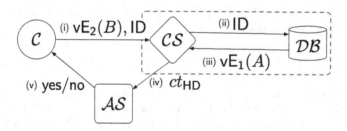

Fig. 2. Authentication phase in the Yasuda et al.'s BAP [23].

3   Combining the BFR and the SHE Schemes

In this section, we describe how to efficiently combine the verifiable computation scheme BFR by Backes et al. [4] with the somewhat homomorphic scheme SHE by Yasuda et al. [23]. We call the resulting scheme BFR+SHE. Our motivation for defining this new scheme is to build a tailored version of BFR that we insert in Yasuda et al.'s biometric authentication protocol to mitigate the template recovery attack of [1].

As a preliminary step, we explain the most challenging point of the combination of the two schemes: the ring range problem. This problem rises because the elements and operations in BFR and SHE are defined over two different rings. This passage is quite mathematical, but it is necessary to guarantee the correctness of our composition BFR+SHE, presented later on in this section.

3.1 The Ring Range Problem

The most significant challenge in combining the Backes et al. VC scheme with the Yasuda et al. SHE scheme is the different range of the base rings. While BFR handles all operations in $\mathbb{Z}_p$, where p is a prime, the operations in the SHE scheme [23] are handled in $\mathbb{Z}_d$, where d is the resultant of two polynomials. Therefore, in our BFR+SHE scheme, we need to tweak the input data if there is a mismatch in the ranges. In the calculation of the secure Hamming distance, there is a constant term equal to -2, which lives in $[-d/2, d/2)$ but not in $[0, p)$. In order to verify and generate proper tags, we can write -2 as $D = (d - 2)$ mod p. Furthermore, we need to check the impact of the range difference to the verification carried out by the client. The first equation in the **Ver** algorithm of BFR is:

$$ct_{\mathsf{HD}} = y_0^{(\mathsf{HD})}, \tag{1}$$

where $ct_{\mathsf{HD}} \in \mathbb{Z}_d$ and $y_0^{(HD)} \in \mathbb{Z}_p$. In our instantiation, the term ct_{HD} corresponds to the encrypted Hamming distance between the fresh and the reference templates, while $y_0^{(\mathsf{HD})}$ is a component of the final authentication tag. As long as $d \neq p$, Eq. (1) is not satisfied even when the computation is carried out correctly.

We present a general solution to this problem. For simplicity, we assume $p < d$ (as the tag size should be ideally small), although the reasoning also applies when $p > d$ by swapping the place of p and d. Our solution relies on keeping track of the dividend. Given a stored template $\alpha \in \mathbb{Z}_d$ and a fresh template $\beta \in \mathbb{Z}_d$, both encrypted, we have that: $\alpha = \alpha' + mp, \alpha' = \alpha \mod p \in \mathbb{Z}_p$, $\beta = \beta' + kp$ and $\beta' = \beta \mod p \in \mathbb{Z}_p$.

Let $\mathsf{SWHD}(x, y)$ be the arithmetic circuit for calculating the encrypted Hamming distance without the final modulo d. Let $c = \mathsf{SWHD}(\alpha, \beta)$ and $c' = \mathsf{SWHD}(\alpha', \beta')$, we can derive: $\mathsf{SWHD}(\alpha, \beta) \mod p = \mathsf{SWHD}(\alpha', \beta') \mod p$; and $c = \ell \cdot p + c'$. The value ℓ is the dividend. In our case of study, we want to perform the comparison between the Hamming distance (of the biometric templates) and the threshold κ which determines if the templates match, i.e., the client is authenticated, or not. To this end, we would track $\ell \mod d$ instead of ℓ directly. The reason is that ℓ contains more information and would lead to a privacy leak. Relating back to Eq. (1) we have: $ct_{\mathsf{HD}} = c \mod d \in \mathbb{Z}_q$ and $y_0^{(\mathsf{HD})} = c' \in \mathbb{Z}_p$. Given $c = \ell \cdot p + c'$, it holds that:

$$
\begin{aligned}
ct_{\mathsf{HD}} = c \mod d &= (\ell * p + c') \mod d \\
&= c' \mod d + (\ell \mod d) \cdot (p \mod d) \\
&= (y_0^{(\mathsf{HD})} \mod d) + (\ell \mod d) \cdot (p \mod d)
\end{aligned}
\tag{2}
$$

Thus, if we define $\ell_d := (\ell \mod d) \cdot (p \mod d)$, the verification equation in (1) becomes $ct_{\mathsf{HD}} = y_0^{(\mathsf{HD})}(\mod d + \ell_d)$, which is satisfied whenever ct_{HD} is computed correctly (as we show in Sect. 3.3).

3.2   Our BFR+SHE Scheme

To facilitate the intuition of how we incorporate BFR+SHE in Yasuda et al.'s BAP we describe the algorithms of BFR+SHE directly in the case the encrypted vectors are biometric templates:

BFR+SHE.**KeyGen**(λ): The key generation algorithm of BFR+SHE runs SHE.**KeyGen**$(\lambda) \to (pk, sk)$ and BFR.**KeyGen**$(\lambda) \to (ek, vk)$. The output is the four-tuple (ek, pk, sk, vk).

BFR+SHE.**Enc**$(pk, A, phase)$: The encryption algorithm takes as input the (encryption) public key pk, a plaintext biometric template $A \in \{0,1\}^{2048}$ and a *phase* $\in \{1, 2\}$ to select the appropriate packing method. It outputs the ciphertext ct computed as $ct = \mathsf{vEnc}_{phase}(A)$, using the type-*phase* packing method of the SHE scheme.

BFR+SHE.**Auth**(vk, L, ct): on input the verification key vk, a ciphertext ct and a multi-label $L = (\Delta, \tau)$, with Δ the *data set identifier* (e.g., the client's ID) and τ the *input identifier* (e.g., "stored biometric template" or "fresh biometric template"); this algorithm outputs $\sigma \leftarrow$ BFR.**Auth**(vk, L, ct), with $\sigma = (y_0, Y_i, 1) = (ct, F_K(\Delta, \tau) \cdot g^{-ct})^{1/\theta})$, where the value θ and the function F_K are defined in vk.

BFR+SHE.**Comp**(pk, ct_1, ct_2): The compute algorithm takes as input the encryption public key pk, and two ciphertexts ct_1, ct_2, which intuitively correspond to the encryptions $\mathsf{vEnc}_1(A)$ and $\mathsf{vEnc}_2(B)$ respectively. The output is the encrypted Hamming distance HD calculated as: $ct_{\mathsf{HD}} = C_2 \cdot \mathsf{vEnc}_1(A) + C_1 \cdot \mathsf{vEnc}_2(B) + (-2 \cdot \mathsf{vEnc}_1(A) \cdot \mathsf{vEnc}_2(B)) \in \mathbb{Z}_d$, where $C_1 := \left[\sum_{i=0}^{n-1} r^i \right]_d$ and $C_2 := [-C_1 + 2]_d$ and r, d are extracted from pk. To solve the ring range problem described in Sect. 3.1 we compute ℓ_d as follows. Let c be the result of the (encrypted) Hamming distance computation without the final modulo d. Then $c' = c \mod p$ and $c = \ell p + c'$, where c' is a component in the authentication tag and ℓ is a dividend. We compute $\ell_d = \ell \mod d = (c - [c \mod p])/p \mod d$. The output is $(ct_{\mathsf{HD}}, \ell_d)$.

BFR+SHE.**Eval**$(ek, pk, \sigma_1, \sigma_2)$: The evaluation algorithm takes as input the evaluation key ek, the ecryption public key pk, and two tags, which intuitively correspond to the authenticators for the two biometric templates, A, B. In our case of study, the function to be evaluated is fixed to be $f = \mathsf{HD}$ the Hamming distance. This algorithm outputs $\sigma_{\mathsf{HD}} = (y_0, Y_1, \hat{Y}_2) \leftarrow$ BFR.**Eval**$(ek, \mathsf{HD}, (\sigma_1, \sigma_2))$.

In details, every input gate accepts either two tags $\sigma_A, \sigma_B \in (\mathbb{Z}_p \times \mathbb{G} \times \mathbb{G}_T)^2$, or one tag and a constant $\sigma, c \in ((\mathbb{Z}_p \times \mathbb{G} \times \mathbb{G}_T) \times \mathbb{Z}_p)$. The output of a gate is a new tag $\sigma' \in (\mathbb{Z}_p \times \mathbb{G} \times \mathbb{G}_T)$, which will be fed into the next gate in the circuit as one of the two inputs. The operation stops when the final gate of f is reached and the resulting tag σ_{HD} is returned. A tag has the format $\sigma^{(i)} = (y_0^{(i)}, Y_1^{(i)}, \hat{Y}_2^{(i)}) \in \mathbb{Z}_p \times \mathbb{G} \times \mathbb{G}_T$ for $i = 1, 2$ (indicating the two input tags), which corresponds respectively to the coefficients of (x^0, x^1, x^2) in a polynomial. If $\hat{Y}_2^{(i)}$ is not defined, it is assumed that it has value $1 \in \mathbb{G}_T$. Next we define the specific operations for different types of gates:

- **Addition.** The output tag $\sigma' = (y_0, Y_1, \hat{Y}_2)$ is calculated as:

$$y_0 = y_0^{(1)} + y_0^{(2)}, \quad Y_1 = Y_1^{(1)} \cdot Y_1^{(2)}, \quad \hat{Y}_2 = \hat{Y}_2^{(1)} \cdot \hat{Y}_2^{(2)}.$$

- **Multiplication.** The output tag $\sigma' = (y_0, Y_1, \hat{Y}_2)$ is calculated as:

$$y_0 = y_0^{(1)} \cdot y_0^{(2)}, \quad Y_1 = Y_1^{(1)} \cdot Y_1^{(2)}, \quad \hat{Y}_2 = e(\hat{Y}_1^{(1)}, \hat{Y}_1^{(2)}).$$

Since the circuit f has maximum degree 2, the input tags to a multiplication gate can only have maximum degree 1 each.

- **Multiplication with constant.** The two inputs are one tag σ and one constant $c \in \mathbb{Z}_p$. The output tag $\sigma' = (y_0, Y_1, \hat{Y}_2)$ is calculated as: $y_0 = c \cdot y_0^{(1)}$, $Y_1 = (Y_1^{(1)})^c$, $\hat{Y}_2 = (\hat{Y}_2^{(1)})^c$.

BFR+SHE.**Ver**$(vk, \mathcal{P}_\Delta, ct_{HD}, \sigma_{HD}, \ell_d)$: The verification algorithm computes $b \leftarrow$ BFR.**Ver**$(vk, sk, \mathcal{P}_\Delta, ct_{HD}, \sigma_{HD}, \kappa)$ to verify the correctness of the outsourced computation. In our case of study, $\mathcal{P}_\Delta$ is a multi-labeled program [4] for the arithmetic circuit for calculating the encrypted HD. The BFR.**Ver** algorithm essentially performs two integrity-checks:

$$ct_{HD} = y_0 \mod (d + \ell_d) \tag{3}$$

$$W = e(g, g)^{y_0} \cdot e(Y_1, g)^\theta \cdot (\hat{Y}_2)^{\theta^2} \tag{4}$$

If the verification output is $b = 0$ the algorithm returns

$$(acc_{VC}, acc_{HD}) = (0, 0).$$

Otherwise, if $b = 1$, it proceeds with the biometric authentication check: it computes $w \leftarrow$ SHE.Dec(ct) to retrieve the actual Hamming distance $w = HD(A, B)$. If $HD(A, B) \leq \kappa$, here κ corresponds to the accuracy of the BAP, the algorithm returns

$$(acc_{VC}, acc_{HD}) = (1, 1).$$

If $HD(A, B) > \kappa$, the output is

$$(acc_{VC}, acc_{HD}) = (1, 0).$$

3.3 Correctness Analysis

In our BFR+SHE scheme the outsourced function is the Hamming distance HD, that can be represented by a bi-variate deterministic quadratic function. Thus, we can avoid using gate-by-gate induction proofs, as done in [4], and demonstrate the correctness in a direct way. In what follows, we adopt the notation in [4], and we prove the correctness of BFR+SHE by walking through the arithmetic circuit of HD step by step.

Figure 3 depicts the arithmetic circuit for calculating the encrypted Hamming distance. A and B denote the encrypted stored and fresh biometric templates respectively. C_1 and C_2 are the constants in the function as defined in the BFR+SHE.**Comp** algorithm. The D letter indicates the -2 in the function, but since -2 is not in the valid range $\mathbb{Z}_p$ required by the original BFR scheme, we need to have an intermediate transformation of $D = d - 2$. All A, B, C_1 and C_2 are in $\mathbb{Z}_d$. Finally, the σs are the outcome tags of the form $\sigma^{(i)} = (y_0^{(i)}, Y_1^{(i)}, \hat{Y}_2^{(i)}) \in \mathbb{Z}_p \times \mathbb{G} \times \mathbb{G}_T$ after each gate operation, and the Rs are values in either $\mathbb{G}$ or $\mathbb{G}_T$, which are used for homomorphic evaluation over bilinear groups (i.e., **GroupEval** in [4]).

We let α and β be $\mathsf{vEnc}_1(A)$ and $\mathsf{vEnc}_2(B)$ and each of them has a tag: $\sigma_\alpha = (y_0^{(A)}, Y_1^{(A)}, 1)$ and $\sigma_\beta = (y_0^{(B)}, Y_1^{(B)}, 1)$. These two tags are generated by the BFR.**Auth** algorithm, which specifies that $y_0^{(A)} = \alpha$ and $Y_1^{(A)} = (R_\alpha \cdot g^{-\alpha})^{1/\theta}$. Similarly, we have $y_0^{(B)} = \beta$ and $Y_1^{(B)} = (R_\beta \cdot g^{-\beta})^{1/\theta}$. To verify the correctness of our BFR+SHE scheme, we need to check that the two equations specified in the BFR.**Ver** algorithm are satisfied if the computation is performed correctly. To this end, let $\sigma_{\mathsf{HD}} = (y_0^{(\mathsf{HD})}, Y_1^{(\mathsf{HD})}, \hat{Y}_2^{(\mathsf{HD})})$ be the final tag (which is equivalent to σ_6 in the arithmetic circuit depicted in Fig. 3).

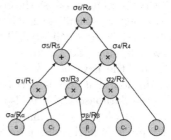

Fig. 3. The arithmetic circuit for calculating the encrypted Hamming distance.

The first step is to derive the tags for the intermediate calculation and eventually the final tag. If we run the SHE.**Eval** algorithm homomorphically through the circuit, we will get the outcome tags $\sigma_1, \ldots, \sigma_6$ (for details see Appendix A). We thus derive σ_{HD} (equivalent to σ_6):

$$
\begin{aligned}
\sigma_{HD} &= (y_0^{(\mathsf{HD})}, Y_1^{(\mathsf{HD})}, \hat{Y}_2^{(\mathsf{HD})}) \\
&= (\ C_2 \cdot y_0^{(A)} + C_1 \cdot y_0^{(B)} + D \cdot y_0^{(A)} \cdot y_0^{(B)}, \\
&\quad (Y_1^{(A)})^{y_0^{(B)} \cdot D + C_2} \cdot (Y_1^{(B)})^{y_0^{(A)} \cdot D + C_1}, \ e(Y_1^{(A)}, Y_1^{(B)})^D).
\end{aligned}
$$

Now we show the proofs for the two verification equations. First we need to prove Eq. (3), i.e., $ct_{\mathsf{HD}} = y_0 \bmod (d + \ell_d)$. The equality holds as for Eq. (2). The end result is: $ct_{\mathsf{HD}} = y_0^{(\mathsf{HD})} \bmod d + (\ell \bmod d) \cdot (p \bmod d)$. As we define $\ell_d = (\ell \bmod d) \cdot (p \bmod d)$, we can derive Eq. (3). Secondly, we need to prove that Eq. (4) holds, i.e., $W = e(R_\alpha^{C_2} \cdot R_\beta^{C_1}, g) \cdot e(R_\alpha, R_\beta)^D$. To this end, we run **GroupEval**(f, R_α, R_β) and execute the bilinear gate operations. Recall that R_α and R_β correspond to R_A and R_B in the notation used in the construction, Denote by R_6 the final result of running **GroupEval** over the circuit of HD. It holds that:

$$R_6 = \mathbf{GroupEval}(f, R_\alpha, R_\beta) = e(R_\alpha^{C_2} \cdot R_\beta^{C_1}, g) \cdot e(R_\alpha, R_\beta)^D$$
$$= \mathbf{GroupEval}(f, R_\alpha, R_\beta) = e(g, g)^{y_0^{\mathsf{HD}}} \cdot e(Y_1^{\mathsf{HD}}, g)^\theta \cdot (\hat{Y}_2^{(\mathsf{HD})})^{\theta^2}$$

By expanding the last expression the desired result (see Appendix A for details). Thus, we have proved the correctness of the BFR+SHE scheme. which are the results of the pseudo-random function F_K in the BFR.**Ver** algorithm.

4   Improving the Yasuda et al. Protocol

In this section, we describe a modified version of the Yasuda et al. [23] protocol that is secure against the recently identified *hill-climbing* attack that can be performed by a malicious computation server $\mathcal{CS}$. It is composed of four distributed parties: a client $\mathcal{C}$ (holding the keys pk, ek and vk), a computation server/database $\mathcal{CS}$ (holding the keys pk and ek), an authentication server $\mathcal{AS}$ (holding the keys pk, sk and vk). In the proposed protocol, we preserve the assumption that $\mathcal{AS}$ is a trusted party and furthermore assume the client $\mathcal{C}$ and the database $\mathcal{DB}$ are also trusted parties. $\mathcal{C}$ is responsible to manage the secret key vk for the verifiable computation scheme and $\mathcal{DB}$ stores the encrypted reference biometric templates with the identities of the corresponding clients. However, $\mathcal{CS}$ can be malicious and cheat with flawed computations. We describe the three main phases of our proposed improvement of Yasuda et al.'s privacy-preserving biometric authentication protocol:

Setup Phase: In this phase the authentication server $\mathcal{AS}$ runs SHE.**KeyGen**(λ) to generate the public key pk and the secret key sk of the somewhat homomorphic encryption (SHE) scheme. $\mathcal{AS}$ keeps sk and distributes pk to both the client $\mathcal{C}$ and the computation server $\mathcal{CS}$.

Enrollment Phase: Upon client registration, the client $\mathcal{C}$ runs BFR. **KeyGen**(λ) to generate the public evaluation key ek and the secret verification key vk. $\mathcal{C}$ distributes ek to $\mathcal{CS}$ and vk to $\mathcal{AS}$. The client $\mathcal{C}$ generates a 2048-bit feature vector A from the client's biometric data, runs BFR+SHE.**Enc**$(pk, A, 0)$ to obtain the ciphertext ct_A. $\mathcal{C}$ authenticates ct_A by running BFR+SHE.**Auth**(vk, L_A, ct_A) and outputs a tag σ_A. Then $\mathcal{C}$ sends the three-tuple $(\mathsf{ID}, ct_A, \sigma_A)$ to the database. This three-tuple serves as the reference biometric template for the specific client with identity ID.

Authentication Phase: The client provides fresh biometric data as a feature vector $B \in \{0,1\}^{2048}$. $\mathcal{C}$ runs BFR+SHE.**Enc**$(pk, B, 1)$ to obtain the ciphertext ct_B and authenticates it by running $\sigma_B \leftarrow$ BFR+SHE.**Auth**(vk, L_B, ct_B). $\mathcal{C}$ sends $(\mathsf{ID}, ct_B, \sigma_B)$ to $\mathcal{CS}$, who extracts the tuple $(\mathsf{ID}, ct_A, \sigma_A)$ corresponding to the client to be authenticated (using the ID as the search key). $\mathcal{CS}$ calculates the encrypted Hamming distance $ct_{\mathsf{HD}} \leftarrow$ BFR+SHE.**Comp**(pk, ct_A, ct_B) and generates a corresponding tag $\sigma_{\mathsf{HD}} \leftarrow$ BFR+SHE.**Eval**$(ek, pk, \sigma_A, \sigma_B)$. Then, $\mathcal{CS}$ sends $(\mathsf{ID}, ct_{\mathsf{HD}}, \sigma_{\mathsf{HD}})$ to the authentication server. $\mathcal{AS}$ runs $(acc_{\mathsf{VC}}, acc_{\mathsf{HD}})$

$\leftarrow$ BFR + SHE.$\mathbf{Ver}(vk, sk, \mathcal{P}_{\Delta}, ct_{\mathsf{HD}}, \sigma_{\mathsf{HD}}, \kappa)$, where κ is the desired accuracy level of the BAP. If either acc_{VC} or acc_{HD} is 0 $\mathcal{AS}$ outputs a no, for authentication rejection. Otherwise, $(acc_{\mathsf{VC}}, acc_{\mathsf{HD}}) = (1, 1)$ and $\mathcal{AS}$ outputs yes, for authentication success.

4.1   Security Analysis of the Proposed BAP

Our primary aim is to demonstrate that our privacy-preserving biometric authentication protocol is not vulnerable to Abidin et al.'s template recovery attack [1]. To this end, we sketch the attack setting in Fig. 4.

Input: the client's identity ID, the public key for SHE scheme pk, the public evaluation key for the BFR scheme ek, the stored reference template and tag $(\mathsf{vEnc}_1(A), \sigma_A)$, the encrypted fresh template $\mathsf{vEnc}_2(B)$ and its tag σ_B.

Output: the inner product $ct_P = \mathsf{vEnc}_1(A) \cdot \mathsf{vEnc}_2(A')$ and the tag σ'_{HD}.

Goals: make the authentication server accept the inner product computation and return yes, and use the hill-climbing strategy recover the reference template A.

Fig. 4. Setting for Abidin et al.'s template recovery attack in [1].

We recall that for this attack, the adversary is a malicious computational server who tries to recover the stored reference biometric template of a client with identity ID. All the other parties of the BAP, are trusted and behave honestly.

In what follows, we show that the malicious $\mathcal{CS}$ cannot forge a tag $\sigma_{\mathsf{HD}'}$ that passes the verification checks performed in BFR. It is possible for the adversary to cheat on the first equality (Eq. (3)) as it only tests that the returned computation result (ct_{HD} or ct_P) aligns with the arithmetic circuit used to generate the tag (σ_{HD} or $\sigma_{\mathsf{HD}'}$). In [1], $\mathcal{CS}$ succeeds by computing the arithmetic circuit for the inner product instead of HD. In this case, it is not possible for the malicious computational server to fool the second integrity check (Eq. (4)). In details, $\mathcal{AS}$ calculates $W = \mathbf{GroupEval}(f, R_{\alpha}, R_{\beta})$, and since $\mathcal{AS}$ is honest, $f = \mathsf{HD}$ is the arithmetic circuit for the Hamming distance. If $\mathcal{CS}$ returns incorrect results, with overwhelming probability the second verification equation does not hold, thus the attack is mitigated.

Other Threats. In what follows, we consider attack scenarios in which one of the participating entities in the BAP is malicious.

Malicious Client. $\mathcal{C}$ is responsible to capture the reference template and the fresh template as well as to perform the encryption. If the client is malicious, the knowledge of the encryption secret key and of the identity ID enables $\mathcal{C}$ to initiate a *center search attack* and recover the stored template A as explained in [17]. Unfortunately, Pagnin et al. [17] show that this class of attacks cannot be detected using verifiable computation techniques, since the attacker is not cheating with the computation.

A new concern with the modified Yasuda et al. protocol is the key generation for BFR. In the protocol, we let the client $\mathcal{C}$ generate the private key vk, the evaluation key ek and the authentication tags because we assume $\mathcal{C}$ is a trusted party. If $\mathcal{C}$ turns malicious, it could give a fake vk to the authentication server $\mathcal{AS}$ and initiate the template recovery attack with the inner product by simulating $\mathcal{CS}$. Since the adversary ($\mathcal{C}$) controls vk, the computation verification step becomes meaningless.

Malicious Computation Server. The main motivation to integrate VC in BAPs is indeed to prevent $\mathcal{CS}$ from behaving dishonestly. Unlike the client $\mathcal{C}$, $\mathcal{CS}$ only has access to the encrypted templates $\mathsf{vEnc}_1(A)$ and $\mathsf{vEnc}_2(B)$ and the user pseudonyms. $\mathcal{CS}$ cannot modify the secret key of the BFR scheme. We have analysed how the template recovery attack conducted by $\mathcal{CS}$ can be countered and hence we shorten the discussion here.

In contrast to the original protocol, $\mathcal{CS}$ needs to calculate an extra value ℓ_d to solve the range issue after integrating BFR. However, ℓ_d is still operated on the ciphertext level and is not involved in the second verification equation. Thus learning ℓ_d does not provide any significant advantage in recovering the templates.

Malicious Authentication Server. A malicious $\mathcal{AS}$ will completely break down the privacy of the BAP since it controls the secret key sk used by the SHE scheme. If $\mathcal{AS}$ successfully eavesdrops and obtains the ciphertext $\mathsf{vEnc}_1(A)$ or $\mathsf{vEnc}_2(B)$, it can recover the plaintext biometric templates.

4.2   Efficiency Analysis

The original BFR scheme in [4] allows alternative algorithms to improve the efficiency of the verifier. Although in our instantiation we did not use these algorithms, the current definition of the multi-labels in BFR+SHE is extensible. Given also that the function to be computed is $f = \mathsf{HD}$ and has a very simple description as arithmetic circuit, running the BFR+SHE.**Ver** algorithm requires $O(|f|)$ computational time. In addition, if the amortized closed-form efficiency functionality is adopted, the verification function will run in time $O(1)$. Nonetheless, the arithmetic circuit of HD has 6 gates only and the saved computation overhead would be relatively small.

5   A Flawed Approach

Privacy and integrity are the two significant properties desired in a privacy-preserving BAP. There are two possible ways to combine VC and homomorphic encryption (HE): running VC on top of HE, and viceversa, running HE on top of VC.

In the first case, the data (biometric template) is first encrypted and then encoded to generate an authentication proof. Our construction of BFR+SHE follows this principle. In this approach, $\mathcal{AS}$ can make the judgement whether the output of $\mathcal{CS}$ is from a correct computation of HD *before* decrypting the ciphertext.

In the second case, the data is first encoded for verifiable computation and then the encoded data is encrypted. This combination is not really straightforward and is prune to security breaches.

In this section, we demonstrate an attack strategy that may lead to information leakage in case the homomorphic encryption scheme (henceforth FHE)[1] is applied on top of a VC scheme. For the sake of generality, we define FHE $= (KeyGen_{FHE}, Enc, Dec, Eval)$. For verifiable computation scheme we adopt the notation of Gennaro et al. [11] and define VC $= (KeyGen_{VC}, ProbGen, Compute, Ver)$, where $KeyGen_{VC}$ outputs the private key sk_{vc} and public key pk_{vc}; $ProbGen$ takes sk_{vc} and the plaintext x as input and outputs the encoded value σ_x; $Compute$ takes the circuit f, the encrypted encoded input and outputs the encoded version of the output; Ver is performed to verify the correctness of the computation given the secret key sk_{vc} and the encoded output σ_y. The main idea of the flawed approach is to first encode the data in plaintext and then encrypt the encoded data. It can be represented by $\hat{x} = Enc(ProbGen(x))$, where $\hat{x}$ is what the malicious server gets access to.

5.1 The Attack

We describe now a successful attack strategy to break the privacy-preservation property of a BAP built with the second composition method: HE on top of VC (or HE *after* VC). The adversary's goal is to recover σ_y, i.e., the encoded value of the computation result. The attack runs in different phases. We show that the privacy-preserving property is broken if $q \geq n$, where q is the number of queries in the learning phase and n is the length of the encoded result σ_y. For simplicity we collect the two entities $\mathcal{C}$ and $\mathcal{AS}$ into a single trusted party $\mathcal{V}$ that we refer to as the Verifier.

The attack is depicted in Fig. 5, a more detailed description follows.

Setup phase: $\mathcal{V}$ generates the keys of the protocol and gives pk_{vc}, pk_{FHE} to $\mathcal{A}$.
Challenge phase: $\mathcal{V}$ generates the encoded version σ_x for the input x. $\mathcal{V}$ encrypts the encoded input and sends $Enc(\sigma_x)$ to $\mathcal{A}$.
Learning phase: $\mathcal{A}$ uses $\mathcal{V}$ as a decryption oracle by sending verification queries, which can be further divided into the following steps:
1. $\mathcal{A}$ performs honest computations and derives the $Enc(\sigma_y)$.
2. $\mathcal{A}$ constructs a vector $A' \in \mathbb{Z}_2^n$ equal in length to σ_y. A' is initialized with the last bit set to 0 and the rest of the bits set to 1. For the ith trial, we

---

[1] The same leakage of information could happen if a SHE scheme is used.

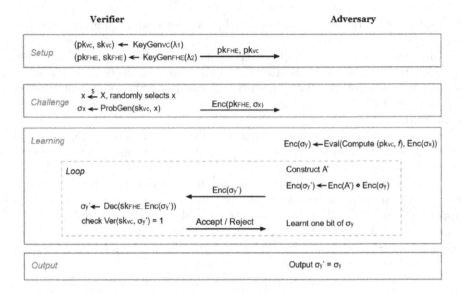

Fig. 5. An attack strategy against the naïve the integration of FHE on top of VC.

    set $A' = (1_1, \ldots, 0_i, 1_{i+1}, \ldots, 1_{n-1}, 1_n)$, i.e., set the ith bit to 0 and the rest bits to 1.

3. $\mathcal{A}$ encrypts the tailored vector A' and reuses the honest result $Enc(\sigma_y)$ from step 1. Then she computes: $Enc(\sigma_{y'}) = Enc(A') \diamond Enc(\sigma_y)$, where $\diamond$ represents the Hadamard product for binary vectors and sends the result to $\mathcal{V}$ for verification.

4. $\mathcal{V}$ decrypts $Enc(\sigma'_y)$. Thanks to the homomorphic property of FHE, $\mathcal{V}$ can derive $\sigma_{y'} = A' \diamond \sigma_y$. $\mathcal{V}$ checks the computation based on the encoded result $\sigma_{y'}$ and returns either *accept* if $Ver(sk, \sigma_{y'}) = 1$ or *reject*, otherwise.

5. The attacker A' acts as a "mask": it copies all the bit values of σ_y into $\sigma_{y'}$ except for the ith bit, which is always set to zero. Consequently, if the output of the verification is *accept*, $\mathcal{A}$ will learn that $\sigma_y = \sigma_{y'}$ as well as $Enc(\sigma_y) = Enc(\sigma_{y'})$, which reveals that the ith bit of σ_y equals to 0. Similarly, if the output of the verification is *reject*, $\mathcal{A}$ learns that the ith bit of σ_y is 1. In both cases, one bit of σ_y is leaked.

Output phase: After $q \geq n$ verification queries, where n equals the length of σ_y, $\mathcal{A}$ outputs $\sigma_{y'}$.

It is trivial to check that that $\sigma_{y'} = \sigma_y$ and thus $Ver(sk, \sigma'_y) = Ver(sk, \sigma'_y) = 1$ and attacker's goal is achieved.

    The attack demonstrates that the order of combining a VC and a (F)HE is very crucial: the verifier must decrypt the ciphertext *before* it can determine whether it is the result of the correct outsourced computation. Adopting such a scheme in a BAP would make $\mathcal{AS}$ a decryption oracle. Leaking information on the Hamming distance may be exploited to perform further attacks that might lead to the full recovery of biometric templates as it has been recently shown [17]. Formally speaking, we can say that the HE on top of VC is not a chosen-ciphertext attack (CCA) secure scheme.

6 Conclusions

Biometric authentication protocols have gained considerable popularity for access control services. Preserving the privacy of the biometric templates is highly critical due to their irrevocable nature. Yasuda et al. proposed a biometric authentication protocol [23] using a SHE scheme. However, a hill-climbing attack [1] has been presented against this protocol that relies on a malicious internal computation server CS that performs erroneous computations and leads to the disclosure of the biometric reference template. We counter the aforementioned attack by constructing a new scheme named BFR+SHE which adds a verifiable computation layer to the SHE scheme. We then describe a modified version of the Yasuda et al. protocol that utilizes our BFR+SHE scheme, and demonstrate that the improved BAP provides higher privacy guarantees. Although employing VC to mitigate hill-climbing attack techniques seems a quite straightforward step, we demonstrate that not all combinations of a VC scheme with a HE one are secure, and show how a naïve combination leads to a drastic private information leakage in BAP.

Acknowledgements. This work was partially supported by the People Programme (Marie Curie Actions) of the European Union's Seventh Framework Programme (FP7/2007-2013) under REA grant agreement no 608743; the VR grant PRECIS no 621-2014-4845 and the STINT grant "Secure, Private & Efficient Healthcare with wearable computing" no IB2015-6001.

A   Details in the Correctness Analysis

In this section, we show the intermediate steps of the calculation.
The derived tags are:

$$\sigma_1 = (C_2 \cdot y_0^{(A)}, (Y_1^{(A)})^{C_2}, 1); \ \sigma_2 = (C_1 \cdot y_0^{(B)}, (Y_1^{(B)})^{C_1}, 1);$$
$$\sigma_3 = (y_0^{(A)} \cdot y_0^{(B)}, (Y_1^{(A)})^{y_0^{(B)}} \cdot (Y_1^{(B)})^{y_0^{(A)}}, e(Y_1^{(A)}, Y_1^{(B)}));$$
$$\sigma_4 = (D \cdot y_0^{(A)} \cdot y_0^{(B)}, (Y_1^{(A)})^{y_0^{(B)} \cdot D} \cdot (Y_1^{(B)})^{y_0^{(A)} \cdot D}, e(Y_1^{(A)}, Y_1^{(B)}))^D;$$
$$\sigma_5 = (C_2 \cdot y_0^{(A)} + C_1 \cdot y_0^{(B)}, (Y_1^{(A)})^{C_2} \cdot (Y_1^{(B)})^{C_1}, 1);$$
$$\sigma_6 = \begin{pmatrix} C_2 \cdot y_0^{(A)} + C_1 \cdot y_0^{(B)} + D \cdot y_0^{(A)} \cdot y_0^{(B)} \\ (Y_1^{(A)})^{y_0^{(B)} \cdot D + C_2} \cdot (Y_1^{(B)})^{y_0^{(A)} \cdot D + C_1} \\ e(Y_1^{(A)}, Y_1^{(B)})^D \end{pmatrix}$$

The homomorphic bilinear map calculation results are:

$$R_1 = R_\alpha^{C_2}; \ R_2 = R_\beta^{C_1}; \ R_3 = e(R_\alpha, R_\beta); \ R_4 = e(R_\alpha, R_\beta)^D;$$
$$R_5 = R_\alpha^{C_2} \cdot R_\beta^{C_1}; \ R_6 = e(R_\alpha^{C_2} \cdot R_\beta^{C_1}, g) \cdot e(R_\alpha, R_\beta)^D.$$

To prove that $W = \mathbf{GroupEval}(f, R_\alpha, R_\beta)$ satisfies Eq. (4), we start by analysing the three factors that made up the righthand of the equation, namely: $e(g,g)^{y_0^{HD}} \cdot e(Y_1^{HD}, g)^\theta \cdot (\hat{Y}_2^{(HD)})^{\theta^2}$. We in turn expand each one of the factors and finally compute the product of the results, evaluating it against W.

The first factor can be expanded as:

$$e(g,g)^{y_0^{HD}} = e(g,g)^{C_2 \cdot y_0^{(A)} + C_1 \cdot y_0^{(B)} + D \cdot y_0^{(A)} \cdot y_0^{(B)}} = e(g,g)^{C_2\alpha + C_1\beta + \alpha\beta D}.$$

The second factor is expanded as:

$$\begin{aligned}
e(Y_1^{HD}, g)^\theta &= e((Y_1^{(A)})^{y_0^{(B)} \cdot D + C_2} \cdot (Y_1^{(B)})^{y_0^{(A)} \cdot D + C_1}, g)^\theta \\
&= e((R_\alpha \cdot g^{-\alpha})^{(\beta D + C_2)/\theta} \cdot (R_\beta \cdot g^{-\beta})^{(\alpha D + C_1)/\theta}, g)^\theta \\
&= e(R_\alpha^{\beta D + C_2} \cdot R_\beta^{\alpha D + C_1} \cdot g^{-2\alpha\beta D - \alpha C_2 - \beta C_1}, g) \\
&= e(R_\alpha, g)^{\beta D + C_2} \cdot e(R_\beta, g)^{\alpha D + C_1} \cdot e(g,g)^{-2\alpha\beta D - \alpha C_2 - \beta C_1}.
\end{aligned}$$

The third factor is expanded as:

$$\begin{aligned}
(\hat{Y}_2^{(HD)})^{\theta^2} &= e(Y_1^{(A)}, Y_1^{(B)})^{D\theta^2} = e((R_\alpha \cdot g^{-\alpha})^{1/\theta}, (R_\beta \cdot g^{-\beta})^{1/\theta})^{D\theta^2} \\
&= e(R_\alpha \cdot g^{-\alpha}, R_\beta \cdot g^{-\beta})^D = e(R_\alpha, R_\beta \cdot g^{-\beta})^D \cdot e(g^{-\alpha}, R_\beta \cdot g^{-\beta})^D \\
&= e(R_\alpha, R_\beta)^D \cdot e(R_\alpha, g)^{-\beta D} \cdot e(R_B, g)^{-\alpha D} \cdot e(g,g)^{\alpha\beta D}.
\end{aligned}$$

Here we need to prove the right hand side is equal to W. We use a temporary variable $P = e(g,g)^{y_0^{HD}} \cdot e(Y_1^{HD}, g)^\theta \cdot (\hat{Y}_2^{(HD)})^{\theta^2}$ to denote the expansion result of the righthand-side. The expression below proves the correctness of the second verification Eq. (4).

$$\begin{aligned}
P &= e(g,g)^{C_2 \cdot \alpha + C_1 \cdot \beta + D \cdot \alpha \cdot \beta} \cdot e(R_\alpha, g)^{\beta D + C_2} \cdot e(R_\beta, g)^{\alpha D + C_1} \cdot e(g,g)^{-2\alpha\beta D - \alpha C_2 - \beta C_1}. \\
&\quad \cdot e(R_\alpha, R_\beta)^D \cdot e(R_\alpha, g)^{-\beta D} \cdot e(R_B, g)^{-\alpha D} \cdot e(g,g)^{\alpha\beta D} \\
&= e(g,g)^0 \cdot e(R_\alpha, g)^{C_2} \cdot e(R_\beta, g)^{C_1} \cdot e(R_a, R_b)^D \\
&= e(R_\alpha^{C_2}, g) \cdot e(R_\beta^{C_1}, g) \cdot e(R_a, R_b)^D \\
&= e(R_\alpha^{C_2} \cdot R_\beta^{C_1}, g) \cdot e(R_a, R_b)^D = W.
\end{aligned}$$

References

1. Abidin, A., Mitrokotsa, A.: Security aspects of privacy-preserving biometric authentication based on ideal lattices and ring-LWE. In: Proceedings of the IEEE Workshop on Information Forensics and Security 2014 (WIFS 2014) (2014)
2. Babai, L.: Trading group theory for randomness. In: Proceedings of STOC 1985, pp. 421–429. ACM, New York (1985)
3. Backes, M., Barbosa, M., Fiore, D., Reischuk, R.M.: ADSNARK: nearly practical and privacy-preserving proofs on authenticated data. In: Proceedings of the 36th IEEE Symposium on Security and Privacy (Oakland) (2015)

4. Backes, M., Fiore, D., Reischuk, R.M.: Verifiable delegation of computation on outsourced data. In: Proceedings of the 2013 ACM SIGSAC Conference on Computer and Communications Security, pp. 863–874. ACM (2013)

5. Barbosa, M., Brouard, T., Cauchie, S., de Sousa, S.M.: Secure biometric authentication with improved accuracy. In: Mu, Y., Susilo, W., Seberry, J. (eds.) ACISP 2008. LNCS, vol. 5107, pp. 21–36. Springer, Heidelberg (2008). https://doi.org/10.1007/978-3-540-70500-0_3

6. Bringer, J., Chabanne, H., Kraïem, F., Lescuyer, R., Soria-Vázquez, E.: Some applications of verifiable computation to biometric verification. In: 2015 IEEE International Workshop on Information Forensics and Security (WIFS), pp. 1–6. IEEE (2015)

7. Bringer, J., Chabanne, H., Patey, A.: Privacy-preserving biometric identification using secure multiparty computation: an overview and recent trends. IEEE Sig. Process. Mag. 30(2), 42–52 (2013)

8. Bringer, J., Chabanne, H., Patey, A.: SHADE: Secure HAmming DistancE computation from oblivious transfer. In: Adams, A.A., Brenner, M., Smith, M. (eds.) FC 2013. LNCS, vol. 7862, pp. 164–176. Springer, Heidelberg (2013). https://doi.org/10.1007/978-3-642-41320-9_11

9. Costello, C., et al.: Geppetto: versatile verifiable computation. In: 2015 IEEE Symposium on Security and Privacy, pp. 253–270. IEEE (2015)

10. Fiore, D., Gennaro, R., Pastro, V.: Efficiently verifiable computation on encrypted data. In: Proceedings of the 2014 ACM SIGSAC Conference on Computer and Communications Security, pp. 844–855. ACM (2014)

11. Gennaro, R., Gentry, C., Parno, B.: Non-interactive verifiable computing: outsourcing computation to untrusted workers. In: Rabin, T. (ed.) CRYPTO 2010. LNCS, vol. 6223, pp. 465–482. Springer, Heidelberg (2010). https://doi.org/10.1007/978-3-642-14623-7_25

12. Gentry, C.: A fully homomorphic encryption scheme. Ph.D. thesis, Stanford University (2009)

13. Goldwasser, S., Micali, S., Rackoff, C.: The knowledge complexity of interactive proof systems. SIAM J. Comput. 18(1), 186–208 (1989)

14. Kolesnikov, V., Sadeghi, A.-R., Schneider, T.: Improved garbled circuit building blocks and applications to auctions and computing minima. In: Garay, J.A., Miyaji, A., Otsuka, A. (eds.) CANS 2009. LNCS, vol. 5888, pp. 1–20. Springer, Heidelberg (2009). https://doi.org/10.1007/978-3-642-10433-6_1

15. Naehrig, M., Lauter, K., Vaikuntanathan, V.: Can homomorphic encryption be practical? In: Proceedings of the 3rd ACM Workshop on Cloud Computing Security Workshop, pp. 113–124. ACM (2011)

16. Pagnin, E.: Authentication under Constraints. Licentiate dissertation, Chalmers University of Technology (2016)

17. Pagnin, E., Dimitrakakis, C., Abidin, A., Mitrokotsa, A.: On the leakage of information in biometric authentication. In: Meier, W., Mukhopadhyay, D. (eds.) INDOCRYPT 2014. LNCS, vol. 8885, pp. 265–280. Springer, Cham (2014). https://doi.org/10.1007/978-3-319-13039-2_16

18. Pagnin, E., Mitrokotsa, A.: Privacy-preserving biometric authentication: challenges and directions. IACR Cryptology ePrint Archive 2017:450 (2017)

19. Parno, B., Howell, J., Gentry, C., Raykova, M.: Pinocchio: nearly practical verifiable computation. In: Proceedings of the 2013 IEEE Symposium on Security and Privacy, pages 238–252. IEEE Computer Society, Washington (2013)

20. Simoens, K., Bringer, J., Chabanne, H., Seys, S.: A framework for analyzing template security and privacy in biometric authentication systems. IEEE Trans. Inf. Forensics Secur. **7**(2), 833–841 (2012)
21. Simoens, K.: A framework for analyzing template security and privacy in biometric authentication systems. IEEE Trans. Inf. Forensics Secur. **7**(2), 833–841 (2012)
22. Stoianov, A.: Cryptographically secure biometrics. In: SPIE 7667, Biometric Technology for Human Identification VII, p. 76670C–12 (2010)
23. Yasuda, M., Shimoyama, T., Kogure, J., Yokoyama, K., Koshiba, T.: Practical packing method in somewhat homomorphic encryption. In: Garcia-Alfaro, J., Lioudakis, G., Cuppens-Boulahia, N., Foley, S., Fitzgerald, W.M. (eds.) DPM/SETOP -2013. LNCS, vol. 8247, pp. 34–50. Springer, Heidelberg (2014). https://doi.org/10.1007/978-3-642-54568-9_3

Towards Attribute-Based Credentials in the Cloud

Stephan Krenn[(⊠)], Thomas Lorünser, Anja Salzer, and Christoph Striecks

AIT Austrian Institute of Technology GmbH, Vienna, Austria
{stephan.krenn,thomas.loruenser,christoph.striecks}@ait.ac.at,
anjasalzer3@gmail.com

Abstract. Attribute-based credentials (ABCs, sometimes also anonymous credentials) are a core cryptographic building block of privacy-friendly authentication systems, allowing users to obtain credentials on attributes and prove possession of these credentials in an unlinkable fashion. Thereby, users have full control over which attributes the user wants to reveal to a third party while offering high authenticity guarantees to the receiver. Unfortunately, up to date, all known ABC systems require access to all attributes in the clear at the time of proving possession of a credential to a third party. This makes it hard to offer privacy-preserving identity management systems "as a service," as the user still needs specific key material and/or dedicated software locally, e.g., on his device.

We address this gap by proposing a new cloud-based ABC system where a dedicated cloud service ("wallet") can present the users' credentials to a third-party *without* accessing the attributes in the clear. This enables new privacy-preserving applications of ABCs "in the cloud."

This is achieved by carefully integrating proxy re-encryption with structure-preserving signatures and zero-knowledge proofs of knowledge. The user obtains credentials on his attributes (encrypted under his public key) and uploads them to the wallet, together with a specific re-encryption key. To prove a possession, the wallet re-encrypts the ciphertexts to the public key of the receiving third party and proves, in zero-knowledge, that all computations were done honestly. Thereby, the wallet never sees any user attribute in the clear.

We show the practical efficiency of our scheme by giving concrete benchmarks of a prototype implementation.

Keywords: Attribute-based credentials
Privacy-preserving authentication · Proxy re-encryption
Implementation

The project leading to this publication has received funding from the European Unions Horizon 2020 research and innovation programme under grant agreement No 653454 (CREDENTIAL).

© Springer Nature Switzerland AG 2018
S. Capkun and S. S. M. Chow (Eds.): CANS 2017, LNCS 11261, pp. 179–202, 2018.
https://doi.org/10.1007/978-3-030-02641-7_9

1   Introduction

Privacy-enhancing attribute-based credential systems—originally envisioned by
Chaum [15,16] and, among others, refined by Brands et al. [7,24] and Camenisch
and Lysyanskaya et al. [8,9,11–13]—are a cryptographic primitive enabling user-
centric identity management. They allow for strong user authentication while
still respecting the user's privacy and giving him full control over the data that
is revealed. In a basic ABC system, a *user* receives a credential from an *issuer*,
certifying certain information such as name or date of birth. Later, when the user
wants to authenticate towards a *service provider*, he can decide which information to reveal or to keep private. This can be done in a way that guarantees to
the user that no two actions can be linked to each other even by colluding service
providers and issuers. The importance of such data minimization efforts has been
recognized by the European Commission [17,18] and the US government [26].

In recent years, services have been experiencing a massive trend towards
cloudification. That is, services should be available ubiquitously, preferably without requiring any dedicated software on the end user's side. However, in all
existing ABC schemes such as [5,7–9,11–13,20,24], presentation of a credential
requires access to the plaintext attributes that have been certified by the issuer.
Thus, a full cloudification would directly subvert the initial motivation of such
systems which aim at protecting the users' privacy. One would therefore need
key material on the user's side and perform (parts of) the computation locally[1]
in order to not leak the attribute values to the cloud provider.

However, this may not be a satisfying solution in different scenarios. From a
computational point of view, just consider a low-cost embedded device such as a
smart card or a sensor that shall be used to authenticate a user or a device to a
service. In this case, it might not be possible to perform expensive computations
(such as pairings or full-length exponentiations, potentially in RSA groups) on
the device with acceptably low delay. Furthermore, from a trust point of view,
it may also not desirable to store key material on the end device that is needed
for authentication, because the device is not fully under the user's control and
therefore not fully trustworthy. For instance, employees may not wish to store
sensitive key material on a company phone which they are also allowed to use
privately. Or as another example, consider restricted access areas which only
residents are allowed to enter: in this case, users might not wish to store their
secret credentials on a rental car.

In all these cases, a fully cloud-based authentication process still maintaining
the user's privacy would be preferable due to usability and trust reasons.

Contribution. In this paper, we propose a solution to the problem sketched
before. That is, we consider a setting where the user obtains credentials from
an issuer, and stores those credentials—potentially together with some auxiliary
data—at some central *wallet*. Authenticating towards a service provider is then

---

[1] Confer, e.g., the experimental service of identity mixer, https://console.ng.bluemix.
net/docs/services/identitymixer/index.html.

done through this central wallet in a way that does not allow the wallet to learn any information about the attributes that are stored and potentially revealed to the service provider.

This seemingly contradictory goal is realized by usage of proxy re-encryption, which allows a semi-trusted party (the *proxy*) to convert a ciphertext encrypted under public key pk_A of party A to a ciphertext encrypted under pk_B of party B without learning any information about the plaintext data whenever A handed a re-encryption key $rk_{A \to B}$ to the proxy. This re-encryption key is used by the wallet to compute presentations without requiring sensitive information on the user's side for this step.

Our construction allows for re-using credentials, i.e., credentials can be used for arbitrarily many presentations without becoming linkable, similar to, e.g., identity mixer [9,11–13]. The proposed scheme is based on Blaze et al.'s [6] proxy re-encryption scheme and Abe et al.'s [1] structure-preserving signature scheme. For the security analysis, we assume a semi-trusted wallet, i.e., we assume that the wallet does not collude with service providers or issuers.[2] We provide formal security definitions inspired by Camenisch et al. [10] with rigorous computational-security proofs.

To enrich our basic scheme with advanced functionalities, we explain how it can be used to support also features such as inspection (i.e., anonymity revocation in case of abuse), revocation (by the issuer, user, or service provider), or multi-credential presentations (e.g., for proving possession of an eID and a valid service subscription by the same user). Furthermore, our scheme guarantees that any outsider (including issuers or service providers) cannot profile users, as from their point of view all actions are unlinkable. However, the wallet learns when a user is authenticating to which service provider, which may leak sensitive metadata even though the wallet does not learn the revealed attributes. We therefore present a straightforward yet practical solution which disables the wallet from learning which service provider is currently accessed, but only leaks that the user is authenticating to *some* service, thereby significantly reducing the level of leaked information.

Finally, we give concrete benchmarks showing the practical efficiency of our main construction.

Related Work. Anonymous credential systems were first envisioned by Chaum [15,16]. Over the last two decades, a large body of protocols and instantiations have been developed, with the two most prominent being Microsoft's U-Prove based on Brands' signatures [7,24] and IBM's identity mixer based on CL-signatures [9,11–13]; other work includes Hanser and Slamanig [20], Camenisch et al. [8], or Belenkiy et al. [5]. All this work has in common that presenting a credential requires access to the plaintext attributes of a user. The most com-

---

[2] Note that is a natural and unavoidable assumption as issuers and service providers are intended to learn (parts of) the attributes, opening a trivial way for the wallet to learn attributes in the case of collusion.

prehensive definitional framework for ABCs has been presented by Camenisch et al. [10].

Proxy re-encryption has been introduced by Blaze et al. [6] and has been a vivid area of research since then, resulting in schemes with different features, security properties, and underlying complexity assumptions, e.g., [3,4,21,23].

Recently, Sabouri [25] suggested a mobile and install-free architecture based proxy re-encryption, without providing a formal security analysis of the construction. Also, this architecture requires the usage of smart cards, which impede the real-world adoption of the system.

Outline. This document is structured as follows. In Sect. 2, we recap the cryptographic background needed for the rest of the paper. Section 3 contains the formal syntax and security model. A generic high-level construction is presented in Sect. 4 and instantiated in Sect. 5. We provide benchmarks of a prototype implementation in Sect. 7, and then briefly conclude in Sect. 8.

2   Preliminaries

Notation. We declare $\lambda \in \mathbb{N}$ to be the security parameter and for $n \in \mathbb{N}$, let $[n] := \{1, \ldots, n\}$. We denote algorithms by sans-serif letters (A, B, ...) and sets by calligraphic letters ($\mathcal{R}, \mathcal{S}, \ldots$). For a finite set $\mathcal{S}$, $s \xleftarrow{\$} \mathcal{S}$ denotes the process of sampling s uniformly from $\mathcal{S}$. For an algorithm A, $y \leftarrow \mathsf{A}(\lambda, x)$ denotes the process of running A, on input λ and x, with access to uniformly random coins and assigning the result to y. We assume that all used algorithms take λ as input and we will sometimes not make this explicit. An algorithm A is probabilistic polynomial time (PPT) if its running time is polynomially bounded in λ. A function f is negligible if $\forall c \exists \lambda_0 \forall \lambda \geq \lambda_0 : |f(\lambda)| \leq 1/\lambda^c$.

Pairings, DDH Assumption, Commitments, and ElGamal Encryption. We provide definitions of these well-known preliminaries in the full version.

Proxy Re-encryption (With Additional Properties). A bidirectional Proxy Re-Encryption (PRE) scheme PRE with message space $\mathcal{M}$ consists of the PPT algorithms (Par, Gen, Enc, ReKey, ReEnc, Dec). $\mathsf{Par}(\lambda)$, on input the security parameter λ, outputs system parameter pp. $\mathsf{Gen}(pp)$, on input pp, outputs a public and secret key pair (pk, sk). $\mathsf{Enc}(pk, M)$, on input pk and a message $M \in \mathcal{M}$, outputs a ciphertext C. $\mathsf{ReKey}(pk, sk, pk', sk')$, on input the public-secret-key pairs (pk, sk, pk', sk'), outputs a re-encryption key rk. $\mathsf{ReEnc}(rk, C)$, on input rk and C, outputs a re-encrypted ciphertext C' or $\bot$. $\mathsf{Dec}(sk, C)$, on input sk and C, outputs $M \in \mathcal{M} \cup \{\bot\}$.

For correctness, we require that for all $\lambda \in \mathbb{N}$, for all $(pk, sk) \leftarrow \mathsf{Gen}(\lambda)$, for all $M \in \mathcal{M}$, for all $C \leftarrow \mathsf{Enc}(pk, M)$, for all $(pk', sk') \leftarrow \mathsf{Gen}(\lambda)$, for all $rk \leftarrow \mathsf{ReKey}(pk, sk, pk', sk')$, for all $C' \leftarrow \mathsf{ReEnc}(rk, C)$, we have that $\mathsf{Dec}(sk, C) = M$ and $\mathsf{Dec}(sk', C') = M$.

Definition 2.1 (PRE-IND-CPA). *We say a PRE scheme* PRE *is PRE-IND-CPA-secure if for any PPT adversary* A, *the advantage function*

$$\mathsf{Adv}_{\mathsf{PRE,A}}^{\mathsf{pre-ind-cpa}}(\lambda) := \left| \Pr\left[\mathsf{Exp}_{\mathsf{PRE,A}}^{\mathsf{pre-ind-cpa}}(\lambda) = 1\right] - \frac{1}{2} \right|$$

is negligible in λ, *where* $\mathsf{Exp}_{\mathsf{PRE,A}}^{\mathsf{pre-ind-cpa}}$ *is defined as:*

Experiment $\mathsf{Exp}_{\mathsf{PRE,A}}^{\mathsf{pre-ind-cpa}}(\lambda)$
$pp \leftarrow \mathsf{Par}(\lambda), (pk, sk) \leftarrow \mathsf{Gen}(pp)$
$(\ell, \mathsf{st}) \leftarrow \mathsf{A}(pp), (pk'_i, sk'_i) \leftarrow \mathsf{Gen}(pp)$, for all $i \in [\ell]$
$rk'_i \leftarrow \mathsf{ReKey}(pk, sk, pk'_i, sk'_i)$, for all $i \in [\ell]$
$(M_0, M_1, \mathsf{st}) \leftarrow \mathsf{A}(\mathsf{st}, pk, (pk'_i, rk'_i)_{i \in [\ell]})$
$b \xleftarrow{\$} \{0, 1\}, b^* \leftarrow \mathsf{A}(\mathsf{st}, \mathsf{Enc}(pk, M_b))$
if $b = b^*$ and $|M_0| = |M_1|$ then return 1, else return 0

Definition 2.2 (Anonymous PRE). *We say a PRE scheme* PRE *is anonymous if for any PPT adversary* A, *the advantage function* $\mathsf{Adv}_{\mathsf{PRE,A}}^{\mathsf{pre-anon}}(\lambda) :=$ $\left|\Pr\left[\mathsf{Exp}_{\mathsf{PRE,A}}^{\mathsf{pre-anon}}(\lambda) = 1\right] - \frac{1}{2}\right|$ *is negligible in* λ, *where* $\mathsf{Exp}_{\mathsf{PRE,A}}^{\mathsf{pre-anon}}$ *is given as:*

Experiment $\mathsf{Exp}_{\mathsf{PRE,A}}^{\mathsf{pre-anon}}(\lambda)$
$pp \leftarrow \mathsf{Par}(\lambda), (pk_0, sk_1) \leftarrow \mathsf{Gen}(pp), (pk_1, sk_1) \leftarrow \mathsf{Gen}(pp)$
$(M, \mathsf{st}) \leftarrow \mathsf{A}(pk_0, pk_1)$
$b \xleftarrow{\$} \{0, 1\}, b^* \leftarrow \mathsf{A}(\mathsf{st}, \mathsf{Enc}(pk_b, M))$
if $b = b^*$ then return 1, else return 0

Definition 2.3 (Re-randomization of PRE ciphertexts). *A PRE scheme* PRE *has the ciphertext re-randomization property if there exists a PPT algorithm* ReRand(pk, C) *which, on input of a PRE public key* pk *and ciphertext* C *under* pk, *outputs a consistent re-randomized ciphertext* $\hat{C}$ *under* pk *such that* $\hat{C}$ *is computationally indistinguishable from a uniformly distributed ciphertext in the image of* Enc. *More formally, we require* $\mathsf{Dist}[C] \stackrel{c}{\approx} \mathsf{Dist}[\hat{C}]$, *for any honestly generated* pk, *any* $M \in \mathcal{M}$, *any* $C \leftarrow \mathsf{Enc}(pk, M)$, *and any* $\hat{C} \leftarrow \mathsf{ReRand}(pk, C)$.

Definition 2.4 (Re-encryption key verification). *A PRE scheme* PRE *has the re-encryption-key verifiability property if there is a PPT algorithm* RKVerify(pk, pk', sk', rk), *on input public keys* (pk, pk'), *secret key* sk', *and a re-encryption key* rk, *outputs a verdict (indicating whether* rk *is a valid re-encryption key from* pk *to* pk'). *We require that for all honestly generated* pk, sk, pk', sk' *and* $rk \leftarrow \mathsf{ReKey}(pk, sk, pk', sk')$, $\mathsf{RKVerify}(pk, pk', sk', rk) = 1$.

Definition 2.5 (Verify honestly generated PRE public keys). *A PRE scheme* PRE *has the property of verifying honestly generated public keys if there exists a PPT algorithm* KVerify(pk, sk), *on input public and secret keys* (pk, sk), *outputs a verdict (indicating whether pk and sk are honestly generated). We require that for all $\lambda \in \mathbb{N}$, all $pp \leftarrow \mathsf{Par}(\lambda)$, all $(pk, sk) \leftarrow \mathsf{Gen}(pp)$ it holds that* $\mathsf{KVerify}(pk, sk) = 1$.

The above Definition 2.5 is a standard assumption to avoid rogue-key attacks and can be realized by appending a zero-knowledge proof (see below) to pk, showing that the keys are indeed a valid output of the key generation Gen.

The BBS PRE Scheme. In the following, we recap the PRE scheme of Blaze, Bleumer, and Strauss (BBS) [6], which we will use within our construction and can be seen as an extension of the ElGamal scheme to the PRE setting. We formally argue that their scheme is correct, PRE-IND-CPA secure, and exhibits the anonymous and re-randomization properties. (We only give claims here and prove them in the full version of the paper due to space constraints.) Further, the BBS scheme naturally satisfies Definition 2.5 as shown below.

Let $\mathsf{PRE}_{BBS} := (\mathsf{Par}, \mathsf{Gen}, \mathsf{Enc}, \mathsf{ReKey}, \mathsf{ReEnc}, \mathsf{Dec})$ be a PRE scheme as follows. $\mathsf{Par}(\lambda)$ outputs system parameter $pp = (g, q)$ and sets $\mathcal{M} = \mathcal{G}_1$. (We assume that $(\mathcal{G}_1, g, q) \leftarrow \mathsf{G}(\lambda)$ are output by some group generator as defined in the full version of the paper.) $\mathsf{Gen}(pp)$ outputs a public and secret key pair $(pk, sk) := ((pp, g^x), x)$, for $x \xleftarrow{\$} \mathbb{Z}_q$. $\mathsf{Enc}(pk, M)$, for $(pp, g^x) := pk$, outputs a ciphertext $C := ((g^x)^y, g^y \cdot M)$, for $y \xleftarrow{\$} \mathbb{Z}_q$. $\mathsf{ReKey}(pk, sk, pk', sk')$ outputs a re-encryption key $rk = sk'/sk$. $\mathsf{ReEnc}(rk, C)$, for $(C_1, C_2) := C$, outputs a re-encrypted ciphertext $C' := (C_1^{rk}, C_2)$. $\mathsf{Dec}(sk, C)$, for $(C_1, C_2) := C$, outputs $M := C_2/C_1^{1/sk}$.

Correctness follows as $\mathsf{Dec}(sk, C) = M = C_2/C_1^{1/sk} = g^y \cdot M/(g^{yx})^{1/x}$. Note that this also holds for re-encrypted ciphertexts C' since re-encryption implicitly "changes" the public key (and only the first part) of the ciphertext, i.e., $C' = ((g^x)^{x'/x}, g^y \cdot M)$ which can be decrypted using the corresponding secret key x'.

Claim (PRE-IND-CPA security of PRE_{BBS}). Under the DDH assumption, PRE_{BBS} is PRE-IND-CPA-secure for any PPT adversaries A and B with $\mathsf{Adv}_{\mathsf{PRE}_{BBS}, A}^{pre-ind-cpa}(\lambda) = \mathsf{Adv}_{G, A'}^{ddh}(\lambda)$.

Claim (Anonymity of PRE_{BBS}). Under the DDH assumption, PRE_{BBS} is anonymous for any PPT adversaries A and B with $\mathsf{Adv}_{\mathsf{PRE}_{BBS}, A}^{pre-anon}(\lambda) = \mathsf{Adv}_{G, A'}^{ddh}(\lambda)$.

Claim (Re-randomizable ciphertext of PRE_{BBS}). The ciphertexts in PRE_{BBS} are re-randomizable.

Regarding $\mathsf{RKVerify}$, it is easy to see that the requested guarantees in Definition 2.5 can be obtained by checking whether $pk^{rk} = pk'$.

Definition 2.6 (Structure-preserving signatures). *A signature scheme SIG with message space $\mathcal{M}$ consists of the four PPT algorithms* $(\mathsf{Par}, \mathsf{Gen}, \mathsf{Sig}, \mathsf{Ver})$ *as follows. $\mathsf{Par}(\lambda)$, on input the security parameter λ, outputs public parameter pp. $\mathsf{Gen}(pp)$, on input pp, outputs a signing and verification key pair (sk, vk).*

Sig(sk, M), *on input sk and a message $M \in \mathcal{M}$, outputs a signature σ on M.*
Ver(vk, M, σ), *on input vk, M, and σ, outputs a verdict $b \in \{0, 1\}$ (i.e., it outputs
1 if the signature σ is valid on M, and 0 otherwise).*

*For correctness, we require for all $\lambda \in \mathbb{N}$, for all $pp \leftarrow$ Par(λ), for all
$(sk, vk) \leftarrow$ Gen(pp), for $M \in \mathcal{M}$, for all $\sigma \leftarrow$ Sig(sk, M), that* Ver(vk, M, σ) = 1
holds. We say, a signature scheme SIG *is structure-preserving if the message
space, the verification key, and the signatures are group elements, and the verification equations use pairing-product equations in the sense of [1]. We prove the
security (i.e., strong EUF-CMA security) in the full version of the paper.*

The AGHO SPS Scheme. The structure-preserving signatures (SPS) scheme
of Abe, Groth, Haralambiev, and Ohkubo (AGHO) is defined (and restricted
to our needs) as follows. Let $\mathsf{SIG}_{AGHO} = (\mathsf{Par}, \mathsf{Gen}, \mathsf{Sig}, \mathsf{Ver})$ be a SPS scheme.
The message space $\mathcal{M}$ and the public parameter pp (as output by Par) are
given by $\mathcal{G}_1^l$ and $(\mathcal{G}_1, \mathcal{G}_2, \mathcal{G}_{\mathrm{T}}, \mathrm{e}, g, h, q, l)$, for q-prime-order groups $\mathcal{G}_1, \mathcal{G}_2, \mathcal{G}_{\mathrm{T}}$, pairing $\mathrm{e} : \mathcal{G}_1 \times \mathcal{G}_2 \to \mathcal{G}_{\mathrm{T}}$, generators $g \in \mathcal{G}_1, h \in \mathcal{G}_2$, and $l \in \mathbb{N}$, respectively.
Gen(λ) samples $w_1, \ldots, w_l, v, z \xleftarrow{\$} \mathbb{Z}_q$ and outputs $vk := (h^{w_1}, \ldots, h^{w_l}, h^v, h^z)$
and $sk := (w_1, \ldots, w_l, v, z)$. Sig($sk, M_1, \ldots, M_l$) samples $r \xleftarrow{\$} \mathbb{Z}_q^*$ and outputs
$\sigma := (R, S, T) := (g^r, g^{z-rv} \prod_i^l M_i^{-w_i}, h^{1/r})$. Ver($vk, M_1, \ldots, M_l, \sigma$), for $vk = (W_1, \ldots, W_l, V, Z)$ and $\sigma = (R, S, T)$, outputs 1 if $\mathrm{e}(R, V)\,\mathrm{e}(S, h) \prod_i^l \mathrm{e}(M_i, W_i) = \mathrm{e}(g, Z)$ and $\mathrm{e}(R, T) = \mathrm{e}(g, h)$ is satisfied, otherwise 0.

The correctness is easy is verify when one considers the exponent equations,
i.e., $rv + (z - rv) - (w_1 + \cdots + w_l) + (w_1 + \cdots + w_l) = z$ and $r/r = 1$. Abe et
al. proved (strong) EUF-CMA-security in the generic group model [1]:

Corollary 2.7 ([1]). SIG_{AGHO} *is strongly EUF-CMA-secure in the generic
group model. (We give the strong EUF-CMA security model in the full version
of the paper.)*

Zero-Knowledge Proofs of Knowledge. Zero-knowledge proofs of knowledge allow one party (the *prover*) to convince another party (the *verifier*) that
he knows some secret piece of information without revealing anything about the
secret except for what has already been revealed by the claim itself.

Definition 2.8 (ZKP). *A zero-knowledge proof (ZKP) system* ZKP *for a language $\mathcal{L}$ (with PPT witness relation R) consists of PPT algorithms* (Gen, Prove,
Verify). Gen(λ), *on input security parameter λ, outputs a common reference
string (CRS) crs.* Prove(crs, x, w), *on input crs, a word $x \in \mathcal{L}$, and a witness w
for x, outputs a proof P.* Verify(crs, x, P), *on input crs, $x \in \mathcal{L}$, and P, outputs
a verdict $b \in \{0, 1\}$ (i.e., it outputs 1 if the proof P is valid on x under crs, and
0 otherwise).*

*For correctness, we require that for all $\lambda \in \mathbb{N}$, all crs $\leftarrow$ Gen(λ), all $x \in \mathcal{L}$,
all $P \leftarrow$ Prove(crs, x, w), it holds that* Verify(crs, x, P) = 1. *We give the zero-
knowledge and (simulation-)soundness properties in the full version.*

In this paper, we use the standard notation introduced by Camenisch and Stadler [14] to specify proof goals. In particular, an expression like $\mathsf{ZKP}\big[(\alpha, \beta, \gamma) : y_1 = g^\alpha h^\beta \wedge y_2 = y_1^\gamma h^\beta\big]$ specifies a zero-knowledge proof of knowledge proving knowledge of values α, β, γ such that the expression on the right-hand side is satisfied. We follow the convention that knowledge of variables denoted by Greek letters is to be proven, while all other values are supposed to be publicly known. (For more details on zero-knowledge proofs, we refer to the full version of the paper.)

3   Encrypted Attribute-Based Credentials

In this section, we introduce and give formal definitions for Encrypted Attribute-Based Credentials (EABCs).

Participants and Attack Model of EABCs. Within EABCs, we consider four (types of) participants: users, issuers, a (central) wallet, and service providers. Each participant except the wallet is in possession of its own public and secret key material, i.e., $(\mathsf{pk_U}, \mathsf{sk_U}), (\mathsf{pk_I}, \mathsf{sk_I}), (\mathsf{pk_S}, \mathsf{sk_S})$, respectively, which are the output of the EABC key generation algorithm. The user possesses attributes $(a_i)_i$ (e.g., name or date of birth) and engages in an issuance protocol with an issuer (e.g., to certify the name or date of birth). At the end of each invocation of the issuance protocol, the user outputs a credential C on the attributes $(a_i)_i$. The intention of the user (associated to $\mathsf{pk_U}$) is to share some of the attributes with a service provider (associated to $\mathsf{pk_S}$) via the wallet in a selectively disclosed and authenticated manner. Therefore, the wallet holds the user-provided credential C on the attributes $(a_i)_i$ and specific transformation information $t_{\mathsf{U} \to \mathsf{S}}$. (We stress that the wallet is not able to learn any attributes in the plain at any time.) The wallet and the service provider engage in a presentation protocol to disclose specific user attributes $(a_i')_i$ (which are a subset of $(a_i)_i$) from a user's credential C to the service provider.

We need that the wallet and the service providers or issuers are not allowed to collude, i.e., both do not share any (secret) material which is only intended to be received by only one of them. Furthermore, we assume that the wallet is semi-trusted in the sense that it follows the protocol specifications. Otherwise, the adversary may behave arbitrarily malicious.

Syntax of EABCs. An EABC system EABC with attribute space $\mathcal{A}$ consists of the PPT algorithms (Par, $\mathsf{Gen_I}$, $\mathsf{Gen_U}$, $\mathsf{Gen_S}$, Trans) and (User, Issuer, Wallet, Service) specified in the following.

Parameter, key, and transformation generation. $\mathsf{Par}(\lambda)$, on input security parameter λ, outputs system-wide parameter sp. $\mathsf{Gen_I}(\mathsf{sp}), \mathsf{Gen_U}(\mathsf{sp})$, and $\mathsf{Gen_S}(\mathsf{sp})$, on input sp, outputs public and secret key pairs $(\mathsf{pk_I}, \mathsf{sk_I}), (\mathsf{pk_U}, \mathsf{sk_U})$, and $(\mathsf{pk_S}, \mathsf{sk_S})$, respectively. $\mathsf{Trans}(\mathsf{pk_U}, \mathsf{sk_U}, \mathsf{pk_S}, \mathsf{sk_S})$, on input public and secret key pairs $(\mathsf{pk_U}, \mathsf{sk_U}, \mathsf{pk_S}, \mathsf{sk_S})$, outputs a transformation information $t_{\mathsf{U} \to \mathsf{S}}$.

Issuance protocol. The issuance protocol

$$\langle \mathsf{User}[\mathsf{pk_U}, \mathsf{sk_U}, \mathsf{pk_I}, (a_i)_{i \in [\ell]}, D], \mathsf{Issuer}[\mathsf{pk_I}, \mathsf{sk_I}, (a_i)_{i \in D}, D]\rangle,$$

for inputs the public and secret keys $(\mathsf{pk_I}, \mathsf{sk_I}, \mathsf{pk_U}, \mathsf{sk_U})$, a list of attributes $(a_i)_{i \in [\ell]}$, with $\ell \leq |\mathcal{A}|$, and index set D, outputs a credential C.

Presentation protocol. The presentation protocol

$$\langle \mathsf{Wallet}[C, \mathsf{pk_U}, \mathsf{pk_S}, \mathsf{pk_I}, t_{U \to S}, R], \mathsf{Service}[\mathsf{pk_S}, \mathsf{sk_S}, \mathsf{pk_I}, R]\rangle,$$

for inputs credential C as well as public and secret keys $\mathsf{pk_U}, \mathsf{sk_U}, \mathsf{pk_S}, \mathsf{sk_S}, \mathsf{pk_I}$, transformation information $t_{U \to S}$, and index set R, outputs a list of attributes $(a)_{i \in R}$ or $\bot$.

We follow the correctness and security definitions from Camenisch et al. [10] aligned to our features and setting. For correctness, we require for all $\lambda \in \mathbb{N}$, all $\mathsf{sp} \leftarrow \mathsf{Par}(\lambda)$, all $(\mathsf{pk_I}, \mathsf{sk_I}) \leftarrow \mathsf{Gen_I}(\mathsf{sp})$, all $(\mathsf{pk_U}, \mathsf{sk_U}) \leftarrow \mathsf{Gen_U}(\mathsf{sp})$, all $(\mathsf{pk_S}, \mathsf{sk_S}) \leftarrow \mathsf{Gen_S}(\mathsf{sp})$, all $t_{U \to S} \leftarrow \mathsf{Trans}(\mathsf{pk_U}, \mathsf{sk_U}, \mathsf{pk_S}, \mathsf{sk_S})$, all $\langle C, \bot \rangle \leftarrow \langle \mathsf{User}[\mathsf{pk_U}, \mathsf{sk_U}, \mathsf{pk_I}, (a_i)_{i \in [\ell]}, D], \mathsf{Issuer}[\mathsf{pk_I}, \mathsf{sk_I}, (a_i)_{i \in D}, D]\rangle$, that $\langle \bot, (a_i')_{i \in R}\rangle = \langle \mathsf{Wallet}[C, \mathsf{pk_U}, \mathsf{pk_S}, \mathsf{pk_I}, t_{U \to S}, R], \mathsf{Service}[\mathsf{pk_S}, \mathsf{sk_S}, \mathsf{pk_I}, R]\rangle$ for $(a_i')_{i \in R} = (a_i)_{i \in D}$ holds.

Unforgeability for EABC. A EABC scheme EABC is unforgeable if any PPT adversary A succeeds in the following experiment only with negligible probability. The experiment generates key material for an honest issuer $(\mathsf{pk_I^*}, \mathsf{sk_I^*})$ and an honest service provider id_S^* as $(\mathsf{pk_S}^*, \mathsf{sk_S}^*)$ and hands the public keys to the adversary A, who is then given access to the following oracles:

Issuer oracle. The oracle $\mathsf{Issuer}'((a_i)_{i=1}^\ell, D, \mathsf{pk_U}, \mathsf{sk_U})$, on input ℓ attributes $(a_i)_{i=1}^\ell$, index set D, and user public and secret key $(\mathsf{pk_U}, \mathsf{sk_U})$, outputs a credential C (and, hence, simulates an issuance protocol for honest issuer associated with $\mathsf{pk_I^*}$ and user associated with $\mathsf{pk_U}$). Before sending its last message in a successful issuance protocol, the oracle adds $(a_i)_{i \in D}$ to the initially empty list L managed by the experiment.

User-credential oracle. $\mathsf{User}'(id_U, (a_i)_{i \in [\ell]}, D)$, on input of user id id_U, attributes $(a_i)_{i \in [\ell]}$, and index list D, will output a handle h_{C, id_U} on user credential C for user id_U. If id_U is entered the first time, the oracle generates a new key pair $(\mathsf{pk_U}, \mathsf{sk_U})$ for the user which it uses in the following, otherwise it reuses the already existing key pair for id_U (where ids and keys are maintained by the oracle). Internally, User' obtains a credential from the honest issuer (with public key $\mathsf{pk_I^*}$) for id_U.

User-transformation information oracle. $\mathsf{Trans}'(id_U, id_S)$, on input user and service-provider ids id_U and id_S, respectively, outputs a transformation information $t_{U \to S}$. (Internally, required public and secret keys $(\mathsf{pk_U}, \mathsf{sk_U})$ or $(\mathsf{pk_S}, \mathsf{sk_S})$ are generated and maintained.)

User-presentation oracle. $\mathsf{User}''(h_{C, id_U}, id_S, R)$, on input a handle h_{C, id_U} on credential C for user id_U, a service provider id id_S, and an index set R, outputs a presentation p_{C, id_U}. If the service provider associated with id_S is first used for the given user id_U, $\mathsf{Trans}'(id_U, id_S)$ is queried internally to obtain the required transformation information $t_{U \to S}$. (Again, required keys for service providers and users are maintained by the oracle).

Service-provider oracle. $\mathsf{Service}'(p_{C,id_U}, \mathsf{pk_I}*, id_S)$, on input p_{C,id_U}, outputs 1 if p_{C,id_U} is a valid presentation under $\mathsf{pk_I^*}$ for id_S (i.e., for all $(a_i)_{i \in D} \in L$ there exists an index $j \in D \cap R^*$ such that $a_j \neq a_j^*$), else output 0.

The experiment ends when the adversary presents a specially labeled presentation p^* for honest issuer associated with $\mathsf{pk_I^*}$ with index set R^* of revealed attributes $(a_i^*)_{i \in R^*}$ to the service-provider oracle $\mathsf{Service}'$. We now require that no PPT adversary can, with more than negligible probability, come up with a valid presentation p^* under $\mathsf{pk_I^*}$ and $\mathsf{pk_S}^*$.

Note that in this definition the adversary canonically takes the role of malicious users as well as the functionality of the wallet for those, and it is thus not necessary to consider the wallet as a separate entity in the game. More formally:

Definition 3.1 (Unforgeability). *We say a EABC scheme* EABC *is unforgeable if for any PPT adversary* A, *the advantage function*

$$\mathsf{Adv}_{\mathsf{EABC},\mathsf{A}}^{\mathsf{unforge}}(\lambda) := \left| \Pr\left[\mathsf{Exp}_{\mathsf{EABC},\mathsf{A}}^{\mathsf{unforge}}(\lambda) = 1 \right] \right|$$

is negligible in λ, *where* $\mathsf{Exp}_{\mathsf{EABC},\mathsf{A}}^{\mathsf{unforge}}$ *is given below:*

Experiment $\mathsf{Exp}_{\mathsf{EABC},\mathsf{A}}^{\mathsf{unforge}}(\lambda)$
$sp \leftarrow \mathsf{Par}(\lambda)$, $(\mathsf{pk_I^*}, \mathsf{sk_I^*}) \leftarrow \mathsf{Gen_I}(sp)$, $(\mathsf{pk_S}^*, \mathsf{sk_S}^*) \leftarrow \mathsf{Gen_S}(sp)$
associate id_S^* with $(\mathsf{pk_S}^*, \mathsf{sk_S}^*)$
$p^* \leftarrow \mathsf{A}^{\mathsf{Issuer}', \mathsf{User}', \mathsf{Trans}', \mathsf{User}'', \mathsf{Service}'}(\mathsf{pk_I^*}, \mathsf{pk_S}^*)$
if $\mathsf{Service}'(p^*, \mathsf{pk_I^*}, id_S^*) = 1$ then return 1, else return 0

Unlinkability. An EABC scheme EABC is unlinkable if any PPT adversary A succeeds in the following experiment only with probability negligibly larger than $1/2$. The experiment samples key material for two target users $(pk^{(0)}, sk^{(0)})$ and $(pk^{(1)}, sk^{(1)})$, and passes the public keys $(pk^{(0)}, pk^{(1)})$ to A. During the experiment, A has access to target-user credential and a presentation oracles as follows:

Credential oracle. $\mathsf{Cred}(id_U^{(\{0,1\})}, (a_i)_{i=1}^{\ell}, D, \mathsf{pk_I})$, on input user id $id_U^{(0)}$ or $id_U^{(1)}$ associated to the target public keys, attributes $(a_i)_{i=1}^{\ell}$, index-set D, and issuer ID id_I, outputs a credential C; while the oracle stores $(id_U, (a_i)_{i=1}^{\ell})$ in an initially empty list L. (If there are no key material associated with the ids, the oracle internally creates it using the key generation algorithm $\mathsf{Gen_I}$.)

Presentation oracle. $\mathsf{Present}(C_{id_U^{(\{0,1\})}}, id_I, id_S, R)$, on input credential C_{id_U} for target-user id $id_U^{(\{0,1\})}$, issuer and service provider IDs (id_I, id_S), and index set R, outputs a presentation $p_{C,id_U^{(\{0,1\})}}$. (Again, if key material for id_S or transformation information $t_{U \to S}$ is needed, the oracle generates it.)

Eventually, the adversary outputs credentials C_0 under $pk^{(0)}$ and C_1 under $pk^{(1)}$, issuer and service-provider ids (id_I, id_S), and index set R such that $(a^*_{0,i})_{i \in R} = (a^*_{1,i})_{i \in R}$ where $(pk^*_0, (a^*_{0,i})^\ell_{i=1})$ and $(pk^*_1, (a^*_{0,i})^\ell_{i=1})$ are in L. The experiment tosses a coin $b \in \{0,1\}$ and queries the presentation oracle with $(C_b, id^*_I, id^*_S, R^*)$. Eventually, the adversary outputs a guess on b. We require that every PPT adversary guesses b^* with probability at most negligibly larger than $1/2$. Note again that in this definition it is not necessary to consider the wallet as a separate entity, but it would be internally simulated in the presentation oracle. More formally:

Definition 3.2 (Unlinkability). *We say a EABC scheme EABC is unlinkable if for any PPT adversary A, the advantage function*

$$\mathsf{Adv}^{\mathsf{unlink}}_{\mathsf{EABC},\mathsf{A}}(\lambda) := \left| \Pr\left[\mathsf{Exp}^{\mathsf{unlink}}_{\mathsf{EABC},\mathsf{A}}(\lambda) = 1 \right] \right|$$

is negligible in λ, where $\mathsf{Exp}^{\mathsf{unlink}}_{\mathsf{EABC},\mathsf{A}}$ is given below:

Experiment $\mathsf{Exp}^{\mathsf{unlink}}_{\mathsf{EABC},\mathsf{A}}(\lambda)$

$sp \leftarrow \mathsf{Par}(\lambda)$, $(pk^{(0)}, sk^{(0)}) \leftarrow \mathsf{Gen}_\mathsf{U}(sp)$, $(pk^{(1)}, sk^{(1)}) \leftarrow \mathsf{Gen}_\mathsf{U}(sp)$

associate $id^{(0)}_\mathsf{U}$ and $id^{(1)}_\mathsf{U}$ to $pk^{(0)}$ and $pk^{(1)}$, respectively

$(C_0, C_1, id_I, id_S, R), \mathsf{st}) \leftarrow \mathsf{A}^{\mathsf{Cred},\mathsf{Present}}(pk^{(0)}, pk^{(1)})$

$b \xleftarrow{\$} \{0,1\}$, $p^* \leftarrow \mathsf{Present}(C_{id^{(b)}_\mathsf{U}}, id_I, id_S, R)$

$b^* \leftarrow \mathsf{A}^{\mathsf{Cred},\mathsf{Present}}(\mathsf{st}, p^*)$

if $b = b^*$ then return 1, else return 0

Wallet-Privacy. An EABC scheme EABC is private towards the wallet if any PPT adversary A succeeds in the following experiment only with probability negligibly larger than $1/2$. The experiment samples target-user key material $(pk^*, sk^*) \leftarrow \mathsf{Gen}_\mathsf{U}(sp)$ and provides the adversary A with the pk_U^*. During the experiment, A has access to a credential oracle:

Credential Oracle. $\mathsf{Cred}((a_i)^\ell_{i=1}, D, id_I)$, on input attributes $(a_i)^\ell_{i=1}$, index set D, and issuer id id_I, outputs a credential C under pk^*. (Implicitly, needed key material will be created.)

Further, A sends attribute-lists $(a^{(0)}_i)^\ell_{i=1}, (a^{(1)}_i)^\ell_{i=1}$ with $(a^{(0)}_i)_{i \in D} = (a^{(1)}_i)_{i \in D}$, an index set D, and an issuer public key id^*_I to the experiment. The experiment tosses a coin $b \xleftarrow{\$} \{0,1\}$ and sends a credential C^* on $(a^{(b)}_i)^\ell_{i=1}$, D, and id^*_I to A. (Implicitly, the experiment engages in an issuance protocol with A.) Eventually, A outputs a guess on b. We require that every PPT adversary guesses b with probability at most negligibly larger than $1/2$. More formally:

Definition 3.3 (Wallet-privacy). *We say a EABC scheme EABC is private with regard to the wallet if for any PPT adversary A, the advantage function*

$$\mathsf{Adv}^{\mathsf{wallet-privacy}}_{\mathsf{EABC},\mathsf{A}}(\lambda) := \left| \Pr\left[\mathsf{Exp}^{\mathsf{wallet-privacy}}_{\mathsf{EABC},\mathsf{A}}(\lambda) = 1 \right] \right|$$

is negligible in λ, where $\mathsf{Exp}_{\mathsf{EABC,A}}^{\mathsf{wallet-privacy}}$ *is given below:*

Experiment $\mathsf{Exp}_{\mathsf{EABC,A}}^{\mathsf{wallet-privacy}}(\lambda)$

$\mathrm{sp} \leftarrow \mathsf{Par}(\lambda), (pk^*, sk^*) \leftarrow \mathsf{Gen_U}(\mathrm{sp})$

$(a_i^{(0)})_{i=1}^{\ell}, (a_i^{(1)})_{i=1}^{\ell}, D, \mathrm{pk}_\mathsf{I}^*, \mathrm{st} \leftarrow \mathsf{A}^{\mathsf{Cred}}(pk^*)$ with $(a_i^{(0)})_{i \in D} = (a_i^{(1)})_{i \in D}$

$b \xleftarrow{\$} \{0,1\}, C^* \leftarrow \Big\langle \mathsf{User}[pk^*, sk^*, \mathrm{pk}_\mathsf{I}^*, ((a_i^{(b)})_{i=1}^{\ell}, D], \mathsf{A}[\mathrm{st}, \mathrm{pk}_\mathsf{I}^*, (a_i^{(b)})_{i \in D}, D] \Big\rangle$

$b^* \leftarrow \mathsf{A}^{\mathsf{Cred}}(\mathrm{st}, C^*)$

if $b = b^*$ then return 1, else return 0

---

4 A Generic Construction

We now define our generic EABC system EABC with PPT algorithms
Par, Gen$_\mathsf{I}$, Gen$_\mathsf{U}$, Gen$_\mathsf{S}$, Trans as well as protocol participants User, Issuer, Wallet,
and Service. The system uses a proxy re-encryption scheme PRE $=$
$(\mathsf{Par_{PRE}}, \mathsf{Gen_{PRE}}, \mathsf{ReKey_{PRE}}, \mathsf{Enc_{PRE}}, \mathsf{ReEnc_{PRE}}, \mathsf{Dec_{PRE}})$, a commitment scheme
COM $=$ $(\mathsf{Par_{COM}}, \mathsf{Com_{COM}}, \mathsf{Open_{COM}})$, and a signature scheme SIG $=$
$(\mathsf{Par_{SIG}}, \mathsf{Gen_{SIG}}, \mathsf{Sig_{SIG}}, \mathsf{Ver_{SIG}})$, all defined as in Sect. 2 and in the full version.
Concerning the message spaces, we assume that PRE's message space $\mathcal{M}_{\mathsf{PRE}}$ is
set to the attribute space $\mathcal{A}$, COM's message space $\mathcal{M}_{\mathsf{COM}}$ includes all possible
public keys of PRE, and SIG's message space $\mathcal{M}_{\mathsf{SIG}}$ includes all possible ℓ-length
PRE ciphertexts (for $\ell \leq |\mathcal{A}|$) and all possible COM commitments, respectively.

Parameter and Key Generation. First, a trusted third party generates the
system parameters $sp \leftarrow \mathsf{Par}(\lambda)$ and the attribute space $\mathcal{A}$ as well as length
parameter $\ell \leq |\mathcal{A}|$ (depending on λ), obtains $pp_{\mathsf{PRE}} \leftarrow \mathsf{Par_{PRE}}(\lambda)$, $pp_{\mathsf{COM}} \leftarrow$
$\mathsf{Par_{COM}}(\lambda)$, and $pp_{\mathsf{SIG}} \leftarrow \mathsf{Par_{SIG}}(\lambda)$, and outputs $\mathrm{sp} := (sp, pp_{\mathsf{PRE}}, pp_{\mathsf{COM}}, pp_{\mathsf{SIG}})$;
which in practice can be realized using multi-party techniques. (We assume that
sp determines $\mathcal{A}$ and is implicitly available to each protocol participant.) Further,
each participant generates its key material using the appropriate key generation
algorithm of EABC as follows. Each user U samples $(\mathrm{pk_U}, \mathrm{sk_U}) \leftarrow \mathsf{Gen_U}(\mathrm{sp})$,
where Gen$_\mathsf{U}$ internally obtains $(pk_{\mathsf{U,PRE}}, sk_{\mathsf{U,PRE}}) \leftarrow \mathsf{Gen_{PRE}}(pp_{\mathsf{PRE}})$ and outputs
$(\mathrm{pk_U}, \mathrm{sk_U}) := ((pk_{\mathsf{U,PRE}}, pp_{\mathsf{COM}}), sk_{\mathsf{U,PRE}})$. Each issuer I generates $(\mathrm{pk_I}, \mathrm{sk_I}) \leftarrow$
$\mathsf{Gen_I}(\mathrm{sp})$, where internally Gen$_\mathsf{I}$ samples $(sk_{\mathsf{SIG}}, vk_{\mathsf{SIG}}) \leftarrow \mathsf{Gen_{SIG}}(pp_{\mathsf{SIG}})$ and outputs $(\mathrm{pk_I}, \mathrm{sk_I}) := (vk_{\mathsf{SIG}}, sk_{\mathsf{SIG}})$. Each service provider S computes $(\mathrm{pk_S}, \mathrm{sk_S}) \leftarrow$
$\mathsf{Gen_S}(\mathrm{sp})$, where Gen$_\mathsf{S}$ internally obtains $(pk_{\mathsf{S,PRE}}, sk_{\mathsf{S,PRE}}) \leftarrow \mathsf{Gen_{PRE}}(pp_{\mathsf{PRE}})$ and
outputs $(\mathrm{pk_S}, \mathrm{sk_S}) := (pk_{\mathsf{S,PRE}}, sk_{\mathsf{S,PRE}})$.

Issuance. The user U encrypts its the attributes $(a_i)_{i=1}^{\ell}$, for integer $\ell \leq |\mathcal{A}|$,
under its PRE public key pk_{PRE} using Enc$_{\mathsf{PRE}}$ to obtain a ciphertext c_i and
r_i, where r_i are the random coins used in Enc. (Note that the output of the
random encryption coins is slightly different to the encryption syntax of PRE,
but straightforward to achieve.) Further, U computes a commitment and an

opening to $pk_{\mathsf{U,PRE}}$ as $(com, w) \leftarrow \mathsf{Com}(pp_{\mathsf{COM}}, pk_{\mathsf{U,PRE}})$. The ciphertexts $(c_i)_{i=1}^{\ell}$ and com are sent to the issuer I. U and Iengage in a ZKP sub-protocol where U shows that he is in possession of honestly generated PRE public and secret keys (π_{U}, ξ) (via $\mathsf{KVerify}(\pi_{\mathsf{U}}, \xi) = 1$), consistent user attributes $(\alpha_i)_{i \notin D}$ in $\mathcal{A}$, consistent random coins used during encryption $(\rho_{c_i})_{i=1}^{\ell}$, and a valid opening o for com (and, thus, to π_{U}) via $\mathsf{Ver}(com, \pi_{\mathsf{U}}, o) = 1$. Further, Icomputes a signature $\sigma \leftarrow \mathsf{Sig}_{\mathsf{SIG}}(\mathsf{sk}_{\mathsf{I}}, (c_i)_{i=1}^{\ell}, com)$ under the issuer's public key pk_{I} on the ciphertexts $(c_i)_{i=1}^{\ell}$ and on the commitment com, and sends σ to U. Then, U outputs the credential $(c_i)_{i=1}^{\ell}, \sigma, com, w$, and auxiliary data aux.

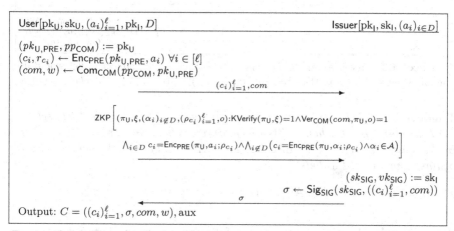

Protocol 4.1: Issuance for the generic construction: an honest user would use the following witnesses in the ZKP: $(pk_{\mathsf{U,PRE}}, \mathsf{sk}_{\mathsf{U}}, (a_i)_{i \notin D}, (r_{c_i})_{i=1}^{\ell}, w)$

Presentation. The presentation phase of EABC is given in Protocol 4.2. The wallet receives as input the credential C, auxiliary input aux, the user and service provider public keys pk_{U} and pks, transformation information $t_{\mathsf{U} \to \mathsf{S}}$, issuer public key pk_{I}, and index set R. The wallet re-randomizes and afterwards re-encrypts the ciphertexts $(c_i)_{i \in R}$ using ReRand and $\mathsf{ReEnc}_{\mathsf{PRE}}$ with $t_{\mathsf{U} \to \mathsf{S}} = rk_{\mathsf{U} \to \mathsf{S}}$, respectively, to obtain $(d_i)_{i \in R}$. (Again, note the output the random coins is slightly different to the encryption syntax of PRE, but straightforward to achieve.) Further, the wallet sends those ciphertexts to the service provider S. Both, the wallet and I agree on an index set R beforehand and the valid re-encryption key is received as the result of a joint re-encryption-key computation by user U and S with $t_{\mathsf{U} \to \mathsf{S}} := rk_{\mathsf{U} \to \mathsf{S}} \leftarrow \mathsf{ReKey}(pk_{\mathsf{U,PRE}}, sk_{\mathsf{U,PRE}}, pk_{\mathsf{S,PRE}}, sk_{\mathsf{S,PRE}})$. The wallet further engages in a ZKP protocol to show that it is in possession of a signature σ on the original ciphertexts $(\gamma_i)_{i=1}^{\ell}$ and on a ψ-commitment, a public key π_{U}, re-randomized ciphertexts $(\hat{c}_i)_{i \in R}$ for random coins $(\rho_{\hat{c}_i})_{i \in R}$, re-encryption coins $(\rho_{d_i})_{i \in R}$ used in $\mathsf{ReEnc}_{\mathsf{PRE}}$, a consistent re-encryption key ξ (via $\mathsf{RKVerify}(\pi_{\mathsf{U}}, \mathsf{pk}_{\mathsf{S}}, \mathsf{sk}_{\mathsf{S}}, \xi) = 1$), and a commitment-opening pair (ψ, o) via $\mathsf{Ver}_{\mathsf{COM}}(\psi, \pi_{\mathsf{U}}, o) = 1$. The service provider decrypts the received ciphertexts $(d_i)_{i \in R}$ and outputs the attributes $(a_i)_{i \in R}$ for $(a_i := \mathsf{Dec}(\mathsf{sk}_{\mathsf{SIG}}, d_i))_{i \in R}$.

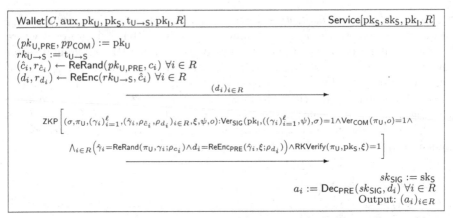

Protocol 4.2: Presentation for the generic construction. An honest wallet would use the following witnesses in the ZKP: $(\sigma, pk_{U,PRE}, (c_i)_{i=1}^{\ell}, (\hat{c}_i, r_{\hat{c}_i}, r_{d_i})_{i \in R}, rk_{U \to S}, com, w)$

Theorem 4.3. *Let* PRE *be an anonymous and re-randomizable PRE scheme,* COM *a commitment scheme,* SIG *a signature scheme, and* ZKP *a ZKP scheme in the sense of Sect. 2, then* EABC *is an EABC scheme in the sense of Sect. 3 against semi-trusted wallets.*

The proof can be found in Appendix A.

Semi-generic Constructions. In the full version, we give a semi-generic construction that uses the BBS PRE scheme [6] and a commitment scheme explicitly.

5   A Concrete Instantiation

In this section, we describe a concrete EABC instantiation $EABC_{BBS,AGHO}$ based on the BBS proxy re-encryption scheme [6] and the structure-preserving signature scheme by Abe et al. [1]. The resulting EABC scheme with multi-show credentials is secure in the generic group model against semi-trusted wallets.

The system parameter are chosen as $sp = (\mathcal{G}_1, \mathcal{G}_2, \mathcal{G}_T, q, e, g, y, z, \lambda, \ell)$, for pairing groups $\mathcal{G}_1, \mathcal{G}_2, \mathcal{G}_T$ with sufficiently large prime order q, pairing e, generators $g, y, z \xleftarrow{\$} \mathcal{G}_1$ such that DDH holds in $\mathcal{G}_1$ (all depending on λ), and number of attributes ℓ. Implicitly, one can see y, z as a parameter of the underlying commitment scheme (where the discrete logarithm $\log_y(z)$ is only known with negligible probability), and g as a parameter for the underlying PRE scheme. Further, the user and the service provider invoke the Gen-algorithm of PRE with the system parameter sp to obtain (pk_U, sk_U) and (pk_S, sk_S), respectively. In the BBS proxy re-encryption scheme, public keys are group elements $pk_U = (g, g^{sk_U})$, for a secret key $sk_U \in \mathbb{Z}_q$.) The issuer runs Gen_{SIG} of Abe et al.'s signature scheme to obtain $pk_I = (V, Z, (W_i)_{i=1}^{2\ell+2})$ and $sk_I = (v, z, (w_i)_{i=1}^{2\ell+2})$.

Issuance. The issuance phase is depicted in Protocol 5.1. The user computes an BBS-encryption of the neutral element in $\mathcal{G}_1$ as $c_0 = (c_{01}, c_{02}) = (g^e, \mathrm{pk_U}^e)$, for $e \xleftarrow{\$} \mathbb{Z}_q$. Further, the user creates ciphertexts by computing $c_i := (c_{01}^{r_i}, a_i c_{02}^{r_i})$, for all attributes $(a_i)_{i=1}^{\ell}$ and random coins $r_i \xleftarrow{\$} \mathbb{Z}_q$, and sends $(c_i)_{i \notin D}, c_0, (r_i)_{i \in D}$ and a ZKP of knowing the secret key $\mathrm{sk_U}$ to the issuer. The issuer can check for all disclosed attributes $(a_i)_{i \in D}$ that the ciphertexts are consistent (since it knows the encryption randomness). Further, the issuer computes an Abe et al.-signature σ on the list of ciphertexts $((c_i)_{i=1}^{\ell}, c_0)$ and sends σ to the user. Eventually, the user outputs the credential $((c_i)_{i=1}^{\ell}, \sigma, c_0, e, \varepsilon)$.

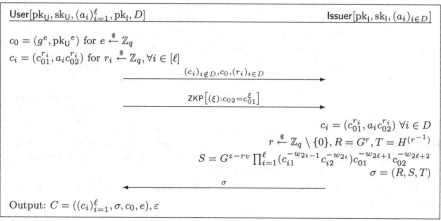

Protocol 5.1: Multi-show credentials from the BBS proxy re-encryption and Abe et al.'s signature schemes: Issuance. An honest user would set $\xi = e$.

Presentation. The presentation phase is depicted in Protocol 5.2. We assume that the user and the service provider have jointly computed a re-encryption key $\mathrm{rk_{U \to S}}$ (via ReKey) in the sense of BBS beforehand. The wallet re-randomizes the ciphertexts consistently for the attributes going to be revealed (i.e., specified by R) and inconsistently for all others. Then, $\mathrm{rk_{U \to S}}$ is used to re-encrypt all consistent ciphertexts to obtain $(d_i)_{i \in R}$. Further, commitments to $\mathrm{rk_{U \to S}}$, $\mathrm{rk_{U \to S}}e$, and $(\mathrm{rk_{U \to S}}f_i)_{i \in R}$ are computed as $k = y^{\mathrm{rk_{U \to S}}} z^b$, $k' = y^{\mathrm{rk_{U \to S}}e} z^{b'}$, and $(k_i = y^{\mathrm{rk_{U \to S}}f_i} z^{b_i})_{i \in R}$, for exponents $b, b', (b_i)_{i \in R} \xleftarrow{\$} \mathbb{Z}_q$, respectively. Finally, $\hat{T} = T^{-x}$, for exponent $x \xleftarrow{\$} \mathbb{Z}_q$, is computed. All re-randomizations $(\hat{c}_i)_{i=1}^{\ell}$, all re-encryptions $(d_i)_{i \in D}$, $k, k', (k_i)_{i \in R}$, and $\hat{T}$ as well as the ZKP are sent to the service provider. The service provider decrypts the ciphertexts $(d_i)_{i \in R}$ and outputs the revealed attributes $(a_i)_{i \in R}$.

Lemma 5.3. $\mathrm{EABC_{BBS,AGHO}}$ *above is an EABC scheme in the sense of Sect. 3 in the generic group model for semi-trusted wallets.*

Wallet$[C, \varepsilon, \mathsf{pk_U}, \mathsf{pk_S}, t_{U \to S}, \mathsf{pk_I}, R]$ Service$[\mathsf{pk_S}, \mathsf{sk_S}, \mathsf{pk_I}, R]$

$\hat{c}_i = (c_{i1}g^{f_i}, c_{i2}\mathsf{pk_U}^{f_i})$ for $f_i \xleftarrow{\$} \mathbb{Z}_q, \forall i \in R$

$\hat{c}_i = (c_{i1}g^{v_i}, c_{i2}g^{w_i})$ for $v_i, w_i \xleftarrow{\$} \mathbb{Z}_q, \forall i \notin R$

$d_i = (\hat{c}_{i1}^{\mathsf{rk}_{U \to S}}, \hat{c}_{i2}), \forall i \in R$

$b \xleftarrow{\$} \mathbb{Z}_q, k = y^{\mathsf{rk}_{U \to S}} z^b$

$b' \xleftarrow{\$} \mathbb{Z}_q, k' = y^{\mathsf{rk}_{U \to S} e} z^{b'}$

$b_i \xleftarrow{\$} \mathbb{Z}_q, k_i = y^{\mathsf{rk}_{U \to S} f_i} z^{b_i}, \forall i \in R$

$\hat{T} = T^{-x}$ for $x \xleftarrow{\$} \mathbb{Z}_q$

$$\xrightarrow{(\hat{c}_i)_{i=1}^{\ell}, (d_i)_{i \in R}, k, k', (k_i)_{i \in R}, \hat{T}}$$

$$\mathsf{ZKP}\Big[\big((\sigma_1, \sigma_2, \chi), (\phi_i, \beta_i, \gamma_i, \theta_i)_{i \in R}, (\nu_i, \omega_i)_{i \notin R}, \alpha, \beta, \beta', \theta', \xi, \eta\big):$$

$$e(\sigma_1, V)\, e(\sigma_2, H) \prod_{i \in R}\Big(e(g, W_{2i-1})^{-\phi_i}\, e(\mathsf{pk_S}, W_{2i})^{-\gamma_i}\Big) \prod_{i \notin R}\Big(e(g, W_{2i-1})^{-\nu_i}\, e(g, W_{2i})^{-\omega_i}\Big)\cdot$$

$$e(g, W_{2l+1})^{\eta}\, e(\mathsf{pk_S}, W_{2l+2})^{\alpha} = e(G, Z)\prod_{i=1}^{\ell}\big(e(\hat{c}_{i1}, W_{2i-1})\, e(\hat{c}_{i2}, W_{2i})\big)^{-1} \wedge$$

$$e(\sigma_1, \hat{T})\, e(G, H)^{\chi} = 1 \wedge k' = y^{\alpha} z^{\beta'} \wedge k' = k^{\eta} z^{\theta'} \wedge k = y^{\xi} z^{\beta} \wedge \bigwedge_{i \in R}\big(k_i = y^{\gamma_i} z^{\beta_i} \wedge k_i = k^{\phi_i} z^{\theta_i} \wedge d_{i1} = \hat{c}_{i1}^{\xi}\big)\Big]$$

$$\xleftarrow{\hspace{7cm}}$$

$\mathsf{sk_{S,PRE}} := \mathsf{sk_I}$

$a_i = \mathsf{Dec}(\mathsf{sk_{S,PRE}}, d_i)\ \forall i \in R$

Output: $(a_i)_{i \in R}$

Protocol 5.2: Multi-show credentials from Blaze's proxy re-encryption and Abe et al.'s signature schemes: Presentation. An honest user would use the following witnesses in the ZKP:

$$((R, S, x), (f_i, b_i, \mathsf{rk}_{U \to S} f_i, b_i - b f_i)_{i \in R}, (v_i, w_i)_{i \notin R}, \mathsf{rk}_{U \to S} e, b, b', b' - b e, \mathsf{rk}_{U \to S}, e)$$

Proof. This lemma readily follows from the properties of the BBS and AGHO schemes from Sect. 3 and Theorem 4.3.

We want to stress that the used re-encryption scheme by Blaze et al. is sufficient for our purposes, and no more advanced schemes (e.g., by Ateniese et al. [4]) are needed here. Namely, the used scheme would become insecure if wallet and service providers colluded, as they could jointly recover the user's secret key $\mathsf{sk_U}$ from $\mathsf{rk}_{U \to S}$ and $\mathsf{sk_S}$. However, this collusion is excluded by definition, as the two parties could also completely break the users privacy without recovering the key but by simply re-encrypting and decrypting the attributes, independent of the scheme being used.

6   Extensions

Our construction can easily be extended to cover most features that are also supported by existing schemes like idemix and U-Prove. In the following, we sketch how some of the more important features can be realized.

Revocation. A core functionality that is needed in almost every practical credential system is revocation. That is, the owner of a credential (and potentially the issuer) should have the possibility to invalidate a credential, e.g., in case of identity theft or abuse. Our scheme can easily be extended to support revocation as follows: during issuance, the user and the issuer agree on a revocation handle $h \in \mathbb{Z}_q$, and the user computes a Pedersen commitment $com_h = y^h z^r$ for some $r \xleftarrow{\$} \mathbb{Z}_q$, which it gets signed by the issuer together with all the ciphertexts and its public key. The user then hands $com_h, h,$ and r to the wallet; note that this does not cause a privacy problem towards the wallet, as the revocation handle does not contain any sensitive information. For presentation, the wallet then gives a re-randomization of com_h to the service provider. Similar to the re-randomization of ciphertexts, it then proves that this computation was done honestly, and that the unrevealed revocation handle contained in the commitment has not been revoked before. The latter can be done very efficiently using existing schemes found in the literature, e.g., Nakanishi et al. [22], where also the algebraic setting would be directly compatible with our construction.

Multi-credential Presentations and Attribute Equality. Sometimes it is necessary to prove possession of multiple credentials in one presentation. For instance, a service provider might request that a user possesses a valid subscription to its service and an electronic identity card proving a certain age. Again, this can directly be achieved by our scheme by simply merging the respective zero-knowledge proofs and proving that the two credentials were issued to the same pk_U. This can be achieved by showing that the same values have been used for ξ and η in the zero-knowledge proof. As during issuance, a user must prove to the issuer knowledge of the secret key corresponding to pk_U, the service provider is then assured that both credentials have indeed be issued to the same user.

Inspection. Anonymity might often lead to abuse by malicious users. It is therefore often required that a dedicated party or judge has the possibility to reveal the identity of an owner of a credential. Similar to related work, this can be done by letting the issuer sign a unique identity of the user, and then extending the zero-knowledge proof for presentations by an encryption of this identity under the judge's public key. This can be realized by letting the user deposit the re-encryption key from his public key to the judge's public key in the wallet. In analogy to the proofs for revealed attributes, the wallet then re-encrypts the identity for the judge and proves that this was done correctly.

6.1 Metadata Privacy

All constructions presented so far prevent the wallet from learning the plain text attributes certified in a credential, and give high unlinkability guarantees to a user. However, they do not hide usage patterns of specific credentials from the wallet. This in turn can be potentially valuable metadata, as information like frequency may reveal sensitive information (e.g., frequent access to an eHealth service may leak other information than frequent access to a video streaming service).

In the following we sketch a solution how to reduce the metadata leaked to the wallet by hiding to which service one is authenticating. The solution works for non-interactive presentations as our concrete instantiation specified above. On a high level, the user lets the wallet compute presentation tokens for defined policies (i.e., set D, etc.) for a set of service providers, and the service provider obtains the correct token running a private information retrieval protocol with the wallet. More precisely, the protocol looks as follows:

1. The user defines presentation policies $pp_1, \ldots, pp_n$ for all service providers among which the presentation should be undistinguishable, where for some j, pp_j is the policy requested by the target service provider. The user sends those policies, as well as a seed s for a pseudo random function PRF to the wallet. Note that for most service providers, the pp_i will be quite stable (many providers always require access to the same attributes), and thus the user can simply use predefined default policies.
2. The wallet computes the non-interactive presentation tokens $pt_1, \ldots, pt_n$ for the defined service providers and policies. For all i, it sends $pt_i' = pt_i \oplus \mathsf{PRF}_s(i)$ to the service provider, where $\oplus$ denotes bitwise XOR.
3. The user sends to the service provider $u = \mathsf{PRF}_s(j)$.
4. The service provider recovers pt_j as $pt_j = pt_j' \oplus u$, and then continues the verification of pt_j as specified in the presentation protocol.

It is easy to see that the service provider can only access the correct presentation token and does not learn any information about the other service providers or presentations as long as the used pseudo random function is secure (note that perfect security could here be achieved by replacing the PRF by a real random string chosen by the user, only at the costs of slightly increasing the communication complexity by a few kilo-bytes on the user's side). From the wallet's perspective, the identity of the target service provider remains perfectly hidden in either case.

We stress that even though the above is natural and straightforward, it still yields practically efficient protocols. Based on the benchmarks presented in the following for single presentations, hiding a presentation among ten service providers can be done in less than two seconds using our prototype implementation, assuming ten attributes per credential.

7   Benchmarks

Figure 1 illustrates the practical efficiency of our main scheme presented in Sect. 5, where all zero-knowledge proofs have been made non-interactive using the Fiat-Shamir heuristic [19]. For two different security parameters (80 and 112, respectively), and two different numbers of attributes (10 and 25, respectively), we show the running times of the issuance and presentation phases for each of the involved parties, where the issuer and service provider learned half of the certified attributes, i.e., the sets D and R in the protocols had cardinality 5 and 12, respectively. For the benchmarks, we pre-computed all pairings with constant inputs on the service provider's and the wallet's sides, which only need to be computed once and can be used for all users. The complexity of this pre-computation is comparable to the service provider's computational efforts.

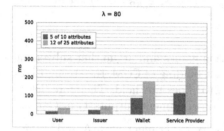 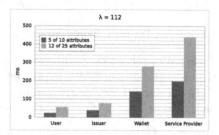

Fig. 1. Execution times for our main scheme for different security parameters and different numbers of attributes of which (around) half of them were opened during presentation. The complexity is roughly linear in the number of attributes, and increases with an increasing fraction of revealed attributes.

The scheme was implemented in Python 2.7 using the Charm cryptographic framework [2] using the concrete PAIRINGGROUPS $MNT159$ and $MNT224$. The benchmarks were performed on an Intel(R) Core(TM) i7-5600 with 2.60 MHz and 16 GB of memory, running Ubuntu 16.04.

8 Conclusion

We presented the first attribute-based credential (ABC) system which can be fully cloudified in a privacy-preserving manner, as presenting credentials does not require access to the plaintext attributes. The basic scheme supports selective disclosure at issuance and presentation, and gives high unlinkability guarantees similar to previous (non-cloudified) ABC systems. We then described various extensions like inspection or revocation, and explained how to protect the user from being profiled by the wallet provider. Furthermore, we showed the practical efficiency of our scheme by means of concrete benchmarks of a prototype implementation.

Future work may further increase metadata privacy, e.g., by also hiding the service providers a user is subscribed to, potentially by splitting the wallet into two non-colluding components.

A   Proof of Theorem 4.3

Proof. We prove the correctness, unforgeability, unlinkability, and wallet-privacy properties of EABC in a sequence of claims:

Correctness is easy to verify. By the correctness of PRE, the re-randomization property of PRE ciphertexts, the correctness of SIG, and correctness of ZKP, correctness of EABC readily follows.

Claim. Under the binding property of COM, the strong EUF-CMA-security of SIG, and the soundness of ZKP, EABC is unforgeable. More concretely, for any PPT adversaries A, A', A'', A''', we have

$$\mathsf{Adv}^{\mathsf{unforge}}_{\mathsf{EABC,A}}(\lambda) \leq q \cdot \mathsf{Adv}^{\mathsf{hiding}}_{\mathsf{COM,A'}}(\lambda) + \mathsf{Adv}^{\mathsf{s-euf-cma}}_{\mathsf{SIG,A''}}(\lambda) + \mathsf{Adv}^{\mathsf{soundness}}_{\mathsf{ZKP,A'''}}(\lambda) + \mathsf{negl}(\lambda), \quad (1)$$

for any $\lambda \in \mathbb{N}$ and polynomial $q = q(\lambda)$.

Proof. We proceed by a sequence of reduction games and argue that subverting the unforgeability of EABC implies either that binding of COM, the strong EUF-CMA-security of SIG, or the soundness of ZKP does not hold. Therefore, let S_i be the event that A wins (i.e., the associated experiment outputs 1) in Game i.

Game 1. Game 1 is the EABC unforgeability experiment with A and, hence, we have $\Pr[S_1] = \mathsf{Adv}^{\mathsf{unforge}}_{\mathsf{EABC,A}}(\lambda)$.

Game 2. Game 2 is identical to Game 1, except that the event F occurs where we have $\mathsf{Open}(pk_{\mathsf{U,PRE}}, com, w_0) = \mathsf{Open}(pk'_{\mathsf{U,PRE}}, com, w_1)$ with $pk_{\mathsf{U,PRE}} \neq pk'_{\mathsf{U,PRE}}$, for some $(com, w_0) = \mathsf{Com}(pp_{\mathsf{COM}}, pk_{\mathsf{U,PRE}})$, $(com, w_1) = \mathsf{Com}(pp_{\mathsf{COM}}, pk'_{\mathsf{U,PRE}})$, i.e., as computation within the issuer or user-credential oracle on input of A. We argue that $\Pr[F] \leq q_O \cdot \mathsf{Adv}^{\mathsf{hiding}}_{\mathsf{COM,A}}(\lambda)$ as the occurring of F directly yields a successful PPT adversary on the binding property of COM, where q is the total number of A-queries to the oracles. Essentially, in a reduction between the binding experiment of COM and unforgeability of EABC, once F occurs with A, the experiment (which has received pp from the binding experiment at the beginning and forwarded pp as part of the public system parameter to A), forwards $(com, pk_{\mathsf{U,PRE}}, w_0, pk'_{\mathsf{U,PRE}}, w_1)$ to the binding experiment which yields a successful PPT adversary A'. Hence, we have $|\Pr[S_2] - \Pr[S_1]| \leq \Pr[F] \leq q \cdot \mathsf{Adv}^{\mathsf{hiding}}_{\mathsf{COM,A'}}(\lambda)$.

Game 3. Game 3 is identical to Game 2, except that the event F' occurs where we have $\mathsf{Ver}(pk_{\mathsf{I}}, ((c_i)_i, com), \sigma^*) = 1$, for $((c_i)_i, com)$ not previously occurred for some potentially already occurred σ^*, i.e., as extracted signature from the A-presentation p^* at the end of the experiment. We argue that $\Pr[F'] \leq q_O \cdot \mathsf{Adv}^{\mathsf{s-euf-cma}}_{\mathsf{SIG,A}}(\lambda)$ as the occurring of F' directly yields a successful PPT adversary on the strong EUF-CMA property of SIG. Essentially, in a reduction between strong EUF-CMA of SIG and unforgeability of EABC, once F' occurs with A, the experiment (which has received pk_{I} from the strong EUF-CMA experiment at the beginning and is able to query signatures under pk_{I}), forwards $(((c_i)_i, com), \sigma^*)$, extracted from the presentation p^* to the strong EUF-CMA experiment which yields a successful PPT adversary A''. Hence, we have $|\Pr[S_3] - \Pr[S_2]| \leq \Pr[F'] \leq \mathsf{Adv}^{\mathsf{s-euf-cma}}_{\mathsf{SIG,A''}}(\lambda)$.

Game 4. Game 4 is identical to Game 3, except that the event F'' occurs where we have a valid A-presentation p^* at the end of the experiment, but the p^* contains values that are not in the language used in the ZKP system. We argue that $\Pr[F''] \leq \mathsf{Adv}^{\mathsf{soundness}}_{\mathsf{ZKP,A}}(\lambda)$ as the occurring of the event directly yields a successful adversary on the soundness property of ZKP. Essentially, in a reduction between the soundness property of ZKP and strong EUF-CMA-security of EABC, once F'' occurs with A, the experiment forwards the values not in the language together with the proof from p^* to the soundness experiment of ZKP which yields a successful PPT adversary A'''. Hence, we have $|\Pr[S_4] - \Pr[S_3]| \leq \Pr[F''] \leq \mathsf{Adv}^{\mathsf{soundness}}_{\mathsf{SIG},A'''}(\lambda)$.

Game 5. Game 5 is identical to Game 4 and we argue that now the adversary has at most negligible advantage (by the perfect correctness of the underlying primitives). Hence, $\Pr[S_4] = \Pr[S_5] \leq \mathsf{negl}(\lambda)$.

Hence, we conclude that Eq. (1) holds. $\qquad\qquad\qquad\qquad\qquad\qquad\qquad\square$

Claim. Assuming the anonymous property of PRE, EABC is unlinkable. More concretely, for any PPT adversaries A, A', A'', we have

$$\mathsf{Adv}^{\mathsf{unlink}}_{\mathsf{EABC},A}(\lambda) \leq q \cdot \mathsf{Adv}^{\mathsf{pre-anon}}_{\mathsf{PRE},A'}(\lambda) + \mathsf{negl}(\lambda), \qquad (2)$$

for any $\lambda \in \mathbb{N}$ and polynomial $q = q(\lambda)$.

Proof. We proceed by a sequence of reduction games and argue that subverting the unlinkability of EABC implies either that the anonymity property of PRE or the soundness property of ZKP does not hold. Therefore, let S_i be the event that A wins (i.e., the associated experiment outputs 1) in Game i.

Game 1. Game 1 is the EABC unlinkability experiment with A and, hence, we have $\Pr[S_1] = \mathsf{Adv}^{\mathsf{unlink}}_{\mathsf{EABC},A}(\lambda)$.

Game 2. Game 2 is identical to Game 1, except that change the way credentials are generated for A. In this game, we do not need the ZKP witness (and, hence, the target-user secret keys) anymore and rely on the zero-knowledge property of ZKP. (That is that we can use a simulator in the sense of ZKP to generate proofs.) This change is purely syntactical. Hence, we have $\Pr[S_2] = \Pr[S_1]$.

Game 3. Game 3 is identical to Game 2, except that we change that the ciphertext in the issuance are generated under an independent and an honestly sampled user public key different to the target-user ciphertext. Hence, A only receives credentials under a different user public key compared to the target public keys. We argue that if A can distinguish under which public keys the ciphertexts are generated, we can directly use A to break the anonymity of the underlying PRE. Essentially, in a reduction between anonymity of PRE and unlinkability of EABC, the experiment (which has received pk_0, pk_1 from the anonymous experiment at the beginning, forwards A's guess to its own challenger which yields a successful PPT adversary A' with probability $1/q$, for q A-queries to Cred. Hence, we have $|\Pr[S_3] - \Pr[S_2]| \leq q \cdot \mathsf{Adv}^{\mathsf{pre-anon}}_{\mathsf{PRE},A'}(\lambda)$.

Game 4. Game 4 is identical to Game 3 and we argue that now the adversary has at most negligible advantage in guessing b. Hence, $\Pr[S_4] = \Pr[S_3] \leq \mathsf{negl}(\lambda)$.

Hence, we conclude that Eq. (2) holds. $\qquad\qquad\qquad\qquad\qquad\qquad\qquad\square$

Claim. Under the IND-CPA security of PRE, EABC is wallet-private. More concretely, for any PPT adversaries A, A′, we have

$$\mathsf{Adv}_{\mathsf{EABC},\mathsf{A}}^{\mathsf{wallet-privacy}}(\lambda) \leq \ell \cdot \mathsf{Adv}_{\mathsf{PRE},\mathsf{A}'}^{\mathsf{pre-ind-cpa}}(\lambda) + \mathsf{negl}(\lambda), \tag{3}$$

for any $\lambda \in \mathbb{N}$ and polynomial $\ell = \ell(\lambda)$.

Proof. We proceed by a sequence of reduction games and argue that subverting the unlinkability of EABC implies that the IND-CPA property of PRE does not hold. Therefore, let S_i be the event that A wins (i.e., the associated experiment outputs 1) in Game i.

Game 1. Game 1 is the EABC unforgeability experiment with A and, hence, we have $\Pr[S_1] = \mathsf{Adv}_{\mathsf{EABC},\mathsf{A}}^{\mathsf{wallet-privacy}}(\lambda)$.

Game 2. Game 2 is identical to Game 1, except that we now do not know the target secret key sk^*. However, sk^* is solely used for the ZKP system within the issuance and, hence, we can use the ZKP zero-knowledge property to provide valid proofs without the witness (where sk^* is part of) using a simulator. This change is purely syntactical. Hence, we have $\Pr[S_3] = \Pr[S_2]$.

Game 3. Game 3 is identical to Game 2, except that we now exchange all ciphertexts with ciphertexts of "0"s. In a reduction between the IND-CPA-security property of PRE and wallet-privacy of EABC, the experiment forwards the answer from A as its own guess to the PRE IND-CPA-security experiment (given the public key from the IND-CPA experiment as target public key for the wallet-privacy adversary). Hence, we have $|\Pr[S_3] - \Pr[S_2]| \leq \ell \cdot \mathsf{Adv}_{\mathsf{PRE},\mathsf{A}'}^{\mathsf{pre-ind-cpa}}(\lambda)$.

Game 4. Game 4 is identical to Game 3 and we argue that now the adversary has at most negligible advantage (by the perfect correctness of the underlying primitives), otherwise, some event occurred which would yield another game. Hence, $\Pr[S_3] = \Pr[S_4] \leq \mathsf{negl}(\lambda)$.

Hence, we conclude that Eq. (3) holds. $\qquad\qquad\qquad\qquad\qquad\qquad\quad\square$

Taken all claims together, this yields the proof. $\qquad\qquad\qquad\qquad\quad\square$

References

1. Abe, M., Groth, J., Haralambiev, K., Ohkubo, M.: Optimal structure-preserving signatures in asymmetric bilinear groups. In: Rogaway, P. (ed.) CRYPTO 2011. LNCS, vol. 6841, pp. 649–666. Springer, Heidelberg (2011). https://doi.org/10.1007/978-3-642-22792-9_37
2. Akinyele, J.A., Green, M., Rubin, A.D.: Charm: a framework for rapidly prototyping cryptosystems. In: NDSS (2012)
3. Aono, Y., Boyen, X., Phong, L.T., Wang, L.: Key-private proxy re-encryption under LWE. In: Paul, G., Vaudenay, S. (eds.) INDOCRYPT 2013. LNCS, vol. 8250, pp. 1–18. Springer, Cham (2013). https://doi.org/10.1007/978-3-319-03515-4_1
4. Ateniese, G., Fu, K., Green, M., Hohenberger, S.: Improved proxy re-encryption schemes with applications to secure distributed storage. In: NDSS (2005)

5. Belenkiy, M., Camenisch, J., Chase, M., Kohlweiss, M., Lysyanskaya, A., Shacham, H.: Randomizable proofs and delegatable anonymous credentials. In: Halevi, S. (ed.) CRYPTO 2009. LNCS, vol. 5677, pp. 108–125. Springer, Heidelberg (2009). https://doi.org/10.1007/978-3-642-03356-8_7

6. Blaze, M., Bleumer, G., Strauss, M.: Divertible protocols and atomic proxy cryptography. In: Nyberg, K. (ed.) EUROCRYPT 1998. LNCS, vol. 1403, pp. 127–144. Springer, Heidelberg (1998). https://doi.org/10.1007/BFb0054122

7. Brands, S.: Rethinking public key infrastructure and digital certificates - building in privacy. Ph.D. thesis, Eindhoven Institute of Technology (1999)

8. Camenisch, J., Dubovitskaya, M., Haralambiev, K., Kohlweiss, M.: Composable and modular anonymous credentials: definitions and practical constructions. In: Iwata, T., Cheon, J.H. (eds.) ASIACRYPT 2015. LNCS, vol. 9453, pp. 262–288. Springer, Heidelberg (2015). https://doi.org/10.1007/978-3-662-48800-3_11

9. Camenisch, J., Van Herreweghen, E.: Design and implementation of the idemix anonymous credential system. In: ACM CCS (2002)

10. Camenisch, J., Krenn, S., Lehmann, A., Mikkelsen, G.L., Neven, G., Pedersen, M.Ø.: Formal treatment of privacy-enhancing credential systems. In: Dunkelman, O., Keliher, L. (eds.) SAC 2015. LNCS, vol. 9566, pp. 3–24. Springer, Cham (2016). https://doi.org/10.1007/978-3-319-31301-6_1

11. Camenisch, J., Lysyanskaya, A.: An efficient system for non-transferable anonymous credentials with optional anonymity revocation. In: Pfitzmann, B. (ed.) EUROCRYPT 2001. LNCS, vol. 2045, pp. 93–118. Springer, Heidelberg (2001). https://doi.org/10.1007/3-540-44987-6_7

12. Camenisch, J., Lysyanskaya, A.: A signature scheme with efficient protocols. In: Cimato, S., Persiano, G., Galdi, C. (eds.) SCN 2002. LNCS, vol. 2576, pp. 268–289. Springer, Heidelberg (2003). https://doi.org/10.1007/3-540-36413-7_20

13. Camenisch, J., Lysyanskaya, A.: Signature schemes and anonymous credentials from bilinear maps. In: Franklin, M. (ed.) CRYPTO 2004. LNCS, vol. 3152, pp. 56–72. Springer, Heidelberg (2004). https://doi.org/10.1007/978-3-540-28628-8_4

14. Camenisch, J., Stadler, M.: Efficient group signature schemes for large groups. In: Kaliski, B.S. (ed.) CRYPTO 1997. LNCS, vol. 1294, pp. 410–424. Springer, Heidelberg (1997). https://doi.org/10.1007/BFb0052252

15. Chaum, D.: Untraceable electronic mail, return addresses, and digital pseudonyms. Commun. ACM **24**, 84–88 (1981)

16. Chaum, D.: Security without identification: transaction systems to make big brother obsolete. Commun. ACM **28**, 1030–1044 (1985)

17. European Parliament and Council of the European Union: Regulation (EC) No 45/2001. Official Journal of the European Union (2001)

18. European Parliament and Council of the European Union: Directive 2009/136/EC. Official Journal of the European Union (2009)

19. Fiat, A., Shamir, A.: How to prove yourself: practical solutions to identification and signature problems. In: Odlyzko, A.M. (ed.) CRYPTO 1986. LNCS, vol. 263, pp. 186–194. Springer, Heidelberg (1987). https://doi.org/10.1007/3-540-47721-7_12

20. Hanser, C., Slamanig, D.: Structure-preserving signatures on equivalence classes and their application to anonymous credentials. In: Sarkar, P., Iwata, T. (eds.) ASIACRYPT 2014. LNCS, vol. 8873, pp. 491–511. Springer, Heidelberg (2014). https://doi.org/10.1007/978-3-662-45611-8_26

21. Libert, B., Vergnaud, D.: Unidirectional chosen-ciphertext secure proxy re-encryption. In: Cramer, R. (ed.) PKC 2008. LNCS, vol. 4939, pp. 360–379. Springer, Heidelberg (2008). https://doi.org/10.1007/978-3-540-78440-1_21

22. Nakanishi, T., Fujii, H., Hira, Y., Funabiki, N.: Revocable group signature schemes with constant costs for signing and verifying. In: Jarecki, S., Tsudik, G. (eds.) PKC 2009. LNCS, vol. 5443, pp. 463–480. Springer, Heidelberg (2009). https://doi.org/10.1007/978-3-642-00468-1_26
23. Nuñez, D., Agudo, I., Lopez, J.: NTRUReEncrypt: an efficient proxy re-encryption scheme based on NTRU. In: ASIA CCS (2015)
24. Paquin, C., Zaverucha, G.: U-prove cryptographic specification v1.1 (revision 2). Technical report, Microsoft Corporation, April 2013
25. Sabouri, A.: A cloud-based model to facilitate mobility of privacy-preserving attribute-based credential users. In: TrustCom/BigDataSE/ISPA (2015)
26. Schmidt, H.A.: National strategy for trusted identities in cyberspace. Cyberwar Resources Guide, Item 163 (2010)

Unlinkable and Strongly Accountable Sanitizable Signatures from Verifiable Ring Signatures

Xavier Bultel[1,2] and Pascal Lafourcade[1,2(✉)]

[1] CNRS, UMR 6158, LIMOS, 63173 Aubière, France
[2] Université Clermont Auvergne, BP 10448, 63000 Clermont-Ferrand, France
pascal.lafoucade@uca.fr

Abstract. An *Unlinkable Sanitizable Signature* scheme (USS) allows a sanitizer to modify some parts of a signed message in such away that nobody can link the modified signature to the original one. A *Verifiable Ring Signature* scheme (VRS) allows the users to sign messages anonymously within a group where a user can prove *a posteriori* to a verifier that it is the author of a given signature. In this paper, we first revisit the notion of VRS: we improve the proof capabilities of the users, we give a complete security model for VRS and we give an efficient and secure scheme called EVeR. Our main contribution is GUSS, a *Generic USS* based on a VRS scheme and an unforgeable signature scheme. We show that GUSS instantiated with EVeR and Schnorr's signature is twice as efficient as the best USS scheme of the literature. Moreover, we propose a stronger definition of *accountability*: an USS is *accountable* when the signer can prove whether a signature is sanitized. We formally define the notion of *strong accountability* where the sanitizer can also prove the origin of a signature. We show that the notion of strong accountability is important in practice. Finally, we prove the security properties of GUSS (including strong accountability) and EVeR under the Decisional Diffie-Hellman (DDH) assumption in the random oracle model.

1 Introduction

Sanitizable Signatures (SS) were introduced by Ateniese *et al.* [1], but similar primitives were independently proposed in [31]. In this primitive, a *signer* allows a proxy (called the *sanitizer*) to modify some parts of a signed message. For example, a magistrate wishes to delegate the power to summon someone to the court to his secretary. He signs the message "*Franz* is summoned to court for an interrogation on *Monday*" and gives the signature to his secretary, where "Franz" and "Monday" are sanitizable and the other parts are fixed. Thus, in order to summon Joseph K. on Saturday in the name of the magistrate, the

This research was conducted with the support of the "Digital Trust" Chair from the University of Auvergne Foundation.

S. Capkun and S. S. M. Chow (Eds.): CANS 2017, LNCS 11261, pp. 203–226, 2018.
https://doi.org/10.1007/978-3-030-02641-7_10

secretary can change the signed message into *"Joseph K.* is summoned to the court for an interrogation on *Saturday"*.

Ateniese *et al.* in [1] proposed some applications of this primitive in privacy of health data, authenticated media streams and reliable routing information. They also introduced five security properties formalized by Brzuska *et al.* in [8]:

Unforgeability: no unauthorized user can generate a valid signature.

Immutability: the sanitizer cannot transform a signature from an unauthorized message.

Privacy: no information about the original message is leaked by a sanitized signature.

Transparency: nobody can say if a signature is sanitized or not.

Accountability: the signer can prove whether a signature is sanitized.

Finally, in [9] the authors point out a non-studied but relevant property called *unlinkability*: a scheme is unlinkable when it is not possible to link a sanitized signature to the original one. The authors give a generic unlinkable scheme based on group signatures. In 2016, Fleischhacker *et al.* [21] give a more efficient construction based on signatures with re-randomizable keys.

On the other hand, ring signature is a well-studied cryptographic primitive introduced by Rivest *et al.* in [29], where some users can sign anonymously within a group of users. Such a scheme is *verifiable* [28] when any user can prove that he is the signer of a given message. In this paper, we improve the properties of VRS by increasing the proof capabilities of the users. We also give an efficient VRS scheme called EVeR and a generic unlinkable sanitizable signature scheme called GUSS that uses a verifiable ring signature. We also show that the definition of accountability is too weak for practical use and we propose a stronger definition.

Contributions: Existing VRS schemes allow any user to prove that he is the signer of a given message. We extend the definition of VRS to allow a user to prove that he is not the signer of a given message. We give a formal security model for VRS that takes into account this property. We first extend the classical security properties of ring signatures to verifiable ring signatures, namely the *unforgeability* (no unauthorized user can forge a valid signature) and the *anonymity* (nobody can distinguish who is the signer in the group). In addition we define the *accountability* (if a user signs a message then he cannot prove that he is not the signer) and the *non-seizability* (a user cannot prove that he is the signer of a message if it is not true, and a user cannot forge a message such that the other users cannot prove that they are not the signers). To the best of our knowledge, it is the first time that formal security models are proposed for VRS. We also design an efficient secure VRS scheme under the DDH assumption in the random oracle model.

The definition of accountability for SS given in [8,9,21] considers that the signer can prove the origin of a signature (signer or sanitizer) by using a proof algorithm such that:

1. The signer cannot forge a signature and a proof that the signature has been forged by the sanitizer.

2. The sanitizer cannot forge a signature such that the proof algorithm blames the signer.

The proof algorithm requires the secret key of the signer. To show that this definition is too weak, we consider a signer who cannot prove the origin of a litigious signature. The signer claims that he lost his secret key because of problems with his hard drive. There is no way to verify whether the signer is lying. Unfortunately, without his secret key, the signer cannot generate the proof for the litigious signature, hence nobody can judge whether the signature is sanitized or not. Depending on whether the signer is lying, there is a risk of accusing the signer or the sanitizer wrongly. To solve this problem, we add a second proof algorithm that allows the sanitizer to prove the origin of a signature. To achieve *strong accountability*, the two following additional properties are required:

1. The sanitizer cannot sanitize a signature σ and prove that σ is not sanitized.
2. The signer cannot forge a signature such that the sanitizer proof algorithm accuses the sanitizer.

The main contribution of this paper is to propose an efficient and generic unlinkable SS scheme called GUSS. This scheme is instantiated by a VRS and an unforgeable signature scheme. It is the first SS scheme that achieves strong accountability. We compare GUSS with the other schemes of the literature:

Brzuska et al. [9]. This scheme is based on group signatures. Our scheme is built on the same model, but it uses ring signatures instead of group signatures. The main advantage of group signatures is that the size of the signature is not proportional to the size of the group. However, for small groups, ring signatures are much more efficient than group signatures. Since the scheme of Brzuska et al. and GUSS uses group/ring signatures for groups of two users, GUSS is much more practical for an equivalent level of genericity.

Fleischhacker et al. [21] This scheme is based on *signatures with re-randomizable keys*. It is generic, however it uses different tools that must have special properties to be compatible with each other. To the best of our knowledge, it is the most efficient scheme of the literature. GUSS instantiated with EVeR and Schnorr's signature is twice as efficient as the best instantiation of this scheme. In Fig. 1, we compare the efficiency of each algorithm of our scheme and the scheme of Fleischhacker et al.

Lai et al. [26] Recently, Lai et al. proposed a USS that is secure in the standard model. This scheme uses pairings and is much less efficient than the scheme of Fleischhacker et al., so this scheme is much less efficient than our scheme. In their paper [26], Lai et al. give a comparison of the efficiency of the three schemes of the literature.

Related Works: *Sanitizable Signatures* (SS) was first introduced by Ateniese et al. [1]. Later, Brzuska et al. gave formal security definitions [8] for *unforgeability*, *immutability*, *privacy*, *transparency* and *accountability*. *Unlinkability* was introduced and formally defined by Brzuska et al. in [9]. In [10], Brzuska et al.

	SiGen	SaGen	Sig	San	Ver	SiProof	SiJudge	Total	pk	spk	sk	ssk	σ	π	Total
[21]	7	1	15	14	17	23	6	**83**	7	1	14	1	14	4	**41**
GUSS	2	1	8	7	10	3	4	**35**	2	1	2	1	12	5	**23**

Fig. 1. Comparison of GUSS and the scheme of Fleischhacker *et al.*: The first six columns give the number of exponentiations of each algorithms of both schemes, namely the key generation algorithm of the signer (SiGen) and the sanitizer (SaGen), the signature algorithm (Sig), the verification algorithm (Ver), the sanitize algorithm (San), the proof algorithm (SiProof) and the judge algorithm (SiJudge). The last six columns give respectively the size of the public key of the signer (pk) and the sanitizer (spk), the size of the secret key of the signer (sk) and the sanitizer (ssk), the size of a signature (σ) and the size of a proof (π) outputted by SiProof. This size is measured in elements of a group $\mathbb{G}$ of prime order. As in [21], for the sake of clarity, we do not distinguish between elements of $\mathbb{G}$ and elements of $\mathbb{Z}_p^*$. We consider the best instantiation of the scheme of Fleischhacker *et al.* given in [21]. In Appendix A, we give a detailed complexity evaluation of our schemes.

introduce an alternative definition of accountability called *non-interactive public accountability* where the capability to prove the origin of a signature is given to a third party. One year later, the same authors propose a stronger definition of unlinkability [11] and design a scheme that is both strongly unlinkable and non-interactively public accountable. However, non-interactive public accountability is not compatible with transparency. In this paper, we focus on schemes that are unlinkable, transparent and interactively accountable. To the best of our knowledge, there are only 3 schemes with these 3 properties, *i.e.* [9,21,26].

Some works focus on other properties of SS that we do not consider here, such as SS with multiple sanitizers [15], or SS where the power of the sanitizer is limited [14]. Finally, there exist other primitives that solve related but different problems such as homomorphic signatures [25], redactable signatures [7] or proxy signatures [22]. The differences between these primitives and sanitizable signatures are detailed in [21].

On the other hand, *Ring Signatures* (RS) [29] were introduced by Rivest *et al.* in 2003. Security models of this primitive were defined in [4]. *Verifiable Ring Signatures* (VRS) [28] were introduced in 2003 by Lv. RS allow the users to sign anonymously within a group, and VRS allow a user to prove that it is the signer of a given message. The authors of [28] give a VRS construction that is based on the discrete logarithm problem. Two other VRS schemes were proposed by Wand *et al.* [32] and by Changlung *et al.* [16]. The first one is based on the Nyberg-Rueppel signature scheme and the second one is a generic construction based on multivariate public key cryptosystems. In these three schemes, a user can prove that he is the signer of a given signature, however, he has no way to prove that he is not the signer, and it seems to be non-trivial to add this property to these schemes. Convertible ring signatures [27] are very close to verifiable ring signatures: they allow the signer of an anonymous (ring) signature to transform it into a standard signature (*i.e.* a *deanonymized* signature). It can be used as a verifiable ring signature because the deanonymized signature can be viewed as a proof that the user is the signer of a given message. However, in this paper we propose a stronger definition of VRS where a user also can prove that he is not the signer of a message, and this property cannot be achieved using convertible signatures.

A *Revocable-iff-Linked Ring Signature* (RLRS) [2] (also called *List Signature* [13]) is a kind of RS that has the following property: if a user signs two messages for the same *event-id*, then it is possible to link these signatures and the user's identity is publicly revealed. It can be used to design a VRS in our model: to prove whether he is the signer of a given message, the user signs a second message using the same event-id. If the two signatures are linked, then the judge is convinced that the user is the signer, else he is convinced that the user is not the signer. However, RLRS requires security properties that are too strong for VRS (linkability and traceability) and it would result in less efficient schemes.

Outline: In Sect. 2, we recall the standard cryptographic tools used in this paper. In Sect. 3 and Sect. 4, we present the formal definition and the security models for verifiable ring signatures and unlinkable sanitizable signatures. In Sect. 5, we present our scheme EvER. Finally, in Sect. 6, we present our scheme GUSS, before concluding in Sect. 7.

2   Cryptographic Tools

We present the cryptographic tools used throughout this paper. We first recall the DDH assumption.

Definition 1 (DDH [5]**).** *Let $\mathbb{G}$ be a multiplicative group of prime order p and $g \in \mathbb{G}$ be a generator. Given an instance $h = (g^a, g^b, g^z)$ for unknown $a, b, z \xleftarrow{\$} \mathbb{Z}_p^*$, the Decisional Diffie-Hellman (DDH) problem is to decide whether $z = a \cdot b$ or not. The DDH assumption states that there exists no polynomial time algorithm that solves the DDH problem with a non-negligible advantage.*

In the following, we recall some notions about digital signatures.

Definition 2 ((Deterministic) Digital Signature (DS)). *A Digital Signature scheme $S = (D.Init, D.Gen, D.Sig, D.Ver)$ is a tuple of four algorithms defined as follows:*

$D.Init(1^k)$: *It returns a setup value* set.
$D.Gen(set)$: *It returns a pair of signer public/private keys* (pk, sk).
$D.Sig(m, sk)$: *It returns a signature σ of m using the key* sk.
$D.Ver(pk, m, \sigma)$: *It returns a bit b.*

S is said to be deterministic when the algorithm $D.Sig$ is deterministic. S is said to be correct when for any security parameter $k \in \mathbb{N}$, any message $m \in \{0,1\}^*$, any set $\leftarrow D.Init(1^k)$ and any $(pk, sk) \leftarrow D.Gen(set)$, $D.Ver(pk, m, D.Sig(m, sk)) = 1$. Moreover, such a scheme is unforgeable when

no polynomial adversary wins the following experiment with non-negligible probability where D.Sig(·, sk) is a signature oracle, q_S is the number of queries to this oracle and σ_i is the ith signature computed by the signature oracle:

$\mathsf{Exp}^{\mathsf{unf}}_{S,\mathcal{A}}(k)$:
$set \leftarrow D.\mathsf{Init}(1^k)$
$(pk, sk) \leftarrow D.\mathsf{Gen}(set)$
$(m_*, \sigma_*) \leftarrow \mathcal{A}^{D.\mathsf{Sig}(\cdot, sk)}(pk)$
$if\ (D.\mathsf{Ver}(pk, m_*, \sigma_*) = 1)\ and\ (\forall\ i \in [\![1, q_S]\!], \sigma_i \neq \sigma_*)$
$then\ return\ 1,\ else\ return\ 0$

As it is mentioned in [21], any DS scheme can be changed into a deterministic scheme without loss of security using a pseudo random function, that can be simulated by a hash function in the random oracle model. The following scheme is the deterministic version of the well-known Schnorr's Signature scheme [30].

Scheme 1 (Deterministic Schnorr's Signature [30]). *This signature is defined by the following algorithms:*

D.$\mathsf{Init}(1^k)$: *It returns a setup value* $set = (\mathbb{G}, p, g, H)$ *where* $\mathbb{G}$ *is a group of prime order* p, $g \in \mathbb{G}$ *and* $H : \{0, 1\}^* \to \mathbb{Z}_p^*$ *is a hash function.*

D.$\mathsf{Gen}(set)$: *It picks* $sk \xleftarrow{\$} \mathbb{Z}_p^*$, *computes* $pk = g^{sk}$ *and returns* (pk, sk).

D.$\mathsf{Sig}(m, \mathsf{sk})$: *It computes the* $r = H(m\|sk)$, $R = g^r$, $z = r + sk \cdot H(R\|m)$ *and returns* $\sigma = (R, z)$.

D.$\mathsf{Ver}(pk, m, \sigma)$: *It parses* $\sigma = (R, z)$, *if* $g^z = R \cdot pk^{H(R\|m)}$ *then it returns 1, else 0.*

This DS scheme is deterministic and unforgeable under the DL assumption in the random oracle model.

A Zero-Knowledge Proof (ZKP) [23] allows a prover knowing a witness w to convince a verifier that a statement s is in a given language without leaking any information. Such a proof is a Proof of Knowledge (PoK) [3] when the verifier is also convinced that the prover knows the witness w. We recall the definition of a non-interactive zero-knowledge proof of knowledge.

Definition 3 (NIZKP). *Let* $\mathcal{R}$ *be a binary relation and let* $\mathcal{L}$ *be a language such that* $s \in \mathcal{L} \Leftrightarrow (\exists w, (s, w) \in \mathcal{R})$. *A* non-interactive ZKP *(NIZKP) for the language* $\mathcal{L}$ *is a couple of algorithms* (Prove, Verify) *such that:*

Prove(s, w). *This algorithm outputs a proof* π.
Verify(s, π). *This algorithm outputs a bit* b.

A NIZKP proof has the following properties:

Completeness. *For any statement* $s \in \mathcal{L}$ *and the corresponding witness* w, *we have that* Verify$(s; \mathsf{Prove}(s, w)) = 1$.

Soundness. *There is no polynomial time adversary* $\mathcal{A}$ *such that* $\mathcal{A}(\mathcal{L})$ *outputs* (s, π) *such that* Verify$(s, \pi) = 1$ *and* $s \notin \mathcal{L}$ *with non-negligible probability.*

Zero-knowledge. *A proof π leaks no information, i.e. there exists a polynomial time algorithm Sim (called the simulator) such that Prove(s, w) and Sim(s) follow the same probability distribution.*

Moreover, such a proof is a proof of knowledge when for any $s \in \mathcal{L}$ and any algorithm $\mathcal{A}$, there exists a polynomial time knowledge extractor $\mathcal{E}$ such that the probability that $\mathcal{E}^{\mathcal{A}(s)}(s)$ outputs a witness w such that $(s, w) \in \mathcal{R}$ given access to $\mathcal{A}(s)$ as an oracle is as high as the probability that $\mathcal{A}(s)$ outputs a proof π such that Verify$(s, \pi) = 1$.

3 Formal Model of Verifiable Ring Signatures

We formally define the Verifiable Ring Signatures (VRS) and the corresponding security notions. A VRS is a ring signature scheme where a user can prove to a judge whether he is the signer of a message or not. It is composed of six algorithms. V.Init, V.Gen, V.Sig and V.Ver are defined as in the usual ring signature definitions. V.Gen generates public and private keys. V.Sig anonymously signs a message according to a set of public keys. V.Ver verifies the soundness of a signature. A VRS has two additional algorithms: V.Proof allows a user to prove whether he is the signer of a message or not, and V.Judge allows anybody to verify the proofs outputted by V.Proof.

Definition 4 (Verifiable Ring Signature (VRS)). *A Verifiable Ring Signature scheme is a tuple of six algorithms defined by:*

V.Init(1^k): *It returns a setup value* set.
V.Gen(set): *It returns a pair of signer public/private keys* (pk, sk).
V.Sig(L, m, sk): *This algorithm computes a signature σ using the key sk for the message m according to the set of public keys L.*
V.Ver(L, m, σ): *It returns a bit b: if the signature σ of m is valid according to the set of public key L then $b = 1$, else $b = 0$.*
V.Proof(L, m, σ, pk, sk): *It returns a proof π for the signature σ of m according to the set of public key L.*
V.Judge(L, m, σ, pk, π): *It returns a bit b or the bottom symbol $\bot$: if $b = 1$ (resp. 0) then π proves that σ was (resp. was not) generated by the signer corresponding to the public key pk. It outputs $\bot$ when the proof is not well formed.*

Unforgeability: We first adapt the unforgeability property of ring signatures [4] to VRS. Informally, a VRS is unforgeable when no adversary is able to forge a signature for a ring of public keys without any corresponding secret key. In this model, the adversary has access to a signature oracle V.Sig$(\cdot, \cdot, \cdot)$ (that outputs signatures of chosen messages for chosen users in the ring) and a proof oracle V.Proof$(\cdot, \cdot, \cdot, \cdot, \cdot)$ (that outputs proofs as the algorithm V.Proof for chosen signatures and chosen users). The adversary succeeds when it outputs a valid signature that was not already generated by the signature oracle.

Definition 5 (Unforgeability). *Let P be a VRS and n be an integer. Let the two following oracles be:*

$\mathsf{V.Sig}(\cdot,\cdot,\cdot)$: *On input (L, l, m), if $1 \leq l \leq n$ then it runs $\sigma \leftarrow \mathsf{V.Sig}(L, sk_l, m)$ and returns σ, else it returns $\perp$.*

$\mathsf{V.Proof}(\cdot,\cdot,\cdot,\cdot,\cdot)$: *On input (L, m, σ, l), if $1 \leq l \leq n$ then this proof oracle runs $\pi \leftarrow \mathsf{V.Proof}(L, m, \sigma, pk_l, sk_l)$ and returns π, else it returns $\perp$.*

P is n-unf secure when for any polynomial time adversary $\mathcal{A}$, the probability that $\mathcal{A}$ wins the following experiment is negligible, where q_S is the number of calls to the oracle $\mathsf{V.Sig}(\cdot,\cdot,\cdot)$ and σ_i is the ith signature outputted by the oracle $\mathsf{V.Sig}(\cdot,\cdot,\cdot)$:

> $\mathbf{Exp}^{n\text{-}\mathsf{unf}}_{P,\mathcal{A}}(k)$:
> $set \leftarrow \mathsf{V.Init}(1^k)$
> $\forall 1 \leq i \leq n, (pk_i, sk_i) \leftarrow \mathsf{V.Gen}(set)$
> $(L_*, \sigma_*, m_*) \leftarrow \mathcal{A}^{\mathsf{V.Sig}(\cdot,\cdot,\cdot), \mathsf{V.Proof}(\cdot,\cdot,\cdot,\cdot,\cdot)}(\{pk_i\}_{1 \leq i \leq n})$
> if $\mathsf{V.Ver}(L_*, \sigma_*, m_*) = 1$ and $L_* \subseteq \{pk_i\}_{1 \leq i \leq n}$ and $\forall\, i \in [\![1, q_S]\!], \sigma_i \neq \sigma_*$
> then return 1, else return 0

P is unforgeable when it is $t(k)$-unf secure for any polynomial t.

Annonymity: We adapt the anonymity property of ring signatures [4] to VRS. Informally, a VRS is anonymous when no adversary is able to link a signature to the corresponding user. The adversary has access to the signature oracle and the proof oracle. During a first phase, it chooses two honest users in the ring, and in the second phase, it has access to a challenge oracle $\mathsf{LRSO}_b(d_0, d_1, \cdot, \cdot)$ that outputs signatures of chosen messages using the secret key of one of the two chosen users. The adversary succeeds if he guesses which user is chosen by the challenge oracle. Note that if the adversary uses the proof oracle on the signatures generated by the challenge oracle then he loses the experiment.

Definition 6 (Anonymity). *Let P be a VRS and let n be an integer. Let the following oracle be:*

$\mathsf{LRSO}_b(d_0, d_1, \cdot, \cdot)$: *On input (m, L), if $\{pk_{d_0}, pk_{d_1}\} \subseteq L$ then this oracle runs $\sigma \leftarrow \mathsf{V.Sig}(L, sk_{d_b}, m)$ and returns σ, else it returns $\perp$.*

P is n-ano secure when for any polynomial time adversary $\mathcal{A} = (\mathcal{A}_1, \mathcal{A}_2)$, the probability that $\mathcal{A}$ wins the following experiment is negligibly close to $1/2$, where $\mathsf{V.Sig}(\cdot,\cdot,\cdot)$ and $\mathsf{V.Proof}(\cdot,\cdot,\cdot,\cdot,\cdot)$ are defined as in Definition 5 and where q_S (resp. q_P) is the number of calls to the oracle $\mathsf{V.Sig}(\cdot,\cdot,\cdot)$ (resp. $\mathsf{V.Proof}(\cdot,\cdot,\cdot,\cdot,\cdot)$), $(L_i, m_i, \sigma_i, l_i)$ is the ith query sent to oracle $\mathsf{V.Proof}(\cdot,\cdot,\cdot,\cdot,\cdot)$ and σ'_j is the jth signature outputted by the oracle $\mathsf{LRSO}_b(d_0, d_1, \cdot, \cdot)$:

$\mathbf{Exp}_{P,\mathcal{A}}^{n\text{-ano}}(k)$:

$set \leftarrow V.Init(1^k)$

$\forall 1 \leq i \leq n, (pk_i, sk_i) \leftarrow V.Gen(set)$

$(d_0, d_1) \leftarrow \mathcal{A}_1^{V.Sig(\cdot,\cdot,\cdot),V.Proof(\cdot,\cdot,\cdot,\cdot,\cdot)}(\{pk_i\}_{1 \leq i \leq n})$

$b \xleftarrow{\$} \{0,1\}$

$b_* \leftarrow \mathcal{A}_2^{V.Sig(\cdot,\cdot,\cdot),V.Proof(\cdot,\cdot,\cdot,\cdot,\cdot),LRSO_b(d_0,d_1,\cdot,\cdot)}(\{pk_i\}_{1 \leq i \leq n})$

$if\ (b = b_*)\ and\ (\forall\ i,j \in [\![1, \max(q_S, q_P)]\!],\ (\sigma_i \neq \sigma'_j)\ or\ (l_i \neq d_0\ and\ l_i \neq d_1))$

$then\ return\ 1,\ else\ return\ 0$

P is anonymous *when it is* $t(k)$-ano *secure for any polynomial* t.

Accountability: We consider an adversary that has access to a proof oracle and a signature oracle. A VRS is accountable when no adversary is able to forge a signature σ (that was not outputted by the signature oracle) together with a proof that it is not the signer of σ. Note that the ring of σ must contain at most one public key that does not come from an honest user, thus the adversary knows at most one secret key that corresponds to a public key in the ring.

Definition 7 (Accountability). *Let* P *be a VRS and let* n *be an integer.* P *is* n-acc *secure when for any polynomial time adversary* $\mathcal{A}$, *the probability that* $\mathcal{A}$ *wins the following experiment is negligible, where* $V.Sig(\cdot,\cdot,\cdot)$ *and* $V.Proof(\cdot,\cdot,\cdot,\cdot,\cdot)$ *are defined as in Definition 5 and where* q_S *is the number of calls to the oracle* $V.Sig(\cdot,\cdot,\cdot)$ *and* σ_i *is the* i^{th} *signature outputted by the oracle* $V.Sig(\cdot,\cdot,\cdot)$:

$\mathbf{Exp}_{P,\mathcal{A}}^{n\text{-acc}}(k)$:

$set \leftarrow V.Init(1^k)$

$\forall 1 \leq i \leq n, (pk_i, sk_i) \leftarrow V.Gen(set)$

$(L_*, m_*, \sigma_*, pk_*, \pi_*) \leftarrow \mathcal{A}^{V.Sig(\cdot,\cdot,\cdot),V.Proof(\cdot,\cdot,\cdot,\cdot,\cdot)}(\{pk_i\}_{1 \leq i \leq n})$

$if\ (L \subseteq \{pk_i\}_{1 \leq i \leq n} \cup \{pk_*\})\ and\ (V.Ver(L_*, \sigma_*, m_*) = 1)\ and$

$\quad (V.Judge(L_*, m_*, \sigma_*, pk_*, \pi_*) = 0)\ and\ (\forall\ i \in [\![1, q_S]\!],\ \sigma_i \neq \sigma_*)$

$then\ return\ 1,\ else\ return\ 0$

P is accountable *when it is* $t(k)$-acc *secure for any polynomial* t.

Non-seizability: We distinguish two experiments for this property: the first experiment, denoted non-sei-1, considers an adversary that has access to a proof oracle and a signature oracle. Its goal is to forge a valid signature with a proof that the signer is another user in the ring.

Definition 8 (n-non-sei-1 experiment). *Let* P *be a SS.* P *is* n-non-sei-1 *secure when for any polynomial time adversary* $\mathcal{A}$, *the probability that* $\mathcal{A}$ *wins the following experiment is negligible, where* $V.Sig(\cdot,\cdot,\cdot)$ *and* $V.Proof(\cdot,\cdot,\cdot,\cdot,\cdot)$ *and where* q_S *is the number of calls to the oracle* $V.Sig(\cdot,\cdot,\cdot)$ *and* (L_i, l_i, m_i) *(resp.* σ_i) *is the* i^{th} *query to the oracle* $V.Sig(\cdot,\cdot,\cdot)$ *(resp. signature outputted by this oracle):*

$\mathsf{Exp}_{P,\mathcal{A}}^{n\text{-non-sei-1}}(k)$:

set $\leftarrow$ V.Init(1^k)

$\forall 1 \leq i \leq n, (pk_i, sk_i) \leftarrow$ V.Gen(set)

$(L_*, m_*, \sigma_*, l_*, \pi_*) \leftarrow \mathcal{A}^{V.Sig(\cdot,\cdot,\cdot), V.Proof(\cdot,\cdot,\cdot,\cdot)}(\{pk_i\}_{1 \leq i \leq n})$

$\pi \leftarrow$ V.Proof($L_*, m_*, \sigma_*, pk, sk$)

if (V.Ver(L_*, σ_*, m_*) = 1) and

$\quad$ (V.Judge($L_*, m_*, \sigma_*, pk_{l_*}, \pi_*$) = 1) and

$\quad$ ($\forall\, i \in [\![1, q_S]\!], (L_i, l_i, m_i, \sigma_i) = (L_*, l_*, m_*, \sigma_*)$))

then return 1, else return 0

The second experiment, denoted non-sei-2, considers an adversary that has access to a proof oracle and a signature oracle and that receives the public key of a honest user as input. The goal of the adversary is to forge a signature σ such that the proof algorithm ran by the honest user returns a proof that σ was computed by the honest user (*i.e.* the judge algorithm returns 1) or a non-valid proof (*i.e.* the judge algorithm returns $\perp$). Moreover, the signature σ must not come from the signature orale.

Definition 9 (Non-seizability). *Let P be a VRS and n be an integer. P is n-non-sei-2 secure when for any polynomial time adversary $\mathcal{A}$, the probability that $\mathcal{A}$ wins the following experiment is negligible, where V.Sig($\cdot, \cdot, \cdot$) and V.Proof($\cdot, \cdot, \cdot, \cdot, \cdot$) are defined as in Definition 5 and where q_S is the number of calls to the oracle V.Sig($\cdot, \cdot, \cdot$) and σ_i is the ith signature outputted by the oracle V.Sig($\cdot, \cdot, \cdot$):*

$\mathsf{Exp}_{P,\mathcal{A}}^{n\text{-non-sei-2}}(k)$:

set $\leftarrow$ V.Init(1^k)

$(pk, sk) \leftarrow$ V.Gen(set)

$(L_*, m_*, \sigma_*) \leftarrow \mathcal{A}^{V.Sig(\cdot,\cdot,\cdot), V.Proof(\cdot,\cdot,\cdot,\cdot,\cdot)}(pk)$

$\pi \leftarrow$ V.Proof($L_*, m_*, \sigma_*, pk, sk$)

if (V.Ver(L_*, σ_*, m_*) = 1) and

$\quad$ (V.Judge($L_*, m_*, \sigma_*, pk_*, \pi_*$) $\neq$ 0) and ($\forall\, i \in [\![1, q_S]\!], \sigma_i \neq \sigma_*$)

then return 1, else return 0

P is non-seizable when it is both $t(k)$-non-sei-1 and $t(k)$-non-sei-2 secure for any polynomial t.

4   Formal Model of Sanitizable Signature

We give the formal definition and security properties of the sanitizable signature primitive. Compared to the previous definitions where only the signer can prove the origin of a signature, our definition introduces algorithms that allow the sanitizer to prove the origin of a signature. Moreover, in addition to the usual security models of [8], we present two new security experiments that improve the accountability definition.

A SS scheme contains 10 algorithms. Init outputs the setup values. SiGen and SaGen generate respectively the signer and the sanitizer public/private keys. As in classical signature schemes, the algorithms Sig and Ver allow the users to sign a message and to verify a signature. However, the signatures are computed using a sanitizer public key and an admissible function ADM. The algorithm San allows the sanitizer to transform a signature of a message m according to a modification function MOD: if MOD is admissible according to the admissible function (*i.e.* MOD(ADM) = 1) this algorithm returns a signature of the message $m' = $ MOD(m).

SiProof allows the signer to prove whether a signature is sanitized or not. Proofs outpoutted by this algorithm can be verified by anybody using the algorithm SiJudge. Finally, algorithms SaProof and SaJudge have the same functionalities as SiProof and SiJudge, but the proofs are computed from the secret parameters of the sanitizer instead of the signer.

Definition 10 (Sanitizable Signature (SS)). *A Sanitizable Signature scheme is a tuple of 10 algorithms defined as follows:*

Init(1^k): *It returns a setup value* set.

SiGen(set): *It returns a pair of signer public/private keys* (pk, sk).

SaGen(set): *It returns a pair of sanitizer public/private keys* (spk, ssk).

Sig(m, sk, spk, ADM): *This algorithm computes a signature σ from the message m using the secret key sk, the sanitizer public key spk and the admissible function* ADM. *Note that we assume that* ADM *can be efficiently recovered from any signature as in the definition of Fleischhacker et al.* [21].

San(m, MOD, σ, pk, ssk): *Let the admissible function* ADM *according to the signature σ. If* ADM(MOD) = 1 *then this algorithm returns a signature σ' of the message $m' = $ MOD(m) using the signature σ, the signer public key pk and the sanitizer secret key ssk. Else it returns $\perp$.*

Ver(m, σ, pk, spk): *It returns a bit b: if the signature σ of m is valid for the two public keys pk and spk then $b = 1$, else $b = 0$.*

SiProof(sk, m, σ, spk): *It returns a signer proof π_{si} for the signature σ of m using the signer secret key sk and the sanitizer public key spk.*

SaProof(ssk, m, σ, pk): *It returns a sanitizer proof π_{sa} for the signature σ of m using the sanitizer secret key ssk and the signer public key pk.*

SiJudge(m, σ, pk, spk, π_{si}): *It returns a bit d or the bottom symbol $\perp$: if π_{si} proves that σ comes from the signer corresponding to the public key pk then $d = 1$, else if π_{si} proves that σ comes from the sanitizer corresponding to the public key spk then $d = 0$, else the algorithm outputs $\perp$.*

SaJudge(m, σ, pk, spk, π_{sa}): *It returns a bit d or the bottom symbol $\perp$: if π_{sa} proves that σ comes from the signer corresponding to the public key pk then $d = 1$, else if π_{sa} proves that σ comes from the sanitizer corresponding to the public key spk then $d = 0$, else the algorithm outputs $\perp$.*

As it is mentioned in Introduction, SS schemes have the following security properties: *unforgeability, immutability, privacy, transparency* and *accountability*. In [8] authors show that if a scheme has the *immutability*, the *transparency* and the *accountability* properties, then it has the *unforgeability* and the *privacy* properties. Hence we do not need to prove these two properties, so we do not recall their formal definitions.

Immutability: A SS is immutable when no adversary is able to sanitize a signature without the corresponding sanitizer secret key or to sanitize a signature using a modification function that is not admissible (*i.e.* ADM(MOD) = 0). To help him, the adversary has access to a signature oracle Sig(., sk, ., .) and a proof oracle SiProof(sk, ., ., .).

Definition 11 (Immutability [8]). *Let the following oracles be:*

Sig(., sk, ., .): *On input* (m, ADM, spk), *this oracle returns* Sig(m, sk, ADM, spk).
SiProof(sk, ., ., .): *On input* (m, σ, spk), *this oracle returns* SiProof(sk, m, σ, spk).

Let P be a SS. P is Immut *secure (or immutable) when for any polynomial time adversary $\mathcal{A}$, the probability that $\mathcal{A}$ wins the following experiment is negligible, where q_{Sig} is the number of calls to the oracle Sig(., sk, ., .), $(m_i, \text{ADM}_i, spk_i)$ is the ith query asked to the oracle Sig(., sk, ., .) and σ_i is the corresponding response:*

$\text{Exp}_{P,\mathcal{A}}^{\text{Immut}}(k)$:
$set \leftarrow Init(1^k)$
$(pk, sk) \leftarrow SiGen(set)$
$(spk_*, m_*, \sigma_*) \leftarrow \mathcal{A}^{Sig(.,sk,.,.),SiProof(sk,.,.,.)}(pk)$
$if\ (Ver(m_*, \sigma_*, pk, spk_*) = 1)\ and\ (\forall\ i \in [\![1, q_{Sig}]\!], (spk_* \neq spk_i)\ or$
$\quad (\forall\ \text{MOD}\ such\ that\ \text{ADM}_i(\text{MOD}) = 1, m_* \neq \text{MOD}(m_i)))$
$then\ return\ 1,\ else\ return\ 0$

Transparency: The transparency property guarantees that no adversary is able to distinguish whether a signature is sanitized or not. In addition to the signature oracle and the signer proof oracle, the adversary has access to a sanitize oracle San(., ., ., ., ssk) that sanitizes chosen signatures and a sanitizer proof oracle SaProof(ssk, ., ., .) that computes sanitizer proofs for given signatures. Moreover the adversary has access to a challenge oracle Sa/Si$(b, pk, spk, sk, ssk, ., ., .)$ that depends on a randomly chosen bit b: this oracle signs a given message and sanitizes it, if $b = 0$ then it outputs the original signature, otherwise it outputs the sanitized signature. The adversary cannot use the proof oracles on the signatures outputted by the challenge oracle. To succeed the experiment, the adversary must guess b.

Definition 12 (Transparency [8]). *Let the following oracles be:*

San(., ., ., ., ssk): *On input* $(m, \text{MOD}, \sigma, pk)$, *it returns* San$(m, \text{MOD}, \sigma, pk, ssk)$.
SaProof(ssk, ., ., .): *On input* (m, σ, pk), *this oracle returns* SaProof(ssk, m, σ, pk).
Sa/Si$(b, pk, spk, sk, ssk, ., ., .)$: *On input* $(m, \text{ADM}, \text{MOD})$, *if* $\text{ADM}(\text{MOD}) = 0$, *this oracle returns* $\perp$. *Else if* $b = 0$, *this oracle returns* Sig$(\text{MOD}(m), sk, spk, \text{ADM})$, *else if* $b = 1$, *this oracle returns* San$(m, \text{MOD}, \text{Sig}(m, sk, spk, \text{ADM}), pk, ssk)$.

Let P be a SS. P is Trans *secure (or transparent) when for any polynomial time adversary $\mathcal{A}$, the probability that $\mathcal{A}$ wins the following experiment is negligible, where Sig(., sk, ., .) and SiProof(sk, ., ., .) are defined as in Definition 11, and where $S_{Sa/Si}$ (resp. $S_{SiProof}$ and $S_{SaProof}$) corresponds to the set of all signatures outputted by the oracle Sa/Si (resp. sent to the oracles SiProof and SaProof):*

$\mathbf{Exp}_{P,\mathcal{A}}^{\text{Trans}}(k):$

$set \leftarrow Init(1^k)$

$(pk, sk) \leftarrow SiGen(set)$

$(spk, ssk) \leftarrow SaGen(set)$

$b \xleftarrow{\$} \{0,1\}$

$b' \leftarrow \mathcal{A}^{\,Sig(.,sk,.,.),San(.,.,.,.,ssk),SiProof(sk,.,.,.)}_{\,SaProof(ssk,.,.,.),Sa/Si(b,pk,spk,sk,ssk,.,.,.)}(pk, spk)$

$if\ (b = b')\ and\ (S_{Sa/Si} \cap (S_{SiProof} \cup S_{SaProof}) = \emptyset)$

$then\ return\ 1,\ else\ return\ 0$

Unlinkablility: The unlinkablility property ensures that a sanitized signature cannot be linked with the original one. We consider an adversary that has access to the signature oracle, the sanitize oracle, and both the signer and the sanitizer proof oracles. Moreover, the adversary has access to a challenge oracle $\mathsf{LRSan}(b, pk, ssk, ., .)$ that depends to a bit b: this oracle takes as input two signatures σ_0 and σ_1, the two corresponding messages m_0 and m_1 and two modification functions MOD_0 and MOD_1 chosen by the adversary. If the two signatures have the same admissible function ADM, if MOD_0 and MOD_1 are admissible according to ADM and if $\text{MOD}_0(m_0) = \text{MOD}_1(m_1)$ then the challenge oracle sanitizes σ_b using MOD_b and returns it. The goal of the adversary is to guess the bit b.

Definition 13 (Unlinkability [8]). *Let the following oracles be:*

$\mathsf{LRSan}(b, pk, ssk, ., .):$ *On input* $((m_0, \text{MOD}_0, \sigma_0)(m_1, \text{MOD}_1, \sigma_1))$, *if for $i \in \{0, 1\}$,* $Ver(m_i, \sigma_i, pk, spk) = 1$ *and* $\text{ADM}_0 = \text{ADM}_1$ *and* $\text{ADM}_0(\text{MOD}_0) = 1$ *and* $\text{ADM}_1(\text{MOD}_1) = 1$ *and* $\text{MOD}_0(m_0) = \text{MOD}_1(m_1)$, *then this oracle returns* $San(m_b, \text{MOD}_b, \sigma_b, pk, ssk)$, *else it returns* 0.

Let P be a SS of security parameter k. P is Unlink secure (or unlinkable) when for any polynomial time adversary $\mathcal{A}$, the probability that $\mathcal{A}$ wins the following experiment is negligibly close to $1/2$, where $Sig(., sk, ., .)$ and $SiProof(sk, ., ., .)$ are defined as in Definition 11 and $San(., ., ., ., ssk)$ and $SaProof(ssk, ., ., .)$ are defined as in Definition 12:

$\mathbf{Exp}_{P,\mathcal{A}}^{\text{Unlink}}(k):$

$set \leftarrow Init(1^k)$

$(pk, sk) \leftarrow SiGen(set)$

$(spk, ssk) \leftarrow SaGen(set)$

$b \xleftarrow{\$} \{0,1\}$

$b' \leftarrow \mathcal{A}^{\,Sig(.,sk,.,.),San(.,.,.,.,ssk)}_{\,SiProof(sk,.,.,.),SaProof(ssk,.,.,.),LRSan(b,pk,spk,.,.)}(pk, spk)$

$if\ (b = b')\ then\ return\ 1,\ else\ return\ 0$

Accountability: Standard defintion of accountability is shared into two security experiments: the sanitizer accountability and the signer accountability. In the sanitizer accountability experiment, the adversary has access to the signature oracle and the signer proof oracle. Its goal is to forge a signature such that the signer proof algorithm returns a proof that this signature is not sanitized. To succeed the experiment, this signature must not come from the signature oracle.

Definition 14 (Sanitizer Accountability [8]). *Let P be a SS. P is SaAcc-1 secure (or* sanitizer accountable*) when for any polynomial time adversary $\mathcal{A}$, the probability that $\mathcal{A}$ wins the following experiment is negligible, where the oracles $Sig(.,sk,.,.)$ and $SiProof(sk,.,.,.)$ are defined as in Definition 11, q_{Sig} is the number of calls to the oracle $Sig(.,sk,.,.)$, the tuple $(m_i, \text{ADM}_i, spk_i)$ is the ith query asked to the oracle $Sig(.,sk,.,.)$ and σ_i is the corresponding response:*

$\mathbf{Exp}_{P,\mathcal{A}}^{\mathsf{SaAcc\text{-}1}}(k):$
$set \leftarrow Init(1^k)$
$(pk, sk) \leftarrow SiGen(set)$
$(spk_*, m_*, \sigma_*) \leftarrow \mathcal{A}^{Sig(.,sk,.,.),SiProof(sk,.,.,.)}(pk)$
$\pi_{si}^* \leftarrow SiProof(sk, m_*, \sigma_*, spk_*)$
$if\ \forall\ i \in [\![1, q_{Sig}]\!],\ (\sigma_* \neq \sigma_i)$
    $and\ (Ver(m_*, \sigma_*, pk, spk_*) = 1)$
    $and\ (SiJudge(m_*, \sigma_*, pk, spk_*, \pi_{si}^*) \neq 0)$
$then\ return\ 1,\ else\ return\ 0$

In the signer accountability experiment, the adversary knows the public key of the sanitizer and has access to the sanitize oracle and the sanitizer proof oracle. Its goal is to forge a signature together with a proof that this signature is sanitized. To succeed the experiment, this signature must not come from the sanitize oracle.

Definition 15 (Signer Accountability [8]). *Let P be a SS. P is SiAcc-1 secure (or* signer accountable*) when for any polynomial time adversary $\mathcal{A}$, the probability that $\mathcal{A}$ wins the following experiment is negligible, where the oracle $San(.,.,.,.,ssk)$ and $SaProof(ssk,.,.,.)$ are defined as in Definition 12 and where q_{San} is the number of calls to the oracle $San(.,.,.,.,ssk)$, $(m_i, \text{MOD}_i, \sigma_i, pk_i)$ is the ith query asked to the oracle $San(.,.,.,.,ssk)$ and σ_i' is the corresponding response:*

$\mathbf{Exp}_{P,\mathcal{A}}^{\mathsf{SiAcc\text{-}1}}(k):$
$set \leftarrow Init(1^k)$
$(spk, ssk) \leftarrow SaGen(set)$
$(pk_*, m_*, \sigma_*, \pi_{si}^*) \leftarrow \mathcal{A}^{San(.,.,.,.,ssk),SaProof(ssk,.,.,.)}(spk)$
$if\ \forall\ i \in [\![1, q_{San}]\!],\ (\sigma_* \neq \sigma_i')$
    $and\ (Ver(m_*, \sigma_*, pk_*, spk) = 1)$
    $and\ (SiJudge(m_*, \sigma_*, pk_*, spk, \pi_{si}^*) = 0)$
$then\ return\ 1,\ else\ return\ 0$

Strong Accountability: Since our definition of sanitizable signature provides a second proof algorithm for the sanitizer, we define two additional security experiments (for signer and sanitizer accountability) to ensure the soundness of the proofs computed by this algorithm. We say that a scheme is strongly accountable when it is signer and sanitizer accountable for both the signer and the sanitizer proof algorithms.

Thus, in our second signer accountability experiment, we consider an adversary that has access to the sanitize oracle and the sanitizer proof oracle. Its goal is to forge a signature such that the sanitizer proof algorithm returns a proof that this signature is sanitized. To win the experiment, this signature must not come from the sanitize oracle.

Definition 16 (Strong Signer Accountability). *Let P be a SS. P is SiAcc-2 secure when for any polynomial time adversary $\mathcal{A}$, the probability that $\mathcal{A}$ wins the following experiment is negligible, where q_{San} is the number of calls to the oracle $San(.,.,.,.,ssk)$, $(m_i, \mathrm{MOD}_i, \sigma_i, pk_i)$ is the ith query asked to the oracle $San(.,.,.,.,ssk)$ and σ'_i is the corresponding response:*

$\mathbf{Exp}_{P,\mathcal{A}}^{\mathsf{SiAcc\text{-}2}}(k)$:
$set \leftarrow Init(1^k)$
$(spk, ssk) \leftarrow SaGen(set)$
$(pk_*, m_*, \sigma_*) \leftarrow \mathcal{A}^{San(.,.,.,.,ssk),SaProof(ssk,.,.,.)}(spk)$
$\pi_{sa}^* \leftarrow SaProof(ssk, m_*, \sigma_*, pk_*)$
$if\ \forall\ i \in [\![1, q_{San}]\!],\ (\sigma_* \neq \sigma'_i)$
$\quad and\ (Ver(m_*, \sigma_*, pk_*, spk) = 1)$
$\quad and\ (SaJudge(m_*, \sigma_*, pk_*, spk, \pi_{sa}^*) \neq 1)$
$then\ return\ 1,\ else\ return\ 0$

P is strong signer accountable *when it is both SiAcc-1 and SiAcc-2 secure.*

Finally, in our second sanitizer accountability experiment, we consider an adversary that knows the public key of the signer and has access to the signer oracle and the signer proof oracle. Its goal is to sanitize a signature with a proof that this signature is not sanitized. To win the experiment, this signature must not come from the signer oracle.

Definition 17 (Strong Sanitizer Accountability). *Let P be a SS. P is SaAcc-2 secure when for any polynomial time adversary $\mathcal{A}$, the probability that $\mathcal{A}$ wins the following experiment is negligible, $Sig(., sk, ., .)$ and $SiProof(sk, ., ., .)$ are defined as in Definition 11, q_{Sig} is the number of calls to the oracle $Sig(., sk, ., .)$, $(m_i, \mathrm{ADM}_i, spk_i)$ is the ith query asked to the oracle $Sig(., sk, ., .)$ and σ_i is the corresponding response:*

$\mathbf{Exp}_{P,\mathcal{A}}^{\mathsf{SaAcc\text{-}2}}(k)$:
$set \leftarrow Init(1^k)$
$(pk, sk) \leftarrow SaGen(set)$
$(spk_*, m_*, \sigma_*, \pi_{sa}^*) \leftarrow \mathcal{A}^{Sig(.,sk,.,.),SiProof(sk,.,.,.)}(spk)$
$if\ \forall\ i \in [\![1, q_{Sig}]\!],\ (\sigma_* \neq \sigma_i)$
$\quad and\ (Ver(m_*, \sigma_*, pk, spk_*) = 1)$
$\quad and\ (SaJudge(m_*, \sigma_*, pk, spk_*, \pi_{sa}^*) = 1)$
$then\ return\ 1,\ else\ return\ 0$

P is strong sanitizer accountable *when it is both SaAcc-1 and SaAcc-2 secure.*

5   An Efficient Verifiable Ring Signature: EVeR

We present our VRS scheme called EVeR (for *Efficient VErifiable Ring signature*). EVeR works as follows: the signer produces an anonymous commitment from his secret key and the message (*i.e.* a commitment that leaks no information about the secret key), then he proves that this commitment was produced with a secret key corresponding to one of the public keys of the group members using a zero-knowledge proof system. Note that the same methodology was used to design several ring signature schemes of the literature [2,13,17,24]. Moreover, to prove that he is (resp. he is not) the signer of a message, the user proves that the commitment was (resp. was not) produced from the secret key that corresponds to his public key using a zero-knowledge proof system. Our scheme is based on the DDH assumption and uses a NIZKP of equality of two discrete logarithms out of n elements. We show how to build this NIZKP: Let $\mathbb{G}$ be a group of prime order p, n be an integer and let the following binary relation be:

$$\mathcal{R}_n = \left\{ (s,w) : \begin{array}{c} s = \{(h_i, z_i, g_i, y_i)\}_{1 \le i \le n}; \\ \exists i \in [\![1,n]\!], (((h_i, z_i, g_i, y_i) \in \mathbb{G}^4) \\ \wedge (w = \log_{g_i}(y_i) = \log_{h_i}(z_i))); \end{array} \right\}.$$

We denote by $\mathcal{L}_n$ the language $\{s : \exists w, (s,w) \in \mathcal{R}_n\}$. Consider the case $n = 1$. In [18], authors present an interactive zero-knowledge proof of knowledge system for the relation $\mathcal{R}_1$. It proves the equality of two discrete logarithms. For example using $(h, z, g, y) \in \mathcal{L}_1$, a prover convinces a verifier that $\log_g(y) = \log_h(z)$. The witness used by the prover is $w = \log_g(y)$. This proof system is a *sigma protocol* in the sense that there are only three interactions: the prover sends a commitment, the verifier sends a challenge, and the prover returns a response.

To transform the proof system of $\mathcal{R}_1$ into a generic proof system of any $\mathcal{R}_n$, we use the generic transformation given in [19]. For any integer n and any relation $\mathcal{R}$, the authors show how to transform a zero-knowledge proof of knowledge of a witness w such that $(s,w) \in \mathcal{R}$ for a given statement s into a zero-knowledge proof of knowledge of a witness w such that there exists $s \in S$ such that $(s,w) \in \mathcal{R}$ for a given set of n statements S, under the condition that the proof is a sigma protocol. Note that the resulting proof system is also a sigma protocol.

The final step is to transform it into a non-interactive proof system. We use the well-known Fiat-Shamir transformation [20]. This transformation changes any interactive proof system that is a sigma protocol into a non interactive one. The resulting proof system is complete, sound and zero-knowledge in the random oracle model. Finally, we obtain the following scheme.

Scheme 2 (LogEq$_n$). *Let $\mathbb{G}$ be a group of prime order p, $H : \{0,1\} \to \mathbb{Z}_p^*$ be a hash function and n be an integer. We define the NIZKP LogEq$_n$ = (LEprove$_n$, LEverif$_n$) for $\mathcal{R}_n$ by:*

LEprove$_n(\{(h_i, z_i, g_i, y_i)\}_{1 \le i \le n}, x)$. *Let $x = \log_{g_j}(y_j) = \log_{h_j}(z_j)$, this algorithm picks $r_j \xleftarrow{\$} \mathbb{Z}_p^*$, computes $R_j = g_j^{r_j}$ and $S_j = h_j^{r_j}$. For all $i \in [\![1,n]\!]$ and $i \ne j$, it* picks $c_i \xleftarrow{\$} \mathbb{Z}_p^*$ and $\gamma_i \xleftarrow{\$} \mathbb{Z}_p^*$, and computes $R_i = g_i^{\gamma_i}/y_i^{c_i}$ and $S_i = h_i^{\gamma_i}/z_i^{c_i}$. It computes $c = H(R_1||S_1||\dots||R_n||S_n)$. It then computes $c_j = c/(\prod_{i=1;i\neq j}^{n} c_i)$ and $\gamma_j = r_j + c_j \cdot x$. It outputs $\pi = \{(R_i, S_i, c_i, \gamma_i)\}_{1 \leq i \leq n}$.

LEverif$_n(\{(h_i, z_i, g_i, y_i)\}_{1 \leq i \leq n}, \pi)$. It parses $\pi = \{(R_i, S_i, c_i, \gamma_i)\}_{1 \leq i \leq n}$. If $H(R_1|| S_1||\dots||R_n||S_n) \neq \prod_{i=1;i\neq j}^{n} c_i$ then it returns 0. Else if there exists $i \in [\![1, n]\!]$ such that $g_i^{\gamma_i} \neq R_i \cdot y_i^{c_i}$ or $h_i^{\gamma_i} \neq S_i \cdot z_i^{c_i}$ then it returns 0, else 1.

Theorem 1. *The NIZKP* LogEq$_n$ *is a proof of knowledge, moreover it is complete, sound, and zero-knowledge in the random oracle model.*

The proof of this theorem is a direct implication of [18,19] and [20]. Using this proof system, we build our VRS scheme called EVeR:

Scheme 3 (Efficient Verifiable Ring Signature (EVeR)). *EVeR is a VRS defined by:*

V.Init(1^k): *It generates a prime order group setup* $(\mathbb{G}, p, g)$ *and a hash function* $H : \{0,1\}^* \to \mathbb{G}$. *It returns the setup* $set = (\mathbb{G}, p, g, H)$.

V.Gen(set): *It picks* $sk \xleftarrow{\$} \mathbb{Z}_p^*$, *computes* $pk = g^{sk}$ *and returns a pair of signer public/private keys* (pk, sk).

V.Sig(L, m, sk): *It picks* $r \xleftarrow{\$} \mathbb{Z}_p^*$, *it computes* $h = H(m||r)$ *and* $z = h^{sk}$, *it runs* $P \leftarrow$ LEprove$_{|L|}(\{(h, z, g, pk_l)\}_{pk_l \in L}, sk)$ *and returns* $\sigma = (r, z, P)$.

V.Ver(L, m, σ): *It parses* $\sigma = (r, z, P)$, *computes* $h = H(m||r)$ *and returns* $b \leftarrow$ LEverif$_{|L|}(\{(h, z, g, pk_l)\}_{pk_l \in L}, P)$.

V.Proof(L, m, σ, pk, sk): *It parses* $\sigma = (r, z, P)$, *computes* $h = H(m||r)$ *and* $\bar{z} = h^{sk}$, *runs* $\bar{P} \leftarrow$ LEprove$_1(\{(h, \bar{z}, g, pk)\}, sk)$ *and returns* $\pi = (\bar{z}, \bar{P})$.

V.Judge(L, m, σ, pk, π): *It parses* $\sigma = (r, z, P)$ *and* $\pi = (\bar{z}, \bar{P})$, *computes* $h = H(m||r)$ *and runs* $b \leftarrow$ LEverif$_1(\{(h, \bar{z}, g, pk)\}, \pi)$. *If* $b \neq 1$ *then it returns* $\perp$. *Else, if* $z = \bar{z}$ *then it returns* 1, *else it returns* 0.

All users have an ElGamal key pair (pk, sk) such that $pk = g^{sk}$, where g is a generator of a prime order group. To sign a message m according to a set of public key L using her key pair (pk, sk), Alice chooses a random r and computes $h = H(m||r)$ and $z = h^{sk}$ where H is an hash function. Alice produces a proof π that there exists $pk_l \in L$ such that $\log_g(pk_l) = \log_h(z)$ using the NIZKP LogEq$_{|L|}$, where $|L|$ denotes the cardinal of L. The signature is the triplet (r, z, π). To verify a signature, it suffices to verify the proof π according to L, m and the other parts of the signature. To prove that she is the signer of the message m, Alice generates a proof that $\log_g(pk) = \log_h(z)$ using the NIZKP LogEq$_1$. By verifying this proof, a judge is convinced that $z = h^{sk}$. Let (r', z', π') be a second signature of a message m' produced from another key pair (pk', sk'). We set $h' = H(m'||r')$, and we recall that $z' = (h')^{sk'}$. To prove that she is not the signer of m', Alice computes $\bar{z}' = (h')^{sk}$ and she generates a proof that $\log_g(pk) = \log_{h'}(\bar{z}')$. Since $\bar{z}' \neq z'$, Alice proves that $\log_g(pk) \neq \log_{h'}(z')$, then she is not the signer of (r', z', π').

Theorem 2. *EVeR is unforgeable, anonymous, accountable and non-seizable under the DDH assumption in the random oracle model.*

We give the intuition of these properties, the proof of the theorem is given in the full version of this paper [12]:

Unforgeability: The scheme is unforgeable since nobody can prove that $\log_g(\mathsf{pk}_l) = \log_h(z)$ without the knowledge of $\mathsf{sk} = \log_h(z)$.

Anonymity: Breaking the anonymity of such a signature is equivalent to breaking the DDH assumption. Indeed, to link a signature $z = h^{\mathsf{sk}}$ with the corresponding public key of Alice $\mathsf{pk} = g^{\mathsf{sk}}$, an attacker must solve the DDH problem on the instance (pk, h, z). Moreover, note that since the value r *randomizes* the signature, it is not possible to link two signatures of the same message produced by Alice.

Accountability: To break the accountability, an adversary must forge a valid signature (*i.e.* to prove that there exists pk_l in the group such that $\log_g(\mathsf{pk}_l) \neq \log_h(z)$) and to prove that he is not the signer (*i.e.* $\log_g(\mathsf{pk}) \neq \log_h(z)$ where pk is the public key chosen by the adversary). However, since the adversary does not know the secret keys of the other members of the group, he would have to break the soundness of LogEq to win the experiment, which is not possible.

Non-seizable: (non-sei-1) no adversary is able to forge a proof that it is the signer of a signature produced by another user since it is equivalent to proving a false statement using a sound NIZKP. (non-sei-2) the proof algorithm run by a honest user with the public key pk returns a proof that this user is the signer of a given signature only if $\log_g(\mathsf{pk}) = \log_h(z)$. Since no adversary is able to compute z such that $\log_g(\mathsf{pk}) = \log_h(z)$ without the corresponding secret key, no adversary is able to break the non-seizability of EvER.

6   Our Unlinkable Sanitizable Signature Scheme: GUSS

We present our USS instantiated by a digital signature (DS) scheme and a VRS.

Scheme 4. (Generic Unlinkable Sanitizable Signature (GUSS)). *Let D be a deterministic digital signature scheme such that $D = (D.\mathsf{Init}, D.\mathsf{Gen}, D.\mathsf{Sig}, D.\mathsf{Ver})$ and V be a verifiable ring signature scheme such that $V = (V.\mathsf{Init}, V.\mathsf{Gen}, V.\mathsf{Sig}, V.\mathsf{Ver}, V.\mathsf{Proof}, V.\mathsf{Judge})$. GUSS instantiated with (D, V) is a sanitizable signature scheme defined by the following algorithms:*

$\mathsf{Init}(1^k)$: *It runs* $\mathsf{set}_d \leftarrow D.\mathsf{Init}(1^k)$ *and* $\mathsf{set}_v \leftarrow V.\mathsf{Init}(1^k)$. *Then it returns the setup* $\mathsf{set} = (\mathsf{set}_d, \mathsf{set}_v)$.

$\mathsf{SiGen}(\mathsf{set})$: *It parses* $\mathsf{set} = (\mathsf{set}_d, \mathsf{set}_v)$, *runs* $(\mathsf{pk}_d, \mathsf{sk}_d) \leftarrow D.\mathsf{Gen}(\mathsf{set}_d)$ *and* $(\mathsf{pk}_v, \mathsf{sk}_v) \leftarrow V.\mathsf{Gen}(\mathsf{set}_v)$. *Then it returns* $(\mathsf{pk}, \mathsf{sk})$ *where* $\mathsf{pk} = (\mathsf{pk}_d, \mathsf{pk}_v)$ *and* $\mathsf{sk} = (\mathsf{sk}_d, \mathsf{sk}_v)$.

$\mathsf{SaGen}(\mathsf{set})$: *It parses* $\mathsf{set} = (\mathsf{set}_d, \mathsf{set}_v)$ *and runs* $(\mathsf{spk}, \mathsf{ssk}) \leftarrow V.\mathsf{Gen}(\mathsf{set}_v)$. *It returns* $(\mathsf{spk}, \mathsf{ssk})$.

$\mathsf{Sig}(m, \mathsf{sk}, \mathsf{spk}, \mathrm{ADM})$: *It parses* $\mathsf{sk} = (\mathsf{sk}_d, \mathsf{sk}_v)$. *It first computes the fixed message part* $M \leftarrow \mathrm{FIX}_{\mathrm{ADM}}(m)$ *and runs* $\sigma_1 \leftarrow D.\mathsf{Sig}(\mathsf{sk}_d, (M \| \mathrm{ADM} \| \mathsf{pk} \| \mathsf{spk}))$ *and* $\sigma_2 \leftarrow V.\mathsf{Sig}(\{\mathsf{pk}_v, \mathsf{spk}\}, \mathsf{sk}_v, (\sigma_1 \| m)))$. *It returns* $\sigma = (\sigma_1, \sigma_2, \mathrm{ADM})$.

San(m, MOD, σ, pk, ssk): *It parses* $\sigma = (\sigma_1, \sigma_2, \text{ADM})$ *and* pk $= (pk_d, pk_v)$. *This algorithm first computes the modified message* $m' \leftarrow \text{MOD}(m)$ *and it runs* $\sigma_2' \leftarrow V.Sig(\{pk_v, spk\}, ssk, (\sigma_1 \| m'))$. *It returns* $\sigma' = (\sigma_1, \sigma_2', \text{ADM})$.

Ver(m, σ, pk, spk): *It parses* $\sigma = (\sigma_1, \sigma_2, \text{ADM})$ *and it computes the fixed message part* $M \leftarrow \text{FIX}_{\text{ADM}}(m)$. *Then it runs* $b_1 \leftarrow D.Ver(pk_d, (M \| \text{ADM} \| pk \| spk), \sigma_1)$ *and* $b_2 \leftarrow V.Ver(\{pk_d, spk\}, (\sigma_1 \| m), \sigma_2)$. *It returns* $b = (b_1 \wedge b_2)$.

SiProof(sk, m, σ, spk): *It parses* $\sigma = (\sigma_1, \sigma_2, \text{ADM})$ *and the key* sk $= (sk_d, sk_v)$. *It runs* $\pi_{si} \leftarrow V.Proof(\{pk_v, spk\}, (m \| \sigma_1), \sigma_2, pk_v, sk_v)$ *and returns it*.

SaProof(ssk, m, σ, pk): *It parses the signature* $\sigma = (\sigma_1, \sigma_2, \text{ADM})$. *It runs* $\pi_{sa} \leftarrow V.Proof(\{pk_v, spk\}, (m \| \sigma_1), \sigma_2, spk, ssk)$ *and returns it*.

SiJudge(m, σ, pk, spk, π_{si}): *It parses* $\sigma = (\sigma_1, \sigma_2, \text{ADM})$ *and* pk $= (pk_d, pk_v)$. *It runs* $b \leftarrow V.Judge(\{pk_v, spk\}, (m \| \sigma_1), \sigma_2, pk_v, \pi_{si})$ *and returns it*.

SaJudge(m, σ, pk, spk, π_{sa}): *It parses* $\sigma = (\sigma_1, \sigma_2, \text{ADM})$ *and* pk $= (pk_d, pk_v)$. *It runs* $b \leftarrow V.Judge(\{pk_v, spk\}, (m \| \sigma_1), \sigma_2, spk, \pi_{sa})$ *and returns* $(1 - b)$.

The signer secret key sk $= (sk_d, sk_v)$ contains a secret key sk_d compatible with the DS scheme and a secret key sk_v compatible with the VRS scheme. The signer public key pk $= (pk_d, pk_v)$ contains the two corresponding public keys. The sanitizer public/secret key pair (spk, ssk) is generated as in the VRS scheme.

Let m be a message and M be the *fixed part* chosen by the signer according to the admissible function ADM. To sign m, the signer first signs M together with the public key of the sanitizer spk and the admissible function ADM using the DS scheme. We denote this signature by σ_1. The signer then signs in σ_2 the full message m together with σ_1 using the VRS scheme for the set of public keys $L = \{pk_v, spk\}$. Informally, he anonymously signs $(\sigma_1 \| m)$ within a group of two users: the signer and the sanitizer. The final sanitizable signature is $\sigma = (\sigma_1, \sigma_2)$. The verification algorithm works in two steps: it verifies the signature σ_1 and it verifies the anonymous signature σ_2.

To sanitize this signature $\sigma = (\sigma_1, \sigma_2)$, the sanitizer chooses an admissible message m' according to ADM (*i.e.* m and m' have the same fixed part). Then he anonymously signs m' together with σ_1 using the VRS for the group $L = \{pk_v, spk\}$ using the secret key ssk. We denote by σ_2' this signature. The final sanitized signature is $\sigma' = (\sigma_1, \sigma_2')$.

Theorem 3. *For any deterministic and unforgeable DS scheme D and any unforgeable, anonymous, accountable and non-seizable VRS scheme V, GUSS instantiated with (D, V) is immutable, transparent, strongly accountable and unlinkable.*

We give the intuition of these properties, the proof of the theorem is given in the full version of this paper [12]:

Transparency: According to the anonymity of σ_2 and σ_2', nobody can guess if a signature comes from the signer or the sanitizer, and since both signatures have the same structure, nobody can guess whether a signature is sanitized or not.

Immutability: Since it is produced by a unforgeable DS scheme, nobody can forge the signature σ_1 of the fixed part M without the signer secret key. Thus the sanitizer cannot change the fixed part of the signatures. Moreover, since σ_1 signs the public key of the sanitizer in addition to M, the other users cannot forge a signature of an admissible message using σ_1.

Unlinkability: An adversary knows (i) two signatures σ^0 and σ^1 that have the same fixed part M according to the same function ADM for the same sanitizer and (ii) the sanitized signature $\sigma' = (\sigma'_1, \sigma'_2)$ computed from σ^b for a given admissible message m' and an unknown bit b. To achieve unlinkability, it must be hard to guess b. Since the DS scheme is deterministic, the two signatures $\sigma^0 = (\sigma_1^0, \sigma_2^0)$ and $\sigma^1 = (\sigma_1^1, \sigma_2^1)$ have the same first part $(i.e.\ \sigma_1^0 = \sigma_1^1)$. As it was shown before, the σ' has the same first part σ'_1 as the original one, thus $\sigma'_1 = \sigma_1^0 = \sigma_1^1$ and σ'_1 leaks no information about b. On the other hand, the second part of the sanitized signature σ'_2 is computed from the modified message m' and the first part of the original signature. Since $\sigma_1^0 = \sigma_1^1$, we deduce that σ'_2 leaks no information about b. Finally, the best strategy of the adversary is to randomly guess b.

(Strong) Accountability: the signer must be able to prove the provenance of a signature. It is equivalent to breaking the anonymity of the second parts σ_2 of this signature: if it was created by the signer then it is the original signature, else it was created by the sanitizer and it is a sanitized signature. By definition, the VRS scheme used to generate σ_2 provides a way to prove whether a user is the author of a signature or not. GUSS uses it in its proof algorithm to achieve accountability. Note that since the sanitizer uses the same VRS scheme to sanitize a signature, he also can prove the origin of a given signature to achieve the strong accountability.

7   Conclusion

In this paper, we revisit the notion of verifiable ring signatures. We improve its properties of verifiability, we give a security model for this primitive and we design a simple, efficient and secure scheme named EvER. We extend the security model of sanitizable signatures in order to allow the sanitizer to prove the origin of a signature. Finally, we design a generic unlinkable sanitizable signature scheme named GUSS based on verifiable ring signatures. This scheme is twice as efficient as the best scheme of the literature. In the future, we aim at finding other applications for the verifiable ring signatures that require our security properties.

Acknowledgement. We acknowledge Dominique Schröder for his helpful comments and feedback on our paper.

A   Algorithms Complexity

In this section, we detail the complexity of the algorithms of our schemes. More precisely, we give the number of exponentiations in a prime order group for each algorithm. Since our schemes are pairing free, this is the main operation. Moreover, we give the size of some values outputted by these algorithms (keys, signatures and proofs). This size is given in the number of elements of a group of prime order p. For the sake of clarity, we do not distinguish between elements of a group $\mathbb{G}$ of prime order p where the DDH assumption is hard and elements of $\mathbb{Z}_p^*$.

In Fig. 2, we give the number of exponentiations of each algorithm of Schnorr's signature, and we give the size of the secret/public keys $\mathsf{sk_{Sh}}$ and $\mathsf{pk_{Sh}}$ and the size of a signature σ_{Sh}.

Schnorr	D.Gen	D.Sig	D.Ver	$\mathsf{sk_{Sh}}$	$\mathsf{pk_{Sh}}$	σ_{Sh}
exp/size	1	1	2	1	1	2

Fig. 2. Complexity analysis of Schnorr (Scheme 1).

In Fig. 3, we give the number of exponentiations of each algorithm of the LogEq_n proof system and the size of a proof π_n^{LE} depending to the number n. The first line corresponds to the general case, the two other lines correspond to the case where $n = 1$ and $n = 2$.

LogEq_n	$\mathsf{LEprove}_n$	$\mathsf{LEverif}_n$	π_n^{LE}
n	$2 + 4 \cdot (n - 1)$	$4 \cdot n$	$4 \cdot n$
$n = 1$	2	4	4
$n = 2$	6	8	8

Fig. 3. Complexity analysis of LogEq (Scheme 2).

In Fig. 4, we give the number of exponentiations of each algorithm of the EVeR verifiable ring signature scheme (first table) and the size the secret/public keys $\mathsf{sk_{EV}}$ and $\mathsf{pk_{EV}}$, the size of a signature σ_n^{EV} and the size of a proof π_n^{EV} (second table). These values depend on the size of the ring n. The first line corresponds to the generic case, where the values depend on the chosen proof system. The second line corresponds to the case where $n = 2$ and where the proof system is LogEq_2.

In Fig. 5, we give the number of exponentiations of each algorithm of the GUSS verifiable ring signature scheme (first table) and the size of the secret/public keys of the signer and the sanitize sk, pk, ssk and spk, the size of a signature σ and the size of a proof π (second table). We ommit the complexity of the algorithms SaProof and SaJudge since these algorithms are similar to SiProof and SiJudge. The first line corresponds to the generic case, where the values depend on the chosen signature scheme and the chosen verifiable ring signature scheme. The second line corresponds to the case where GUSS is instantiated with Schnorr and EVeR.

EVeR	V.Gen	V.Sig$_n$	V.Ver$_n$	V.Proof	V.Judge
n (generic)	1	$1 + \mathsf{LEprove}_n$	$\mathsf{LEverif}_n$	$1 + \mathsf{LEprove}_1$	$\mathsf{LEverif}_1$
$n = 2$ (with LogEq_n)	1	7	8	3	4

EVeR	sk$_{EV}$	pk$_{EV}$	σ_n^{EV}	π_n^{EV}
n (generic)	1	1	$2 + \pi_n^{LE}$	$1 + \pi_1^{LE}$
$n = 2$ (with LogEq_n)	1	1	10	5

Fig. 4. Complexity analysis of EVeR (Scheme 3).

GUSS	SiGen	SaGen	Sig	San	Ver	SiProof
generic	D.Gen + V.Gen	V.Gen	D.Sig + V.Sig$_2$	V.Sig$_2$	D.Ver + V.Ver$_2$	V.Proof
EvER and Schnorr	2	1	8	7	10	3

GUSS	SiJudge	sk	pk	ssk	spk	σ	π
generic	V.Judge	sk$_{EV}$ + sk$_{Sh}$	pk$_{EV}$ + pk$_{Sh}$	sk$_{EV}$	pk$_{EV}$	$\sigma_2^{EV} + \sigma^{Sc}$	π_2^{EV}
EvER and Schnorr	4	2	2	1	1	12	5

Fig. 5. Complexity analysis of GUSS (Scheme 4).

References

1. Ateniese, G., Chou, D.H., de Medeiros, B., Tsudik, G.: Sanitizable signatures. In: di Vimercati, S.C., Syverson, P., Gollmann, D. (eds.) ESORICS 2005. LNCS, vol. 3679, pp. 159–177. Springer, Heidelberg (2005). https://doi.org/10.1007/11555827_10

2. Au, M.H., Susilo, W., Yiu, S.-M.: Event-oriented k-times revocable-iff-linked group signatures. In: Batten, L.M., Safavi-Naini, R. (eds.) ACISP 2006. LNCS, vol. 4058, pp. 223–234. Springer, Heidelberg (2006). https://doi.org/10.1007/11780656_19

3. Bellare, M., Goldreich, O.: On defining proofs of knowledge. In: Brickell [6], pp. 390–420

4. Bender, A., Katz, J., Morselli, R.: Ring signatures: stronger definitions, and constructions without random oracles. In: Halevi, S., Rabin, T. (eds.) TCC 2006. LNCS, vol. 3876, pp. 60–79. Springer, Heidelberg (2006). https://doi.org/10.1007/11681878_4

5. Boneh, D.: The decision Diffie-Hellman problem (Invited paper). In: Buhler, J.P. (ed.) ANTS 1998. LNCS, vol. 1423, pp. 48–63. Springer, Heidelberg (1998). https://doi.org/10.1007/BFb0054851

6. Brickell, E.F. (ed.): CRYPTO 1992. LNCS, vol. 740. Springer, Heidelberg (1993)

7. Brzuska, C., et al.: Redactable signatures for tree-structured data: definitions and constructions. In: Zhou, J., Yung, M. (eds.) ACNS 2010. LNCS, vol. 6123, pp. 87–104. Springer, Heidelberg (2010). https://doi.org/10.1007/978-3-642-13708-2_6

8. Brzuska, C., et al.: Security of sanitizable signatures revisited. In: Jarecki, S., Tsudik, G. (eds.) PKC 2009. LNCS, vol. 5443, pp. 317–336. Springer, Heidelberg (2009). https://doi.org/10.1007/978-3-642-00468-1_18

9. Brzuska, C., Fischlin, M., Lehmann, A., Schröder, D.: Unlinkability of sanitizable signatures. In: Nguyen, P.Q., Pointcheval, D. (eds.) PKC 2010. LNCS, vol. 6056, pp. 444–461. Springer, Heidelberg (2010). https://doi.org/10.1007/978-3-642-13013-7_26

10. Brzuska, C., Pöhls, H.C., Samelin, K.: Non-interactive public accountability for sanitizable signatures. In: De Capitani di Vimercati, S., Mitchell, C. (eds.) EuroPKI 2012. LNCS, vol. 7868, pp. 178–193. Springer, Heidelberg (2013). https://doi.org/10.1007/978-3-642-40012-4_12

11. Brzuska, C., Pöhls, H.C., Samelin, K.: Efficient and perfectly unlinkable sanitizable signatures without group signatures. In: Katsikas, S., Agudo, I. (eds.) EuroPKI 2013. LNCS, vol. 8341, pp. 12–30. Springer, Heidelberg (2014). https://doi.org/10.1007/978-3-642-53997-8_2

12. Bultel, X., Lafourcade, P.: Unlinkable and strongly accountable sanitizable signatures from verifiable ring signatures. Cryptology ePrint Archive, Report 2017/605 (2017). http://eprint.iacr.org/2017/605

13. Canard, S., Schoenmakers, B., Stam, M., Traor, J.: List signature schemes. Discret. Appl. Math. **154**(2), 189–201 (2006)

14. Canard, S., Jambert, A.: On extended sanitizable signature schemes. In: Pieprzyk, J. (ed.) CT-RSA 2010. LNCS, vol. 5985, pp. 179–194. Springer, Heidelberg (2010). https://doi.org/10.1007/978-3-642-11925-5_13

15. Canard, S., Jambert, A., Lescuyer, R.: Sanitizable signatures with several signers and Sanitizers. In: Mitrokotsa, A., Vaudenay, S. (eds.) AFRICACRYPT 2012. LNCS, vol. 7374, pp. 35–52. Springer, Heidelberg (2012). https://doi.org/10.1007/978-3-642-31410-0_3

16. Changlun, Z., Yun, L., Dequan, H.: A new verifiable ring signature scheme based on Nyberg-Rueppel scheme. In: 2006 8th International Conference on Signal Processing, vol. 4 (2006)

17. Chase, M., Lysyanskaya, A.: On signatures of knowledge. In: Dwork, C. (ed.) CRYPTO 2006. LNCS, vol. 4117, pp. 78–96. Springer, Heidelberg (2006). https://doi.org/10.1007/11818175_5

18. Chaum, D., Pedersen, T.P.: Wallet databases with observers. In: Brickell [6], pp. 89–105

19. Cramer, R., Damgård, I., Schoenmakers, B.: Proofs of partial knowledge and simplified design of witness hiding protocols. In: Desmedt, Y.G. (ed.) CRYPTO 1994. LNCS, vol. 839, pp. 174–187. Springer, Heidelberg (1994). https://doi.org/10.1007/3-540-48658-5_19

20. Fiat, A., Shamir, A.: How to prove yourself: practical solutions to identification and signature problems. In: Odlyzko, A.M. (ed.) CRYPTO 1986. LNCS, vol. 263, pp. 186–194. Springer, Heidelberg (1987). https://doi.org/10.1007/3-540-47721-7_12

21. Fleischhacker, N., Krupp, J., Malavolta, G., Schneider, J., Schröder, D., Simkin, M.: Efficient unlinkable sanitizable signatures from signatures with re-randomizable keys. In: Cheng, C.-M., Chung, K.-M., Persiano, G., Yang, B.-Y. (eds.) PKC 2016. LNCS, vol. 9614, pp. 301–330. Springer, Heidelberg (2016). https://doi.org/10.1007/978-3-662-49384-7_12

22. Fuchsbauer, G., Pointcheval, D.: Anonymous proxy signatures. In: Ostrovsky, R., De Prisco, R., Visconti, I. (eds.) SCN 2008. LNCS, vol. 5229, pp. 201–217. Springer, Heidelberg (2008). https://doi.org/10.1007/978-3-540-85855-3_14

23. Goldwasser, S., Micali, S., Rackoff, C.: The knowledge complexity of interactive proof systems. SIAM J. Comput. **18**(1), 186–208 (1989)

24. Hoshino, F., Kobayashi, T., Suzuki, K.: Anonymizable signature and its construction from pairings. In: Joye, M., Miyaji, A., Otsuka, A. (eds.) Pairing 2010. LNCS, vol. 6487, pp. 62–77. Springer, Heidelberg (2010). https://doi.org/10.1007/978-3-642-17455-1_5

25. Johnson, R., Molnar, D., Song, D., Wagner, D.: Homomorphic signature schemes. In: Preneel, B. (ed.) CT-RSA 2002. LNCS, vol. 2271, pp. 244–262. Springer, Heidelberg (2002). https://doi.org/10.1007/3-540-45760-7_17

26. Lai, R.W.F., Zhang, T., Chow, S.S.M., Schröder, D.: Efficient sanitizable signatures without random oracles. In: Askoxylakis, I., Ioannidis, S., Katsikas, S., Meadows, C. (eds.) ESORICS 2016. LNCS, vol. 9878, pp. 363–380. Springer, Cham (2016). https://doi.org/10.1007/978-3-319-45744-4_18

27. Lee, K.C., Wen, H.A., Hwang, T.: Convertible ring signature. IEE Proc. - Commun. **152**(4), 411–414 (2005)

28. Lv, J., Wang, X.: Verifiable ring signature. In: DMS Proceedings, pp. 663–665 (2003)

29. Rivest, R.L., Shamir, A., Tauman, Y.: How to leak a secret. In: Boyd, C. (ed.) ASIACRYPT 2001. LNCS, vol. 2248, pp. 552–565. Springer, Heidelberg (2001). https://doi.org/10.1007/3-540-45682-1_32

30. Schnorr, C.P.: Efficient identification and signatures for smart cards. In: Brassard, G. (ed.) CRYPTO 1989. LNCS, vol. 435, pp. 239–252. Springer, New York (1990). https://doi.org/10.1007/0-387-34805-0_22

31. Steinfeld, R., Bull, L., Zheng, Y.: Content extraction signatures. In: Kim, K. (ed.) ICISC 2001. LNCS, vol. 2288, pp. 285–304. Springer, Heidelberg (2002). https://doi.org/10.1007/3-540-45861-1_22

32. Wang, S., Ma, R., Zhang, Y., Wang, X.: Ring signature scheme based on multivariate public key cryptosystems. Comput. Math. Appl. **62**(10), 3973–3979 (2011)

Web Security

Out of the Dark: UI Redressing and Trustworthy Events

Marcus Niemietz[✉] and Jörg Schwenk

Chair for Network and Data Security, Horst Görtz Institute for IT Security,
Ruhr-University Bochum, Bochum, Germany
{marcus.niemietz,joerg.schwenk}@rub.de

Abstract. Web applications use *trustworthy events* consciously triggered by a human user (e.g., a left mouse click) to authorize security-critical changes. Clickjacking and UI redressing (UIR) attacks trick the user into triggering a trustworthy event unconsciously. A formal model of Clickjacking was described by Huang et al. and was later adopted by the W3C UI safety specification. This formalization did not cover the target of these attacks, the *trustworthy events*.

We provide the first extensive investigation on this topic and show that the concept is not completely understood in current browser implementations. We show major differences between widely-used browser families, even to the extent that the concept of trustworthy events itself becomes unrecognizable. We also show that the concept of *trusted events* as defined by the W3C is somehow orthogonal to *trustworthy events*, and may lead to confusion in understanding the security implications of both concepts. Based on these investigations, we were able to circumvent the concept of trusted events, introduce three new UIR attack variants, and minimize their visibility.

1 Introduction

UI Redressing attacks are powerful attacks which can be used to circumvent browser security mechanisms like sandboxing and the Same-Origin Policy (SOP). They are far less intrusive than, for example, Phishing mails because the user thinks he performs a legal action on an innocent-looking web page. In 2008 Grossmann et al. had to cancel their OWASP talk about a new attack technique called Clickjacking [10]: it turned out that they were able to bypass a major protection mechanism of Adobe's Flash – Clickjacking allowed the attacker's website to automatically get access to the camera and microphone of the victim without any explicit permission. According to Adobe, Clickjacking had the "highest level of damage potential that any exploit can have" [26].

In contrast to Clickjacking that is usually associated with left-click mouse events only, the broader term UIR also covers events from the keyboard and even touch gestures [12,30]. In the past years, many attacks and defense mechanisms were published by the industry as well as the academic community (e.g., [3,4, 28,33], and [18]).

© Springer Nature Switzerland AG 2018
S. Capkun and S. S. M. Chow (Eds.): CANS 2017, LNCS 11261, pp. 229–249, 2018.
https://doi.org/10.1007/978-3-030-02641-7_11

Formal Definition of UIR. Huang et al. [12] defined Clickjacking to be an attack that violates the integrity of either the *visual context* or the *temporary context* of a trustworthy user action on a sensitive element of the web application. *Visual context integrity* may either be violated by making the sensitive element invisible (e.g., by placing it in fully transparent mode above some other element), or by hiding the fact that the user is actually clicking on such an element (e.g., by modifying the image of the mouse pointer, also referred to as Cursorjacking [15]). *Temporal context integrity* can be violated by replacing a non-sensitive element, just before the user clicks on it, by the sensitive element.

The definitions from Huang et al. [12] can easily be extended to the broader class or UIR attacks. However, the treatment of trustworthy events becomes more complex because in addition to left-click events, also right-click-and-select, keyboard and inter alia touch events must be taken into account.

Events in Web Applications. Events can be triggered by humans (e.g., by clicking on a button or moving the mouse pointer), by network operations, or automatically with the help of scripts. From the network, events like load or the status change events in XMLHttpRequest queries can be triggered. Purely script based are, for example, those triggered by the setTimeout() or setIntervall() method. For human interaction, a distinction must be made between events that the user consciously starts (e.g., click or keydown), and events that he may not notice (e.g., mouseover). Event-handlers are procedures with an on-prefix; they are called when the corresponding event occurs. For example, the onclick event-handler is called whenever a click event occurs.

Events are managed in the event system of the browser and there exist many differences across browsers. For example, the event wheel will only be executed on the event system of Internet Explorer (IE) when the method addEventListener() is used. The event system of Google Chrome (GC) will recognize this event with the same conditions when the event-handler onwheel is used. To foster interoperability, there exists a working draft of an UI event specification designed by the World Wide Web Consortium (W3C) [13]. The specification describes event systems and subsets of different event types.

Trusted vs. Trustworthy Events. Trusted events are defined by the W3C as follows: "Events that are generated by the user agent, either as a result of user interaction, or as a direct result of changes to the DOM, are trusted by the user agent with privileges that are not afforded to events generated by script through the createEvent() method, modified using the initEvent() method, or dispatched via the dispatchEvent() method. The isTrusted attribute of trusted events has a value of *true*, while untrusted events have a isTrusted attribute value of *false*." ([40], Sect. 3.4).

This definition is very broad and therefore not suitable for a distinction between events that may be allowed to cause security critical changes, and those that may not. For example, the mouseover and click events are both "trusted" according to the W3C definition when caused by a human user; however, displaying a pop-up window or sending the contents of an HTML form simply because the mouse pointer crossed over a certain area of the browser window (`mouseover`) seems far too permissive. Our definition is more specific: *a trustworthy event is an event that is triggered by a conscious user action* (e.g., by left-click, right-click, or keystroke).

Unreliability of `isTrusted`. To mark trusted actions, the DOM Level 3 specification of the W3C mentions a read-only property called `isTrusted`, which returns a boolean value depending on the dispatched state [39]. In Sect. 4 we show that this property cannot be used to distinguish *trustworthy* events from other events, since pop-ups are blocked even if `isTrusted=true`, and are allowed even if `isTrusted=false`.

Trustworthy Event Scenarios. Trustworthy events are used in different security critical scenarios. *User consent* in activating potentially dangerous browser features (e.g., activating the webcam via Adobe's Flash) was the main target in previously described UIR attacks. *Pop-up windows* are usually blocked when there was no former click with the mouse pointing device. One reason is that pop-up windows are used by the advertisement industry and thus they might disturb the user or they may even trick him to install malware. The *clipboard* should only be accessible by user initiated keyboard or mouse pointing events. If the clipboard would be accessible by JavaScript code only, an attacker's website could steal saved data like passwords stored in a password manager (paste action). *Drag-and-drop* is a scenario where a user is able to move data cross-origin. Again, if this feature was accessible by JavaScript code only, the SOP could be circumvented. *Additional scenarios:* in Firefox (FF), the deprecated XML User Interface Language (XUL) handlers and commands can only be triggered by trustworthy events like click and touch [23]. In modern browsers such as GC, forms can be filled out automatically by using the autofill feature that could be activated by trustworthy events like keystrokes and left-clicks [34,42].

Investigation of Trustworthy Events. We study *(1)* all mouse events including (a) different left-clicks (`click`, `dblclick`, `mousedown`, `mouseup`), (b) right-click, (c) mouse movements (`mouseover`), (d) drag (`drag`, `dragstart`) and (e) wheel; *(2)* the keyboard events `keydown`, `keyup`, `keypress`, and *(3)* combinations of mouse and keyboard events. We show that many of these user-triggered actions have different interpretations as trustworthy or non-trustworthy events in the different browser families.

We investigate the three lesser-researched application areas of trustworthy events: pop-up windows bypassing pop-up blockers, escaping the browser sandbox via copy-and-paste to and from the clipboard, and bypassing the SOP via drag-and-drop.

Research questions. In this work we investigate the following questions: *Which events are recognized as trustworthy by a modern web browser? How is trustworthy event handling implemented in modern web browsers? Could the knowledge of these implementations lead to new UIR variants?*

Contribution. The contributions of this paper are as follows:

- We systematically evaluate trustworthy events in web applications originating in a mouse device, the keyboard, or a combination thereof, and describe differences in modern browsers implementations.
- We thoroughly analyze three security critical trustworthy event scenarios (pop-up windows, drag-and-drop, and clipboard), both same and cross-origin.
- We introduce and discuss three new UIR attack variants by making use of particularities of trustworthy event implementations in modern browsers.

2   UI Redressing

The initial Clickjacking attack of Grossman et al. raised a lot of attention due the hijacking possibilities of the webcam and microphone, but they also discovered a general security problem. As listed by Niemietz et al. [28], UIR is a set of attacks that include Clickjacking as a subset. Next to Classic Clickjacking there are other attacks like Sharejacking and Likejacking (e.g., to attack Facebook [35]), and inter alia Cursorjacking [7,15]. UIR does not only cover clicks, it also covers drag operations (drag-and-drop attacks [38]), keystrokes (Strokejacking [43]) and even maskings (SVG-based attacks [27]).

In a classic Clickjacking attack illustrated in Fig. 1, the victim has opened the attacker's website, which consists of two Iframes. The first Iframe ("Funny Kittens") is loading a visible HTML document to lure the victim into clicking on the `More` button. The second Iframe loads the target "Account Setting" website, but this frame is rendered invisibly (e.g., with the help of the property `opacity=0`) on top of the visible frame. Because of invisible Iframe's position above the `Funny Kittens` Iframe, the victim will actually click on `Delete` instead on `More`.

UIR Contexts. According to Huang et al., the definition of UIR is that "an attacker application presents a sensitive UI element of a target application out of context to a user and hence the user gets tricked to act out of context" [12]. This definition describes the root cause of UIR.

Visual Context. This context defines what the user sees. It does not include actions (e.g., clicking) on sensitive elements (e.g., buttons). To ensure target *display integrity*, sensitive elements must be fully visible to the user. In contrast, *pointer integrity* requires that input mechanisms and their resulting actions are fully visible to the user.

Temporal Context. The timing of a user's action is known as the temporal context. To ensure *temporal integrity*, the user's action is actually intended by the user. To compromise temporal integrity, a visible button could be replaced

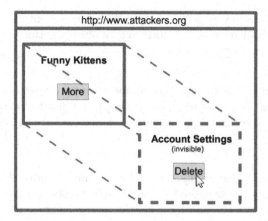

Fig. 1. Illustration for a classic Clickjacking attack.

by the attacker right before the victim is clicking on it (e.g., with a Facebook *Like* button).

These context definitions provide an important insight on how UIR attacks work in the important case that the user does simple events such as a single left-click. However, in reality there exists a much broader set of user events (e.g., keystroke, right-click, and a chain of left-clicks). This could lead to new attack variants and therefore different events must be considered (shown in Sect. 6).

3   Events in Web Applications

Browser events can be divided into different event types according to the W3C working drafts for handling browser events [13,14]. In the following, we map common event types into different event type groups. To the best of our knowledge, we completely cover all commonly used user interactions.

All event types can be either triggered by *user* or *script* actions. To name one example, a user can consciously trigger a `click` event by explicitly clicking on a button with the event-handler `onclick`. In addition, a script can also trigger this event automatically by using the DOM's `click()` method (e.g., `document.getElementById ("button").click()`).

Resource Events. These are frame or object events that are triggered by HTTP events. Examples for resource events are `error` (failed to load), `load` (finished loading), and `unload` (unloading of a document or depending resource).

Mouse Events. Consciously created mouse events are usually left and right clicks. In addition, mouse events can also be generated unconsciously when the pointer is moved or when drag-and-drop actions are done. The most deeply nested element is always the target of a mouse event. Except for user interactions on a virtual keyboard, touch events act similar to mouse events and are thus included in the mouse event set. Examples for mouse events are `click` (button has been pressed and released), `mousemove` (moved pointing device), and `drag` (dragged element or text).

Keyboard Events. This event type is for example triggered when a user is pressing (`keydown`) or releasing a key (`keyup`). Virtual keyboards, from input devices like touch screens, trigger keyboard events and are therefore also in this even type group.

Multiple Events. Some events cannot be assigned to only the mouse or keyboard; they can also be triggered by both variants. As an example, a user can select text in an `input` element by using the mouse cursor (click and mark) and also the keyboard (shift and arrow keys).

Based on these event types, we provide a definition for trustworthy events:

Definition 1. *An event is called* trustworthy *when it was triggered by a* conscious *user action.*

---

4   DOM Property IsTrusted

The W3C specification ([40], Sect. 3.4) describes a boolean attribute `isTrusted`: "The `isTrusted` attribute of trusted events has a value of *true*, while untrusted events have a `isTrusted` attribute value of *false*." We investigate this attribute in detail and show that it is not related to *trustworthy* events.

Different isTrusted Implementations. According to the W3C, the DOM property `event.isTrusted` only returns `true` when an event was dispatched by the user agent [39]. According to the Mozilla Developer Network (MDN), the property is defined as *true* "when the event was generated by a user action, and *false* when the event was created or modified by a script or dispatched via dispatchEvent" [24]. IE is an exception because all events are *true* except they are created with `createEvent()`. This JavaScript feature can be used to create an event object and simulate an event type such as a mouse event (e.g., an automatically fired click on a button for testing web applications).

isTrusted=false, but Pop-Ups are Allowed. Listing 1.1 contains a button and a hyperlink. If the button is clicked by the user, the `onclick` event-handler calls `document.getElementByID("test").click()`, and this JavaScript function selects the hyperlink (which has `id="test"`), and performs a script-generated click event on it. Consequently, the value of `isTrusted`, which is shown in the `alert()` window, is `false`, as described in the W3C specification. Nevertheless, `window.open()` is executed, and a pop-up window is displayed.

```
1 <button onclick="document.getElementByID("test").click()">
    </button>
2 <a href="#" id="test" onclick="alert('isTrusted: '+
    event.isTrusted); window.open('http://example.org', 'rub
    ','height=200,width=200');">Trusted Click</a>
```

Listing 1.1. Pop-ups are not blocked although isTrused is `false`.

isTrusted=true, but Pop-Ups are Blocked. Listing 1.2 provides an example with the `<video>` element introduced with HTML5. It contains an `onloadstart` event-handler, which executes code when the browser starts looking for the video file given in line 2. Thus, JavaScript code will be directly executed without any real user interaction. Due to this reason, the JavaScript code generated pop-up will be blocked. The alert-window with `event.isTrusted` displays `true` on all browsers although the only user interaction was an initial opening of the page (e.g., FF, GC, and Edge).

```
1     <video onloadstart="window.open('http://example.org',
        null, 'height=200,width=400,status=yes,toolbar=no,
        menubar=no,location=no');alert(event.isTrusted);">
2     <source src="movie.mp4" type="video/mp4">
3     </video>
```

Listing 1.2. Pop-ups are blocked although isTrusted is `true`.

Inheritance of Trustworthiness. Our evaluation of the behavior of `isTrusted` and the displaying of pop-up windows shows an interesting result; events occurring within a delay of one second after an initial trustworthy event are also treated as trustworthy events, although they may be triggered purely by JavaScript.

More formally: let $P_t = true$ denote the fact that the pop-up window opened at time t was *not* blocked by the pop-up-blocker. Let $iT = t_0$ denote the fact that a trustworthy event was initiated by the user at time t_0. Then we have:

$$P_t := \begin{cases} true, & if\ (iT = t_0) \wedge (|t - t_0| \leq 1\ sec) \\ false, else \end{cases}$$

The interesting discovery is that a pop-up window will not be blocked in the event that there was *once* a (real) user's click in the chain of events. This behavior was observed for the tested versions of FF and Safari (SA).

5   Trustworthy Scenarios

The W3C UI Events specification [40] does not recommend actions that are allowed after a trustworthy event. As shown by Huang et al. [12], a missing formal definition could lead to different browser implementations and thus to browser bugs and vulnerabilities.

Next to our trustworthy event definition, we address this issue by providing a description of three different trustworthy scenarios. We believe that the scientific community and browser vendors will get a valuable overview about this currently not examined area and thus derive new attack variants and countermeasures (cf. Sect. 6).

5.1   Pop-Up Scenario

Need of Trustworthy Events. In the past, JavaScript code was able to automatically open pop-up windows when the user simply opened a website. The advertising industry used this feature to show unwanted ads to the user and thus modern browsers distinguish between wanted and unwanted pop-up windows: a pop-up window should only be shown when a trustworthy event (e.g., click) was used to call the required JavaScript pop-up-code (e.g., window.open).

Evaluation. Table 1 lists four different types of events with each event type containing different events. Each event type includes different events. The test cases for these events were executed in four different browsers: IE 11, FF 47, GC 54, Opera (OP) 41, and SA 10. Our test function for pop-ups is given in Listing 1.3. It tries to create up to five pop-up windows in case that the code is indeed called. If this is the case, all five pop-ups are displayed in FF and SA; in contrast, only one pop-up with a warning window in IE, GC, and OP.

```
1    <script>
2    function createPopups(){
3      for (i=1;i<6;i++) {
4        window.open('//evil.org', i, 'width=50,height=50');
5      }
6    }
7    </script>
```

Listing 1.3. Our test function for pop-ups.

In the first event type group, *resource* events are given. These events are inter alia triggered by loading the browser's window or by simply reloading it. The user does not use an input device like a mouse or a keyboard and thus pop-up windows are not displayed.

Mouse events are the second type of events. Our test cases cover left-clicks, right-clicks, mouse movements, dragging actions, and the usage of the mouse wheel. In the event of a left-click, pop-ups will be shown. A right-click only leads to pop-up windows in IE. Mouse movements and dragging actions do not let the tested browser open pop-up windows. The event `wheel` is triggered when the wheel rolls up or down over an HTML element; it does not lead to the displaying of pop-up windows in FF, GC, and OP. Furthermore, this event is not supported in IE.

With the third defined type called *keyboard events*, only GC and OP act in a pop-up scenario. IE and FF behave differently, pop-ups will be blocked.

The fourth type called *multiple events* consists of events that can be triggered in different ways like keyboard actions and left-clicks. It shows that there are events which act different across browsers; only some browsers allow access to the pop-up scenario and IE only in case of a left-click in combination with the event `select`. In IE 11 and FF 47, a left-click in combination with `focus` or `blur` does not lead to a pop-up execution. As another example, FF grants access when an `input` event in combination with a right-click for copy-and-paste is used. This is not the case when this event is used in combination with a keyboard action. GC and OP act exactly in the opposite way.

5.2   Clipboard Scenario

Need of Trustworthy Events. Clipboard data may contain sensitive information that should not be shared with an arbitrary website. For example, password managers usually save stored passwords into the clipboard such that they could be inserted into login forms (e.g., for banking or shopping). Therefore, JavaScript code that is able to automatically read clipboard data could copy the password

Table 1. Events and their triggered pop-up windows. ✓ indicates that the pop-up was shown, ✗ that it was blocked. For the category of multiple events, "keyboard" denotes all events of type "Keyboard", and (✓, ✗) means that a keyboard event did result in a pop-up, whereas the mentioned click event did not.

Events	Type	IE 11	FF 47	GC 54	OP 41
Load, error, unload	Resource	✗			
Click, dblclick, mousedown, mouseup (left-click)	Mouse	✓			
Contextmenu (right-click)		✓	✗		
Mouseenter, mouseleave, mousemove, mouseout, mouseover (movement)		✗			
Drag, dragstart (dragging)		✗			
Wheel		✗			
Keydown, keyup, keypress	Keyboard	✗		✓	
Search (keyboard, left-click)	Multiple	–		(✗, ✓)	
Select (keyboard, left-click)		(✗, ✓)		✓	
Input (keyboard, right-click paste)		✗	(✗, ✓)	(✓, ✗)	
Focus (keyboard, left-click)		✗		✓	
Focusin, focusout (keyboard, left-click)		✗	–	✓	
Blur (keyboard, left-click)		✗		✓	(✗, ✓)
Scroll (keyboard, wheel)		✗			

from the clipboard and send it to the attacker. For this reason, browsers should only allow access to clipboard data after a conscious user action, i.e. after a trustworthy event. A moderate security problem arises in the event of copy and cut operations to the clipboard; a website should not overwrite clipboard data without an explicit permission of the user.

Evaluation. As shown in Table 2, the clipboard always allows copy, cut, and paste operations with the help of a keyboard or mouse pointing device (no script execution). In the event of automatically executed scripts, it is usually not possible to access the user's clipboard. IE is an exception as it allows access to copy, cut, and paste operations (see Listing 1.4) by showing the user a confirmation window which only gives access when the user explicitly clicks on `Allow access`.

```
1    //read data of type ``Text'' from clipboard
2    window.clipboardData.getData("Text");
3    //write data of type ``Text'' to the clipboard
4    var input = "This text is written to the clipboard";
5    window.clipboardData.setData("Text",input);
```

Listing 1.4. JavaScript functions to access the clipboard.

By looking at the results from the pop-up scenario (cf. Table 1), JavaScript code can act on a higher privileged authorization level in case that the script was triggered by a trustworthy event. We found that the clipboard copy and cut capabilities are also enabled when a trustworthy event calls JavaScript code. To name an example, a listener on the event `click` can be used to copy data into the clipboard via `clipboardData.setData`. Except IE, event handlers

Table 2. Clipboard handling. ✓ denotes that the text is copied, ✗ that it is not copied. (✓) denotes that the text is copied, but a warning is displayed. The reference to Table 1 means that any trustworthy event that could be used to trigger a pop-up in FF 47, GC 54, or OP 41 can be used, in combination with the JavaScript code given in Listing 1.4 write text to the clipboard.

Action via	IE 11	FF 47	GC 54	OP 41
Copy/cut				
Right mouse-click then copy/cut	✓			
Keyboard: `Ctrl+C`	✓			
Script	(✓)	✗		
Trustworthy event and then script		cf. Table 1		
Paste				
Right mouse-click then paste	✓			
Keyboard: `Ctrl+V`	✓			
Script	(✓)	✗		
Trustworthy event and then script		(✗)		

which are able to open pop-up windows are also able to access the clipboard API with copy and cut capabilities within a delay of one second (e.g., via the `EventTarget.addEventListener()` method) [25]. Thus, our pop-up definition with P_t (cf. Sect. 4) also applies to these kinds of clipboard API access.

Paste operations can only be accessed with the help of JavaScript code when the user triggers a trustworthy `paste` event via `Ctrl+V` and `Edit->Paste`. This clipboard API [37] paste event behavior is important from the security perspective (discussed in Sect. 6.3).

5.3 Drag-and-Drop Scenario

Need of Trustworthy Events. Drag-and-drop operations can be done same-origin or cross-origin. Thus, the usual access limitations of the SOP in the HTML context does not apply in this scenario. Modern browsers like GC even allow the user to drag content from the desktop into the browser's website (e.g., for file uploads). Without trustworthy events, arbitrary data from another window and environment could be stolen automatically with the help of JavaScript code.

JavaScript DOM Access. An example for transferring data via drag-and-drop is given in Table 3. In this table, the host document (HD) shown in Listing 1.5 includes the embedded document (ED) displayed in Listing 1.6.

The first part of Table 3 illustrates that the code of Listing 1.5 can be used, in the same-origin case, to copy the word `Test` into the input field of Listing 1.6. This is possible because we select this word by using the ID `HDt` and afterwards we copy it into the input field with the ID `EDi`. To do this, one must select the embedding element with the ID `EDf`. In the cross-origin case, the browser does not allow the copy-action.

Table 3. A HD wants to transfer data to the Iframe's web page (✓ access, ✗ no access).

Iframe access	IE 11	FF 47	GC 54	OP 41
JavaScript				
Same-Origin (SO)	✓			
Cross-Origin (CO)	✗			
Mouse Events				
Click calls function (SO)	✓			
Click calls function (CO)	✗			
Drag & Drop (SO, CO)	✓		✗	

From an attacker's perspective, it is interesting to know whether it is possible to do actions which are restricted by the SOP [5, 31]:

1. We trigger the JavaScript function of Listing 1.5 by dragging the content of <div> to trigger the JavaScript function copy() with the help of the ondragstart event-handler. In this case, only same-origin access from the HD to the ED is allowed.
2. Cross-origin drag-and-drop operations are allowed in two browsers: IE 11 and FF 47. Trustworthy events like selecting the text test with the mouse, dragging it into the Iframe's input field and dropping the selected text into this field allows to do actions that are (cross-origin) restricted with JavaScript code. GC and OP also allowed these actions in former versions (cf. Sect. 6).

```
1   <i id="HDt">Test</i><br>
2   <iframe id="EDf" src="http://example.org/form.html">
        </iframe>
3   <div draggable="true" ondragstart="copy()">Drag me
        </div>
4   <script> function copy() {
5     document.getElementById("EDf").
          contentDocument.getElementById("EDi").value =
          document.getElementById("HDt").innerHTML;
6   } </script>
```

Listing 1.5. The HD executes JavaScript code when a **dragstart** event occurs.

```
1   <form action="action.php">
2   <input type="text" id="EDi"><br>
3   </form>
```

Listing 1.6. HTML code of the ED.

6   New UIR Attack Variants

Based on the described trustworthy scenarios, we demonstrate that known UI redressing techniques in combination with trustworthy events can be used to derive attacks with a higher attack surface. We construct three new attack variants and evaluate their practicability on modern browsers.

6.1   Optimized Drag-and-Drop Attack

In 2010 Stone published a Clickjacking attack that makes use of the HTML5 drag-and-drop API [38]. In a proof of concept, he showed a website with a frog and a blender. By using social engineering, he lured the victim intro dragging the frog into the blender. What the victim actually does is a cross-origin-drag of attacker defined content into another website. This bypasses protection mechanisms against Cross-Site Request Forgery and could be used in webmail application, document editors, or even to set passwords as shown by Niemietz et al. [21,29].

Drag-and-drop across windows was supported between browsers and therefore an attractive feature which could be abused by attackers. Nowadays, this feature is disabled in modern browsers like GC, SA, and OP; it still works in IE, Edge, and FF. In the following, we derive an attack variant which highlights the importance of different UIR contexts. It points out that trustworthy events play an indispensable role in browser security.

In Stone's initial attack of dragging a frog into a blender, the victim had to clearly visible move the mouse cursor a certain distance (frog to blender) such that the victim might know that it initialized a drag action. The following described attack shrinks the cursor distance to a minimum (e.g., two pixels). Thus, the victim might not notice that any drag actions occurred.

Attack Summary. By looking on the left-hand side of Fig. 2, the attacker's website without any user action is displayed (cf. Listing 1.8). This website could be opened by the user due to a click on a link in a phishing mail. What the victim does is that it slightly moves the button causing a *cross-origin* drag-and-drop injection to occur (cf. Listings 1 .7 and 1.9). For demonstration purposes, an alert-window generated by JavaScript code appears with the injected content (cf. Listing 1.9).

Fig. 2. Attacker defined content can be cross-origin injected.

Attack Structure. With the help of Listing 1.8, there is a web page shown, making use of social engineering techniques. By showing an image with a button that should be moved, attacker defined content will be dragged but not the selected image. By dragging the image, the function *hover* of Listing 1.7 will also be called. This function places an invisible Iframe directly under the mouse cursor such that an drop action attempts to put the attacker defined content into the Iframe's document.

```
1 function hover(e){
2   var x=document.getElementsByTagName("iframe")[0].style;
```

```
3   x.left=(e.clientX-60)+"px";
4   x.top=(e.clientY-10)+"px";
5   x.display='inline';
6   x.opacity='0.0';
7 }
```

Listing 1.7. JavaScript code of the HD (scenario: drag-and-drop attack).

```
1     <h3>Show picture</h3>
2     <iframe src="a.html" style="position:fixed;
          display:none"></iframe>
3     <div id="d" style="background-image:url('evil.png');
          height:1px; width:127px; opacity:0"></div>
4     <img src="s.png" draggable="true" ondragstart="
          this.src=''; event.dataTransfer.setData('text/
          plain','malicious code'); hover(event); var
          d=document.getElementById('d').style; d.height='57
          px'; d.opacity='1'">
```

Listing 1.8. HTML code of the HD (scenario: drag-and-drop attack).

The Iframe's content is shown in Listing 1.9. It only consists of an input area and JavaScript code which shows an alert-window on the condition that the attacker defined content is dropped. Thus, the alert-window only appears in case that the proof-of-concept functions as expected. In a real world application, there could be a search engine in the background which automatically looks up the dropped user input by pulling XMLHttpRequest leading to a code injection, and thus to Cross-Site Scripting.

```
1 <script>
2 var t = setInterval(function() {
3   if (document.getElementsByTagName("input")[0].value) {
4     alert('Cross-origin injection succesful! Value: '+
          document.getElementsByTagName("input")[0].value);
5     clearInterval(t);
6   }
7 }, 500);
8 </script>
9 <input type="text" style="position:absolute; top:0px;
      left:0px">
```

Listing 1.9. HTML and JavaScript code of the ED (scenario: drag-and-drop attack).

6.2   Multiple Pop-Up Attack

As shown in Table 1, a pop-up window can be generated with a trustworthy event like a `click` within a delay which is shorter than one second. For FF and SA, we evaluated that more than one pop-up window will not be blocked once a single pop-up is generated. In contrast, GC, OP, IE, and Edge show one pop-up window and an additional warning window as an information about the blocking of the other pop-up windows.

```
1 <script>
2 function makePopups(){
3   for (i=1;i<1000;i++) {
4     window.open('x.html',i,'width=500,height=500');
5   }
6 }
7 </script>
8 <a href="#" onclick="makePopups()">Spam</a>
```

Listing 1.10. HTML and JavaScript code of the ED (scenario: multiple pop-up attack).

An example is given in Listing 1.10. After a click on Spam the trustworthy event click is triggered and thus the function makePopups() is called. The function includes a for-loop which generates 1,000 windows that could be either pop-ups (this example) or new tabs (by removing the third parameter with width and height). In FF and SA, all of these windows are shown to the user. This behavior leads to a heavy memory consumption and thus heavily slows down the underlying system's speed. It is likely that a victim will close all browser windows simultaneously and for this reason, it may also lose existing browser sessions (e.g., in other tabs). Another use case is click-fraud by creating multiple pop-ups with advertisements; an attempt to close these unwanted windows could lead to an unintended click and thus a successfully clicked advertisement.

The behavior of FF unexpected due to browser settings that are reachable via about:config. Firstly, the property dom.popup_maximum (maximum number of pop-up windows) has a default value of 20. We are clearly able to generate more windows with trustworthy events. Secondly, the property dom.popup_allowed_events (events that spawn pop-ups) has the value change click dblclick mouseup notificationclick reset submit touchend.

As shown in Table 1, we could also use other events like a left-click triggered select (not listed within dom.popup_allowed_events). Therefore, there is a lack of handling pop-up windows properly. We have reported these problems to Mozilla.

6.3 Hijacking Clipboard Data

In contrast to browsers like FF, GC, and even Edge, IE allows full access to the clipboard after a confirmation on a warning window (cf. Table 2). Clickjacking can be used to attack an IE user and thus to get access to the saved clipboard data that may contain sensitive data like a password.

We introduce two new attack sub-variants to steal clipboard data. Firstly by stealing the second click from a double-click scenario which was described by Huang et al. [12]. Secondly by just using a single click; this highlights the importance to look on different trustworthy events.

The first variant is displayed in Listing 1.11. With the help of social engineering, the attackers lures a user to make a double click on the displayed button. The first click of the double click triggers the onclick event-handler, which shows the accessed clipboard data in an alert window (as a proof-of-concept). For the clickjacking attack, the second click of the double-click actually occurs on the `Allow access` button of the confirmation window. To ensure that a user always hits the `Allow access` button, the `Double Click` button will always be positioned in the middle of the screen (with slide adjustments).

The second variant is targeting an impatient user. It consists nearly of the same code and displayed Figure, except for two changes. The `Double Click` button is named `DL in X` where X is a counter with a number in seconds which decreases until zero. An impatient user will wait until the button's counter reaches zero to download a file, and thus the click will be correctly timed. The attacker will therefore show the confirmation dialog 300ms before the button's counter reaches zero, such that the click will be successfully hijacked.

The limitation of both attack variants is that the confirmation window must be visible for at least 300 milliseconds; this is the lower bound we measured. The Human Benchmark Project[1] recorded over 51 million clicks and measured that the average reaction time of a human is 282 milliseconds (where the user was aware of being timed). Therefore, it is very likely that a user is not able to cancel the hijacked click on the confirmation window.

```
1 <style> button {position: fixed; top: 50%; left: 50%;
2   margin-top: 15px; margin-left: -20px;} </style>
3 <button onclick="if (window.clipboardData.getData('Text').
      length > 0) {alert('Hijacked Clipboard data: '+
      window.clipboardData.getData('Text'));}">
4 Double Click</button>
```

Listing 1.11. HTML and JavaScript code of the ED (scenario: clipboard attack).

7   Defenses Discussion

We have evaluated that trustworthy events are implemented differently across browsers. Our formal definition of trustworthy events and the thereby derived descriptions of three different scenarios might help browser vendors to minimize the high number of event handling differences.

An approach to help browser vendors to avoid bugs and features that may lead to security vulnerabilities is to compare their browser result with the result of the majority of other modern browsers. For example, it may be suspicious if just one out of seven tested browsers allows access (or a particular interaction) after a trustworthy event; for clarification reasons, the set of browsers could be extended (e.g., by considering more browsers like Brave and Chromium).

Drag-and-Drop Attack. Drag-and-drop actions are known since the introduction of web browsers, which still allow restricted draggings of for example text elements (selected text), images (image URL), and anchor-elements (anchor URL). Moreover, HTML5 has introduced a drag-and-drop API [41] that is nowadays integrated in all modern web browsers.

---

[1] http://www.humanbenchmark.com/tests/reactiontime/statistics.

We constructed a drag-and-drop attack variant that can be executed in three (IE 11, Edge 20, and FF 47) tested browsers. A simple but effective countermeasure is to prohibit drag-and-drop frame attacks by *disallowing drag operations with data across frames with different origins*. Browser vendors like Google and Opera allowed cross-frame drag-and-drop operations in the past; nowadays, this is not anymore possible due to security reasons (cf. Sect. 6.1)

Pop-Up Attack. FF is the only tested browser which allows creating hundreds of pop-ups after a trustworthy event like a left-click within the measured delay of one second. All other tested browsers disallow the execution of multiple pop-ups and therefore the user will not be annoyed when, for example, they appear unintentionally. The majority of our tested browser behavior results can therefore be used to derive a countermeasure for FF; this browser should *only show one pop-up window after a trustworthy event*.

Clipboard Data. Our clipboard data attack variant on IE showed that a user should not get an unlimited control over the whole clipboard data by just executing JavaScript code. For this reason, there are different access types (copy, cut, paste) that are implemented in modern browsers due to the W3C clipboard API [37]. However, the behavior of IE underlined that read access should only be allowed with a trustworthy event like a keystroke combination (e.g., STRG+V).

The countermeasure of disallowing clipboard read access is very strict and it might be more convenient to get only read access if the user explicitly gives the permission by *showing a clipboard permission window for a time that is significantly higher than the human response time*; this should be longer than the short display time of the IE permission window (cf. Sect. 6.3).

According to the Human benchmark project, only a negligible amount of the measurements ($<0.1\%$) have a longer human response time than 500 ms. As a consequence, a browser implementation should only activate the Allow access button of the permission window after a trustworthy event and a delay of at least half a second. This ensures with a high probability that the second click will not be hijacked by an attacker.

8 Related Work

Definitions and Specifications. Huang et al. [12] discussed UIR attacks and defenses with a definition of UIR. They developed a defense called in InContext to mitigate UIR attacks. The W3C created a UI safety specification [20] that is based on the ideas of InContext. Similar UI contexts are mentioned in the W3C UI security and visibility API [14]. These foundations of describing trusted events do not consider conscious user actions, which we define as trustworthy events. Without these events, UIR attacks could not be executed.

By looking at the concept of zones and scenarios, IE includes predefined zones like *Internet, Local Intranet,* and *Trusted Sites* [22]. This concept is partially adopted between browsers by explicitly white-listing trusted sites [11]. Trusted site lists can be used to manage whether certain actions should be automatically executed (e.g., generate cryptographic keys, play Flash files, and show pop-ups).

Attacks and Countermeasures. Grossman et al. [10] introduced Clickjacking as an attack which is nowadays considered as a class of attacks which relies on the broader set of UIR attacks. Although the attack on Flash received high media attention and several bugfixes since 2008 [2], it was successfully attacked years later (e.g., in 2011 [1]). Next to JavaScript-based frame busters [33], the HTTP Header X-Frame-Options [8,16], and nowadays even the Content-Security-Policy [36] can be used to defend against many types of UI redressing. In an evaluation about different JavaScript-based UIR protection mechanisms, Rydstedt et al. [33] pointed out that there exist attacks which can be used to attack protection mechanism and thus disable them. Balduzzi et al. [4] designed and implemented an automated system to analyze Clickjacking attacks. Niemietz et al. [29] evaluated the security of home routers and found that none of them are protected against UIR. Rydstedt et al. [32] published a paper about UIR on mobile sites and also on home routers.

Lekies et al. [17] presented bypasses for Clickjacking defense tools like NoScript's ClearClick. Furthermore, they introduced a new attack technique called nested Clickjacking. By showing that UI time delays as defense mechanisms are not sufficient to protect the user, Akhawe et al. [3] created examples which bypass the W3C UI safety specification [20].

Mobile Devices. Lin et al. [19] published Screenmilker, which analyzes the user interface of an Android device. By using the Android debug bride (ADB), they showed that Screenmilker is able to make screenshots during user interactions and they were able to steal secrets like passwords. Bianchi et al. [6] published a study on Android-based graphical user interface confusion attacks [128]. These attacks concentrate on phishing and privacy violations. Niemietz et al. enumerated different UIR attacks [27] and their countermeasures. Furthermore, they provide a Tapjacking attack to compromise Android devices [28]. Based on this work, Fratantonio et al. [9] created malicious apps that completely control the UI feedback loop. They furthermore showed with a user study that none of the created attacks could be detected by a user.

9   Conclusions

In this paper, we provide a definition of *trustworthy events*, which are the target of UI Redressing attacks. We show that this concept is significantly different from the concept of *trusted events* as defined by the W3C. Interpretations of events as being trustworthy differ significantly between browser families, and by a non-documented inheritance mechanism trustworthiness may be transferred, within the time frame of one second, from a trustworthy event to a sequence of events triggered by JavaScript. This, for example, allowed us to circumvent the FF pop-up blocker.

We investigated three scenarios where trustworthy events play a major role in protecting the security of web applications: pop-ups, drag-and-drop, and copy-and-paste. In all three scenarios, differences in the interpretation of trustworthy events could be shown. We refined one new example attack variant in each scenario, based on a more detailed investigation of these scenarios. Finally, we discuss defense mechanisms by analyzing the causes of our trustworthy event attacks. With the definition and description of *trustworthy events*, we hope that this paper will contribute to a better understanding of UIR attacks, and thus improved web application security.

References

1. Aboukhadijeh, F.: Spy on the webcams of your website visitors, October 2011. http://feross.org/webcam-spy/
2. Aharonovsky, G.: Malicious camera spying using clickjacking, October 2008. http://blog.guya.net/2008/10/07/malicious-camera-spying-using-clickjacking/
3. Akhawe, D., He, W., Li, Z., Moazzezi, R., Song, D.: Clickjacking revisited: a perceptual view of UI security. In: 8th USENIX Workshop on Offensive Technologies (WOOT 2014). USENIX Association, San Diego, August 2014. https://www.usenix.org/conference/woot14/workshop-program/presentation/akhawe
4. Balduzzi, M., Egele, M., Kirda, E., Balzarotti, D., Kruegel, C.: A solution for the automated detection of clickjacking attacks. In: Proceedings of the 5th ACM Symposium on Information, Computer and Communications Security, ASIACCS 2010, pp. 135–144. ACM, New York (2010). https://doi.org/10.1145/1755688.1755706
5. Barth, A.: The Web Origin Concept. IETF, RFC 6454, December 2011. http://tools.ietf.org/html/rfc6454, http://tools.ietf.org/html/rfc6454
6. Bianchi, A., Corbetta, J., Invernizzi, L., Fratantonio, Y., Kruegel, C., Vigna, G.: What the app is that? Deception and countermeasures in the android user interface. In: IEEE Symposium on Security and Privacy. Department of Computer Science, University of California, Santa Barbara (2015)
7. Bordi, E.: Cursorjacking proof of concept. http://static.vulnerability.fr/noscript-cursorjacking.html (August 2010)
8. Braun, F., Heiderich, M.: X-Frame-Options: All about Clickjacking? (2013) https://cure53.de/xfo-clickjacking.pdf
9. Fratantonio, Y., Qian, C., Chung, S., Lee, W.: Cloak and dagger: from two permissions to complete control of the UI feedback loop. In: Proceedings of the IEEE Symposium on Security and Privacy (Oakland), San Jose, CA, May 2017
10. Hansen, R., Grossman, J.: Clickjacking attack, December 2008. http://www.sectheory.com/clickjacking.htm
11. Help, G.C.: Allow or block content settings for certain sites, March 2017. https://support.google.com/chrome/answer/3123708?hl=en

12. Huang, L.S., Moshchuk, A., Wang, H.J., Schecter, S., Jackson, C.: Clickjacking: attacks and defenses. In: Presented as part of the 21st USENIX Security Symposium (USENIX Security 2012), pp. 413–428. USENIX, Bellevue (2012). https://www.usenix.org/conference/usenixsecurity12/technical-sessions/presentation/huang

13. Kacmarcik, G., Leithead, T.: UI events - W3C working draft, August 2016. https://www.w3.org/TR/uievents/

14. Kaminsky, D., Huang, D.L.S., Maone, G.: W3C - user interface security and the visibility API, June 2016. https://www.w3.org/TR/UISecurity/

15. Kotowicz, K.: Cursorjacking again, January 2012. http://blog.kotowicz.net/2012/01/cursorjacking-again.html

16. Lawrence, E.: Combating clickjacking with x-frame-options, March 2010. http://blogs.msdn.com/b/ieinternals/archive/2010/03/30/combating-clickjacking-with-x-frame-options.aspx

17. Lekies, S., Heiderich, M., Appelt, D., Holz, T.: On the fragility and limitations of current browser-provided clickjacking protection schemes. In: Presented as Part of the 6th USENIX Workshop on Offensive Technologies. USENIX, Berkeley (2012). https://www.usenix.org/conference/woot12/workshop-program/presentation/Lekies

18. Lekies, S., Heiderich, M., Appelt, D., Holz, T., Johns, M.: On the fragility and limitations of current browser-provided clickjacking protection schemes. In: USENIX Workshop on Offensive Technologies (WOOT 2012) (2012)

19. Lin, C.C., Li, H., Zhou, X., Wang, X.: Screenmilker: how to milk your android screen for secrets. In: Network and Distributed System Security (NDSS) Symposium 2014 (2014)

20. Maone, G., Huang, D.L.S., Gondrom, T., Hill, B.: W3C - user interface security directives for content security policy, June 2014. https://dvcs.w3.org/hg/user-interface-safety/raw-file/tip/user-interface-safety.html

21. Mayer, A., Niemietz, M., Mladenov, V., Schwenk, J.: Guardians of the clouds: when identity providers fail. In: The ACM Cloud Computing Security Workshop, CCSW 2014 (2014)

22. Microsoft: How to use security zones in internet explorer, June 2012. https://support.microsoft.com/en-us/help/174360/how-to-use-security-zones-in-internet-explorer

23. Needham, K.: The future of developing firefox add-ons, August 2015. https://blog.mozilla.org/addons/2015/08/21/the-future-of-developing-firefox-add-ons/

24. Network, M.D.: Event.istrusted, February 2017. https://developer.mozilla.org/en-US/docs/Web/API/Event/isTrusted

25. Network, M.D.: Web apis - document.execcommand(), January 2017. https://developer.mozilla.org/de/docs/Web/API/Document/execCommand

26. Niemietz, M.: Clickjacking und UI-Redressing - Vom Klick-Betrug zum Datenklau: Ein Leitfaden für Sicherheitsexperten und Webentwickler. dpunkt-Verlag (2012)

27. Niemietz, M.: UI Redressing: Attacks and Countermeasures Revisited. In: CON-Fidence, May 2011

28. Niemietz, M., Schwenk, J.: UI Redressing Attacks on Android Devices, December 2012. https://media.blackhat.com/ad-12/Niemietz/bh-ad-12-androidmarcus_niemietz-WP.pdf

29. Niemietz, M., Schwenk, J.: Owning your home network: router security revisited. In: Web 2.0 Security & Privacy 2015, San Jose (2015). http://ieee-security.org/TC/SPW2015/W2SP/papers/W2SP_2015_submission_9.pdf

30. Roesner, F., Kohno, T., Moshchuk, A., Parno, B., Wang, H., Cowan, C.: User-driven access control: Rethinking permission granting in modern operating systems. In: 2012 IEEE Symposium on Security and Privacy (SP), pp. 224–238, May 2012
31. Ruderman, J.: The same origin policy (2008). http://www-archive.mozilla.org/projects/security/components/same-origin.html
32. Rydstedt, G., Bursztein, E., Boneh, D.: Framing attacks on smart phones and dumb routers: tap-jacking and geo-localization. In: in USENIX Workshop on Offensive Technologies (wOOt 2010) (2010). http://seclab.stanford.edu/websec/framebusting/tapjacking.pdf
33. Rydstedt, G., Bursztein, E., Boneh, D., Jackson, C.: Busting frame busting: a study of clickjacking vulnerabilities at popular sites. In: IEEE Oakland Web 2.0 Security and Privacy (W2SP 2010) (2010). http://seclab.stanford.edu/websec/framebusting/framebust.pdf
34. Sherman, I.: Making form-filling faster, easier and smarter, January 2012. https://webmasters.googleblog.com/2012/01/making-form-filling-faster-easier-and.html
35. Sophos: Facebook worm - "likejacking", May 2010. http://nakedsecurity.sophos.com/2010/05/31/facebook-likejacking-worm/
36. Stamm, S., Sterne, B., Markham, G.: Reining in the web with content security policy. In: Proceedings of the 19th International Conference on World Wide Web, WWW 2010, pp. 921–930. ACM, New York (2010). https://doi.org/10.1145/1772690.1772784
37. Steen, H.R.M.: W3C - clipboard API and events, December 2016. https://www.w3.org/TR/clipboard-apis/
38. Stone, P.: Next generation clickjacking - new attacks against framed web pages, April 2010. https://www.contextis.com/documents/5/Context-Clickjacking_white_paper.pdf
39. W3C: W3C DOM4: Dom event istrusted, November 2015. https://www.w3.org/TR/dom/
40. W3C: UI events, January 2016. https://w3c.github.io/uievents/
41. WHATWG: Html, living standard - drag and drop, November 2013. http://www.whatwg.org/specs/web-apps/current-work/multipage/dnd.html#dnd
42. WHATWG: form control infrastructure, July 2017. https://html.spec.whatwg.org/multipage/form-control-infrastructure.html
43. Zalewski, M.: Strokejacking, June 2010. http://lcamtuf.blogspot.de/2010/06/curse-of-inverse-strokejacking.html

A Paged Domain Name System
for Query Privacy

Daniele E. Asoni[✉], Samuel Hitz, and Adrian Perrig

Network Security Group, Department of Computer Science, ETH Zürich,
Zurich, Switzerland
{daniele.asoni,samuel.hitz,adrian.perrig}@inf.ethz.ch

Abstract. The lack of privacy in DNS and DNSSEC is a problem that
has only recently begun to see widespread attention by the Internet and
research communities, and the solutions proposed so far only look at a
narrow slice of the design space. In this paper we investigate a new approach for a privacy-preserving DNS mechanism that hides query information from root name servers and TLD registries. Our architecture lets
TLD registries group the DNS records in their zones together into *pages*.
Resolvers cache all pages locally, and retrieve only small incremental
updates to optimize performance. We show that this strategy is particularly effective given the relatively static nature of TLD zone records. We
analyze the privacy guarantees to assess the potential and limitations of
our approach; we also evaluate the memory overhead for a resolver, and
obtain feasibility guarantees through a prototype implementation of the
new functionalities for resolvers and registries.

1 Introduction

The Domain Name System (DNS) [29,30] is a fundamental building block of
the Internet, providing host name to IP address translation. Its design has been
sufficiently scalable to cope with the Internet's growth, but among its deficiencies
is the lack of privacy protection. The DNS security extensions (DNSSEC), which
are still far from widespread adoption, have addressed some of DNS's flaws, but
privacy has explicitly remained a non-goal in the design of DNSSEC [5]. While
the Internet Engineering Task Force (IETF) has recently started considering
privacy concerns for DNS more seriously [8], so far only minor improvements
have been proposed [9].

Users with very high privacy requirements will resort to an anonymous communication system (ACS) such as Tor [16], which will anonymize not only the
DNS lookups, but also the subsequent communications with the hosts whose
addresses are obtained through the lookup. This is necessary, for instance, if a
user wishes to hide from its own Internet service provider (ISP) what hosts it
communicates with. However, communication over Tor comes with harsh performance penalties, so for clients which have some degree of trust in their ISP,
a more lightweight solution is desirable. In particular, we identify the main privacy threat in this scenario to be large-scale information collection at the highest

S. Capkun and S. S. M. Chow (Eds.): CANS 2017, LNCS 11261, pp. 250–273, 2018.
https://doi.org/10.1007/978-3-030-02641-7_12

levels of the DNS hierarchy: the name servers of the root and of the top-level domain (TLD) registries. These are centralized observation points that are ideally suited for surveillance by a nation state actor. While a recent proposal [9] would, if accepted, hide query information from the root name servers, there seems to be no possibility in current DNS to hide sensitive information from TLD registries.

Although users gain some privacy by relying on the recursive resolver provided by their ISP, this method is not secure against an adversary who is able to correlate multiple queries through timing. For instance, the adversary may observe that a certain set of domains is always queried together in a short time interval, and thus infer that the same user is responsible: if one of the domains identifies the user (e.g., because it is the user's own website, which is otherwise scarcely visited), then the adversary is able to deanonymize all the other queries as well. Furthermore, users may wish to conceal the fact that a certain domain is being queried at all, a property akin to private information retrieval (PIR) [11,31], which, incidentally, cannot be achieved even if DNS lookups are performed over an ACS.

In this paper we propose a system whose goal is to prevent information leakage to the DNS root and TLD registries, including the information of what domains are actually queried. Our system requires changes to the TLD registries and to the recursive resolvers, but is transparent to clients and to second level domain (SLD) authoritative name servers, which continue to use the traditional DNS protocol. The core idea of the system is to group records in the TLD zones into fixed sets we call *pages*, which are created and maintained jointly by the TLD registries. Recursive resolvers query for entire pages, rather than single records, which provides a basic amount of privacy (see Fig. 1 for a high-level overview). We improve the performance of this basic mechanism with optimizations such as full page caching on the recursive resolvers, and we improve its privacy with enhancements such as cover page queries from the recursive resolver to the TLD registries.

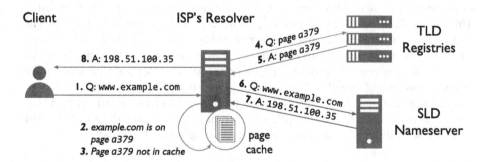

Fig. 1. High-level overview of the PageDNS architecture. Steps 2–5 are specific to PageDNS, while the others are as in DNS. In the example shown, the entire page a379 is retrieved from the TLD registries, but in practice the page would be cached at the resolver, and at most a (much smaller) incremental update would be retrieved.

1.1   Overview

Our *Paged Domain Name System* (PageDNS) introduces a new way in which recursive resolvers can obtain records from TLD registries in a privacy-preserving manner. The TLD registries collaborate to group the records of all their zones into 2^l ($\simeq 10^5$) sets of records which we call *pages*. Each page contains n records, with $n \simeq 10^4$, assuming a total number of records of around one billion (see Sect. 4.1). The overwhelming majority of these records are name server records of second level domains (SLDs), i.e., records providing the IP addresses of the name servers authoritative for various SLDs (e.g., `example.com`). To spread the records uniformly across the pages in a way that allows resolvers to easily determine which page stores which record, each page is given a unique l-bit identifier, and each record is assigned to the page whose identifier matches the first l bits of the hash value of the record's domain name.

A recursive resolver with PageDNS support should, for performance reasons, cache all pages locally, de facto mirroring all TLD zones. These cached copies are kept indefinitely, in particular beyond their expiration time. When queried for a domain name (say `www.example.com`), the recursive resolver proceeds as follows. First, it determines which page should store the corresponding SLD record (the NS record for `example.com`). Second, it checks whether it has an up-to-date copy of that page in its cache. If not, the resolver sends a page query to one of the TLD registries, specifying the *version number* of the locally cached copy. The registry then replies with a list of records that have been changed, added, or removed since the specified version. Because the name servers for SLDs are relatively static, as we show in Sect. 4.1, the size of these *incremental updates* will typically be small. Finally, once the recursive resolver has obtained the NS record for the SLD, it completes the iterative lookup as in DNS, and returns the response to the client.

Replication of Popular Records. Requesting a page instead of a single domain protects query privacy, because an adversary observing a page request cannot directly determine which domain among the $\sim 10^4$ domains in the requested page is accessed by a client. However, domains have very different popularity, meaning that the probability that a certain page is requested because it contains, e.g., the record for `twitter.com` is much higher than the probability that the page was accessed due to some obscure domain that has its record on the same page. To counter this problem and provide privacy protection even for popular domains we *replicate* records of popular domains across multiple pages: the higher the popularity of a domain, the higher the replication degree of its record. In Sect. 3.1, we show in particular that an optimal solution is to replicate the $\sim 0.01\%$ most popular domains, with the most popular one being replicated on all pages.

2   Design

In this section we describe the details of our Paged Domain Name System (PageDNS). We begin by describing our threat model and privacy goals, and then define the structure of pages, name resolution process, and other protocol aspects.

2.1   Threat Model

We want to prevent (government-level) monitoring which targets DNS query information. In particular, we want to prevent linkability between clients and the queries they make, and, to the extent possible, we also try to hide the fact that a certain name is being queried, i.e., that there is interest by *some* client for a certain name. We exclude the case of a compromised client ISP, which would be able to observe not only the DNS queries of its clients, but also the communications after that, requiring the use an anonymous communication system (ACS) to achieve anonymity. For analogous reasons, we do not consider other types of in-network adversaries. Instead, we aim to provide privacy with respect to the DNS root and TLD registries, which constitute a centralized point which is ideal for mass surveillance.

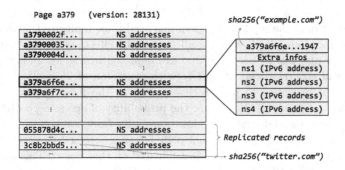

Fig. 2. Structure of a PageDNS page. The highlighted detail on the right shows the structure of a single record for the SLD example.com.

2.2   Page Structure

PageDNS pages consist of a sequence of records, plus information such as the page's length, expiration time, and version number. Each page is identified by an *l*-bit hash prefix, and all records whose domains' hash values match the prefix are contained in the page. Additionally, each page has a small separate section which contains the records of replicated domains (whose hashes will typically not match the prefix, see below). Besides the hash of the domain name, each record contains a set of addresses of the domain's name servers, and additional information such as the type of the addresses (e.g., IPv6 or IPv4). The reasons for indexing records by hash instead of including the domain name are twofold: first, it fixes the length of the identifier; second, it provides a degree of protection against zone enumeration [26]. In Fig. 2 the high-level structure is depicted for one sample page, and the details for the record of domain example.com are shown. This sample record contains four IPv6 addresses, which in terms of size we consider to be a reasonable upper bound.

Since one IPv6 address is 16 bytes long, and assuming a hash size of 32 bytes[1], we obtain 96 bytes, excluding the additional information, so overall we round the size of a record to 100 bytes for the scope of this discussion. Considering a limit of 1 MB on page size and a total of 10^9 domain names (see Sect. 4), this implies a total of 10^5 pages, each containing around 10^4 records. We estimate the variance of the size of pages in Appendix B, and we find that the probability of a deviation of the size of over 10% is negligible.

Record Replication. We observe that the popularity of domains has the potential to heavily affect the privacy of page queries. For instance, a query for a page which contains the record of a very popular domain is likely to be due to an access to that domain. To mitigate this problem we adopt record replication for popular domains. In Appendix A, we show analytically that the optimal replication is proportional to the popularity of the domain, assuming that the popularity of domains follows a Zipf distribution.[2] In particular this means that approximately only the top 200,000 domains need to be replicated, with the most popular domain being replicated on all pages (the effects of replication become clearer in our analysis in Sect. 3.1, and are depicted in Fig. 3). This replication has an overall size overhead of 0.23% (with a median of 25 replicated records per page), but it allows to hide accesses even to the most popular domains. Furthermore, our analysis shows that the probability of accessing a certain page because of interest in a domain with low popularity (any non-replicated domain) is lower than the probability that that page is being accessed because of interest in the most popular domain (which is replicated on that page). This means that replication provides effective and relatively uniform privacy protection for less popular domains. These replicas are placed separately on a page (e.g., at the end, see Fig. 2) and are identified by their hash value.

Assigning Records to Pages. Mapping records to the $m = 2^l$ pages is not entirely trivial. For non-replicated domains, we use the first l bits of the hash value of the domain name to determine the page identifier to which the domain is assigned. However, for replicated records another scheme is needed. We propose a scheme based on a pseudo-random permutation (PRP), keyed with the hash value of the domain name: this PRP has as domain the set of all page identifiers (i.e., all integers from 1 to m).[3] For a replicated domain d, the PRP maps all integers

---

[1] We consider SHA-256 as a reasonable choice for the hash function. While the size could be reduced to 16 bytes while still retaining a negligible collision probability in a non-adversarial setting, a larger size is necessary if we want to have a negligible probability even in a scenario where the adversary actively tries to find a domain name which will result in a collision.

[2] In practice, popularity will vary on a regional basis. We envision that replication may be made region-specific (the non-replicated part of each page would remain the same). We leave a more detailed analysis of these aspects to future work.

[3] PRPs of small domain size can be implemented using format-preserving encryption (FPE) schemes, there exist suitable encryption modes that use standard AES block ciphers as a primitive and achieve FPEs of arbitrary domain size.

between 1 and the replication degree of d, $r(d)$, to the page identifiers on which the replicas should be stored. Since the PRP is keyed with the hash of d, the mapping will be independent from the mapping of replicas for other domains.

Note that this method ensures minimal modifications to the pages as the replication of a domain changes: assuming a domain previously replicated r_{old} times increases its popularity and has to be replicated r_{new} times ($r_{new} > r_{old}$), the first r_{old} replicas will remain the same, and only $r_{new} - r_{old}$ additional pages have to be changed to include the replica. Similarly, if $r_{new} < r_{old}$, the first r_{new} replicas will remain the same, and only $r_{old} - r_{new}$ pages have to be changed to remove the extra replicas.

We point out that for this mechanism to work, the resolvers need to be aware of the replication degree of the domains they look up. To that end, the TLD registries create a special *meta-page* for replication, which lists all the most popular domains (by their hash value), and for each of them it provides the replication degree. This meta-page has a size of about 7 normal pages (7 MB), and is updated less frequently. We show how the TLD registries determine the popularity of domains in Sect. 2.4. In the next section, we show how clients can resolve a name, including more details about how the case of replicated domains is handled.

2.3   Resolving a Name

Algorithm 1 shows the steps a recursive resolver performs when resolving a fully qualified domain name (FQDN), e.g., www.example.com. First, the algorithm splits the FQDN into the SLD (example.com) and the remaining part (typically the host name, www). On a high-level, the algorithm then retrieves the address of an authoritative name server for example.com using PageDNS (Lines 3–30) before completing the lookup for www.example.com using traditional DNS. The resolver starts by identifying the replication degree of the SLD. To obtain this information, it retrieves the meta page from the registries which contains all the replicated domains together with their replication degree. As for ordinary pages, the resolvers also keeps the meta page in its cache, and will therefore usually only need to retrieve an incremental update of the meta page. If a domain is not replicated it has a replication degree of 1.

Then the algorithm checks whether it already has a cached copy of a page that contains the record for the domain. To that end, the algorithm computes the possible page identifiers that could contain the record by calling CalcIdentifier (Algorithm 2) for all $i \in \{1..ReplDeg\}$ and checks if the cache contains any of these pages. A possible optimization for this step would be to keep track, for the more highly replicated records that are requested, of the cached pages that contain them, in order to avoid the computation of tens of thousands of hashes. This step can also be optimized when all pages are cached by the resolver.

The page identifier calculation depends on the chosen replica ID (k): if k is 1, the original record for the domain is chosen, and the page identifier determined as the l-bit prefix of the hash of the domain name (Algorithm 2, line 3). If $k > 1$,

Algorithm 1. FQDN resolution on the recursive resolver.

```
 1: procedure RESOLVEFQDN(FQDN, MetaPage, Cache, Registry)
 2:     Host, Domain ← DomainSplit(FQDN)
 3:     if Domain ∈ MetaPage then
 4:         ReplDeg ← MetaPage[Domain].ReplDeg
 5:     else
 6:         ReplDeg ← 1
 7:     end if
 8:     for all i ∈ RandomShuffle({1..ReplDeg}) do
 9:         PageId ← CalcIdentifier(Domain, i)
10:         if PageId ∈ Cache then
11:             Page ← Cache[PageId]
12:             break
13:         end if
14:     end for
15:     if not Page or HasExpired(Page) then
16:         if not Page then
17:             k ← RandomChoice({1..ReplDeg})
18:             PageId ← CalcIdentifier(Domain, k)
19:             Page ← Query(Registry, PageId)
20:             Page.Registry ← Registry
21:         else
22:             Page ← Query(Page.Registry, PageId, Page.Version)
23:         end if
24:         Key ← PubKey(Registry)
25:         if not Verify(Page.MerkleRoot.Sig, Key) then
26:             abort()
27:         end if
28:         Cache[PageId] ← Page
29:     end if
30:     NS ← BinSearch(Page, SHA256(Domain))
31:     IP ← CompleteLookup(NS, Host)
32:     return IP
33: end procedure
```

then the page is determined by applying a PRP with domain $\{1, \ldots, 2^l\}$ keyed with the hash of the domain name to k (line 5).

If the cached page containing the domain has expired, or if no page containing the domain is cached by the resolver, a page query has to be send out. In case a cached page is available but outdated, the resolver can perform an incremental query by attaching the version number of the cached page to the request. Note that we avoid querying multiple registries for the same page (line 22). If, on the other hand, no page is cached, the algorithm needs to download an entire page. First, a replica ID is chosen uniformly at random from the set $\{1..ReplDeg\}$. Then, a page containing the chosen replica is determined using CalcIdentifier, and the registry is queried for that page.

Algorithm 2. Calculating the page identifier for a (possibly replicated) domain.

```
1: procedure CALCIDENTIFIER(Domain, k)
2:     if k == 1 then
3:         PageId ← SHA256(Domain)[0:l]
4:     else
5:         PageId ← PRP_{2^l}(SHA256(Domain); k)
6:     end if
7:     return PageId
8: end procedure
```

Once the page is obtained, the algorithm verifies the page's integrity. For this, the resolver obtains a signed root of a Merkle hash tree (not show in the algorithm). This hash tree is computed by the registries, for every version number, over all the pages. The resolver verifies the signature of the root (using the standard Web PKI), and verifies that the obtained page is in the tree. This mechanism allows resolvers to use gossiping protocols to ensure that the same hash tree root provided by the queried registry for a specific version is seen by all resolver, and across all registries.

If the verification is successful the page is accepted and updated in/added to the cache. The lookup for the record of interest on the page can be done efficiently by a binary search over the domain name hashes. Communication has to be done over TCP or another reliable transport protocol, given the size of the data returned by the registry (similar to what is done today in DNS for large responses [15]); this has the advantage of preventing reflection and amplification attacks [34]. To complete the lookup and obtain the address of the actual host (Algorithm 1, line 31), the resolver sends an ordinary DNS query directly to the name server whose address was obtained through PageDNS.

2.4   Keeping Pages Updated

Popularity Estimation for Replication. As explained in Sect. 2.2, popular domains are replicated on multiple pages, according to their popularity. TLD registries have to determine the approximate popularity for all these domains, and replicate the records across the pages accordingly. To determine the popularity, we assume that a large fraction of the resolvers can authenticate themselves to the registries (possibly through some out-of-band mechanism), and then provide, at regular time intervals, the approximate number of requests received for the most popular domains (using some randomization to hide the exact numbers). Based on the reports by the resolvers, the registries can then assess the overall popularity of the most popular records and update the pages and meta page accordingly. We point out that strong fluctuation in the popularity are possible (the so-called slashdot effect) and would require updating a high number of pages, which is expensive for the registries. For efficiency we therefore allow registries to consider an averaged popularity, computed for example as a moving average, which obviates the need for rapid and expensive updates of a large number of pages. It is important to note that this comes with some privacy cost for accesses to domains whose popularity has recently increased, which become more identifiable. Similarly, regional differences in popularity can also impact the identifiability of queries.

Page Updates and Authentication. At regular intervals, TLD registries will issue new versions of the pages which need to change as a consequence of updates, insertions, and removals. To authenticate the updated pages (the new versions), each registry constructs a Merkle tree over all the updated pages, and signs the root. When resolvers query for a page, the registries will also provide the signed root of the tree, as well as a proof (which consists of a list of hashes) that the page is part of the tree. We use a binary tree with the pages sorted according to their identifier (every level determines one additional bit): this makes it impossible for a registry to include two pages with the same identifier.

3 Privacy and Security Analysis

In this section we analytically model the privacy guarantees of PageDNS. We start by identifying the ideal replication degree of every domain across pages, depending on their popularity. We then derive the analytic expression of the probability that an adversary is able to correctly guess the *target domain* (i.e., the domain the client is accessing) depending on the domain's popularity, given that the adversary is able to observe the page requests made by the recursive resolver queried by the client. We use this probability distribution as the main metric for the efficacy of PageDNS, and we analyze the impact of replication, *page fingerprints* (one website access causing multiple PageDNS page requests) and of *cover* page queries (retrieving additional pages to provide extra privacy). Cover page queries also model the fact that the DNS queries by other clients of the same resolver cause additional page requests, as well as the fact that the resolver can autonomously update expired pages when idle.

3.1 Replication

Our goal in PageDNS is to hide a client's target domain, and ideally we would like to hide it independently of its popularity. As discussed in Sect. 2.2, to hide target domains with high popularity we have to replicate their records across multiple pages. To determine how much each record should be replicated depending on its popularity, we try to optimize for two goals. First, we want to keep the total number of replicas to a minimum. Second, denoting with $\mathcal{I}_x$ the *identifiability* of a domain x, i.e., how easily x can be guessed based on a page request (we provide a formal definition in Sect. 3.2), we want to minimize the maximum ratio $\mathcal{I}_x/\mathcal{I}_y$ for any two domains x and y. We solve this problem analytically in Appendix A under the assumption that domain popularity follows a Zipf distribution with parameter s (Jung et al. [23] show that this is the case, and that $s = 0.91$). We obtain the following *replication function* that maps a domain's rank $k \in \{1, \ldots, N\}$ to the replication degree of that domain:

$$r(k) = max\{1, Rk^{-s}\} \tag{1}$$

where R is the maximum replication degree. Note that $r(k) = 1$ means that only one record exists for the domain of rank k. Evidently $R \leq m$, where m denotes the total number of pages; the optimal choice in terms of privacy is $R = m$.

3.2 Identifiability

When accessing a certain domain for web browsing, it is likely that a number of additional domains have to be looked up by the client: web pages contain external content from CDNs, from advertisement providers, or from user tracking sites (e.g., `google-analytics.com` [1]). While some of these might be safely blocked by the browser, others are necessary for correctly displaying a page.

For simplicity in our analysis we consider the case where each domain corresponds to a single website (e.g., this can be the index `www` webpage). We define the *page fingerprint* of a domain as the set of pages which are requested due to an access to the website corresponding to that domain. Different domains may have the same page fingerprint; we also note that, owing to replication, the same target domain may have many different possible fingerprints. Intuitively, the size of a fingerprint and the replication degree of the domains in the fingerprint determine its uniqueness: a relatively unique fingerprint can undermine the protection provided by PageDNS. In this section we investigate how identifiable queries are according to the popularity of the domain, depending on the replication degree, on page fingerprinting, and the amount of cover page queries.

To measure the privacy risk of domain queries, we analytically determine the probability of an adversary correctly guessing a target domain with a certain rank, given that the adversary is able to observe the page fingerprint resulting from the client's access to the target domain. We assume that domains are distributed according to a perfect Zipf distribution with parameter $s = 0.91$ [23]. This implies that every domain has a unique rank, and we will therefore often use the rank of a domain to refer to the domain itself. For instance, we use (lowercase) k to indicate a specific domain of rank k, where the set of possible values of k is $\mathcal{K} = \{1, \ldots, N\}$.

We define a random variable K indicating the rank of a domain chosen according to the Zipf distribution. We also define a stochastic process F that maps each domain k to its possible fingerprints. More precisely, F maps each domain k to a random variable that has as possible values all the sets of pages that can be k's fingerprint—we assume that for any replicated domain the resolver chooses one of the possible pages uniformly at random. We will also, as a slight abuse of notation, consider the application of F to the random variable K, $F(K)$: this represents another stochastic process, the possible outcomes of which are determined by first drawing a domain k from K, and then applying F to k. Finally, we also consider a random variable $T = T(k)$, which represents the choice of the cover page queries when querying for domain k.[4] For ease of notation we will often write the argument of the stochastic processes as subscript, e.g., F_K for $F(K)$ or T_k for $T(k)$.

We can now provide the definition of the *identifiability* of domain k, which denotes the probability of the adversary correctly guessing k having observed one of k's fingerprints.

Definition 1 (Identifiability). *We define the identifiability of a domain k as follows:*

$$\mathcal{I}_k = \Pr(K = k \mid F_K \cup T_K = F_k \cup T_k) \tag{2}$$

Intuitively, this models a rational adversary that has no prior information about the preferences of the client. The adversary observes a set of page requests coming from a resolver, and assumes that they are due to an access to an unknown domain K chosen by the client according to the Zipf distribution. The adversary then determines, for all $k' \in \mathcal{K}$, the probability that $K = k'$ given the observed set of pages. This probability for $k' = k$ (where k is the domain actually accessed by the client) is the identifiability of k.

We now show how the identifiability can be expressed in a form that allows us to compute it. For the definition of the basic notation see Sect. 3.2. First, we apply Bayes theorem.

$$
\begin{aligned}
\mathcal{I}_k &= \Pr(K = k \mid F_K \cup T_K = F_k \cup T_k) \\
&= \frac{\Pr(F_K \cup T_K = F_k \cup T_k \mid K = k)\Pr(K = k)}{\sum_{k' \in \mathcal{K}} \Pr(F_K \cup T_K = F_k \cup T_k \mid K = k')\Pr(K = k')}
\end{aligned} \tag{3}
$$

From Appendix A, $\Pr(K = k) = f(k)$ (Zipf distribution). We can rewrite Eq. 3 as follows, where for a random variable X we use notation X' to indicate another random variable with the same distribution.

$$\mathcal{I}_k = \frac{\Pr(F_k \cup T_k = F'_k \cup T'_k)f(k)}{\sum_{k' \in \mathcal{K}} \Pr(F_{k'} \cup T_{k'} = F_k \cup T_k)f(k')} \tag{4}$$

Denoting with A the event $F_k \cup T_k = F'_k \cup T'_k$ and with B the event $F_{k'} \cup T_{k'} = F_k \cup T_k$ (for $k' \neq k$), we rewrite the equation as follows.

$$\mathcal{I}_k = \frac{\Pr(A)f(k)}{\Pr(A)f(k) + \sum_{k' \in \mathcal{K} \setminus \{k\}} \Pr(B)f(k')} \tag{5}$$

If we consider random variable $L_k = F_k \cup T_k$, it can be seen that the values it assumes (sets of pages) are all equiprobable, since all pages in F_k are chosen uniformly at random among the possible replications of each domain, and the cover pages in T_k are chosen uniformly at random among the remaining pages.

---

[4] The pages in T are chosen uniformly at random; the only dependency that T has from k is for its size. For instance, $|T|$ may be chosen such that the total number of page requests is higher than or equal to a given minimum.

Therefore, denoting with $d(X)$ the possible values (range) of a random variable X, we have that $\Pr(A) = 1/d(L_k)$. With the assumption that the size of the fingerprint is equal to constant q for all domains, and assuming also a constant number of cover pages t, we can rewrite the probability as follows.[5]

$$\Pr(A) = \frac{1}{d(L_k)} = \frac{1}{r(k)^q m^t} \tag{6}$$

Note that $r(k)$ is the replication degree of k (Eq. 1). There is an important assumption behind this equation, which is that all pages in the fingerprint of a domain have the same popularity as the domain itself. We call this *popularity inheritance*, and the rationale behind it is that if a domain is very popular and requires access to another domain, then the other domain will be requested at least as often. However, this means that for non-popular domains we might be overestimating the identifiability, since non-popular domains may very well include contents from popular domains. We leave it to future work to make the calculation of the identifiability with fingerprints for unpopular domains more realistic.

To compute probability $\Pr(B)$, we note that it is actually independent of the value of k', as long as $k' \neq k$. It can be shown[6] that $\Pr(B)$ is simply equal to the probability of guessing a randomly chosen set of pages of size $q + t$ (lottery-winning probability). We rewrite the probability as follows.

$$\Pr(B) = \frac{1}{\binom{m}{q+t}} \simeq \frac{1}{m^{q+t}/(q+t)!} = \frac{(q+t)!}{m^{q+t}} \tag{7}$$

Finally we rewrite Eq. 5 as follows:

$$\mathcal{I}_k \simeq \frac{\frac{1}{r(k)^q m^t} f(k)}{\frac{1}{r(k)^q m^t} f(k) + (1 - f(k)) \frac{(q+t)!}{m^{q+t}}} \tag{8}$$

Limitations of the Identifiability Metric. Because of how the identifiability is defined, our results are in a sense *averaged* over all possible assignments of records to pages (i.e., over all possible hash functions). This means that in a concrete instantiation, there might be pages which are worse for privacy. In Appendix B we determine the distribution of the number of ordinary records and of the number of replicas per page. Our results show that the number of ordinary records will in the worst case be less than 10% below the average, which would not significantly impact the identifiability. However, the number of replicas will be 7 at the minimum with high probability, the median being 25, so this could have an

---

[5] We are slightly approximating the exact value in Eq. 6, ignoring the fact that the pages in T are chosen from the set of all m pages *excluding* those that are already part of the fingerprint.

[6] To formally show this step, one needs to average out the probability over all possible replica sets that could be assumed by all domains, i.e., over all possible hash functions (or all possible sets of domain names of size N).

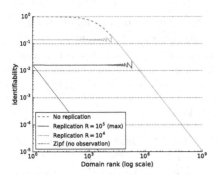

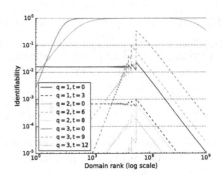

Fig. 3. Identifiability of domains according to their rank, depending on the maximum replication allowed. The figure also shows the Zipf distribution, which is equal to the identifiability *prior* to any observation of pages by the adversary. We point out that the sawtooth pattern is due to rounding in the replication function.

Fig. 4. Identifiability of domains according to their rank, depending on the size of the fingerprint q and on the number of cover queries t. As can be seen, fingerprints can be highly effective for identifying a website access. However, with a number of cover queries between twice and three times the size of the fingerprint, strong privacy can still be guaranteed.

impact: in particular, a page with few replicas would cause higher identifiability for the popular records on it (closer to the cases with little or no replication in Fig. 3). For the non-replicated records, the worst case happens for a somewhat popular record to be on a page with few replicas and few other records of similar or higher popularity. Even in such an (unlikely) case, the replicas alone will still provide privacy protection.

3.3   Results

We use identifiability (Definition 1) as a metric to measure the effectiveness of PageDNS, showing in particular how the *identifiability curve* (obtained from all possible values of k) varies depending on the replication degree, on the use of cover page queries, and on page fingerprinting.

Identifiability and Replication. Considering the basic scenario without fingerprinting or cover queries, we study the effectiveness of the basic mechanism of PageDNS. Figure 3 shows the identifiability curves for two maximum replication degrees, $R = 10^4$ and $R = m = 10^5$, the latter being the highest possible replication degree where the most popular record is replicated on all m pages. The figure also shows for comparison the case where no replication is used, and it can be seen how replication is indeed able to achieve its goal of hiding requests to popular domains. We have plotted also the prior knowledge of the adversary, i.e., the identifiability of domains when the adversary does not observe the requested pages (this is simply the Zipf distribution).

Fingerprinting and Cover Queries. In Fig. 4 we show the impact on identifiability of page fingerprinting and of cover queries, considering small fingerprint sizes. We consider a fingerprint size of q, including the page of the main domain, and an amount of cover queries t. As can be seen, without cover queries the use of page fingerprinting by an adversary can be very effective, leading to complete privacy loss in many cases. Fortunately, we find that cover queries are sufficient to compensate for this loss. In particular, an amount of cover queries of three times the size of the fingerprint appears to be enough to provide a privacy level lower than the basic one obtained for $q = t = 0$ (even for $q > 3$, which is not shown in the figure).

We have also analyzed the size of fingerprints in practice: we have logged all SLDs that appear in HTTP GET/POST requests for 10,971[7] out of the 20,000 most popular domains (according to Alexa [35]) by automatically loading all these pages in a browser and relaying all requests made by the browser through a custom proxy. This gave us a list of unique SLDs for each domain loaded by the browser.

The results, reported in Table 1, show that indeed many websites require external content from a large number of domains, but also that many of these domains are common across different websites. Indeed, if we discard the most popular 100 SLDs, the median number of additional DNS queries performed is 4, and it drops to only 2 when discarding the top 1,000. Still, there appear to be certain websites which require an exceptionally large number of SLD lookups. While we expect that in most cases these accesses could be hidden due to the large number of queries being constantly performed by the resolver of a medium-large ISP, and due to the fact that a number of page requests can be avoided as fresh page copies are still in the resolver's cache, we cannot in general provide strong guarantees for such websites. To be secure, clients would need to be made privacy-aware, and restrict the number of SLD queries per page (or perhaps space them over a longer period of time). We leave a more detailed investigation of these possibilities to future work.

Table 1. Distribution of the fingerprint size q as observed loading 10,971 web-pages. The total number of SLDs seen over all page loadings is 20,777. The table shows how most domains in the fingerprints are common (e.g., advertisement, analytics, social media) by showing how the fingerprint size is reduced when we exclude the most common 100 or 1,000 SLDs.

	Min	Median	95th	Max
Considering all SLDs	1	12	47	238
Without top 100	1	4	24	211
Without top 1,000	1	2	8	172

---

[7] The number of accessed domains is almost half of the 20,000 we consider: this is because many of them did not have a www host, and also due to some restrictions we imposed on the loading time.

3.4   Security Against Active Attacks

In previous sections we have analyzed the privacy guarantees of PageDNS with the assumption that the adversary is able to see the page queries made by recursive resolvers, but does not perform any active attack (i.e., deviating from the protocol). However, it is possible that an adversary may try to improve its ability to identify the domains accessed by a client by the use of active attacks. In particular, the adversary could modify the records for some domains he wishes to monitor to point to a honeypot server under his control: he could use a different server for every connecting resolver, and would therefore be able to link accesses to one of honeypot servers to a resolver, revealing that one of the resolver's clients has an interest in the monitored domain (this type of attack is sometimes called *split world attack*).

In PageDNS this attack is prevented by requiring registries to authenticate every new set of page versions they create through a Merkle hash tree, and by having resolvers gossip about the roots of the trees obtained from all the registries that they contact. Additionally, domain owners can monitor the pages distributed by PageDNS to ensure that the information contained in them is correct. This public auditability property also provides significantly higher integrity guarantees than those achieved in plain DNS. Since we assume that the adversary wishes to avoid detection, this scheme ensures that active attacks of this kind are prevented.

4   Evaluation

In this section we present an evaluation of the computational overhead for maintaining the PageDNS, as well as the memory overhead for registries using PageDNS.

4.1   Cost of Maintaining the PageDNS

Frequency of Authoritative NS Changes. Records in PageDNS pages only contain name servers of SLDs (Sect. 2.2) to limit the total number of records in PageDNS, but also to ensure that pages will not change too frequently. The frequency of page changes affects both TLD registries (cost of creating the updated pages) and resolvers (cost of downloading incremental updates in order to keep the local cache updated).

Table 2. Number of changes of authoritative name servers for the 3000 domains that had at least one change over the monitored 25 days.

Total	Min	Max	Mean	Median	95th
16205	1	32	5.40	4	30

To evaluate how frequently authoritative name servers of SLDs change, we monitored the authoritative name servers of the 100,000 most popular domains (according to the Alexa Top 1M domains list [35]) over 25 days in July 2017. Out of the 100,000 monitored domains, we could resolve the authoritative name server for 75,622 domains[8]. 72,622 of these, or 96%, did not change their authoritative name servers over the 25 days. For the remaining 3,000 (4%) domains, Table 2 shows statistics about the number of name server changes. From these results we can calculate the expected number of changes C to the name servers for each domain per day:

$$E[C] = \frac{16205}{75622} \cdot \frac{1}{25} \approx 0.0086 \qquad (9)$$

Thus, there are expected $10,000 \cdot 0.0086/24 \simeq 3.6$ updates per page per hour. If new page versions are created, e.g., every 4 h, less than 15 records would need to be changed per page on average between two versions. Out of our monitored domains which where updated, only around 5% had, right before the change, a TTL lower than 4 h, so only a small number of domains might suffer from higher inconsistencies than with DNS's caching. Furthermore, for all planned updates, domain owners could schedule an update with their registrar, ensuring that the update will be included by the registry at a specified version, at a specified time.

Update Costs per Registry. According to Verisign's Domain Name Industry Brief [37] (cf. also ICANN's monthly report for .com [22]), the DNS has reached a size of 335M domain names across all TLDs with an increase of about 15 million per year over the last few years ($\sim$5%). This is considering only the higher level domains to which TLDs delegate, e.g., example.org or example.co.uk, excluding subdomains like cs.example.com. We therefore set, for our analyses in this paper, the number of SLDs $N = 10^9$.[9]

Given the total number of SLDs and the update frequency we calculated above (Eq. 9), we find that the TLD registries need to perform, overall, around 100 page updates per second, assuming as above that each page is updated every 4 h. We expect this to be well within the capacity of TLD registries; furthermore, we note that if PageDNS were widely used, the root name servers would see a significantly reduced query load, meaning that their space resources could also be spent to assist the registries (this would be particularly easy in cases where the same company manages both some root name servers and TLDs, which is the case for instance for Verisign). Still, we assess the feasibility of these update with our prototype implementation in Sect. 4.3, and find that even with low-end hardware this update frequency can be sustained.

---

[8] The reason almost 25% of domains were not resolved is that for our monitoring we kept low timeouts, and excluded the domains which frequently resulted in time-outs.

[9] With a growth rate of $\sim$5% the number of domains and thus the number of pages doubles approximately every 14 years. We expect that the available bandwidth and computing power can easily keep up with the growth of PageDNS.

4.2   Memory Overhead for Resolvers

We assume that resolvers will locally cache the set of all pages, corresponding to all records in the TLD zones. Given our assumptions (Sect. 4.1) of 10^9 SLD records, distributed across 10^5 pages of about 1 MB of size each, we have that the total storage requirement for a resolver is 100 GB (excluding optimizations such as compression). While this could entail non-negligible upgrading costs for the ISPs managing the resolvers, in particular for ISPs of small size, we believe that by the time PageDNS would reach widespread deployment, the cost of this memory upgrade would be bearable even for smaller ISPs. Initially, ISPs will still have an incentive for adoption as PageDNS would allow them to provide a privacy-preserving lookup service to their clients, and it would in all likelihood also be a faster lookup service than in todays DNS, since a significant fraction of queries made by clients would be for up-to-date pages, and thus the resolver can directly query the SLD name server.

4.3   Prototype Implementation

We have implemented a prototype of PageDNS in Python to obtain some preliminary performance results and assess the feasibility of our system. Our code defines both a TLD registry and a resolver, for a total of almost 6K LOCs. Our evaluation of this prototype was made running a registry instance and a resolver instance on two Amazon AWS instances (in Ireland and in Germany, respectively), each with an Intel Xeon CPU with 2.53 GHz and 2 GB of RAM.

The registry is implemented as a server providing pages over HTTP using a RESTful API. The pages are represented as plain text for human readability, and for our evaluation we have not implemented optimizations to compress the size of pages. In our implementation, the registry lazily computes incremental updates between the cached version of the resolver, specified in the query, and the last available version. These incremental updates are cached by the registry until the new page versions are generated.

We use this setting to evaluate the latency of a page query, both in the case of cold cache, in which the entire page has to be downloaded (this should happen only in exceptional cases), and in the case of the download of an incremental update. The results are averaged over 1000 queries. The time to download an entire page consisting of 10,000 records is 789 ms on average, while for an incremental update of one version the required time was 41 ms. This last value is in the same order of magnitude as a request to Google Public DNS for NS records.

We have also implemented and evaluated the page-updating functionality of registries, offering a RESTful API for domain owners to communicate their updates to the registry. The time needed for one page update was on average 68ms, meaning that a registry can perform around 14 page updates per second using one low-end AWS instance. In practice, we expect registries to deploy significantly more powerful machines. We leave a more comprehensive evaluation of the registries' performance using PageDNS to future work.

5   Related Work

In response to the revelations about the NSA's mass surveillance programs [2], the IETF took a stance considering such surveillance practices as an attack [17], and began analyzing the problem of how to defend against it [6]. One of the identified threat vectors is DNS [8], but the countermeasures proposed so far are relatively weak. The simplest (but also weakest) of these proposals calls for query minimization [9], which would only hide some information from root and TLD servers.[10] Another proposal (a now expired IETF draft) aims to extend DNS with the option to encrypt queries between the recursive resolver and the authorities [39]. Recently also an academic proposal by Zhu et al. called T-DNS [41] was submitted as an RFC [21]: their suggestion is to use TLS between clients and recursive resolvers (and possibly between recursive resolvers and authorities) to protect against network eavesdroppers.

Outside the IETF other solutions have been devised which are similar in scope. DNSCurve [7] allows clients or recursive resolvers to establish secure channels to the authoritative resolvers using efficient cryptography. A related system is DNSCrypt [13], which offers similar guarantees. Both of these systems have seen some adoption, e.g., they are supported by OpenDNS [3]. These systems, as well as the RFCs currently under examination at the IETF, are easily deployable, and could be used complementarily to PageDNS, since their threat model is orthogonal to ours.

Other researchers aimed to protect against stronger adversaries. Zhao et al. [40] suggest a simple approach called *range queries*, which consists in the client sending extra "dummy" queries to the resolver, in order to hide the real query. Federrath et al. [18], however, show that range queries are vulnerable to semantic intersection attacks. In this same paper, the authors propose another system, based on a combination of broadcasting and sending the queries over an anonymous communication system (ACS): popular records are broadcast to all clients, while in order to retrieve less popular entries the clients have to query a resolver through a mixnet or an onion routing system. It is unclear however how these systems can effectively guarantee privacy against a malicious ISP, which will inevitably see the communications following the lookup, thus apparently nullifying the efforts to anonymize the queries. We believe that, to protect against a malicious ISP, an ACS necessarily has to be used.

Our goal in PageDNS of hiding the information of what records are being queried is analogous to that of private information retrieval (PIR) [10,25], which leverages either multiple non-colluding servers or computationally expensive cryptography to significantly reduce communication costs. Unfortunately, even with recent improvements [4,14], PIR has remained too costly in terms of computation to be used for a critical application like DNS. In PageDNS we do not try to trade off lower communication costs for additional computation: instead, we show how domain-specific aspects, such as the low variability of TLD zones,

---

[10] It appears that Verisign, Inc. was able to obtain a patent [28] on this technology, and it is unclear what this will mean for its adoption.

paired with extensive caching, allow us to achieve privacy properties close to those of PIR, but without its prohibitive performance penalties. Another related direction regarding weak PIR was taken by Toledo et al. [36], who shown how privacy guarantees can be increased through the use of anonymous communication systems, or by leveraging multiple servers.

Other related projects aim to push DNS entries to multiple entities, such as recursive resolvers, across the Internet. For instance, Cohen and Kaplan [12] propose proactive caching of records, and similarly Handley and Greenhalgh [20] also advocate for pushing records to thousands of name servers for higher robustness. Kangasharju and Ross [24] take an even more radical stance, proposing a new design for DNS involving distributed servers storing the complete DNS database. This is perhaps the closest to PageDNS, although without the goal of privacy. However, in our system we put a much stronger emphasis on efficiency and deployability: in particular, PageDNS pages only contain TLD zones, not the entire DNS database, which would be orders of magnitude larger, and change more frequently.

Researchers have also investigated new approaches to name resolution that are fundamentally different from DNS, based on distributed architectures that are not structured hierarchically, which can provide privacy. One such approach by Lu and Tsudik [27], called PPDNS, adds privacy on top of CoDoNS [33], an alternative naming system based on distributed hash tables (DHTs). The scheme also uses computational PIR to reduce communication overhead, but the ensuing computational costs strongly limit the size of the range and thus also the privacy guarantees, and leave the system vulnerable to denial of service attacks. The GNU Name System (GNS) [19,38] is another scheme based on DHTs, but because it uses a fully peer-to-peer approach it does not provide global naming consistency, and is thus quite different from today's DNS. Pappas et al. [32] have analyzed more generally DHT-based designs for DNS, and arrived at the conclusion that compared to the current DNS they are inferior in terms of performance and availability, except in terms of protection against some specific denial-of-service attacks.

6   Conclusions

We explore the design space of the solutions to the scarcely studied problem of privacy-preserving DNS lookups, and we identify a yet unexplored but promising direction. We propose an architecture, PageDNS, which aims to hide query information from root name servers and TLD registries. PageDNS lets TLD registries group together the name server records in their zones into pages; recursive resolvers retrieve entire pages rather than single records, which provides a first level of privacy protection. Additionally, we design a number of optimizations and enhancements to make the architecture more efficient, such as full caching of pages at the resolver, and incremental updates, which reduce the overhead. PageDNS requires significant changes to resolvers and TLD registries, and a certain memory overhead for resolvers, but it provides privacy properties close to those of PIR, and it may even speed up the average DNS query, since effectively the resolvers will be caching all TLD zones. Furthermore, name is incrementally deployable by TLD registries, and does not need to be adopted by all resolvers. Since PageDNS is orthogonal to other privacy solutions, it can be combined with other approaches to achieve different tradeoffs in efficiency and privacy. These are interesting directions for future work.

Acknowledgments. We thank Jinank Jain for his help with the prototype implementation.

A   Replication Function

Let $\mathcal{P}$ be a page of PageDNS containing n records, and let k' be the rank of a record in $\mathcal{P}$. For ease of notation, we write $k' \in \mathcal{P}$, and in general we will often use a domain's rank to refer to the domain. We assume a total of N domains, thus $k' \in \{1, \ldots, N\}$, where $k' = 1$ is the highest rank. We consider a random variable K indicating the rank of a domain chosen at random (by a generic client) according to a Zipf distribution with parameter $s = 0.91$, i.e., such that:

$$\Pr(K = k) = f(k; s, N) = \frac{1}{k^s H_{N,s}} \quad \text{with } H_{N,s} = \sum_{k=1}^{N} \frac{1}{k^s} \tag{10}$$

To simplify the notation, we will write $f(k)$ to mean $f(k; s, N)$ and H for $H_{N,s}$.

Now we try to analytically express the probability that an adversary would assign to k' being the target domain having observed a request to $\mathcal{P}$, which is equal to the probability that $K = k'$ given that K is restricted to $\mathcal{P}$. By applying Bayes theorem, we obtain the following equation:

$$\Pr(K = k' \mid K \in \mathcal{P}) = \frac{\Pr(K \in \mathcal{P} \mid K = k') \Pr(K = k')}{\sum_{k \in \mathcal{P}} \Pr(K \in \mathcal{P} \mid K = k) \Pr(K = k)} \tag{11}$$

Probability $\Pr(K \in \mathcal{P} \mid K = k)$ is equal to 1 if k is not replicated, since we are assuming that $k \in \mathcal{P}$. More generally, if k has a replication degree of $r(k)$ (i.e., the record for domain k exists on $r(k)$ pages), then the probability of choosing the replica in $\mathcal{P}$ is $1/r(k)$. We can therefore rewrite Eq. 11 as follows:

$$\Pr(K = k' \mid K \in \mathcal{P}) = \frac{f(k')/r(k')}{\sum_{k \in \mathcal{P}} f(k)/r(k)} \tag{12}$$

Now let k'' be another domain on the same page, i.e., $k'' \in \mathcal{P}$. Ideally, we would like the replication function to be such that $\Pr(K = k' \mid K \in \mathcal{P}) = \Pr(K = k'' \mid K \in \mathcal{P})$ for all possible choices of k' and k''. Unfortunately, it is possible to see that the only scenario where this could theoretically be achieved is one where the number of pages is equal to the number of domains, and the cost of replication would be excessive (the total size of all PageDNS pages would increase by almost a hundredfold). Instead, we try to get the ratio of those probabilities as close to 1 as possible. Since we also want to minimize the cost of replication, we do not replicate the least popular domain (i.e., $r(N) = 1$): replication should only help to reduce the probability in Eq. 11 for high-rank domains, to get it closer to the probability of the more unpopular domains. It is reasonable therefore for the ratio to be at its maximum when $k' = 1$ and $k'' = N$.

$$\rho_{MAX} = \frac{\Pr(K = 1 \mid K \in \mathcal{P})}{\Pr(K = N \mid K \in \mathcal{P})} = \frac{f(1)/r(1)}{f(N)/r(N)} \tag{13}$$

Denoting with R the replication degree of the most popular domain ($r(1) = R$), and since $r(N) = 1$, Eq. 13 becomes the following:

$$\rho_{MAX} = \frac{f(1)/R}{f(N)} = \frac{H^{-1}/R}{N^{-s}H^{-1}} = \frac{N^s}{R} \tag{14}$$

All other domains should be replicated in order not to increase this ratio further. From this requirement, we obtain the following bound $\forall k$.

$$\frac{\Pr(K = k \mid K \in \mathcal{P})}{\Pr(K = N \mid K \in \mathcal{P})} \leq \rho_{MAX} \tag{15}$$

$$\implies \quad \frac{f(k)/r(k)}{f(N)/r(N)} = \frac{k^{-s}H^{-1}/r(k)}{N^{-s}H^{-1}} = \frac{N^s}{k^s r(k)} \leq \frac{N^s}{R} \tag{16}$$

$$\implies \quad r(k) \geq \frac{R}{k^s} \tag{17}$$

We derived the bound in Eq. 17 for the worst case of the page containing both the most and the least popular domains, so by applying it generally to the replication for all k-s we ensure that on no page there will be two domains for which the ratio of their identification probabilities (Eq. 11) exceeds ρ_{MAX}. Furthermore, with the approximation that the denominator in Eq. 12, $\sum_{k \in \mathcal{P}} f(k)/r(k)$, has the same value for all pages, the bound in Eq. 17 actually guarantees the following for any two pages $\mathcal{P}$, $\mathcal{P}'$:

$$\forall k \in \mathcal{P}, \forall k' \in \mathcal{P}' \quad \frac{\Pr(K = k \mid K \in \mathcal{P})}{\Pr(K = k' \mid K \in \mathcal{P}')} \leq \rho_{MAX} \tag{18}$$

Since we desire to minimize the cost of replication, we try to match the bound of Eq. 17 as closely as possible (rounding it to the nearest integer), with the additional constraint that replication of any domain be at least 1. Thus the replication function we use is the following:

$$r(k) = max\{1, round(Rk^{-s})\} \tag{19}$$

In Fig. 5 we plot the function. Note how only the most popular domains with rank from 1 to k^* are replicated: these will all have approximately (because of rounding) the same identification probability, while for less popular domains the probability will be lower. We also point out that in our scenario $R \leq m$, where m is the number of pages, and that the best (lowest) probabilities are obtained for the equality: in this case, we have the most popular domain replicated on all pages. For realistic values ($m = 10,000$ and $s = 0.91$), we obtain $k^* \simeq 200,000$.

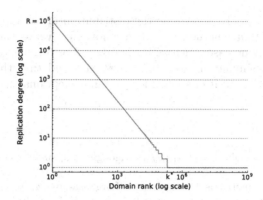

Fig. 5. Replication degree of domain names according to their rank.

B   Page-Size Variance

We can think of the size of a page P as the sum of N random variables X_k, each assuming value 1 if the k-th domain is assigned by the hash function to page P, and value 0 otherwise. The size of page P is thus $X = \sum_{k=1}^{N} X_k$. Assuming that the hash function behaves as a random function, and considering a set of m pages, we can easily compute the expected value of the size of the generic page P as follows:

$$\mu = \mathrm{E}\left[X\right] = \sum_{k=1}^{N} \mathrm{E}\left[X_k\right] = \sum_{k=1}^{N} \frac{1}{m} = \frac{N}{m} \tag{20}$$

Since X is the sum of independent random variables with values in the set $\{0, 1\}$, we can apply the multiplicative Chernoff bound to estimate the probability that the size of a specific page will deviate from the expected value μ by a certain factor $(1 - \delta)$ (we aim to find a lower bound). The bound has the following form.

$$\Pr(X \leq (1 - \delta)\mu) \leq e^{-\frac{\delta^2 \mu}{2}} \tag{21}$$

Considering for the parameters the values $N = 10^9$ and $m = 10^5$, as we have done throughout the paper, we obtain from Eq. 20 that $\mu = 10^4$. Setting $\delta = 0.1$ for a deviation of at least 10% from the mean, Eq. 21 yields the following bound:

$$\Pr(X \leq 0.9\mu) \leq e^{-\frac{10^{-2} 10^4}{2}} = e^{-50} \simeq 2 \cdot 10^{-22} \tag{22}$$

We see from these numbers that the probability of having pages significantly smaller than the average is clearly negligible. Another Chernoff bound can be used to find similar limitations for the probability of pages to be 10% larger than the average.

Replicas Distribution. To determine the distribution of the number of replicas per page, we find that Chernoff bounds are not effective, as they do not allow us to rule out extreme cases such as having only 2 or 3 replicas on some page. Instead, we use a simulation over 10^6 pages, and find that the median is 25 records per page, though it can be as low as 7 in exceptional cases.

References

1. Google Analytics Solutions. https://www.google.com/analytics. Accessed 22 Sept 2017
2. NSA Spying on Americans. https://www.eff.org/nsa-spying. Accessed 22 Sept 2017
3. OpenDNS. https://www.opendns.com/. Accessed 22 Sept 2017
4. Aguilar-Melchor, C., Barrier, J., Fousse, L., Killijian, M.-O.: XPIR: private information retrieval for everyone. In: PETS (2016)
5. Arends, R., Austein, R., Larson, M., Massey, D., Rose, S.: DNS security introduction and requirements. RFC 4033 (2005)
6. Barnes, R., et al.: Confidentiality in the face of pervasive surveillance: a threat model and problem statement. RFC 7624 (2015)
7. Bernstein, D.J.: DNSCurve: usable security for DNS. https://dnscurve.org/. Accessed 22 Sept 2017
8. Bortzmeyer, S.: DNS privacy considerations. RFC 7626 (2015)
9. Bortzmeyer, S.: DNS query name minimisation to improve privacy. RFC 7816 (2016)
10. Chor, B., Goldreich, O., Kushilevitz, E., Sudan, M.: Private information retrieval. In: IEEE FOCS (1995)
11. Chor, B., Goldreich, O., Kushilevitz, E., Sudan, M.: Private information retrieval. J. ACM **45**(6) (1998)
12. Cohen, E., Kaplan, H.: Proactive caching of DNS records: addressing a performance bottleneck. In: IEEE/IPSJ International Symposium on Applications and the Internet (SAINT) (2001)
13. Denis, F., Fu, Y.: DNSCrypt (2011). https://dnscrypt.org/. Accessed 22 Sept 2017
14. Devet, C., Goldberg, I., Heninger, N.: Optimally robust private information retrieval. In: USENIX Security (2012)
15. Dickinson, J., Dickinson, S., Bellis, R., Mankin, A., Wessels, D.: DNS transport over TCP - implementation requirements. RFC 7766 (2016)
16. Dingledine, R., Mathewson, N., Syverson, P.: Tor: the second-generation onion router. In: USENIX Security (2004)
17. Farrell, S., Tschofenig, H.: Pervasive monitoring is an attack. RFC 7258 (2014)
18. Federrath, H., Fuchs, K.-P., Herrmann, D., Piosecny, C.: Privacy-preserving DNS: analysis of broadcast, range queries and mix-based protection methods. In: ESORICS (2011)
19. Grothoff, C., Wachs, M., Emert, M., Appelbaum, J.: NSA's MORECOWBELL: knell for DNS. Technical report, GNUnet e.V. (2015)
20. Handley, M., Greenhalgh, A.: The case for pushing DNS. In: HotNets (2005)
21. Hu, S., Zhu, L., Heidemann, J., Mankin, A., Wessels, D., Hoffman, P.: Specification for DNS over Transport Layer Security (TLS). RFC 7858 (2016)
22. ICANN: .com Monthly Registry Reports. https://www.icann.org/resources/pages/com-2014-03-04-en. Accessed 22 Sept 2017

23. Jung, J., Sit, E., Balakrishnan, H., Morris, R.: DNS performance and the effectiveness of caching. IEEE/ACM TON **10**(5), 589–603 (2002)
24. Kangasharju, J., Ross, K.W.: A replicated architecture for the domain name system. In: IEEE INFOCOM (2000)
25. Kushilevitz, E., Ostrovsky, R.: Replication is not needed: single database, computationally-private information retrieval. In: IEEE FOCS (1997)
26. Laurie, B., Sisson, G., Arends, R., Blacka, D.: DNS security (DNSSEC) hashed authenticated denial of existence. RFC 5155 (2008)
27. Lu, Y., Tsudik, G.: Towards plugging privacy leaks in the domain name system. In: IEEE P2P (2010)
28. McPherson, D., Osterweil, E.: Providing privacy enhanced resolution system in the domain name system. US Patent 8,880,686 B2 (2014)
29. Mockapetris, P.: Domain names - concepts and facilities. RFC 1034 (1987)
30. Mockapetris, P.: Domain names - implementation and specification. RFC 1035 (1987)
31. Ostrovsky, R., Skeith III, W.E.: A survey of single-database PIR: techniques and applications. In: PKC (2007)
32. Pappas, V., Massey, D., Terzis, A., Zhang, L.: A comparative study of the DNS design with DHT-based alternatives. In: IEEE INFOCOM (2006)
33. Ramasubramanian, V., Sirer, E.G.: The design and implementation of a next generation name service for the Internet. In: ACM SIGCOMM (2004)
34. Rossow, C.: Amplification hell: revisiting network protocols for DDoS abuse. In: NDSS (2014)
35. Alexa the Web Information Company. Alexa Top 500 Global Sites (2016). http://www.alexa.com/topsites
36. Toledo, R.R., Danezis, G., Goldberg, I.: Lower-cost ϵ-private information retrieval. PoPETS **2016**(4), 184–201 (2016)
37. Verisign, Inc.: The domain name industry brief, vol. 14, no. 2 (2017). https://www.verisign.com/assets/domain-name-report-Q12017.pdf. Accessed 22 Sept 2017
38. Wachs, M., Schanzenbach, M., Grothoff, C.: A censorship-resistant, privacy-enhancing and fully decentralized name system. In: International Conference on Cryptology and Network Security (CANS) (2014)
39. Wijngaards, W., Wiley, G.: Confidential DNS. Internet Draft draft-wijngaards-dnsop-confidentialdns-03 (2015)
40. Zhao, F., Hori, Y., Sakurai, K.: Analysis of privacy disclosure in DNS query. In: International Conference on Multimedia and Ubiquitous Engineering (MUE) (2007)
41. Zhu, L., Hu, Z., Heidemann, J., Wessels, D., Mankin, A., Somaiya, N.: Connection-oriented DNS to improve privacy and security. In: IEEE Symposium on Security and Privacy (2015)

Bitcoin and Blockchain

A New Approach to Deanonymization
of *Unreachable* Bitcoin Nodes

Indra Deep Mastan[1](✉) and Souradyuti Paul[2]

[1] Indian Institute of Technology Gandhinagar, Gandhinagar, India
immastan@gmail.com
[2] Indian Institute of Technology Bhilai, Raipur, India
souradyuti.paul@gmail.com

Abstract. Mounting deanonymization attacks on the *unreachable* Bitcoin nodes – these nodes do not accept incoming connections – residing
behind the NAT is a challenging task. Such an attack was first given by
Biryukov, Khovratovich and Pustogarov based on their observation that
a node can be uniquely identified in a *single session* by their directly-
connected neighbouring nodes (ACM CCS'15). However, the BKP15
attack is less effective across *multiple sessions*. To address this issue,
Biryukov and Pustogarov later on devised a new strategy exploiting certain properties of address-cookies (IEEE S&P'15). Unfortunately, the
BP15 attack is also rendered ineffective by the present modification to
the Bitcoin client.

In this paper, we devise an efficient method to link the sessions of
unreachable nodes, even if they connect to the Bitcoin network over the
Tor. We achieve this using a new approach based on organizing the *block-
requests* made by the nodes in a *Bitcoin session graph*. This attack also
works against the modified Bitcoin client. We performed experiments on
the *Bitcoin main network*, and were able to link consecutive sessions with
a *precision* of 0.90 and a *recall* of 0.71. We also provide counter-measures
to mitigate the attacks.

1 Introduction

Bitcoin works in a peer-to-peer network over which users create transactions that
are stored in a distributed ledger known as the Blockchain [9]. All transactions
in the Bitcoin network can be publicly viewed and analyzed. One of the most
important properties of Bitcoin is its anonymity. If an adversary is able to link
the transactions to its owner, then she has broken the anonymity property; this
event is known as deanonymization.

Let $U = \{U_1, U_2, \ldots, U_n\}$ be the set of addresses of the user u (note that
Bitcoin allows a user to have multiple addresses). If an adversary is able to
find U, then she will be able to get the full transaction history of u from the
Blockchain (publicly available distributed ledger). In this setting, there are two
main problems associated to deanonymization. First, how to link the bitcoin
addresses (or transactions) to determine U? Next, how to link U to the real
identity u? Our work is in the direction of the first problem.

© Springer Nature Switzerland AG 2018
S. Capkun and S. S. M. Chow (Eds.): CANS 2017, LNCS 11261, pp. 277–298, 2018.
https://doi.org/10.1007/978-3-030-02641-7_13

Motivation. Deanonymization attack is performed mainly in two ways: *transaction graph* analysis and *Bitcoin network* analysis. Several papers show how to perform *transaction graph analysis* to link the transactions (and thus Bitcoin addresses) of users [8,11]. Some papers even link real world entities such as *mtgox*, *silkroad* with their Bitcoin addresses [8,10]. Biryukov, Khovratovich and Pustogarov observed that deanonymization using *transaction graph analysis* is less effective when a user makes multiple transactions using *distinct* bitcoin addresses, since such transactions might not have any relations in the graph [4].

Now we concretely discuss the main challenges in deanonymization with respect to Bitcoin network analysis. Suppose, a victim node v having the public IP address IP creates a transactions tx. Even if the adversary discovers that tx is related to IP, he is still not sure about the owner of tx, because there could be multiple Bitcoin nodes (or *users*) having the same public IP IP behind the NAT. Therefore, an adversary first needs to distinguish the nodes behind the NAT, before linking the transactions. Another challenge in the deanonymization is when a user connects to Bitcoin over Tor: the user can change its onion address in a new session, making it difficult for the attacker to trace his activities.[1]

Many of the deanonymization problems in Bitcoin network have already been solved in [4,5,7]. These attacks have the following limitations: attacks given in [7] do not apply to unreachable Bitcoin nodes; attacks given in [4] are not performed on the Bitcoin *main network*, and can not deanonymize nodes across the sessions; the *address-cookies* method given in [5] is ineffective in the *updated* version of the Bitcoin client. In short, it is not satisfactorily solved as to how to deanonymize Bitcoin nodes, when the victim nodes behind the NAT are unreachable or are using Bitcoin over Tor.

In this paper we solve this issue using a novel technique based on analysing the sequence of block-header hashes (block-ids) requested across multiple sessions by the unreachable nodes.

Related Work. Following are various attempts at deanonymization attacks and their limitations, in chronological order.

2014: Koshy et al. have shown how to perform deanonymization by analyzing the relay patterns of transactions [7]. They collected data over 5 months, and used statistical analysis to determine the source IP addresses of the transactions. They were able to deanonymize 1162 bitcoin addresses of *reachable* Bitcoin nodes. However, in [4], it was pointed out that their attack applied to only *reachable* nodes that constituted only 10% of all nodes.

2014: Biryukov, Khovratovich and Pustogarov gave the first attack to deanonymize transactions of nodes that are *unreachable* and hidden behind the NAT [4]. They performed experiments on the *Bitcoin Testnet*. Their attack was based on the observation that a node can be uniquely identified by its direct connections (or entry-nodes). They gave a strategy to learn the entry-nodes;

---

[1] Tor is a circuit-based communication service which provides anonymity by relaying traffic through routers as proxies (see Sect. 2).

however, their solution has an inherent limitation that the entry-nodes change in a new Bitcoin session. Thus, their attack can not relate the transactions created in the multiple sessions. This shows that the ability to identify a node across the sessions is an essential step of any deanonymization attack.

2015: Biryukov and Pustogarov gave a fingerprinting technique for identification of the nodes across the sessions [5]. The technique is as follows: adversary sends unique *address-cookies* to his peers in ADDR messages, the peers store IP addresses contained in *address-cookies* into their IP address tables (address-cookies are created into peers IP address tables); after some time, adversary sends GETADDR to query IP addresses in the peers IP address tables; the peers respond using ADDR; now the adversary can analyse the responses received, and check if it matches some *address-cookies*. The present modification to Bitcoin's inbuilt fingerprinting protection[2] makes *unreachable* nodes ignore the GETADDR; thus, the fingerprinting attack is prevented for *unreachable* nodes.

Our Contribution. Our main contribution is launching a deanonymization attack on *unreachable* Bitcoin nodes, even if they are behind the NAT, in both direct connection and proxy connection settings (running Bitcoin over Tor). Most importantly, unlike the previous attacks, our technique works against the new version of Bitcoin client [3]. Our attack is fairly generic, and does not seem to exploit any rectifiable mistake in the Bitcoin implementation. The crux of the attack is an observation that a Bitcoin node requests for blocks following a specific pattern, in particular, in the increasing order of Blockchain height. This pattern is observable even in the following scenarios: when the node is connected via Tor; and when the node is connected in multiple sessions with or without Tor. Using this observation we linked the consecutive sessions (sessions that follow each other continuously) of an unreachable node, which could then be used to link majority of the sessions. Linked sessions help in linking the transactions created in the different sessions.

The above attack has been experimentally verified in the Bitcoin main network. We have performed experiments by running Bitcoin nodes on Amazon EC2.

The main objective of the first experiment is to give a concrete measure of the quality of the attack when the victim nodes are connected to Bitcoin network directly. We ran eight sessions with the four *unreachable* nodes (therefore, a total of 32 sessions). In this experiment, we link the consecutive sessions with a *precision* of 0.90 and a *recall* of 0.71.

The objective of the second experiment is to show the performance of the attack when victim nodes connect to Bitcoin network with Tor and without Tor in different sessions. We ran six sessions with four nodes (therefore, a total of 24 sessions), where, in the first three sessions, nodes are connected to Bitcoin network over Tor, and, in the next three sessions the nodes are connected directly to

---

[2] The modification that are done to provide inherent fingerprinting protection to Bitcoin network.

Bitcoin network. In the experiment, we link consecutive sessions with a *precision* of 1.0 and a *recall* of 0.75. To thwart this attack, we propose a counter-measure, where the blocks are requested in a random fashion.

2   Background

Here we first give necessary background of Bitcoin network, nodes and Bitcoin protocol messages. Next, we describe the Tor network.

2.1   Bitcoin Network

Bitcoin network is a peer-to-peer network. A node in the Bitcoin network can have at most 8 outgoing and 117 incoming connections. The nodes get connected to each other by establishing a TCP connection.

Nodes. There are mainly two classes of Bitcoin nodes: (1) first one is based on the *reachability* criterion, and (2) the second one is based on the existence of proxy nodes in the connection. The first class of nodes is further divided into two types: (1a) nodes that accept incoming connections, we call them *reachable* nodes, and (1b) nodes that do not accept incoming connections, we call them *unreachable* nodes. Both *reachable* and *unreachable* nodes can make outgoing connections. The second class of nodes is also subdivided into three categories: (2a) nodes that connect to Bitcoin network directly, (2b) nodes that connect to Bitcoin network using Tor anonymity system, and (2c) nodes that connect to Bitcoin network sometimes with Tor and sometimes without Tor.

Bitcoin Protocol Messages. Bitcoin protocol uses a large number of application layer messages that are exchanged between the nodes for various purposes. Below we describe the important messages.

- VERSION & VERACK: In the beginning of a connection, two nodes exchange VERSION and VERACK messages. The VERSION message contains information of the Bitcoin version, best-height[3] of the sender, a random nonce, network addresses of the sender and receiver, etc. The VERACK message sent from the receiver denotes acknowledgement of VERSION message.
- GETADDR & ADDR: Every node normally holds a list of IP addresses that are working as active nodes in the network in the recent past. Using GETADDR message, a node X can request another node Y for the list of those IP addresses. This is done to help X find the potential active nodes in the network. In response, Y returns the requested list of IP addresses using ADDR message.
- PING & PONG: The PING is sent to check the status of the connection is alive. The PONG is sent in response to a PING message.

---

[3] The best-height of a node is the height of the Blockchain of the node.

- INV: A node advertises suitably chosen transactions and blocks it possesses to its peers using INV messages. It is a tuple *(count, inventory)*, where count is the size of *inventory* and *inventory* is a list of inventory vectors. Each inventory vector is a tuple *(type, hash)*, where *type* identifies the object type; i.e. transaction or block, and *hash* denote transaction-id or block-id (block header hash). The INV message can be issued by a node unsolicited.

- GETDATA: The GETDATA message is sent in the response to an INV message. It is a request to retrieve the full content of specific transactions or blocks[4]. Similar to INV, the GETDATA is also a tuple *(count, inventory)*.

- TX: It describes a bitcoin transaction. When a Bitcoin user create a new transaction, it broadcasts the transaction-id in INV message. The peers connected to the user receive the INV messages and get the transaction-id of new transaction, next, they request the transaction by sending the GETDATA message for it. Then user sends transaction in TX message.

- BLOCK: It describes a block. The BLOCK message sent for two different reasons: (1) sent as response to the GETDATA message, and (2) sent by miners to broadcast newly-mined blocks.

- GETBLOCKS: A GETBLOCKS message is exchanged between peers to tell each other the block-ids of the top block on their Blockchain, this helps in updating their Blockchain. For example, suppose node X and Y have exchanged GETBLOCKS message, and Blockchain height of node X is more than Y. Since GETBLOCKS sent from Y contains the block-id of the block at the top of Blockchain of Y, X will determine the set of blocks that Y needs in-order to update his Blockchain. Next, X sends INV message containing upto 500 block-ids to Y, and then Y can request the desired blocks using GETDATA message. This way of synchronizing Blockchain is called Blocks-First Sync.

- GETHEADERS & HEADERS: These messages are exchanged between peers to update the block headers[5]. The GETHEADERS message contains the block-id from where the sender wants to receive the headers. When a peer sends GETHEADERS message, it gets a HEADERS message as a response. The HEADERS message contains up to 2000 block headers. Note that similar to GETBLOCKS, the peer with higher Blockchain height sends HEADERS message. This way of synchronizing Blockchain is called Headers-First Sync. After updating the block headers a node can request the GETDATA message for the blocks.

2.2   The Onion Router (Tor)

Tor is a circuit-based communication service which provides anonymity by relaying traffic through routers in the Tor network as proxies [6]. Tor network is a

---

[4] Node X advertises blocks and transactions to node Y using INV, where INV contains block header hashes (block-ids) or transaction-ids. Then Y requests specific transactions or blocks from X using GETDATA; such a communication is called *pull-based communication*.

[5] Each block has a 80-byte block header, which contains important information such as the hash value of the previous block, the time of creation of the block, a nonce, number of transactions etc.

distributed overlay network, which consists of approximately 7,000 volunteer-operated routers or *onion routers* (ORs) or *relays*.

When a user runs Tor client, it creates a *Tor circuit* to route the traffic through the Tor network by choosing three relays – namely entry, middle and exit – and establishes a session key with each relay. Suppose, Alice is using a Tor client to connect to Bob. When Alice starts the Tor client in her machine, it creates the following three-hop circuit.

Alice ↔ entry relay ↔ middle relay ↔ exit relay.

Next, the Tor client sends *data* to Bob by encapsulating it in three layers of encryption, using the session keys established with the relays. Each relay in the circuit removes its layer by performing decryption, and finally Bob receives data in the unencrypted form.

Each relay in the circuit knows the IP addresses of its predecessor and successor. The entry relay knows the IP address of Alice, but does not know the *data* Alice is sending. Exit relay knows the *data* sent by Alice, but does not know the IP address of Alice. Therefore, none of the relays (as well as Bob) can relate the *data* with the IP address of Alice.

The three-hop Tor circuit does not provide anonymity to Bob, because the IP address of Bob is known to Alice; however, using *Tor Hidden Service* (THS) Bob can hide its IP address from Alice while offering a TCP service, e.g. a *Bitcoin server*.

The THS is accessed through its *onion* address rather than IP address, which is of the form "x.onion", where x is the base-32 encoded THS identifier.[6] The onion address – as opposed to IP address – does not reveal geographical information. The Tor client routes data *to and from* THS using the onion address. For example, suppose Bob is running a THS, and suppose Alice wants to use it. They construct the following circuit to exchange data, where *RP* is a Tor relay, also known as *Rendezvous Point* [6].

Alice ↔ Relay ↔ Relay ↔ RP ↔ Relay ↔ Relay ↔ Relay ↔ Bob

RP connects Alice's circuit to Bob's circuit; it does not know the IP address of Alice and Bob and the *data* they exchange.

3   Peer-Representations and Sessions

Here we give important definitions required to formalise our deanonymization attack. Our main focus is to give a strategy to link the sessions of an identical node. Linking sessions of a node enables the attackers to trace the activities of the user and monitor its transactions.

In our attack model, victim nodes are assumed to be inside the NAT, whereas the adversarial nodes are outside of it. Nodes inside the NAT connect to Bitcoin

---

[6] Base-32 encoding is done using 32-character: twenty-six letters A to Z and six digits 2 to 7.

network mainly in two ways: Direct connection and Proxied connection (Bitcoin over Tor). The directly connected nodes share the same public IP, making distinguishing difficult outside the NAT. The Bitcoin nodes over Tor may not share the same onion address; however, they could change their onion addresses (e.g. opening new Bitcoin sessions), making tracing difficult.

Let $\mathcal{A} = \{a_i\}_{i \in [n]}$ and $\mathcal{V} = \{v_i\}_{i \in [m]}$ denote the sets of adversarial and victim nodes. Two nodes are called *peers* of each other, if they are connected. In the Bitcoin network, an attacker node a_j identifies a peer victim node v_i by assigning it a *peer-id* p_{ij}. Note that a victim node may be assigned different peer-ids by different attacker nodes, making it harder for the attacker to determine whether the peer-ids actually are of a single node or of multiple nodes behind the NAT.

To represent the victim by a single representation outside the NAT in one session, we provide a *peer-representation* based on the time at which a victim is disconnected. Let T_v denote the set of all *disconnect times* of the victim v from the Bitcoin network (Suppose, v comes online k times; therefore, $T_v = \{t_v^1, t_v^2, \ldots, t_v^k\}$). Also, let $addr(v)$ be the set of public addresses by which v is identified outside the NAT. (Suppose, IP is the public IP address of the victim node v. Also, suppose that $o_v^1, o_v^2 \ldots o_v^k$ are the onion addresses of v. Therefore, $addr(v) = \{IP, o_v^1, o_v^2 \ldots o_v^k\}$.)

The set of all *peer-representations* of all victim nodes is defined as follows.

$$\mathcal{A}.peers = \{(a,t) : v \in \mathcal{V}, t \in T_v, a \in addr(v), a \ disconnects \ at \ t\}.$$

Note that $\mathcal{A}.peers$ contains the *peer-representations* different from *peer-ids*; however, to compute $\mathcal{A}.peers$, adversary needs *peer-ids*. The technical details of how $\mathcal{A}.peers$ is computed is provided in Appendix A. In the Fig. 1 a pictorial representation of relation between $\mathcal{A}.peers$ and $\mathcal{V}$ is given.

A *session* (or Bitcoin session) is the collection of the Bitcoin protocol messages exchanged between the times a node connects to Bitcoin network and disconnect from it. Suppose, a victim represented by $x \in \mathcal{A}.peers$ comes online and exchanges various Bitcoin protocol messages, let S_x be the set of messages exchanged with x; therefore, S_x is a *session*. We define $\mathcal{A}.data$ as follows:

$$\mathcal{A}.data = \{S_x : x \in \mathcal{A}.peers\}.$$

The set $\mathcal{A}.data$ contains all the *sessions* of the victims. For example, the set $S_{(IP, t_{v_1}^1)}$ contains Bitcoin protocol messages exchanged with victim v_1; thus, $S_{(IP, t_{v_1}^1)}$ is a *session* of v_1. It is easy to establish a bijection $S : \mathcal{A}.peers \rightarrow \mathcal{A}.data$, which shows that for a victim $x \in \mathcal{A}.peers$, its *session* $S(x)$ is contained in $\mathcal{A}.data$. We shall use $S(x)$ and S_x synonymously. The technical details of how S_x is computed is provided in Appendix A.

4 A New Form of Deanonymization: Linking the Sessions

Here we give the necessary background for our deanonymization attack using the definitions of the *peer-representation* and *session* as described in Sect. 3. First we describe why linking the *sessions* is a deanonymization attack, and then we outline the major steps.

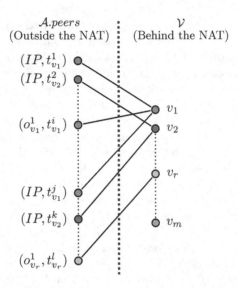

Fig. 1. Representations of victim nodes behind and outside the NAT. For example, the node v_1 is identified as $(IP, t_{v_1}^1)$ in one session (direct connection) and $(o_{v_1}^1, t_{v_1}^j)$ in another session (a Bitcoin over Tor connection), here IP and $o_{v_1}^1$ denote the public IP address and onion address of v_1; $t_{v_1}^1$ and $t_{v_1}^j$ denote the disconnect times.

Why Linking Sessions is a Deanonymization Attack. The first deanonymization attack of nodes behind the NAT was given in [4], where the adversary logs the first 10 nodes broadcasting the transaction-ids in `INV` messages, and then assigns the transactions to a node behind the NAT; however, their attack fails to link transactions created in different sessions.

We now give an example of how linking of sessions helps in deanonymizing transactions. Suppose, a user creates transactions T_1 and T_2 in sessions s_1 and s_2. After creating the transactions, he broadcasts them to his peers. Suppose, the user is connected to adversarial nodes; therefore, they receive T_1 and T_2. The adversary determines that T_1 was first broadcast in s_1, and T_2 in s_2. Next, the adversary checks if s_1 and s_2 are of identical victim. If so, she then concludes that T_1 and T_2 were created by the same user; this way he is able to link transactions created in different sessions.

Let $\gamma_i \subseteq \mathcal{A}.data$ contains the *sessions* (sets of Bitcoin protocol messages) of a victim v_i. If adversary is able to link *sessions* of v_i and compute γ_i, then he will be able to link transactions of v_i.

Linking of *sessions* also gives an additional interesting result. If adversary is able to compute γ_i, then he can also get the set of *peer-representations* α_i of the victim v_i outside the NAT. For example, in Fig. 1, the node v_1 is represented by the set of *peer-representations* $\alpha_1 = \{(IP, t_{v_1}^1), (o_{v_1}^1, t_{v_1}^i), (IP, t_{v_1}^j)\}$. Suppose, the adversary is able to link the sessions of v_1, and to compute γ_1, where

$$\gamma_1 = \{S(IP, t_{v_1}^1), S(o_{v_1}^1, t_{v_1}^i), S(IP, t_{v_1}^j)\}.$$

Each *session* $S(x) \in \gamma_1$ is a set of Bitcoin protocol messages, that contains information on IP address or onion address of the victim, and also the time of disconnect (see Sect. 2); thus, the adversary can determine *peer-representations* x from the Bitcoin protocol messages in $S(x)$. The adversary can compute α_1 using γ_1. Hence, by linking *sessions* of v_1, it is also possible to achieved identification of v_1 when it is connected to Bitcoin network directly, and when it is connected to Bitcoin network over Tor.

Major Steps. In our attack, we analyse the GETDATA messages sent by the victims. Below are the major steps of the attack.

1. *Extracting the Block-ids:* We compute the block-ids requested by the victims in each session using the GETDATA messages sent by them.
2. *Linking consecutive sessions:* We take two sequences of block-ids, and determine if they are requested in the consecutive sessions of a node (linking consecutive sessions).
3. *Linking all the sessions:* To link all the sessions of a victim, we define a *Bitcoin session graph*, where each vertex represents a sequence of block-ids requested by the victim in a session; and two vertices have an edge if they are related to the consecutive sessions. The vertices of the maximally connected component of the graph gave the sequences of block-ids requested by the victim; which in turn gives all the sessions of the victim node.

In what follows, we describe above steps in detail.

4.1   Step 1: Extracting the Block-ids

Here we describe the first step of our attack. We focus on the analysing the block-ids (block header hashes) requested by the victims. We first give the motivation for extracting block-ids, and then show how to extract them.

Motivation. The Bitcoin protocol messages sent by a victim contain GETDATA messages. A GETDATA message contains the list of block-ids the victim *does not* have at a specific time (see Sect. 2). Each block-id is associated with a unique block, and each block has a unique height in the Blockchain. By analysing the block-ids in GETDATA messages issued by the victim, adversary can get two pieces of important information: First, estimate of the Blockchain height of victim at a specific time; second, the block-ids of the blocks that the victim has updated into its Blockchain. Below we describe how they are useful for adversary.

The estimate of Blockchain height of a node can help in linking the consecutive sessions of the node: if adversary gets the Blockchain height of the victim v_i when it disconnects, then, in the new session, v_i starts requesting blocks from the height achieved in previous session; the height achieved in previous session and starting height of new session of v_i are equal. Therefore, the height of the blocks requested by v_i in the beginning of a new session will be close to the height of requested blocks, when v_i disconnected in previous session; thus, adversary can compare the block-requests, and identify victim v_i in the new session.

The block-ids requested by the victims can help in distinguishing their sessions: a node does not request blocks after they are updated into its Blockchain; however, two nodes can request same blocks; thus, by comparing the block-ids requested in the two sessions, adversary can determine if sessions correspond to a single (or two different) victim node(s).

Extracting Block-ids. The set S_x contains the GETDATA messages issued by a victim whose peer-representation is x. When a victim sends GETDATA message to retrieve a block, the adversary sends BLOCK message in the response. A GETDATA message may contain a maximum of 50,000 entries for blocks or transaction ids [1]. Let E denote the algorithm that, given the element $S \in \mathcal{A}.data$, outputs the multiset $E(S)$ containing the block-ids of the blocks requested in S. We define the set $\mathcal{A}.SessionBid$ as follows:

$$\mathcal{A}.SessionBid = \{E(S) : S \in \mathcal{A}.data\}. \tag{1}$$

Another way of formalization is:

$$\mathcal{A}.SessionBid = \{E(S_x) : x \in \mathcal{A}.peers\}$$
$$= \{E(S_x) : x \in \bigcup_{i \in [m]} \alpha_i\}.$$

Here, $\alpha_1 | \alpha_2 | \ldots | \alpha_m$ is a disjoint partition of $\mathcal{A}.peers$, where α_i is the set of *peer-representations* of victim v_i. Let $E(S_x)$ be denoted by β_x. One should observe, if there are multiple GETDATA messages for an identical block, then there are multiple entries of one block-id in β_x (therefore, β_x is a multiset!). A node sends multiple GETDATA messages for an identical block, if the response of GETDATA is not received.

The set $\mathcal{A}.SessionBid$ can be computed inside the NAT (from the nodes in $\mathcal{V}$) because the GETDATA messages are known to both victims (nodes in $\mathcal{V}$) and the adversary (who is the set $\mathcal{A}$). For example, suppose α_i is the set of *peer-representations* of a victim v_i outside the NAT, when a GETDATA message sent by v_i comes out of NAT, it appears that it is sent by a *peer-representation* contained in α_i; therefore, the block-ids requested by node v_i is same as those requested by the set of peers in α_i. Let $\hat{\beta}_{v_i}$ contain the sets of block-ids requested in various sessions by victim v_i. We define $\hat{\beta}_{v_i}$ as follows:

$$\hat{\beta}_{v_i} = \{\beta_x : x \in \alpha_i\}.$$

Putting the β_{v_i} in the definition of $\mathcal{A}.SessionBid$ we get

$$\mathcal{A}.SessionBid = \bigcup_{v_i \in \mathcal{V}} \hat{\beta}_{v_i}. \tag{2}$$

We have computed the set $\mathcal{A}.SessionBid$ from the files contained in the Bitcoin's application data folder of victim nodes.[7]

4.2   Step 2: Linking Consecutive Sessions

After extracting the block-ids, our next step is to link the *consecutive sessions*. The consecutive sessions of a node follow each other continuously. The phrase "consecutive sessions" is *only* meaningful for a single node.

Let us take two sequences of block-ids β_x and β_y in $\mathcal{A}.SessionBid$. If the adversary determines that they are requested in the consecutive sessions of a victim node; then using the inverse of *extract operation*, she can determine that S_x and S_y are consecutive sessions, where $E^{-1}(\beta_x) = S_x$ and $E^{-1}(\beta_y) = S_y$.[8] Since S_x and S_y contain information of IP address or onion address of the victim, and also the time of disconnect, the adversary can determine *peer-representations* x and y of the victim in two consecutive sessions.

We run the Algorithm 1 (also called `consecutive`), which returns *True* iff the inputs β_x and β_y are requested in consecutive sessions. The algorithm uses a fixed parameter th, which is a threshold of the number of common block-requests sent in the consecutive sessions. The correctness of the algorithm and the parameter threshold th are described in Appendices B and C. The function $H(b)$ used in `consecutive` returns the Blockchain height of the input block-id b.

4.3   Step 3: Linking All Sessions

In Sect. 4.2, we already described how to link the *consecutive sessions*. In this section, we describe how to link all the sessions – not necessarily consecutive – of a victim. We achieve it by constructing a *Bitcoin session graph*, and, finally, extracting the connected components in it. First, we define the *Bitcoin session graph*, and then we show how it will help in linking all the sessions of nodes.

Bitcoin Session Graph. A *Bitcoin session graph* $G(\mathcal{S}, \mathcal{E})$ is defined as follows: (1) $\mathcal{S} = \{E(S) : S \in \mathcal{A}.data\}$, where E is an algorithm, which, given a session S, outputs certain data; (2) for all $(a, b) \in \mathcal{S} \times \mathcal{S}$, there is an undirected edge between a and b, *iff* a and b are data contained in the consecutive sessions of an identical node.

In our setting, $E(S)$ is a sequence of block-ids requested in `GETDATA` contained in the session $S \in \mathcal{A}.data$; therefore, $\mathcal{S} = \mathcal{A}.SessionBid$; two vertices a

---

[7] Bitcoin's application data folder: A set of data files containing the following information of the Bitcoin client: Private keys, Peer IP addresses, and various information related to the current Blockchain.

[8] E is a bijection from $\mathcal{A}.data$ to $\mathcal{A}.SessionBid$. It shows that the sequence of block-ids requested in a session is unique, which we found to be true in our experiments (see Sect. 5).

```
1  consecutive(βₓ, β_y)
2      h_x^s = min{H(b) : b ∈ βₓ}
3      h_x^e = max{H(b) : b ∈ βₓ}
4      h_y^s = min{H(b) : b ∈ β_y}
5      h_y^e = max{H(b) : b ∈ β_y}
6      if |βₓ ∩ β_y| < th then
7          if  max{|h_y^e − h_x^s|, |h_x^e − h_y^s|} < |βₓ| + |β_y| then
8              return True

9      return False
```

Algorithm 1: consecutive(β_x, β_y) determines, if block-ids in β_x and β_y are requested in consecutive sessions.

and b have an edge, if consecutive(a,b) = *True*, where consecutive is the Algorithm 1.

Linking *All* Sessions of Victim Nodes. Here we describe a procedure that links all the sessions of the victim nodes and finally gives the set of *peer-representations* of a victim. Since the set $\mathcal{A}.SessionBid$ contains the sequences of block-ids extracted from *all* the sessions of *all* the victim nodes, therefore, there exists a subset of vertices in $\mathcal{S}$ corresponding to the sequences of block-ids requested by a victim in all its sessions.

A path in *Bitcoin session graph* is a sequence of edges, where each edge gives information of two vertices related to a victim; therefore, if two vertices have path between them, then they correspond to an identical victim. To determine the vertices related to a victim, adversary can compute the set of vertices of a maximally connected component of the *Bitcoin session graph*.[9] The graph $G(\mathcal{S}, \mathcal{E})$ can have more than one maximally connected component, where each of them gives the set of vertices related to a victim node.

Let $\mathbb{M}$ contain the sets of vertices of the maximally connected components in the *Bitcoin session graph* $G(\mathcal{S}, E)$. The details of the constructions of the sets γ_i and α_i for $i \in [|\mathbb{M}|]$ are as follows:

The following are the *sessions* and *peer-representations* of a victim node.

$$\gamma_i = \{E^{-1}(\beta_x) : \beta_x \in c_i; c_i \in \mathbb{M}\}, \tag{3}$$

$$\alpha_i = \{S^{-1}(E^{-1}(\beta_x)) : \beta_x \in c_i; c_i \in \mathbb{M}\}. \tag{4}$$

---

[9] A maximally connected component of a graph $G = (V, E)$ is a subgraph $C = (V', E')$ such that: C is connected, and, for all vertices $u \in V \setminus V'$, there is no vertex $v \in V'$ such that $(u, v) \in E$.

5 Experiments

Precision and Recall. We ran experiments to evaluate the performance of Algorithm 1 (a.k.a `consecutive`). In particular, we compute two parameters, namely, *precision* and *recall*, whose definitions are given below. Suppose, $G = (\mathcal{S}, \mathcal{E})$ is a Bitcoin session graph (see Sect. 4.3 for definition). Let $G^* = (\mathcal{S}^*, \mathcal{E}^*)$ denote the Bitcoin session graph obtained from our experiment.

1. *precision*: This captures a measure of correct linking of the vertices.

$$precision = \frac{|\mathcal{E} \bigcap \mathcal{E}^*|}{|\mathcal{E}^*|}$$

   (The *precision* of 1 menas that the edges we have guessed are all correctly linked.)
2. *recall*: It captures a measure of how much the result is close to the best case of linking all the vertices related to consecutive sessions.

$$recall = \frac{|\mathcal{E} \bigcap \mathcal{E}^*|}{|\mathcal{E}|}$$

   (The *recall* of 1 means that, for each victim node, we have linked all its consecutive sessions. Note that, unlike *precision*, *recall* does not capture the scenario of linking sessions of two different nodes.)

Details of the Experimental Set-Up. We have done two sets of experiments to measure the performance of `consecutive` procedure. Experiment 1 had 4 victim bitcoin nodes, and each node had 8 sessions (this implies that the experiment included a total of $4 \times 8 = 32$ sessions); Experiment 2 had 4 victim nodes and each node had 6 sessions, implying that we experimented with a total of $4 \times 6 = 24$ sessions. In Experiment 1, the victims are directly connected to Bitcoin network, and in the Experiment 2, victims connect to Bitcoin network with (and without) Tor in different sessions. We believe that 32 and 24 sessions are good enough to demonstrate the proof of concept, which is the main purpose of this paper. Also, we would like to point out that, in order to obtain more realistic results, our experiments were performed on the *Bitcoin main network*, rather than on the *Bitcoin Testnet* (unlike the attack in [4]). We constructed Bitcoin session graph $G_1^* = (\mathcal{S}_1^*, \mathcal{E}_1^*)$ in Experiment 1, and $G_2^* = (\mathcal{S}_2^*, \mathcal{E}_2^*)$ in Experiment 2. Before giving our experimental results, below we provide the technical details of the experimental set-up.

In our experiment, we have used the following components: (a) Amazon Elastic Compute Cloud (*a.k.a* Amazon EC2), (b) a software container platform *Docker* [2], (c) a Bitcoin client *Bitcoind*, and (d) Tor client. See Fig. 2 for the layers in which the components reside. Amazon EC2 is a web service that provides a cloud server on which *Ubuntu* runs. In the cloud server, we ran *Docker* to create multiple instances of the container, and each container had *Bitcoind* and Tor client in it. We used Docker because it allowed us to run instances of

Bitcoind in the same EC2 machine, so that we had the same connection speed for all the Bitcoin nodes. The nodes running at the same speed are less vulnerable to fingerprinting attack (more on that later). The Tor client is used to connect to Bitcoin network over Tor.

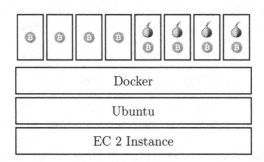

Fig. 2. Setup for Experiment 1 and 2

After we ran Bitcoin nodes, they connect to the Bitcoin network and started requesting blocks from other running peers to update their Blockchain. The session timings vary from 5 min to 160 min. Figure 3 shows the height of Blockchain at the start of each session we ran, we can see that the height of the nodes are close to each other. The nodes were running in the same EC2 instance for approximately the same time in each experiment, making their Blockchain growth very close to each other. This way of running Bitcoin nodes is a challenging scenario for deanonymization. After checking the block-ids requested in the sessions, we found that each victim requests a unique sequence of block-ids in its sessions.

Experiment 1. The main objective of the first experiment is to measure the performance when victim nodes are connected to Bitcoin network directly. We ran eight sessions for each victim in $\mathcal{V}_1 = \{v_1, v_2, v_3, v_4\}$ (total 32 sessions), then constructed the *Bitcoin session graph* to see if block-ids requested by a victim in different sessions can be linked. The major steps are as follows:

1. We extract sequence of block-ids of the blocks requested in each session to compute $\mathcal{A}.SessionBid = \{\beta_1, \beta_2, \ldots, \beta_{32}\}$. (see Sect. 4.1).
2. We ran `consecutive`(β_i, β_j) for each $\beta_i, \beta_j \in \mathcal{A}.SessionBid$. If it returns *True*, then the inputs β_i and β_j are related to the consecutive sessions of a victim node. (see Sect. 4.2).
3. We construct a *Bitcoin session graph* $G_1^* = (\mathcal{S}_1^*, \mathcal{E}_1^*)$ using the output we got by running `consecutive` procedure, see Fig. 4.
   (a) For each β_i in $\mathcal{A}.SessionBid$, we have a vertex in $\mathcal{S}_1^*$; thus, $\mathcal{S}_1^* = \{\beta_1, \beta_2, \ldots, \beta_{32}\}$. (see Sect. 4.3).
   (b) The edges $\{\beta_i, \beta_j\} \in \mathcal{E}_1^*$, if `consecutive`$(\beta_i, \beta_j) = $ *True*.

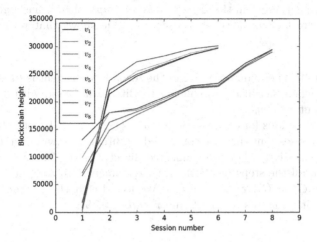

Fig. 3. This graph shows how the blockchain (best) height of a victim node (y-axis) varies with session number (x-axis). We ran nodes $\{v_1, v_2, v_3, v_4\}$ in Experiment 1 and nodes $\{v_5, v_6, v_7, v_8\}$ in Experiment 2.

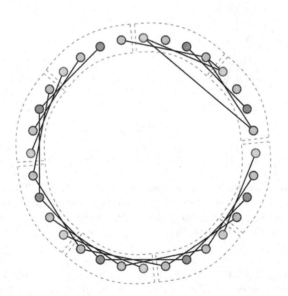

Fig. 4. This figure shows the *Bitcoin session graph* $G_1^* = (\mathcal{S}_1^*, \mathcal{E}_1^*)$. There are 32 vertices representing 32 sessions we ran. The vertices related to an identical victim node have the same color. (Color figure online)

We got a *precision* of 0.90 and *recall* of 0.71. The high *precision* value shows that if an edge is present in the *Bitcoin session graph*, then it has a good chance of being a correct edge; however, the *recall* value shows that we have missed some edges.

Experiment 2. The main objective of the second experiment is to measure the performance when victim nodes connect to Bitcoin network, with and without Tor, in different sessions.

We ran six sessions for each victim in $V_2 = \{v_5, v_6, v_7, v_8\}$ (total 24 sessions); in the first three sessions they are connected to Bitcoin netowrk over Tor, and, in the next three sessions, they are connected directly to the Bitcoin network. We have executed all the steps mentioned in Experiment 1 to compute the *Bitcoin session graph* $G_2^* = (\mathcal{S}_2^*, \mathcal{E}_2^*)$, see Fig. 5. We found that the consecutive links the sessions with a *precision* of 1.0 and *recall* of 0.75.

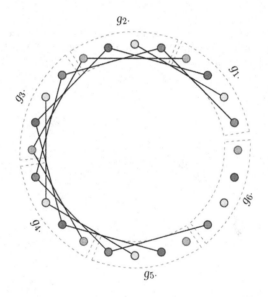

Fig. 5. This figure shows the *Bitcoin session graph* $G_2^* = (\mathcal{S}_2^*, \mathcal{E}_2^*)$. There are 24 vertices representing 24 sessions we ran. The vertices related to an identical victim node have the same color. (The set of vertices is partitioned into subsets $g_1|g_2|\ldots|g_6$, see Experiment 2 for more details.) (Color figure online)

To get more insight of the graph $G_2^* = (\mathcal{S}_2^*, \mathcal{E}_2^*)$, we have partitioned $\mathcal{S}_2^*$ according to the six sessions of each victim, $\mathcal{S}_2^* = g_1|g_2|\ldots|g_6$ (shown by dotted regions in the figure), where g_i contains the sequence of block-ids requested by the victims in i^{th} session. For example, g_1 contains the sequences of block-ids requested by victims in their first session.

The edges between the vertices contained in $g_1 \bigcup g_2 \bigcup g_3$ show that we could link the Bitcoin over Tor sessions, and the edges between the vertices contained in $g_4 \bigcup g_5 \bigcup g_6$ show that we could link the sessions when victims connect to Bitcoin network directly.

An edge between a vertex from g_3 to a vertex in g_4 is important, because g_3 contains the sequences of block-ids requested when the victims connect to the Bitcoin network over Tor, and g_4 contains the sequences of block-ids requested when they connect to Bitcoin network directly. An edge between a vertex from g_3 and a vertex from g_4 confirms that we can link the sessions of victims when they ran Bitcoin over Tor and Bitcoin without Tor (direct connection).

Discussion. We got a high *precision*, because initially the nodes request blocks at a faster rate; thus, in a short period of time, the differences of *best-heights* of the nodes become significant, allowing the attacker to distinguish between the sessions of different nodes. The *recall* value is also high because the heights of the consecutive sessions achieved at the end of the first session and at the beginning of the other session are close.

Another interesting observation is that when nodes are running Bitcoin over Tor then Algorithm 1 performed better. We found the rate of blockchain update is comparatively slower for proxied connections than for direct connections, it is because of the additional number of proxy nodes (onion routers) between the sender and the receiver. We believe that a slower rate of blockchain update could be one of the reasons why nodes did not catch up heights close to each other.

One might ask how *precision* and *recall* change when we increase the numbers of nodes and sessions. Theoretically, increasing the numbers of victim nodes and sessions could result in the following situation: suppose s_1 and s_2 are the sessions of two distinct nodes, such that the blockchain height at the end of session s_1 is *close* (i.e., the height difference is less than the threshold th described in Appendix C) to the starting blockchain height of the session s_2. In this scenario, our algorithm will incorrectly link s_1 and s_2, thereby, will decrease the experimentally obtained *precision*. However, we emphasize that such events will be infrequent, even in large-scale experiments. For concreteness, such event occurred only once in our 56 sessions conducted across two experiments. Moreover, it is worth noting that a node requests for blocks from his peers in the increasing order of blockchain height, and such a pattern does not change even in large-scale experiments. Therefore, our algorithm that crucially relies on the aforementioned pattern will still be able to correctly link them; as a result, the *recall* is unlikely to be significantly affected. We leave it as a future work as to how to appropriately design a large-scale experimental set-up to test Algorithm 1 that also takes into consideration the *ethical issues*.

6   Countermeasure

If the victim nodes request blocks in an unordered fashion, then it will not be possible for an adversary to estimate their Blockchain height, and, therefore, he cannot link sessions. Below we describe it in more detail.

Nodes exchange information on their Blockchains using VERSION, GETBLOCKS, and GETHEADERS messages (see Sect. 2). Using this information, the peers compare their Blockchain heights. A peer with higher height sends INV or HEADERS to the one with lower height.[10] Instead of sending the actual information of Blockchain, if a node sends height-values chosen uniformly from 1 to *best-height*, and if the block-id is chosen uniformly, then the other peers will not be able to deduce the Blockchain height. Similarly, if a node chooses a point uniformly from the entire Blockchain, and starts requesting the GETDATA from there in every session, then it will result in requesting blocks which are already updated into the Blockchain. Therefore, the number of common block-requests of consecutive sessions might not be bonded by a threshold *th*. As a result, adversary would not be able to get a fixed value *th* to determine the consecutive sessions, ruling out the attack. However, due to requesting blocks already updated into the Blockchain, the node's Blockchain growth will become slow.

7   Conclusion

In this paper, we have shown that, in the Bitcoin main network, linking of the *consecutive sessions* is possible by analysing the block-requests the victim makes. Our approach relies on the observation that a node requests blocks in the increasing order of height; this observation leads to linking consecutive sessions of the node. Once consecutive sessions are linked, all others could be linked as well. We were able to link (consecutive) sessions with a high success rate in three settings: (1) when nodes connect directly; (2) when nodes connect using the Tor; and (3) when nodes connect with Tor and then without Tor. We have also suggested countermeasures against our attacks.

Acknowledgments. The authors are grateful to the reviewers of CANS 2017 for their constructive comments. We also thank Sherman Chow for his insightful remarks regarding the experimental setup used in our work; the paper has benefitted immensely from them.

A   Peer Representation and Session

Suppose a victim node v_i is connected to adversarial nodes $\{a_j\}_{j \in [k]}$. The node v_i gets assigned multiple peer-ids p_{ij} outside the NAT making it difficult for a_j's to determine if they are connected to a single or multiple victim(s). The attacker nodes can check the *time of disconnect* of v_i to relate all the peer-ids assigned to v_i, and get a single representation (a, t), where a is the public address of v_i and t is the time of disconnect.[11]

---

[10] The Blocks-First Sync and Headers-First Sync methods are two ways to update the Blockchain as described in Sect. 2.

[11] The public address can be either public IP address or the onion address.

The representation (a, t) is achieved as follows: the *ping* and *pong* protocol messages are exchanged periodically between peers to get the status of the connection; after node v_i goes offline at time t, it is not able to respond to the *ping* messages sent by a_j's; as a result, all the a_j's will detect at the same time t that *pong* messages are not coming from peer p_{ij} (a_j continues to identify v_i by p_{ij} even when it is disconnected); therefore, they conclude a victim is disconnected at time t enabling them to associate $p_{i1}, p_{i2}, \ldots, p_{ik}$ with a victim node, and get the *peer-representation* (a, t). Note that (a, t) is a peer-representation of the victim v_i outside the NAT.

After getting a single peer-representation (a, t) of v_i, the adversary can combine the Bitcoin protocol messages exchanged with the peers in $\{p_{i1}, p_{i2}, \ldots, p_{ik}\}$ to compute the session $S(a, t)$.

B   Correctness of Algorithm 1

B.1   Background

Let $\beta_x, \beta_y \in \mathcal{A}.SessionBid$. The pair β_x and β_y can be related to three types: (1) consecutive sessions of an identical node, (2) non-consecutive sessions of an identical node, and (3) two different nodes. Below we describe the three cases in detail.

Related to Consecutive Sessions of an Identical Node. A Bitcoin node continuously sends GETDATA messages to update its Blockchain. When the node disconnects, it misses response (BLOCK messages) of some GETDATA messages. Then in the new session, the node starts from the Blockchain height achieved at the end of the previous session, and again sends the GETDATA for the blocks it has missed in the previous session. Thus, there are repeated block-requests in the consecutive sessions of an identical node. This is an important observation for our attack, because it helps in defining a threshold th, below which if two sessions have the common block-requests, then they could correspond to consecutive sessions of an identical node.[12] More formally, if $|\beta_x \cap \beta_y| \leq th$, then they may correspond to an identical node.

Related to Non-consecutive Sessions of an Identical Node. A node requests and receives blocks in each session; thus, the sessions between the two non-consecutive sessions update blocks into the Blockchain. Therefore, if we combine the block-ids requested in two non-consecutive sessions, the output will not have ids of blocks that lie between the heights achieved at the end of first session and at the start of the other session; these are the blocks updated between the two non-consecutive sessions. Since a node does not request the blocks after it is updated into the Blockchain, it requests disjoint sets of blocks in the non-consecutive sessions. More formally, if $|\beta_x \cap \beta_y| = 0$, then they may correspond to non-consecutive sessions.

---

[12] In our experiments, we take the maximum number of repeated block-requests in the consecutive sessions to be the threshold th (see Appendix C for more details).

Related to Two Different Nodes. Two nodes can request common blocks or disjoint sets of blocks depending upon their Blockchain height. Therefore, if β_x and β_y are from two different nodes, then we could get the following cases:

1. $|\beta_x \cap \beta_y| = 0$. This happens when the sessions related to β_x and β_y contain block-requests for disjoint sets of blocks.
2. $0 < |\beta_x \cap \beta_y| \leq th$. When the sessions related to β_x and β_y have Blockchain heights similar to consecutive sessions, then the numbers of common block-requests contained in them could be bounded by the threshold th.
3. $|\beta_x \cap \beta_y| > th$. The sessions related to β_x and β_y can contain the numbers of common block-requests greater than the threshold th. It happens when the first node achieves the height of the other in some session and then requests common blocks greater than th.

To determine if two sessions are of identical node or two different nodes, the case 3 as described above could be useful. This is explained using the following example. Suppose, two nodes v_i and v_j, where height of v_j is significantly large (greater than th) before they start the session. In the new session, the node v_i will make block requests for the blocks which are already requested and received by v_j, whereas v_j will not make repeated block requests for the blocks it has. Therefore, the previous session of v_j and the current session of v_i could have common block-requests much greater than th, whereas two sessions of v_j have common block-requests bounded by a threshold th. This shows that, if two sessions have common block-requests more than th then they are of different nodes, otherwise of an identical node. More formally, if $|\beta_x \cap \beta_y| \geq th$, then β_x and β_y are of two different nodes.

B.2   On the Correctness of Algorithm 1

We note that Algorithm consecutive requires two checks (two *if conditions*) to conclude whether the inputs are of consecutive sessions of an identical node. Below we describe the two *conditions*.

1. We compare the intersection of β_x and β_y with a value th, where th is the threshold of the number of common block-ids a node requests in the consecutive sessions (see Appendix B.1). If we have number of common block-ids in β_x and β_y greater than than the threshold th, then they are related to different nodes and the algorithm return *False* (see Appendix B.1).
2. Let us denote the sessions related to β_x and β_y by S_x and S_y. As we can see, the second *if* condition makes use of parameters h_x^s, h_x^e, h_y^s, and h_y^e. Since the blocks are requested in the increasing order of height, h_x^s is close to the Blockchain height at the *start* of the session S_x and h_x^e is close to the height at *end* of S_x, same holds for h_y^s and h_y^e. Following are the reasons why the condition is true for consecutive sessions.
- Without loss of generality, assume that session S_x happens before S_y. If S_x and S_y are consecutive, then we have $|h_y^e - h_x^s|$ as the output of *max*. Since the consecutive sessions have common block-requests from the same blocks (see

Appendix B.1), we get $|\beta_x| + |\beta_y|$ greater than $|h_y^e - h_x^s|$. Thus, the algorithm returns *True*.

• Without loss of generality, assume that session S_x happens before S_y. If S_x and S_y are non-consecutive, then we have $|h_y^e - h_x^s|$ as the output of *max*. The sessions between S_x and S_y update blocks into Blockchain (see Appendix B.1); thus, there are blocks between heights h_x^e and h_y^s whose block-ids are not contained in $\beta_x \bigcup \beta_y$. Thus, $|h_y^e - h_x^s|$ is greater then the value $|\beta_x| + |\beta_y|$ and the algorithm returns *False*.

• The third case is when S_x and S_y are of different nodes but have common block-requests less than th. This could be further divided into two cases $|\beta_x \bigcap \beta_y| = 0$ and $0 < |\beta_x \bigcap \beta_y| \leq th$ (see Appendix B.1). When intersection is empty then correctness is proved as follows: without loss of generality assume that the output of *max* is $|h_y^e - h_x^s|$; the output of *max* is greater than $|\beta_x| + |\beta_y|$ because of missing requests for the blocks between the heights h_y^e and h_x^s; thus, the algorithm returns *False*. (When intersection is non-empty and less than th, the correctness holds because it is unlikely that two different nodes have heights similar to consecutive sessions.)

C   Determining Threshold th

The value th is the threshold for the number of common block-requests sent in the consecutive sessions. It is used in the Algorithm 1, namely, `consecutive`, to determine: if two sets of block-ids correspond to consecutive sessions.

We have set th at 200 in our experiments, because we found the value 200 to be close to the number of common block-ids exchanged in the consecutive sessions. We now describe how the th is related to `consecutive` procedure.

Difficulty in Setting th Very High. Let us define a random variable $X_{ij} = |\beta_i \bigcap \beta_j|$ (that denote the number of common block-requests). Note that X_{ij} is uniformly distributed over all integers between zero and the current *best-height* of Blockchain, because we *assume* that the block-requests issued by a node inside the NAT is independent of the block-requests made by others. Using Markov inequality we get.

$$Pr(X_{ij} \geq th) \leq \frac{\mathbb{E}(X_{ij})}{th}. \tag{5}$$

Let us define a random variable `consecutive`(β_i, β_j) as follows:

$$\texttt{consecutive}(\beta_i, \beta_j) = \begin{cases} 1 \ if \ X_{ij} < th. \\ 0 \ if \ X_{ij} \geq th. \end{cases}$$

From the definition above, we get

$$Pr\big(\texttt{consecutive}(\beta_i, \beta_j) = 0\big) = Pr(X_{ij} \geq th).$$

By putting values from Eq. 5, we get

$$Pr(\texttt{consecutive}(\beta_i, \beta_j) = 0) \leq \frac{\mathbb{E}(X_{ij})}{th}.$$

Let p be the probability that consecutive(β_i, β_j) *returns* "1", we get

$$p = 1 - Pr(\text{consecutive}(\beta_i, \beta_j) = 0)$$
$$\geq 1 - \frac{\mathbb{E}(X_{ij})}{th}$$

The expected number of common block requests $\mathbb{E}(X_{ij})$ is a constant. Therefore, if the value of th is set to be high, then the probability that consecutive(β_i, β_j) returns "1" increases; thus, the attack might end up linking the sessions which correspond to different nodes (wrong linking).

References

1. Bitcoin Wiki (2017). https://en.bitcoin.it/wiki/
2. Docker Project Code (2017). https://github.com/docker/docker
3. v0.13.2, Bitcoin Code Project (2017). https://github.com/bitcoin/bitcoin
4. Biryukov, A., Khovratovich, D., Pustogarov, I.: Deanonymisation of clients in Bitcoin P2P network. In: Proceedings of the 2014 ACM SIGSAC Conference on Computer and Communications Security. ACM (2014)
5. Biryukov, A., Pustogarov, I.: Bitcoin over tor isn't a good idea. In: 2015 IEEE Symposium on Security and Privacy, pp. 122–134. IEEE (2015)
6. Dingledine, R., Mathewson, N., Syverson, P.: Tor: the second-generation onion router. Technical report, DTIC Document (2004)
7. Koshy, P., Koshy, D., McDaniel, P.: An analysis of anonymity in bitcoin using P2P network traffic. In: Christin, N., Safavi-Naini, R. (eds.) FC 2014. LNCS, vol. 8437, pp. 469–485. Springer, Heidelberg (2014). https://doi.org/10.1007/978-3-662-45472-5_30
8. Meiklejohn, S., et al.: A fistful of Bitcoins: characterizing payments among men with no names. In: Proceedings of the 2013 Conference on Internet Measurement Conference, pp. 127–140. ACM (2013)
9. Nakamoto, S.: Bitcoin: a peer-to-peer electronic cash system (2008)
10. Reid, F., Harrigan, M.: An analysis of anonymity in the Bitcoin system. In: Altshuler, Y., Elovici, Y., Cremers, A., Aharony, N., Pentland, A. (eds.) Security and Privacy in Social Networks, pp. 197–223. Springer, New York (2013). https://doi.org/10.1007/978-1-4614-4139-7_10
11. Ron, D., Shamir, A.: Quantitative analysis of the full Bitcoin transaction graph. In: Sadeghi, A.-R. (ed.) FC 2013. LNCS, vol. 7859, pp. 6–24. Springer, Heidelberg (2013). https://doi.org/10.1007/978-3-642-39884-1_2

Towards a Smart Contract-Based, Decentralized, Public-Key Infrastructure

Christos Patsonakis[1(✉)], Katerina Samari[1], Mema Roussopoulos[1],
and Aggelos Kiayias[2]

[1] Department of Informatics and Telecommunications, National and Kapodistrian
University of Athens, University Campus, 15784 Ilisia, Athens, Greece
{c.patswnakis,ksamari,mema}@di.uoa.gr
[2] University of Edinburgh, 10 Crichton St., Edinburgh EH8 9AB, UK
Aggelos.Kiayias@ed.ac.uk

Abstract. Public-key infrastructures (PKIs) are an integral part of the security foundations of digital communications. Their widespread deployment has allowed the growth of important applications, such as, internet banking and e-commerce. Centralized PKIs (CPKIs) rely on a hierarchy of trusted Certification Authorities (CAs) for issuing, distributing and managing the status of *digital certificates*, i.e., unforgeable data structures that attest to the authenticity of an entity's public key. Unfortunately, CPKI's have many downsides in terms of security and fault tolerance and there have been numerous security incidents throughout the years. Decentralized PKIs (DPKIs) were proposed to deal with these issues as they rely on multiple, independent nodes. Nevertheless, decentralization raises other concerns such as what are the incentives for the participating nodes to ensure the service's availability.

In our work, we leverage the scalability, as well as, the built-in incentive mechanism of blockchain systems and propose a smart contract-based DPKI. The main barrier in realizing a smart contract-based DPKI is the size of the contract's state which, being its most expensive resource to access, should be minimized for a construction to be viable. We resolve this problem by proposing and using in our DPKI a *public-state* cryptographic accumulator with constant size, a cryptographic tool which may be of independent interest in the context of blockchain protocols. We also are the first to formalize the DPKI design problem in the Universal Composability (UC) framework and formally prove the security of our construction under the strong RSA assumption in the Random Oracle model and the existence of an ideal smart contract functionality.

1 Introduction

Public key, or asymmetric, cryptography is a critical building block for securing important communications across the Internet, such as e-commerce and internet banking. To enable such applications, public-key infrastructures (PKIs) are essential because they provide a verifiable mapping from an entity's name to its

S. Capkun and S. S. M. Chow (Eds.): CANS 2017, LNCS 11261, pp. 299–321, 2018.
https://doi.org/10.1007/978-3-030-02641-7_14

corresponding public key. In essence, a PKI is a system that allows the creation, revocation, storage, and distribution of *digital certificates*, i.e., unforgeable data structures that attest to the authenticity of an entity's public key.

In a centralized PKI (CPKI), a Certification Authority (CA), is responsible for issuing, distributing and managing the status of digital certificates. Two assumptions must be made when deploying a CPKI. These are: (1) everyone knows the CA's (correct) public key and, (2) statements signed by the CA's private key are valid, i.e., everyone trusts the CA. In a CPKI, registration is handled in two phases. In the first phase, the user proves her claim on an identity to a Registration Authority (RA). Assuming the RA validates the claim, it forwards the user's request to the CA. In the second phase, the user receives her digital certificate, which is signed by the CA's private key, thus, attesting its validity. CAs *periodically* publish signed data structures that contain revoked certificates, e.g., a certificate revocation list (CRL). Distribution of certificate-related information is handled either by the CA (online CA), or, it is delegated to online, publicly accessible directories (offline CA).

While predominant in use, CPKIs have several shortcomings. A CA constitutes a single point of failure, both in terms of security and availability. There have been several incidents where CAs have been hacked that led to the issuance of false certificates for domains of high-profile corporations, such as Google [3]. Other prominent examples are the Symantec [4] and TrustWave [8] incidents, as well as the growing concern of governments and private organizations being able to issue false certificates for surveillance, thus, violating the privacy of end-users [38]. In practice, there exist multiple CAs, which are linked with well-defined, parent-child relationships, based on trust and other policies. The most notable example of this architecture is the SSL/TLS certificate chain. This hierarchical, tree-like, certification model is designed to increase the system's scalability and fault-tolerance. However, root, or even, subordinate CA compromises are still catastrophic [20].

In a decentralized PKI (DPKI), multiple, independent nodes cooperate and deliver the same set of services, without relying on one, or more, trusted third parties (TTPs). DPKIs have been proposed because, as distributed systems, they have the potential to offer a number of desirable properties that CPKIs cannot offer, such as scalability, fault-tolerance, load balancing and availability. Researchers have proposed DPKIs based on various distributed primitives, such as distributed hash tables (DHTs) (e.g., [9]). To account for malicious nodes and provide increased security, they employ secret sharing, threshold and byzantine agreement protocols (e.g., [10,18]). These techniques, while more complex to design and implement correctly, lead to systems that do not exhibit single points of failure. Unfortunately, prior DPKIs do not provide incentives for the participating nodes to ensure that the offered service remains available in the long term, e.g., they fail to address the *free-riding* problem [27].

Blockchain protocols (e.g., Bitcoin [31]), feature a reward mechanism that incentivizes parties to engage in the protocol. The rewards come in the form of a digital currency that compensates its participants, thus, creating a counter-incentive to free-riding, while still retaining a highly scalable, free-entry system.

In this work, we present the design of a DPKI on top of a smart contract platform, a new generation of blockchains that allow the development of smart contracts, i.e., stateful agents that "live" in the blockchain and can execute arbitrary state transition functions. The main barrier in realizing a smart contract-based DPKI is the size of the smart contract's state which, being its most expensive resource to access, should be minimized for a construction to be considered viable. Previous blockchain-based solutions, such as Namecoin [6] and Emercoin [2], fall short on this part as their state is linear to the number of registered entities. Fromknecht et al. [21] improve on this by harnessing the power of cryptographic accumulators, i.e., space-efficient data structures that allow for (non) membership queries. However, we believe that they do not exploit, sufficiently, their potential for the following reasons: (1) their system's state is still of logarithmic complexity, due to the use of a Merkle tree-based accumulator and, (2) their construction recomputes accumulator values to handle deletions of elements, i.e., each deletion (revocation) has a linear computational complexity. We resolve these inefficiencies by presenting a construction whose state is constant and avoids recomputing accumulator values. Our main building block is a public-state, additive, universal accumulator, based on the strong RSA assumption in the Random Oracle model, which, among others, has the following nice properties: (1) the accumulator and the structures for proving (non) membership (referred to as *witnesses*) have constant size and, (2) all of its operations can be performed efficiently by having access only to the accumulator's public key.

In short, the contributions of this paper are as follows:

- We propose the design of a DPKI on top of a smart contract platform. Due to the interoperability of smart contracts, our system provides a generic mechanism for on-blockchain authentication that, up to this point, was handled in an ad-hoc manner. Furthermore, the programmable nature of these platforms allows us to evolve our system with more efficient primitives, when such become available, without the need for a fork in the blockchain, which is the case for specialized PKI blockchains.
- We resolve the main barrier of realizing a viable smart contract-based DPKI by providing a construction that has the "constant-ness" property, i.e., both the smart contract's state, as well as, the structures for proving (non) membership, have constant size. We stress the importance of this property as it guarantees, in addition to efficiency, uniform digital currency costs for any given operation across all users, i.e., fairness in terms of costs.
- Our construction is based on a public-state, additive, universal accumulator, a cryptographic tool which may be of independent interest for protocols that employ blockchains for verifying, efficiently, the validity of information.
- We are the first to formalize the DPKI design problem in the Universal Composability (UC) framework [16] and we formally prove the security of our construction under the strong RSA assumption in the Random Oracle model

and the existence of an ideal smart contract functionality. Due to space limitations, we provide a part of the proof in Appendix A.

- Even though our envisioned application is a PKI, we specifically model our service as a generic "Naming Service". Thus, our design can be ported to implement, efficiently, other services that reside in this paradigm, e.g., a distributed domain name system (DDNS).

2   Related Work

Several previously proposed systems utilize the same underlying primitive, each in its own unique way, to decentralize the services of a PKI. In the interest of space, we focus on full-fledged DPKIs, i.e., systems that implement registration, revocation, certificate storage and retrieval. Thus, we will not be concerned with certification systems (e.g., [28]), which do not offer revocation, hybrid approaches, e.g., coupling CAs with structured overlays (e.g., [37]), or, even PGP [40], whose operation relies on centralized servers. We also review related work regarding cryptographic accumulators, which form the basis of our construction.

Researchers have proposed DPKIs based on the replicated state machine (RSM) paradigm [34,39] to enforce a global, consistent view of the system's state. This is achieved by having nodes participate in an authenticated agreement protocol and typically assume: (1) a threshold t of faulty nodes, (2) *join()* and *leave()* protocols for nodes wishing to enter, or leave, a replica group, to adjust the system's threshold parameter and, (3) nodes are able to authenticate any (potential) participant. In RSM-based PKIs, registration requires one to perform an "out-of-band" negotiation with multiple administrative domains, which is cumbersome for the user. In addition, non-determinism, e.g., time-stamping, is a key difficulty of consistent replication since it can lead to replica state-divergence, thus, compromising fault-tolerance. However, time-stamping is essential in a PKI for tracking certificate lifetime. Blockchain-based systems, on the other hand, do not suffer from this issue and they have already been used for the implementation of time-stamping services (e.g., [25]). Furthermore, they employ a different form of agreement which is based on computation. This, alternative, agreement algorithm has the nice property of being adaptable as nodes freely join and leave the system. Experience has illustrated, that the blockchain approach has been highly favored by both the research community, as well as, the industry, due to its highly scalable, adaptive and non-restrictive nature [1,5,7].

Structured overlays have also been proposed to distribute the services of a PKI [10,18]. These are, by design, scalable, load-balanced and provide for efficient storage and retrieval of data. Unfortunately, these systems do not defend against Sybil attacks. Douceur [19] has proved that to defend against the Sybil attack, distributed systems must employ either authentication, or, computational power. However, the aforementioned DHT-based systems do not employ either, thus, they are insecure. Blockchains, on the contrary, are resilient to the Sybil attack since their operation inherently depends on computational power.

In all of the above systems, nodes are expected to participate in resource-intensive protocols. Unfortunately, these systems do not incentivize node participation, nor, enforce correct behavior of participating nodes.

The initial approach of constructing a blockchain-based PKI is based on the observation that there is an inherent similarity between the services of a DNS and a PKI, respectively. Both essentially map identity names to some value (be it an IP address, or, a public key). One of the biggest *altcoins*, Namecoin [6], provides a distributed DNS as its main function. In Namecoin, the blockchain is used both for storing, as well as, verifying/querying DNS records. Several DPKIs follow this approach, the most notable example of which is Emercoin [2]. Unfortunately, this approach is inefficient as it forces each user to store an entire copy of the blockchain and traverse its contents every time she needs to validate a mapping. This limits the system's applicability significantly; for example, storing the entire blockchain on a smartphone is prohibitive. Moreover, validating mappings, which is the most frequent operation, requires an increasing amount of computation as more blocks are appended to the blockchain.

A modern, more involved approach, is to employ cryptographic accumulators, which were first introduced in the work of Benaloh et al. [14] as a decentralized alternative to digital signatures. These are space-efficient data structures that allow for membership queries. Their initial construction was refined in the work of Barić et al. [13] by strengthening the original security notion to that of *collision freeness*. Camenisch et al. [15] extended previous works and presented the first accumulator scheme that allowed for elements to be dynamically added/deleted. In this scheme, membership witnesses can be updated by utilizing only the accumulator's public key, i.e., no trapdoor information is required. Following this work, Li et al. [29] introduced *universal accumulators*, i.e., accumulators that support non-membership witnesses as well. All of the aforementioned accumulator schemes are RSA-based. Other proposed accumulator schemes are based on Merkle trees (e.g., [33]), bilinear pairings (e.g., [32]) and lattices (e.g., [26]).

Cryptographic accumulators provide a number of benefits. First, their compact (or even constant) size makes them suitable candidates for storage-limited devices (e.g., smartphones). Second, (non) membership verifications have constant computational cost, regardless of the number of accumulated values (accumulation is the addition of an element to the accumulator). Third, their security properties are based on standard hardness assumptions, thus, making them suitable for critical security infrastructures. An additional benefit of accumulator-based constructions is that they do not employ the blockchain for enforcing consensus on the entire set of (identity, public-key) mappings, as is the case for Namecoin and Emercoin. Instead, the consensus object is the accumulator value(s), which has the following benefits. First, users are not required to perform a complete retrieval and verification of the entire transaction history, i.e., downloading and validating the entire blockchain. Instead, an outdated, or, new client, can download and validate only block headers to update her state, which is far more efficient both in terms of communication and computation. Second, it allows the introduction of an unreliable component that users can query to efficiently obtain, among others, a more compact version of the entire history of operations, compared to the full transaction history. Due to the verifiable nature of cryptographic accumulators, this increased efficiency comes at no cost.

Certcoin [21] is a blockchain-based PKI which deals with the aforementioned inefficiencies of Namecoin and Emercoin by decoupling information storage from its verification. It employs an authenticated DHT for storing digital certificates, based on Kademlia [30]. These networks facilitate storage and retrieval queries in logarithmic complexity, i.e., they are very efficient. Furthermore, its authenticated nature makes it secure against the Sybil attack [19]. To facilitate the verification of (identity, public-key) mappings, Certcoin maintains two cryptographic accumulators in the blockchain. Certcoin's first accumulator is based on the strong RSA assumption and accumulates identity names. Thus, users can infer if an identity has been registered in the system. The second accumulator is based on Merkle trees (formally presented in [35]) and accumulates (identity, public-key) mappings. This allows clients to validate the authenticity of any mapping retrieved from the DHT network by downloading and validating the latest block and by performing the appropriate membership queries.

In the following, we highlight the differences of our design compared to Certcoin. We open the discussion with issues regarding accumulators. First, our design employs two RSA-based accumulators, which have constant size and small public-keys. On the contrary, Certcoin employs one RSA and one Merkle-based accumulator. The size of Merkle-based accumulators increases as more elements are accumulated, i.e., it is not constant. This is an issue in the blockchain world as miners prefer small blocks which can be hashed faster and with reduced operational costs to increase their profits. Therefore, it would be difficult to incentivize miners to support Certcoin's blockchain whose blocks are of variable size. Moreover, in blockchain-based systems, transaction execution costs are a function of their size. Thus, Certcoin does not guarantee *fairness* in terms of transaction costs. Our construction does not face these issues due to its constant state. Second, while Merkle-based accumulators have the nice property that elements can be deleted without the knowledge of trapdoor information, the same does not hold for RSA-based accumulators. Consequently, when an (identity, public-key) mapping is revoked, thus making the identity available again, Certcoin recomputes its RSA accumulator from scratch, which is, inefficient. To deal with the fact that deletions in RSA accumulators require access to their secret key, which, if publicly known, can break their security, we employ a trick that is presented in the work of Baldimtsi et al. [12]. Essentially, we use tags to mark elements as "added" (during registration) or "deleted" (during revocation). Thus, in contrast to Certcoin, we do not recompute any accumulator. Third, Certcoin tightly couples the process of mining with the blockchain's actual application, which is to deliver the services of a DPKI. These two issues are orthogonal to each other, in terms of the system's architecture and, we believe, should be addressed at different layers. Instead, we are the first to propose a DPKI that is based on a smart contract platform, i.e., programmable blockchains that decouple the blockchain's consensus protocol from the applications' functionalities that run on top of it.

This key difference allows us to evolve our system with more efficient primitives, when such become available, without the need for a hard fork in the blockchain, which is not the case for application-specific blockchains, such as Certcoin. Additionally, in these platforms, contracts can interact with each other, thus, creating an ecosystem of applications that can interoperate. By leveraging this feature, our system can provide a generic mechanism for on-blockchain authentication that, up to this point, was handled in an ad-hoc manner. Fourth, Certcoin has no security model for the PKI it implements nor a proof that it provides the claimed service. In contrast, we are the first to formalize the DPKI design problem in the UC framework [16] and we formally prove the security of our construction under the strong RSA assumption in the Random Oracle model.

3   Public-State Accumulator

In this section, we present the main building block of our naming service. Specifically, in Sect. 3.1, we provide the definition of a public-state, additive, universal accumulator and, in Sect. 3.2, we present a construction for such an accumulator under the strong RSA assumption in the Random Oracle model.

3.1   Definition of a Public-State, Additive, Universal Accumulator

Informally, we consider an accumulator as *public-state*, if one can perform all of its operations by only having access to its public-key, i.e., no trapdoor knowledge is required. According to the terminology presented in [12], an accumulator is *additive*, if it only allows for addition of elements, and *universal*, if it allows for both membership and non-membership witnesses. In the following, we present the definition of a public-state, additive, universal accumulator. Our definition employs two trusted parties. The first one, T, runs the key-generation algorithm $(\mathsf{KeyGen}(1^\lambda))$ and publishes the accumulator's public-key. The second one, the "accumulator manager" T_{acc}, is responsible for maintaining the accumulator.

Definition 1 (public-state, additive, universal accumulator). Let D be the domain of the accumulator's elements, and X, the current accumulated set. A public-state, additive, universal accumulator consists of the following algorithms:

- KeyGen: On input a security parameter λ, it generates a key pair (pk, sk) and outputs pk. This algorithm is run by T.
- InitAcc: On input pk and the empty accumulated set $X = \emptyset$, it outputs an accumulator value c_0. This algorithm is run by T_{acc}.
- Add: On input pk, an element $x \in D$ to be added and an accumulator value c, it outputs (c', W), where c', is the new value of the accumulator, and W, is a membership witness for x.
- MemWitGen: On input pk, X, c and $x \in X$, it outputs a membership witness W for x.
- NonMemWitGen : On input pk, X, c, x, where, $x \in D$ and $x \notin X$, it outputs a non-membership witness W for x.

- UpdMemWit: On input pk, x, y, W, where, W is a membership witness for x, it outputs an updated membership witness W' for x. This algorithm is run after $(c', W_y) \leftarrow \mathsf{Add}(pk, y, c)$, where W_y is a membership witness for y.
- UpdNonMemWit : On input pk, x, y, W, where, $x, y \in D$, $x \neq y$ and W is a non-membership witness for x, it outputs an updated non-membership witness W' for x. This algorithm is run after $(c', W_y) \leftarrow \mathsf{Add}(pk, y, c)$.
- VerifyMem : On input $pk, x \in D, W$ and c, it outputs 1 or 0.
- VerifyNonMem : On input $pk, x \in D, W$ and c, it outputs 1 or 0.

The correctness and security properties for a cryptographic accumulator can be defined with the aid of a security game between a challenger $\mathcal{C}$ and an adversary $\mathcal{A}$. Informally, an accumulator is *correct* if, for any honestly produced membership witness, the membership verification algorithm outputs 1, and, for any honestly produced non-membership witness, the non-membership verification algorithm outputs 1. Furthermore, we consider a universal accumulator as *secure* if no p.p.t. adversary can produce a valid non-membership witness for a member of the accumulated set, nor, a valid membership witness for an element which is not a member of the accumulated set. The security property for an accumulator can be met as *collision-freeness*, or, *soundness* in the literature. A formal definition for security is given below (Definition 2). Due to lack of space, we omit the formal definition of correctness.

Definition 2. We say that an accumulator is *secure* if, for any p.p.t. adversary $\mathcal{A}$ interacting with a challenger $\mathcal{C}$, as illustrated in the security game of Fig. 1, it holds that $\Pr[G(1^\lambda) = 1] \leq \varepsilon$, where ε is a negligible function of λ.

G: On input 1^λ,

- $\mathcal{C}$ runs $\mathsf{KeyGen}(1^\lambda)$, generates (pk, sk) and gives pk to the adversary $\mathcal{A}$.
- $\mathcal{A}$ first makes an InitAcc query to $\mathcal{C}$. $\mathcal{C}$ initializes a set $X \leftarrow \emptyset$, runs InitAcc and returns the value c_0 to $\mathcal{A}$.
- When $\mathcal{A}$ performs an Add query for an element x, $\mathcal{C}$ sets $X \leftarrow X \cup \{x\}$ and computes $(c', W) \leftarrow \mathsf{Add}(pk, x, c)$. Then, $\mathcal{C}$ returns the pair (c', W) to $\mathcal{A}$.
- $\mathcal{A}$ outputs (x^*, W^*).

The game returns 1 if at least one of the following conditions holds:

1. $x^* \notin X$ and $\mathsf{VerifyMem}(pk, x^*, W^*, c) = 1$,
2. $x^* \in X$ and $\mathsf{VerifyNonMem}(pk, x^*, W^*, c) = 1$.

Fig. 1. The security game between the adversary $\mathcal{A}$ and a challenger $\mathcal{C}$, where $\mathcal{C}$ plays the roles of both T and T_{acc}.

3.2   Construction

In Fig. 2, we present a construction of a public-state, additive, universal accumulator. We aim to accumulate identities, or, (identity, public-key) pairs, i.e., arbitrary strings, thus, the accumulator's domain is $D = \{0,1\}^*$. This construction is a combination of the RSA-based universal accumulator of Li et al. [29], accompanied with a procedure which maps arbitrary strings to prime numbers, as suggested in [24]. Namely, for any algorithm run on input an element $z \in \{0,1\}^*$, the party who runs the algorithm, first, executes a procedure Map which maps z to a prime number, e.g., z_p, and then, proceeds by running the same algorithm as in the accumulator of Li et al. [29] for the prime number z_p.

Mapping Arbitrary Strings to Primes. Gennaro et al. [24] describe a procedure that utilizes a universal hash function family U of functions [17] which map strings of $3k$ bits to strings of k bits, with the additional property that, for any $y \in \{0,1\}^k$ and given $f \in U$, one can efficiently sample uniformly from the set $\{x \in \{0,1\}^{3k} : f(x) = y\}$. On input $z \in \{0,1\}^*$, it first computes $h(z)$, where $h : \{0,1\}^* \to \{0,1\}^k$ is a collision-resistant hash function. It then samples from the set $\{x \in \{0,1\}^{3k} : f(x) = h(z)\}$ for a prime number. This procedure is collision-resistant if h is collision resistant. The authors illustrate that, for all but a $(1/2^k)$-fraction of functions $f \in U$ and for any $y \in \{0,1\}^k$, at least $1/ck$ elements in the set $\{x \in \{0,1\}^{3k} : f(x) = y\}$ are primes, for a small constant c. Thus, an algorithm which samples from the set $\{x \in \{0,1\}^{3k} : f(x) = h(z)\}$ at least $ck \log^2 k$ times will find a prime number, except with negligible probability.

In our construction (Fig. 2), we employ a deterministic version of the aforementioned Map procedure. Specifically, we utilize a labeled hash function $h : \{0,1\}^* \times \{0,1\}^* \to \{0,1\}^k$, which we model as a Random Oracle, and pick two labels, i.e., $label_0, label_1 \in \{0,1\}^*$. Then, Map, on input $z \in \{0,1\}^*$, computes $h(label_0, z)$ and fixes the coins for sampling from the set $\{x \in \{0,1\}^{3k} : f(x) = h(label_0, z)\}$ to depend on z. This is accomplished by computing $G(h(label_1, z))$, where $G : \{0,1\}^k \to \{0,1\}^{p(k)}$ is a pseudorandom generator and $p(k)$ is some polynomial in k. Then, Map samples a prime from the set $\{x \in \{0,1\}^{3k} : f(x) = h(label_0, z)\}$ using the bits of $G(h(label_1, z))$ as randomness.

We stress that Map is collision-resistant in the Random Oracle model. Informally, if Map is not collision-resistant, then, there is a p.p.t. adversary which finds $z_1, z_2 \in \{0,1\}^*$, such that, $\mathsf{Map}(z_1) = \mathsf{Map}(z_2)$. If this holds, then there are two possible cases: (i) the adversary breaks the collision-resistance of the hash function h, which is a contradiction, or, (ii) the adversary breaks the security of the pseudorandom generator G. Namely, given that h is collision-resistant and sampling with a truly random function gives us collisions only with negligible probability, then, by the security of G, a p.p.t. adversary finds collisions in the Map procedure only with negligible probability.

We elaborate by giving a simple example as to why the Map procedure has to be deterministic in our construction. First, assume that we used the procedure of [24] without the suggested modification and that an element $x \in \{0,1\}^*$

The domain of the accumulator is $D = \{0,1\}^*$.

- KeyGen : On input 1^λ, it generates a pair of safe primes p, q of equal length, such that, $p = 2p' + 1$, $q = 2q' + 1$ and p', q' are also primes. It computes $n = pq$ and chooses g randomly from QR_n. It sets $\ell = \lfloor \lambda/2 \rfloor - 2$ and chooses a deterministic procedure Map (as described in the previous paragraph), which receives as input an arbitrary string and outputs a prime number less than $2^{\lfloor \lambda/2 \rfloor - 2}$. It sets $pk = (n, g, \mathsf{Map})$, $sk = (p, q)$, and outputs pk.
- InitAcc : On input pk, it outputs $c_0 = g$.
- Add : On input pk, $x \in \{0,1\}^*$ and c, it invokes Map on input x and receives a prime number x_p. Then, it computes $c' = c^{x_p} \bmod n$, sets $W = c$ and outputs (c', W).
- MemWitGen : On input pk, X and $x \in \{0,1\}^*$, it computes and outputs $W = g^{\prod_{x_i \in X \setminus \{x\}} \mathsf{Map}(x_i)}$.
- NonMemWitGen : On input $pk, X, c, x \notin X$, it invokes Map on input x and receives a prime number x_p. Then, it computes $u = \prod_{x_i \in X} \mathsf{Map}(x_i)$. Since $gcd(x_p, u) = 1$, it runs the extended Euclidean algorithm and computes $a', b' \in \mathbb{Z}$, such that, $a'u + b'x_p = 1$. Then, since x_p is a positive integer in $\mathbb{Z}_{2\ell}$, it finds $k \in \mathbb{Z}$, such that, $a' + kx_p \in \mathbb{Z}_{2\ell}$, and sets $a = a' + kx_p$ and $b = b' - ku$. Finally, it outputs $W = (a, d)$, where $d = g^{-b} \bmod n$.
- UpdMemWit: On input pk, x, y, W, it invokes Map on input x and receives a prime number x_p. Then, it computes and outputs $W' = W^{x_p} \bmod n$.
- UpdNonMemWit : On input $pk, x \notin X, y \in X, c$ and $W = (a, d)$, it invokes Map on inputs x, y and receives the prime numbers x_p, y_p respectively. Since $y_p \neq x_p$, it runs the extended Euclidean algorithm and computes $a_0, r_0 \in \mathbb{Z}$, such that, $a_0 y_p + r_0 x_p = 1$. Then, it multiplies both sides by a, i.e., $aa_0 y_p + ar_0 x_p = a$ and computes $a' = a_0 a \bmod x_p$. Then, it finds $r \in \mathbb{Z}$, such that, $a' y_p = a + rx_p$, computes $d' = dc^r \bmod n$ and outputs $W' = (a', d')$.
- VerifyMem : On input pk, x, W, c, it invokes Map on input x and receives a prime number x_p. Then, it outputs 1, if $W^{x_p} = c \bmod n$, otherwise, it outputs 0.
- VerifyNonMem : On input pk, x, W, c, where, $W = (a, d)$, it invokes Map on input x and receives a prime number x_p. Then, it outputs 1, if $c^a = d^{x_p} g \bmod n$, otherwise, it outputs 0.

Fig. 2. Construction of a public-state, additive, universal accumulator.

was added in our accumulator. This means that T_{acc} first produces a prime x_p and then adds x_p in the underlying RSA accumulator. Then, an adversary can produce a non-membership witness W for x simply by producing a different prime $x_p' \neq x_p$ for the element x and then running the non-membership witness generation algorithm for x_p'. Therefore, the security property of the accumulator, as defined in the security game of Fig. 1, would not hold, since the adversary can output (x, W), such that, $x \in X$ and VerifyNonMem$(pk, x, W, c) = 1$.

The security of our accumulator is derived by the security of the accumulator of Li et al. [29], which is proven secure under the strong RSA assumption, and the collision-resistance of Map in the Random Oracle model. The strong RSA assumption [13] states that, given an RSA modulus n and a random element $x \in \mathbb{Z}_n^*$, no p.p.t. adversary can find (y, e), such that, $y^e = x \mod n$, except with negligible probability.

Theorem 1. *The accumulator of Fig. 2 is secure according to Definition 1 under the strong RSA assumption and the collision-resistance of Map in the Random Oracle model.*

Proof sketch. Assume that there is a p.p.t. adversary $\mathcal{A}$ which breaks the security of the accumulator of Fig. 2. Then, according to Definition 1, $\mathcal{A}$ can output (x^*, W^*), such that: (1) $x^* \notin X$ and VerifyMem$(pk, x^*, W^*, c) = 1$, or, (2) $x^* \in X$ and VerifyNonMem$(pk, x^*, W^*, c) = 1$. Suppose that (1) holds. Then, there are two possible cases: (a) $\mathcal{A}$ comes up with x, x^*, such that, Map$(x) =$ Map(x^*) and $x \in X$, thus, breaking the collision-resistance of the Map procedure, or, (b) $\mathcal{A}$ computes a valid membership witness W^* for a prime x_p^*, where Map$(x^*) = x_p^*$ and $x^* \notin X$. In the latter case, we can construct a p.p.t adversary $\mathcal{B}$ which breaks the strong RSA assumption. We refer for further details to the proof of Li et al. [29]. We follow a similar argument for the case where (2) holds.

Constructing a Universal Accumulator from an Additive, Universal Accumulator [12]. Assume that ACC_U^{add} is an additive, universal accumulator, which accumulates elements of the form (x, i, op), where, x is the element to be added, i, is an index, and op, is either a or d. We construct a universal accumulator ACC_U, from ACC_U^{add}, as follows. When an element x is added to ACC_U for the first time, T_{acc} adds the value $(x, 1, a)$ to ACC_U^{add}. Otherwise, it adds (x, i, a), where the index i indicates that this is the i-th time that x is added to ACC_U^{add}. When an element x is deleted from ACC_U, T_{acc} adds (x, i, d) to ACC_U^{add}. In order to prove membership of x in ACC_U, one should find an index i, such that, $\big((x, i, a) \in ACC_U^{add} \big) \wedge \big((x, i, d) \notin ACC_U^{add} \big)$. Accordingly, to prove that $x \notin X$, one should either prove that $(x, 1, a) \notin ACC_U^{add}$, or, find an index i, such that, $\big((x, i-1, d) \in ACC_U^{add} \big) \wedge \big((x, i, a) \notin ACC_U^{add} \big)$.

4   Defining a Naming Service Functionality

In this section, we describe the security of a naming service in the UC framework [16] by defining it as an ideal functionality $\mathcal{F}_{ns}$ (Fig. 3). $\mathcal{F}_{ns}$ interacts with n clients, m servers, a party T, which is responsible for setup, and an adversary $\mathcal{S}$. It stores (identity, public-key) pairs and supports a number of operations. During setup, the trusted party T, specifies a relation R, which defines under which condition a public key can be revoked. In practice, this relation might be a verification algorithm for a NIZK, or, a signature on a randomly selected message. After the setup phase, a client can register an (identity, public-key) pair, assuming the identity is available, and, can revoke an (identity, public-key) pair,

$\mathcal{F}_{ns}$:

- On input (sid, Init) by a server S_i, $\mathcal{F}_{ns}$ sets $\leftarrow Y \cup \{S_i\}$ (where Y is initialized as $Y = \emptyset$). When all servers have sent a message (sid, Init), $\mathcal{F}_{ns}$ sends $(sid, \mathsf{ServersInitialized})$ to S. If S returns allow, then $\mathcal{F}_{ns}$ sets flag = start. Also, S corrupts a number of clients. We denote as C_{cor} the set of corrupted clients.
- On input (sid, Setup, R) by T, $\mathcal{F}_{ns}$ checks if flag = start and forwards (sid, Setup, R) to S. If S returns allow, $\mathcal{F}_{ns}$ stores R, initializes a set $X \leftarrow \emptyset$, sets flag = manage and returns success to T.
- On input $(sid, \mathsf{Register}, id, pk)$ by a client C, $\mathcal{F}_{ns}$ checks if flag = manage and forwards $(sid, \mathsf{Register}, id, pk)$ to S.
  - If S returns allow, $\mathcal{F}_{ns}$ checks if there is $(id, \cdot) \in X$. If there is no $(id, \cdot) \in X$, it sets $X \leftarrow X \cup (id, pk)$ and returns success to C via public delayed output, otherwise, it returns fail to C via public delayed output.
  - If S returns fail, then $\mathcal{F}_{ns}$ returns fail to C via public delayed output.
  - If S returns $(sid, \mathsf{Register}, id', pk', C)$, $\mathcal{F}_{ns}$ checks if $C \in C_{cor}$ and if there is $(id', \cdot) \in X$. If there is no $(id', \cdot) \in X$, it sets $X \leftarrow X \cup (id', pk')$ and returns success to C via public delayed output, otherwise, it returns fail to C via public delayed output.
- On input $(sid, \mathsf{Revoke}, id, pk, \mathsf{aux})$ by C, $\mathcal{F}_{ns}$ checks if flag = manage and forwards $(sid, \mathsf{Revoke}, id, pk, \mathsf{aux})$ to S.
  - If S returns allow, $\mathcal{F}_{ns}$ checks whether $R(pk, \mathsf{aux}) = 1$ and $(id, pk) \in X$. If both conditions hold, $\mathcal{F}_{ns}$ computes $X \leftarrow X \setminus (id, pk)$ and returns success to C via public delayed output, otherwise, it returns fail to C via public delayed output.
  - If S returns fail, then $\mathcal{F}_{ns}$ returns fail to C via public delayed output.
  - If S returns $(sid, \mathsf{Revoke}, id', pk', \mathsf{aux}', C)$, $\mathcal{F}_{ns}$ checks if $C \in C_{cor}$, $R(pk', \mathsf{aux}') = 1$ and $(id', pk') \in X$. If so, $\mathcal{F}_{ns}$ computes $X \leftarrow X \setminus (id', pk')$, and returns success to C via public delayed output, otherwise, it returns fail to C via public delayed output.
- On input $(sid, \mathsf{Retrieve}, id)$ by C, if flag = manage, $\mathcal{F}_{ns}$ forwards this message to S. If S returns allow, then, if there is a pair $(id, pk) \in X$, for some pk, $\mathcal{F}_{ns}$ returns pk to C via public delayed output, otherwise, it returns $\perp$. If S returns fail to $\mathcal{F}_{ns}$, then $\mathcal{F}_{ns}$ returns fail to C via public delayed output.
- On input $(sid, \mathsf{VerifyID}, id)$ by a client C, if flag = manage, $\mathcal{F}_{ns}$ forwards this message to S. If S returns allow, then, if there is a pair $(id, pk) \in X$, for some pk, $\mathcal{F}_{ns}$ returns 1 to C via public delayed output, otherwise, it returns 0. If S returns fail to $\mathcal{F}_{ns}$, then $\mathcal{F}_{ns}$ returns fail to C via public delayed output.
- On input $(sid, \mathsf{VerifyMapping}, id, pk)$ by a client C, if flag = manage, $\mathcal{F}_{ns}$ forwards this message to S. If S returns allow, then, if there is a pair $(id, pk) \in X$, $\mathcal{F}_{ns}$ returns 1 to C via public delayed output, otherwise, it returns 0. If S returns fail to $\mathcal{F}_{ns}$, then $\mathcal{F}_{ns}$ returns fail to C via public delayed output.

Fig. 3. The naming service functionality $\mathcal{F}_{ns}$ interacts with a set of n clients, a set of m servers, a trusted party T and the simulator S. It allows clients to register, revoke, retrieve and verify (id, pk) mappings.

assuming her public key satisfies relation R. Furthermore, she is able to retrieve the public key of a registered identity and check, whether an identity, or, an (identity, public-key) pair, is registered or not. Our model considers only static corruptions, thus, we assume that the simulator specifies the set of corrupted clients, C_{cor}, before setup.

5 Naming Service Implementation

At a high-level, the service must allow the storage, retrieval and deletion of (identity, public-key) pairs. The main barrier in realizing a smart contract-based DPKI is the size of its state, which, being its most expensive resource to access, must be minimized. We resolve this issue in a twofold manner. First, we separate storage from the process of verifying the validity of mappings by maintaining two public-state, universal accumulators (as presented in Sect. 3) as our smart contract's state. Second, our accumulators are RSA-based and have constant size. The public-state property is required as the contract's state is publicly accessible. To circumvent the issue that deletions require access to an accumulator's private key, we employ the methodology that we presented at the end of Sect. 3.

$\mathcal{F}_{TP}$:

- On input (sid, Init) by a server S_i, $\mathcal{F}_{TP}$ sets $S_{init} \leftarrow S_{init} \cup \{S_i\}$ (initialized as $S_{init} \leftarrow \emptyset$). If all servers have sent a message (sid, Init), $\mathcal{F}_{TP}$ sends $(sid, \mathsf{ServersInitialized})$ to $\mathcal{A}$ and if $\mathcal{A}$ returns allow, $\mathcal{F}_{TP}$ sets flag = ready.
- On input $(sid, \mathsf{Install}, P)$ by T, if flag = ready, send $(sid, \mathsf{Install}, P)$ to $\mathcal{A}$. If $\mathcal{A}$ returns allow, set flag = start, store P, set $state \leftarrow \varepsilon$, where ε is the empty string, and return success to T.
- On input (sid, x) by a client C, if flag = start, send (sid, x) to $\mathcal{A}$. If $\mathcal{A}$ returns allow, run P on input $(x, state)$ and output $(y, state')$. Set $state \leftarrow state'$ and return $(y, state')$ to C via public delayed output. If x is an invalid input for program P, return $\perp$ to C via public delayed output.

Fig. 4. The functionality $\mathcal{F}_{TP}$ captures the role of the smart contract. It interacts with a trusted party T, a set of n clients, a set of t servers and the adversary $\mathcal{A}$.

In Fig. 4, we define the functionality $\mathcal{F}_{TP}$, which captures the role of the smart contract in our protocol. This functionality interacts with a party T, a set of n clients and a set of m servers, some of which may be corrupted by the adversary prior to the initialization phase. $\mathcal{F}_{TP}$ is initialized by a trusted party T by receiving as input a program P. The state of $\mathcal{F}_{TP}$ is updated after a call to the program P and the output is received by the calling party. Note that the implementation of $\mathcal{F}_{TP}$ requires an honest majority of servers, along the lines of [11,22,23]. The adversary has always full knowledge of all the computations

1. On input (Setup, *params*), where *params* is of the form (pk_1, pk_2, R), run InitAcc on input pk_1 and on input pk_2 and compute $c_{0,1}, c_{0,2}$ respectively. This procedure initializes two public-state, additive, universal accumulators c_1 and c_2 by setting $c_1 \leftarrow c_{0,1}, c_2 \leftarrow c_{0,2}$. It sets $state \leftarrow (params, c_1, c_2)$ and returns $state$.

2. On input (Register, id, pk, i, W_1, W_2),
   (a) If $i = 1$, check if VerifyNonMem$(pk_2, (id, 1, a), W_1, c_2) = 1$.
   (b) If $i \geq 2$, check if VerifyNonMem$(pk_2, (id, i, a), W_1, c_2) = 1$ and VerifyMem$(pk_2, (id, i-1, d), W_2, c_2) = 1$.

   If all the above checks succeed, run $(c_1', W_1') \leftarrow$ Add$(pk_1, (id, pk, i, a), c_1)$ and $(c_2', W_2') \leftarrow$ Add$(pk_2, (id, i, a), c_2)$. Update $state$ by setting $c_1 \leftarrow c_1'$ and $c_2 \leftarrow c_2'$, and return $((c_1', W_1'), (c_2', W_2'))$. Otherwise, return fail.

3. On input (Revoke, $id, pk, i, W_1, W_2, W_3, $aux),
   (a) Check if $R(pk, aux) = 1$.
   (b) Check if VerifyMem$(pk_2, (id, i, a), W_1, c_2) = 1$, VerifyNonMem$(pk_2, (id, i, d), W_2, c_2) = 1$ and VerifyMem$(pk_1, (id, pk, i, a), W_3, c_1) = 1$.

   If none of the above verifications fail, run $(c_2', W_2') \leftarrow$ Add$(pk_2, (id, i, d), c_2)$ and $(c_1', W_1') \leftarrow$ Add$(pk_1, (id, pk, i, d), c_1)$. Update $state$ by setting $c_1 \leftarrow c_1'$ and $c_2 \leftarrow c_2'$ and return $((c_1', W_1'), (c_2', W_2'))$. Otherwise, return fail.

4. On input RetrieveState, return $state \leftarrow (params, c_1, c_2)$.

Fig. 5. The program P which is input to $\mathcal{F}_{TP}$, during initialization, in our construction.

performed, and may still interfere by either aborting, or, allowing, an execution of P at will, however, it is restricted from modifying the output. Implementing $\mathcal{F}_{TP}$ using a blockchain protocol has the servers acting as "miners" and T and the clients interacting with the blockchain by posting transactions. Installing a program P is a special transaction that includes P in the blockchain, and subsequently executing P requires running P by all miners and recording the state update in the blockchain as well. The security properties of the underlying blockchain, specifically related to persistence of transactions, cf. [11,22,23], would imply the security of $\mathcal{F}_{TP}$'s realization.

Furthermore, we assume that all operations are completed in a synchronous, atomic fashion. In practice, some time is required for an operation (transaction) to be validated, i.e., to be recorded in the blockchain. Nevertheless, blockchains enforce a total ordering of transactions and execute them serially, which has the same net result. In our protocol, $\mathcal{F}_{TP}$ is input the program P of Fig. 5, thus, $\mathcal{F}_{TP}$ essentially maintains the aforementioned accumulators as its state, i.e., it acts as the accumulator manager T_{acc}. To simplify the description and security analysis of our design, we assume a trusted setup phase that establishes the relation R and generates the accumulators' keys. This assumption does not introduce a single point of failure in our design as it can be replaced, in a practical implementation, with distributed protocols for generating parameters (e.g., [36]).

$\mathcal{F}_{UDB}$:

- On input (sid, Init) by a server S_i', $\mathcal{F}_{UDB}$ sets $S_{init}' \leftarrow S_{init}' \cup \{S_i'\}$ (initialized as $S_{init}' \leftarrow \emptyset$). If all servers have sent a message (sid, Init), $\mathcal{F}_{UDB}$ sends $(sid, \mathsf{ServersInitialized})$ to $\mathcal{A}$. If $\mathcal{A}$ returns allow, then $\mathcal{F}_{UDB}$ sets flag = ready, $DBstate \leftarrow \emptyset$ and $p \leftarrow 0$.
- On input (sid, Post, x) by a client C, if flag = ready, forward $(sid, \mathsf{Post}, x, C)$ to $\mathcal{A}$. If $\mathcal{A}$ sends allow, set $p \leftarrow p + 1$, $DBstate[p] \leftarrow x$ and return success to C.
- On input $(sid, \mathsf{RetrieveDB})$ by a client C, if flag = ready, forward $(sid, \mathsf{RetrieveDB}, C)$ to $\mathcal{A}$. If $\mathcal{A}$ returns allow, return $DBstate$ to C.
- On input $(sid, \mathsf{ChangeDBstate}, DBstate')$ by $\mathcal{A}$, set $DBstate \leftarrow DBstate'$.

Fig. 6. The functionality $\mathcal{F}_{UDB}$ models an unreliable database and interacts with a set of n clients, a set of ℓ servers and the adversary $\mathcal{A}$.

In Fig. 6, we introduce the functionality $\mathcal{F}_{UDB}$, which handles the storage of information that are relevant to our protocol, e.g., (identity, public-key) pairs. $\mathcal{F}_{UDB}$ interacts with n clients and a set of ℓ servers and the adversary. This functionality models an "unreliable database", i.e., the adversary may tamper with its contents. Its involvement in our protocol is twofold. First, a client queries this functionality to retrieve all the necessary information that will allow her to, subsequently, interact with $\mathcal{F}_{TP}$. Second, following the completion of an interaction with $\mathcal{F}_{TP}$, the client stores in $\mathcal{F}_{UDB}$, among others, information that were output by the smart contract and reflect the new state of the system. We elaborate more on the information that clients query/store from/to $\mathcal{F}_{UDB}$ later on in this section where we provide a high-level description of each operation. A practical realization of $\mathcal{F}_{UDB}$ is out of the scope of this paper. However, an authenticated DHT network comprised of nodes that have registered in our PKI would be a suitable candidate, both in terms of security (i.e., it is Sybil resilient), as well as efficiency, due to its logarithmic message complexity.

In our scheme, we accumulate (id, pk, i, op) tuples in c_1, where, $\mathsf{op} = a$ or $\mathsf{op} = d$ mark an element as "added" or "deleted", respectively. This allows clients to infer if an (identity, public-key) mapping is valid. In c_2, we accumulate (id, i, op) tuples, which allows clients to infer if an identity is registered in the system. In the following, and due to space limitations, we present a high-level description of the Register, Revoke, Retrieve, VerifyID, and VerifyMapping operations.

Informally, a client that is interested in registering an (identity, public-key) mapping must prove to the smart contract that the identity is, currently, available. To achieve this, she generates two witnesses. First, a membership witness of the tuple (id, i, d) for c_2, which proves that the i-th instance of this identity has been marked as deleted. Second, a non-membership witness of the tuple $(id, i+1, a)$ for c_2, which proves that the $(i+1)$-th instance of this identity, i.e., the one she is interested in registering, has not been marked as added. If both of the aforementioned conditions hold, she can convince the smart contract to accumulate her mapping in c_1. These witnesses are constructed by, first, querying $\mathcal{F}_{UDB}$ for the history of operations and, second, locating records regarding id, in an attempt to find the proper value for index i. Following a successful registration, the client posts a (Register, id, pk, i, W_1, W_2, W_3) record to $\mathcal{F}_{UDB}$. The witnesses W_1, W_2, W_3 facilitate queries from future clients regarding, e.g., the validity of the $(i+1)$-th instance of her (identity, public-key) mapping.

To revoke an (identity, public-key) mapping, a client generates the following proofs. First, a proof of ownership of the corresponding secret key, which is captured by the relation R. Second, a membership witness of the tuple (id, i, a) for c_2, which proves that this identity has been marked as added for index i. Third, a non-membership witness of the tuple (id, i, d) for c_2, which proves that this identity has not been marked as deleted for index i. Fourth, a membership witness of the tuple (id, pk, i, a) for c_1, which proves that the identity is indeed mapped to the same public-key that satisfies the relation R. Assuming that witnesses are generated honestly, the client convinces the contract to revoke her mapping and, then, she proceeds on posting (Revoke, id, pk, i) to $\mathcal{F}_{UDB}$.

To retrieve an identity's public-key, the client queries $\mathcal{F}_{UDB}$ to check whether the last record related to this identity is a registration record. If so, the client updates the witnesses W_1, W_2, W_3 stored in the retrieved registration record and, subsequently, invokes the smart contract to validate the (identity, public-key) mapping. VerifyID and VerifyMapping follow the same procedure as Retrieve to verify if an identity, or, an (identity, public-key) mapping, is registered.

In Fig. 7, we present the formal description of protocol π, which realizes the functionality $\mathcal{F}_{ns}$. Recall that the entities that participate in the protocol are: (1) a trusted party T, which is used for setup purposes, (2) a functionality $\mathcal{F}_{TP}$, (3) n clients $C_1, \ldots, C_n$ and, (4) a functionality $\mathcal{F}_{UDB}$. We denote with X_1 and X_2 the sets of accumulated elements of c_1 and c_2, respectively. These sets are constructed by the client as follows. For any record of the form (Register, $id, pk, i, \cdot$), a client adds (id, pk, i, a) to X_1 and (id, i, a) to X_2. For any record of the form (Revoke, id, pk, i), a client adds (id, pk, i, d) to X_1 and (id, i, d) to X_2.

For ease of presentation, we have described our protocol using two accumulators. We can achieve the same net result using only one accumulator since both c_1 and c_2 accumulate arbitrary strings. Thus, we are able to accumulate both types of tuples, i.e., (id, i, op) and (id, pk, i, op), in one accumulator, while still being able to generate the (non) membership witnesses required in our protocol. To achieve this, we modify the Register and Revoke operations of program P as follows. First, the second call to Add, in either operation, receives as parameter the accumulator value that is returned from the first call to Add. Second, we invoke UpdMemWit after the second call to Add, to update the membership witness that was returned by the first call to Add. This approach, cuts down in half the contract's state, but, increases the computation of both Register and Revoke by one exponentiation and one invocation of the Map procedure.

1. On input (sid, Setup, R), the party T runs $\mathsf{KeyGen}(1^\lambda)$ twice, sets $params = (pk_1, pk_2, R)$ and sends $(sid, \mathsf{Install}, P)$ to $\mathcal{F}_{TP}$, where P is the program of Fig. 5. Then, T sends $(sid, (\mathsf{Setup}, params))$ to $\mathcal{F}_{TP}$, which executes the program P on input $(\mathsf{Setup}, params)$ (if $\mathcal{A}$ returns allow to $\mathcal{F}_{TP}$). Therefore, if $\mathcal{F}_{TP}$ returns $state \leftarrow (params, c_1, c_2)$ to T, T returns success.

2. On input $(sid, \mathsf{Register}, id, pk)$, C sends $(sid, \mathsf{RetrieveDB})$ to $\mathcal{F}_{UDB}$. Upon receiving $DBstate$, C checks for records regarding id.

   (a) If there is at least one record, it finds the last record regarding id and checks if it is of the form $(\mathsf{Revoke}, id, pk, i)$. If it is not, C returns fail. If it is, she computes a non-membership witness W_1 for $(id, i+1, a)$ and a membership witness W_2 for (id, i, d) by running $\mathsf{NonMemWitGen}(pk_2, (id, i + 1, a), X_2, c_2)$ and $\mathsf{MemWitGen}(pk_2, (id, i, d), X_2, c_2)$ respectively.

   (b) If no record is found, C computes a non-membership witness W_1 for $(id, 1, a)$ by running $\mathsf{NonMemWitGen}(pk_2, (id, 1, a), X_2, c_2)$, sets $W_2 = \bot$ and $i = 0$.

   Then, C sends $(sid, \mathsf{Register}, id, pk, i + 1, W_1, W_2)$ to $\mathcal{F}_{TP}$, which runs P on this input. If $\mathcal{F}_{TP}$ returns $((c'_1, W'_1), (c'_2, W'_2), state)$, where W'_1 is a membership witness for $(id, pk, i + 1, a)$ in c'_1 and W'_2 is a membership witness for $(id, i + 1, a)$ in c'_2, C computes a non-membership witness W'_3 for $(id, i + 1, d)$ by running $\mathsf{NonMemWitGen}((id, i + 1, d), X_2, c'_2)$ and sends $(sid, \mathsf{Post}, (\mathsf{Register}, id, pk, i + 1, W'_1, W'_2, W'_3))$ to $\mathcal{F}_{UDB}$. If $\mathcal{F}_{UDB}$ returns success, C outputs success. Otherwise, C outputs fail.

3. On input $(sid, \mathsf{Revoke}, id, pk, \mathsf{aux})$, C sends $(sid, \mathsf{RetrieveDB})$ to $\mathcal{F}_{UDB}$. Upon receiving $DBstate$, C searches for records that precede her registration record. Assuming ℓ such records, and depending if an encountered record is of the form $\mathsf{Register}$, or, Revoke, C, on each iteration, updates her witnesses as follows:

   (a) C encounters a $(\mathsf{Register}, id', pk', j, W_1^{id'}, W_2^{id'}, W_3^{id'})$. She updates W'_1 by running $W'_1 \leftarrow \mathsf{UpdMemWit}(pk_1, (id, pk, i, a), (id', pk', j, a), W'_1)$. W'_2, W'_3 are updated accordingly.

   (b) C encounters a $(\mathsf{Revoke}, id', pk', j)$. She updates W'_1 by running $W'_1 \leftarrow \mathsf{UpdMemWit}(pk_1, (id, pk, i, a), (id', pk', j, d), W'_1)$. W'_2, W'_3 are updated accordingly.

   Then, C sends $(sid, \mathsf{Revoke}, id, pk, i, W'_1, W'_2, W'_3, \mathsf{aux})$ to $\mathcal{F}_{TP}$. If $\mathcal{F}_{TP}$ returns $((c'_1, W'_1), (c'_2, W'_2), state)$, C sends $(sid, \mathsf{Post}, (\mathsf{Revoke}, id, pk, i))$ to $\mathcal{F}_{UDB}$. If $\mathcal{F}_{UDB}$ returns success, C outputs success. Otherwise, C outputs fail.

4. On input $(sid, \mathsf{Retrieve}, id)$, C sends $(sid, \mathsf{RetrieveDB})$ to $\mathcal{F}_{UDB}$. If there is no record related to id, C outputs fail, otherwise, C:

   (a) Checks if the last record related to id is of the form $\mathsf{Register}$. If so, she retrieves W'_1, W'_2, W'_3 from the record and runs Steps 3a and 3b to compute the updated witnesses W'_1, W'_2, W'_3. Otherwise, C outputs fail.

   (b) Sends $(sid, \mathsf{RetrieveState})$ to $\mathcal{F}_{TP}$. If $\mathcal{F}_{TP}$ returns $state$ then C runs $\mathsf{VerifyMem}(pk_1, (id, pk, i, a), W'_1, c_1)$, $\mathsf{VerifyMem}(pk_2, (id, i, a), W'_2, c_2)$ and $\mathsf{VerifyNonMem}(pk_2, (id, i, d), W'_3, c_2)$. If all algorithms output 1, C outputs pk as the retrieved public key, otherwise, C outputs $\bot$.

5. On input $(sid, \mathsf{VerifyID}, id)$, C runs Step 4. If Step 4 outputs some pk, C outputs 1, otherwise, C outputs 0.

6. On input $(sid, \mathsf{VerifyMapping}, id, pk)$, C runs Step 4. If Step 4 outputs pk, C outputs 1, otherwise, C outputs 0.

Fig. 7. Description of the protocol π built upon the program P of Fig. 5.

Lastly, we show that our construction is secure by proving Theorem 2. Due to space limitations, we provide a part of the proof for this theorem in Appendix A.

Theorem 2. *The protocol π of Fig. 7 securely realizes the functionality $\mathcal{F}_{ns}$ of Fig. 3 in the $(\mathcal{F}_{TP}, \mathcal{F}_{UDB})$-hybrid world under the strong RSA assumption in the Random Oracle model.*

A   Proof of Theorem 2

We construct a simulator $\mathcal{S}$ (Fig. 8) which emulates an execution of the protocol π in the hybrid world, in the presence of an adversary $\mathcal{A}$. $\mathcal{S}$ plays the role of T, $\mathcal{F}_{TP}$, $\mathcal{F}_{UDB}$ and acts on behalf of a number of honest clients and on behalf of servers in the emulation of π. We show that an environment can distinguish between the executions in the hybrid and the ideal world only by influencing the way the membership or non-membership tests take place in the hybrid world, i.e., it is reduced to the security of the accumulators utilized in protocol π. Note that relation $R(pk, \mathsf{aux})$ does not provide an opportunity for distinguishing since it is the same in both worlds. Furthermore, our proof would also work if the functionalities $\mathcal{F}_{TP}$ and $\mathcal{F}_{UDB}$ were global, however, for simplicity, we have the simulator $\mathcal{S}$ playing their role in the simulation. Assuming that all servers have sent a (sid, Init) message, we consider the case where an environment $\mathcal{Z}$ sends a $(sid, \mathsf{Register}, id, pk)$ message to a client C. We examine the output of C in the hybrid and the ideal world, by examining the following cases:

1. $\mathcal{A}$ has not sent $(sid, \mathsf{ChangeDBstate}, DBstate')$ until $(sid, \mathsf{Register}, id, pk)$ is sent to the client C. We consider two different sub-cases:
   (i) **The identity id is *not* registered:** In the hybrid world, an *honest* client C sends $(sid, \mathsf{RetrieveDB})$ to $\mathcal{F}_{UDB}$ and, if $\mathcal{A}$ returns allow, then C receives $DBstate$ and computes the witnesses W_1, W_2 according to Steps 2a, 2b of Fig. 7. Then, C sends $(sid, (\mathsf{Register}, id, pk, i+1, W_1, W_2))$ to $\mathcal{F}_{TP}$. If $\mathcal{A}$ returns allow to $\mathcal{F}_{TP}$, $\mathcal{F}_{TP}$ returns $((c_1', W_1'), (c_2', W_2'), state)$ and C computes the witness W_3'. Then, C sends $(sid, \mathsf{Post}, (\mathsf{Register}, id, pk, i+1, W_1', W_2', W_3'))$ to $\mathcal{F}_{UDB}$ and, if $\mathcal{A}$ returns allow, C outputs success in the hybrid world. In the ideal world, C also outputs success because $\mathcal{S}$, who acts as C and $\mathcal{F}_{TP}$, returns allow to $\mathcal{F}_{ns}$ (Step 3, Fig. 8). Finally, $\mathcal{F}_{ns}$ verifies that id is not registered and sends success to C.
   (ii) **The identity id is currently registered:** In the hybrid world, an *honest* client C sends $(sid, \mathsf{RetrieveDB})$ to $\mathcal{F}_{UDB}$ and, when C receives $DBstate$, checks that id is registered and returns fail. In the ideal world, $\mathcal{S}$ simulates C and $\mathcal{F}_{UDB}$ and, since id is registered, $\mathcal{S}$ returns fail to $\mathcal{F}_{ns}$. Next, $\mathcal{F}_{ns}$ returns fail to C. A *malicious* client C (on behalf $\mathcal{A}$) in the hybrid world may try to convince $\mathcal{F}_{TP}$ that id is **not** registered. Then, C should either provide a non-membership witness W_1 for (id, j, a), for some $j \geq 2$, such that $\mathsf{VerifyNonMem}(pk_2, (id, j, a), W_1, c_2) = 1$ and a membership witness for W_2 for $(id, j-1, d)$, such that $\mathsf{VerifyMem}(pk_2, (id, j-1, d), W_2, c_2) = 1$,

---

Simulator $\mathcal{S}$:

1. Upon receiving $(sid, \mathsf{ServersInitialized})$ by $\mathcal{F}_{ns}$, $\mathcal{S}$ playing the role of $\mathcal{F}_{TP}$ and $\mathcal{F}_{UDB}$, sends $(sid, \mathsf{ServersInitialized})$ to $\mathcal{A}$. If $\mathcal{A}$ returns allow as a response to both $\mathcal{F}_{TP}$ and $\mathcal{F}_{UDB}$, then $\mathcal{S}$ sends allow to $\mathcal{F}_{ns}$.

2. Upon receiving (sid, Setup, R) by $\mathcal{F}_{ns}$, $\mathcal{S}$, on behalf of the party T in the hybrid world, runs $\mathsf{KeyGen}(1^\lambda)$ twice (according to Step 1 of Fig. 7) and then, playing the role of $\mathcal{F}_{TP}$ in the hybrid-world protocol, sends $(sid, \mathsf{Install}, P)$ to $\mathcal{A}$. If $\mathcal{A}$ returns allow, then $\mathcal{S}$ sends $(sid, \mathsf{Setup}, params)$ to $\mathcal{A}$. If $\mathcal{A}$ returns allow, $\mathcal{S}$ runs P on input $(\mathsf{Setup}, params)$ and sends allow to $\mathcal{F}_{ns}$.

3. Upon receiving $(sid, \mathsf{Register}, id, pk)$ by $\mathcal{F}_{ns}$, $\mathcal{S}$, playing the role of an honest client C and the role of $\mathcal{F}_{UDB}$ in the hybrid-world protocol π, sends $(sid, \mathsf{RetrieveDB}, C)$ to $\mathcal{A}$. If $\mathcal{A}$ returns allow, $\mathcal{S}$ runs Step 2a of Fig. 7. If the last record related to id is a register record, $\mathcal{S}$ sends fail to $\mathcal{F}_{ns}$, otherwise, it proceeds by running Step 2b if necessary. Then $\mathcal{S}$, playing the role of the $\mathcal{F}_{TP}$, sends $(sid, (\mathsf{Register}, id, pk, i + 1, W_1', W_2'))$ to $\mathcal{A}$. If $\mathcal{A}$ returns allow, then $\mathcal{S}$ runs P on input $(\mathsf{Register}, id, pk, i + 1, W_1', W_2')$. If P outputs fail, $\mathcal{S}$ sends fail to $\mathcal{F}_{ns}$, otherwise, P returns $((c_1', W_1'), (c_2', W_2'), state)$ and $\mathcal{S}$ computes a non-membership witness W_3' for $(id, i + 1, d)$. Then, on behalf of $\mathcal{F}_{UDB}$, $\mathcal{S}$ sends $(sid, \mathsf{Post}, (\mathsf{Register}, id, pk, i, W_1', W_2', W_3'))$ to $\mathcal{A}$. If $\mathcal{A}$ returns allow then $\mathcal{S}$ sends allow to $\mathcal{F}_{ns}$ and updates $DBstate$ by storing $(\mathsf{Register}, id, pk, i, W_1', W_2', W_3')$, as $\mathcal{F}_{UDB}$ does in Fig. 6.

4. Upon receiving $(sid, \mathsf{Revoke}, id, pk, \mathsf{aux})$ by $\mathcal{F}_{ns}$, $\mathcal{S}$, playing the role of an honest client C and the role of $\mathcal{F}_{UDB}$, sends $(sid, \mathsf{RetrieveDB}, C)$ to $\mathcal{A}$. If $\mathcal{A}$ returns allow, $\mathcal{S}$ runs Steps 3a, 3b of Fig. 7 and computes the updated witnesses W_1', W_2', W_3'. Then, $\mathcal{S}$, on behalf of $\mathcal{F}_{TP}$, sends $(sid, \mathsf{Revoke}, id, pk, i, W_1', W_2', W_3', \mathsf{aux})$ to $\mathcal{A}$. If $\mathcal{A}$ returns allow, $\mathcal{S}$ runs P on input $(\mathsf{Revoke}, id, pk, i, W_1', W_2', W_3', \mathsf{aux})$ and if it outputs $(fail, state)$ then $\mathcal{S}$ sends fail to $\mathcal{F}_{ns}$. Otherwise, $\mathcal{S}$ returns allow to $\mathcal{F}_{ns}$ and updates $DBstate$ by storing $(\mathsf{Revoke}, id, pk, i)$.

5. Upon receiving $(sid, \mathsf{Retrieve}, id)$ by $\mathcal{F}_{ns}$, $\mathcal{S}$, on behalf of an honest client C and playing the role of $\mathcal{F}_{UDB}$, sends $(sid, \mathsf{RetrieveDB})$ to $\mathcal{A}$. If $\mathcal{A}$ returns allow, $\mathcal{S}$ runs Step 4a of Fig. 7. If Step 4a returns fail, then, $\mathcal{S}$ returns fail to $\mathcal{F}_{ns}$, otherwise $\mathcal{S}$ runs Step 4b and simulating $\mathcal{F}_{TP}$, sends $(sid, \mathsf{RetrieveState})$ to $\mathcal{A}$. If $\mathcal{A}$ returns allow and all the algorithms at Step 4b return 1, then, $\mathcal{S}$ sends allow to $\mathcal{F}_{ns}$, otherwise it sends fail.

6. Upon receiving $(sid, \mathsf{VerifyID}, id)$ or $(sid, \mathsf{VerifyMapping}, id, pk)$ by $\mathcal{F}_{ns}$, $\mathcal{S}$ runs the simulation similarly to Step 5.

7. Upon receiving $(sid, \mathsf{ChangeDBstate}, DBstate')$ by $\mathcal{A}$, $\mathcal{S}$ sets $DBstate \leftarrow DBstate'$.

8. Upon receiving $(sid, \mathsf{Register}, id, pk)$ or $(sid, \mathsf{Revoke}, id, pk)$ by $\mathcal{F}_{ns}$ for a corrupted client C, $\mathcal{S}$ waits for the actions of $\mathcal{A}$, and simulates $\mathcal{F}_{TP}$ and $\mathcal{F}_{UDB}$ as in previous cases. If $\mathcal{S}$ receives $(\mathsf{Register}, id', pk', i', W_1', W_2')$ by $\mathcal{A}$, checks if $id \neq id'$ or/and $pk \neq pk'$. If the program P does not return $(fail, state)$ on this input then $\mathcal{S}$ sends $(\mathsf{Register}, id', pk', C)$ to $\mathcal{F}_{ns}$. $\mathcal{S}$ runs similarly when it receives $(\mathsf{Revoke}, id', pk', i', W_1', W_2', W_3')$.

Fig. 8. The simulator $\mathcal{S}$ that emulates the execution of protocol π in the hybrid world.

or, C should provide a valid non-membership witness W_1 for $(id, 1, a)$. We show that, by the security of the c_2, such an attack takes place only with negligible probability. Recall that since id is currently registered, it holds that either: (1) there is $\ell \geq 2$ such that, $(id, \ell, a) \in X_2$ and $(id, \ell, d) \notin X_2$, where X_2 is the accumulated set of c_2, or, (2) $(id, 1, a) \in X_2$. Starting with (1), we consider the cases where $1 < j \leq \ell$, and $j > \ell$. If $1 < j \leq \ell$, then $(id, j, a) \in X_2$ and $(id, j - 1, d) \in X_2$. By the security of the c_2, C can produce a valid non-membership witness W_1 for (id, j, a) only with negligible probability. If $j > \ell$, then $(id, j, a) \notin X_2$ and $(id, j - 1, d) \notin X_2$. By the security of the accumulator c_2, C cannot produce a valid membership witness W_2 for $(id, j - 1, d)$. Regarding (2), similarly, C can produce a valid non-membership witness W_1 for $(id, 1, a)$ only with negligible probability. Therefore, C returns fail in the hybrid world. In the ideal world, C would also return fail, because S sends fail to $\mathcal{F}_{ns}$, which sends fail to C.

2. $\mathcal{A}$ has sent $(sid, \mathsf{ChangeDBstate}, DBstate')$ until $\mathcal{Z}$ sends $(sid, \mathsf{Register}, id, pk)$, such that $DBstate' \neq DBstate$ and the set X_2' derived by $DBstate'$ is different from the set X_2 accumulated in c_2[1]. We consider the following sub-cases:

   (i) **The identity id is not registered but the last record including id in $DBstate'$ is of the form** $(\mathsf{Register}, id, pk, j, W_1, W_2, W_3)$: In the hybrid world, an *honest* C sends $(sid, \mathsf{RetrieveDB})$ to $\mathcal{F}_{UDB}$ and when C receives $DBstate'$, checks that id is registered and returns fail. In the ideal world, S, simulating C and $\mathcal{F}_{UDB}$, sends fail to $\mathcal{F}_{ns}$, which returns fail to C. If a *malicious* C sends $(sid, \mathsf{Register}, id', pk', j', W_1', W_2')$ to $\mathcal{F}_{TP}$ and $\mathcal{F}_{TP}$ returns $(\mathsf{fail}, state)$, then in the ideal world, S sends fail to $\mathcal{F}_{ns}$. Even if $\mathcal{F}_{TP}$ returns $((c_1', W_1'), (c_2', W_2'), state)$ and $\mathcal{F}_{UDB}$ returns success to C after receiving a message of the form $(sid, \mathsf{Post}, \cdot)$, this means that in the ideal world, S returns allow to $\mathcal{F}_{ns}$. Then, $\mathcal{F}_{ns}$, verifying that id is not registered, sends success to C. In both cases, C returns consistent outputs in the hybrid and ideal world.

   (ii) **The identity id is not registered and the last record including id in $DBstate'$ is of the form** $(\mathsf{Revoke}, id, pk, j)$, **or there is no record for id** : The analysis for this case is similar to the same with 2(i) except that an honest client interacts with $\mathcal{F}_{TP}$ after receiving $DBstate'$ from $\mathcal{F}_{UDB}$.

   (iii) **The identity id is registered and the last record including id in $DBstate'$ is of the form** $(\mathsf{Revoke}, id, pk, j)$, **or there is no record for id**: In the hybrid world, an *honest* C sends $(sid, \mathsf{RetrieveDB})$ to $\mathcal{F}_{UDB}$ and when C receives $DBstate'$, computes the witnesses W_1, W_2 according to Steps 2a, 2b and sends $(sid, \mathsf{Register}, id, pk', j + 1, W_1, W_2)$ to $\mathcal{F}_{TP}$. Following the same reasoning with the case 1(ii), $\mathcal{F}_{TP}$ returns $(\mathsf{fail}, state)$ except with negligible probability. If an honest C, given the accumulated

---

[1] As we explained in Sect. 5, a set X_2' is derived by $DBstate'$ in the following way: For any record of the form $(\mathsf{Register}, id, pk, i, W_1, W_2, W_3)$, (id, i, a) is added to X_2 and for any record of the form $(\mathsf{Revoke}, id, pk, i)$, (id, i, d) is added to X_2.

set X_2' could produce a valid non-membership witness W_1 for $(id, j + 1, a)$ and a valid membership witness W_2 for (id, j, d), with non-negligible probability then the security of c_2 would break. As a result, C returns fail, both in the hybrid and ideal world. A *malicious* C, in the hybrid world may send $(sid, \mathsf{Register}, id, pk^*, \ell, W_1^*, W_2^*)$ to convince $\mathcal{F}_{TP}$ that id is not registered. The analysis for this case is also the same as 1(ii).

(iv) **The identity id is registered but the last record including id in** $DBstate'$ **is of the form** $(\mathsf{Register}, id, pk, i+1, W_1, W_2, W_3)$**:** The analysis for this case is the same with 2(iii) except that an honest client returns fail after receiving $DBstate'$ from $\mathcal{F}_{UDB}$.

3. $\mathcal{A}$ has sent $(sid, \mathsf{ChangeDBstate}, DBstate')$ until $(\mathsf{Register}, id, pk)$ sent by $\mathcal{Z}$, such that $DBstate' \neq DBstate$ but the set X_2' derived by $DBstate'$ is the same as the set X_2 accumulated in c_2. In this case, our reasoning is similar to case 1, where the adversary has not sent such a message, since an honest client is able to compute the witnesses W_1, W_2 utilizing a correct accumulated set.

We proved that when an environment $\mathcal{Z}$ sends a message $(\mathsf{Register}, id, pk)$ to a client C, $\mathcal{Z}$ can distinguish between the executions in the hybrid and ideal world only with negligible probability, relying on the security of the accumulator c_2. We argue that, following similar arguments, the same holds for the cases where $\mathcal{Z}$ sends $(\mathsf{Revoke}, id, pk, \mathsf{aux})$, $(\mathsf{Retrieve}, id)$, $(\mathsf{VerifyID}, id)$, $(\mathsf{VerifyMapping}, id, pk)$. However, due to lack of space, a complete proof, including the above cases, will be provided in the full version of the paper. □

References

1. ASCAP, PRS and SACEM join forces for blockchain copyright system. https://tinyurl.com/y7aruwlw. Accessed 06 July 2017
2. Emercoin - distributed blockchain services for business and personal use. http://www.emercoin.com. Accessed 30 Sept 2010
3. Final report on diginotar hack shows total compromise of ca servers. https://tinyurl.com/hnmuahc. Accessed 07 Apr 2017
4. Google takes symantec to the woodshed for mis-issuing 30,000 HTTPS certs. https://tinyurl.com/kwkvfur. Accessed 07 Apr 2017
5. IBM pushes blockchain into the supply chain. https://tinyurl.com/yazgt9pk. Accessed 06 July 2017
6. Namecoin. https://namecoin.org/. Accessed 07 Apr 2017
7. Swiss industry consortium to use Ethereum's blockchain. https://tinyurl.com/zlbfmnt. Accessed 06 July 2017
8. Trustwave admits it issued a certificate to allow company to run man-in-the-middle attacks. https://tinyurl.com/ycfv6kfs. Accessed 07 Apr 2017
9. Aberer, K.: P-grid: a self-organizing access structure for P2P information systems. In: Batini, C., Giunchiglia, F., Giorgini, P., Mecella, M. (eds.) CoopIS 2001. LNCS, vol. 2172, pp. 179–194. Springer, Heidelberg (2001). https://doi.org/10.1007/3-540-44751-2_15

10. Avramidis, A., Kotzanikolaou, P., Douligeris, C., Burmester, M.: Chord-PKI: a distributed trust infrastructure based on P2P networks. Comput. Netw. **56**, 378–398 (2012)

11. Badertscher, C., Maurer, U., Tschudi, D., Zikas, V.: Bitcoin as a transaction ledger: a composable treatment. In: Katz, J., Shacham, H. (eds.) CRYPTO 2017. LNCS, vol. 10401, pp. 324–356. Springer, Cham (2017). https://doi.org/10.1007/978-3-319-63688-7_11

12. Baldimtsi, F., et al.: Accumulators with applications to anonymity-preserving revocation. In: EuroS&P (2017)

13. Barić, N., Pfitzmann, B.: Collision-free accumulators and fail-stop signature schemes without trees. In: Fumy, W. (ed.) EUROCRYPT 1997. LNCS, vol. 1233, pp. 480–494. Springer, Heidelberg (1997). https://doi.org/10.1007/3-540-69053-0_33

14. Benaloh, J., de Mare, M.: One-way accumulators: a decentralized alternative to digital signatures. In: Helleseth, T. (ed.) EUROCRYPT 1993. LNCS, vol. 765, pp. 274–285. Springer, Heidelberg (1994). https://doi.org/10.1007/3-540-48285-7_24

15. Camenisch, J., Lysyanskaya, A.: Dynamic accumulators and application to efficient revocation of anonymous credentials. In: Yung, M. (ed.) CRYPTO 2002. LNCS, vol. 2442, pp. 61–76. Springer, Heidelberg (2002). https://doi.org/10.1007/3-540-45708-9_5

16. Canetti, R.: Universally composable security: a new paradigm for cryptographic protocols. IACR Cryptology ePrint Archive 2000:67 (2000)

17. Carter, L., Wegman, M.N.: Universal classes of hash functions. J. Comput. Syst. Sci. **18**(2), 143–154 (1979)

18. Datta, A., Hauswirth, M., Aberer, K.: Beyond "web of trust": enabling P2P e-commerce. In: CEC 2003, pp. 303–312 (2003)

19. Douceur, J.R.: The sybil attack. In: Druschel, P., Kaashoek, F., Rowstron, A. (eds.) IPTPS 2002. LNCS, vol. 2429, pp. 251–260. Springer, Heidelberg (2002). https://doi.org/10.1007/3-540-45748-8_24

20. Ellison, C., Schneier, B.: Ten risks of PKI: what you're not being told about public key infrastructure (2000)

21. Fromknecht, C., Velicanu, D., Yakoubov, S.: A decentralized public key infrastructure with identity retention. IACR (2014)

22. Garay, J., Kiayias, A., Leonardos, N.: The bitcoin backbone protocol: analysis and applications. In: Oswald, E., Fischlin, M. (eds.) EUROCRYPT 2015. LNCS, vol. 9057, pp. 281–310. Springer, Heidelberg (2015). https://doi.org/10.1007/978-3-662-46803-6_10

23. Garay, J., Kiayias, A., Leonardos, N.: The bitcoin backbone protocol with chains of variable diculty. IACR Cryptology ePrint Archive (2016)

24. Gennaro, R., Halevi, S., Rabin, T.: Secure hash-and-sign signatures without the random oracle. In: Stern, J. (ed.) EUROCRYPT 1999. LNCS, vol. 1592, pp. 123–139. Springer, Heidelberg (1999). https://doi.org/10.1007/3-540-48910-X_9

25. Gipp, B., Meuschke, N., Gernandt, A.: Decentralized trusted timestamping using the crypto currency bitcoin. CoRR, abs/1502.04015 (2015)

26. Jhanwar, M.P., Safavi-Naini, R.: Compact accumulator using lattices. In: Chakraborty, R.S., Schwabe, P., Solworth, J. (eds.) SPACE 2015. LNCS, vol. 9354, pp. 347–358. Springer, Cham (2015). https://doi.org/10.1007/978-3-319-24126-5_20

27. Karakaya, M., Korpeoglu, I., Ulusoy, Ö.: Free riding in peer-to-peer networks. IEEE Internet Comput. **13**(2), 92–98 (2009)

28. Lesueur, F., Me, L., Tong, V.V.T.: An efficient distributed PKI for structured P2P networks. In IEEE P2PC (2009)

29. Li, J., Li, N., Xue, R.: Universal accumulators with efficient nonmembership proofs. In: Katz, J., Yung, M. (eds.) ACNS 2007. LNCS, vol. 4521, pp. 253–269. Springer, Heidelberg (2007). https://doi.org/10.1007/978-3-540-72738-5_17

30. Maymounkov, P., Mazières, D.: Kademlia: a peer-to-peer information system based on the XOR metric. In: Druschel, P., Kaashoek, F., Rowstron, A. (eds.) IPTPS 2002. LNCS, vol. 2429, pp. 53–65. Springer, Heidelberg (2002). https://doi.org/10.1007/3-540-45748-8_5

31. Nakamoto, S.: Bitcoin: a peer-to-peer electronic cash system. http://bitcoin.org/bitcoin.pdf. Accessed 07 Apr 2017

32. Nguyen, L.: Accumulators from bilinear pairings and applications. In: Menezes, A. (ed.) CT-RSA 2005. LNCS, vol. 3376, pp. 275–292. Springer, Heidelberg (2005). https://doi.org/10.1007/978-3-540-30574-3_19

33. Nyberg, K.: Fast accumulated hashing. In: Gollmann, D. (ed.) FSE 1996. LNCS, vol. 1039, pp. 83–87. Springer, Heidelberg (1996). https://doi.org/10.1007/3-540-60865-6_45

34. Reiter, M.K.: Franklin, M.K., Lacy, J.B., Wright, R.N.: The ω key management service. In: CCS 1996 (1996)

35. Reyzin, L., Yakoubov, S.: Efficient asynchronous accumulators for distributed PKI. In: Zikas, V., De Prisco, R. (eds.) SCN 2016. LNCS, vol. 9841, pp. 292–309. Springer, Cham (2016). https://doi.org/10.1007/978-3-319-44618-9_16

36. Sander, T.: Efficient accumulators without trapdoor extended abstract. In: Varadharajan, V., Mu, Y. (eds.) ICICS 1999. LNCS, vol. 1726, pp. 252–262. Springer, Heidelberg (1999). https://doi.org/10.1007/978-3-540-47942-0_21

37. Wouhaybi, R.H., Campbell, A.T.: Keypeer: a scalable, resilient distributed public-key system using chord (2008)

38. Yüce, E., Selçuk, A.A.: Server notaries: a complementary approach to the web PKI trust model. IACR Cryptology ePrint Archive 2016:126 (2016)

39. Zhou, L., Schneider, F.B., Van Renesse, R.: COCA: a secure distributed online certification authority. ACM Trans. Comput. Syst. 20, 329–368 (2002)

40. Zimmermann, P.: Pretty good privacy. https://philzimmermann.com

Embedded System Security

Secure Code Updates for Smart Embedded Devices Based on PUFs

Wei Feng[1(✉)], Yu Qin[1(✉)], Shijun Zhao[1], Ziwen Liu[2(✉)], Xiaobo Chu[1], and Dengguo Feng[1]

[1] Trusted Computing and Information Assurance Laboratory, Institute of Software Chinese Academy of Sciences, Beijing, China
vonwaist@gmail.com
[2] School of Software Engineering, South China University of Technology, Guangzhou, China
ziwenliu@scut.edu.cn

Abstract. Code update is a very useful tool commonly used in low-end embedded devices to improve the existing functionalities or patch discovered bugs or vulnerabilities. If the update protocol itself is not secure, it will only bring new threats to embedded systems. Thus, a secure code update mechanism is required. However, existing solutions either rely on strong security assumptions, or result in considerable storage and computation consumption, which are not practical for resource-constrained embedded devices (e.g., in the context of Internet of Things). In this work, we first propose to use intrinsic device characteristics (i.e., Physically Unclonable Functions or PUF) to design a practical and lightweight secure code update scheme. Our scheme can not only ensure the freshness, integrity, confidentiality and authenticity of code update, but also verify that the update is installed correctly on a specific device without any malicious software. Cloned or counterfeit devices can be excluded as the code update is bound to the unpredictable physical properties of underlying hardware. Legitimate devices in an untrustworthy software state can be restored by filling suspect memory with PUF-derived random numbers. After update installation, the initiator of the code update is able to obtain the verifiable software state from device, and the device can maintain a sustainable post-update secure check by enforcing a secure call sequence. To demonstrate the practicality and feasibility, we also implement the proposed scheme on a low-end MCU platform (TI MSP430) by using onboard SRAM and Flash resources.

Keywords: Firmware update · Secure code update
Physically Unclonable Function (PUF) · Remote attestation
Embedded security

1 Introduction

With the rise of new trends like the Internet of Things (IoT), Industry 4.0, or Industrial Internet, smart embedded devices are being increasingly used in various scenarios, such as industrial control, smart home, wireless sensor networks,

© Springer Nature Switzerland AG 2018
S. Capkun and S. S. M. Chow (Eds.): CANS 2017, LNCS 11261, pp. 325–346, 2018.
https://doi.org/10.1007/978-3-030-02641-7_15

etc. Firmware or code update is an important mechanism for these scenarios as it offers many benefits [31,35]: fix bugs or vulnerabilities that have been disclosed in the deployed devices; add new features or functionalities to system; enable or disable product functionality in the field; reduce the number of product returns to be handled. For example, as recently reported, Dyn DNS DDoS attack[1] is caused by a large number of IoT botnet nodes infected with the Mirai malware. Code update mechanisms may be used to repair these embedded nodes without having to recall or destroy these devices. However, if the update process itself is vulnerable, it can be exploited by attackers to compromise the security of embedded systems. As low-end embedded devices are resource-constrained and often lack the security capabilities of general purpose computing platforms, it's difficult and challenging to design a secure code update mechanism for them.

A secure code update scheme for embedded systems should not only consider a protocol for secure downloading, but also ensure that the newly downloaded code is installed properly and its memory can be verified with the confidence that no attacker or malicious code is involved. Ideally, a secure code update should provide the following security attributes [31,35]: (1) **Freshness**, the downloaded code is newest, not a simple replay or downgrading; (2) **Integrity**, the update code installed on device is expected and unmodified; (3) **Authenticity**, the update code comes from an authorized source and is loaded onto an authorized device (cloning can also be detected), i.e., mutual authentication is needed; (4) **Confidentiality**, the code may be an important intellectual property, which should not be revealed to other parties; (5) **Feasibility**, the scheme is applicable to existing commodity low-end embedded devices based on existing resources; (6) **Verifiability**, after update installation, the software state of the updated device should be verified and the verification result should be eventually fed back to the source who issues the update; (7) **Restorability**, secure code update is able to restore the software state of a compromised device; and (8) **Secure Call**, only trustworthy code can be called and executed on the device after the update process is complete, which aims to alleviate TOCTOU attack [14].

Currently, there are few solutions that can satisfy all these attributes. By pointing out the inadequacies of existing techniques (hardware and software-based attestation), Perito and Tsudik [42] introduced a new notion called Proofs of Secure Erasure (PoSE) for secure code update, in which new code is downloaded onto an embedded device after secure erasure of all its prior state. PoSE meets the integrity, feasibility and restorability attributes. However, other security attributes are not supported. Furthermore, PoSE relies on strong security assumptions [42], e.g., the adversary maintains complete communication silence during attestation, and it also results in considerable energy and time overhead. The follow-up researches [16,32] of PoSE all focus on reducing the communication and computation overhead, and rarely consider to improve the assumptions or strengthen security guarantees. Recently, Kohnhauser et al. [35] proposed a novel secure code update scheme for mesh networked embedded devices, which achieves much stronger security guarantees and satisfies most of the secu-

---

[1] https://en.wikipedia.org/wiki/2016_Dyn_cyberattack.

rity attributes. Their method relies on three hardware security requirements: immutable code, secure storage and uninterruptible execution. Nevertheless, their method has the following flaws: (1) it uses the traditional secure storage technology (like EEPROM, BBRAM or eFuse) for device secret or private keys, which is expensive, inflexible and unsafe [39,53]; (2) it uses public-key cryptography, which results in apparent storage consumption (66KB for signature) and increased running time; and (3) it is vulnerable to device cloning attack and TOCTOU attack.

Contributions. In this paper, we propose the first secure code update scheme for current commodity low-end embedded devices by using Physically Unclonable Functions (PUFs). Firstly, our scheme reserves the design of secure erasure from PoSE; however, the prover does not need to download random data as large as its own memory from the verifier. As an improvement, we fill the prover's memory with high entropy data derived from PUF. Additionally, we don't rely on the strong security assumption like communication silence. Secondly, as opposed to the latest method in ESORICS 2016 [35], we use PUF-based secure key generation to replace traditional secure key storage, and use symmetric cryptography and message authentication code (MAC) instead of public-key cryptography to achieve confidentiality and authenticity. Thirdly, we design a secure code update protocol based on reverse fuzzy extractor, which satisfies all the security attributes mentioned above. To illustrate this, we conclude eight possible security threats that may break these security attributes and show how our scheme can be used to address them. Finally, we implement and evaluate our protocol building blocks in a low-cost and general-purpose MSP430 MCU. The evaluation results demonstrate the feasibility and validity of PUF-based secure code update in low-end embedded devices.

Outline. In Sect. 2, we conclude the security threats and present some background knowledge about PoSE and PUF. In Sect. 3, we first give the system requirements and adversary model, and then introduce our new proposal for secure code update by using PUF. In Sect. 4, we implement and evaluate the building blocks of our novel scheme using a MSP430 device. In Sect. 5, we overview the related work, and we conclude the paper in Sect. 6.

2   Background and Preliminaries

2.1   Security Threats of Code Update

Code update involves a verifier V and a prover P. P is a generic embedded device with constrained resources, e.g. a medical instrument, a wearable device or an industrial control device. V is a more powerful computing device, e.g. a smartphone, a laptop or a cloud platform. Secure code update can be viewed as a means to ensure that a code update issued by a trusted V has been securely distributed and correctly installed on P. Specifically, for secure code update, we aim to provide measures to solve the following security threats [21,31,35]:

Threat-1 Code Alteration. The binary code (or firmware image) distributed by V is modified by attackers during the update process.

Threat-2 Code Reverse Engineering. Attackers intercept the binary image code, and use the reverse engineering technique to analyze the functionality and contents of the update image code.

Threat-3 Loading Unauthorized Code. The update binary code may be created by an unauthorized party, and P is cheated to install the unauthorized or malicious code.

Threat-4 Loading Code onto an Unauthorized Device. The code intended for one device is installed on another, or the code generated by the product manufacturer is loaded onto an unauthorized device.

Threat-5 Code Downgrading. An attacker in possession of an old code package may resend it to the device reverting it to a previous, possibly vulnerable, state in order to exploit it.

Threat-6 Incomplete Update. A compromised device may simply deny the execution of code update or execute it inappropriately without restoring software integrity. And at the same time, V is cheated with a response indicating a successful update.

Threat-7 TOCTOU (Time Of Check to Time Of Use). After a complete update, the update code stored in the device may have been tampered with when it's called to run a specific embedded task.

Threat-8 Device Cloning. Attackers may simply copy the memory contents (including code, data, secrets or keys, and other intellectual property) and create a cloned device to replace the original one.

2.2 Proofs of Secure Erasure (PoSE)

While hardware-based attestation [8,41] is not practical for low-cost embedded systems and software-based attestation [47] offers unclear security guarantees, Perito and Tsudik [42] proposed a new technique called *Proofs of Secure Erasure* (PoSE) for low-end embedded devices.

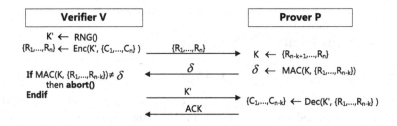

Fig. 1. Secure code update protocol based on PoSE

According to [42], PoSE can be used to implement a secure code update protocol, which we conclude in Fig. 1. Suppose the size of P's memory (all writable storage on the embedded device) is n, the verifier V first encrypts the update code using a random key K'. Upon receiving the ciphertext blocks $\{R_1, ..., R_n\}$, P uses the last k-blocks of randomness as the key to compute a MAC (Message Authentication Code) and sends the MAC to V. V verifies the MAC to ensure that P's memory is reliably erased with the high entropy data (ciphertext) sent by V. If MAC verified correctly, V sends the encryption key K' to P in order for P to decrypt the ciphertext into the new code $\{C_1, ..., C_{n-k}\}$. Both the performance and security of the PoSE mechanism are not optimistic, and a detailed analysis is given in the full version [18]. Our approach attempts to improve performance and security without relying on any hardware modification.

2.3   Physically Unclonable Function and Reverse Fuzzy Extractor

A physically unclonable function (PUF) is an entity that uses manufacturing variation to generate a device-specific output, which can be seen as the *fingerprint* of a device [11]. Specifically [44], when queried with a challenge C_i, a PUF generates a response $R_i = PUF(C_i)$ that depends on both, C_i and the unique IC intrinsic physical properties of the device containing PUF. The tuples (C_i, R_i) are thereby termed the challenge-response pairs (CRPs) of the PUF. PUFs are inherently noisy and their responses are not uniformly random, thus some mechanisms are needed to correct noise and extract randomness from the PUF responses. Depending on the computing power of the prover device, there are two different mechanisms [10,23]: fuzzy extractor (FE) and reverse fuzzy extractor (RFE). In the full version of the paper [18], we provide a detailed description about FE and RFE. In our update protocol, we adopt RFE to extract reconstructible random keys from PUF, in which $FE.Gen$ (the generation procedure of FE) and $FE.Rep$ (the reproduction procedure of FE) are represented as follows:

$$(K, h) \leftarrow FE.Gen(R'):$$
$$\{r \leftarrow RNG(), CW \leftarrow Encode(r), h \leftarrow R' \oplus CW, K \leftarrow Ext(R')\}$$
$$K \leftarrow FE.Rep(R, h):$$
$$\{CW' \leftarrow R \oplus h, r \leftarrow Decode(CW'), CW \leftarrow Encode(r), K \leftarrow Ext(CW \oplus h)\}$$

Encode and *Decode* are two procedures included in the error correction. The random extractor (Ext) is used to obtain a full-entropy key K from PUF response. A random number generator RNG is used to choose a random codeword (CW), and CW only serves for error correction.

3   Secure Code Update Based on PUF

3.1   System Requirements and Adversary Model

Our system consists of (at least) two players: a verifier V and a resource-constrained prover P. We denote the adversary with $\mathcal{A}$. The main goal is to allow V to update the application code of P, while providing effective measures to mitigate all kinds of security threats mentioned above.

Verifier V. We assume V to be trusted. Further, V initializes and deploys P in a secure environment, extracts adequate (at least two) challenge-response pairs (CRPs) from the PUF of P and stores them securely. V also keeps a copy of the update's binary code generated by the product manufacturer of P.

Prover P. We assume P to be equipped with a root of trust (RoT), which contains *a robust and unpredictable PUF, a reverse fuzzy extractor, a random number generator, a symmetric cryptographic algorithm, a secure one-way hash function* and *a message authentication code*. We also assume P has a static non-volatile write-protected memory region $\mathcal{R}$, which can be implemented based on Flash memory with dedicated lock bits as described in [35]. We assume the RoT code is stored in the protected region $\mathcal{R}$ isolated from the application code, and once the RoT code in $\mathcal{R}$ gets executed, it cannot be interrupted until the control flow intentionally leaves $\mathcal{R}$. The difference from [35] is that $\mathcal{R}$ here doesn't rely on a traditional secure storage, which is replaced by PUF-based key generation. We also assume the protection on $\mathcal{R}$ can be temporarily removed by RoT during the update and is restored immediately after the update, which can also be implemented on existing commercial embedded devices as described in [30]. It is worth noting that the update of RoT code itself (infrequently) should be offline in a secure environment.

Adversary $\mathcal{A}$. We assume that $\mathcal{A}$ has complete control over the communication channel between V and P. This means that $\mathcal{A}$ can eavesdrop, manipulate and reroute all messages sent between V and P. We assume $\mathcal{A}$ cannot clone or tamper the PUF feature of P. Following the typical assumptions on PUF-based key generation (like [10,23]), we assume that $\mathcal{A}$ cannot access the challenge-response interface of PUF and cannot obtain temporary data (such as PUF-derived key information) stored in registers or on-device RAM during the update protocol. The temporary data can be erased by RoT immediately after the update protocol. In addition, we assume that $\mathcal{A}$ can be physically present and introduce additional (cloned) prover device. Finally, we assume $\mathcal{A}$ cannot bypass any of the hardware protections and cryptographic algorithms used in P. Data remanence attacks and physical attacks are not considered in our mechanism. We assume RoT code is immune from vulnerabilities, but the application code may be vulnerable. The device debug interfaces are disabled after deployment.

3.2   Update Protocol

Our new code update protocol is described in Fig. 2, and the memory layout of P during the protocol execution is illustrated in Fig. 3. V prepares a code update package *cupkg*, which includes (at least) the binary update code (*cupkg.code*), the current package version number (*cupkg.ver*), and the hash values over the expected memory contents for a successful update (*cupkg.hash*). P stores the RoT code and the expected integrity data (consisting of *cupkg.ver*

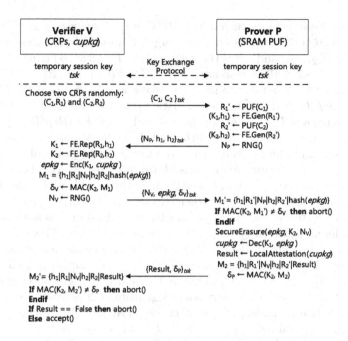

Fig. 2. Secure code update protocol based on PUF

and *cupkg.hash*) in the protected region $\mathcal{R}$, and the integrity data can also be updated securely by RoT during the protocol. Two PUF CRPs are used in the protocol, one for encryption and the other for mutual authentication.

Before each update protocol, we assume a temporary session key (tsk) is established between V and P by using a key exchange protocol (e.g., Diffie-Hellman or ECDH). Liu et al [37] have presented an efficient implementation of ECDH key exchange for MSP430 devices. tsk is mainly used to build a secure channel, and $\{M\}_{tsk}$ denotes that a message M is encrypted with tsk. All the exchanged messages are encrypted with tsk by using a symmetric encryption algorithm. Specifically, the key features of the protocol can be summarized as follows:

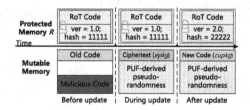

Fig. 3. Illustration of prover's memory layout during protocol execution

(1) Key Generation based on PUF and Reverse Fuzzy Extractor without Relying on Secure Storage. V randomly chooses two CRPs (C_1, R_1) and (C_2, R_2), and sends the challenges to P. After receiving C_1 and C_2, P reads the physical PUF responses $R_1' \leftarrow PUF(C_1)$ and $R_2' \leftarrow PUF(C_2)$, and generates the secret key and helper data as $(K_1, h_1) \leftarrow FE.Gen(R_1')$, $(K_2, h_2) \leftarrow FE.Gen(R_2')$. The helper data h_1 and h_2 are sent to V, and V uses them to recover $K_1 \leftarrow FE.Rep(R_1, h_1)$ and $K_2 \leftarrow FE.Rep(R_2, h_2)$. In this way, P doesn't need to store keys with the help of NVM-based secure storage, and can generate random keys on demand every time the protocol is started.

(2) Mutual Authentication based on K_2 and MAC. Based on the reproducibility property of PUF, V and P share the same keys K_1 and K_2 now. We use K_2 and MAC to achieve authentication. As the correct CRPs are only known to the trusted V and the physical PUF embedded in P is unclonable and unpredictable, no other party (e.g., $\mathcal{A}$) can forge a valid key. Thus, the authentication can be mutual. In detail, P generates a random nonce N_P and sends it to V. Once V receives N_P, it uses K_2 to create an authenticated message $\delta_V \leftarrow MAC(K_2, M_1)$ where M_1 contains the nonce N_P and other exchanged messages between V and P. δ_V serves as a signature, and prevents any modifications to the exchanged messages since P checks $MAC(K_2, M_1')? = \delta_V$. Similarly, δ_P is an authenticated message created by P, and verified by V.

(3) Encryption Transmission and Secure Code Erasure based on PUF. The code update package *cupkg* is encrypted by using K_1 and symmetric cryptography. Only P with a valid K_1 can decrypt the encrypted package. After P receives *epkg*, it first checks the authenticated message. If δ_V passes the verification, P believes that the messages come from an authorized V. Then P performs a secure code erasure (Algorithm 1): the encrypted package *epkg* is used to overwrite the memory occupied by the old code, and the extra memory space is filled with PUF-derived pseudorandom noises. The parameters K_2 and N_V assure that the secure code erasure is device-specific and protocol-specific, and no attackers can predict a valid memory layout. The use of *cnt* (inspired by [51]) is convenient for V to reconstruct the prover's memory layout and compute expected integrity values in advance. Secure code erasure can also eliminate possible malicious codes and restore P to a clean environment.

Algorithm 1: SecureErasure(*epkg*, K_2, N_V).

Variables:
> The counter value, *cnt*;
> The extra memory range, $[Mem_{Start} : Mem_{End}]$.

1   $Mem(OldCode, size) \leftarrow epkg$;
2   $cnt = 0$;
3   **for** $i = Mem_{Start}; i < Mem_{End}; i++$ **do**
4      $prandom \leftarrow Hash(K_2, N_V, cnt)$;
5      $Mem[i] \leftarrow prandom$;
6      $cnt++$;
7   **end**

---

(4) Local Code Integrity Attestation. After a secure code erasure, P can decrypt *epkg* and finish the installation of the update binary code. In order to attest an untampered and up-to-date software state, the *RoT* code in the protected region $\mathcal{R}$ triggers a local attestation routine. As illustrated in Algorithm 2, the attestation routine uses *cupkg* to perform three checks: (1) check whether the version number contained in *cupkg* is higher than the version number stored in $\mathcal{R}$, (2) check whether the hash values in *cupkg* are different from the values stored in $\mathcal{R}$, and (3) check whether the hash values over all memory regions match the expected integrity reference values specified in *cupkg.hash* (denoted by *CheckCodeIntegrity()*). If all checks pass, the verification of code update and software integrity is successful. Upon a successful verification, *RoT* disables the protection on $\mathcal{R}$ and writes the newest integrity reference values (*cupkg.ver* and *cupkg.hash*) into $\mathcal{R}$. As the prover device has just performed secure code erasure and integrity attestation, no malicious values can be written into $\mathcal{R}$ at this moment. Once $\mathcal{R}$ is updated, *RoT* enables the write protection immediately.

Algorithm 2: LocalAttestation(*cupkg*).

---

if $(cupkg.ver \leq \mathcal{R}.ver) \vee (cupkg.hash == \mathcal{R}.hash) \vee \neg CheckCodeIntegrity(cupkg.hash)$
then
    return *False* ;
else
    Disable protection on $\mathcal{R}$;
    $UpdateR(cupkg.ver, cupkg.hash)$;
    Enable protection on $\mathcal{R}$;
    return *True* ;
end

---

(5) Verification Result Feedback and Secure Call. The result of local integrity attestation is included in the computation of δ_P to ensure integrity, and it is sent back to V along with δ_P. If δ_P is verified successfully, V can ensure that the result comes from the correct P as no attackers can forge K_2. According to the feedback result, V knows whether P is in an up-to-date and unmodified software state. After a successful update, *RoT* code in P will enforce a strict white list policy to ensure a secure code call: the entry point of the update binary application code is hardcoded in $\mathcal{R}$, and each time the control flow is passed to the application code only when *CheckCodeIntegrity()* returns *True*.

Since SRAM is used for PUF in implementation (Sect. 4), a reboot is needed for each update protocol. In the experiment, we turn off the power manually to implement a full power cycle to collect SRAM PUF data. The initial SRAM values are used as R_1 and R_2 for each reboot, and *RoT* uses these values to generate K_1 and K_2. *RoT* is always executed after device reset, and the whole update process is handled by *RoT*. After the update, the keys are immediately erased by *RoT*. *RoT* also decides if the application code can be executed. Thus, we define a standard secure call sequence for P in Fig. 4.

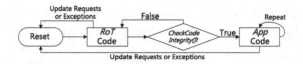

Fig. 4. Secure call sequence of P

Memory Integrity Check. If P's memory space is relatively large, we can divide it into multiple small sections and use hash tree (or Merkle tree) [19] to implement memory integrity check ($CheckCodeIntegrity()$). Figure 5 illustrates an example of a binary hash tree, in a setting where the memory is divided into four sections, denoted by S_1, S_2, S_3 and S_4. The hash values of these sections are the leaves, and a parent is the hash of the concatenation of its children. Only h_{root} (the root of the tree) is stored in the protected region $\mathcal{R}$. Before each update, V must decide the size of each section, and prepare a bran-new hash tree as the integrity reference value. During the update, all the sections should be checked, i.e., RoT should compute:

$$Hash(Hash(Hash(S_1), Hash(S_2)), Hash(Hash(S_3), Hash(S_4))).$$

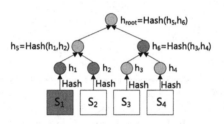

Fig. 5. A binary hash tree. Hash values of each memory section are aggregated to the root of the tree.

All intermediate values during the computation should match the hash values in the tree (including all nodes). After a succssful check, the root value h_{root} is written into $\mathcal{R}$, and other hash values are stored in the mutable memory along with the code. As the check of the entire memory is time-consuming, the hash tree method also supports to check the integrity of a specific memory section (e.g., the memory section containing the application code). For example, to check the integrity of S_1, RoT only needs to read S_1, h_2 and h_6, and the resultant aggregation value $Hash(Hash(Hash(S_1), h_2), h_6)$ is compared to h_{root}.

3.3   Analysis

The security of reverse fuzzy extractor is described in [23]. In this section, we mainly focus on the analysis of security threats (Sect. 2). We also give some comparisons and discussions about our method.

Our mechanism can defend against all mentioned security threats, and the specific analysis is as follows:

(1) **Code Alteration.** For each update, a local attestation is used to check the code integrity and any changes to the update binary code will be found.

(2) **Code Reverse Engineering.** It is almost impossible to absolutely guarantee the code confidentiality. Our main goal is to prevent code extraction during the network transmission and increase the difficulty of breaking the prover device P. As shown in Fig. 2, the communication channel only discloses $epkg$, which is encrypted with PUF-based key K_1. As we assume PUF is secure, $\mathcal{A}$ cannot decrypt $epkg$. Moreover, secure code erasure can remove any malicious code in P during the update, and at other times, RoT maintains a secure code call by enforcing a strict white list policy. Thus, it's difficult for $\mathcal{A}$ to break P and extract the update code.

(3) **Loading Unauthorized Code.** The update binary code is included in $epkg$, which is sent to P along with δ_V. $epkg$ is created based on K_1 and δ_V is generated based on K_2. Since K_1 and K_2 originate from the PUF of the same prover device P, it can be inferred that $epkg$ is from an authorized V if δ_V is verified successfully by P. If an unauthorized $epkg$ (created randomly or using a malicious key) arrives at P, its decryption is meaningless and cannot pass the verification of a local integrity attestation.

(4) **Loading Code onto an Unauthorized Device.** Due to the uniqueness and unpredictability of PUF, an unauthorized device cannot derive a correct decryption key K_1 and thus cannot install a update code intended for another device.

(5) **Code Downgrading.** An ascending version number $cupkg.ver$ is included in each code update package $cupkg$, the attestation routine will check the version number.

(6) **Incomplete Update.** Firstly, the result of $Local Attestation$ is included in δ_P, and thus V can ensure the integrity and authenticity of the feedback result. Secondly, RoT resides in the protected region $\mathcal{R}$ which is write-protected and execution-uninterruptible, the only entry to RoT is reset, and the only chance to write $\mathcal{R}$ is after a secure code erasure and an integrity check $CheckCodeIntegrity()$. Since the feedback result and δ_P are created by RoT, the result $True$ indeed indicates a complete update and the result $False$ illustrates the other situations.

(7) **Alleviating TOCTOU.** It's difficult to completely prevent TOCTOU, e.g., runtime attacks may break our system easily, which are not discussed here. Our mechanism uses the post-update defense to alleviate the TOCTOU attack, which is not considered in previous update mechanisms. During each update, the code is checked in the local integrity attestation routine and the newest reference values are written to $\mathcal{R}$. After update (post-update defense), RoT checks the integrity of application code by using the newest reference values to run $CheckCodeIntegrity()$ each time the application code is called. If the code has been tampered with, RoT will never give the system control to the code. In this case, RoT could trigger a new update protocol.

(8) **Device Cloning.** Even if $\mathcal{A}$ obtains all the memory contents (including *RoT* code) of an authorized prover device, it cannot copy or clone a similar device to pretend to be a legitimate P because $\mathcal{A}$ cannot clone a physical PUF or predict the responses of a particular PUF.

(9) **Control-flow Attack.** Our system provides no control flow integrity, and we assume RoT code is immune from vulnerabilities. But the application code may be compromised, we need to prevent application code from jumping to the RoT code arbitrarily. We can achieve this by enforcing a single well-defined entry point to RoT code in the ARMv8-M architecture [52]. Or in other devices, we can use software fault isolation [45] to sandbox the application code.

(10) **Physical Adversary.** Due to the unclonability and unpredictability of PUF, a physical clone or replacement of a valid prover device will be found. However, we cannot defend against other physical attacks, such as reprogramming the whole flash memory, data remanence of SRAM, or invasive attacks with micro-probing. Possible solutions to mitigate physical adversary contain the heartbeat protocol in DARPA [27].

Comparison with PoSE [42] and [35]. Our comparison with recently proposed update mechanisms mainly covers five aspects: the dependent assumptions, the supported security attributes, the ability to resist all mentioned security threats, the main communication and computation costs. As shown in Table 1, our mechanism has the following advantages: (1) Don't rely on a traditional secure storage; (2) Resist all 8 security threats by providing security attributes like mutual authentication, confidentiality (or secrecy), integrity, unclonability and secure call; (3) The message transmitted from V to P is the size of the update binary code, and the extra memory of P is filled with PUF-derived pseudorandom numbers; and (4) Use symmetric cryptography and MAC instead of public-key cryptography and signature, which is more suitable for low-end embedded systems.

Comparison with Remote Attestation. Remote attestation mechanisms are mainly used for verifying the software integrity of a remote device. Our update mechanism not only verifies the integrity of a remote device after an update installation, but also needs to ensure the correctness, freshness, confidentiality and authenticity of code update. Schulz et al. [44] gave a lightweight remote attestation by combing software-based attestation and PUF. PUFatt [36] implemented Schulz's idea by presenting a novel PUF design (called ALU PUF) based on the delay difference in two different arithmetic and logic units (ALUs). These works mainly focused on remote attestation, and did not consider secure code update. Furthermore, ALU PUF needs to change the microprocessor of device and is not available in current embedded devices. Researches (like SMART [17], Sancus [40], TyTAN [13], etc.) all tried to propose lightweight secure architecture for embedded devices, which can be used to implement remote attestation (also called hybrid attestation by [1]). In our opinion, these architecture can be easily extended to implement secure code update although none of them mentioned

Table 1. Comparison.

	Our mechanism	[42] (ESORICS 10)	[35] (ESORICS 16)
Assumptions	Immutable Code, uninterruptible execution and a robust and unpredictable PUF	Immutable code and secure communication (P only communicates with V and no other party)	Immutable code, secure storage and uninterruptible execution
Security Attributes Supported	Freshness, Integrity, Authenticity, Confidentiality, Feasibility, Verifiability, Restorability and Secure Call	Integrity, Feasibility and Restorability	Freshness, Integrity, Authenticity, Feasibility, Verifiability, and Restorability
Resisting Security Threats	Threat-1, 2, 3, 4, 5, 6, 7, 8	Only Threat-1	Threat-1, 3, 5, 6
Communication costs	The size of $cupkg$	The size of P's writable memory	The size of $cupkg$
Computation costs	Symmetric cryptography, MAC, Hash, RFE	Symmetric cryptography, MAC, Hash	Symmetric and Asymmetric cryptography, Signature and verification, Hash

this. However, all hybrid attestation schemes need some hardware modifications, which are not available commercially. Our secure code update mechanism can be applicable using existed resources in current commodity embedded devices.

Limitation. Firstly, our method requires that the prover device must have enough SRAM space, meeting the memory requirements for PUF and program variables at the same time. For low-end embedded devices, we may consider increasing the size of SRAM memory or exploring new PUF primitives (like Flash-based PUF [50]). Secondly, the scalability of our scheme is not good. To update multiple devices in a large network, V has to establish an update protocol for each individual device. Even if all devices have the same configuration (that is, the same $cupkg$), V must prepare different hash reference values and different encryption package $epkg$ for different devices. Our future work will be focused on the design and implementation of a scalable and lightweight secure code update mechanism based on PUF. A preliminary idea is to combine PUF physical properties with attribute-based encryption (ABE) [2], where PUF responses can be viewed as specific attributes associated with a decryption key.

Discussion. Helfmeier et al. [22] used a Focused Ion Beam (FIB) circuit edit (CE) to successfully produce a physical clone of a SRAM PUF. Although we assume a 'good' PUF in the adversary model, it's better to strengthen SRAM PUF with synthesized logic as recommended in [22] or adopt other PUF instances (like Flash-based PUF [50]) for high-security applications. Recently, data remanence attack [4] brought a new threat to SRAM PUF, but the attack needs a harsh condition (low-temperature between $-110\,°C$ and $-40\,°C$). Verifying the temperature using the sensors within embedded devices before each update may mitigate this attack. Note that, our work is not to design an ideal PUF, but to use PUF to design a secure code update mechanism. Actually, any PUF instances can be used in our update protocol. In addition, we adopt SRAM PUF because SRAM is ubiquitous in various computing devices and there are no modeling attacks currently found against weak PUFs. But we have to assume $\mathcal{A}$ cannot access the challenge-response interface of the PUF and cannot obtain temporary data stored in volatile memory during the update protocol. Although this is a strong assumption (the assumption is also used in other literatures like [10,23]), it is necessary because no secure execution environment (like TEE) exists in current embedded devices. However, this assumption can be improved by forcing memory access control based on a Memory Protection Unit (MPU) [13,34] or using other techniques such as obfuscation and white-box cryptography. We adopt reverse FE due to less performance overhead, actually any FEs (like a computationally secure FE [15]) can be used if they are more effective. Finally, our work mainly focuses on providing security without changing hardware for legacy devices. However, in many embedded scenes, modifying hardware is necessary to provide strong security, and we think ARM TrustZone technology in ARMv8-M architecture will be a good choice.

4   Implementation and Performance Considerations

Setup. We implement and evaluate our proposed secure code update scheme on a MSP-EXP430G2 LaunchPad Development Board. The board is a complete USB-based development and experimenter tool from Texas Instrument with a MSP430G2553 MCU by default. The key features of the MSP430G2553 MCU include [29]: ultralow-power, von-Neumann architecture; 16-bit RISC CPU (up to 16 MHz); 16 KB of programmable Flash; 512 bytes of SRAM.

We use the on-board SRAM as the source of entropy to implement the *PUF* and random number generator (*RNG*). For reverse fuzzy extractor (*RFE*), we adopt the BCH error correction code to eliminate noises and use a hash function as an entropy accumulator to generate unpredictable random keys. We implement the hash function using SHA256, while the symmetric algorithm uses 128-bit AES. The MAC computation is implemented by using the construct of HMAC-SHA256. As no hardware acceleration is supported in MSP430G2553, all of the cryptographic algorithms are implemented in software based on [28]. As 512B SRAM is relatively small, our implementation is based on the following guidelines: (1) Use more constants and Flash space; (2) Use fewer variables and RAM space; (3) Initial SRAM values are written to Flash used for *PUF* and *RNG*, and the actual SRAM space is reserved for global and local variables (.bss,

.data and .stack). Our time performance is measured in clock cycles. As we set the clock frequency to 1MHz, m cycles are equal to $m/1,000,000$ seconds. Our evaluation code (in python) and data for PUF are uploaded to the Github[2].

SRAM PUF and SRAM RNG. We collect the startup SRAM values from two different MSP430G2553 devices, each measured over 50 power cycles. Based on these data, we first evaluate the robustness, uniqueness and randomness of SRAM PUF by analyzing the min-entropy and Hamming distance. For robustness, we compute the intra-chip Hamming distance (HD_{Intra}) between repeated measurements of SRAM cells from the same chip. The resulting HD_{Intra} is 260 ($260/4096 = 6.3\%$) at average, and 743 ($743/4096 = 18\%$) at worst. For uniqueness, we compute the inter-chip Hamming distance (HD_{Inter}) and min-entropy over the measurements from different chips. The average ratio for HD_{Inter} is 42.3%, and the min-entropy rate is 87% which means the average min-entropy per bit is 0.87. For randomness, we compute the min-entropy over 50 repeatedly measured SRAM values from the same chip, which gives an average min-entropy rate of 7.76%. This means that we need at least $1/7.76\% = 12.88$ SRAM cells to obtain one random bit. These evaluated results show a well-featured PUF.

4096-bit (=512B) SRAM space is allocated as follows: 2628 bits are used to generate two PUF CRPs, and the remaining 1468 bits are used to derive random numbers. The address spaces are separated to avoid direct correlation between *PUF* and *RNG*. As only two CRPs can be used in each device, C_1 and C_2 needs not to be transmitted over the network. Using multiple CRPs corresponds to storing multiple session keys. It means that we have two default session keys. Additionally, we use 256 bits SRAM to derive a 16-bit random nonce, which is achieved by XORing adjacent bytes 16 times. Thus, 5 (1468/256) random numbers can be used for each power cycle. Aysu et al. [10] showed that the SRAM data can pass all experiments in the NIST statistical Test Suite after 8-fold XORing, thus our 16-fold XORing is random enough. *RNG* is implemented in assembly by using only two registers (one for the start address of SRAM RNG and the other for the *xor* result). The code size of RNG is 56 bytes and it takes 44 clock cycles to output one random number. Theoretically, a random extractor should be used instead to generate RNG, we choose XOR due to low overhead and Aysu's experience in [10].

Reverse Fuzzy Extractor. A BCH($n, k, d = 2t + 1$) [39] code allows to correct errors up to t-bit within a n-bit block. We customize a BCH(127,15,53) based on the open source code[3], which can correct up to 20.5% noisy bits (greater than the worst SRAM noise level of 18%). As the average min-entropy rate for uniqueness is 87%, 1314 (2628/2) bits SRAM data contains 1143 (1314×0.87) bits entropy. We use 1143 bits PUF entropy in 9 blocks of a BCH(127,15,53) code, and 1008(=(127−15)×9) bits are leaked in the helper data. The remaining entropy is 135 (=1143−1008) bits, which are enough for a 128-bit key. We use SHA256 to hash the PUF response, and the 256-bit result is 2-XORed to obtain

---

[2] https://github.com/vonwaist/PUFRNG.
[3] http://www.eccpage.com/.

a 128-bit key. We assume that a single bit flips with a probability of $P_{error} = 7\%$ (greater than the average HD_{Intra}), then the probability that 27 bits or more will flip in a 127-bit block is $P_{block} = \sum_{i=27}^{n=127} \binom{127}{i} P_{error}^i (1 - P_{error})^{(127-i)} \approx 1.87 \times 10^{-7}$, and thus the error cannot be corrected in this case. For 9 blocks of a BCH(127,15,53) code, the probability that a key can be fully reconstructed is $P_{correct} = (1 - P_{block})^9 > 1 - 1.69 \times 10^{-6}$.

The *PUF* and *RFE.Gen* are implemented in C with a code size of 3274 bytes, and it also uses 768 bytes constant space and 426 bytes variable space. To save RAM, we pre-compute the coefficients of the generator polynomial, log table and antilog table of the Galois field $GF(2^m)$, and store these parameters as the constants in the flash memory. The implementation contains four steps: read SRAM values to generate a 1314-bit PUF response (it takes 1471 cycles); use SHA256 and 2-XORing to generate a 128-bit key (it takes 290,951 clock cycles); BCH Encoder for 9 blocks (it takes 585,504 clock cycles); write the result to Flash (132,982 cycles).

Symmetric Algorithm, Hash and MAC. There is a decryption operation for each update protocol, and we adopt 128-bit AES algorithm. The code size of *Dec* is 910 bytes, and the memory requirements for its constants and variables are 522 bytes and 119 bytes, respectively. To decrypt a 128-bit cipher text, *Dec* takes about 23,487 CPU cycles. The hash function is SHA256, and its implementation costs 1530 bytes of code size, 288 bytes of constant space and 271 bytes of variable space. The performance of SHA256 depends on the specific input size, e.g., 96,617 cycles for 50-byte input, 291,040 cycles for 150-byte. HMAC is implemented based on SHA256, and its code size is 2348 bytes. For a 16-byte message, HMAC-SHA256 takes about 392,174 clock cycles. As many MCUs support cryptographic hardware security[4], the performance can be improved further.

Secure Erasure and Local Attestation. Two algorithms *SecureErasure()* and *LocalAttestation()* are both implemented based on SHA256. The code size of *SecureErasure()* is 2,568 bytes, and the number of clock cycles it takes to erase a 512B flash section is 1,615,880. The main time consumption of *SecureErasure()* is caused by SHA256 computation and Flash write operation. The primary role of *LocalAttestation* is *CheckCodeIntegrity()*, which is also the most time-consuming part. *CheckCodeIntegrity()* computes the hash value of a given memory block and compares it with the reference value, and it takes about 292,422 clock cycles for a 128-byte application program code.

Protected Memory. In our method, write-protection is needed for storing the version, reference hashes and RoT code, and we use existing hardware resources in embedded devices to implement a static non-volatile write-protected region $\mathcal{R}$. In MSP430G2553, the hardware resources are Flash memory. According to [29], the Flash memory of MSP430G2553 is partitioned into main and information memory sections. The information memory has four 64-byte segments, and the main memory has multiple 512-byte segments. The information memory can be locked separately from the main memory with a LOCKA bit. When LOCKA is

---

[4] http://www.ti.com/ww/en/embedded/security/index.shtml.

set, the information memory is protected and cannot be written or erased. Thus, *RoT* code can be stored in the information memory. As the size (256-byte) of information memory in MSP430G2553 is smaller than the size of our *RoT* code, our evaluation described above uses the main memory. However, this does not affect the evaluation results because there are no other differences between the information and main memory except for the lock bit.

In MSP430FR family [30], the protected hardware resources are FRAMs similar for MPU. An FRAM is a non-volatile memory that can be read and written like a standard SRAM. An MPU can be used to divide the device's main memory into three variable-sized segments with configurable read, write and execute access. Furthermore, the protection of the second segment can be temporarily removed when necessary by the bootloader, which can be used to store and update the integrity reference hash values. Bootloader (similar to our *RoT*) locks the MPU settings before jumping to the application, preventing the application from corrupting or overwriting the protected area. For the security of PUF, we propose to allow only the *RoT* code to access the start-up values at boot time, and after that the SRAM space is erased by *RoT*.

For uninterruptible execution, we suggest to disable the interrupt during the execution of *RoT* code. Before the control is handed over to the applicaton code, *RoT* enables the interrupt and at the same time checks the integrity of application code and all interrupt handlers.

Comparison with Public-Key Cryptography. As the MSP430G2553 device does not have enough resources to implement and run a ECC/RSA algorithm, we compare our PUF-based AES encryption (with 128-bits key) with a RSA encryption (with 2048-bits key) in a host environment. RSA is implemented based on the open-source mbed TLS library[5]. For a 100-bytes plain message, we test the two encryption operations 1000 times respectively. The min, max and average runtime for PUF-based AES encryption are 0.023 ms, 1.927 ms, and 0.0549 ms; and the runtime for RSA are 1.076 ms, 16.07 ms, and 1.37 ms. Obviously, our method is more lightweight and more suitable for tiny embedded devices. In the future, we plan to purchase a more rich embedded development board (e.g., MSP430FR family) to make a more comprehensive comparison.

5   Related Work

Remote Attestation. Remote attestation can be categorized in three main branches: hardware-based attestation, software-based attestation and hardware-software co-design with minimum hardware requirements. Hardware-based attestation relies on strong hardware features, such as TCG's TPM [8,41], ARM TrustZone [6] and Intel SGX [5], which are not supported on low-cost commodity embedded devices. Software-based attestation [7,26,46,48] does not require secure hardware and thus is well suitable for constrained embedded systems.

---

[5] https://tls.mbed.org/.

However, its security guarantee is weak. Between the two mechanisms, hardware-software co-design [13,17,34,36,40] aims to build a dynamic trust anchor in a low-end embedded device with minimal changes to existing MCUs. The trust anchor established can be further used to design a scalable collective attestation protocol (SEDA [9] and SANA [3]), meeting the global security requirements of large groups of interconnected smart devices. In our opinion, all remote attestation mechanisms can be used to strengthen secure updates, e.g., to verify the code integrity after update. But a complete secure code update is more than a remote attestation mechanism.

Secure Code Updates For Embedded Devices. SCUBA [47] is a secure code update mechanism by using software-based attestation to ensure indisputable code execution (ICE) on a remote sensor node. PoSE [42] is a different approach that can enable a prover device to convince a verifier that it has erased all its memory. As the overhead of PoSE is relatively high, some researchers try to explore effective skills to reduce the overhead including uncomputable hash function [16], invert-hash PoSE and graph-based PoSE [33], and All or Nothing Transforms [32]. Recently, Kohnhauser and Katzenbeisser [35] presented a novel code update scheme which verifies and enforces the correct installation of code updates on all commodity low-end embedded devices in a mesh network. To address the security threats involved with the in-field firmware updates process, Texas Instruments [21,30,31] proposes to integrate cryptographic algorithms and security mechanisms into the bootloader of its ultra-low-energy MCUs.

SRAM PUF. The SRAM PUF was first introduced in 2007 by Holcomb et al. [24,25] and Guajardo et al. [20] concurrently and independently. Holcomb et al. [24,25] proposed to use SRAM physical fingerprints for identification and generation of true random numbers in RFID tag circuits, while Guajardo et al. [20] used initial SRAM values to design new protocols for IP protection on FPGAs. To provide a viable alternative to costly protected non-volatile memory (NVM), Maes et al. [38] presented a low-overhead implementation of helper data algorithm for SRAM PUFs using soft decision information. The SRAM PUF was implemented and evaluated on a microcontroller in [12]. Researchers from intrinsic-ID showed the construction of a FIPS 140-3 compliant random bit generator based on SRAM PUF in [49], and presented a comparative analysis of several types of SRAM memories from different technology nodes and demonstrated the reliability and uniqueness of all the tested SRAMs when used as PUFs in [43]. Aysu et al. showed in [10] that SRAM PUF can be used to design and implement a provably secure protocol that supports privacy-preserving mutual authentication.

6   Conclusion

In this paper, we presented a novel secure code update scheme for commodity low-end embedded devices by combing the advantages of secure erasure and physically unclonable function. We concluded eight security threats that may happen in secure code updates from the existing literature, and showed how our scheme can be used to prevent or mitigate these threats. Our scheme doesn't rely on secure storage or secure communication. By using the symmetric cryptography and lightweight construction of a reverse fuzzy extractor, our approach offers acceptable communication and computation overhead. Finally, we also eliminate the gap from the world of protocol theory to concrete realization through evaluating all protocol components in a single TI MSP430 device. Our implementation uses only on-board SRAM and the protected memory resources without requiring any hardware modifications, which is applicable to a broad range of popular low-end embedded systems.

Acknowledgments. The work has been supported by the National Natural Science Foundation of China (No. 61602455 and No. 61402455). We thank anonymous reviewers for their helpful comments. We specially thank Aurlien Francillon for his suggestions on improving the paper.

References

1. Abera, T., et al.: Invited: things, trouble, trust: on building trust in IoT systems. In: 53nd ACM/EDAC/IEEE Design Automation Conference (DAC), pp. 1–6 (2016)
2. Ambrosin, M., Anzanpour, A., Conti, M., Dargahi, T., Moosavi, S.R., Rahmani, A.M., Liljeberg, P.: On the feasibility of attribute-based encryption on internet of things devices. IEEE Micro **36**(6), 25–35 (2016)
3. Ambrosin, M., Conti, M., Ibrahim, A., Neven, G., Sadeghi, A.-R., Schunter, M.: SANA: secure and scalable aggregate network attestation. In: Proceedings of the 2016 ACM SIGSAC Conference on Computer and Communications Security, CCS 2016, pp. 731–742. ACM, New York (2016)
4. Anagnostopoulos, N.A., Katzenbeisser, S., Rosenstihl, M., Schaller, A., Gabmeyer, S., Arul, T.: Low-temperature data remanence attacks against intrinsic SRAM PUFs. Cryptology ePrint Archive, Report 2016/769 (2016). http://eprint.iacr.org/2016/769
5. Anati, I., Gueron, S., Johnson, S.P., Scarlata, V.R.: Innovative technology for CPU based attestation and sealing. In: Proceedings of the 2nd International Workshop on Hardware and Architectural Support for Security and Privacy, vol. 13 (2013)
6. ARM. Arm security technology: Building a secure system using trustzone technology. Technical report, ARM Technical White Paper (2009)
7. Armknecht, F., Sadeghi, A.-R., Schulz, S., Wachsmann, C.: A security framework for the analysis and design of software attestation. In: Proceedings of the 2013 ACM SIGSAC Conference on Computer and Communications Security, CCS 2013, pp. 1–12. ACM, New York (2013)
8. Arthur, W., Challener, D.: A Practical Guide to TPM 2.0: Using the Trusted Platform Module in the New Age of Security. Apress, Berkely (2015)
9. Asokan, N., et al.: SEDA: scalable embedded device attestation. In: Proceedings of the 22nd ACM SIGSAC Conference on Computer and Communications Security, CCS 2015, NY, USA, pp. 964–975 (2015)
10. Aysu, A., Gulcan, E., Moriyama, D., Schaumont, P., Yung, M.: End-to-end design of a PUF-based privacy preserving authentication protocol. In: Güneysu, T., Handschuh, H. (eds.) CHES 2015. LNCS, vol. 9293, pp. 556–576. Springer, Heidelberg (2015). https://doi.org/10.1007/978-3-662-48324-4_28

11. Bhm, C., Hofer, M.: Physical Unclonable Functions in Theory and Practice. Springer, Heidelberg (2012). https://doi.org/10.1007/978-1-4614-5040-5
12. Bohm, C., Hofer, M., Pribyl, W.: A microcontroller SRAM-PUF. In: 5th International Conference on Network and System Security (NSS), pp. 269–273, September 2011
13. Brasser, F., El Mahjoub, B., Sadeghi, A.-R., Wachsmann, C., Koeberl, P.: TyTAN: tiny trust anchor for tiny devices. In: Proceedings of the 52nd Annual Design Automation Conference, DAC 2015, pp. 34:1–34:6. ACM, New York (2015)
14. Bratus, S., D'Cunha, N., Sparks, E., Smith, S.W.: TOCTOU, traps, and trusted computing. In: Lipp, P., Sadeghi, A.-R., Koch, K.-M. (eds.) Trust 2008. LNCS, vol. 4968, pp. 14–32. Springer, Heidelberg (2008). https://doi.org/10.1007/978-3-540-68979-9_2
15. Canetti, R., Fuller, B., Paneth, O., Reyzin, L., Smith, A.: Reusable fuzzy extractors for low-entropy distributions. In: Fischlin, M., Coron, J.-S. (eds.) EUROCRYPT 2016. LNCS, vol. 9665, pp. 117–146. Springer, Heidelberg (2016). https://doi.org/10.1007/978-3-662-49890-3_5
16. Dziembowski, S., Kazana, T., Wichs, D.: One-time computable self-erasing functions. In: Ishai, Y. (ed.) TCC 2011. LNCS, vol. 6597, pp. 125–143. Springer, Heidelberg (2011). https://doi.org/10.1007/978-3-642-19571-6_9
17. Eldefrawy, K., Francillon, A., Perito, D., Tsudik, G.: SMART: secure and minimal architecture for (establishing a dynamic) root of trust. In: 19th Annual Network and Distributed System Security Symposium, NDSS 2012, San Diego, USA, 5–8 February (2012)
18. Feng, W., Qin, Y., Zhao, S., Feng, D.: Secure code updates for smart embedded devices based on PUFs. Cryptology ePrint Archive, Report 2017/991 (2017). http://eprint.iacr.org/2017/991
19. Gassend, B., Edward Suh, G., Clarke, D., van Dijk, M., Devadas, S.: Caches and hash trees for efficient memory integrity verification. In: Proceedings of the 9th International Symposium on High-Performance Computer Architecture, HPCA 2003, Washington, DC, USA, p. 295 (2003)
20. Guajardo, J., Kumar, S.S., Schrijen, G.-J., Tuyls, P.: FPGA intrinsic PUFs and their use for IP protection. In: Paillier, P., Verbauwhede, I. (eds.) CHES 2007. LNCS, vol. 4727, pp. 63–80. Springer, Heidelberg (2007). https://doi.org/10.1007/978-3-540-74735-2_5
21. Guillen, O., Nisarga, B., Reynoso, L., Brederlow, R.: Crypto-bootloader secure in-field firmware updates for ultra-low power MCUs. Texas Instruments Incorporated (2015)
22. Helfmeier, C., Boit, C., Nedospasov, D., Seifert, J.P.: Cloning physically unclonable functions. In: 2013 IEEE International Symposium on Hardware-Oriented Security and Trust (HOST), pp. 1–6, June 2013
23. Van Herrewege, A.: Reverse fuzzy extractors: enabling lightweight mutual authentication for PUF-enabled RFIDs. In: Keromytis, A.D. (ed.) FC 2012. LNCS, vol. 7397, pp. 374–389. Springer, Heidelberg (2012). https://doi.org/10.1007/978-3-642-32946-3_27
24. Holcomb, D.E., Burleson, W.P., Fu, K.: Initial SRAM state as a fingerprint and source of true random numbers for RFID tags. In: Proceedings of the Conference on RFID Security, vol. 7 (2007)
25. Holcomb, D.E., Burleson, W.P., Fu, K.: Power-up SRAM state as an identifying fingerprint and source of true random numbers. IEEE Trans. Comput. 58(9), 1198–1210 (2009)

26. Horsch, J., Wessel, S., Stumpf, F., Eckert, C.: SobTra: a software-based trust anchor for ARM cortex application processors. In: Proceedings of the 4th ACM Conference on Data and Application Security and Privacy, pp. 273–280. ACM (2014)
27. Ibrahim, A., Sadeghi, A.-R., Tsudik, G., Zeitouni, S.: DARPA: device attestation resilient to physical attacks. In: Proceedings of the 9th ACM Conference on Security and Privacy in Wireless and Mobile Networks, WiSec 2016, pp. 171–182. ACM, New York (2016)
28. Texas Instruments Incorporated. C implementation of cryptographic algorithms, SLAA547A-July 2013 (2013)
29. Texas Instruments Incorporated. MSP430x2xx family user's guide, SLAU144J-December 2004, Revised July 2013
30. Texas Instruments Incorporated. Crypto-bootloader (CryptoBSL) for MSP430FR 59xx and MSP430FR69xx MCUs, user's guide, SLAU657-November 2015 (2015)
31. Texas Instruments Incorporated. Secure in-field firmware updates for MSP MCUs, application report, SLAA682-November 2015 (2015)
32. Karame, G.O., Li, W.: Secure erasure and code update in legacy sensors. In: Conti, M., Schunter, M., Askoxylakis, I. (eds.) Trust 2015. LNCS, vol. 9229, pp. 283–299. Springer, Cham (2015). https://doi.org/10.1007/978-3-319-22846-4_17
33. Karvelas, N.P., Kiayias, A.: Efficient proofs of secure erasure. In: Abdalla, M., De Prisco, R. (eds.) SCN 2014. LNCS, vol. 8642, pp. 520–537. Springer, Cham (2014). https://doi.org/10.1007/978-3-319-10879-7_30
34. Koeberl, P., Schulz, S., Sadeghi, A.-R., Varadharajan, V.: TrustLite: a security architecture for tiny embedded devices. In: Proceedings of the Ninth European Conference on Computer Systems, EuroSys 2014, pp. 10:1–10:14. ACM, New York (2014)
35. Kohnhäuser, F., Katzenbeisser, S.: Secure code updates for mesh networked commodity low-end embedded devices. In: Askoxylakis, I., Ioannidis, S., Katsikas, S., Meadows, C. (eds.) ESORICS 2016. LNCS, vol. 9879, pp. 320–338. Springer, Cham (2016). https://doi.org/10.1007/978-3-319-45741-3_17
36. Kong, J., Koushanfar, F., Pendyala, P.K., Sadeghi, A.-R., Wachsmann, C.: PUFatt: embedded platform attestation based on novel processor-based PUFs. In: Proceedings of the 51st Annual Design Automation Conference, DAC 2014, pp. 109:1–109:6. ACM, New York (2014)
37. Liu, Z., Seo, H., Hu, Z., Hunag, X., Grosschadl, J.: Efficient implementation of ECDH key exchange for MSP430-based wireless sensor networks. In: Proceedings of the 10th ACM Symposium on Information, Computer and Communications Security, ASIACCS 2015, pp. 145–153. ACM, New York (2015)
38. Maes, R., Tuyls, P., Verbauwhede, I.: Low-overhead implementation of a soft decision helper data algorithm for SRAM PUFs. In: Clavier, C., Gaj, K. (eds.) CHES 2009. LNCS, vol. 5747, pp. 332–347. Springer, Heidelberg (2009). https://doi.org/10.1007/978-3-642-04138-9_24
39. Maes, R., Van Herrewege, A., Verbauwhede, I.: PUFKY: a fully functional PUF-based cryptographic key generator. In: Prouff, E., Schaumont, P. (eds.) CHES 2012. LNCS, vol. 7428, pp. 302–319. Springer, Heidelberg (2012). https://doi.org/10.1007/978-3-642-33027-8_18
40. Noorman, J., et al.: Sancus: low-cost trustworthy extensible networked devices with a zero-software trusted computing base. In: Proceedings of the 22nd USENIX Conference on Security, SEC 2013, Berkeley, CA, USA, pp. 479–494 (2013)
41. Parno, B., McCune, J.M., Perrig, A.: Bootstrapping trust in commodity computers. In: 2010 IEEE Symposium on Security and Privacy, SP 2010, pp. 414–429. IEEE Computer Society, May 2010

42. Perito, D., Tsudik, G.: Secure code update for embedded devices via proofs of secure erasure. In: Gritzalis, D., Preneel, B., Theoharidou, M. (eds.) ESORICS 2010. LNCS, vol. 6345, pp. 643–662. Springer, Heidelberg (2010). https://doi.org/10.1007/978-3-642-15497-3_39

43. Schrijen, G.-J., van der Leest, V.: Comparative analysis of SRAM memories used as PUF primitives. In: Proceedings of the Conference on Design, Automation and Test in Europe, DATE 2012, pp. 1319–1324. EDA Consortium, San Jose (2012)

44. Schulz, S., Sadeghi, A.-R., Wachsmann, C.: Short paper: lightweight remote attestation using physical functions. In: Proceedings of the Fourth ACM Conference on Wireless Network Security, WiSec 2011, pp. 109–114. ACM, New York (2011)

45. Sehr, D., et al.: Adapting software fault isolation to contemporary CPU architectures. In: Proceedings of the 19th USENIX Conference on Security, USENIX Security 2010, p. 1. USENIX Association, Berkeley (2010)

46. Seshadri, A., Perrig, A., van Doorn, L., Khosla, P.: SWATT: software-based attestation for embedded devices. In: Proceedings of 2004 IEEE Symposium on Security and Privacy, pp. 272–282, May 2004

47. Seshadri, A., Luk, M., Perrig, A., van Doorn, L., Khosla, P.: SCUBA: secure code update by attestation in sensor networks. In: Proceedings of the 5th ACM Workshop on Wireless Security, WiSe 2006, pp. 85–94. ACM, New York (2006)

48. Seshadri, A., Luk, M., Shi, E., Perrig, A., van Doorn, L., Khosla, P.: Pioneer: verifying code integrity and enforcing untampered code execution on legacy systems. In: Proceedings of the Twentieth ACM Symposium on Operating Systems Principles, SOSP 2005, pp. 1–16. ACM, New York (2005)

49. van der Leest, V., van der Sluis, E., Schrijen, G.-J., Tuyls, P., Handschuh, H.: Efficient implementation of true random number generator based on SRAM PUFs. In: Naccache, D. (ed.) Cryptography and Security: From Theory to Applications. LNCS, vol. 6805, pp. 300–318. Springer, Heidelberg (2012). https://doi.org/10.1007/978-3-642-28368-0_20

50. Wang, Y., Yu, W., Wu, S., Malysa, G., Edward Suh, G., Kan, E.C.: Flash memory for ubiquitous hardware security functions: true random number generation and device fingerprints. In: Proceedings of the 2012 IEEE Symposium on Security and Privacy, SP 2012, pp. 33–47. IEEE Computer Society, Washington (2012)

51. Yang, Y., Wang, X., Zhu, S., Cao, G.: Distributed software-based attestation for node compromise detection in sensor networks. In: 26th IEEE International Symposium on Reliable Distributed Systems, SRDS 2007, pp. 219–230, October 2007

52. Yiu, J.: White paper: ARMv8-M architecture technical overview (2015)

53. Zhao, S., Zhang, Q., Hu, G., Qin, Y., Feng, D.: Providing root of trust for arm trustzone using on-chip SRAM. In: Proceedings of the 4th International Workshop on Trustworthy Embedded Devices, TrustED 2014, pp. 25–36. ACM, New York (2014)

A Privacy-Preserving Device Tracking System Using a Low-Power Wide-Area Network

Tomer Ashur[1], Jeroen Delvaux[1], Sanghan Lee[2], Pieter Maene[1(✉)],
Eduard Marin[1], Svetla Nikova[1], Oscar Reparaz[1], Vladimir Rožić[1],
Dave Singelée[1], Bohan Yang[1], and Bart Preneel[1]

[1] imec-COSIC KU Leuven, Leuven, Belgium
{tomer.ashur,jeroen.delvaux,pieter.maene,eduard.marin,
svetla.nikova,oscar.reparaz,vladimir.rozic,dave.singelee,bohan.yang,
bart.preneel}@esat.kuleuven.be
[2] The Attached Institute of ETRI, 1559, Yuseong-daero, Yuseong-gu,
Daejeon 34044, South Korea
freewill71@nsr.re.kr

Abstract. This paper presents the design and implementation of a low-power privacy-preserving device tracking system based on *Internet of Things* (IOT) technology. The system consists of low-power nodes and a set of dedicated beacons. Each tracking node broadcasts pseudonyms and encrypted versions of observed beacon identifiers over a *Low-Power Wide-Area Network* (LPWAN). Unlike most commercial systems, our solution ensures that the device owners are the only ones who can locate their devices. We present a detailed design and validate the result with a prototype implementation that considers power and energy consumption as well as side-channel attacks. Our implementation uses *Physically Unclonable Function* (PUF) technology for secure key-storage in an innovative way. We build and evaluate a complete demonstrator with off-the-shelf IoT nodes, *Bluetooth Low Energy* (BLE) beacons, and LoRa long distance communication (LPWAN). We validate the setup for a bicycle tracking application and also estimate the requirements for a low-cost ASIC node.

1 Introduction

The *Internet of Things* (IOT) will transform our society. Predictions about the scale and speed of this transformation vary – Gartner claims that there will be 26 billion IOT devices by 2020, Cisco predicts 50 billion, Intel 200 billion and IDC 212 billion – but there is no doubt that this development will have a major impact on our lives.

One of the main benefits of the IoT is that it links the physical world to the online world. A key application is the localization and tracking of objects. The simplest way to track objects is by adding a barcode. However, those are limited to the line of sight of the reader and store little information. *Radio-Frequency*

© Springer Nature Switzerland AG 2018
S. Capkun and S. S. M. Chow (Eds.): CANS 2017, LNCS 11261, pp. 347–369, 2018.
https://doi.org/10.1007/978-3-030-02641-7_16

IDentification (RFID) tags overcome these limitations; they were mostly considered for deployment in warehouses or shops, but with the appropriate back-office infrastructure, they could be used for tracking during the complete supply chain. Unfortunately, RFID tags require external power, have only limited computational capabilities, and can only be used for short-range communications. Another development is the tracking of more sophisticated mobile devices such as smartphones, tablets, and laptops. The first two typically have *Global Positioning System* (GPS) receivers and all of them can detect nearby WiFi networks. Using an Internet connection, it is rather straightforward to make them 'phone home' so that legitimate owners can recover them. A similar feature is being added to cars, in order to assist users to find their car in large parking lots, or to help track lost or stolen cars. Multiple location services allow users to locate their friends or spouses using smartphones. Another fascinating development created by modern technologies is participatory sensing: while the original goal was to crowdsource data on weather or mobility patterns, they could also be used to track objects or users. Dedicated IOT trackers are also arriving on the market: low-cost nodes broadcast information that helps owners locate their keys, wallets, or bicycles. Some of these devices such as Tile [6] and TrackR [7] use a combination of *Bluetooth Low Energy* (BLE) with crowdsourcing (GPS location on smartphones); others such as SemTech exploit the fact that – unlike RFID tags – modern IOT nodes can communicate over long distances with a *Low-Power Wide-Area Network* (LPWAN) such as LoRa. For several devices, vulnerabilities have been reported, e.g., in [4].

With the deployment of these applications, there has been an increased interest in the privacy aspects of tracking, and in a broader sense, in the privacy of location-based services. At first sight, it seems to be impossible to reconcile these properties: devices should be tracked by their legitimate owners, yet one wants to offer privacy against third parties and service providers. Cryptographic techniques can overcome this paradox. In particular in the area of RFID, several solutions have been proposed that offer a broad range of tradeoffs between security, privacy, and scalability. However, low-cost IoT devices have capabilities that go well beyond those of RFID tags. While one can repurpose some of the ideas, long distance communications bring new threats but also new opportunities to enhance security and privacy. For devices that are connected to the Internet, more sophisticated solutions have been developed that offer advanced privacy features (e.g., the Adeona and Eddystone systems described in Sect. 2). We show how similar goals can be achieved with low-energy and low-cost nodes and for devices that are not inherently connectable.

Contributions. We present a system that allows users to track the location of their devices such that the intermediate entities learn neither the device's location nor to whom it belongs. Our privacy-preserving tracking system is based on a small, low-cost tracking device, also known as *a tag*, that is attached to any user's personal belongings (e.g., a bicycle), and a set of dedicated beacons. A schematic overview of the system is depicted in Fig. 1. Our protocol provides message confidentiality and integrity as well as entity authentication, unlinkability, and forward privacy. Furthermore, it has built-in resistance to side-channel attacks. Our prototype uses commercial components and offers secure key storage based on a PUF. The latency of our prototype, as well as its power and energy consumption, is evaluated, and we also outline an ASIC design for the tag.

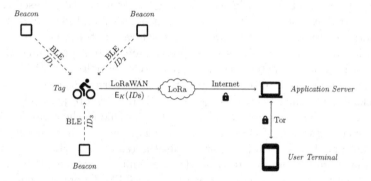

Fig. 1. The proposed tracking system showing all components and communication channels involved.

Notation. Binary vectors are denoted by a bold-faced, lowercase character, e.g., $\mathbf{x} = (x_1\, x_2)$. All vectors are row vectors. The all-zeros vector is denoted by $\mathbf{0}$. Binary matrices are denoted by a bold-faced, uppercase character, e.g., $\mathbf{X}$. A random variable is denoted by a regular uppercase character, e.g., X. The set of all n-bit vectors is denoted by $\{0,1\}^n$. Procedure names are printed in a sans-serif font, e.g., $\mathsf{Func}(\cdot)$. The operator $\|$ denotes concatenation.

Organization. The remainder of this paper is organized as follows. Section 2 gives an overview of the related work. Section 3 details the threat model and assumptions about the building blocks. Section 4 describes the design requirements and the proposed protocol, whereas the practical aspects and the implementation of our system are discussed in Sect. 5. The security and privacy are analyzed in Sect. 6. An evaluation of our prototype implementation is given in Sect. 7. Finally, Sect. 8 presents concluding remarks.

2   Related Work

Over the last years, the problem of providing privacy in tracking systems has been extensively studied in RFID systems. See for instance the series of papers by Avoine et al. [14,15,17–20], Juels et al. [37,38], Weis et al. [54], Molnar et al. [47], Henrici et al. [33–35], Saito et al. [50], Alomair et al. [12,22] and Spiekermann et al. [52].

Variants of this problem arise in other technologies such as electronic passports [16] or Bluetooth-based localization services. Google's Eddystone [32] is a modern example of a BLE-based tracking system featuring some privacy enhancements. Eddystone allows "ephemeral identifiers" that rotate periodically to enhance privacy. Only trusted parties can resolve the ephemeral identifier to the actual identity by performing an expensive, yet feasible, computation. Other parties that attempt to deanonymize ephemeral identifiers are faced with a computational problem of prohibitive cost. Apple's iBeacon is a BLE-based tracking system but it does not support any equivalent of ephemeral identifiers [9].

In addition to BLE, key technologies for location tracking are GPS and RFID. The former is a satellite-based positioning system providing 3D coordinates with up to five-meter-precision anywhere on Earth. This high accuracy, however, comes at a cost in terms of power, with modern GPS receivers drawing up to tens of mA when tracking objects [8]. RFID tags are passively powered circuits that are typically used to identify objects. Due to their passive nature, the power consumption of these tags is extremely low, with read-outs requiring only around $10\,\mu W$ of input power from the reader [11]. However, this also implies that tags can only be read out from a limited distance, therefore offering a shorter range and a lower accuracy than GPS. In our system, the tag determines its position using BLE, which has a current draw on the order of hundreds of μA when scanning for beacons [10]. Our solution can therefore be positioned between GPS and RFID in terms of range, accuracy, and power.

Several systems have been proposed that allow tracking of objects such as bicycles, while using existing IOT technologies and preserving the user's privacy. Bochem et al. [25] presented a privacy-preserving tracking system that relies on GPS-enabled devices and base stations distributed over the tracking area. When a device is reported as stolen by its legitimate user, base stations broadcast its ID in order to activate its tracking module. This solution, however, presents several limitations. First, broadcasting the unique ID of a stolen device can pose significant privacy risks. Second, GPS may not be suitable for resource-constrained tracking modules since it incurs high energy costs. Third, this solution requires the service provider to be trusted. A rogue service provider could, for example, pretend that a given device has been reported as stolen and start broadcasting its ID. Fourth, the device needs to support downlink communication with the base stations. This process can be expensive in terms of energy since the device needs to periodically scan for messages sent by base stations. We address all four limitations: tags encrypt beacon IDs and use pseudonyms, localization is achieved using BLE, the service provider is trusted only for availability reasons, and tags only use uplink communication.

The Adeona system by Ristenpart et al. [49] presents an efficient solution that can be used to track Internet-connected devices. However, Adeona has several limitations. Firstly, it only supports the tracking of devices that are connected to the Internet. Secondly, the location is determined using IP geolocation services. Thirdly, it requires devices and users to have a synchronized clock. Our solution addresses all these shortcomings. By using BLE and LoRa technologies, objects that are isolated from the Internet can be tracked without the need for a synchronized clock.

3   Threat Model and Assumptions

Threat Model. Our privacy-preserving tracking system considers the presence of both passive adversaries, who can eavesdrop on any given communication channel in Fig. 1, and active adversaries, who can additionally jam, replay, modify, inject, delay, and forge messages. We do not consider fingerprinting attacks that uniquely identify a device based on its physical characteristics [28]. Adversaries can obtain physical access to a tag and perform simple side-channel attacks such as *Simple Power Analysis* (SPA) and *Differential Power Analysis* (DPA). More sophisticated side-channel attacks such as template attacks require extensive profiling of the device and are not considered. We do not claim any resistance to adversaries who can perform physically invasive attacks on a tag. This implies that data stored in volatile memory, e.g., registers, and *Non-Volatile Memory* (NVM), e.g., flash, can be neither read nor modified. In Sect. 4.4, we extend our design such that the assumption of secure NVM can be relaxed. Debugging and programming interfaces are irreversibly disabled, i.e., internal components cannot be accessed through software tools. Adversaries can detach tags from their assets; for a bicycle, this could be made more difficult by hiding tags inside the frame. Finally, note that an adversary could always track an object in a limited region with relatively modest means, e.g., a set of cameras.

Assumptions. While some of the beacons can act maliciously, e.g., by broadcasting the ID of another beacon, we assume that the majority of the beacons is legitimate. This enables the user terminal to detect inconsistencies in the received location data. The LoRa infrastructure is trusted only for availability reasons; LoRa provides built-in device authentication which allows to protect against *Denial-of-Service* (DOS) attacks such as flooding. However, the current LoRa protocol does not provide any location privacy since all messages contain the unique *Media Access Control* (MAC) address of the sender. Our solution targets a future, privacy-preserving version of LoRaWAN that would support privacy features such as MAC randomization. Furthermore, while we choose LoRa to build our prototype, any similar networking technology could be used.

The application server is trusted for availability reasons and is assumed to be honest-but-curious. In other words, it follows the protocol specifications but may try to learn information about the users and the tags. The application server communicates with both the LoRa infrastructure and the user terminal over a secure channel, e.g., using *Transport Layer Security* (TLS). To preserve anonymity, users can establish a connection over the Internet with, e.g., Tor [30]. Moreover, users should regularly delete their tag's location history from their terminal, thereby limiting forward privacy loss if the terminal were to be compromised.

4   Design

Section 4.1 describes the main entities and communication channels of our proposed system. The requirements of the corresponding protocol are listed in Sect. 4.2, and followed by its specification in Sect. 4.3. To allow for more cost-effective NVM technologies with relaxed assumptions on physical security, we extend our design with a PUF-based key generator in Sect. 4.4.

4.1   System Model

Our privacy-preserving tracking system is shown in Fig. 1 and can be summarized as follows: a tag scans the BLE channel at fixed time intervals and stores the IDs of all beacons it sees. In addition, the tag periodically sends an encrypted list of observed beacon IDs to the application server via LoRa. LoRa is currently one of the most promising wireless technologies, because it allows for long-range, low-power, and low-cost communication. This makes LoRa suitable for tracking devices that need to transmit small amounts of data a few times per day over long distances. It is important to note, however, that other LPWAN technologies could be used as well.

To prevent unauthorized parties from linking LoRa messages to tags, the use of a public and static tag ID should be avoided. In our proposal, tags reveal their identity through a one-time pseudonym that is refreshed with every message and that can only be regenerated by the user terminal. By querying the application server with a valid pseudonym, the user terminal can retrieve the location of its tracking device.

Below, we describe the five main entities of our system in more detail. This includes tags, BLE beacons, the LoRa infrastructure, the application server, and the user terminal.

Tags. Tags are small, inexpensive, low-energy, and self-powered devices that are attached to an object, e.g., a bicycle, for tracking purposes. They consist of a low-end microcontroller and the following two network interfaces: BLE and LoRaWAN. The former is used to collect the unique ID of all beacons that are in close proximity to the tag, whereas the latter is used to regularly send the encrypted version of the beacon IDs to the application server.

Beacons. Battery-powered beacons are provisioned in tracking-enabled areas. Each beacon continuously transmits its unique ID over BLE.

LoRa Infrastructure. The LoRa infrastructure includes gateways and servers that are managed by a network provider. The main function of the LoRa infrastructure is to act as a channel between the tags and the application server. More specifically, the LoRa infrastructure uses gateways to gather the messages sent by tags over LoRa, stores them on internal servers, and then forwards them to the corresponding application server.

Application Server. The application server collects the encrypted versions of beacon IDs, which are sent by the tags through the LoRa infrastructure, and provides them upon request to the authorized users.

User Terminal. Users download the location history of their tags from the application server. The user terminal can be a smartphone or a laptop.

Beacons communicate with tags wirelessly using a unidirectional BLE channel, whereas the communication between tags and the LoRa infrastructure takes place over a LoRa uplink. The application server communicates with the LoRa infrastructure and the user terminal over the Internet, respectively using a secure channel (e.g. TLS) and through an anonymous network (e.g. Tor).

4.2 Design Requirements

Our privacy-preserving tracking system should satisfy the following functional (F), security (S) and privacy (P) requirements.

(F1) Energy Usage: A tag's energy usage for both transmissions and computations should be as low as possible.

(F2) Cost: The cost of tags after adding the security and privacy mechanisms should remain as low as possible.

(F3) Efficiency: Users should be able to retrieve the location data from the application server in an efficient manner.

(S1) Confidentiality of Location Data: Only authorized users should be able to access the location history of their tags.

(S2) Message Integrity: Users should be assured that the received messages are fresh and have not been altered during transit.

(S3) Tag Authentication: Users should be assured that the received messages are sent by the tags attached to their personal belongings.

(P1) Tag Identity Privacy: No unauthorized entity should be able to learn the identity of a tag from its transmissions.

(P2) Message Unlinkability: No entity should be able to link a transmission to a tag, or even associate two transmissions to the same source.

(P3) Forward Privacy: Adversaries who compromise a tag or its corresponding user terminal should not be able to learn the tag's past locations.

4.3 Protocol

We now specify a protocol that allows users to track their belongings in a privacy-preserving manner. At constant time intervals, a tag wakes up from sleep mode and stores all beacon IDs it can receive over BLE. The geographical coordinates of each beacon are known and hence allow users to estimate the location of their tags. For this purpose, each tag periodically transmits an encrypted version of the collected beacon IDs to the application server over LoRa, using a key that is shared with the legitimate user only. Each message also contains an ID of the tag and can hence be retrieved by the user. Given that adversaries, including the application server, would be able to link messages if tag IDs are static, this comprises a dynamic, one-time pseudonym that can be reconstructed by the user terminal only. To prevent linkage through predicable transmission times, each message is sent at a random time within a pre-determined interval.

Following the manufacturing of a tag, a state stored in physically secure NVM is initialized with a tag-specific, symmetric master key $\mathbf{k}_{\text{Tag}}$. The programming interface is irreversibly disabled afterwards. The value of $\mathbf{k}_{\text{Tag}}$ is also printed in the form of a machine-readable barcode, i.e., a *Quick Response* (QR) code [36], and given to the user when the tag is bought. To prevent the key $\mathbf{k}_{\text{Tag}}$ from leaking through the supply chain, the QR code is stored inside the tracker's tamper-evident packaging. The user terminal is initialized by scanning the QR code and hence shares $\mathbf{k}_{\text{Tag}}$ with the tag. After the initialization, the QR code should be either destroyed or stored in, e.g., the user's home safe. To facilitate an implementation resistant to side-channel attacks (cf. Sect. 6.2), a tag updates the value of its key after every transmission. The first session key $\mathbf{k}_0$ is obtained by feeding $\mathbf{k}_{\text{Tag}}$ into a cryptographic hash function H, thereby overwriting the state stored in NVM. As depicted in Fig. 2, the same update mechanism applies to all subsequent session keys $\mathbf{k}_i$.

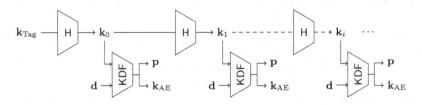

Fig. 2. A one-time session key $\mathbf{k}_i$ is obtained by hashing the previous session key $\mathbf{k}_{i-1}$, with $\mathbf{k}_0 = \mathsf{H}(\mathbf{k}_{\text{Tag}})$. Next, the pseudonym $\mathbf{p}$ and encryption key $\mathbf{k}_{\text{AE}}$ are derived from $\mathbf{k}_i$. Here H is a cryptographic hash function and KDF$(\mathbf{d}, \mathbf{k})$ a Key Derivation Function with key domain separator $\mathbf{d}$.

To protect a user's privacy, we designed our system such that tags only transmit a single LoRa message per hour and the LoRa network provider does not learn to whom a tag belongs. For this purpose, a tag generates a unique pseudonym $\mathbf{p}$ and a one-time encryption key $\mathbf{k}_{\text{AE}}$ for every transmission. Both are derived from the session key $\mathbf{k}_i$ using a key derivation function KDF$(\mathbf{d}, \mathbf{k})$, where $\mathbf{d}$ denotes the key separation domain and $\mathbf{k}$ denotes the key. To be precise, $\mathbf{p} = \mathsf{KDF}(0, \mathbf{k}_i)$ and $\mathbf{k}_{\text{AE}} = \mathsf{KDF}(1, \mathbf{k}_i)$.

Messages consist of a pseudonym $\mathbf{p}$ and an encrypted payload. The latter, in turn, consists of a ciphertext $\mathbf{c}$ and a message authentication tag $\mathbf{t}$, i.e., $(\mathbf{c}, \mathbf{t}) = \mathsf{AE}_{\mathbf{k}_{\text{AE}}}(\mathbf{n} = 0, \mathbf{a} = \mathbf{p}, \mathbf{m} = \mathbf{b})$. Here, $\mathsf{AE}_{K_{\text{AE}}}$ denotes authenticated encryption with associated data using key $\mathbf{k}_{\text{AE}}$, where the first argument is the nonce $\mathbf{n}$, the second argument the associated data $\mathbf{a}$ (which is authenticated but not encrypted) and the third argument is the plaintext $\mathbf{m}$. Note that the nonce $\mathbf{n}$ can be set to zero as a unique key is used for every encryption. The list of beacon IDs discovered during a scan interval is denoted by $\mathbf{b}$; it always contains q entries, where q is a constant that is constrained by the bandwidth of the LoRa connection. If less than q beacons were detected during the scan interval, the list is padded with zeroes. If more than q beacons are discovered, the oldest ones are discarded.

When users download the location history data from the application server through their user terminal, the following procedure applies. First, they regenerate the current pseudonym $\mathbf{p}$ from the previous session key $\mathbf{k}_i$ by hashing it, where $\mathbf{k}_0 = H(\mathbf{k}_{\mathrm{Tag}})$. Subsequently, they send a request to the application server for data $(\mathbf{c}, \mathbf{t})$ corresponding to pseudonym $\mathbf{p}$, issuing an individual request for each pseudonym. Finally, the user terminal decrypts the ciphertext $\mathbf{c}$, verifies the message authentication tag $\mathbf{t}$, and processes the data items.

Since the networking service may not always be guaranteed, it is possible that messages sent by the tag are not received by the application server. In this event, the session keys $\mathbf{k}_i$ at the user terminal and tag would be out of sync. Furthermore, when the user terminal has not been used for some period of time, its hash chain would also become desynchronized. In order to resynchronize, the user terminal advances the hash chain and queries the application server for the pseudonym derived from it, until the server again replies with a valid message. The tag's transmission interval determines an upper bound on the number of attempted requests. On initialization, the user terminal performs a similar exhaustive search to recover the session key of the first message stored on the server. However, in order to prevent timing attacks where the application server would link requests for different pseudonyms to the same user terminal, and consequently the same tag, the user terminal should issue these requests with random delays between them. As we assume the user terminal features powerful processing and high-bandwidth network connectivity, the time required for such searches is on the order of seconds, when the randomization delays are not considered.

4.4 PUF-Based Key Generation

For tags to remain functional after an occasional reboot, the value of the last session key $\mathbf{k}_i$ previously had to be stored in physically secure NVM. We now extend our design with a PUF-based key generator such that this assumption is partially mitigated. To be precise, the attacker is allowed to have read-access, and for the most part also write-access, to the NVM contents. Volatile memory, however, is still assumed to be inaccessible to the attacker. Moreover, the extended design only requires the NVM to be one-time programmable, thereby eliminating the higher manufacturing cost of multiple-time programmable NVM. Flash, for example, requires floating-gate transistors and hence additional photomasks and processing steps with respect to *Complementary Metal–Oxide–Semiconductor* (CMOS)-compatible fuses.

Each tag implements an oversize PUF such that $u > 1$ independent keys $\mathbf{k}_{\mathrm{Tag}}$ can be extracted from its long response $\mathbf{x}$ [21]. For this purpose, the response $\mathbf{x}$ is subdivided into u equally-sized partitions. A *fuzzy extractor* [31] ensures that each $\mathbf{k}_{\mathrm{Tag}}$ can be reproduced in a reliable manner despite the inherently noisy evaluation of the PUF. This requires the storage of public helper data $\mathbf{h}$ in NVM. Each $\mathbf{k}_{\mathrm{Tag}}$ is used for the derivation of at most v session keys $\mathbf{k}_i$, as previously depicted in Fig. 2. After v iterations, or whenever a device reboots, we move to the next partition of response $\mathbf{x}$. A monotonic counter $c \in \{1, 2, \cdots, u\}$, which can be implemented with u fuses, points to the currently active partition. The physical protection of this counter should primarily prohibit write-access: a rollback of c would breach forward privacy. Read-access is of limited importance, given that for large-scale systems with numerous tags, c is only a weak identifier. Moreover, the need for repeated physical access defeats the purpose of this attack: mostly, a user's belongings are inherently identifiable with the naked eye, and if not, it is would be easier to add a low-tech, nearly-invisible marking.

The use of $u > 1$ partitions not only enables storing the session key hash chain in volatile memory, as a new chain is started when the device reboots, it also prevents side-channel attacks where a tag is repeatedly rebooted. We suggest using $u = 100$ keys $\mathbf{k}_{\text{Tag}} \in \{0, 1\}^{128}$, given that 12.8 kbit still fits the storage offered by a single QR code [36]. To support a device lifetime of, for example, 5 years, $v = 438$ session keys $\mathbf{k}_i$ hence would have to derived from each $\mathbf{k}_{\text{Tag}}$. An occasional reboot therefore shortens the lifetime with at most 18.25 days. The ESP32 microcontroller has plenty of *Static Random-Access Memory* (SRAM), i.e., the core component of our PUF, and NVM available, and hence induces no extra cost for the extraction of multiple keys $\mathbf{k}_{\text{Tag}}$. On an ASIC, however, additional resources would have to allocated. Finally, we note that if the keys $\mathbf{k}_{\text{Tag}}$ are recovered by the bootloader, the boot process should be protected, e.g., by placing it in ROM.

5   Implementation

In Sect. 4, we proposed a system consisting of five entities. We now present its prototype on commercial hardware, thereby implementing a tag (Sect. 5.1), the application server (Sect. 5.2), and the user terminal (Sect. 5.3). Since the beacons do not require any custom functionality, we used commercial *Proximity Beacons* by Estimote [2]. They were configured to broadcast iBeacon packets, which contain a *Universally Unique Identifier* (UUID) [9]. The beacons have a configurable range of up to 70 m. To increase their battery life, the range is set to approximately 50 m. The use of the crowd-sourced LoRa infrastructure, managed by The Things Network [5], is not discussed in detail.

5.1   Tag

We selected the Pycom LoPy development board [48] to build a prototype implementation of the tag. This board features an Espressif ESP32 chipset which interfaces with Bluetooth, LoRa, and Wi-Fi radios. This chipset includes a dual-core microcontroller running at 240 MHz, and has 512 KB of *Dynamic Random-Access Memory* (DRAM) and 4 MB of flash memory. It also has hardware accelerators for the *Advanced Encryption Standard* (AES) and the *Secure Hash Algorithm* (SHA) family. The board runs MicroPython, which is an implementation of the Python 3 language for microcontrollers, offering high-level APIs for a lot of its functionality (e.g., the wireless radios and crypto accelerators).

When a device boots, it connects to the LoRa network and configures two internal timers. At the firing of the first clock with period t_{sci}, the Bluetooth radio is enabled and the tag starts scanning during a time t_{scd} for iBeacon advertisement frames. When a beacon is discovered, the upper $m = 4$ bytes of its UUID are stored. This enables the deployment of up to $2^{32} \approx 4.3$ billion beacons in a given area. As the beacons broadcast continuously, advertisements are received multiple times during each scan interval. Discovered identifiers are therefore only added to the transmission buffer once during each scan interval. After the scan has finished, the Bluetooth radio is turned off again. At most $q_1 = 3$ IDs are stored during each scan. In our current prototype, the scan interval t_{sci} is set to five minutes, and the scan duration t_{scd} to one minute.

At the rate of the second, slower clock with period $t_{txi} = q_2 \cdot t_{sci}$, where q_2 is a positive integer, an encrypted version $\mathbf{c}$ of the collected IDs is transmitted in addition to a pseudonym $\mathbf{p}$ and its tag $\mathbf{t}$. The latter quantities are computed from the current session key $\mathbf{k}_i$ stored in flash memory. Afterwards, $\mathbf{k}_{i+1}$ is computed and overwrites the previous session key. To prevent an attacker from linking transmissions from the same tag if they were all sent at the end of their corresponding transmission intervals, each message $(\mathbf{p}, \mathbf{c}, \mathbf{t})$ is sent at a random time within its transmission interval. An alternative solution requires all tags to be perfectly synchronized in order to transmit at the exact same global time, regardless of whether reboots occur. In our current prototype, the transmission interval t_{txi} to one hour, which implies $q_2 = 12$. Thus, at most $q = q_1 \cdot q_2 = 36$ IDs need to be stored by a tag.

The size of pseudonym $\mathbf{p}$ is determined by the probability that a collision occurs between pseudonyms that belong to either the same or different users. Since the user terminal only has access to the master key $\mathbf{k}_{Tag}$ associated with its own tag, it cannot generate session keys $\mathbf{k}_i$ other than its own. Therefore, only messages belonging to the associated tags can be decrypted; as a consequence, collisions between pseudonyms do not leak information but increase the amount of data transmitted to and processed by the terminal. Our prototype uses eight-byte pseudonyms. Consequently, the MACs $\mathbf{t}$ should be at least 64 bits (eight bytes) to protect the messages' integrity against online attacks and to filter out such dummy traffic relating to other users. These choices result in messages of 160 bytes: $(\mathbf{p} \to 8$ bytes$) + (\mathbf{c} \to q \cdot m = 144$ bytes$) + (\mathbf{t} \to 8$ bytes$)$.

Since the LoPy board features AES acceleration, we selected AES-COLM [13] as the authenticated encryption scheme. This algorithm is a candidate in the CAESAR competition and supersedes the COPA and ELmD proposals. Aside from the AES algorithm, its building blocks consist of XORs, multiplications with constants in $GF(2^{128})$, and a linear mixing function. This makes it suited for both software and hardware implementations. Furthermore, the availability of hardware support for the SHA family of hash functions determined our choice of a hash-based key derivation function, i.e., $\mathsf{KDF}(\mathbf{d}, \mathbf{k}) = \mathsf{SHA256}(\mathbf{d}\|\mathbf{k})$. Note that our protocol depends in no way on the algorithms selected for the prototype, and that they can all be replaced with functionally identical alternatives.

The tag connects to a nearby LoRa gateway registered with The Things Network [5]. This is a global community which is building a LoRaWAN network by crowdsourcing gateways. Anyone with a compatible device can add their node as a gateway, while the organization provides all other backend infrastructure (e.g., servers to handle the messages received by the LoRa gateways).

SRAM PUF. The extended version of our protocol requires the implementation of a PUF. As pointed out by Layman et al. [43], a *Static Random-Access Memory* (SRAM) can be adopted as a PUF. Its initial state $\mathbf{x} \in \{0,1\}^n$ after power-up provides a device-unique fingerprint. For our proof-of-concept implementation on an ESP32 microcontroller, an internal SRAM is readily available. The state $\mathbf{x}$ is read-out by listing addresses in counter mode and concatenating the corresponding 32-bit words.

Secure Sketch. A *secure sketch* [31] provides an *information-theoretically secure* mechanism to transform the noisy state X into a stable secret Y. During the enrollment, public helper data $\mathbf{h}$ is generated for a reference response $\mathbf{x}$ and is subsequently stored in flash memory. In the field, the helper data $\mathbf{h}$ allows the reference response $\mathbf{x}$, or a related variable $\mathbf{y}$, to be recovered from a newly generated response $\tilde{\mathbf{x}}$ that is sufficiently close to $\mathbf{x}$.

Instances of a secure sketch are most frequently based on a binary $[n, k, d]$ block code, where n is the codeword length, k is the message length, and d is the minimum distance. Seven code-based constructions are known to be equivalent in terms of min-entropy loss [29]. For a uniformly distributed reference input X, i.e., $\mathbb{H}_\infty(X) = n$, it holds that $\tilde{\mathbb{H}}_\infty(Y|H) = k$. More generally, for potentially non-uniformly distributed inputs X, it holds that $\tilde{\mathbb{H}}_\infty(Y|H) \geq \mathbb{H}_\infty(X) - (n - k)$. Unfortunately, the latter bound is not very tight if the non-uniformities are major. For our prototype, we simply assume that the state X is uniformly distributed, given that neither the distribution of X nor the min-entropy $\mathbb{H}_\infty(X)$ can precisely be determined from experimental data.

Out of seven constructions, we opt for the proposal of Kang et al. [39] because of three reasons: the size of the helper data is $(n - k)$ bits rather than n bits, the size of $\mathbf{y}$ is k bits rather than n bits, and the enrollment does not require a random number. The generator matrix is required to be in standard form, i.e., $\mathbf{G} = (\mathbf{I}_k\,\mathbf{P})$, where $\mathbf{I}_k$ is the $k \times k$ identity matrix. During the enrollment, the helper data is computed as follows: $\mathbf{h} \leftarrow (x_1\,x_2\,\cdots\,x_k)\mathbf{P} \oplus (x_{k+1}\,x_{k+2}\,\cdots\,x_n)$. In the field, the reference output $\mathbf{y} = (x_1\,x_2\,\cdots\,x_k)$ is recovered as follows: $\hat{\mathbf{y}} \leftarrow \mathsf{Decode}(\tilde{\mathbf{x}} \oplus (\mathbf{0}\|\mathbf{h}))$.

We instantiate the binary code with the proposal of Van der Leest et al. [53]. To be precise, the response $\mathbf{x}$ is subdivided into z partitions and the concatenation of an $[n_1 = 24, k_1 = 12, d_1 = 8]$ Golay code and an $[n_2, k_2 = 1, d_2 = n_2]$ repetition code is applied to each partition. Globally, this can be understood as an $[n = z\,n_1\,n_2, k = z\,k_1, d = d_1\,d_2]$ linear code. We choose $z = 11$ so that

$\tilde{\mathbb{H}}_\infty(Y|H) = 132$ exceeds the key size 128. The only remaining degree of freedom, n_2, is chosen in conformity with the error rate of the PUF. The repetition decoder outputs, in addition to a presumed message $\hat{y}_1 = x_1$, a level of confidence that decreases with the number of errors $\in \{0, 1, \cdots, (n_2 - 1)/2\}$ that has presumably been corrected. Subsequently, we iterate over all 2^{12} Golay codewords and retain the one that achieves maximum likelihood. For a bit error rate of 15% and $n_2 = 3$, i.e., a total of $n = 792$ response bits, a Monte Carlo experiment shows that the failure rate for reconstructing the key $\mathbf{k}_{\text{Tag}}$ is approximately 10^{-5}.

Robust Fuzzy Extractor. To detect helper data manipulation, which might otherwise allow an attacker to recover the master key through failure statistics [24, 29], and to ensure that the key has been recovered correctly, we apply an integrity scheme [27]. First, we check whether the Hamming distance between the regenerated response $\tilde{\mathbf{x}}$ and the recovered response $\hat{\mathbf{x}}$ does not exceed a certain threshold. Subsequently, it is checked whether a precomputed hash value $\mathbf{h}_* = \mathsf{H}_1(\mathbf{y}, \mathbf{h})$, which is stored in either fuses or flash, can be reproduced. Only if the latter two checks succeed, the master key is computed as $\mathbf{k}_{\text{Tag}} = \mathsf{H}_2(\mathbf{y}, \mathbf{h})$.

ASIC. Finally, in addition to the LoPy-based design, we also propose a system-level architecture mapping our design to ASIC. Since tags are high-volume, low-cost, battery-powered devices, the primary design objectives are low area and low energy consumption. In the presented application scenario, a tag spends most of its lifetime in sleep mode and only occasionally performs computations or transmits messages. Therefore, the power consumption of the *System-on-Chip* (SOC) without communication blocks is dominated by the stand-by currents and hence not by dynamic switching behavior. Typically, in standard-cell-based designs, high area results in high static power consumption.

To avoid the high area and energy costs associated with a SHA-256 implementation, we recommend using AES-COLM for all cryptographic operations. The key $\mathbf{k}_{Tag}$ and the integrity data $\mathbf{h}_*$ are computed as the tag output $\mathbf{t}$ of the authenticated encryption module, using the helper data $\mathbf{h}$ and the reference output $\mathbf{y}$ of the secure sketch as associated data $\mathbf{a}$. To be precise, $(-, \mathbf{k}_{Tag}) = \mathsf{AE}_{\mathbf{k}=\mathbf{0}}(\mathbf{n} = \mathbf{0}, \mathbf{a} = \mathbf{y}\|\mathbf{h}, \mathbf{m} = \mathbf{0})$ and $(-, \mathbf{h}_*) = \mathsf{AE}_{\mathbf{k}=\mathbf{0}}(\mathbf{n} = \mathbf{1}, \mathbf{a} = \mathbf{y}\|\mathbf{h}, \mathbf{m} = \mathbf{0})$. Ciphertexts $\mathbf{c}$ are not used and hence do not have to be computed. The values of nonce $\mathbf{n}$ can be chosen arbitrarily, but should differ for both instances. Similarly, the session key $\mathbf{k}_{i+1}$, the pseudonym $\mathbf{p}$, and the encryption key $\mathbf{k}_{\text{AE}}$ are derived using three different values of the nonce $\mathbf{n}$. To be precise, $(-, \mathbf{k}_{i+1}) = \mathsf{AE}_{\mathbf{k}_i}(\mathbf{n} = \mathbf{1}, \mathbf{a} = \mathbf{0}, \mathbf{m} = \mathbf{0})$, $(-, \mathbf{p}) = \mathsf{AE}_{\mathbf{k}_i}(\mathbf{n} = \mathbf{2}, \mathbf{a} = \mathbf{0}, \mathbf{m} = \mathbf{0})$, and $(-, \mathbf{k}_{\text{AE}}) = \mathsf{AE}_{\mathbf{k}_i}(\mathbf{n} = \mathbf{3}, \mathbf{a} = \mathbf{0}, \mathbf{m} = \mathbf{0})$.

Figure 3 shows the architecture of a tag. The system contains an embedded SRAM to store the beacon identifiers and the encrypted message before the transmission. Its power-up state serves as the PUF response $\mathbf{x}$. Each PUF response is generated using 132 bytes of raw SRAM data. A 128 kb SRAM block is sufficient for generating 100 PUF responses. In addition, the tag uses a one-time programmable NVM to store the helper data and the integrity data.

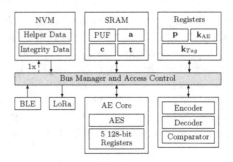

Fig. 3. System-level design of a prospective ASIC.

The required capacity of this memory is 10.6 kB. A AES-COLM core consists of the round-based AES-128 module, data paths for linear operations and five 128-bit registers for storing intermediate results. Flip-flop-based registers are used for storing the session key, the pseudonym and the encryption key. Dedicated data paths are used for encoding and decoding the PUF responses and for the integrity check. The bus manager controls bus access and communication between the cores and networking blocks. A *Finite State Machine* (FSM) is used to execute the protocol. In order to reduce the leakage power, the system should also contain power management logic for switching off the cores during sleep mode.

5.2   Application Server

After receiving a message from the tag through one of its gateways, the LoRa infrastructure provider forwards the payload to the application server over HTTP. The application server therefore runs a simple web service that accepts the messages and stores them in a database. It also provides an endpoint where the user terminal can retrieve the message corresponding to a specific pseudonym p. In addition, the application server stores the geographical coordinates of each beacon, which can be requested from a second endpoint.

The application server was built as a REST-based web service, which can be queried through a JSON API. The service was implemented in Go, with all data being stored in Bolt [1], a high-performance key/value store. Since both the messages and beacon coordinates are always accessed through their keys, i.e., the pseudonym and beacon identifier respectively, this database was preferred over more traditional relational databases. In order to reduce storage cost, the server periodically removes messages from its database (e.g., every month).

5.3   User Terminal

We implemented a prototype of the terminal as an iOS app written in Swift. Both AES-COLM and our KDF based on SHA-256 could be easily implemented for this platform. The performance of these operations is of lesser concern here, as our messages are relatively short and smartphones have significantly more processing power than tags. When the app is started for the first time, access is requested to the phone's camera to scan a QR code containing the tag key k_{Tag}, after which it is stored in iOS' keychain. Recall that this process links the terminal to the tag, allowing it to reconstruct the sequence of pseudonyms which can be used to retrieve messages from the server. The app will then synchronize its session key k_i by repeatedly querying the server until no new message is found. In order to preserve forward privacy, the terminal will always update the iOS keychain with the last synchronized session key, as this prevents an attacker from reconstructing the hash chain if they manage to breach the app. As mentioned in Sect. 3, protecting forward privacy if the terminal is compromised, requires that the user sets limits on the history kept by the user terminal. Configuring the user terminal to remove data immediately after displaying it would fully guarantee forward privacy, at the cost of not having access to location history.

When launched, the app retrieves all messages from the application server since its previous run: it repeatedly advances the session key hash chain and calculates the corresponding pseudonym p until the server responds that no data is available. The app will also retrieve the public list of beacons and their geographical location from the server. After decrypting each message, the beacon identifiers it contains are matched to the coordinate list and the tag's position during the scan interval is calculated through interpolation. Its location during the last 24 h is then shown on a map. In addition, the user can view the tag's calculated location during each scan interval.

6 Analysis

We first present a security and privacy analysis of our design, referencing the requirements identified in Sect. 4.2. Next, we discuss how our design intrinsically protects against side-channel attacks (Sect. 6.2).

6.1 Security and Privacy

Confidentiality and Authenticity of Location Data *(Satisfies S1, S2, and S3)*. A tag's location data is always sent in encrypted form using a secret key k_{AE} that is known only to the tag and an authorized user terminal. This offers end-to-end confidentiality, as it prevents intermediate entities from learning anything about the tag's location. Due to the use of an authenticated encryption scheme, unauthorized modifications are detected with extremely high probability.

Tag Identity Privacy *(Satisfies P1)*. To hide its identity, a tag appends a pseudonym p and hence not a public identifier to each message. This pseudonym p is generated using a KDF and a session key k_i that is known only to the tag and the authorized user terminal. Since p is relatively short, an occasional collision could imply that user A inadvertently downloads the encrypted location data of user B. Note, however, that user A cannot successfully decrypt user B's data.

Message Unlinkability *(Satisfies P2)*: Due to the freshness of the pseudonyms **p**, an unauthorized entity cannot link two or more messages that have been sent by the same tag. Furthermore, all LoRa messages have the same size and are sent at random intervals to prevent meta-data and traffic analysis attacks. An anonymous channel, e.g., Tor, between each user terminal and the application server prevents the latter from linking pseudonyms **p** of which a single terminal requests the location data.

Forward privacy *(Satisfies P3)*: Since only the last session key k_i is stored in the secure NVM, if adversaries compromise a tag or a user terminal, they cannot compute old session keys to learn the tag's past locations.

6.2   Side-Channel Security

An adversary who obtains physical access to a tag could mount a side-channel attack in an attempt to recover its cryptographic keys. This involves measuring the tag's instantaneous power consumption (or any related magnitude such as electromagnetic field intensity) while it performs a cryptographic operation. We focus on two common side-channel attacks: SPA and DPA. The protocol described in Sect. 4.3 features an intrinsic countermeasure. By construction, the protocol limits the number of cryptographic executions that an adversary may ask the device to perform under the same key. This limits the information leaked by each key, so that if the key is changed often enough, it becomes unfeasible for an adversary to reconstruct the whole key. This is the main working principle of "fresh re-keying" systems, initiated independently in the seminal patent of Kocher [41] and [26, § 6.6], and continued in [44,45].

More concretely, fresh re-keying guarantees resistance to SPA and DPA based on the following two assumptions [45]. Firstly, the key update mechanism must be resistant to DPA and SPA. In our case, the key update mechanism is $k_{i+1} = H(k_i)$. This makes DPA impossible, since each value of the hash chain is handled only once. Given that session key k_i is stored in NVM, a reboot does not allow an attacker to remeasure traces. The resistance to SPA must be guaranteed by constant-time, constant-flow code and a noisy enough hardware platform.

Secondly, the pseudonym and encryption key generation $G(k_i)$ must be resistant to SPA. This is again guaranteed by constant-time, constant-flow code and a noisy enough hardware platform. Note that this makes the protocol "somewhat stateful", since the receiving end must keep a minimal state as the last seen key from the hash chain. Nevertheless, we believe that this is an attractive property of the protocol since it protects forward privacy: key breaches do not allow to retroactively de-anonymize previous messages (and thus learn the past locations of the device). Thus, we fulfill design requirement P3 from Sect. 4.2.

For the extended version of our protocol, it is expected that the fuzzy extractor is resistant to first-order DPA attacks as well. Not only is the SRAM partition regularly updated, the monotonic counter prevents traces from being remeasured after a reboot. The masking-based countermeasure of Merli et al. [46], which precludes DPA attacks on a secure sketch, is therefore omitted. More sophisticated attacks such as template attacks [40] are ruled out since they require a characterization of the device leaking behaviour.

7 Evaluation

We evaluate our prototype on Pycom's LoPy board (Sect. 5) in terms of latency in Sect. 7.1. Subsequently, we quantify its power/energy usage in Sect. 7.2 through a custom-designed measurement board for *Universal Serial Bus* (USB)-powered devices. In Sect. 7.3, we quantify the randomness and stability of its SRAM PUF. Finally, we estimate the energy usage of our ASIC design in Sect. 7.4.

7.1 Latency

The energy usage, which is the dominant constraint in our application, scales to some extent proportionally with the latency of the cryptographic operations on our microcontroller. First, we benchmarked the derivation of the pseudonym $\mathbf{p}$ and the encryption key $\mathbf{k}_{AE}$: this operation takes on average 855.21µs. Next, our COLM implementation averages 96.13 ms when encrypting a single 144-byte list $\mathbf{b}$ of beacon IDs (Sect. 5.1).

7.2 Power and Energy

In addition to the tag's performance, we evaluated its power and energy usage. To this end, we designed a USB pass-through board, powering the LoPy through this board, rather than directly from the power source. A 1 Ω resistor was placed between the source's ground line and the Lopy's ground, allowing us to measure the current drawn by the board by sampling the voltage over this resistor. The measurement points were connected to an MSP430F5529 development board, which features a 12-bit *Analog-to-Digital Converter* (ADC) (cf. Fig. 4). As the LoPy is rated to draw at most 400 mA, the microcontroller's REF module was used to generate a 1.5 V reference voltage, yielding a measurement resolution of 3.66 mA, as the measured voltage is equal to current due to the use of a 1Ω resistor. Since a USB port is used as the source, the board is powered at 5 V.

Before measuring the energy usage of our overall application, we looked at the board's power usage in four different instances. First, we measured that the device draws 39.25 mA when idling with all radios asleep. Placing the board in deep sleep mode reduced the average current to 19.87 mA, which is still much higher than expected due to a silicon bug in our version of the LoPy.[1] Next, we programmed the board to perform one million AES encryptions of a single block in ECB mode and to hash 16 bytes one million times, measuring respective average energy usages of 35.58 µJ per encryption and 44.17 µJ per hash.

---

[1] https://forum.pycom.io/topic/1022/root-causes-of-high-deep-sleep-current.

Fig. 4. Measurement setup for the power usage of our prototype implementation. In addition to the USB pass-through board connected to an MSP430F5529 (bottom left) and LoPy (bottom right), two Estimote beacons are shown.

Lastly, we investigated the current usage when our protocol implementation is running on the board. During a one-hour interval with a single LoRa transmission, 874.29 J is consumed on average. As can be seen from Fig. 5, this is mainly caused by the Bluetooth scanning, with the board drawing 88.16 mA during the scan interval, because the current ESP32 SDK only supports enabling both the BLE and Classic modes of the Bluetooth radio. Note that our implementation does not place the board in deep sleep, resulting in high average power usage. However, it can be seen from the graph that the device is only active for short bursts, and the required energy to run our protocol would therefore drop significantly when less current is drawn in between. Extending the prototype implementation to make use of deep sleep functionality is left as future work. Due to the high Bluetooth current usage caused by the ESP32 SDK, our prototype implementation runs for about 103 h when used with a 5000 mA h battery. However, note that the battery life of our application can be improved by tweaking the scan duration and scan interval. For example, the energy usage over the one-hour interval drops to 718.55 J when t_{sci} and t_{scd} are configured to 15 min and 15 s respectively. Finally, the development of an ASIC, rather than a general-purpose chip, greatly reduces current consumption, enabling the use of coin-sized batteries (Sect. 7.4).

The listed issues with the current prototype complicate direct energy comparison with GPS and RFID (Sect. 2). Furthermore, the evaluated prototype is a SOC featuring several components in addition to the BLE receiver. Additionally, as shown by our analysis, the energy usage very much depends on the exact implementation and the activity of each component. However, while we did not evaluate the tracking accuracy and range of our design, it performs better than RFID in these areas through its use of BLE, but is outperformed by GPS.

7.3 SRAM PUF

Additionally, we have evaluated whether the LoPy's SRAM could be used for PUF-based key generation. The nominal environment, during which the enrollment takes place, is defined as $T = 25\,°C$. The error rate between two read-outs, and averaged over $n = 10^6$ response bits, is 6.1%. In the field, however, different temperatures might apply, thereby deteriorating the reproducibility of the state **x**. Using an ESPEC SH-662 temperature chamber, we evaluated the PUF

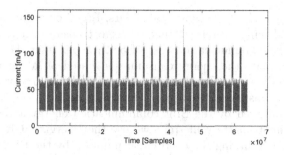

Fig. 5. The LoPy's current usage (20-140 mA) when running the prototype of the protocol implementation during two hours of operation, with the graph starting halfway through the scan interval.

of a single device under an extremely cold, i.e., $T = -40\,°C$, and an extremely hot, i.e., $T = 85\,°C$, environment. The error rates with respect to the nominal environment are 16.8% and 11.2% respectively. Although the latter two error rates are higher than for the 65 nm CMOS ASIC of Koeberl et al. [42], we emphasize that the ESP32 is only a prototype device. A *temporal majority vote*, which could have lowered the error rates at the cost of multiple read-outs, was not implemented.

Given that an SRAM consists of ideally autonomous cells, the distribution of X has the potential to be fairly uniform over $\{0,1\}^n$. For a set of 10 devices and $n = 10^6$, we apply statistical tests that can detect non-uniformities. To be precise, we estimate several probabilities through a 95% confidence interval and check whether their ideal value 50% is enclosed. The interval for the probability that a given response bit of a given device equals 1 is $[50.01\%, 50.07\%]$. The interval for the probability that two neighboring response bits in a given word of a given device are equal is $[50.44\%, 50.50\%]$. The interval for the probability that a given bit in two consecutive words of a given device is equal is $[49.71\%, 49.77\%]$. The interval for the probability that two devices produce the same value for a given response bit is $[50.24\%, 50.27\%]$. We conclude that minor non-uniformities are likely to exist.

7.4   ASIC

In order to estimate the battery lifespan for the future ASIC implementation we look into the energy consumption of the cryptographic core and communication blocks. The energy cost of a round-based AES-128 implementation is 350.7pJ per encryption using the STM 90 nm CMOS technology [23]. Every hour, 35 AES encryptions are required for preparing the message and updating the key, which costs 12.3 nJ. According to a study presented in [51], the energy cost of a 128 kb SRAM using 32-bit words and manufactured in 90nm CMOS technology is less than 10 pJ per access.

For SRAM using 32-bit word lengths, the presented protocol requires approximately 760 memory accesses per hour, which consumes less than 7.6 nJ. The transmission of each LoRa message takes $2\,\mu s$. Based on the experiments using the LoPy, the driving current during message transmission is $55.2\,mA$. This results in a LoRa communication cost of $55.2\,\mu J$ per hour. For BLE communication, we estimate the energy consumption based on a Texas Instruments Bluetooth controller [10]. From the number of scanning intervals we approximate that BLE consumes $2.8\,mJ$ every hour. The energy consumption of the chip will be dominated by the BLE communication energy and static leakage. In order to estimate battery lifetime, we use the Energizer CR1620 coin-sized battery [3]. Approximating the energy consumption by the BLE leakage power in shutdown mode and the transmission power, we end up with a rough battery lifetime approximation of 74 months.

8   Conclusion

This paper presented the design and a prototype implementation of a privacy-friendly tracking system that builds upon the following IOT technologies: inexpensive tags, *Bluetooth Low Energy* (BLE), and LoRa communications. Only legitimate tag owners can track their devices, while no information is leaked to network operators or service providers. The system was evaluated through a prototype based on an off-the-shelf IOT node. We have also produced estimates for the energy and power consumption of a possible ASIC taking the role of the tag. We have demonstrated that secure key storage using an SRAM PUF and side-channel resistance is achievable within realistic area and energy budgets. Future work should focus on a detailed ASIC design and a prototype for a privacy-friendly variant of the LoRa protocol with changing MAC addresses. The automated detection of compromised beacons could be added to the demonstrator. A large-scale field test would be helpful to obtain a more thorough experimental validation of the resistance against traffic analysis.

Acknowledgements. We would like to thank the anonymous reviewers for their feedback, as well as Patrick Tague for acting as our shepherd. This work is the result of collaborative research partially funded by the Attached Institute of ETRI. It was also supported in part by the KU Leuven Research Council through C16/15/058, the European Union's Horizon 2020 research and innovation programme under grant agreements No 644052 HECTOR and No 644371 WITDOM, ERC Advanced Grant 695305. Pieter Maene is an SB PhD fellow at Research Foundation - Flanders (FWO).

References

1. Bolt. https://github.com/boltdb/bolt
2. Estimote. https://estimote.com
3. Product Datasheet Energizer CR1620. http://data.energizer.com/pdfs/cr1620. pdf. Accessed 01 July 2017
4. RAPID7. https://community.rapid7.com/community/infosec/blog/2016/10/25/ multiple-bluetooth-low-energy-ble-tracker-vulnerabilities
5. The Things Network. https://thethingsnetwork.org

6. Tile. https://www.thetileapp.com
7. TrackR. https://thetrackr.com/bravo
8. A2235-H Stack-up Antenna SiRFstarIV Integrated Solution. Datasheet, Maestro (2012)
9. Proximity Beacon Specification. Specification, Apple (2015)
10. CC256x Dual-Mode Bluetooth Controller (Rev. E). Datasheet (2016)
11. SL3S1214 UCODE 7m Rev. 3.3. Datasheet, NXP Semiconductors (2016)
12. Alomair, B., Clark, A., Cuellar, J., Poovendran, R.: Scalable RFID systems: a privacy-preserving protocol with constant-time identification. IEEE Trans. Parallel Distrib. Syst. **23**(8), 1536–1550 (2012)
13. Andreeva, E., et al.: COLM v1 (2016). https://competitions.cr.yp.to/round3/colmv1.pdf
14. Avoine, G.: Privacy Issues in RFID Banknote Protection Schemes. In: Quisquater, J.J., Paradinas, P., Deswarte, Y., El Kalam, A.A. (eds.) 6th International Conference on Smart Card Research and Advanced Applications. IFIP International Federation for Information Processing, vol. 153, pp. 33–48. Springer, Boston (2004). https://doi.org/10.1007/1-4020-8147-2_3
15. Avoine, G.: Privacy challenges in RFID. In: Garcia-Alfaro, J., Navarro-Arribas, G., Cuppens-Boulahia, N., de Capitani di Vimercati, S. (eds.) DPM/SETOP -2011. LNCS, vol. 7122, pp. 1–8. Springer, Heidelberg (2012). https://doi.org/10.1007/978-3-642-28879-1_1
16. Avoine, G., Beaujeant, A., Hernandez-Castro, J., Demay, L., Teuwen, P.: A survey of security and privacy issues in ePassport protocols. ACM Comput. Surv. **48**(3), 47:1–47:37 (2016)
17. Avoine, G., Bingöl, M.A., Carpent, X., Yalcin, S.B.O.: Privacy-friendly authentication in RFID systems: on sublinear protocols based on symmetric-key cryptography. IEEE Trans. Mob. Comput. **12**(10), 2037–2049 (2013)
18. Avoine, G., Coisel, I., Martin, T.: Untraceability model for RFID. IEEE Trans. Mob. Comput. **13**(10), 2397–2405 (2014)
19. Avoine, G., Oechslin, P.: A scalable and provably secure hash-based RFID protocol. In: 3rd IEEE Conference on Pervasive Computing and Communications Workshops, pp. 110–114 (2005)
20. Avoine, G., Oechslin, P.: RFID traceability: a multilayer problem. In: Patrick, A.S., Yung, M. (eds.) FC 2005. LNCS, vol. 3570, pp. 125–140. Springer, Heidelberg (2005). https://doi.org/10.1007/11507840_14
21. Aysu, A., Gulcan, E., Moriyama, D., Schaumont, P., Yung, M.: End-to-end design of a PUF-based privacy preserving authentication protocol. In: Güneysu, T., Handschuh, H. (eds.) CHES 2015. LNCS, vol. 9293, pp. 556–576. Springer, Heidelberg (2015). https://doi.org/10.1007/978-3-662-48324-4_28
22. Lazos, L., Alomair, B., Poovendran, R.: Securing low-cost RFID systems: an unconditionally secure approach (2010)
23. Banik, S., et al.: Midori: a block cipher for low energy. In: 21st International Conference on Advances in Cryptology, pp. 411–436 (2015)
24. Becker, G.T.: Robust fuzzy extractors and helper data manipulation attacks revisited: theory vs practice. Cryptology ePrint Archive, Report 2017/493 (2017). http://eprint.iacr.org/2017/493
25. Bochem, A., Freeman, K., Schwarzmaier, M., Alfandi, O., Hogrefe, D.: A privacy-preserving and power-efficient bicycle tracking scheme for theft mitigation. In: 2nd IEEE International Conference on Smart Cities, pp. 1–4 (2016)

26. Borst, J.: Block Ciphers: Design, Analysis and Side-Channel Analysis. Ph.D. thesis, Katholieke Universiteit Leuven (2001). Bart Preneel and Joos Vandewalle (promotors)
27. Boyen, X., Dodis, Y., Katz, J., Ostrovsky, R., Smith, A.: Secure remote authentication using biometric data. In: Cramer, R. (ed.) EUROCRYPT 2005. LNCS, vol. 3494, pp. 147–163. Springer, Heidelberg (2005). https://doi.org/10.1007/11426639_9
28. Danev, B., Zanetti, D., Capkun, S.: On physical-layer identification of wireless devices. ACM Comput. Surv. 45(1), 6:1–6:29 (2012)
29. Delvaux, J.: Security Analysis of PUF-Based Key Generation and Entity Authentication. Ph.D. thesis, KU Leuven, June 2017
30. Dingledine, R., Mathewson, N., Syverson, P.F.: Tor: the second-generation onion router. In: 13th USENIX Security Symposium, pp. 303–320 (2004)
31. Dodis, Y., Ostrovsky, R., Reyzin, L., Smith, A.: Fuzzy extractors: how to generate strong keys from biometrics and other noisy data. SIAM J. Comput. 38(1), 97–139 (2008)
32. Hassidim, A., Matias, Y., Yung, M., Ziv, A.: Ephemeral identifiers: mitigating tracking & spoofing threats BLE beacons (2016)
33. Henrici, D., Götze, J., Müller, P.: A hash-based pseudonymization infrastructure for RFID systems. In: 2nd International Workshop on Security, Privacy and Trust in Pervasive and Ubiquitous Computing, pp. 22–27 (2006)
34. Henrici, D., Müller, P.: Hash-based enhancement of location privacy for radio-frequency identification devices using varying identifiers. In: 2nd IEEE Conference on Pervasive Computing and Communications Workshops, pp. 149–153 (2004)
35. Henrici, D., Müller, P.: Providing security and privacy in RFID systems using triggered hash chains. In: 6th Annual IEEE International Conference on Pervasive Computing and Communications, pp. 50–59 (2008)
36. Information - Automatic identification and data capture techniques - QR Code barcode symbology specification. Standard, International Organization for Standardization, vol. 2 (2015)
37. Juels, A., Pappu, R.: Squealing euros: privacy protection in RFID-enabled banknotes. In: Wright, R.N. (ed.) FC 2003. LNCS, vol. 2742, pp. 103–121. Springer, Heidelberg (2003). https://doi.org/10.1007/978-3-540-45126-6_8
38. Juels, A., Rivest, R.L., Szydlo, M.: The blocker tag: selective blocking of RFID tags for consumer privacy. In: 10th ACM Conference on Computer and Communications Security, pp. 103–111 (2003)
39. Kang, H., Hori, Y., Katashita, T., Hagiwara, M., Iwamura, K.: Cryptographic key generation from PUF data using efficient fuzzy extractors. In: 16th International Conference on Advanced Communication Technology, pp. 23–26. IEEE, February 2014
40. Karakoyunlu, D., Sunar, B.: Differential template attacks on PUF enabled cryptographic devices. In: 2nd Workshop on Information Forensics and Security (WIFS 2010), pp. 1–6. IEEE, December 2010
41. Kocher, P.: Leak-resistant Cryptographic Indexed Key Update (2003). US Patent 6,539,092
42. Koeberl, P., Maes, R., Rožić, V., van der Leest, V., Van der Sluis, E., Verbauwhede, I.: Experimental evaluation of physically unclonable functions in 65 nm CMOS. In: 38th European Conference on Solid-State Circuits, pp. 486–489, September 2012
43. Layman, P.A., Chaudhry, S., Norman, J.G., Thomson, J.R.: Electronic fingerprinting of semiconductor integrated circuits, May 2004. US Patent 6738294

44. Medwed, M., Petit, C., Regazzoni, F., Renauld, M., Standaert, F.-X.: Fresh Re-keying II: securing multiple parties against side-channel and fault attacks. In: Prouff, E. (ed.) CARDIS 2011. LNCS, vol. 7079, pp. 115–132. Springer, Heidelberg (2011). https://doi.org/10.1007/978-3-642-27257-8_8

45. Medwed, M., Standaert, F.-X., Großschädl, J., Regazzoni, F.: Fresh Re-keying: security against side-channel and fault attacks for low-cost devices. In: Bernstein, D.J., Lange, T. (eds.) AFRICACRYPT 2010. LNCS, vol. 6055, pp. 279–296. Springer, Heidelberg (2010). https://doi.org/10.1007/978-3-642-12678-9_17

46. Merli, D., Stumpf, F., Sigl, G.: Protecting PUF error correction by codeword masking. Cryptology ePrint Archive, Report 2013/334 (2013). http://eprint.iacr.org/2013/334

47. Molnar, D., Wagner, D.A.: Privacy and security in library RFID: issues, practices, and architectures. In: 11th ACM Conference on Computer and Communications Security, pp. 210–219 (2004)

48. Pycom. LoPy. https://www.pycom.io/product/lopy/

49. Ristenpart, T., Maganis, G., Krishnamurthy, A., Kohno, T.: Privacy-preserving location tracking of lost or stolen devices: cryptographic techniques and replacing trusted third parties with DHTs. In: 17th USENIX Security Symposium, pp. 275–290 (2008)

50. Saito, J., Ryou, J.-C., Sakurai, K.: Enhancing privacy of universal re-encryption scheme for RFID tags. In: Yang, L.T., Guo, M., Gao, G.R., Jha, N.K. (eds.) EUC 2004. LNCS, vol. 3207, pp. 879–890. Springer, Heidelberg (2004). https://doi.org/10.1007/978-3-540-30121-9_84

51. Sharma, V., Cosemans, S., Ashouie, M., Huisken, J., Catthoor, F., Dehaene, W.: Ultra low-energy SRAM design for smart ubiquitous sensors. IEEE Micro **32**(5), 10–24 (2012)

52. Spiekermann, S., Berthold, O.: Maintaining privacy in RFID-enabled environments. In: Robinson, P., Vogt, H., Wagealla, W. (eds.) Privacy, Security and Trust within the Context of Pervasive Computing. The International Series in Engineering and Computer Science, vol. 380, pp. 137–146. Springer, Boston (2005). https://doi.org/10.1007/0-387-23462-4_15

53. van der Leest, V., Preneel, B., van der Sluis, E.: Soft decision error correction for compact memory-based PUFs using a single enrollment. In: Prouff, E., Schaumont, P. (eds.) CHES 2012. LNCS, vol. 7428, pp. 268–282. Springer, Heidelberg (2012). https://doi.org/10.1007/978-3-642-33027-8_16

54. Weis, S.A., Sarma, S.E., Rivest, R.L., Engels, D.W.: Security and privacy aspects of low-cost radio frequency identification systems. In: Hutter, D., Müller, G., Stephan, W., Ullmann, M. (eds.) Security in Pervasive Computing. LNCS, vol. 2802, pp. 201–212. Springer, Heidelberg (2004). https://doi.org/10.1007/978-3-540-39881-3_18

Anonymous and Virtual Private Networks

Oh-Pwn-VPN! Security Analysis of OpenVPN-Based Android Apps

Qi Zhang, Juanru Li, Yuanyuan Zhang$^{(\boxtimes)}$, Hui Wang, and Dawu Gu

Shanghai Jiao Tong University, Shanghai, China
yyjess@sjtu.edu.cn

Abstract. Free VPN apps have gained popularity among millions of users due to their convenience, and have been massively used for accessing blocked sites and preventing network eavesdropping. As a popular open-source VPN solution, OpenVPN is widely used by developers to implement their own VPN services. Despite the prevalence of OpenVPN, it can be insecurely customized and deployed by developers in lack of security guide.

In this paper, we perform a systematic security analysis of 84 popular OpenVPN-based apps on the Google Play store. We analyze the deployment security of OpenVPN on Android from the aspects of client profile, code implementation, and permission management. Our experiment reveals three types of misconfigurations that exist in several apps: insecure customized protocols, weak authentication at the client side, and incorrect file permissions on Android. The misconfigurations found by us can lead to some serious attacks, such as VPN traffic decryption and Man-in-the-Middle attacks, endangering millions of users' privacy. Our work shows that, although OpenVPN protocol itself has withstood security analysis, insecure custom modification and configuration can still compromise the security of VPN apps. We then discuss potential causes of these misconfigurations and make practical recommendations for developers to securely deploy OpenVPN services.

Keywords: OpenVPN · Android apps · Security assessment

1 Introduction

Security concerns about network communications of Android apps have been raised in recent years. A straightforward protection approach is to use a virtual private network (VPN) as a secure connection between the device and VPN server over the Internet. VPN services are useful for securely accessing sensitive content in a public network and are commonly used to circumvent censorship.

This work was partially supported by the Key Program of National Natural Science Foundation of China (Grants No. U1636217), the Major Project of the National Key Research Project (Grants No. 2016YFB0801200), and the Technology Project of Shanghai Science and Technology Commission under Grants No. 15511103002.

S. Capkun and S. S. M. Chow (Eds.): CANS 2017, LNCS 11261, pp. 373–389, 2018.
https://doi.org/10.1007/978-3-030-02641-7_17

On Android, mobile app developers can use native support to create VPN clients through the Android VPN Service class [2]. Thus many apps legitimately use the VPN permission to offer online anonymity by intercepting and taking full control of the network traffic on device.

However, the use of VPN within an Android app is a new scenario for most developers. Previous researches [22,25,26] have revealed several privacy issues and security flaws in implementations of these VPN services and applications. The most serious security flaw found is the usage of insecure VPN tunneling protocols. Various VPN tunneling protocols are used among different Android VPN-based apps. Despite promising online anonymity and security to their users, many VPN apps still implement unencrypted tunneling protocols. Since implementing a secure VPN tunneling protocol from scratch is sophisticated, a group of VPN apps utilize OpenVPN, the most popular open-source VPN solution [4,17], to build their own VPN services. Because OpenVPN is open-source and has been tested by security analysts over a long period of time, it is generally considered as a secure VPN solution and is widely used on both desktop and mobile platforms.

Although OpenVPN-based apps (in short, OpenVPN apps) are believed to guarantee better security and anonymity compared to those apps with home-brewed tunneling protocols, unfortunately, real world OpenVPN apps are not always secure. On Android platform, how OpenVPN should be incorporated and deployed in these VPN apps is not regulated. Android developers may misuse OpenVPN or modify the original execution flow of it and thus lead to an insecure VPN service.

In this paper, we conduct an in-depth misuse analysis on widely used Android OpenVPN apps. To unveil those misuses, we focus on the variation of OpenVPN apps' tunnel implementations and deployment policies. Our analysis finds that due to three kinds of misuses, the security of the VPN tunnel is weakened or even completely broken. The first one is misuse of modified OpenVPN protocol. Developers add custom operation to the standard OpenVPN protocol implementation, as we called *custom obfuscation*, for the purpose of obfuscating the VPN traffic. The VPN connection is configured to replace the standard OpenVPN encryption with custom obfuscation, and finally leads to an insecure VPN tunnel. The second one is weak authentication at the client side, which leaves the identity of server insecurely validated and finally induces a Man-in-the-Middle attack. The third one is incorrect usage of native library, which assigns an improper privilege to the management interface and finally causes a Denial-of-Service threat. More seriously, the implementation and deployment of these apps cannot be modified by users. Users are generally unaware of relevant security flaws, and can be easily attacked if using such apps to protect their network communications.

We analyzed 84 popular free OpenVPN apps in Android market and found that such misuses widely exist. Among them, 11 apps replace the standard encryption of OpenVPN with custom obfuscation. Due to vulnerable key agreement of the custom obfuscation, VPN traffic can be completely decrypted by attackers. Seven of the apps are susceptible to Man-in-the-Middle (MITM) attacks as a result of weak authentication at the client side. Four of the apps suffer from Denial-of-Service attacks by reason of unprotected management interface. Our study indicates that even if the VPN apps adopt an robust VPN library, situations of insecure deployment are still common and severely threaten users' security and privacy.

The main contributions of our work are summarized as follows:

- We summarize how OpenVPN is incorporated and utilized by Android VPN apps. We conclude the typical usage of OpenVPN on Android and spot developers' customizations by analyzing popular OpenVPN apps and auditing source code of forked OpenVPN projects.
- We conduct an in-depth analysis of OpenVPN misuses. Our assessment methodology is able to find misconfigurations of OpenVPN apps in the aspects of client profile, code implementation, and permission management.
- We uncover a typical previously unknown security issue in OpenVPN apps. Specifically, we find that some apps add a new tunnel protocol into OpenVPN following the security-by-obscurity policy: these implementations of tunnel modify the original protocol to hide the feature and evade network censorship technologies such as deep packet inspection (DPI). However, our study demonstrates that the modified protocols often adopt vulnerable key agreement that leads to complete insecure communications.

2 Background

2.1 OpenVPN Security Mechanisms

The goal of a VPN system is to provide private communications. To secure the network traffic, OpenVPN has implemented many features for authentication, encryption and management. OpenVPN utilizes SSL as the underlying cryptographic layer for authentication and encryption. There are two channels in OpenVPN: the control channel for authentication and key exchange, and the data channel for traffic encryption. Moreover, OpenVPN also provides an interface for managing the VPN process.

Authentication. In the control channel, OpenVPN has two modes of authentication [16]: *(a)* **Static key mode** static keys are pre-shared by client and server. All traffic are encrypted by the same static key, thus this mode cannot provide perfect forward security. *(b)* **SSL/TLS mode** A mutual authentication is established inside an SSL session. Most security related features are implemented in this mode.

In SSL/TLS mode, the identity of server is validated by its certificate the same way as in HTTPS. Meanwhile, multiple ways of client authentication are provided by OpenVPN. The client can be authenticated by the traditional username/password mechanism, by client certificate, or by the combination of these two types. To mitigate possible vulnerabilities in the TLS handshake, e.g., the famous *Heartbleed* bug [7], additional HMAC of TLS control channel packets can be required by enabling *tls-auth* option. After the authentication, session keys are generated by Diffie-Hellman key exchange and updated periodically.

Encryption. After authentication and session key generation, OpenVPN uses data channel to tunnel the actual network traffic. *Encrypt-then-Mac* scheme is used to protect data channel packets. Specifically, the encryption and HMAC algorithm are determined by option *cipher* and *auth* in the configuration file. The cipher algorithm used for data encryption must be specified at both the client and the server side.

Management. OpenVPN provides a management interface [11] that allows itself to be administratively controlled from an external program via a TCP or UNIX domain socket. Control commands such as *setting proxy address, providing passwords* or *suspending VPN service* can be transmitted through the management interface.

2.2 OpenVPN on Android

Since Android version 4.0, the *VpnService* API [2] is provided for developers to build their own VPN solutions. This API returns a descriptor of a virtual network interface (the *tun* interface) for apps to read and modify all the network traffic on device. While the *VpnService* API makes it convenient for developers to build VPN services, malicious apps may use this API to eavesdrop network activity of other apps. Android system takes several actions to prevent the abuse of VPN Service API. To obtain the VPN interface by using this API, apps have to request the *BIND_VPN_SERVICE* permission, and the first time a VPN connection is created, Android alerts users by displaying system dialog and notification.

Typically, OpenVPN is ported to Android platform as a shared library. For instance, *ics-openvpn* [10], a popular open-source OpenVPN app, implements OpenVPN as an ELF shared library *libopenvpn.so*. The shared library is invoked by a native process on Android (the native layer). Other functions of the app are usually implemented at the Java layer. To handle the Inter-Process Communication (IPC) between the Java layer (i.e., the UI thread) and the native layer (i.e., the OpenVPN process), UNIX domain socket is adopted to implement the management interface.

The execution flow of an OpenVPN app is depicted in Fig. 1 and divided into four steps: *profile assembly, VPN initiation, management interaction* and *VPN connection*. The client profile for the OpenVPN app is retrieved in various ways (step 1). Based on the configuration file, OpenVPN process is initiated at the native layer (step 2), then the Java layer controls the OpenVPN process via the management interface (step 3) and the VPN tunnel is established by the OpenVPN process (step 4). Details of these steps are described as follows:

Step 1: The VPN client assembles a configuration profile for connecting to a remote VPN server. Note that this step can be implemented differently by VPN providers. The client can directly obtain the configuration from the APK file, or the client retrieves VPN server address from a server (the profile

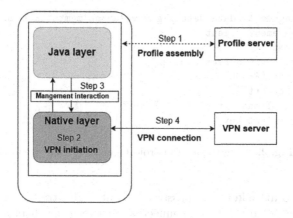

Fig. 1. A typical workflow of android OpenVPN apps

server), and then assembles a complete configuration at the client side, or the configuration is fully downloaded from the profile server. Other options such as file location of the management interface and protocol used (TCP or UDP) are also included in the configuration profile. An example of Android OpenVPN client profile is shown in Fig. 2.

Step 2: The OpenVPN library is loaded and invoked by a native process. Based on the client profile, the OpenVPN process is initiated. Network parameters of the *tun* interface such as IP address, DNS server are pushed from the VPN server. As shown in Fig. 2, the *management-client* option is enabled, thus OpenVPN process acts as the client and the management interface is created by the Java layer. The management interface is created in the app's private directory (e.g., */data/data/pkg.name/cache/*) and is waiting for connection.

Step 3: The OpenVPN process connects to the management interface and then it is controlled by the UI thread. To utilize the Android *VpnService* API, the Java layer sends several commands to the OpenVPN process to gather network parameters of the *tun* interface. After the call of *VpnService*, the descriptor of the *tun* interface is sent from the Java layer to the native layer and the descriptor of the link to VPN server is sent in the reverse direction.

Step 4: The OpenVPN process has obtained two descriptors for controlling the traffic between the device and the remote server. After that, network traffic on device is tunneled inside the VPN connection.

3 Attacking OpenVPN Apps

3.1 Adversary Model

Our adversary model consists of two types of attackers:

```
management /data/data/pkg.name/cache/mgmtsocket unix
management-client
client
remote vpn.server.address
cipher BF-CBC
ca  ca.crt
cert client.crt
key client.key
```

Fig. 2. An example of android OpenVPN client profile

1. **A network attacker** can passively monitor the traffic, or can actively intercept and modify network connections between the client and OpenVPN server. Mobile devices are commonly used under different network environments. Users may connect to a free public Wi-Fi for convenient Internet access, and then protect network activity by using a VPN app. The public Wi-Fi could be controlled by a network attacker, thus the VPN traffic can be observed and manipulated by the attacker.

2. **A malicious app** that attempts to attack OpenVPN apps is installed on the user device. Apps installed on the user device cannot be all trusted. Users may install apps from third-party app markets, where the attacker can repackage malicious payload into popular apps and distribute them.

3.2   Vulnerabilities and Attacks

From the attacker's perspective, the profile distribution (step 1) is the critical step for discovering vulnerabilities in the execution flow of OpenVPN apps. Most security related information can be found from the VPN client profile, which can be obtained after step 1. The client profile provides all the prerequisites for attacking the management interaction and the VPN connection procedure, such as the address of VPN server, cipher algorithm, authentication types and file location of the management interface. Without these critical information, it is impossible for attackers to find vulnerabilities in other steps of the OpenVPN workflow.

Free VPN apps indeed expose VPN client profiles to attackers, which makes conducting a certain attack feasible. Most free VPN apps do not require user registration, or some even provide same private key for different users [29]. Any user can obtain a valid client profile, by just connecting to the VPN servers in these apps. The attacker can utilize the profile distribution step of these apps on his own device to collecting the configuration profiles of VPN clients. Except client credentials like certificates and private keys which may be user-unique, the attacker can obtain the same configuration as other normal users due to the same client implementation and server logic. After that, the attacker can explore configuration profiles and client implementations to find vulnerabilities and attack specific VPN apps.

Based on our adversary model and the leakage of client profile, we present three types of attacks against OpenVPN apps, which compromise the confidentiality, authenticity and availability of the OpenVPN service. These attacks are caused by insecure customization and deployment of OpenVPN apps, not by OpenVPN protocol itself.

1. **Traffic Decryption.** Some VPN service providers claim that they use some proprietary VPN protocols or Anti-DPI [9, 21] technology to prevent VPN traffic from being identified or blocked. Also, a few custom OpenVPN patches intended to obfuscate the OpenVPN traffic and bypass firewalls have been proposed in the OpenVPN community [19] and GitHub [20]. We identify a typical misuse that developers disable the encryption of OpenVPN and use custom obfuscation to replace the standard encryption. These custom obfuscations are commonly implemented by scramble operations such as XOR, and adopt vulnerable key agreements (e.g., hard-coded keys). Thus the misconfiguration of replacing standard encryption with custom obfuscation will lead any passive network attacker to completely decrypt the VPN traffic. Details of custom obfuscation and its misconfiguration are discussed in Sect. 5.1.
2. **Man-in-the-Middle Attack.** The publicly available client profiles of these free VPN apps may lead to possible MITM attacks. This MITM attack happens when the client certificate is signed by the same CA of server certificate and the usage of server's certificate is not verified at the client side. A valid client certificate and private key are sufficient to conduct this MITM attack if the OpenVPN app is misconfigured. An active network attacker can truncate the connection request from the client, then claim to be the server by using a valid client certificate. OpenVPN provides several ways to defend this attack [13], however, developers may not enable these security features, leaving their apps vulnerable to this attack.
3. **Denial of Service.** Besides network attackers, threats can also come from a malicious app at the client side. Since Android is a multi-app platform, improper permission of the management interface may allow other apps on the same device to control the OpenVPN process, prevent the normal connection and cause a Denial-of-Service attack.

4 Methodology

This section describes our approach of analyzing the deployment security of Android OpenVPN apps in consideration of the three attacks we proposed. Figure 3 illustrates the procedure of our analysis. In detail, our approach consists of three phases: OpenVPN identification, profile collection, and security assessment. Most prior studies on security and privacy of VPN services focus on the network traffic. However, security flaws in code implementation and permission management can also break the security of OpenVPN. We propose a comprehensive assessment methodology that evaluates OpenVPN apps from three aspects: client profile, code implementation and permission management.

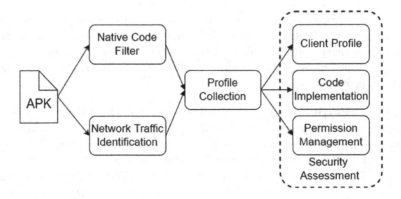

Fig. 3. An overview of how we analyze the security of OpenVPN apps

4.1 OpenVPN Identification

Given a set of Android VPN permission-enabled apps, we need to determine the tunneling protocols used by them. We propose two general methods to identify the usage of OpenVPN among these VPN apps: native code filter and network traffic identification.

1. **Native Code Filter.** This method filters OpenVPN apps by inspecting the symbol table of native libraries. Since cryptographic operations inside VPN service are CPU-intensive, VPN protocols on Android are commonly implemented in native code. If the OpenVPN library is incorporated in the native libraries of the app, function names and other symbol information from OpenVPN source code are preserved in the binary code. By searching meaningful strings like function name *openvpn_encrypt* in the symbol table of shared library files, we can quickly determine whether OpenVPN is used in the native code. This static method is efficient, and is fully automatic. If symbols in the library are stripped, or developers intentionally obfuscate the binary code, this method cannot detect the usage of OpenVPN, thus we need runtime traffic identification.

2. **Network Traffic Identification.** This method focuses on investigating network activity of apps and treats APK files as a black box. After capturing the network traffic of VPN apps, the tunneling protocol can be identified by using protocol parsers such as Bro [3]. The accuracy of this method depends on the precision of the protocol parser. As reported in the work by Ikram et al. [25], a large portion of the tunneling protocols used by VPN apps cannot be recognized by protocol parsers. If a VPN app has obfuscated its traffic by modifying the protocol implementation, common protocol analysis tools are incapable of identifying its network traffic. While this method cannot detect custom obfuscations and needs manual interference to establish VPN connection in apps, it helps us to find the usage of typical OpenVPN implementation regardless of the binary code information.

In this step our goal is to identify OpenVPN apps as many as possible, therefore we combine these two methods. We adopt native code filter as the main detection method, which narrows the assessment scope automatically, and is capable of detecting custom OpenVPN implementations. Then we utilize network analysis tools for those apps that are not identified by our native code filter.

4.2 Profile Collection

As discussed before, the profile distribution of different apps can vary from each other, therefore it is complicated to collect client profiles from profile servers. Instead, for each OpenVPN app, we gather the runtime arguments of the Open-VPN process to figure out the client profile.

OpenVPN allows options to be provided either by the command line arguments or by a configuration file. Actually the configuration file is used as a command line option *–config*. Therefore, by inspecting the app's native process and its command line arguments (i.e. */proc/PID/cmdline*), we are able to extract the client connection configuration of VPN apps.

We build a semi-automatic tool for collecting client profiles of OpenVPN apps. While the VPN service is running, this tool automatically parses the arguments of OpenVPN process belong to each VPN app, then it records all the configurations or directly extracts the configuration file on device. The necessary manual part is that for each app we need to actively connect to the VPN server and approve the VPN connection in the Android system dialog.

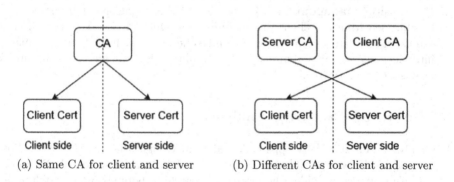

(a) Same CA for client and server   (b) Different CAs for client and server

Fig. 4. Two types of CA trust model

4.3 Security Assessment

1. **Client profile.** We implement a parser of OpenVPN profile to extract all the options from configuration files which we collect from different apps. Then, we perform a statistical survey of the usage of security related options. Particularly, we inspect the cipher algorithm used in OpenVPN apps (*cipher*),

whether the client is authenticated by passwords (*auth-user-pass*), by certificates (*cert*), or additional TLS authentication is used (*tls-auth*). When client and server are authenticated by certificates, there are two types of CA trust model, as shown in Fig. 4. The certificate of client and server can be signed by the same CA (Fig. 4a), or by different CAs (Fig. 4b). Under the CA model in Fig. 4a, if the usage of certificate is not checked at the client side, an active network attacker can impersonate a valid server by using another client certificate retrieved from the OpenVPN app. By checking whether the client certificate (*cert*) is signed by the CA (*ca*), we determine the CA model of authentication in the OpenVPN app. If the certificate of client and server is signed by the same CA, we then examine whether options for MITM prevention (*remote-cert-tls*, *ns-cert-type*) are applied.

2. **Code implementation.** We focus on evaluating the implementation of custom features added for the purpose of obfuscating OpenVPN traffic. By selecting options that exist in these configuration profiles but not in the official manual page of OpenVPN [12], we are able to filter the custom features. In order to understand these custom obfuscation behaviors, we perform reverse engineering on the modified OpenVPN library. To target the obfuscation operations, we concentrate on the code implementation located after the original cryptographic procedure in OpenVPN, and before the actual network send/receive logic, e.g., *process_outgoing_link* or *read_incoming_link* functions in OpenVPN source code [17]. Utilizing the source code of OpenVPN and comparing it with the decompiled code generated by IDA Pro [8], we identify custom obfuscations added in these OpenVPN apps. After that, the obfuscation key is examined in the client profile.

3. **Permission management.** File location of the management interface can be obtained from the client profile. While the OpenVPN service is running, We use a script to automatically examine the permission of its management interface. The permission management is vulnerable if the management interface is world-accessible.

5 Result and Security Analysis

We analyzed top 200 VPN apps collected from Google Play in May 2017. We utilize *google-play-scraper* [6] and *gplaycli* [5] to select and download VPN apps. *Google-play-scraper* provides the feature of searching popular apps matching a certain term, and *gplaycli* is able to automatically download a list of APK files from the Google Play store. Among the top 200 VPN apps on Google Play, 111 apps using OpenVPN were identified by our methods. We successfully analyzed the client deployment status of 84 OpenVPN-based apps. The remaining 27 apps were not evaluated due to the need of in-app purchase or the failure of server connection. All the apps were tested on a rooted Moto G with Android 7.1.

The main vulnerabilities we found are summarized in Table 1. 11 of the analyzed apps replace the standard OpenVPN encryption with custom obfuscation, thus the VPN traffic can be decrypted by network attackers. There are seven apps vulnerable to MITM attacks due to their lack of certificates usage validation or using the static key mode. Four of the OpenVPN apps leave the management interface unprotected, which may lead to Denial-of-Service attacks. The maximum number of installs among the apps belong to each misconfiguration type is also listed in Table 1.

Table 1. Main vulnerabilities we found in OpenVPN apps

Category	Vulnerability type	# of apps	Max installs
Cipher	Replacing encryption with obfuscation	11	1M
	Encryption disabled	2	1M
Auth	Lacking cert usage validation	6	1M
	Static key mode	1	100K
Management	Unprotected interface	4	1M

5.1 Insecure Encryption

Insecure Cipher Algorithm. We found that 30 OpenVPN apps use the default cipher algorithm BF-CBC. As a 64-bit block cipher, Blowfish is vulnerable to the SWEET32 attack [23], thus it is no longer recommended. Despite the publication of this attack, Blowfish is still the default cipher algorithm of OpenVPN [12]. This *insecure-by-default* setting may influence the security of OpenVPN deployment.

Meanwhile, two of the analyzed apps explicitly set cipher to *none*, which disables encryption and transfers all traffic in plain text. When using option *cipher none*, OpenVPN has a warning in its standard output. However, since Android users cannot notice this warning, they will be unaware of this insecure setting.

Replace Encryption with Obfuscation. The usage of custom obfuscation patches in the 84 OpenVPN apps is described in Table 2. Obfuscation is realized by adding extra encryptions of the OpenVPN packet data, and the key for obfuscation needs to be configured the same at both client and server side. We notice that 13 apps use RC4 to obfuscate the OpenVPN traffic, and the key of RC4 is set to the IP address of VPN server. Obfuscation itself does not weaken the security of OpenVPN. However, 11 of them **use custom obfuscation to replace standard encryption by setting cipher to *none***, which completely breaks the security of OpenVPN. Besides, nine apps use XOR-based obfuscation while four of them choose the same obfuscation key. There is only one app that uses two's complement to obfuscate the traffic.

Furthermore, we perform a thorough analysis of the most commonly used custom obfuscation option *antidpi*. Before OpenVPN's network send/receive logic,

Table 2. The usage of custom obfuscation patches

Obfuscation type	Option name	Obfuscation key	# of apps
RC4	*antidpi, antidpi_remote*	Server's IP address	13
XOR	*obsecure_key, scramble, link-key*	Random string	9
Two's complement	*ob-key*	-	1

a custom RC4 encryption/decryption of the whole OpenVPN packet is added. The key of RC4 is determined by the argument of *antidpi_remote*. The crux deployment security problem is that 11 apps disable the standard encryption of OpenVPN and set a poor key for the RC4 encryption. OpenVPN's own encryption is disabled by setting *cipher none* and the key of RC4 is set to server's IP address by setting *antidpi_remote*. In this case, users and developers who believe network data is protected will be in fact fully exposed to threats. Any passive attacker that obtains network traffic of these apps can completely decrypt the data and recover users' online activity, e.g., the attacker can learn all the HTTP traffic.

We investigated several forked OpenVPN projects to analyze potential causes of the misconfiguration of custom obfuscation. A custom obfuscation patch called *xorpatch* in OpenVPN community forum claims that encryption of the scramble patch is secure thus *'it is OK to use cipher none'* [19]. Due to neglecting the importance of encryption, the patch author gives an insecure demonstration of configuration, which may mislead other developers to make the same mistake. In addition, we audited some OpenVPN patches on GitHub [14,15,20]. All of them are in lack of guide about how to set the key for obfuscation. Users may not know how to set the correct patched options with OpenVPN's original features, thus weak arguments like using IP address as the key may occur.

5.2 Weak Authentication

Table 3 presents the usage of different client authentication methods of Open-VPN apps. Different security levels are provided by these authentication methods. *Password+Certificate+TLS-auth* is the most secure method, which is

Table 3. The usage of client authentication methods

Password	Certificate	TLS-auth	# of apps
√	√	√	18
×	√	√	4
√	×	√	8
√	√	×	29
×	√	×	10
√	×	×	14

adopted by 18 apps, while 14 apps use less secure *Password only* authentication. Besides, one app is found to use static key mode. In this mode OpenVPN connection can be decrypted and MITM-ed as the key is shared by different users.

We observe that *Password+Certificate* is the most used type. In our results, 61 apps use certificate-based authentication and 39 of them use the same CA trust model. It is convenient to use the CA trust model in Fig. 4a since the CAs deployed at client and server side are the same. However, six apps lack the validation of server certificate usage, which means a rogue client can conduct MITM attacks on other clients.

The trust model in Fig. 4b is immune to this attack due to different CAs are used. To prevent this attack against the same CA model, extra validation of server's certificate usage (i.e., certificate for server only) is needed. OpenVPN has provided several options like *remote-cert-tls* to require an explicit key usage of peer certificates. These security protections are not enabled by default since in most case private keys are not leaked. While for most free OpenVPN apps, client certificates and private keys are publicly available, thus developers must apply these options to prevent MITM attacks.

Due to lack of security awareness, developers of these free VPN apps usually make their VPN client profiles public, or even provide the same client credential for different users. While providing convenient VPN services, these VPN apps leak the client profiles at the same time.

The *insecure-by-default* policy in OpenVPN may also cause this misconfiguration. OpenVPN provides various of options, some are required for enabling the basic VPN function, some are for security hardening purpose. Developers may omit these complicated security protection options, leaving their OpenVPN service insecure by default.

5.3 Unprotected Management Interface

In the circumstance of OpenVPN on Android, the management interface handles the communication between native and Java layer. The problem is that, the management interface itself does not have any authentication mechanism, thus the file permission of the interface must be correctly set. Otherwise it can can be exploited by other apps on the same device.

In our experiment, four of the analyzed apps set the permission of management interface world-accessible, i.e., *srwxrwxrwx*. Because the file location of management interface can be inferred from the client profile, a malicious app on the same device can exploit the insecure permission of the interface, and access to it before the normal connection. The implementation of OpenVPN management interface does not support multiplex, thus the first connection will block others from accessing the interface. After the malicious app connects to the management interface, the normal connection is blocked and this eventually leads to a Denial-of-Service attack.

The misuse of Android UNIX domain socket has been analyzed by [28], here we focus on OpenVPN and explore the causes of unprotected file permission. We observe that *management-client* option is not used by the four apps, while all other apps enable this option. The typical execution flow of OpenVPN is modified by disabling this option. When *management-client* is disabled, the native process, not the Java process, acts as the server of UNIX domain socket. The management interface is created at the native layer and the OpenVPN process listens on the UNIX domain socket. We find that the default file permission of the UNIX domain socket in OpenVPN is world-accessible because *umask(0)* is used by OpenVPN [17]. To protect the UNIX domain socket and only allow specific user to access the interface, developers need to enable the *management-client-user* option, which specifies the file permission of the management interface.

On the other side, if *management-client* is enabled, the Java layer is responsible for creating the management interface. At Java layer the security model is supplied by Android Java VM and file is created with correctly protected default permission. In a word, the different default file permissions of native layer and Java layer result in this vulnerability.

6   Recommendations

Don't Use Custom Obfuscation to Replace Encryption. Custom obfuscation is commonly implemented by simple scramble operations thus it is not secure enough to replace the standard encryption of OpenVPN. The purpose of obfuscation is to hide protocol metadata, not to protect the payload. For bypassing network censorship, the OpenVPN team disapproves of custom patches and suggests to use *obfsproxy* [18]. Another approach is to tunnel the VPN traffic in common secure protocols like TLS or SSH.

Deploy Countermeasures Against MITM. OpenVPN provides different ways to avoid the Man-in-the-Middle attack from an authorized client. Certificates can be assigned with specific key usage and extended key usage. Options like *remote-cert-tls server* or *ns-cert-type server* make OpenVPN clients accept server-only certificates. Signing certificates for server and client with different CAs can also prevent this MITM attack.

Set Secure File Permissions on Android. Since Android is a multi-app platform, developers should protect their own files from being tempered by other apps on the same device. File permission at Java layer is correctly protected by default. However, at the native layer, developers should take their own responsibility and use *umask* and *chmod* to securely protect their files.

Securely Distribute Client Profiles. Developers should harden the client configuration, protect the distribution of client profiles (e.g., transmit them via email) and securely store them at the client side. Unique client credentials should be generated for different users to prevent the abuse of public client profiles. To achieve a better security level, VPN profiles can be encrypted or stored in Android *Keystore* [1].

7   Related Work

Several studies have been working on the privacy and security of VPN services. Appelbaum et al. [22] are the first to uncover the VPN traffic leakage problem caused by misconfiguration of route tables. Perta et al. [26] extend their work and analyze popular commercial VPN services. Their results reveal that the majority of VPN services suffer from IPv6 leakage and DNS hijack attacks. Ikram et al. [25] conduct a comprehensive privacy and security analysis of Android VPN permission-enabled apps. Their study mainly focuses on investigating VPN apps' manipulation of TLS traffic and behavior of tracking user privacy. Instead of concentrating on network analysis, our work evaluates the security of VPN apps from the aspects of security related configuration and code implementation at the client side. Recently OpenVPN 2.4.0 has been audited and several security issues have been found [27]. Our work reveals that, in addition to the flaws in official implementations, developers' custom modification and configuration in VPN applications can also lead to severe security vulnerabilities.

There are some studies about the security of custom VPN protocols and misuses of UNIX domain sockets on Android. Peter [24] gives a classic cryptographic audience of the weakness of some custom VPN protocols. He suggests to use standard-protocol-based VPN, such as OpenVPN and IPsec, while we demonstrate that misuses of OpenVPN can still threaten the VPN communication. Shao et al. [28] conduct a systematic study of the misuses of Android UNIX domain sockets. We analyze the causes of insecure permission of OpenVPN management interface based on their work.

8   Conclusion

In this work, we focus on the client side deployment security of Android OpenVPN apps. After summarizing the procedure of client deployment and VPN connection, we present a security assessment methodology by evaluating the security of client profile, code implementation and permission management. The configuration status of 84 popular OpenVPN-based apps on Google Play are analyzed. To our best knowledge, we are the first to identify a typical misuse of insecure custom obfuscation in several OpenVPN apps. Our experiment also shows that MITM vulnerability and Denial-of-Service problem due to misconfigurations still exists in these apps. The misconfigurations are either due to patch authors' wrong advices and lacking of document, the 'insecure-by-default' OpenVPN configuration, or due to developers' incorrect file permission setting on Android. Finally we develop some practical recommendations for securing the OpenVPN deployment.

References

1. Android keystore system. https://developer.android.com/reference/java/security/KeyStore.html
2. Android vpn service documentation. https://developer.android.com/reference/android/net/VpnService.html
3. Bro network security monitor. https://www.bro.org
4. Detailed vpn comparison chart. https://thatoneprivacysite.net/vpn-comparison-chart/
5. Google play downloader via command line. https://github.com/matlink/gplaycli
6. Google-play-scraper. https://github.com/facundoolano/google-play-scraper
7. The heartbleed bug. http://heartbleed.com/
8. Ida pro. https://www.hex-rays.com/products/ida/
9. Nvpn antidpi. http://www.nvpn.net/. Accessed 21 July 2017
10. Openvpn for android source code. https://github.com/schwabe/ics-openvpn
11. Openvpn management interface. https://openvpn.net/index.php/open-source/documentation/miscellaneous/79-management-interface.html
12. Openvpn manual page. https://community.openvpn.net/openvpn/wiki/Openvpn24ManPage
13. Openvpn mitm protection. https://openvpn.net/index.php/open-source/documentation/howto.html#mitm
14. Openvpn obfuscation patch. https://github.com/siren1117/openvpn-obfuscation-release/
15. Openvpn patch from tunnelblick. https://github.com/Tunnelblick/Tunnelblick/tree/master/third_party/sources/openvpn/openvpn-2.4.3/patches
16. Openvpn security overview. https://openvpn.net/index.php/open-source/documentation/security-overview.html
17. Openvpn source code. https://github.com/OpenVPN/openvpn
18. Openvpn traffic obfuscation guide. https://community.openvpn.net/openvpn/wiki/TrafficObfuscation
19. Xorpatch in openvpn forum. https://forums.openvpn.net/viewtopic.php?f=15&t=12605&hilit=openvpn_xorpatch
20. Xorpatch source code. https://github.com/clayface/openvpn_xorpatch
21. Zpn antidpi. https://zpn.im/blog/total-anonymity-connectivity-antidpi. Accessed 21 July 2017
22. Appelbaum, J., Ray, M., Koscher, K., Finder, I.: vpwns: virtual Pwned networks. In: 2nd USENIX Workshop on Free and Open Communications on the Internet. USENIX Association (2012)
23. Bhargavan, K., Leurent, G.: On the practical (in-) security of 64-bit block ciphers: collision attacks on HTTP over TLS and OpenVPN. In: Proceedings of the 2016 ACM SIGSAC Conference on Computer and Communications Security, pp. 456–467. ACM (2016)
24. Peter, G.: Linux's answer to MS-PPTP. https://www.cs.auckland.ac.nz/~pgut001/pubs/linux_vpn.txt
25. Ikram, M., Vallina-Rodriguez, N., Seneviratne, S., Kaafar, M.A., Paxson, V.: An analysis of the privacy and security risks of android VPN permission-enabled apps. In: Proceedings of the 2016 ACM on Internet Measurement Conference, pp. 349–364. ACM (2016)
26. Perta, V.C., Barbera, M.V., Tyson, G., Haddadi, H., Mei, A.: A glance through the VPN looking glass: IPv6 leakage and DNS hijacking in commercial VPN clients. Proc. Priv. Enhanc. Technol. **2015**(1), 77–91 (2015)

27. Quarkslab: Security assessment of openvpn. https://blog.quarkslab.com/security-assessment-of-openvpn.html. Accessed 21 July 2017
28. Shao, Y., Ott, J., Jia, Y.J., Qian, Z., Mao, Z.M.: The misuse of android unix domain sockets and security implications. In: Proceedings of the 2016 ACM SIGSAC Conference on Computer and Communications Security, pp. 80–91. ACM (2016)
29. White, K.: Most VPN services are terrible. https://gist.github.com/kennwhite/1f3bc4d889b02b35d8aa. Accessed 21 July 2017

Two Cents for Strong Anonymity: The Anonymous Post-office Protocol

Nethanel Gelernter[1], Amir Herzberg[3,2], and Hemi Leibowitz[2(✉)]

[1] Department of Computer Science, College of Management Academic Studies,
Rishon LeZion, Israel
[2] Department of Computer Science, Bar Ilan University, Ramat Gan, Israel
Leibo.hemi@gmail.com
[3] Department of Computer Science, University of Connecticut, Mansfield, CT, USA

Abstract. We introduce the *Anonymous Post-Office Protocol (Anon-PoP)*, a practical strongly-anonymous messaging system. Its design effectively combines known techniques such as (synchronous) mix-cascade and constant sending rate, with several new techniques including *request-pool, bad-server isolation* and *per-epoch mailboxes*. AnonPoP offers *strong anonymity against strong, globally-eavesdropping adversaries*, that may also control *multiple* servers, including all-but-one servers in a mix-cascade. Significantly, AnonPoP's anonymity holds even when clients may occasionally disconnect, which is essential for supporting mobile clients.

AnonPoP is *affordable*, with monthly costs of 2 cents per client. It is also *efficient* with respect to latency, communication, and energy, making it suitable for mobile clients. We developed an API that allows other applications to use AnonPoP for adding strong anonymity. We evaluated AnonPoP in several experiments, including a 'double-blinded' usability study, a cloud-based deployment, and simulations.

1 Introduction

The growing awareness for the importance of anonymous communication, has resulted in many efforts to develop, analyze, and deploy anonymous communication protocols and systems. Specifically, the Tor anonymous network [1] is widely used. That said, the road to strongly-secured anonymous communication is still paved with many challenges.

For instance, Tor is designed for low-latency services, which leaves it vulnerable to globally-eavesdropping adversaries. Several works showed that Tor is also vulnerable to other (weaker) attackers, e.g., off-path attackers [2] and malicious servers/clients [3,4].

Tor provides a popular *communication channel* but is lacking a complete *messaging* system. A complete messaging system should also provide 'mailbox' facilities to keep messages until users pick them up; this is also needed to prevent detection of a pair of users that frequently communicate with each other

© Springer Nature Switzerland AG 2018
S. Capkun and S. S. M. Chow (Eds.): CANS 2017, LNCS 11261, pp. 390–412, 2018.
https://doi.org/10.1007/978-3-030-02641-7_18

and may get disconnected. A naive mailbox solution, where Tor is used to communicate with a mailbox server, would allow the server to de-anonymize users. This can be done by exploiting Tor's design, e.g., eavesdropping on a particular (suspect) user, and then correlating between messages sent/received by this user and messages received to this mailbox or sent from this mailbox.

Clearly, Tor's popularity indicates that it provides a valuable service to many users. Nevertheless, there are many scenarios that require stronger anonymity properties, even at the cost of somewhat higher latency and overhead.

Several works proposed protocols for stronger guarantees of anonymity, as compared to those of Tor. However, existing research is mostly impractical. Many seem to believe that it is infeasible to ensure strong anonymity properties in a practical system for many users, especially with acceptable overhead and efficiency. Disappointingly, neither Tor nor any other practical (existing or proposed) system allows *strongly-anonymous messaging*. Messaging is used more and more for business and personal communication, and anonymity is often *required* - for reasons ranging from whistle-blowing to consulting on sexual harassment. Strongly-anonymous messaging is *feasible*, since the volume of (text) messages is not very large, and reasonable delays are acceptable. This makes it all the more frustrating that such a system is not yet operative.

In this paper, we present the *Anonymous Post-Office Protocol (AnonPoP)*, a practical anonymous messaging system, designed to ensure strong anonymity, even against strong attackers. AnonPoP is designed with the scalability required to support millions of users, because *anonymity loves company* [5]. AnonPoP uses efficient cryptographic primitives and has acceptable energy consumption, making it appropriate for use on mobile devices. Furthermore, to the best of our knowledge, AnonPoP is the *only* proposed anonymous messaging protocol to support client disconnections, a feature that is essential for mobile clients.

To measure and confirm AnonPoP's low operating costs, we implemented and installed AnonPoP's servers in the cloud, and tested it on hundreds of thousands of clients, communicating anonymously with each other using AnonPoP. We found that the cost of supporting such a large number of clients is less than 25 *cents per user, per year*, or 2 *cents per month*.

We provide an API for messaging applications, making it possible to easily add an option for strong anonymity using AnonPoP. With this API, clients of different applications can form one large anonymity set.

Due to the complexity of providing strong anonymity, we acknowledge that we cannot completely address *all* of the challenges in this paper. Therefore, in this paper, we focus on explaining the challenges, rationale, and system design of AnonPoP, and present a variety of analyses of AnonPoP. We also provide an extended technical report [6] which contains elaborated discussions on selected topics.

Our Contributions. This work makes the following contributions:

1. Design, development, and evaluation of a practical strongly-anonymous messaging protocol, secure against strong adversaries.
2. Support for mobile environments, including energy-saving considerations.

3. Features such as the *request-pool* mechanism and *per-epoch mailboxes*, to handle clients' disconnections and to limit exposure due to disconnections or active tagging attacks.
4. A *bad-server isolation* mechanism, allowing the isolation of a corrupt server that is involved in aggressive (non-stealthy) tagging attacks.
5. An open-source prototype of AnonPoP, including API for applications, and an Android messaging application that uses this API.

Paper Layout and Organization. We start by explaining the model of the system in Sect. 2, and then present a high-level overview of AnonPoP in Sect. 3. After that, we delve into AnonPoP's mechanisms in detail, in Sects. 4 and 5. We then analyze AnonPoP in Sect. 6, and evaluate it on mobile devices and in the cloud in Sects. 7 and 8. We conclude by surveying related work in Sect. 9.

2   Model and Preliminaries

2.1   System Model

In AnonPoP, clients relay messages using two types of servers: *Post-Office (PO)* servers and *timed mixes* (see Fig. 1). These servers are expected to operate continuously, while clients may disconnect from time to time. AnonPoP supports multiple PO servers, where each client can select a PO server; it also supports an arbitrary number of mix servers. For simplicity, our discussion and figures refer to a single PO. As we describe later, AnonPoP allows the detection of tagging attacks by corrupt servers, in which case we expect clients to move to different available servers.

Mixes operate in synchronized *slots* of τ seconds. Each mix collects all the packets received in a slot, shuffles them, decrypts or encrypts them, and forwards them to their next hop so they are all received in the subsequent slot.

Mix Selection. For simplicity, we assume that AnonPoP clients and servers use a trusted, reliable *directory server* for path selection. This is merely a simplification, as it is straightforward to implement such a directory service in a distributed manner, avoiding a single point of failure. The directory maintains a list of all AnonPoP servers, with their public keys and addresses. Clients pick a cascade of mixes uniformly among all paths consisting of pairs of connected mixes, ending in the desired PO.

Mailbox Setup. AnonPoP automatically assigns clients to random mailboxes and generates proper keys. In this paper, we assume that clients have a secure communication channel on which to perform the initial key exchange. Anonymous key-exchange (e.g., in [7]) is a further challenge, beyond the scope of this paper.

2.2 Adversary Model

As in previous works, we focus on probabilistic polynomial time attackers. This is essential, since our design uses cryptographic mechanisms, which are only secure assuming probabilistic polynomial time attackers (e.g., encryption schemes [8]).

The AnonPoP design assumes the attacker has *global eavesdropping* abilities. In other words, it can instantly observe all communication sent between any of the parties in the system. We also consider *additional* attacker capabilities, mainly, control of the PO and/or some of the mixes, thereby allowing complex powerful attacks. We make the reasonable assumptions that the number of adversarial servers is limited. First, we assume a known upper bound f for the number of malicious (faulty) mix servers. In addition, we assume that honest mixes behave as expected and that the adversary can neither drop messages nor delay a message to more than a slot between honest mixes. We also consider communication to be between trusting peers (sender and recipient). In fact, we are already working on advanced extensions that will offer defense against malicious peers, misbehaving honest mixes, and adversaries who can drop messages between honest mixes. However, these extensions are beyond the scope of this paper.

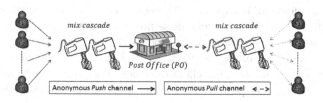

Fig. 1. System architecture of AnonPoP. The PO maintains anonymous mailboxes; clients send/receive messages to/from the mailbox anonymously via mix-cascades. All communication channels (represented by arrows) use fixed rates.

3 High-Level Overview

When Alice wishes to send an anonymous message to Bob, her AnonPoP client pads and packs the message into a fixed-sized packet with Bob's (pseudo)random mailbox address. Her packet is then relayed through a cascade of mixes, until it reaches the PO. The PO then looks at the destination of the packet and delivers it to Bob's mailbox. We call this packet a *push request*, and the cascade it travels a *push channel*.

When Bob wishes to retrieve messages from his mailbox, his client crafts a designated packet with proof of ownership over the specific mailbox, and sends the packet through a cascade of mixes to the PO. The PO verifies Bob's ownership and sends Bob a message from the mailbox. This message is called a *pull request* and the cascade used is called a *pull channel*.

All packets of the same type are padded to the same size[1]. Long messages are fragmented and re-assembled by the clients. The packets are *onion-encrypted* [9,10]. They are encrypted using the public key of the PO, and then consecutively, using the public keys of the mixes. Additionally, messages are encrypted for the destination. When the PO decrypts the final onion layer, it finds only a mailbox identifier and an encrypted message. Furthermore, all the communication between every pair of adjacent entities (adjacent mixes, last mix and PO, or client and first mix) is authenticated and encrypted.

The request packets that travel from clients to the PO are layer decrypted in each mix in the cascade. The response packets traveling from the PO back to the clients are encrypted at each mix; this is done through the same cascade as the requests, but in reverse order. We use authenticated-encryption [11], allowing the client to validate that the response was sent by the PO, over the specified sequence of mixes. The PO sends the response upon receiving the corresponding request from the client. To facilitate the authenticated-encryption of responses, clients include the authenticated-encryption key to be applied to the response in each onion-encryption layer. For a detailed illustration, see Fig. 2.

AnonPoP clients maintain a fixed rate of transmission for the packets, independent of the actual pattern of users. This is accomplished by queuing outgoing messages if their rate exceeds the fixed rate, and sending *dummy* packets [12], if no outgoing messages are queued. Namely, clients send one push and one pull request in every round. If a client does not have a real message to push, she sends a dummy request, which is indistinguishable from a real push request. The PO responds to dummy requests as it does for real requests, but using dummy responses. Similarly, if a client does not have a message to pull[2], the PO sends back a dummy response. All dummy packets are indistinguishable from real packets.

In the following sections, we dive deeper into AnonPoP, presenting and discussing the complex challenges and how AnonPoP faces them.

4   Anti-tagging Defenses

Onion-encryption and padding cannot fully protect anonymity against a rogue PO. There are a few ways in which a rogue PO, possibly colluding with some mixes, can 'tag' a request and/or the corresponding response, allowing it to link between a client and a mailbox. In this section, we present AnonPoP's defenses against such tagging attacks. For an illustration of the possible attacks and how AnonPoP deals with them, see Fig. 3.

---

[1] To further increase the anonymity set, at the small price of extra bandwidth, it is possible to pad all types to be of the same size.

[2] Checking if a mailbox is not empty could be done anonymously and efficiently via [13].

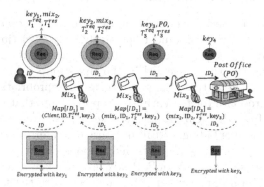

Fig. 2. Onion-encryption with cascades of mixes, as used by both push and pull channels. The circles above the straight lines mark the route of the request from the client to the PO. The response route is illustrated by squares below dashed curves. The senders encode in each onion layer a key (key_i) that Mix_i uses to authenticate and encrypt the response, and timestamps T_i^{req}, T_i^{res} specifying when the request and the response are expected to arrive (see Sect. 4.1). The sender and mixes each select a random identifier ID (for sender) or ID_i (for i^{th} mix), and store the relevant parameters (key_i, T_i^{res} and received ID) in table Map indexed by the chosen ID.

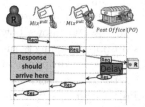

(a) Delaying the response arrival. Only the first pull mix is honest.

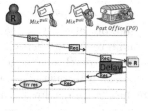

(b) Defense by returning an error report indistinguishable from a real response at the expected response time.

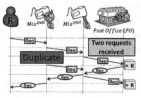

(c) Sending a duplicated request in a subsequent slot. Only one non-first pull mix is honest.

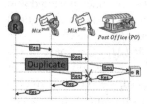

(d) Defense by dropping the duplicated request that arrived in the wrong slot.

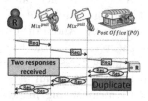

(e) Sending a duplicated response in the same slot. Only one non-first pull mix is honest.

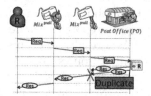

(f) Defense by dropping the second response received.

Fig. 3. Attacks to correlate recipients and their mailboxes, and their defenses

4.1 Timestamps, Anti-duplication, and Anti-tampering

The basic anti-tagging mechanism in AnonPoP is to include *timestamps* in every layer of the onion; the request (and respectively, response) timestamp for the ith mix is denoted T_i^{req} (T_i^{res}). Non-corrupt mixes always return an encrypted response at *exactly* the time specified in the timestamp field. If the expected response is not received on time, the mix returns an appropriate error report. A response received too late (or too early) is dropped. Note that error reports are indistinguishable from the 'real' responses. The adversary cannot learn whether an encrypted response is hiding a 'real' response or an 'error report'.

To further detect *duplicate* requests and responses that are received at the same (correct) time slot, each AnonPoP mix uses the key it receives key_i, as a unique identifier. If a mix receives multiple requests with the same key (for the same slot), then it discards all but one of them. It sends back an error-report containing the plaintext and randomness, allowing the previous mix to validate the collision of keys. Similarly, a mix discards all but one response for each forwarded request. (Random collisions occur with negligible probability.)

To prevent tagging via tweaked packet values or other packet tampering, any corruption of the packet's authentication in any layer is immediately detected and the tampered packet is discarded.

4.2 Bad Server Isolation

An attacker who controls both the first mix and the PO can drop, delay, or corrupt requests and/or responses to correlate between clients and mailboxes. The first mix knows the originators of every request, and the PO knows how many messages reach each mailbox. Even when clients do not disconnect, the PO and the first mix may try to match clients to mailboxes by dropping or delaying requests and/or responses; this may allow an intersection attack.

The *bad-server isolation* mechanism, allows AnonPoP to efficiently deter active attacks involving a rogue first mix and the PO. Previous works (e.g., [14]) discussed the complexity of achieving such a goal. Suppose mix M detects that it did not receive the expected response at its specified time from the 'next server' (mix or PO), denoted X. Next, M encrypts and sends back a signed and time-stamped *problem report*, stating the relevant identities (M, X); the (signed) problem report is also deposited in the AnonPoP directory.

As a result, *all* clients and servers will avoid the pair (M, X) as part of the AnonPoP routes. Namely, a rogue, active server, 'loses' one of its edges to other servers, for each slot in which it uses such an 'aggressive, detectable' tagging attack. Note, the link between a misbehaving/malicious mix and an honest mix is dropped, as opposed to a specific mix. Furthermore, because the adversary controls up to f servers, it cannot single handedly cause the exclusion of an honest mix. We emphasize that if a malicious server deliberately sends a fake error report in an attempt to incriminate an honest mix, they in fact achieve the opposite effect: the honest mix is no longer connected to a malicious mix - and that is actually the goal of the mechanism.

An attacker might attempt to abuse the mechanism by intentionally issuing false error reports in order to disconnect as many links as possible between honest mixes and malicious mixes. Since paths are chosen uniformly among all the reliable paths, this would exclude many paths where at least one mix is honest. This increases the probability of choosing a path where all the mixes are malicious. Nevertheless, as long as the fraction of malicious mixes is low, the advantage gained by the attacker is not significant; see Appendix A. Moreover, by doing so, the attacker also increases the probability of choosing paths where all mixes are honest.

Because the attacker loses links on every attack and there are many available paths which do not contain malicious mixes, launching a DoS attack is limited and could be mitigated through known techniques, e.g., forward error correction. Attempting to overload honest mixes to cause congestion can be mitigated by limiting the amount of work each mix processes.

5   Handling Disconnections

Mobile clients often disconnect from the network. Such disconnections may be observable or sometimes even controlled by the attacker. This can allow an attacking PO to correlate between clients and the mailboxes they pull from, using *intersection and correlation attacks* [15–18]. In particular, if a pull request reaches some mailbox, an eavesdropping PO can learn that all the clients who were offline when the request arrived are not the owners of that mailbox. By repeating this procedure over time, the adversary can correlate a single recipient with his mailbox.

We focus on pull-requests, where the defense works better, and exploit the fact that pull requests do not depend on the mailbox status. We can prepare pull-requests in advance, with each request pulling the 'next' message from a mailbox. Specifically, each client prepares and sends to the first pull-mix a 'pool' of pull requests to be used in future rounds, even in rounds where the client is disconnected. Unlike pull requests that are sent by each client to her own mailbox at a fixed rate and can be prepared in advance, push requests are sent to specific mailboxes according to the current needs of the user. Therefore, Anon-PoP cannot precisely predict push requests in advance. Consequently, we do not use a request-pool for push requests. When clients disconnect, the mechanisms described so far do not protect sender-anonymity against intersection and correlation attacks. For these cases, we suggest *Per-Epoch Mailboxes (PEM)*.

We first explain the *request-pool*, a technique that allows AnonPoP to extend its recipient-anonymity defenses to the case of (reasonably-limited) client disconnections. Then, we present PEM for sender anonymity.

5.1   Request-Pool

When a client is connected, the first pull-mix maintains a 'pool' of μ pull requests, prepared in advance, for the μ next rounds. As long as the client remains connected, in every round, one pull request is used to retrieve a message; the client provides a new pull request, thereby maintaining μ requests in the 'pool'. Namely, the client sends the pull request that will be used μ rounds after the current round. This 'pool' allows the first pull-mix to send a pull request for the client, even in rounds in which the client is disconnected (up to μ consecutive rounds). The mix also holds all the encrypted responses received from the PO; the mix does not know whether the responses are real or dummy.

When a client reconnects after being disconnected for $x \leq \mu$ rounds, it contacts the first pull-mix to retrieve the messages kept by the mix from the previous x rounds. The client also sends to the mix $x + 1$ new pull requests, to be used in future rounds; this replenishes the 'pool' of μ requests. A small μ value would mean that users who disconnect for more than μ rounds would lose the anonymity guarantees, where a large μ value would require appropriate resources (e.g., storage).

5.2   Anonymity of Disconnecting Senders

In practice, when there are many clients, it can take a considerable amount of time to learn information about the sender; however, this is still feasible, hence AnonPoP does not ensure sender anonymity. For example, the adversary can choose two sender-permuted scenarios in which only a single client receives requests. In the first scenario, only client a sends the messages, and in the second scenario, only client b sends. Obviously, if one of the scenarios is simulated, and the adversary observes that only a or only b are online, she can detect the identity of the sender. This is done by simply checking whether any message reached some mailbox or not (the adversary controls the PO). In this extreme case of a single sender, the adversary can simply correlate the incoming messages with the single sender because she knows that no other messages were sent by other clients. In reality, when there are always many clients online, and when messages are sent by many of them, it is significantly harder to detect the sender.

AnonPoP implements *per-epoch mailboxes (PEM)* to heuristically defend sender-anonymity, even in the case of disconnections. Namely, clients change their mailboxes every fixed number δ of rounds, referred to as an *epoch*. PEM does not completely ensure sender-anonymity, but it decreases the amount of data learned by the PO. Furthermore, it ensures the anonymity guarantees of AnonPoP in an epoch, among all the clients that stay online within that time. PEM improves the resistance to sender-mailbox intersection attacks.

We simulated AnonPoP with and without PEM, and empirically found that PEM strengthens the resistance to intersection and correlation attacks. Figure 4 shows that without PEM, even with 25,000 clients, the adversary succeeded in completing the attack for a significant fraction of senders; this was done in a significant but not prohibitive amount of time. Figure 5 depicts the distribution of the anonymity set after 250, 500, 1000, and 1500 slots. It is possible to see that the size of the anonymity set decreases quickly. After 1000 slots, the size of the anonymity set is about a tenth of the number of clients. With PEM, the adversary was unable to complete the attack in a single epoch. For a detailed explanation about our simulations and results, see [6].

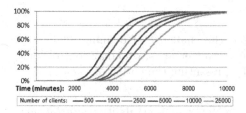

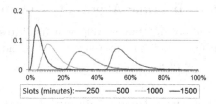

Fig. 4. Percentage of the attacks that were completed over time (x axis) for different numbers of AnonPoP clients.

Fig. 5. The anonymity set distribution after different slots for AnonPoP without PEM. The anonymity set is presented as a fraction of the number of AnonPoP clients.

6   Analysis

Before analyzing AnonPoP's anonymity properties, it is imperative to explain why anonymity analysis is not trivial, and in-fact, imposes a significant challenge. Intuitively, *anonymous communication* means the inability to identify specific communicating entities among the set of potentially communicating entities. Multiple variants were considered by researchers and practitioners, such as unobservability and sender and/or recipient anonymity. One widely-used interpretation [19] is that *sender (recipient) anonymity* refers to the inability of an attacker to identify the sender (recipient) of a message among a set of potential senders (recipients). *Unobservability* refers to inability of an attacker to know whether there was any communication at all. These are useful, intuitive notions; however, they are not sufficiently formal to allow rigorous proofs of security.

Transforming such informal, intuitive notions into precise, well-defined, formal definitions, is a non-trivial challenge. There have been multiple attempts to present appropriate formal definitions, including [20–30]. These definitions differ in multiple aspects, for example in the capabilities of the adversary and what constitutes a successful attack.

Unfortunately, these definitions are not suitable for analyzing AnonPoP, as they fail to satisfy or comprehend all of AnonPoP's goals. None of the existing formal definitions capture AnonPoP's abilities to target *active, adaptive adversaries*, support limited-duration client disconnections, and detect attacks or isolates attackers when complete prevention is not possible.

Since our focus in this work is on system design, we decided to follow 'the spirit' of the existing definitions of anonymity, and to use intuitive notions of anonymity instead of formal definitions. Future work should extend the existing definitions to provide a well-defined notion of practical anonymity. That said, we do provide elaborated definitions and arguments in [6].

6.1   Informal Anonymity Notions

Our notions follow [30], which addressed the challenge of defining anonymity properties in the presence of active, adaptive adversaries, who may control some of the protocol participants. We begin by presenting the notion of *unobservability*.

Notion 1 (Unobservability). *A protocol achieves* unobservability *against a globally eavesdropping attacker that controls a set S of the servers, if the adversary cannot (with significant advantage and in efficient time) distinguish between any pair of communication scenarios.*

To present the (slightly weaker) notions of sender and recipient anonymity, we first present *sender/recipient permuted pairs*. Consider two scenarios σ_0, σ_1, where all the recipients receive the same messages in σ_0 and in σ_1, but the senders in σ_0 are a constant permutation (chosen by the attacker) of the senders in σ_1. Namely, given a permutation π chosen by the attacker, when an honest client i needs to send a message to recipient r in σ_0, the honest client $\pi(i)$ needs to do the same in σ_1. We say that σ_1 and σ_0 are a *sender-permuted pair*. Similarly, the term *recipient-permuted pair* refers to two scenarios where the *recipients* in one scenario are a permutation of the recipients in the second scenario, and the senders are identical.

Notion 2 (Anonymity). *A protocol achieves* sender (recipient) anonymity *against a globally eavesdropping attacker that controls a set S of AnonPoP's servers, if the adversary cannot (with significant advantage and in efficient time) distinguish between any sender (recipient) permuted pair of scenarios.*

These notions pose great challenges because they also consider extreme scenarios that may not occur in reality. For example, for unobservability, the adversary must be unable to distinguish between any two scenarios; this includes a scenario where all parties send messages versus a scenario where nobody sends any message. These are strong anonymity requirements. Surely they do not hold for a low-latency solution such as Tor, since the adversary can easily distinguish between the scenarios. Furthermore, we also allow the corruption of different subsets including most of the servers, as described below.

Detection/Isolation. Existing formal definitions of anonymity (e.g., [29,30]) require the complete prevention of attacks. However, sometimes prevention is infeasible, hard, or expensive. In this cases, a *detection/isolation* approach may be sufficient to deter attackers and hence to ensure anonymity. For example, in AnonPoP and possibly in other efficient strong-anonymity solutions, the PO may 'signal' the use of a particular mailbox, by intentionally dropping its responses or ignoring requests; such a 'signal' seems almost unavoidable for the model where the PO keeps mailboxes. We show that every such abuse is detected and isolated as a specific rogue entity (e.g., the PO).

Our goal is to ensure that every 'bit' of information collected by the adversary has a 'high price'. We present an intuitive notion, which is somewhat tailored to the AnonPoP model.

Notion 3. *Let $X(f), Y(f)$ be two functions (in the number of corrupt servers f), and let x_c be the number of bits that the adversary can learn on (honest) user c communicating only with honest peers, and $y_c = \max\{0, x_c - X(f)\}$. A protocol achieves (X, Y)-attacker-isolation if, with a high probability, $\sum_c y_c \leq Y(f)$.*

Intuitively, the attacker can learn up to $X(f)$ bits 'per client', plus up to $Y(f)$ bits additional (for all clients); in AnonPoP, $X(f) = f$ and $Y(f) = f(f+1)$, as we show below.

6.2 Anonymity Properties

We first consider the anonymity properties of AnonPoP, as a function of the malicious servers along the paths between the senders and recipients. This is summarized in Table 1.

Table 1. Anonymity properties achieved by AnonPoP against a globally-eavesdropping attacker, who controls all servers along the path, except as indicated in the 'honest server' column.

Honest server	Anonymity property	Attack: Defense	
		Without disconnections	With disconnections
Only *first* push mix	Sender anonymity	Passive : Prevent Active : Prevent	Heuristic defense, see Sect. 5.2
Only *non-first* push mix		Passive : Prevent Active : Detect	
Only *first* pull mix	Recipient anonymity	Passive : Prevent Active : Prevent	Passive : Prevent Active : Detect
Only *non-first* pull mix		Passive : Prevent Active : Detect	
Only PO	Unobservability	Passive and Active: Prevent	

Claim. AnonPoP ensures *unobservability* against a global-eavesdropping adversary that further controls any subset of mixes and users, and has the ability to disconnect users, as long as the PO is not corrupted.

Argument: This follows by reduction to the indistinguishability property of (1) the public key encryption scheme $\mathcal{E}_{\mathcal{PK}}$, used to encrypt requests to the PO, and of (2) the shared-key encryption scheme $\mathcal{E}_{\mathcal{SK}}$, used by the PO to encrypt responses. Namely, assume some (efficient) adversary A is able to distinguish between two scenarios of message sending S_0 and S_1. We first check if A also distinguishes between scenarios S_0' and S_1', which are the same as the corresponding S_0 and S_1 except that responses from the PO are all (encryptions of) some fixed message m. If A succeeds, we use it as oracle to distinguish $\mathcal{E}_{\mathcal{PK}}$; otherwise, we use A to create an oracle to distinguish $\mathcal{E}_{\mathcal{SK}}$ (using the fact that A fails to distinguish between S_0' and S_1', yet succeeds in distinguishing between S_0 and S_1). The reduction works because the pattern of transmissions is fixed and independent of input messages.   □

Claim. When clients are always connected and some push (pull) mix is non-corrupted, AnonPoP ensures *sender (recipient) anonymity* against passive attackers.

Argument: We present the argument only for sender anonymity; the argument for recipient anonymity follows the same logic. The argument is by reduction to the indistinguishability of the public key encryption scheme $\mathcal{E}_{\mathcal{PK}}$, used to encrypt requests to the mixes, and in particular, to the non-corrupt mix. Assume some efficient adversary A is able to distinguish between two scenarios of message sending S_0, S_1, where the number and length of messages sent to each mailbox (recipient) are identical (sender anonymity). Due to AnonPoP's padding mechanisms, the pattern of transmissions is fixed and independent of input messages. Moreover, due to the operation of the (non-corrupt) mix, the order of requests arriving at the PO is random. Hence, A provides an oracle that allows an efficient distinguisher for $\mathcal{E}_{\mathcal{PK}}$.   □

Claim. AnonPoP ensures *sender (recipient) anonymity* against active attackers, provided that the first push (pull) mix is honest and that clients are always connected.

Argument: Due to the padding mechanism, requests are sent exactly once a round, with a fixed size. The (honest) first mix shuffles these requests, hence, the subsequent mixes and the PO cannot link between the client and a specific request from the first mix. The traffic *from* the first mix *back* to the client is also fixed, since the mix returns a response at *exactly* the time specified in the time-stamp field; due to the anti-duplication mechanisms, only a single such response is sent and only in that slot.

The encryption applied to messages ensures that eavesdroppers and other mixes cannot link between the sender and the (encrypted) requests. Additionally, in every two scenarios that are different only in the senders (recipients), the same number of messages is pushed (pulled) to (from) mailboxes that differ only by their pseudonym, so the PO cannot distinguish between the two scenarios. In this case, delaying or blocking an encrypted message can be done only when all the messages are already shuffled by the honest first mix; hence, such active attacks are not helpful and Notion 2 holds.   □

Claim. When clients are always connected, AnonPoP achieves $(f, f \cdot (f + 1))$-attacker isolation.

Argument: A server reported by $f + 1$ servers, where f is a bound on the number of malicious servers, is definitely malicious and not used in any channel. Note, the attacker cannot *frame* an honest server. Also, since we assume that at least one mix along the path is honest, it follows that the PO and first mix can only signal each other in the absence of a request/response, i.e., *one bit per round*; they cannot, for example, use content-based signaling. For each such learned bit, a disconnection occurs between one malicious mix and another mix, either due to false report by the attacker or due to a real report by the honest mix. Beyond that, each of the mixes can operate as a first mix for each of the clients; they can drop the message to tag the user without blaming the next mix. In this case, the first mix will not be disconnected from another mix, but the client knows for sure that the first mix is malicious and disconnects from it. Theoretically, this allows each of the f mixes to tag each of the users once. The number of bits can be learned according to each of the cases and therefore satisfies Notion 3.   □

Claim. AnonPoP ensures *recipient anonymity* against passive attackers when some pull mix is honest, even when clients may disconnect, provided clients do not go offline for more than μ consecutive rounds.

Argument: Since the adversary is passive, the traffic from/to the first pull mix to/from the PO is fixed, as though there were no disconnections. There might be a peak in traffic between the first pull mix and the client immediately after the client reconnects; in this case, the rate of traffic between the client and first mix is not completely fixed. However, traffic is still independent of the actual number of messages sent and received, and depends only on the clients' connectivity. Since the connectivity is known to the eavesdropping attacker, this mechanism exposes no additional information. Hence, there is no information leakage.

Since clients are not offline for more than μ rounds, recipient anonymity is achieved against passive attackers according to Notion 2, provided that (at least) one pull mix is honest. This is because all the pull requests and the responses for the requests arrive at an honest mix that forwards them shuffled. Consequently, the adversary cannot correlate incoming messages to outgoing mixed messages.
                                                                                    □

Claim. AnonPoP achieves $(f, f \cdot (f + 1))$-attacker isolation, even when clients may disconnect, provided $f << n$ and clients do not go offline for more than μ consecutive rounds.

Argument: As stated by the previous claim, while clients do not disconnect for more than μ consecutive rounds, and the traffic reaches the servers intact on time, recipient anonymity is ensured. Additionally, the mechanism described in Subsect. 4.2 allows the detection of active attacks. Under the same conditions, while the rate of pull requests remains fixed, Notion 3 is satisfied with regard to recipient anonymity against active adversaries.   □

Finally, we note that AnonPoP can also ensure *forward secrecy and proactive security* by using existing schemes and methods, e.g., [31] and [32].

7   Mobile Environments

Support for mobile clients is critical for the success of anonymous messaging, but also involves serious challenges. In particular, users of mobile devices are reluctant to use energy-hungry applications. We briefly describe one of our energy-optimizations and our experimental evaluation of energy requirements.

7.1   Saving Energy with Lazy Pulling

In a naive implementation, clients would maintain an open TCP connection to the first mix until the response arrives back at the first mix. However, the open connection prevents the device from moving to energy-saving 'sleep' mode. To reduce energy consumption, AnonPoP uses *lazy pulling*, where clients use only short connections. In lazy pulling, the first mix in every channel acts as a proxy for the PO's responses. The client sends requests to the first mixes in each of the push/pull channels, and immediately retrieves the responses for the requests of the previous round. Although it may not be obvious, lazy pulling results in the same average latency as with immediate pulling (for a detailed explanation, see [6]).

7.2   Evaluating Energy Consumption

We briefly describe a user study we conducted to test the impact of AnonPoP on the user-experience of Android phone users. The topology of the network in the experiment included three mixes in each channel and one PO. The user study was conducted with the participation of 20 smartphone users. We wanted to test how the different implementations (using asymmetric or symmetric cryptography) affect the user experience. We created an Android application that runs one of three states: (1) using *asymmetric* cryptography ('real' AnonPoP), (2) using *symmetric* cryptography, and (3) with *cryptography disabled*.

During every installation, the application randomly chose one of the three states. The experiment participants reinstalled the application every week for eight weeks, to change the state randomly. At the end of every week, the participants were asked to rate their user experience with a focus on the battery life, compared to the previous week. The experiment was conducted *double-blindly*; both the authors and the participants did not know which states were assigned to each of them during the course of the experiment. At the end of the experiment, we compared the real changes in the states and the feedback by the users. The experimental results serve to strengthen our hypothesis: AnonPoP overhead does not create a significant degradation in usability for smart-phone users. For more information about the experiment and its results, see [6].

8   Implementation and Evaluation

In this section, we first describe our implementation, focusing on AnonPoP servers and the cryptographic primitives we used. We then show that the Anon-PoP implementation is practical, by evaluating it under real-world conditions, including a cost analysis of the system using commercial cloud services.

8.1   Implementation

We implemented AnonPoP in Java, because of its portability to different platforms. Our implementation for the push and pull channels uses a simple four-layer onion for each request, using a hybrid encryption scheme. For shared-key encryption, we used a simple authenticated encryption scheme with AES/CBC/PKCS5 padding. The key size was 48 bytes, consisting of 128-bit AES key and 256-bit HMAC-SHA key. For the public key encryption scheme, we used RSA with a 1024-bit key. For the push and pull request onions, the cryptographic overhead was slightly more than 1KB. The overhead for the push and the pull response onions was 256 bytes. We used 128-bit tokens as unique identifiers for messages and mailboxes. We expect further improvements in performance by moving to elliptic-curve cryptography, using the efficient and compact Sphinx [33] design.

To decouple any dependency in AnonPoP we developed an API that relieves any direct interaction with AnonPoP. The API autonomously maintains the connectivity, sends dummy messages when needed, handles the encryption/decryption, and generally acts as a friendly intermediary between the application and AnonPoP's infrastructure. The bottom line is that any application or service can use AnonPoP's API as a "carrier" to deliver the data anonymously.

Moreover, a crucial obstacle towards adoption of anonymous communication systems is the fact that users need to migrate to a dedicated application, and cannot continue to use their preferred messaging service. Our energy consumption results, cloud evaluations, and API, suggest that AnonPoP could be integrated as a layer of anonymity for any service, while users continue to use their preferred applications.

8.2   Evaluation in the Cloud

We used Amazon's cloud services c4.8xlarge Linux machines with 36 virtual CPUs and 60 GB of memory. Our evaluation was done on the simple topology of three mixes in each channel and a single PO, with extra machines that simulated the clients. We configured the instances such that every pair of communicating machines will be located on different continents, to emulate worst-case scenarios.

We experimented with slots of $\tau = 30\,$s, rounds of $\lambda = 10$ slots, and epochs of 3 h. We began to run the protocol against 100, 000 concurrent clients. We repeated the experiment, gradually increasing the number of clients until the failure rate was higher than 0.001%. Our implementation was able to support up to 500, 000 concurrent users with only sporadic failures due to clients who were unable to open a connection with the first mix.

Our AnonPoP implementation uses a 1 KB message size and round length of 5 min. The 1 KB size is suitable for most textual messaging services, especially regarding mobile communication. The round length was selected to trade-off latency with energy consumption, which is critical for mobile devices (short rounds could significantly increase energy consumption and bandwidth). Figure 6 demonstrates the effect of payload size and round length in terms of costs, which shows that our choices were sensible.

8.3   Costs Evaluation

Running AnonPoP servers in the cloud is not expensive. The cost for each Amazon instances depends on several variables: location, type of payment, and bandwidth usage. Significant discounts are received for reserving instances for long periods. Reserving c4.8xlarge instances for the first and third mixes in the US and for the second mixes and the PO in Europe has a yearly cost of $60K\$$.

In addition to machine costs, there is a payment for the traffic generated by the machines. There is no need to pay for traffic coming from the Internet, but there is a changing cost for outgoing traffic. The cost starts at 0.09$ per 1 GB and goes down as the amount of outgoing traffic increases.

When a client sends push and pull requests to the first mixes, there is no cost for the system. However, each of these messages travels through the mixes and PO, generating outbound traffic of around 14.7 KB per client per round. The maximal communication volume in the system for a client is 1.47 GB.

Calculating the yearly cost of the system involves two factors: (1) the yearly cost of the instances, and (2) the yearly cost of the traffic for all the clients together. While the first factor does not directly depend on the number of clients and can be referred to as a constant, the second factor depends on the number of clients because the cost per GB decreases as the total amount of traffic increases. Both of the components reflect the yearly cost of running AnonPoP's servers. We divided the yearly cost for the machines and the traffic by the number of clients to get the yearly cost per client.

Using the instances we chose, the yearly cost starts at 1.4$ per client for $50K$ clients, and decreases rapidly to less than 25 cents per client for $500K$ clients. Figure 7 depicts the yearly cost per client as a function of the number of clients. Note, the calculation was based on using strong and relatively expensive instances even for a low number of clients. In practice, for fewer clients, weaker and cheaper machines can be used to further decrease the cost.

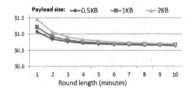

Fig. 6. Yearly cost ($) per client as a function of the payload size and round length, using c4.8xlarge machines with 100,000 concurrent users.

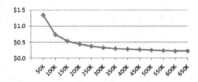

Fig. 7. Yearly cost ($) per client as a function of the number of clients using AnonPoP.

9   Related Work

This section briefly discuss other works dealing with anonymous communication. We focus on works whose goal, like AnonPoP, is to provide anonymity against adversaries with eavesdropping capabilities. Like AnonPoP, most of these works focus on applications with potential to suffer from significant latency, such as messaging. This excludes the many works like Tor and other low-latency systems, which, unlike AnonPoP, are vulnerable to eavesdropping adversaries.

AnonPoP continues the line of mix-based mechanisms whose goal is to provide strong anonymity for messaging or email, with relatively high latency. Other examples include Mixminion [34] and previous proposals, e.g., Babel [35], Mix-Master and Reliable [36]. Mixminion introduced new ideas such as *single-use reply blocks*, which allows anonymity for recipients, and techniques to deal with tagging and replay attacks; some of these techniques are used by AnonPoP. However, Mixminion is vulnerable to long-term intersection, does not provide unobservability, and has latency that can be excessive for messaging applications.

Other proposals for strong anonymous messaging were not really designed for practical deployment, as they are neither efficient nor appropriate for many users. The Busses protocol [37] ensures strong anonymity - even unobservability - by having each message sent through all possible destinations. The Drunk Motorcyclist (DM) design achieves similar properties, e.g., strong recipient anonymity, by sending each message randomly through the network, making it highly likely to reach the destination. These are elegant designs that ensure strong anonymity, but result in excessive overhead, which is inappropriate for real-life applications.

Verdict [38] and Dissent [39] follow the DC-net [40] design, to ensure sender anonymity. The computational overhead for both clients and servers is relatively high, although it was shown to be practical for up to thousands of users.

Riposte [41] is a recent DC-net proposal, which achieves sender anonymity against globally-eavesdropping adversaries for large anonymity sets; however, only a small proportion of the users send messages. In Riposte, many clients write into a shared database, maintained by a small set of servers. To reduce the bandwidth overhead for n clients from $O(n)$ to $O(\sqrt{n})$, Riposte uses private information retrieval (PIR) [42]. However, PIR schemes, even optimized (e.g., [43]), have significant latency and bandwidth overheads. These overheads increase as a function of the number of clients using the system, making them impractical for large-scale messaging. Pynchon Gate [44] is another design using PIR, in this case, to retrieve pseudonymous mail. Again, due to the use of PIR, it suffers from high communication and computation overhead, making it impractical for use in mobile devices and for systems with many users.

Nipane et al. presented Mix-In-Place [45], an architecture based on Secure Function Evaluation (SFE), supporting messaging with a single proxy. SFE is more computationally-intensive than PIR, thus, the system is not practical when considering many users and mobile devices.

Aqua [46] is another related system; although it has a higher overhead (cf. to AnonPoP), it is much more efficient than the systems discussed above. However, the goal of Aqua is file-sharing applications such as BitTorrent. Aqua ensures k-anonymity [47], using onion routing [10] with dummy traffic via multiple paths to resist traffic analysis. It does not provide anonymity against corrupt servers and does not support disconnecting clients.

Vuvuzela [48] is a scalable mix-based anonymous messaging system, with quite similar goals to AnonPoP. Vuvuzela shares several design decisions, but differs considerably as outlined in Appendix B. AnonPoP and Vuvuzela were designed concurrently [49,50] yet independently.

Resisting Intersection Attacks. Buddies [51] offers a mechanism to keep the publisher of a message on a shared board. The publisher remains anonymous within a set of k participants [47] for a long time, to avoid intersection attacks [15–18] by a global eavesdropper. However, Buddies does not mask the communication; instead, it *prevents* its clients from publishing messages when this might cause an exposure of their identity. By requiring many cooperating clients ('buddies') online to create a large anonymity set, Buddies uses significant overhead and latency. Hence, Buddies is unable to efficiently achieve long-term resistance to intersection attacks (see Sect. 5.6 in [51]).

10   Conclusions and Future Work

AnonPoP demonstrates practical anonymous messaging service with defenses against powerful attackers, including those with global eavesdropping capabilities and the ability to control some of the servers. AnonPoP achieves this with low overhead and operational costs and in a scalable manner, practical for mobile clients due to its low energy requirements and support for temporary disconnections. We evaluated AnonPoP using a user study and experiments, including commercial cloud services and mobile devices.

Future work should address significant remaining challenges. One significant challenge is *secure, usable and anonymous key management*; note that even without anonymity, this issue is a long outstanding challenge of usable security, even in popular messaging applications claiming end-to-end security [52].

One of AnonPoP's major contributions is the idea of bad server isolation. However, this design has not been fully analyzed. Completing and refining this mechanism is a significant challenge for future work. Initial steps toward this goal are taken in [53].

Solutions are also required to additional system issues, in particular, controlling traffic to prevent overload of honest servers, clock synchronization, resource reservation or allocation etc.

Acknowledgments. We are grateful to George Danezis, Yossi Gilad, Hezi Moriel, Roee Shlomo, Bogdan Carbunar and the anonymous reviewers for their helpful and constructive feedback. This work was supported by the Israeli Ministry of Science and Technology.

Appendix

A   Probability of Compromised Channel

When the PO is corrupt, AnonPoP's sender (recipient) anonymity may fail, if *all* mixes in the push (resp., pull) channel are malicious (1). We now show that, under the reasonable assumption that $f << n$, the probability of such 'all bad' channel is small.

To increase the probability of 'all bad' channel, the attacker may decrease the number of possible channels where at least one mix is honest, by disconnecting up to f honest servers from each malicious mix, abusing the 'bad server isolation' mechanism. However, as we show, this abuse does not significantly improve the probability of 'all bad' channel. Assume, for simplicity, that the attacker can cancel *every* connection between malicious and honest mixes; for simplicity, assume three mixes in a channel. Hence, there are $3! \cdot \binom{f}{3}$ 'all bad' channels, and $3! \cdot \binom{n-f}{3}$ 'all honest' channels. The probability of choosing an 'all bad' channel is therefore only: $\frac{\binom{f}{3}}{\binom{f}{3}+\binom{n-f}{3}}$.

B   AnonPoP and Vuvuzela

In this appendix, we briefly discuss some of the differences between AnonPoP and Vuvuzela.

Vuvuzela allows communication only between connected (online) users, where AnonPoP aims to provide defense to users who may disconnect. AnonPoP's motivation for this decision is to provide protection for its users from attacks that takes advantage of disconnections to infer information about the users. Furthermore, AnonPoP also aims to have a built-in support for mobile users, and mobile users sometimes disconnect.

AnonPoP's goal to provide support for mobile users is also exhibited in its attempt to minimize the communication overhead requirements to be suitable for the low energy and low bandwidth requirements of usable mobile environments. In Vuvuzela, at each 'dial round' (currently set at 10 min), every Vuvuzela user downloads and decrypts all 'invitations' sent to her invitation dead drop, shared with many other users and determined as the hash of the user's public key. Even with only three servers, this is 7MB per (10-min) dialing round.

AnonPoP presents the bad-server isolation mechanism, which actively takes measures against misbehaving servers, to deter rogue servers from performing active attacks against AnonPoP users.

References

1. Dingledine, R., Mathewson, N., Syverson, P.F.: Tor: the second-generation onion router. In: USENIX Security Symposium, USENIX, pp. 303–320 (2004)

2. Gilad, Y., Herzberg, A.: Spying in the dark: TCP and tor traffic analysis. In: Fischer-Hübner, S., Wright, M. (eds.) PETS 2012. LNCS, vol. 7384, pp. 100–119. Springer, Heidelberg (2012). https://doi.org/10.1007/978-3-642-31680-7_6

3. Bauer, K., McCoy, D., Grunwald, D., Kohno, T., Sicker, D.: Low-resource routing attacks against tor. In: Proceedings of the 2007 ACM Workshop on Privacy in Electronic Society, pp. 11–20. ACM (2007)

4. Borisov, N., Danezis, G., Mittal, P., Tabriz, P.: Denial of service or denial of security? In: Proceedings of the 14th ACM Conference on Computer and Communications Security, pp. 92–102. ACM (2007)

5. Dingledine, R., Mathewson, N.: Anonymity loves company: usability and the network effect. In: WEIS (2006)

6. Gelernter, N., Herzberg, A., Leibowitz, H.: Two cents for strong anonymity: the anonymous post-office protocol. Cryptology ePrint Archive, Report 2016/489 (2016) http://eprint.iacr.org/2016/489

7. Farb, M., Burman, M., Chandok, G., McCune, J., Perrig, A.: SafeSlinger: an easy-to-use and secure approach for human trust establishment. Technical report, Technical Report CMU-CyLab-11-021, Carnegie Mellon University (2011)

8. Bellare, M., Rogaway, P.: Asymmetric encryption. http://cseweb.ucsd.edu/~mihir/cse207/w-asym.pdf

9. Chaum, D.: Untraceable electronic mail, return addresses, and digital pseudonyms. Commun. ACM **24**(2), 84–90 (1981)

10. Goldschlag, D., Reed, M., Syverson, P.: Onion routing. Commun. ACM **42**(2), 39–41 (1999)

11. Bellare, M., Namprempre, C.: Authenticated encryption: relations among notions and analysis of the generic composition paradigm. In: Okamoto, T. (ed.) ASIACRYPT 2000. LNCS, vol. 1976, pp. 531–545. Springer, Heidelberg (2000). https://doi.org/10.1007/3-540-44448-3_41

12. Pfitzmann, A., Pfitzmann, B., Waidner, M.: ISDN-MIXes: untraceable communication with very small bandwidth overhead. GI/ITG Conf. Commun. Distrib. Syst. **267**, 451–463 (1991)

13. Piotrowska, A., Hayes, J., Gelernter, N., Danezis, G., Herzberg, A.: AnNotify: a private notification service. In: Workshop on Privacy in the Electronic Society (WPES 2017) (2017)

14. Dingledine, R., Syverson, P.: Reliable MIX cascade networks through reputation. In: Blaze, M. (ed.) FC 2002. LNCS, vol. 2357, pp. 253–268. Springer, Heidelberg (2003). https://doi.org/10.1007/3-540-36504-4_18

15. Berthold, O., Federrath, H., Köhntopp, M.: Project "anonymity and unobservability in the internet". In: Proceedings of the Tenth Conference on Computers, Freedom and Privacy: Challenging the Assumptions, pp. 57–65. ACM (2000)

16. Berthold, O., Langos, H.: Dummy traffic against long term intersection attacks. In: Dingledine, R., Syverson, P. (eds.) PET 2002. LNCS, vol. 2482, pp. 110–128. Springer, Heidelberg (2003). https://doi.org/10.1007/3-540-36467-6_9

17. Mathewson, N., Dingledine, R.: Practical traffic analysis: extending and resisting statistical disclosure. In: Martin, D., Serjantov, A. (eds.) PET 2004. LNCS, vol. 3424, pp. 17–34. Springer, Heidelberg (2005). https://doi.org/10.1007/11423409_2

18. Wright, M.K., Adler, M., Levine, B.N., Shields, C.: Passive-logging attacks against anonymous communications systems. ACM Trans. Inf. Syst. Secur. (TISSEC) **11**(2), 3 (2008)

19. Pfitzmann, A., Hansen, M.: A terminology for talking about privacy by data minimization: anonymity, unlinkability, undetectability, unobservability, pseudonymity, and identity management, 34 (2010). http://dud.inf.tu-dresden.de/literatur/Anon_Terminology_v0
20. Hughes, D., Shmatikov, V.: Information hiding, anonymity and privacy: a modular approach. J. Comput. Secur. **12**(1), 3–36 (2004)
21. Halpern, J., O'Neill, K.: Anonymity and information hiding in multiagent systems. J. Comput. Secur. **13**(3), 483–514 (2005)
22. Pashalidis, A.: Measuring the effectiveness and the fairness of relation hiding systems. In: IEEE Asia-Pacific Services Computing Conference, APSCC 2008, pp. 1387–1394. IEEE (2008)
23. Tsukada, Y., Mano, K., Sakurada, H., Kawabe, Y.: Anonymity, privacy, onymity, and identity: a modal logic approach. In: International Conference on Computational Science and Engineering, CSE 2009, vol. 3, pp. 42–51. IEEE (2009)
24. Bohli, J., Pashalidis, A.: Relations among privacy notions. ACM Trans. Inf. Syst. Secur. (TISSEC) **14**(1), 4 (2011)
25. Goriac, I.: An epistemic logic based framework for reasoning about information hiding. In: 2011 Sixth International Conference on Availability, Reliability and Security (ARES), pp. 286–293. IEEE (2011)
26. Veeningen, M., de Weger, B., Zannone, N.: Modeling identity-related properties and their privacy strength. In: Degano, P., Etalle, S., Guttman, J. (eds.) FAST 2010. LNCS, vol. 6561, pp. 126–140. Springer, Heidelberg (2011). https://doi.org/10.1007/978-3-642-19751-2_9
27. Backes, M., Goldberg, I., Kate, A., Mohammadi, E.: Provably secure and practical onion routing. In: 2012 IEEE 25th Computer Security Foundations Symposium (CSF), pp. 369–385. IEEE (2012)
28. Feigenbaum, J., Johnson, A., Syverson, P.: Probabilistic analysis of onion routing in a black-box model. ACM Trans. Inf. Syst. Secur. **15**(3), 14:1–14:28 (2012)
29. Hevia, A., Micciancio, D.: An indistinguishability-based characterization of anonymous channels. In: Borisov, N., Goldberg, I. (eds.) PETS 2008. LNCS, vol. 5134, pp. 24–43. Springer, Heidelberg (2008). https://doi.org/10.1007/978-3-540-70630-4_3
30. Gelernter, N., Herzberg, A.: On the limits of provable anonymity. In: Proceedings of the 12th Annual ACM Workshop on Privacy in the Electronic Society, WPES 2013 (2013)
31. Canetti, R., Halevi, S., Katz, J.: A forward-secure public-key encryption scheme. In: Biham, E. (ed.) EUROCRYPT 2003. LNCS, vol. 2656, pp. 255–271. Springer, Heidelberg (2003). https://doi.org/10.1007/3-540-39200-9_16
32. Canetti, R., Halevi, S., Herzberg, A.: Maintaining authenticated communication in the presence of break-ins. J. Cryptol. **13**(1), 61–105 (2000)
33. Danezis, G., Goldberg, I.: Sphinx: a compact and provably secure mix format. In: 2009 30th IEEE Symposium on Security and Privacy, pp. 269–282. IEEE (2009)
34. Danezis, G., Dingledine, R., Mathewson, N.: Mixminion: design of a type iii anonymous remailer protocol. In: Proceedings of 2003 Symposium on Security and Privacy, pp. 2–15. IEEE (2003)
35. Gülcü, C., Tsudik, G.: Mixing email with Babel. In: Ellis, J.T., Neuman, B.C., Balenson, D.M. (eds.) NDSS, pp. 2–16. IEEE Computer Society (1996)
36. Díaz, C., Sassaman, L., Dewitte, E.: Comparison between two practical mix designs. In: Samarati, P., Ryan, P., Gollmann, D., Molva, R. (eds.) ESORICS 2004. LNCS, vol. 3193, pp. 141–159. Springer, Heidelberg (2004). https://doi.org/10.1007/978-3-540-30108-0_9

37. Beimel, A., Dolev, S.: Buses for anonymous message delivery. J. Cryptol. **16**(1), 25–39 (2003)
38. Corrigan-Gibbs, H., Wolinsky, D.I., Ford, B.: Proactively accountable anonymous messaging in Verdict. In: Proceedings of the 22nd USENIX Conference on Security, pp. 147–162. USENIX Association (2013)
39. Wolinsky, D.I., Corrigan-Gibbs, H., Ford, B., Johnson, A.: Dissent in numbers: making strong anonymity scale. In: 10th OSDI (2012)
40. Chaum, D.: The dining cryptographers problem: unconditional sender and recipient untraceability. J. Cryptol. **1**(1), 65–75 (1988)
41. Corrigan-Gibbs, H., Boneh, D., Mazires, D.: Riposte: an anonymous messaging system handling millions of users. In: IEEE Symposium on Security and Privacy, pp. 321–338. IEEE Computer Society (2015)
42. Chor, B., Kushilevitz, E., Goldreich, O., Sudan, M.: Private information retrieval. J. ACM (JACM) **45**(6), 965–981 (1998)
43. Demmler, D., Herzberg, A., Schneider, T.: RAID-PIR: practical multi-server PIR. In: Proceedings of the 6th edition of the ACM Workshop on Cloud Computing Security, pp. 45–56. ACM (2014)
44. Sassaman, L., Cohen, B., Mathewson, N.: The Pynchon gate: a secure method of pseudonymous mail retrieval. In: Proceedings of the 2005 ACM Workshop on Privacy in the Electronic Society, pp. 1–9. ACM (2005)
45. Nipane, N., Dacosta, I., Traynor, P.: "mix-in-place" anonymous networking using secure function evaluation. In: Zakon, R.H., McDermott, J.P., Locasto, M.E. (eds.) ACSAC, pp. 63–72. ACM (2011)
46. Le Blond, S., Choffnes, D., Zhou, W., Druschel, P., Ballani, H., Francis, P.: Towards efficient traffic-analysis resistant anonymity networks. In: Proceedings of the ACM SIGCOMM 2013 Conference on SIGCOMM, pp. 303–314. ACM (2013)
47. von Ahn, L., Bortz, A., Hopper, N.J.: K-anonymous message transmission. In: Proceedings of the 10th ACM Conference on Computer and Communications Security, pp. 122–130. ACM (2003)
48. van den Hooff, J., Lazar, D., Zaharia, M., Zeldovich, N.: Vuvuzela: scalable private messaging resistant to traffic analysis. In: SOSP, pp. 137–152. ACM (2015)
49. Gelernter, N., Herzberg, A.: AnonPoP old anonymous technical report (before the system implementation). Anonymised Technical report, August 2014. https://sites.google.com/site/anonymoustechreports/home
50. Gelernter, N., Herzberg, A.: Hide from the NSA: achieving strong anonymity against strong adversaries. In: 2014 IEEE International Conference on Software Science, Technology and Engineering (SWSTE), Doctoral Symposium (2014)
51. Wolinsky, D.I., Syta, E., Ford, B.: Hang with your buddies to resist intersection attacks. In: Proceedings of the 2013 ACM SIGSAC Conference on Computer & Communications Security, CCS 2013, pp. 1153–1166. ACM, New York (2013)
52. Herzberg, A., Leibowitz, H.: Can Johnny finally encrypt? Evaluating E2E encryption in popular IM applications. In: ACM Workshop on Socio-Technical Aspects in Security and Trust (STAST) (2016)
53. Leibowitz, H., Piotrowska, A., Danezis, G., Herzberg, A.: No right to remain silent: isolating malicious mixes. Cryptology ePrint Archive, Report 2017/1000 (2017). http://eprint.iacr.org/2017/1000

Wireless and Physical Layer Security

Practical Evaluation of Passive COTS Eavesdropping in 802.11b/n/ac WLAN

Daniele Antonioli[1]([⊠])(ID), Sandra Siby[2](ID), and Nils Ole Tippenhauer[1](ID)

[1] Singapore University of Technology and Design (SUTD), Singapore, Singapore
{daniele_antonioli,nils_tippenhauer}@sutd.edu.sg
[2] Ecole Polytechnique Federale de Lausanne (EPFL), Lausanne, Switzerland
sandra.siby@epfl.ch

Abstract. In this work, we compare the performance of a passive eavesdropper in 802.11b/n/ac WLAN networks. In particular, we investigate the downlink of 802.11 networks in infrastructure mode (e. g. from an access point to a terminal) using Commercial-Of-The-Shelf (COTS) devices. Recent 802.11n/ac amendments introduced several physical and link layer features, such as MIMO, spatial diversity, and frame aggregation, to increase the throughput and the capacity of the channel. Several information theoretical studies state that some of those 802.11n/ac features (e. g. beamforming) should provide a degradation of performance for a passive eavesdropper. However, the real impact of those features has not yet been analyzed in a practical context and experimentally evaluated. We present a theoretical discussion and a statistical analysis (using path loss models) to estimate the effects of such features on a passive eavesdropper in 802.11n/ac, using 802.11b as a baseline. We use Signal-to-Noise-Ratio (SNR) and Packet-Error-Rate (PER) as our main metrics. We compute lower and upper bounds for the expected SNR difference between 802.11b and 802.11n/ac using high-level wireless channel characteristics. We show that the PER in 802.11n/ac increases up to 98% (compared to 802.11b) at a distance of 20 m between the sender and the eavesdropper. To obtain a PER of 0.5 in 802.11n/ac, the attacker's maximal distance is reduced by up to 129.5 m compared to 802.11b. We perform an extensive set of experiments, using COTS devices in an indoor office environment, to verify our theoretical estimations. The experimental results validate our predicted effects and show that every amendment add extra resiliency against passive COTS eavesdropping.

Keywords: WLAN · 802.11 · Eavesdropping · MIMO · Beamforming

1 Introduction

In the last decade, wireless network communication has grown tremendously mainly due to standards such as UMTS (3G) and LTE (4G) for cellular networks and IEEE 802.11 (WLAN) for wireless networks. Cisco estimated that in 2017, 68% of all Internet traffic will be generated by wireless devices [5]. As a result,

© Springer Nature Switzerland AG 2018
S. Capkun and S. S. M. Chow (Eds.): CANS 2017, LNCS 11261, pp. 415–435, 2018.
https://doi.org/10.1007/978-3-030-02641-7_19

it can be expected that a majority of sensitive communication services, such as mobile banking and online payments will involve wireless networks. Indeed, it is paramount to secure the broadcast wireless channel against eavesdroppers to protect the confidentiality and integrity of the information.

In this work, we present a theoretical discussion, a numerical analysis (using path loss models), and a practical evaluation of passive eavesdropping attacks targeting several 802.11 (WLAN) networks. Recent 802.11n/ac amendments introduced interesting physical layer and link layer features such as Multiple-Input-Multiple-Output (MIMO), spatial diversity (e. g. CSD, TxBF, STBC) , spatial multiplexing (e. g. MU-TxBF), dual-band antennas[1] and frame aggregation [14]. It is believed that some of those features, that were developed mainly to increase the robustness and throughput of the channel might also *degrade* the performance of a passive eavesdropper. We would like to investigate this claim and experimentally measure whether this degradation happens or not in practice in a simple but yet realistic scenario (e. g. eavesdropping WLAN networks with COTS devices).

Several theoretical discussions have already been presented about passive and active eavesdropping in the wireless channel. The seminal work by Wyner [31] started the wiretap channel research track that has been extended to Gaussian [16], fading [10], and MIMO [20] channels. This set of papers studies asymptotic conditions that very rarely happen in practice. Recently, special attention was given to MIMO and beamforming as a defense mechanism against passive eavesdropping [22,25,32]. However, those works do not focus on 802.11 and they consider only a subset of the 802.11 features. There are also some alternative techniques already proposed against passive eavesdropping including multi-user cooperative diversity and the use of artificial noise [8,19,33]. However, those techniques are neither listed in any 802.11 standards nor implemented in any COTS device.

In this paper, we investigate the disadvantages that a passive eavesdropper has to face when attacking the downlink of an 802.11n/ac (MIMO) network versus an 802.11b (SISO) network. We focus on 802.11 networks in infrastructure mode (e. g. an access point connecting several laptops to the Internet) that use Commercial-Of-The-Shelf (COTS) devices. In particular, we compare *three* of the most widely used 802.11 amendments: b, n, and ac. We look at the downlink (e. g. traffic from the access point to the terminals) because it is the link that supports most of the advanced features of 802.11n/ac (e. g. spatial diversity and spatial multiplexing). We use 802.11b as a baseline. Our attacker model choice is explained in detail in Sect. 3.1, and a brief discussion about a stronger attacker model is presented in Sect. 4.5.

In our theoretical discussion, we estimate lower and upper bounds for the expected Signal-to-Noise-Ratio (SNR) disadvantage of an eavesdropper in 802.11n and ac compared to 802.11b. We numerically derive the expected Packet-Error-Rate (PER) of the intended receiver and the eavesdropper with respect to their distances to the sender. Finally, we present an 802.11b/n/ac downlink

---

[1] In this work we always use the word *antennas* rather than *antennae*.

empirical evaluation using COTS devices. After the experiments, we are able to confirm that in 802.11n/ac networks, the PER of the eavesdropper increases with respect to her distance to the sender, given a minimum distance between the attacker and the intended receiver.

We summarize our contributions as follows:

– We derive the theoretically expected eavesdropper's SNR disadvantage (in dB), for attacks using COTS radios, in 802.11b/n/ac downlinks.
– We discuss how the theoretical SNR disadvantage translates to practical constraints (e. g. reduced range, higher PER) for the attacker.
– We perform a series of experiments to validate that the expected disadvantage is experienced in practice and that its effects were correctly predicted.

The structure of this work is as follows: in Sect. 2 we provide the required wireless communications background. In Sect. 3, we present the system and attacker models, we compare passive eavesdropping 802.11b and 802.11n/ac downlinks, and we estimate the SNR and PER disadvantages for a passive eavesdropper in 802.11n/ac. In Sect. 4, we present our results from a series of eavesdropping experiments that validate our predicted impediments. We summarize related work in Sect. 5, and conclude our paper in Sect. 6.

2 Background

We now provide a summary of the important concepts used in this work: the fading wireless channel, the 802.11b/n/ac amendments, and three wireless communication metrics (SNR, BER, and PER). For additional details, we refer to influential books such as [9, 23].

2.1 The Fading Wireless Channel

The progression of wireless communication systems evolved around two main metrics: *robustness* and *throughput*. Those metrics are severely influenced by channel fading. Fading can be described as a random process affecting the quality of the transmitted wireless signal, by means of attenuation and distortion over time and frequency. There are three additive phenomena contributing to fading: path loss, shadowing, and multipath.

Path loss is a large-scale fading event due to the propagation nature of the electromagnetic waves (that are carrying the useful signal). There are different path loss models according to the system parameters and the channel environment. For example, in the Free Space Path Loss (FSPL) model the transmitted power decays quadratically with the distance from the transmitter to the receiver. Shadowing is another large-scale fading event due to the presence of obstacles between the transmitter and the receiver. There are different ways to model shadowing such as using a log-normal random variable. Multipath is a small-scale fading phenomenon that takes into account constrictive and/or destructive interference at the receiver between direct, reflected and scattered electromagnetic waves.

There are two well-known fading models that take into account all three fading aspects: *Rayleigh fading* for non-line-of-sight (NLOS) environments, and *Rician fading* for line-of-sight (LOS) environments. In both cases, each channel coefficient h is modeled with a complex random number. Each channel coefficient is providing random attenuation (change in amplitude) and distortion (change in phase). In the Rayleigh fading model, the real and imaginary parts of h are modeled with independent identically-distributed (IID) Gaussian random variables with 0 mean and equal variances and the amplitude of h is Rayleigh distributed. In the (more generic) Rician fading model, the amplitude of h is Rice distributed.

2.2   IEEE 802.11 Standard (WLAN)

802.11 is a family of IEEE standards that regulates wireless local area networks (WLAN) [7]. The standards define the physical layer (PHY), and the link layer specifications. An example of physical layer specification is the modulation and coding scheme (MCS) table that lists the supported modulation types, spatial streams, coding rates, bandwidths and data rates of a given PHY. An example of link layer specification is the medium access control (MAC) protocol that governs how the nodes share the wireless medium.

Table 1. Relevant 802.11b/n/ac physical layer specifications. f_c is the carrier frequency, λ is the wavelength, s_{dr} is the theoretical maximum throughput of the channel, n_S is the number of maximum independent data streams, TxBF indicates support for single-user (SU) or multi-user (MU) transmit-beamforming, d_i and d_o are the expected ranges for indoor and outdoor communications.

	Technology	Modulation	f_c [GHz]	λ [cm]	s_{dr} [Mbit/s]	n_S	TxBF	d_i	d_o
b	SISO	DSSS	2.4	12.5	11	N/A	N/A	35	140
n	SU-MIMO	OFDM	2.4, 5	12.5, 6	135	4	SU	70	250
ac	MU-MIMO	OFDM	5	6	780	8	MU	35	N/A

Table 1 lists some relevant physical layer specifications for 802.11b, n, and ac [14]. 802.11b uses Single-Input-Single-Output (SISO) scheme with direct-sequence spread spectrum (DSSS) modulation techniques. In contrast, 802.11n and 802.11ac are Multiple-Input-Multiple-Output (MIMO) schemes, based on orthogonal frequency division multiplexing (OFDM) modulation techniques. Single user MIMO is supported by 802.11n, while 802.11ac supports multi-user MIMO. The major advantage in terms of throughput and robustness of the channel from b to n/ac is given by the usage of multiple radios and antennas that allows transmitting different independent symbol at the same time (spatial multiplexing) or the same symbol on multiple antennas at the same time (spatial diversity).

In particular, 802.11n/ac support transmit-beamforming (TxBF) at the downlink for single user (n) and multiple users (ac). By using TxBF, an access point can optimize the transmission of the symbols to a device located in a particular region of space, given an estimate of the condition of the downlink channel. For a more detailed comparison among the three 802.11 amendments please refer to [13,21].

2.3   Wireless Communications Metrics

Here we present the three wireless communication metrics used in our paper:

- The *Signal-to-Noise-Ratio (SNR)* is the ratio between the power of the useful signal denoted with P, and the noise power σ^2. It is typically expressed in decibel dB, and it convertible from logarithmic to linear scale using: $10 \log_{10} \text{SNR} = \text{SNR}_{dB}$.
- The *Bit-Error-Rate (BER)* is the expected probability of error while decoding 1-bit at the receiver. The BER is not an exact quantity. It can be modeled and estimated according to different factors such as the modulation/coding schemes, the fading model and the number of antennas. Typically, 10^{-6} is considered a reasonable BER value, i.e. 1-bit error per Mbit.
- The *Packet-Error-Rate (PER)* is directly proportional to the BER, and it is computed as: $\text{PER} = 1 - (1 - \text{BER})^N$, where N is the average packet size in bits. In this work, we assume that one or more bit errors in a packet will lead to an incorrect link layer checksum. Packets with an incorrect checksum are not acknowledged by the (legitimate) receiver, and retransmitted by the sender.

3   Passive 802.11 Downlink Eavesdropping

We start this section introducing the system and attacker models. Then we present a theoretical discussion and a numerical analysis (based on 802.11 path loss models) to estimate the SNR and PER disadvantages of a passive eavesdropper in an 802.11n/ac (MISO) downlink, compared to an 802.11b (SISO) downlink.

3.1   System and Attacker Model

Our system model focuses on the *downlink* of indoor 802.11b/n/ac networks in infrastructure mode (e.g. access point that communicates with several wireless terminals), using Commercial-Of-The-Shelf (COTS) devices. The access point is equipped with multiple antennas. The intended receiver and the attacker are equipped either with a single or multiple antennas according to the scenario. We are looking at the ratio of packets that the attacker successfully eavesdrop on the physical layer and we are agnostic to any encryption scheme used at the link layer or above. Attacks on those schemes are possible, but out of the scope of this work [3,26]).

The attacker is assumed to be *equipotent* to the intended receiver in terms of hardware and software capabilities. In particular, both use COTS devices, with a similar chipset, driver, feature set, and maximum throughput. With COTS devices we refer to wireless radios either built into laptops, smartphones, access point or USB dongles. We do not consider an attacker equipped with a software-defined-radio (SDR) or similar devices. We focus on a *passive* eavesdropper who wants to capture the downlink packets in real-time using her wireless card in monitor mode. We are not considering an attacker who is recording and post-processing the traffic offline. We assume an attacker that is static and we evaluate her eavesdropping performance at different distances from the sender. If the sender is using beamforming, we assume that the attacker is outside the beamforming region.

The effectiveness of the attacker is assessed from the Signal-to-Noise-Ratio (SNR) and the Packet-Error-Rate (PER) at her receiver. We chose PER as metric because we are mainly interested in the relative performance of eavesdropping on 802.11b vs. 802.11n/ac. As our passive attacker is unable to request retransmissions, the only chance to recover from bit errors would be to find the offending bit(s) and correct it using a checksum (possibly by brute force). We note that such corrections are expected to have significant cost for increasing number of flipped bits, and that the number of flipped bits is expected to quickly increase with distance. We plan to further investigate this in future work.

Without loss of generality and to simplify our discussion, we are considering an attacker focused on eavesdropping the downlink channel of one pair of transmitter and intended receiver. We understand that our attacker model is relatively weak (e. g. a single attacker, no SDR), however, given the lack of related experimental work and the number of involved moving parts, we decided to start with a simple scenario that is easy to evaluate (e. g. worst-case scenario for the passive eavesdropper). We look forward to investigate more complex attacker models in future work.

Finally, we present the notation used in our paper. The access point is referred as Alice (the transmitter), the victim as Bob (the intended receiver), and the attacker as Eve (passive eavesdropper). We will use A, B, and E subscripts to identify quantities related to Alice, Bob, and Eve respectively. We use x to denote Alice's transmitted symbol, h for complex channel coefficients, and n for the noise at a specific receiver. The relative distances between Alice, Bob, and Eve are written as: d_{AB}, d_{BE}, d_{AE}. Alice is equipped with L antennas and L radios.

3.2   SISO and MISO Channels Eavesdropping

In this section, we analyze and compare two different eavesdropping scenario: (i) 802.11b SISO downlink, (ii) 802.11n/ac MISO downlink. and we derive two essential conclusions about passive eavesdropping in SISO vs. MIMO 802.11 downlinks.

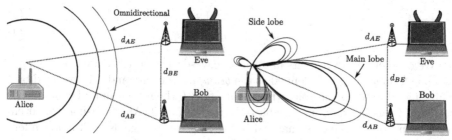

(a) *Omnidirectional radiation (L = 1).* (b) *Transmit-beamforming (L > 1). Eve's*
Eve's success depends on d_{AE}.   *success depends also on d_{BE} and L.*

Fig. 1. 802.11b SISO (left) vs. 802.11 n/ac MISO (right) passive eavesdropping. Bob
and Eve have one antenna. Dashed lines represent distances. Black circles and lobes represent omnidirectional and directional transmission ranges. Circles and lobes decreasing thickness represent the transmission power decay with respect to distance from the transmitter. Both channels are affected by random noise and fading.

802.11b SISO Downlink. Figure 1a shows Eve trying to intercept the communication from Alice to Bob in an 802.11b SISO network. We can represent the signals received by Eve and Bob as:

$$y_E = x \cdot h_E + n_E \tag{1}$$

$$y_B = x \cdot h_B + n_B \tag{2}$$

Intuitively, it is possible to represent Alice's two-dimensional transmission coverage with concentric circles. In free space, the greater is the distance from the transmitter the higher is the transmitted power decay. While one might assume that every receiver inside these circles will be "in range" and receive all transmissions by Alice, this is not the case in practice. If circles are shown around transmitters, their radius commonly refers to a distance in which the average received signal strength is above a certain threshold. However, due to random deep fading (mostly due to multipath), the instantaneous received power will constantly vary. In other words, it is possible to "miss transmissions" while being in the outer circle, or even receive transmissions just outside the outer circle. In this case, Eve's success rate depends on her distance to Alice (d_{AE}) regardless of her distance to Bob (d_{BE}), and random channel characteristics. The SISO wireless channel is providing some sort of resiliency against eavesdropping that an attacker can compensate with other means (e.g.: increase receiver sensitivity, use directional antenna).

802.11n/ac MISO Downlink. Figure 1b shows Eve attempting to intercept the communication from Alice to Bob in an 802.11n/ac MISO network. Alice is equipped with L antennas and uses transmit-beamforming. In this scenario, beamforming has been theoretically proven to provide resiliency against passive eavesdropping [12]. The received signals by Eve and Bob are as follows:

$$y_E = x \cdot g_E + n_E \tag{3}$$
$$y_B = x \cdot g_B + n_B \tag{4}$$

We can derive two benefits in terms of eavesdropping resiliency, one from g_B, and one from g_E. $\|g_B\|^2$ is defined as the *beamforming gain* and it is modeled by a Chi-squared random variable, with parameter $2L$ (being the sum of squared IID standard Gaussian random variables). Indeed, if $L = 2$ (Alice is using two antennas), then Bob's received signal will be the sum of two signals with independent fading paths. The correspondent beamforming gain is computed as:

$$\|g_B\|^2 = \|h_{B1}\|^2 + \|h_{B2}\|^2 \tag{5}$$

and this quantity is certainly greater (or equal) to $\|h_{B1}\|^2$ and $\|h_{B2}\|^2$. The net result is a better SNR at Bob's receiver with respect to the SISO case.

The second benefit arising from transmit-beamforming is encapsulated by g_E. Eve's ability to eavesdrop depends on two more factors with respect to the SISO case. Firstly, her distance from Bob (d_{BE}), and secondly the number of antennas used by Alice (L). This is a consequence of transmit-beamforming employed by Alice (the beamformer) towards Bob (the beamformee). Figure 1b shows Alice beamforming in the direction of Bob (e. g. inside the main lobe) while Eve is outside the main and the side lobes. This results in a smaller SNR at her receiver compared to the one of Fig. 1a (given the same relative distances). Even if we decrease the distance between Eve and Alice, the disadvantage will still hold until Eve is outside the beamforming region. Furthermore, Eve's SNR will be inversely proportional to L because the more antennas are used by Alice to beamform, the more Alice can focus the beam towards a narrower but longer region in space [29].

3.3   Eavesdropper's Theoretical SNR Disadvantage in 802.11n/ac

In the previous section we argued that MISO beamforming from Alice to Bob will degrade Eve's eavesdropping performance according to d_{AE}, d_{BE}, and L. In this section, we will quantify the expected disadvantage of Eve in an 802.11n/ac network compared to an 802.11b network. We will estimate upper and lower bounds for the SNR at Eve's receiver with respect to L. We will provide numerical results for $L = 4$ to match the experimental setup of Sect. 4.1. We note that the bounds we are providing are not supposed to be *strict*—the actual SNR disadvantage will depend on many factors. Nevertheless, we compute the bounds based on the modeling assumptions to provide an intuition about the theoretically expected disadvantage.

Upper Bound. We start comparing high-level wireless channel characteristics of SISO and MISO channels. Table 2 lists the closed-form expressions for the SNR and the BER of SISO and MISO networks using BPSK modulation scheme. In general, we note that the number of antennas deployed by Alice (L) is playing a central role. If we fix the expected BER to 10^{-6}, then we can compute the

Table 2. SNR and BER of 802.11b (SISO) and 802.11n/ac (MISO transmit-beamforming with L antenna) using BPSK modulation scheme.

Metric	SISO	MISO beamforming	
SNR	$\|h\|^2 \frac{P}{\sigma^2}$	$\|g\|^2 \frac{P}{\sigma^2}$	
BER	$\frac{1}{2}(1-\lambda)$	$\left(\frac{1-\lambda}{2}\right)^L \sum_{i=0}^{L-1} \binom{L+i-1}{i} \left(\frac{1+\lambda}{2}\right)^i$	$\lambda = \sqrt{\frac{\text{SNR}}{2+\text{SNR}}}$
DO	1	L	

minimum SNR for the SISO (57 dB) and the MISO case with $L = 4$ (16 dB). There is a notable difference in SNR of 41 dB between the SISO and the MISO cases. We use 41 dB as an upper bound for the SNR disadvantage of Eve with respect to Bob.

Lower Bound. For the lower bound of Eve's SNR disadvantage, we use a standard formula to compute the beamforming gain in a MISO channel where Alice is using Cyclic Delay Diversity (CDD) with L antennas [17]. In this case, the beamforming gain in dB can be computed as follows:

$$\|g\|^2 = 10\log_{10}(L) \quad dB \tag{6}$$

Assuming a COTS access point with 4 antennas and a single receiving antenna, Bob's beamforming gain is 6 dB. As Eve's COTS radio will not benefit from the beamforming gain (being outside the main lobe) Eve's SNR disadvantage lower bound is thus 6 dB with respect to Bob.

Summary. We estimate that an 802.11n/ac downlink that is using transmit-beamforming with four antennas provides an reduction in the SNR of a passive eavesdropper (outside the main lobe, using a COTS receiver) that is bounded *between 6 dB and 41 dB*. The reduction in SNR at Eve's receiver depends on a deterministic and measurable factors: d_{AE} (distance between Alice and Eve) and L (number of antennas used by the Alice). We note that Eve's SNR variation depends also on channel (Rayleigh) fading, however this factor is not considered in our discussion because it randomly affects both Bob and Eve, providing no deterministic disadvantage to Eve. Given this theoretically expected disadvantage, the question now is: *"How does the eavesdropper SNR disadvantage translate to practical constraints on 802.11 passive eavesdroppers?"*

3.4 Numerical Path Loss Analysis

In this section, we present a numerical analysis using three indoor path loss models for 802.11 networks. The models includes both the 2.4 and 5 GHz bands and they are taken from [23]. We now describe their relevant parameters. In particular, d_{BP} is defined as the breakpoint distance between the transmitter and the receiver and it determines the cutoff span between LOS and NLOS channel condition. σ_{SF} represents the standard deviation in dB of the log-normal random variable that models the shadowing term of the path loss. s_{PL} represents the path loss slope before and after d_{BP}. Comma-separated values in the following list indicate values before and after the breakpoint distance:

- **Model B**: Residential (e.g. intra-room, room-to-room).
  - $d_{BP} = 5$ m
  - $\sigma_{SF} = 3, 4$ dB
  - $s_{PL} = 2, 3.5$

- **Model D**: Office (e.g. large conference room, sea of cubes).
  - $d_{BP} = 10$ m
  - $\sigma_{SF} = 3, 5$ dB
  - $s_{PL} = 2, 3.5$

- **Model E**: Large office (e.g. multi-storey building).
  - $d_{BP} = 20$ m
  - $\sigma_{SF} = 3, 6$ dB
  - $s_{PL} = 2, 3.5$

Figure 2 shows the setup used for our numerical analysis and for the experiments. Bob is placed at a fixed distance away from Alice, Eve is placed at different (stationary) distances d_i from Alice, and Alice is constantly sending traffic to Bob. In a two-dimensional plane, Bob and Eve distance vectors are perpendicular to avoid Eve being in the main lobe when Alice is using transmit-beamforming. We note that in an indoor environment multipath is playing a major role than visual of RF line-of-sight conditions that is why we decided to keep altitude and angle constant and vary only the distance between Alice and Eve [6].

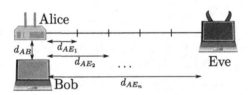

Fig. 2. Setup used for our numerical analysis and for the experiments: Bob is at a fixed distance away from Alice, Alice is sending 802.11 traffic and Eve is passively eavesdropping from different (stationary) distances on a line perpendicular to Bob.

The path loss model function L_P is constructed considering the sum of a free-space loss component (L_{FS}), a shadowing log-normal component due to obstacles (S_F), and a post breakpoint distance component. All terms vary according to the distance d between the transmitter and the receiver. We used the following equations from [23]:

$$L_P(d) = \begin{cases} L_{FS}(d) + S_F(d) & \text{if } d \leq d_{BP} \\ L_{FS}(d_{BP}) + S_F(d) + 35 \log_{10}\left(\dfrac{d}{d_{BP}}\right) & \text{otherwise} \end{cases} \tag{7}$$

$$L_{FS}(d) = 20 \log_{10}(d) + 20 \log_{10}(f) - 147.5 \tag{8}$$

$$S_F(d) = \frac{1}{\sqrt{2\pi}\sigma_{SF}} \exp\left(-\frac{d^2}{2\sigma_{SF}^2}\right) \tag{9}$$

Figures 3 and 4 shows the predicted BER and PER for model B (Residential) vs. distance between the transmitter and the receiver. Solid lines represent results for 2.4 GHz and dash-dotted lines represent results for 5.0 GHz. Red lines represent Eve's expected BER and PER. The other lines represent Bob's expected BER and PER when Alice is using transmit beamforming with two (green lines) and four (blue lines) antennas. If we focus on the solid lines of Fig. 4, then we note that a distance between Alice and Eve d_{AE} of 12.5 m is sufficient to drop Eve's expected PER from 0 to 0.5 (50% chance of decoding). Furthermore a d_{AE} of 20 m is sufficient to increase Eve's PER to 0.98 (0.2% chance of decoding). On the other hand, a d_{AB} of 142 m is required to experience a PER of 0.5 at Bob's receiver when Alice is using four antennas (L = 4).

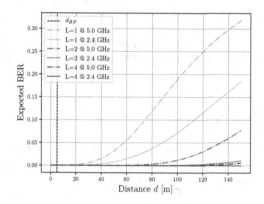

Fig. 3. 802.11n Model B (Residential) expected BER estimation using BPSK. Red lines represent Eve. Green and Blue lines represent Bob when L = 2 and L = 4. (Color figure online)

3.5 Eavesdropping Analysis Summary

In this section, we argued that in 802.11n/ac downlink a passive eavesdropper (Eve) using a COTS radio will have a disadvantage in terms of SNR compared to an eavesdropper in an 802.11b downlink. This disadvantage is due to different features provided by recent 802.11n/ac such as MIMO, and spatial diversity. This disadvantage can be expressed in an SNR decrease at the eavesdropper receiver of 6–41 dB (depending on the chosen scenario). We also express this disadvantage in terms of the distance that the eavesdropper has to be closer to the sender to achieve the same PER as a legitimate receiver, which can reach up to 129.5 m. In contrast, there is no such distance disadvantage for the eavesdropper in 802.11b. Furthermore, we can express the disadvantage in terms of PER at the eavesdropper receiver compared to her distance from the transmitter (d_{AE}). For example, if d_{AE} is 12.5 m, then the PER of Eve is increased to 50%, and if d_{AE} is 20 m, then the PER of Eve is increased to 98%.

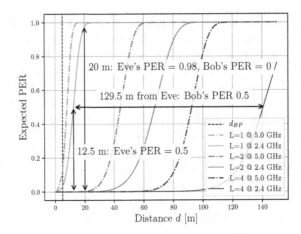

Fig. 4. 802.11n Model B (Residential) expected PER estimation using BPSK. Red lines represent Eve. Green and Blue lines represent Bob when L = 2 and L = 4. (Color figure online)

4   Experimental Validation

In this section, we present an experimental evaluation of COTS passive eavesdropping in 802.11b/n/ac downlink networks. The presented results are in line with the theoretical estimations from Sect. 3.

4.1   Experimental Methodology

We focus our experiments on SNR and PER measurements at Eve's receiver using the setup presented in Fig. 2. We keep a ninety-degree angle between Bob and Eve to ensure that when beamforming is used Eve is outside the beamforming region. We vary the distance from Bob to Eve (d_{BE}) while keeping the distance from Alice to Bob (d_{AB}) constant. Table 3 lists the parameters that we fix for our experiments with a short description. As stated in Sect. 3.1 we are not using link-layer encryption (which does not influence our measurements). Figure 5 shows the layout of the indoor office environment where we conducted the experiments.

Our setup consists of an open access point (Alice) and a laptop (Bob) associated to it. The access point is a Linksys WRT3200 ACM device, equipped with four antennas and supporting 802.11a/b/g/n/ac. We installed the OpenWrt [28] operating system on the access point to have more configuration options at our disposal. For the 802.11b/n experiments (at 2.4 GHz), Bob's laptop runs Ubuntu 16.04 and has a TP-Link TL-WN722N wireless adapter. The adapter has a single antenna and supports 802.11b/g/n. Eve's laptop runs Ubuntu 16.04, and it uses the same TP-Link TL-WN722N wireless adapter. Eve's adapter is not associated with the access point and it tries to record the traffic from Alice and Bob, in

Table 3. Parameters used for the experiments.

Parameter	Value	Description
P_A [dBm]	23	Alice's transmitted power
N_0 [dBm]	−91	Mean noise power at the receivers
$Ch_{b/n/ac}$	11, 11, 36	Channels used for 802.11 b/n/ac
d_{AB} [m]	2	Fixed distance from Alice to Bob
d_{AE} [m]	$[2.5, 5.0, \ldots, 20]$	Eight distances from Alice to Eve

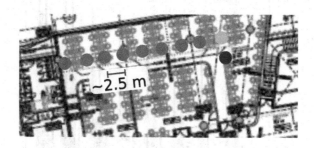

Fig. 5. The layout of the indoor office environment where we conducted the experiments. The green and blue dots indicate the location of Alice and Bob. The red dots indicate the positions of Eve. (Color figure online)

monitor mode using `tcpdump`. Eve listens to the same channel used by Alice and Bob (channel 11 for b and n, channel 36 for ac).

For the 802.11ac experiments (at 5 GHz), Bob's laptop runs Ubuntu 16.04 and uses an Asus USB-AC68 wireless adapter. The adapter has a 3x4:3 antenna configuration and supports 802.11a/b/g/n/ac. Eve's laptop is a MacBook Pro with an inbuilt adapter with 3x3:3 configuration compatible with 802.11a/b/g/n/ac. We use a different adapter for Eve because the Asus adapter could not be put into monitor mode due to some issues with its driver. The other parameters remain the same as in the 802.11b/n experiments.

For all the experiments, we vary Eve's distance from Bob and we obtain pcap traces of the packets transferred from Alice to Bob. The distance between Alice and Bob (d_{AB}) is fixed at 2 m. We used `iperf` to generate UDP downlink traffic. We decided to use UDP to avoid retransmissions at the transport layer. The PER is computed based on the number of received UDP packets with a valid UDP checksums. We acknowledge that this approach slightly underestimates the actual PER, as packets with a valid UDP checksum but incorrect link-layer checksum (FCS) might be included in this calculation. The transmission power of Alice is set to 23 dBm. From the experiments, we are able to obtain the traces from Eve at d_{AE} between 2.5 m and 20 m, using increments of 2.5 m. We do not change the orientation of Eve with respect to Alice in our tests to better compare the results. All the devices have the same fixed elevation, without a visual line-of-sight path between them. The information about the recorded

traffic is obtained from the 802.11 PHY radiotap headers. In the subsequent
section we will compare the experimental results with our estimations from the
path loss model D (Office). Figure 9 shows the predicted BER and PER curves at
Eve's receiver (red curves), and at Bob's receiver when Alice is using transmit-
beamforming with two (green curves) and four antennas (blue curves).

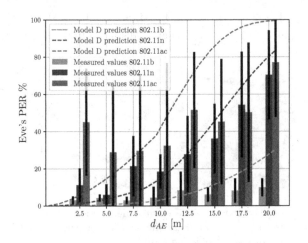

Fig. 6. Eve's measured PER (bars) vs. Model D predicted PER (dashed lines). (Color
figure online)

4.2   Comparison Between 802.11b/n/ac Networks

For the comparison between 802.11b/n/ac networks, we set a 2.4 GHz band for
802.11b/n and a 5 GHz band for 802.11ac. To extract the results we capture
packets both from Eve and Alice. We measured two parameters—the PER of
the passive eavesdropper, and her SNR. We compute Eve's PER by comparing
her pcap traces with the ones from Alice. We compute the SNR by dividing the
extracted signal strength values by the average channel noise. We computed the
average channel noise using noise measurements from the access point, and it
resulted in −91 dBm. We repeat the same experiments with the same distances
30 times and we average the results to obtain mean SNR and PER values, and
related errors (standard deviations).

Figure 6 shows Eve's PER measurements and estimated values for d_{AE} varying from 0 m to 20 m. The red/blue/green bars indicate the experimental results
for 802.11b/n/ac, respectively. The dotted lines indicate the predicted estimates
(from model D). It can be observed that Eve's PER is almost always *increasing* from b to n and from n to ac. In particular, the PER starts to increase
significantly when d_{AE} is greater than 15 m. While such (relatively small-scale)
experiments will hardly produce the exact same results as our theoretical analysis, we observe that the increase in PER that was predicted by us, for even

Table 4. Results from 802.11n and 802.11ac experiments. d_{AE} is the distance from Alice to Eve in meters. n_r is the total number of runs. μ_p is the average number of UDP packets sent by Alice per run. μ_{PER} and σ_{PER} are the Eve's PER means and standard deviations measured in our experiments for 802.11n (n subscript) and 802.11ac (ac subscript).

$d_{AE}[m]$	n_r	μ_p	μ_{PER_n}	σ_{PER_n}	$\mu_{PER_{ac}}$	$\sigma_{PER_{ac}}$
2.5	30	894.00	11.13	8.56	45.07	28.25
5.0	30	894.00	6.02	5.06	28.94	35.13
7.5	30	894.00	21.39	15.57	29.64	40.86
10.0	30	894.00	18.52	8.63	32.33	43.88
12.5	30	894.00	27.79	19.97	51.52	30.55
15.0	30	894.00	36.08	18.16	45.23	33.07
17.5	30	894.00	54.33	27.79	50.20	36.80
20.0	30	894.00	70.32	23.46	77.01	28.80

relatively short distances of around 20 m, can be observed in practice. In particular, our D model predicted a PER for Eve in an 802.11n downlink of around 78% when $d_{AE} = 20$ m, and in our experiments the average PER was around 70%. For convenience, we tabulate in Table 4 the numerical results of Fig. 6.

Figure 7 shows Eve's mean SNR varying her distance (d_{AE}) from Alice for 802.11b (red bars), 802.11n (blue bars) and 802.11ac (green bars). It can be observed that Eve's SNR in 802.11n/ac is always *smaller* than in 802.11b—an effect that we assumed to be caused by advanced 802.11n/ac physical and link layer features (such as TxBF).

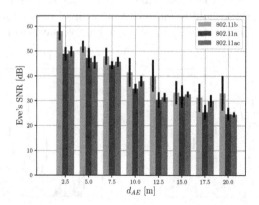

Fig. 7. Eve's measured SNR with respect to d_{AE}. (Color figure online)

4.3   Bob vs. Eve in 802.11n/ac

We conducted a second set of experiments targeting Bob in order to compare his SNR and PER with respect to Eve's SNR and PER in 802.11n/ac networks. In this case, we increased Bob's distance from Alice. As in the previous experiments, we start from 2.5 m and we end at 20 m, with increments of 2.5 m. Bob is placed at the same location that Eve was placed in the previous case. In this scenario, we are expecting that Bob would benefit from 802.11n/ac features. We are not showing the plot for Bob's PER compared to the one Eve experienced in Fig. 6. This is because we observed that Bob's PER is very low (less than 1%), and yet not comparable with Eve's PER. This confirms our assumption that the intended receiver experiences significantly *lower* PER than a passive eavesdropper in 802.11n/ac networks.

Interestingly, as we can see from Fig. 8a, the mean SNR of Bob and Eve at various distances are relatively close. In particular, Bob's SNR in 802.11n is always higher than Eve's SNR (as expected). However in the 802.11ac case, we measure a higher SNR for Eve than Bob. We assume that this is an artifact resulting from the fact that Eve's SNR is reported only for successfully received packets.

4.4   Eve's PER and PER Thresholds

We note that even a small decrease in PER could affect a passive eavesdropper depending on the type of exchanged traffic. That is why we decided to analyze Eve's PER compared to different PER thresholds and distances d_{AE}. Table 5 shows the results of our analysis for 802.11b/n/ac. For example, if we fix the threshold to 15%, then Eve's PER in 802.11ac is above the threshold in at least 33% of all cases. The same holds for 802.11n except for the 5 m measurement. With regards to 802.11b, fixing the same 15% threshold, we note that Eve's PER does not exceed the threshold in more than 16% of all cases. This is another way to confirm our predictions about 802.11n/ac passive eavesdropping.

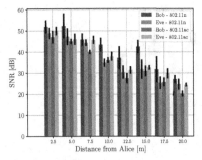
(a) *802.11n and 802.11ac SNR comparison between Bob and Eve at different distances from Alice.*

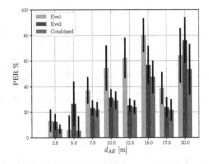
(b) *802.11n PER of Eve using two COTS radios. The green bars represent combined PER.*

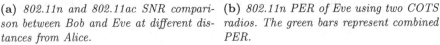

Fig. 8. Experimental results from Sect. 4.3 (a) and Sect. 4.5 (b). (Color figure online)

4.5 Eve with Two COTS Radios in 802.11n

We argued earlier that attackers with COTS radios will not be able to benefit from advanced 802.11n/ac physical layer and link layer features, and discussed an attacker with a single COTS radio. We now discuss a passive eavesdropper with multiple COTS radios in an 802.11n downlink. The attacker aggregates the eavesdropped packets to reduce the number of packets lost (e. g. due to deep fading). In Fig. 8b, we show the PER for an attacker with *two* COTS radios. The radios are placed at a distance of 50 cm from each other (to avoid mutual coupling). Note that we used a different data set from the previous experimental section, and we repeated this experiment 30 times. It can be observed that such a scheme *reduces* the number of lost packets for the attacker (as expected). However, the PER in the aggregated case is still higher than the 802.11b one, especially at distances greater than 5 m. For a threshold PER of 15%, the PER for the aggregated case is higher than the threshold in about 23% of the runs, compared to 6% for 802.11b.

4.6 Summary of 802.11b/n/ac Experiments

Overall, we were able to experimentally confirm our main findings: (a) there is a significant increase of the PER of a passive eavesdropper attacking 802.11n/ac networks compared to 802.11b ones. In our experiments, the difference was approximately 60% increased PER for 802.11n and 70% increased PER for 802.11ac at 20 m distance. In addition, the PER rises from around 12.5 m onward, similar to our predictions based on the theoretical analysis. We also confirmed that the PER experienced by the attacker is related to the non-cooperating Alice. In particular, legitimate receivers at the same locations were able to receive traffic with close to zero PER.

5 Related Work

There are several empirical studies for 802.11 networks. Most of them focus on specific link layer [18] or physical layer [27] features. There are also more generic empirical studies, for example about enterprise WLAN [4], intrusion detection [15], denial of service [2] co-existence [11] and signal manipulation [24] Anyway, those studies neither focuses on wireless security nor compares end experimentally evaluate eavesdropping in various 802.11 networks.

An interesting aspect of eavesdropping is to study how to optimally place a set of antennas in a multiple users scenario to obtain the maximum amount of private information. In [30] Wang et al. compare co-located vs. distributed eavesdropping schemes performance with respect to Eve's number of antennas and the presence of a guard zone. The de-facto standard countermeasure against eavesdropping (complementary to physical layer security) is cryptography. Several studies were done to secure [1] and break [3,26] cryptographic systems used by 802.11 such as WEP and WPA.

6   Conclusions

In this work, we investigated the impact of novel 802.11n/ac features over a passive eavesdropper using COTS devices. We focused on downlink networks in infrastructure mode. We performed a theoretical discussion, a numerical simulation and several experiments comparing the Signal-to-Noise-Ratio and Packet-Error-Rates of the eavesdroppers in 802.11b/n/ac. We showed that theoretically the eavesdropper's effective SNR is decreased by 6–41 dB in 802.11n/ac networks with four antennas ($L = 4$), which translates to a Packet-Error-Rate increase of up to 98% at a distance of 20 m between sender and eavesdropper. To obtain same Packet-Error-Rates as in a legitimate receiver, the attacker's maximal distance has to be reduced by 129.5 m in the case of 802.11n. In our practical experiments, we showed that the predicted effects occur in practice (although we were not able to exactly reproduce the theoretic predictions). Eve's PER for n was at least 20% higher than for b, and more than 30% for ac (with increasing impact over distances greater than 10 m).

We conclude that the evolution of the 802.11 standard actually introduced several physical and link layer features, such as MIMO and spatial diversity, that might degrade the performance of a passive eavesdropper. If properly exploited those features could be used as a part of a defense-in-depth strategy as a complement to well-known eavesdropping defense mechanism. Nevertheless, we understand that further investigations are necessary to characterize the benefits against stronger attacker models and in more complex scenarios. We leave those discussions to future work.

A   Appendix

Figure 9 shows the result of our BER and PER analysis using model D. Figure 10 shows the result of our BER and PER analysis using model E. Figure 11 shows expected BER and PER for a free-space path-loss model.

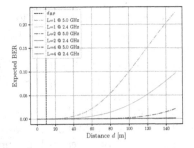

(a) *Expected BER vs. Distance.*

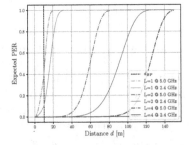

(b) *Expected PER vs. Distance.*

Fig. 9. 802.11n Model D (office) BER/PER using BPSK. Red lines represent Eve. Green and Blue lines represent Bob when L = 2 and L = 4. (Color figure online)

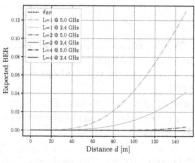
(a) *Expected BER vs. Distance.*

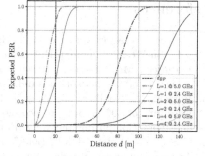
(b) *Expected PER vs. Distance.*

Fig. 10. 802.11n Model E (Large office) BER/PER using BPSK. Red lines represent Eve. Green and Blue lines represent Bob when L = 2 and L = 4. (Color figure online)

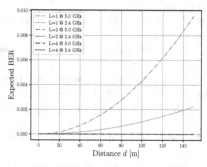
(a) *Expected BER vs. Distance.*

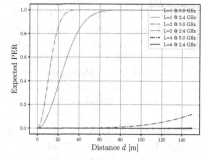
(b) *Expected PER vs. Distance.*

Fig. 11. Free Space Path Loss (LOS) BER/PER using BPSK. Red lines represent Eve. Green and Blue lines represent Bob when L = 2 and L = 4. (Color figure online)

Table 5. Eve's PER vs. PER thresholds vs. distances. Columns represent different distances from Eve to Alice (d_{AE}). Rows represent different PER thresholds. Comma-separated values represent the rounded-down percentage of experimental runs where Eve's PER was above the threshold for 802.11b, n, and ac.

	5.0 [m]	7.5 [m]	10.0 [m]	12.5 [m]	15.0 [m]	17.5 [m]
5%	33, 36, 50	10, 100, 33	20, 100, 33	36, 100, 90	43, 100, 80	60, 100, 96
10%	0, 26, 40	0, 73, 33	6, 83, 33	30, 90, 83	16, 96, 70	30, 100, 70
15%	0, 3, 36	0, 56, 33	6, 53, 33	16, 66, 76	0, 90, 63	13, 100, 60
20%	0, 0, 33	0, 43, 33	3, 36, 33	13, 53, 56	0, 76, 56	6, 96, 53
25%	0, 0, 33	0, 30, 33	3, 26, 33	10, 40, 53	0, 66, 56	0, 83, 53
30%	0, 0, 33	0, 20, 33	0, 13, 33	6, 30, 50	0, 60, 43	0, 73, 53
35%	0, 0, 30	0, 13, 30	0, 3, 33	3, 30, 43	0, 56, 43	0, 63, 50
40%	0, 0, 30	0, 10, 30	0, 0, 33	0, 23, 43	0, 40, 43	0, 53, 46
45%	0, 0, 26	0, 10, 30	0, 0, 33	0, 16, 43	0, 26, 43	0, 46, 46
50%	0, 0, 23	0, 6, 26	0, 0, 33	0, 16, 33	0, 16, 36	0, 43, 46

References

1. Arbaugh, W.A., et al.: Real 802.11 Security: Wi-Fi Protected Access and 802.11 i. Addison-Wesley Longman Publishing Co., Inc., Boston (2003)
2. Bernaschi, M., Ferreri, F., Valcamonici, L.: Access points vulnerabilities to dos attacks in 802.11 networks. Wirel. Netw. (2008)
3. Borisov, N., Goldberg, I., Wagner, D.: Intercepting mobile communications: the insecurity of 802.11. In: Proceedings of the 7th Annual International Conference on Mobile Computing and Networking. ACM (2001)
4. Cheng, Y.-C., Bellardo, J., Benkö, P., Snoeren, A.C., Voelker, G.M., Savage, S.: Jigsaw: solving the puzzle of enterprise 802.11 analysis. In: Proceedings of Conference on Applications, Technologies, Architectures, and Protocols for Computer Communications (SIGCOMM) (2006)
5. Cisco: Cisco's visual networking index forecast projects nearly half the world's population will be connected to the internet by 2017 (2013). https://newsroom.cisco.com/press-release-content?articleId=1197391
6. Coleman, D.D., Westcott, D.A.: CWNA: Certified Wireless Network Administrator Official Study Guide: Exam CWNA-106. Sybex (2014)
7. Crow, B.P., Widjaja, I., Kim, L.G., Sakai, P.T.: IEEE 802.11 wireless local area networks. IEEE Commun. Mag. (1997)
8. Dong, L., Han, Z., Petropulu, A.P., Poor, H.V.: Improving wireless physical layer security via cooperating relays. IEEE Trans. Sig. Process. **58**, 185–1888 (2010)
9. Goldsmith, A.: Wireless Communications. Cambridge University Press, Cambridge (2005)
10. Gopala, P.K., Lai, L., El Gamal, H.: On the secrecy capacity of fading channels. IEEE Trans. Inf. Theory **54**, 4687–4698 (2008)
11. Gummadi, R., Wetherall, D., Greenstein, B., Seshan, S.: Understanding and mitigating the impact of RF interference on 802.11 networks. ACM SIGCOMM Comput. Commun. Rev. **37**, 385–396 (2007)
12. Hero, A.: Secure space-time communication. IEEE Trans. Inf. Theory **49**, 3235–3249 (2003)
13. Hiertz, G.R., Denteneer, D., Stibor, L., Zang, Y., Costa, X.P., Walke, B.: The IEEE 802.11 universe. IEEE Commun. Mag. **48**, 62–70 (2010)
14. IEEE: IEEE standard for information technology-telecommunications and information exchange between systems local and metropolitan area networks-specific requirements - part 11: Wireless LAN medium access control (MAC) and physical layer (PHY) specifications (2016). http://standards.ieee.org/getieee802/download/802.11-2016.pdf
15. Kolias, C., Kambourakis, G., Stavrou, A., Gritzalis, S.: Intrusion detection in 802.11 networks: empirical evaluation of threats and a public dataset. IEEE Commun. Surv. Tutor. **18**, 184–208 (2016)
16. Leung-Yan-Cheong, S.K., Hellman, M.E.: The Gaussian wire-tap channel. IEEE Trans. Inf. Theory **24**, 451–456 (1978)
17. Martin, S.: Directional Gain of IEEE 802.11 MIMO Devices Employing Cyclic Delay Diversity (2013)
18. Mishra, A., Shin, M., Arbaugh, W.: An empirical analysis of the IEEE 802.11 MAC layer handoff process. ACM SIGCOMM Comput. Commun. Rev. **33**, 93–102 (2003)
19. Mukherjee, A., Swindlehurst, A.L.: Robust beamforming for security in MIMO wiretap channels with imperfect CSI. IEEE Trans. Sig. Process. **59**, 351–361 (2013)

20. Oggier, F., Hassibi, B.: The secrecy capacity of the MIMO wiretap channel. In: IEEE Transactions on Information Theory (2011)
21. Ong, E.H., Kneckt, J., Alanen, O., Chang, Z., Huovinen, T., Nihtilä, T.: IEEE 802.11 ac: enhancements for very high throughput WLANs. In: 2011 IEEE 22nd International Symposium on Personal Indoor and Mobile Radio Communications (PIMRC). IEEE (2011)
22. Peppas, K.P., Sagias, N.C., Maras, A.: Physical layer security for multiple-antenna systems: a unified approach. IEEE Trans. Commun. **64**, 314–328 (2016)
23. Perahia, E., Stacey, R.: Next Generation Wireless LANs: 802.11 n and 802.11 ac. Cambridge University Press, Cambridge (2013)
24. Pöpper, C., Tippenhauer, N.O., Danev, B., Capkun, S.: Investigation of signal and message manipulations on the wireless channel. In: Atluri, V., Diaz, C. (eds.) ESORICS 2011. LNCS, vol. 6879, pp. 40–59. Springer, Heidelberg (2011). https://doi.org/10.1007/978-3-642-23822-2_3
25. Prabhu, V.U., Rodrigues, M.R.: On wireless channels with M-antenna eavesdroppers: characterization of the outage probability and-outage secrecy capacity. IEEE Trans. Inf. Forensics Secur. **6**, 853–860 (2011)
26. Robyns, P., Bonné, B., Quax, P., Lamotte, W.: Exploiting WPA2-enterprise vendor implementation weaknesses through challenge response oracles. In: WiSec. ACM (2014)
27. Sheth, A., Doerr, C., Grunwald, D., Han, R., Sicker, D.: MOJO: a distributed physical layer anomaly detection system for 802.11 WLANs. In: Proceedings of the 4th International Conference on Mobile Systems, Applications and Services. ACM (2006)
28. OD Team: OpenWRT wireless freedom. https://openwrt.org/
29. Van Veen, B., Buckley, K.: Beamforming: a versatile approach to spatial filtering. IEEE ASSP Mag. **5**, 4–24 (1988)
30. Wang, J., Lee, J., Quek, T.Q.S.: Best antenna placement for eavesdroppers: distributed or co-located? IEEE Commun. Lett. **20**, 1820–1823 (2016)
31. Wyner, A.D.: The wiretap channel. Bell Syst. Tech. J. **54**, 1355–1387 (1975)
32. Yang, N., Yeoh, P.L., Elkashlan, M., Schober, R., Collings, I.B.: Transmit antenna selection for security enhancement in MIMO wiretap channels. IEEE Trans. Commun. **64**, 144–154 (2013)
33. Zou, Y., Zhu, J., Wang, X., Leung, V.C.M.: Improving physical-layer security in wireless communications using diversity techniques. IEEE Netw. **29**, 42–48 (2015)

A Novel Algorithm for Secret Key Generation in Passive Backscatter Communication Systems

Mohammad Hossein Chinaei[1]([⊠]), Diethelm Ostry[2], and Vijay Sivaraman[1]

[1] University of New South Wales, Sydney, Australia
{m.chinaei,vijay}@unsw.edu.au
[2] Data61, CSIRO, Sydney, Australia
diet.ostry@data61.csiro.au

Abstract. The extreme asymmetry of passive backscatter communications systems such as passive Wi-Fi, while allowing significant reduction of node power consumption for communications, imposes severe resource limitations on implementing secure communications. Target applications for this technology are typically driven by the promise of low power consumption, up to four orders of magnitude lower than commercial Wi-Fi chipsets. Industry standard security approaches using encryption technology are problematic in this power regime, particularly as the potential low complexity and size of passive nodes will encourage application to high-density networks of very small, energy-poor devices. Generation of shared symmetric keys through reciprocal channel measurements, for example of received signal strength (RSS), is a natural approach in this situation. However previous work in this area has focused on the symmetric case where base station and nodes communicate at the same radio frequency. Backscatter communications uses two frequencies, typically a pilot beacon transmitted by a base station on one frequency, and response on a shifted frequency. This paper describes a protocol for RSS-based shared key generation for this architecture and reports the results of an experimental implementation using software radio emulation of backscatter communication.

Keywords: Physical layer security · Secret key generation
Passive sensors · Backscatter communication

1 Introduction

Power consumption remains a key limiting constraint in achieving long-lived networks of wireless sensor nodes, and communications is typically a major component of their power budget. The appearance of many applications requiring small low-power sensors in areas such as the Internet of Things (IoT), wearable devices, and implantable medical sensors, has attracted a great deal of research interest in techniques able to achieve low-power communications. The most extreme

© Springer Nature Switzerland AG 2018
S. Capkun and S. S. M. Chow (Eds.): CANS 2017, LNCS 11261, pp. 436–455, 2018.
https://doi.org/10.1007/978-3-030-02641-7_20

approaches to date employ backscatter technologies which can reduce power consumption by orders of magnitude through transfering as much as possible of the power-consuming transmitter functionality of the wireless communications system out of the nodes and into the base station. Instead of implementing an active wireless transmitter, with correspondingly large power consumption, a node or "tag" employing backscatter communications uses relatively simple RF circuitry to receive, modulate, and reflect either ambient wireless transmissions or beacon signals provided by a base station or "reader".

For example, the authors in [1] presented a "Wi-Fi backscatter" approach as a practical technology for wireless communication for passive sensors. A Wi-Fi backscatter tag is able to send data at a rate of a few kbps to a commodity receiver over a range of 2 meters by modulating ambient Wi-Fi communications packets and thereby influencing the channel state visible to the reader. In [2] the authors propose a similar technique to modulate ambient Bluetooth low power packets and extended the range to over 9 meters. In [3–6] the authors extend the idea of backscatter communication using Wi-Fi signals using different approaches. Interference cancellation is proposed in [3] so that the same frequency can be used by both beacon signal from reader to tag and the reflected signal from tag to reader. In [4–6] the authors use dual frequencies to achieve compatibility with current commercial Wi-Fi devices.

In "passive Wi-Fi" [5], the reader (which can employ standard Bluetooth and Wi-Fi chipsets) sends out a continuous wave (CW) beacon on a Bluetooth frequency. The passive tag modulates its information on the received beacon, shifts its frequency and reflects a normal Wi-Fi (802.11b) packet back to the reader. This technology can provide in principle up to 11 Mbps at 10^{-4} times lower power than current active Wi-Fi chipsets. All these reported technologies have been implemented and tested under real world conditions and for some of them IC implementations have also been designed. It appears likely that many novel applications will become feasible with these new ultra-low-power passive technologies based on backscatter communication.

One of the attractive new application areas is wearable devices, for example for physiological and medical monitoring purposes. Such devices are ideally small and lightweight which restricts their battery capacity and so makes them ideal candidates for using the ultra-low-power backscatter communications technologies. However the communications system in this and other applications may carry sensitive information, e.g. commercial, personal and medical data and so a security capability is often mandatory. In view of the limited computational capabilities of the devices, their deployment in perhaps not-easily-accessible locations, and potentially in large numbers, it is challenging to devise practical security mechanisms to protect their data. Cryptographic means of implementing data confidentiality require the secure distribution of keys between the communicating devices and this is a power-intensive task in a wireless system, making it problematic for ultra-low-power devices.

The use of the wireless channel itself (often termed the physical (PHY) layer for convenience) as a source of shared key material has been studied extensively in recent years [7,8]. From physical principles, the channel is intrinsically reciprocal, i.e. both parties in a wireless communication see the same propagation parameters to within a constant factor, and an eavesdropper in a sufficiently removed location cannot determine those shared parameters. What makes this appealing in the low-power regime is that measurements of the channel parameters can often be made as part of the usual communications protocol without incurring the power overhead of a cryptographic key-exchange protocol. Regular re-keying is generally required in practice for maintaining security and so the secret key generation rate is also a matter of concern in some applications. Researchers have explored high rate key bit extraction in [9–11]. However, in [12], it is shown that for IoT and wearable sensors where high bit rate is not a critical issue, low-complexity algorithms provide benefits in overall device energy consumption.

All the previous work on using the wireless channel to generate shared keys has addressed active symmetric communications where the two communicating parties alternately transmit to each other and make independent measurements of channel properties at the same frequency. However, the inherent asymmetry of passive backscatter communication makes this approach inapplicable. In the "passive WiFi" scenario for example, the reader emits a beacon at one frequency and the tag reflects a WiFi signal at a different frequency. A new approach is needed to generate shared keys from wireless channel properties at two frequencies and in a way which is secure from eavesdropping.

In the remainder of this paper we describe such an approach. Our specific contributions are:

1. We describe a straightforward secret key generation scheme modified for use by passive sensors which implement asymmetric backscatter communication. We develop a three-step protocol to measure received signal strength (RSS) at dual frequencies, allowing a reader and tag to establish a secret shared key with high agreement in principle. We use the universal software radio peripheral (USRP) platform to test the approach experimentally.
2. We identify a specific problem with key generation based on wireless channel parameters caused by the dual frequency operation inherent in many practical backscatter schemes like "passive WiFi".
3. We propose an enhanced of the basic algorithm and device design to allow secure shared key generation in the backscatter communications system.

The rest of the paper is organised as follows: in Sect. 2, the basic system architecture is outlined and the protocol for secret key generation is developed. In Sect. 3, an experimental evaluation of the protocol is presented. A theoretical and practical analysis is carried out to establish the security risks of the proposed protocol. In Sect. 4, an enhancement of the protocol and device is described and evaluated. The paper is concluded in Sect. 5 with directions to future work.

2   Basic System Architecture

2.1   System Model

A data communications transmitter typically comprises a digital baseband processor which constructs an analog signal at a convenient low (baseband) frequency and an RF section which shifts the signal to the final frequency and amplifies it to the required power level. To achieve adequate transmit power, the RF section usually uses an architecture which consumes far more power than the baseband processor. Backscatter communication eliminates the power consuming analog RF part of the transmitter and effectively offloads its function to the reader device (which could be a smart phone in practice). Passive tags do not have the usual active transmitter function but instead essentially piggyback information on ambient communications signals or a reader–generated beacon. Their RF circuitry is far less complex and requires far less power than an active transmitter. Of course also it generates far less RF signal power and so there is a corresponding cost in range reduction. It is convenient to shift the frequency of the tag's reflected signal so that the reader's beacon signal can be at a placed at a non-interfering frequency (e.g. in the Bluetooth band for a backscattered signal in the WiFi band).

PHY layer secret key generation relies on reciprocity of electromagnetic propagation, i.e. two communicating parties under general conditions will independently see identical channel properties. In other words, if two parties, reader and tag, consecutively exchange signals with each other (in less than channel coherence time, i.e. the time during which the channel is effectively constant) so that each can estimate the channel they see, their estimates would match. However, there are two major differences in the backscatter scenario. First, the tag is not able to make an independent transmission but can only reflect the reader's beacon signal. Second, the tag and reader receive signals at different frequencies which may be sufficiently separated to have different channel propagation properties (for example in passive Wi-Fi they are 11 MHz apart).

2.2   Asymmetric Channel Measurements

There are a number of channel characteristics on which key generation can be based, for example the spectrum of multipath components, complex (magnitude and phase) link gain, and received signal strength (RSS). RSS is by far the easiest to implement and measure, particularly by resource constrained devices and so is best suited to backscatter nodes. Our basic system model includes a reader, a tag and an eavesdropper (Eve). The tag is assumed to have two different operation modes, reflecting and listening. In reflecting mode, the tag can only retransmit the modified beacon signal back to the reader. However, in the listening mode, the tag can listen to the reader and measure the RSS of its beacon signals.

The channel characteristics (e.g. RSS) between the legitimate parties, reader and tag, are assumed to fluctuate sufficiently for key generation. The fluctuation

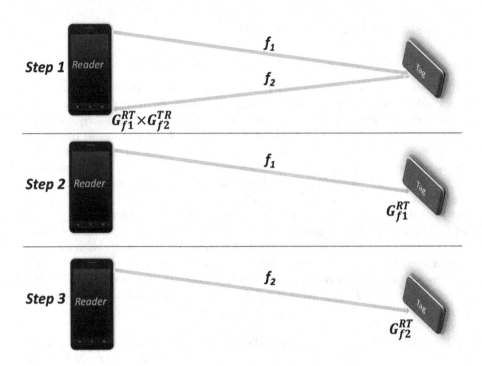

Fig. 1. Asymmetric channel measurement steps

may be due to tag motion for example or the motion of other nearby entities which change the multipath environment. Eve is presumed to be a passive attacker who is able to measure the RSS of different signals transmitted from the reader or reflected from the tag but does not transmit any signals.

The reader transmits a beacon signal at f_1 and receives the tag's reflection at f_2 which causes the RSS at the reader to be influenced by the channel gains at both f_1 and f_2. The reader uses this composite RSS to generate its key. Accordingly, the tag also needs to be provided with the same channel gain information at the same two frequencies. To achieve this, our basic protocol [13] uses the following three-step scheme (see Fig. 1):

1. The reader transmits a CW beacon signal at f_1 to the tag (which is in the reflection mode at this step). The tag reflects the received signal at the second frequency, f_2. The reader measures the RSS of the reflected signal which is the product of channel gains at the two frequencies, i.e. $G_{f_1}^{RT} G_{f_2}^{TR}$.
2. The tag listens at frequency f_1. The reader transmits a CW beacon at f_1. The tag measures the RSS giving it an estimate of channel gain at f_1, i.e. $G_{f_1}^{RT}$.
3. The tag listens at frequency f_2. The reader transmits a beacon at f_2. The tag measures the RSS to estimate the channel gain at f_2, i.e. $G_{f_2}^{RT}$.

The three steps take place consecutively and one set of three steps forms a single round of the key generation protocol. At the end of each round, the tag and reader have independent estimates of the same channel properties. The tag multiplies the gains it measured at steps 2 and 3 to form an estimate of $G_{f_1}^{RT} G_{f_2}^{RT}$. The reader measures its estimate of $G_{f_1}^{RT} G_{f_2}^{TR}$ in step 1. If the channel has remained essentially constant over the duration of a protocol round the tag's and reader's estimates of the gain product would be expected to be in high agreement because of channel reciprocity and so can be used as a source of shared entropy for key generation.

The gain product $G_{f_1}^{RT} G_{f_2}^{RT}$ is used by the reader and tag to generate their keys. The eavesdropper Eve is able to make two different estimates of this product after every round of the protocol. The first estimate is available at step 1, where she receives the reflection signal from the tag, the product of two channel gains: reader-tag link at frequency f_1, $(G_{f_1}^{RT})$ and tag-Eve link at frequency f_2, $(G_{f_2}^{TE})$. This estimate will be termed Eve's reflection estimate: $G_{f_1}^{RT} G_{f_2}^{TE}$. Eve's second estimate, termed the product estimate, is generated by multiplying the RSS values she sees at steps 2 and 3 (channel gains of reader-Eve link at f_1 and f_2), i.e. $G_{f_1}^{RE} G_{f_2}^{RE}$.

2.3 Quantisation Process

After a sequence of protocol rounds in which consecutive channel measurements are made, both reader and tag apply a quantisation process to convert the raw channel measurements into a key bit string. We use the level crossing quantisation technique first proposed in [14]. In this method, an adaptive sliding window of length W_Q is defined to select a block of consecutive raw measurements. In each block upper and lower levels are defined as follows:

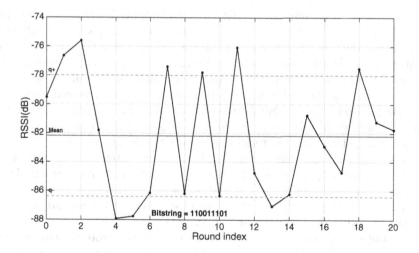

Fig. 2. Level crossing quantisation technique.

$$q+ = \mu + \alpha\sigma$$

$$q- = \mu - \alpha\sigma$$

where μ is the measurement mean, σ is the standard deviation, and $0 \leq \alpha \leq 1$ is a parameter which can be adjusted to trade off key bit rate against key bit agreement. Each of the RSS measurements inside the window produces a key bit value of 1 if it is greater than the upper quantisation level ($q+$) and 0 if it is smaller than the lower level ($q-$). Measurements which fall between two levels of quantisation are discarded (see Fig. 2). The quantisation parameters are the same for both legitimate sides of the communication and we assume Eve knows the quantisation algorithm.

When the parameter α is small, most of the RSS measurements contribute key bits, leading to a higher key bit generation rate although the key bit agreement is likely to decrease drastically due to deriving key bits from uncorrelated noise. On the other hand, for α near 1 and higher, many usable RSS measurements are discarded, thereby reducing the key bit generation rate, but also reducing key discrepancies due to noise. In our target application, higher key agreement is desirable in order to minimise the cost of any subsequent key reconciliation process [12].

2.4 Security Considerations

Threat Model: In this work we are concerned with the threat posed by an eavesdropper who is able to detect all key-generation communications. An eavesdropper is assumed to have full knowledge of the system protocols and is not limited substantially in computational power or receiver capability (e.g. she can receive at multiple frequencies simultaneously). However we restrict an eavesdropper to be passive, i.e. is unable to generate spoofing signals.

Secrecy: A crucial assumption is that any eavesdropper is sufficiently far from the legitimate parties that her radio channel characteristics are uncorrelated with those of the legitimate parties. A half-wavelength separation is theoretically adequate in a multipath–saturated environment, but in practice considerably greater distances may be required [8] and in general the extent of eavesdropper–exclusion zones must be established through measurement or propagation modeling. Backscatter systems have relatively short range due to their passive nature reducing the tag's signal power and this may make an eavesdropper who is close enough to detect the backscattered signal more physically evident.

Although two frequencies are used in backscatter systems, we do not rely on their being uncorrelated. In fact the enhanced algorithms we introduce in Sect. 4 attempt to remove the effect of propagation variations at one of the frequencies. Under our system assumptions, the channel variation at the second frequency can be assumed to be adequate for key generation.

We note that in our protocols, the reader drives the protocol sequence and can therefore in principle monitor the radio environment and detect some spoofing attacks, e.g. by listening for false beacon signals or signal collisions which would indicate attempts to inject counterfeit backscatter signals over the legitimate backscatter signals. The true backscatter signal is returned essentially instantaneously apart from propagation delays, making it more difficult for an active attacker with powerful transmitter (to operate at a standoff for example) to impersonate a legitimate backscatter node. It would also be difficult for an active attacker to ensure that her signal levels at the reader were such that they did not reveal she could not be a backscatter device.

3   Evaluation and Analysis

3.1   Evaluation

System Model and Channel Measurements. We implemented and evaluated the performance of our proposed protocol on USRP software–defined radios. The three nodes in our scenario are represented by three different USRPs, each connected to a PC running LabView software as the control interface. The dual frequencies chosen were $f_1 = 2.171$ GHz and $f_2 = 2.182$ GHz, different from standard Wi-Fi and Bluetooth frequencies because of equipment limitations, but 11 MHz apart as in passive Wi-Fi. The corresponding wavelength is about 14 cm.

Since we are extracting keys from wireless channel characteristics, the channel is required to fluctuate sufficiently to provide key generation at an adequate rate [12,14] and this is achieved in our experiments by moving the tag through a sequence of positions. In the configuration shown in Fig. 3, the reader and eavesdropper Eve are stationary and the tag moves randomly about 5 cm after each 5 rounds of the protocol. The distance between the two legitimate parties (reader and tag) is varied between 150 cm to 190 cm. In each experiment, Eve is located at a different distance from the reader. We chose the configuration where Eve is close to the reader as the worst case since her channels are then most geometrically similar to the tag–reader channels used for key generation.

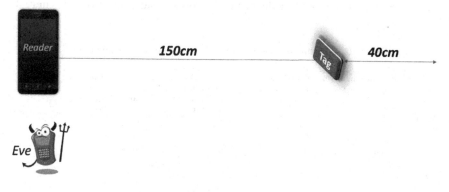

Fig. 3. System model

In the first experiment (Fig. 4), Eve is located 42 cm, about 3λ, from the reader. In subsequent experiments, this distance is increased to 52 cm (4λ) and 84 cm (6λ), shown in Figs. 5 and 6, respectively. In each figure, we have four different curves corresponding to reader's RSS measurement of the backscattered signal, the tag's estimation, which is the product of its RSS measurements at steps 2 and 3, Eve's reflection estimation, based on what she measures at step 1, and Eve's product estimation, based on the product of her RSS measurements at steps 2 and 3. Each experiment comprises 250 protocol rounds lasting about 12 min. The Pearson correlation coefficients between measurements at different nodes are shown in Table 1. The correlation coefficient always lies in the range [−1, 1], where 1, 0, and −1 represents perfect correlation, no correlation and anti-correlation respectively.

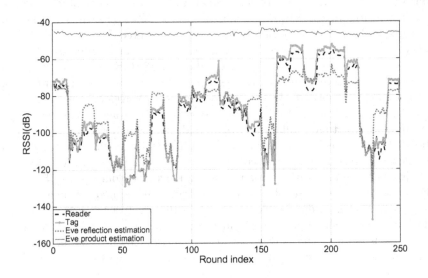

Fig. 4. RSS measurements when the distance between reader and Eve is 42 cm ≈ 3λ.

Table 1. Correlation coefficient between different node signals

Distance between reader and Eve	Correlation between reader and tag	Correlation between reader and Eve reflection estimation	Correlation between reader and Eve product estimation
42 cm ≈ 3λ	0.99	0.90	0.16
56 cm ≈ 4λ	0.99	0.90	0.02
84 cm ≈ 6λ	0.99	0.80	0.16

Secret Key Generation: As outlined in Sect. 1, we use a level crossing quantiser to generate key bits from the RSS measurements of reader and tag. All of the nodes in the experimental scenario record the round index of a successful

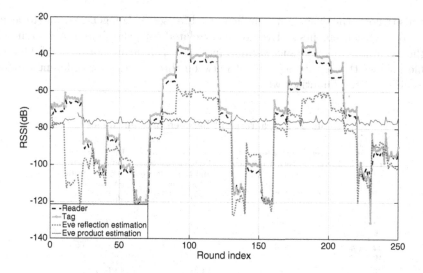

Fig. 5. RSS measurements when the distance between reader and Eve is 56 cm ≈ 4λ.

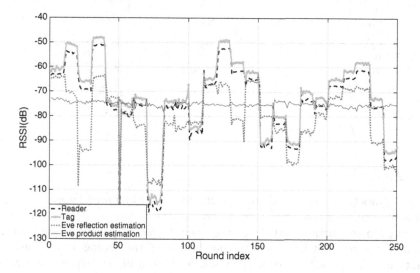

Fig. 6. RSS measurements when the distance between reader and Eve is 84 cm ≈ 6λ.

measurement, i.e. one which produced a key bit. The actual shared key bit string is established through subsequent communication between reader and tag over a public channel. The reader and tag agree on a key bit sequence by exchanging their successful round indices. They discard any bits which do not correspond to a successful round index for both. In this way the reader and tag only keep the key bits formed from successful round indices at both sides. Since a public channel is used by the legitimate parties to exchange successful round indexes,

Eve also knows the exact round indices used for secret key generation. However, if the RSS measurements at Eve are uncorrelated with the shared measurements at the tag and reader, the successful round indices alone are not be helpful to her. Table 2 shows the key bit agreement between three parties for different window sizes (W_Q) of the quantiser with $\alpha = 1$.

Table 2. Key agreement between different nodes

Distance between reader and Eve	Key agreement between reader and tag	Key agreement between reader and Eve reflection estimation	Key agreement between reader and Eve product estimation
42 cm $\approx$ 3λ, $W_Q = 5$	96.96%	81.31%	46.96%
42 cm $\approx$ 3λ, $W_Q = 10$	100%	83.64%	51.40%
42 cm $\approx$ 3λ, $W_Q = 20$	100%	83.96%	52.35%
56 cm $\approx$ 4λ, $W_Q = 5$	92%	74.71%	49.41%
56 cm $\approx$ 4λ, $W_Q = 10$	100%	81.73%	49.13%
56 cm $\approx$ 4λ, $W_Q = 20$	100%	88.43%	49.25%
84 cm $\approx$ 6λ, $W_Q = 5$	93.87%	75.50%	45.91%
84 cm $\approx$ 6λ, $W_Q = 10$	100%	77.81%	47.57%
84 cm $\approx$ 6λ, $W_Q = 20$	100%	76%	47.27%

3.2 Analysis

Table 1 shows that when the three measurement steps in a round were completed in less than channel coherence time, the correlation between RSS measurements at the legitimate parties was 0.99, giving a very high key agreement (see Table 2). The actual agreement level depends on the size of the sliding window used in quantisation process. Greater window size leads to a higher agreement level at the cost of larger memory size and more complexity in the hardware, but requires channels with only slowly changing means. For our experimental implementation, 100% key agreement was reached with a sliding window size of 10 samples.

The product estimation Eve generates by multiplying the measured RSS at steps 2 and 3 is almost constant for all of the experiments (Figs. 4, 5, and 6) with the reader and Eve fixed in position during each experiment. When Eve uses this estimate she has only around a 50% chance of deriving the legitimate key (i.e. no better than a coin toss), and the agreement level does not change significantly with sliding window size. On the other hand, Eve's reflection estimate is in close agreement with the measurements at the reader and this results a near 80% agreement between Eve's key based on the reflection estimate and the legitimate key. This level of agreement is a serious problem which jeopardises the security of our first proposed protocol.

Problem Statement: In active channel measurement scenarios where the nodes communicate symmetrically and alternately, an eavesdropper Eve located more than a half-wavelength away from the legitimate nodes could not in principle form a valid measurement of the legitimate channel (the channel between the reader and tag). However, in the passive backscatter case, Eve's estimate of the RSS gain product based on the reflected signal is strongly correlated to the reader's measurements. Our experiments show that even when Eve is 6λ away from the reader the correlation coefficient between the reader's measurements and Eve's measurements is 0.80, which leads to near 75% key agreement. Here we analyse the reflection behaviour of the tag in detail to identify the underlying problem which causes the unacceptably high key agreement for an eavesdropper situated at even relatively large ranges from tag and reader.

Referring to Sect. 2.2, in the first step of protocol the reader transmits a beacon at f_1, the tag shifts its received signal to f_2 and reflects it back to the reader. The reflected signal is measured by both the reader and Eve. As a result, measurements of the reflected gains at the reader and Eve are:

$$\text{Reader reflection measurement} = G_{f_1}^{RT} G_{f_2}^{TR} \tag{1}$$

$$\text{Eve reflection measurement} = G_{f_1}^{RT} G_{f_2}^{TE} \tag{2}$$

where $G_{f_1}^{RT}$, $G_{f_2}^{TR}$, and $G_{f_2}^{TE}$ are channel gains for reader to tag link at frequency f_1, tag to reader link at frequency f_2, and tag to eve link at frequency f_2, respectively. All of the gain terms in Eqs. (1) and (2) are positive random variables as they represent an attenuation factor. But as shown below, even when the three gain terms are statistically independent, the RSS measurements of the reflected signal by Eve and the reader are not necessarily uncorrelated.

Assume X, Y, and Z are statistically independent random variables with means μ_X, μ_Y, and μ_Z and variances σ_X^2, σ_X^2, and σ_X^2 respectively. (In our case, X and Y will represent the RSS values from tag to reader and tag to Eve, and Z the RSS from reader to tag.) Here we are interested in the correlation coefficient between products such as ZX and ZY under the assumption of statistical independence. The correlation coefficient ρ_{XY} of the processes X and Y is defined as

$$\rho_{XY} = \frac{\sigma_{XY}}{\sigma_X \sigma_Y} = \frac{E[(X - \mu_X)(Y - \mu_Y)]}{\sigma_X \sigma_Y}$$

If X and Y are independent, $\rho_{XY} = 0$, and if they are linearly dependent, $|\rho_{XY}| = 1$. Now consider the product random variables ZX and ZY. Their correlation coefficient is:

$$\rho_{ZX,ZY} = \frac{E[(Z^2 XY - \mu_Z^2 \mu_X \mu_Y]}{\sqrt{var(ZX) var(ZY)}}$$

Now for independent Z, X, and Y:

$$var(ZX) = E[(ZX - \mu_{ZX})^2] = E[Z^2 X^2] - \mu_{ZX}^2$$
$$= \sigma_Z^2 \sigma_X^2 + \mu_Z^2 \sigma_X^2 + \mu_X^2 \sigma_Z^2$$
$$\text{and } var(ZY) = \sigma_Z^2 \sigma_Y^2 + \mu_Z^2 \sigma_Y^2 + \mu_Y^2 \sigma_Z^2$$

So

$$\rho_{ZX,ZY} = \frac{\mu_X \mu_Y (E[Z^2] - \mu_Z^2)}{\sqrt{(\sigma_Z^2 \sigma_X^2 + \mu_Z^2 \sigma_X^2 + \mu_X^2 \sigma_Z^2)(\sigma_Z^2 \sigma_Y^2 + \mu_Z^2 \sigma_Y^2 + \mu_Y^2 \sigma_Z^2)}}$$
$$= \frac{\mu_X \mu_Y}{\sqrt{(\mu_X^2 + \sigma_X^2(1 + (\frac{\mu_X}{\sigma_Z})^2))(\mu_Y^2 + \sigma_Y^2(1 + (\frac{\mu_Y}{\sigma_Z})^2))}}. \tag{3}$$

This shows that when X and Y have non-zero means and Z is not a constant ($\sigma_Z^2 \neq 0$), the correlation of the products ZX and ZY is in general non-zero, even though X, Y, and Z are statistically independent.

Returning to our passive tag scenario with RSS given in Eqs. (1) and (2), this result shows that because the reader and Eve reflection signals both contain the common randomly varying factor $G_{f_1}^{RT}$, their correlation is unlikely to be zero and so their derived key strings will likely have an unacceptable agreement, borne out by the experimental measurements in Tables 1 and 2.

Correlation Analysis: From the experimental measurements we can derive the gains of the individual signal paths comprising the reflected signals measured by the reader and Eve. Since we have used USRPs for our experiments (rather than actual passive tags), the tag is able to measure the gain $G_{f_1}^{RT}$ at the first step of a protocol round. We can derive the RSS corresponding to $G_{f_2}^{TR}$ and $G_{f_2}^{TE}$ from the reflection signal at the reader and Eve and the RSS measurements at the tag in the first step.

$$G_{f_2}^{TR} = \frac{\text{Reader backscatter measurement}}{\text{Tag measurement at } f_1}$$

$$G_{f_2}^{TE} = \frac{\text{Eve backscatter measurement}}{\text{Tag measurement at} f_1}$$

Using this approach we are able to derive the correlation between the different links and the corresponding RSS measurements at Eve and reader for various Eve locations (see Table 3). Our results show that the channel gains of tag to Eve and tag to reader at f_2 become less correlated with greater separation. However, this does not lead to lower correlation in Eve and reader measurements, as none of them are zero-mean random variables (see Eq. 3).

The theoretical (Eq. 3) and experimental (Table 3) analysis in this section shows that in contrast to secret key generation using bidirectional active communications where a separation of more than half a wavelength between Eve and

Table 3. Correlation coefficient between RSS measurements at different links

Distance between Reader and Eve	Correlation between RSS at $G_{f_1}^{RT}$ and $G_{f_2}^{TR}$	Correlation between RSS at $G_{f_2}^{TR}$ and $G_{f_2}^{TE}$	Correlation between reader and Eve reflection estimation
$42\,\text{cm} \approx 3\lambda$	0.80	0.41	0.90
$56\,\text{cm} \approx 4\lambda$	0.80	0.43	0.90
$70\,\text{cm} \approx 5\lambda$	0.81	0.34	0.93
$84\,\text{cm} \approx 6\lambda$	0.77	0.24	0.80
$98\,\text{cm} \approx 7\lambda$	0.75	0.16	0.90

legitimate nodes theoretically results in uncorrelated channel measurements, the common beacon signal in the passive backscatter case causes high correlation between measurements at Eve and reader.

4 Enhanced Algorithm for Secret Key Generation

The evident agreement between the RSS of the tag reflected signal as measured by the reader and eavesdropper makes it unsuitable for generating a secret key. As discussed in the previous section, the tag reflects the beacon back to the reader and eavesdropper (with frequency translation). Although the reflected signal subsequently goes through uncorrelated channels to the reader and eavesdropper, the first channel traversed by the beacon, from reader to tag, is common to both reader and eavesdropper and causes a correlation between the reader and eavesdropper's measurements of RSS.

One approach to removing this correlation and blinding the eavesdropper to the reader–tag channel is to modify the effect of this common channel. In this section, we will discuss two algorithms to achieve this, one at the reader side and one at the tag side. We show that the enhanced algorithm at the reader side can be easily attacked by Eve. On the other hand, an enhanced algorithm at the tag side can remove the common random factor $(G_{f_1}^{RT})$ from the reflection signal and result in nearly uncorrelated RSS measurements at the reader and eavesdropper.

4.1 Reader–Side Enhanced Algorithm

One capability an enhanced reader might use is to modify the key generation process by controlling the power level of the beacon signals used in different steps of the key–generation protocol. Note that Eve's best estimate of the secret key is based on her measurement at the first step of the protocol. Beacons sent by the reader at step 2 and 3 are used by the tag for key generation but Eve's estimation based on her calculation of their product is uncorrelated to the tag's and reader's measurements. Hence the reader can best prevent Eve from measuring the actual channel gains by interfering with her estimate made in step 1 of the protocol.

In the reader–side enhanced algorithm, the reader sends out the beacon in step 1 at a random power level to falsify Eve's estimate of the reflected RSS. Since the random power level is chosen by reader itself, it can easily extract the true channel gain from the reflection signal RSS. On the other hand, this does not affect the step 2 and 3 RSS measurements at the tag and the high correlation between measurements at the reader and tag can be expected to remain unchanged. Table 4 shows the key agreement between the reader and tag, and Eve when the beacon power is randomised in this way at the reader side.

Table 4. Key agreement between different nodes for reader–side enhanced algorithm

Distance between reader and Eve	Key agreement between reader and tag	Key agreement between reader and Eve reflection estimation	Key agreement between reader and Eve product estimation
$42\,\mathrm{cm} \approx 3\lambda$, $W_Q = 5$	96.96%	47.47%	46.96%
$42\,\mathrm{cm} \approx 3\lambda$, $W_Q = 10$	100%	48.59%	51.40%
$42\,\mathrm{cm} \approx 3\lambda$, $W_Q = 20$	100%	46.22%	52.35%
$56\,\mathrm{cm} \approx 4\lambda$, $W_Q = 5$	92%	52.87%	49.41%
$56\,\mathrm{cm} \approx 4\lambda$, $W_Q = 10$	100%	46.52%	49.13%
$56\,\mathrm{cm} \approx 4\lambda$, $W_Q = 20$	100%	57.08%	49.25%
$84\,\mathrm{cm} \approx 6\lambda$, $W_Q = 5$	93.87%	50.51%	45.91%
$84\,\mathrm{cm} \approx 6\lambda$, $W_Q = 10$	100%	42.23%	47.57%
$84\,\mathrm{cm} \approx 6\lambda$, $W_Q = 20$	100%	46.93%	47.27%

Potential Attack: The reader transmits the beacon signal at f_1 with a random sequence of amplitudes, say $\alpha_0, \alpha_1, \alpha_2,...$ If the reader and Eve are both stationary so that $G_{f_1}^{RE}$ is constant for a time, Eve will receive these beacon signals with amplitudes $s_0 = \alpha_0 G_{f_1}^{RE}$, $s_1 = \alpha_1 G_{f_1}^{RE}$, $s_2 = \alpha_2 G_{f_1}^{RE}$ and so on. If she takes ratios of the signals, e.g. $\frac{s_1}{s_0} = \frac{\alpha_1}{\alpha_0}$, $\frac{s_2}{s_0} = \frac{\alpha_2}{\alpha_0}$... she can estimate the α_i to within a scale factor (α_0 in this case). This estimate would then allow her to correct her reflected estimation of $\alpha_i G_{f_1}^{RT} G_{f_2}^{TE}$ signal and find $G_{f_1}^{RT} G_{f_2}^{TE}$ to within a constant scale factor, which does not affect key bit quantisation, and so she can discover the key bits.

4.2   Tag–Side Enhanced Algorithm

In this section, we propose an algorithm at the tag side instead to eliminate the effect of the common reader–to–tag channel. We assume an ideal tag which can accurately control the strength of its reflection signal. However a passive tag is not able to amplify the received beacon but only reduce the amplitude of its reflection. As explained in [4], the tag is in principle able to change its reflection characteristics by altering the impedance load on its antenna and so control the power level of the reflection signal. The power level of the reflection signal can stated in the form:

$$P_{Reflection} = P_{Beacon}\frac{\mid \Gamma_1^* - \Gamma_2^* \mid^2}{4} \tag{4}$$

where Γ_1^* and Γ_2^* are the complex conjugates of the reflection coefficients corresponding to the two impedance states. The backscattered signal can be reflected at different power levels corresponding to the range $[0, P_{Beacon}]$. If the tag can keep the reflected power at some constant level in step 2 of successive rounds of the key generation process, the damaging effects of the common random factor $(G_{f_1}^{RT})$ can be eliminated. The reader and eavesdropper now see just the single channel gains $G_{f_2}^{TR}$ and $G_{f_2}^{TE}$ respectively, and these channel gains are uncorrelated given our assumptions.

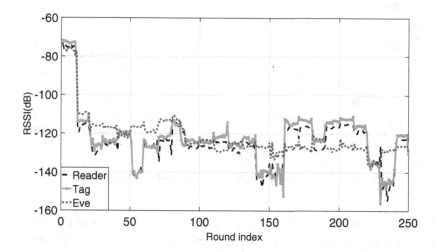

Fig. 7. RSS for tag–side enhanced algorithm, reader and Eve are 42 cm $\approx 3\lambda$ apart.

To implement the tag–side power management algorithm, we need to swap the order of step 1 and step 2 in each protocol round so that the tag can estimate the reader-tag channel gain as a first step. So in the new sequence, the tag is in the

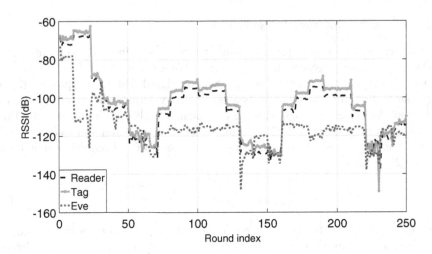

Fig. 8. RSS for tag–side enhanced algorithm, reader and Eve are 56 cm ≈ 4λ apart.

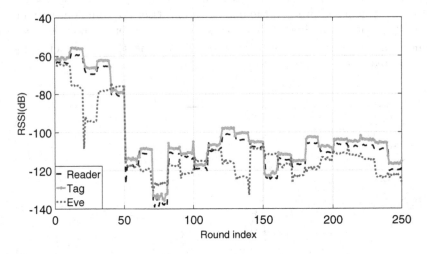

Fig. 9. RSS for tag–side enhanced algorithm, reader and Eve are 84 cm ≈ 6λ apart.

listening mode in step 1 and measures $G_{f_1}^{RT}$. In step 2 the tag is in the reflection mode and controls the reflected power to remove the effects of $G_{f_1}^{RT}$ from the reflection signals $G_{f_1}^{RT} G_{f_2}^{TR}$ and $G_{f_1}^{RT} G_{f_2}^{TE}$ seen by the reader and eavesdropper respectively.

For a simple proof-of-concept demonstration of the approach, we consider time epochs in which the tag adjusts its reflection according to the factor $k(i)$ (with i the protocol round index within an epoch):

Table 5. Key agreement between different nodes for tag–side enhanced algorithm

Distance between reader and Eve	Key agreement between reader and tag	Key agreement between reader and Eve
42 cm ≈ 3λ, W_Q=5	82%	51.28%
42 cm ≈ 3λ, W_Q=10	89%	50%
42 cm ≈ 3λ, W_Q=20	97%	51%
56 cm ≈ 4λ, W_Q=5	83.33%	51.19%
56 cm ≈ 4λ, W_Q=10	94.11%	55.39%
56 cm ≈ 4λ, W_Q=20	98.34%	60%
84 cm ≈ 6λ, W_Q=5	86%	54%
84 cm ≈ 6λ, W_Q=10	100%	55.51%
84 cm ≈ 6λ, W_Q=20	100%	59.48%

$$k(i) = \begin{cases} 1 & i = 1, \\ \dfrac{min(G_{f_1}^{RT}(1),...,G_{f_1}^{RT}(i))}{G_{f_1}^{RT}(i)} & i > 1, \end{cases} \tag{5}$$

where $G_{f_1}^{RT}(i)$ is the RSS at f_1 in round i. Eve's and the reader's reflection measurements in round i are then:

$$\text{Reader reflection measurement} = k(i)G_{f_1}^{RT}(i)G_{f_2}^{TR}(i)$$

$$\text{Eve reflection measurement} = k(i)G_{f_1}^{RT}(i)G_{f_2}^{TE}(i)$$

In step 3 of a protocol round, the reader sends a beacon at f_2 to the tag (which operates in the listening mode) and the tag measures $G_{f_2}^{RT}$. In the enhanced tag–side key generation algorithm, the reader and Eve generate a key based on their measurements at step 2, while the tag multiplies its measurements at step 1 and 3 to compute its product term (from which its key is derived) as:

$$\text{Tag product measurement} = k(i)G_{f_1}^{RT}(i)G_{f_2}^{RT}(i)$$

The effect of the factor $k(i)$ is to reduce the variability of the reader–tag channel at f_1 to a piecewise constant since $k(i)G_{f_1}^{RT}(i) = min(G_{f_1}^{RT}(1),\ldots,G_{f_1}^{RT}(i))$. Even though in the ideal case the effect of the common reader–tag channel at f_1 has been largely removed and so does not contribute to key generation, the remaining component at f_2, i.e. $G_{f_2}^{RT}(i)$ is sufficient for random key generation under our system assumptions (as in symmetric non-backscatter systems).

In order to emulate an ideal tag, we have applied the tag–side enhanced algorithm to the channel gain measurements made in the experiments described in Sect. 3 above. The results are shown in Figs. 7, 8 and 9 and Table 5 and demonstrate that the correlation effects due to the common reader–tag channel have been considerably reduced, with key agreement rates between Eve and the legitimate parties now approaching the desired 50% levels. An investigation of improved forms for the factor $k(i)$ in Equation (5) is left for future work.

5 Conclusion

In this paper, we proposed a novel algorithm for generating shared secret keys in passive backscatter communications systems by measuring wireless channel characteristics at dual frequencies. Restricted capabilities and severe power limitations are typical of passive backscatter sensors and make shared secret key generation based on reciprocal channel characteristics an attractive approach. Previous work on physical–layer secret key generation has focused on the symmetric case where both parties use comparable active transceivers to exploit the symmetric channel characteristics at a single frequency. However, passive backscatter systems operate at dual frequencies, and are asymmetric, with only the reader device being able to transmit arbitrary signals.

A simple RSS–based key generation approach modified for dual frequency operation has been implemented on USRP software–defined radios acting as an emulation of the reader–tag backscatter system and shows good key agreement between the legitimate parties. However the reflection signal from the passive backscatter tag contains a beacon component common to both the tag and an eavesdropper and this compromises the secrecy of the shared key. To overcome the effect of the common beacon component we have described an enhanced algorithm based on giving the tag the additional capability of being able to control its reflected power. The enhanced algorithm was demonstrated using USRP emulation and showed significant improvement in restricting an eavesdropper's ability to derive the secret key by intercepting the communications of the legitimate parties.

Acknowledgements. The authors would like to thank Prof. Sherman Chow and the anonymous reviewers, whose valuable suggestions greatly improved the manuscript. We also would like to express our very great appreciation to Ms Samira Saadatpour for her valuable technical assistance in experimental implementations.

References

1. Kellogg, B., Parks, A., Gollakota, S., Smith, J.R., Wetherall, D.: Wi-fi backscatter: internet connectivity for RF-powered devices. ACM SIGCOMM Comput. Commun. Rev. **44**(4), 607–618 (2015)
2. Ensworth, J.F., Reynolds, M.S.: Every smart phone is a backscatter reader: modulated backscatter compatibility with Bluetooth 4.0 Low Energy (BLE) devices. In: 2015 IEEE International Conference on RFID (RFID), pp. 78–85. IEEE (2015)
3. Bharadia, D., Joshi, K.R., Kotaru, M., Katti, S.: BackFi: high throughput WiFi backscatter. ACM SIGCOMM Comput. Commun. Rev. **45**(4), 283–296 (2015)

4. Kellogg, B., Talla, V., Gollakota, S., Smith, J.R.: Passive Wi-Fi: bringing low power to Wi-Fi transmissions. In: 13th USENIX Symposium on Networked Systems Design and Implementation (NSDI 2016), pp. 151–164 (2016)
5. Iyer, V., Talla, V., Kellogg, B., Gollakota, S., Smith, J.: Inter-technology backscatter: towards internet connectivity for implanted devices. In: Proceedings of the 2016 Conference on ACM SIGCOMM 2016 Conference, pp. 356–369. ACM (2016)
6. Zhang, P., Bharadia, D., Joshi, K.R., Katti, S.: HitchHike: practical backscatter using commodity WiFi. In: SenSys, pp. 259–271 (2016)
7. Wang, T., Liu, Y., Vasilakos, A.V.: Survey on channel reciprocity based key establishment techniques for wireless systems. Wirel. Netw. **21**(6), 1835–1846 (2015)
8. Zou, Y., Zhu, J., Wang, X., Hanzo, L.: A survey on wireless security: technical challenges, recent advances, and future trends. Proc. IEEE **104**(9), 1727–1765 (2016)
9. Premnath, S.N., et al.: Secret key extraction from wireless signal strength in real environments. IEEE Trans. Mob. Comput. **12**(5), 917–930 (2013)
10. Jana, S., Premnath, S.N., Clark, M., Kasera, S.K., Patwari, N., Krishnamurthy, S.V.: On the effectiveness of secret key extraction from wireless signal strength in real environments. In: Proceedings of the 15th Annual International Conference on Mobile Computing and Networking, pp. 321–332. ACM (2009)
11. Patwari, N., Croft, J., Jana, S., Kasera, S.K.: High-rate uncorrelated bit extraction for shared secret key generation from channel measurements. IEEE Trans. Mob. Comput. **9**(1), 17–30 (2010)
12. Ali, S.T., Sivaraman, V., Ostry, D.: Eliminating reconciliation cost in secret key generation for body-worn health monitoring devices. IEEE Trans. Mob. Comput. **13**(12), 2763–2776 (2014)
13. Chinaei, M.H., Sivaraman, V., Ostry, D.: An experimental study of secret key generation for passive wi-fi wearable devices. In: IEEE 18th International Symposium on a World of Wireless, Mobile and Multimedia Networks (WoWMoM), pp. 1–9. IEEE (2017)
14. Mathur, S., Trappe, W., Mandayam, N., Ye, C., Reznik, A.: Radio-telepathy: extracting a secret key from an unauthenticated wireless channel. In: Proceedings of the 14th ACM International Conference on Mobile Computing and Networking, pp. 128–139. ACM (2008)

Short Papers

A Provably-Secure Unidirectional Proxy Re-encryption Scheme Without Pairing in the Random Oracle Model

S. Sharmila Deva Selvi, Arinjita Paul$^{(\boxtimes)}$, and Chandrasekaran Pandurangan

Theoretical Computer Science Lab,
Department of Computer Science and Engineering,
Indian Institute of Technology Madras, Chennai, India
{sharmila,arinjita,prangan}@cse.iitm.ac.in

Abstract. Proxy re-encryption (PRE) enables delegation of decryption rights by entrusting a proxy server with special information, that allows it to transform a ciphertext under one public key into a ciphertext of the same message under a different public key, without learning anything about the underlying plaintext. In Africacrypt 2010, the first PKI-based collusion resistant CCA secure PRE scheme without pairing was proposed in the random oracle model. In this paper, we point out an important weakness in the security proof of the scheme. We also present a collusion-resistant pairing-free unidirectional PRE scheme which meets CCA security under a variant of the computational Diffie-Hellman hardness assumption in the random oracle model.

Keywords: Proxy re-encryption · Random oracle model
Chosen ciphertext security · Provably secure · Unidirectional

1 Introduction

Proxy re-encryption is an important cryptographic primitive that allows a third party termed as *proxy server*, to transform the ciphertext of a user into a cipher-text of another user without learning anything about the underlying message. As pointed out by Mambo and Okamoto in [7], this is a common situation in practice where a data encrypted under PK_{Alice} is required to be encrypted under PK_{Bob}, such as applications like encrypted email forwarding, distributed file systems and outsourced filtering of encrypted spam. Here, Alice provides a secret information to the proxy called Re-Encryption Key (but not her private key

S. Sharmila Deva Selvi—Postdoctoral researcher supported by Project No. CCE/CEP/ 22/VK&CP/CSE/14-15 on Information Security & Awareness(ISEA) Phase-II by Ministry of Electronics & Information Technology, Government of India.
A. Paul and C. Pandurangan—Work partially supported by Project No. CCE/CEP/ 22/VK&CP/CSE/14-15 on ISEA-Phase II.

© Springer Nature Switzerland AG 2018
S. Capkun and S. S. M. Chow (Eds.): CANS 2017, LNCS 11261, pp. 459–469, 2018.
https://doi.org/10.1007/978-3-030-02641-7_21

SK_{Alice}) allowing it to transform $E_{PK_{Alice}}(m)$ to $E_{PK_{Bob}}(m)$ without learning anything about m or SK_{Alice}.

Most of the proxy re-encryption schemes in the literature are based on costly bilinear pairing operation [1,3,6]. Despite recent advances in implementation techniques, bilinear pairing takes more than twice the time taken by modular exponentiation computation and is an expensive operation. As stated by Chow et al. [4], removing pairing operations from PRE constructions is one of the open problems left by [3]. Weng et al. [5] proposed the first CCA secure pairing-free PRE scheme, which was however shown to be vulnerable to *collusion attack* [10]. Collusion resistance, also termed as *delegator secret security* is a desirable property in many practical scenarios such as secure cloud services, which prevents a colluding proxy and malicious delegatees from recovering the private key of the delegator. In 2010, Chow et al. [4] proposed the first construction of a collusion-resistant CCA secure pairing-free PRE scheme. However, in our work, we point out a major weakness in the security proof of the scheme by Chow et al. We also provide a construction of a CCA-secure collusion-resistant pairing-free unidirectional single-hop proxy re-encryption scheme under the Computational Diffie-Hellman (CDH) and the Divisible Computational Diffie-Hellman (DCDH) hardness assumptions in the random oracle model. Prior to our work, Canard et al. [2] exposed a similar flaw in the security proof of the scheme due to Chow et al. [4], and provided a fix to the scheme using NIZKOE (Non-interactive Zero-Knowledge proofs with Online Extractors). We show that our scheme is more efficient than the modified scheme due to Canard et al., providing an efficient pairing-free unidirectional collusion-resistant PRE scheme.

Complexity Assumptions

We define the complexity assumptions used in the security proof of our scheme. Let $\mathbb{G}$ be a cyclic multiplicative group of prime order q.

Definition 1. Computational Diffie Hellman Assumption (CDH): *The CDH problem in $\mathbb{G}$ is, given* $(g, g^a, g^b) \in \mathbb{G}^3$, *compute* g^{ab}, *where* $a, b \leftarrow \mathbb{Z}_q^*$.

Definition 2. Divisible Computational Diffie Hellman Assumption (DCDH): *The DCDH problem in $\mathbb{G}$ is, given* $(g, g^a, g^b) \in \mathbb{G}^3$, *compute* $g^{b/a}$, *where* $a, b \leftarrow \mathbb{Z}_q^*$.

Definition 3. Discrete Logarithm Assumption (DL): *The DL problem in $\mathbb{G}$ is, given* $(g, g^a) \in \mathbb{G}^2$, *compute* a, *where* $a \leftarrow \mathbb{Z}_q^*$.

2   Analysis of a PRE Scheme by Chow et al. [4]

We review the scheme due to Chow et al. [4] and point out the weakness in the security proof of the scheme in this section.

2.1   Review of the Scheme

- **Setup**(λ): Choose two primes p and q such that $q|p-1$ and the security parameter λ defines the bit-length of q. Let $\mathbb{G}$ be a subgroup of $\mathbb{Z}_q^*$ with order q and let g be a generator of the group $\mathbb{G}$. Choose four hash functions: $H_1 : \{0,1\}^{l_0} \times \{0,1\}^{l_1} \rightarrow \mathbb{Z}_q^*, H_2 : \mathbb{G} \rightarrow \{0,1\}^{l_0+l_1}, H_3 : \{0,1\}^* \rightarrow \mathbb{Z}_q^*, H_4 : \mathbb{G} \rightarrow \mathbb{Z}_q^*$. The hash functions H_1, H_2, H_3 are modelled as random oracles in the security proof reduction. Here l_0 and l_1 are security parameters determined by λ, and the message space $\mathcal{M}$ is $\{0,1\}^{l_0}$. Return the public parameters $PARAM = (q, \mathbb{G}, g, H_1, H_2, H_3, H_4, l_0, l_1)$.
- **KeyGen**($U_i, PARAMS$): To generate the private key (SK_i) and the corresponding public key (PK_i) of user U_i:
  - Pick $x_{i,1}, x_{i,2} \in_R \mathbb{Z}_q^*$ and set $SK_i = (x_{i,1}, x_{i,2})$.
  - Compute $PK_i = (PK_{i,1}, PK_{i,2}) = (g^{x_{i,1}}, g^{x_{i,2}})$.
- **ReKeyGen**($SK_i, PK_i, PK_j, PARAMS$): On input of the private key SK_i and public key PK_i of user U_i and user j's public key PK_j, generate the re-encryption key $RK_{i\rightarrow j}$ as shown:
  - Pick $h \in_R \{0,1\}^{l_0}, \pi \in_R \{0,1\}^{l_1}$.
  - Compute $v = H_1(h, \pi)$, $V = PK_{j,2}^v$ and $W = H_2(g^v) \oplus (h||\pi)$.
  - Define $RK_{i\rightarrow j}^{\langle 1 \rangle} = \frac{h}{x_{i,1}H_4(PK_{i,2})+x_{i,2}}$.
  - Return $RK_{i\rightarrow j} = (RK_{i\rightarrow j}^{\langle 1 \rangle}, V, W)$.
- **Encrypt**($m, PK_i, PARAMS$): To encrypt a message $m \in \mathcal{M}$ under PK_i:
  - Pick $u \in_R \mathbb{Z}_q^*, \omega \in_R \{0,1\}^{l_1}$.
  - Compute $D = \left(PK_{i,1}^{H_4(PK_{i,2})} PK_{i,2}\right)^u$.
  - Compute $r = H_1(m, \omega)$.
  - Compute $E = \left(PK_{i,1}^{H_4(PK_{i,2})} PK_{i,2}\right)^r$ and $F = H_2(g^r) \oplus (m||\omega)$.
  - Compute $s = u + r \cdot H_3(D, E, F) \mod q$.
  - Output the ciphertext $\sigma_i = (D, E, F, s)$.
- **ReEncrypt**($\sigma_i, PK_i, PK_j, RK_{i\rightarrow j}, PARAMS$): On input of an original ciphertext $\sigma_i = (D, E, F, s)$ encrypted under the public key of the delegator PK_i, the public key of the delegatee PK_j, the re-encryption key $RK_{i\rightarrow j}$, re-encrypt σ_i into a ciphertext $\hat{\sigma}_j$ under PK_j as follows:
  - Check if the following condition holds to satisfy the well-formedness of ciphertexts, otherwise return $\bot$:

$$\left(PK_{i,1}^{H_4(PK_{i,2})} PK_{i,2}\right)^s \overset{?}{=} D \cdot E^{H_3(D,E,F)} \tag{1}$$

  - If the condition holds, compute $\hat{E} = E^{RK_{i\rightarrow j}^{\langle 1 \rangle}} = g^{rh}$.
  - Output $\hat{\sigma}_j = (\hat{E}, F, V, W)$.
- **Encrypt**$_1$($m, PK_i, PARAMS$): To generate a non-transformable ciphertext under public key PK_i of a message $m \in \mathcal{M}$:
  - Pick $h \in_R \{0,1\}^{l_0}$ and $\pi \in_R \{0,1\}^{l_1}$.
  - Compute $v = H_1(h, \pi)$, $V = PK_{j,2}^v$ and $W = H_2(g^v) \oplus (h||\pi)$.

- Pick $\omega \in_R \{0,1\}^{l_1}$ and compute $r = H_1(m, \omega)$.
- Compute $\hat{E} = (g^r)^h$ and $F = H_2(g^r) \oplus (m||\omega)$.
- Output the non-transformable ciphertext $\hat{\sigma}_j = (\hat{E}, F, V, W)$.

- **Decrypt**$(\sigma_i, PK_i, SK_i, PARAMS)$: On input of a ciphertext σ_i, public key PK_i and private key $SK_i = (x_{i,1}, x_{i,2})$, decrypt according to two cases:
  - Original Ciphertext $\sigma_i = (D, E, F, s)$:
    * If Eq. (1) does not hold, return $\perp$.
    * Otherwise, compute $(m||\omega) = F \oplus H_2(E^{\frac{1}{x_{i,1}H_4(PK_{i,2})+x_{i,2}}})$.
    * Return m if $E \stackrel{?}{=} (PK_{i,1}^{H_4(PK_{i,2})}PK_{i,2})^{H_1(m,\omega)}$ holds; else return $\perp$.
  - Transformed /Non-transformable Ciphertext $\hat{\sigma}_i = (\hat{E}, F, V, W)$:
    * Compute $(h||\pi) = W \oplus H_2(V^{1/SK_{i,2}})$ and $(m||\omega) = F \oplus H_2(\hat{E}^{1/h})$.
    * Return m if $V \stackrel{?}{=} PK_{i,2}^{H_1(h,\pi)}$, $\hat{E} \stackrel{?}{=} g^{H_1(m,\omega)\cdot h}$ holds; else return $\perp$.

2.2 Weakness in the Security Proof of Chow *et al.*

In this section, we point out the weakness of the security proof for the PRE scheme by Chow *et al.* [4]. We show that the simulation of the oracles defined in the security proof of the scheme is not consistent with the real algorithm. This allows the adversary to distinguish the simulation run by the challenger from the real system. We demonstrate this flaw by considering the validity of the ciphertexts with respect to the *ReEncrypt* and *Decrypt* algorithm in the real system and the simulation (re-encryption oracle $OReE$ and decryption oracles $ODec$ respectively). To make it simple, we consider PK_T as the public key of the target user in the challenge phase and the attack is posed in **Phase-II** after the challenge phase is over. We re-encrypt a ciphertext σ_T under PK_T into a ciphertext $\hat{\sigma}_j$ under PK_j (PK_j is corrupt) and further decrypt $\hat{\sigma}_j$. All the computations hereafter are done using PK_T and PK_j.

First, we encrypt a message m under PK_T. Let us consider two forms of ciphertext $\sigma_{Real} = \langle D_{Real}, E_{Real}, F_{Real}, s_{Real} \rangle$ and $\sigma_{Fake} = \langle D_{Fake}, E_{Fake}, F_{Rand}, s_{Fake} \rangle$. σ_{Real} is the ciphertext obtained from the encryption algorithm **Encrypt** (i.e., encryption of m under PK_T by executing the **Encrypt**$(m, PK_T, PARAMS)$). σ_{Fake} is a cooked-up ciphertext that can pass the verification tests of *ReEncrypt* algorithm but not the *Decrypt* algorithm. We denote the algorithm for the construction of σ_{Fake} as **Encrypt**$_{Fake}(m, PK_T)$, which is as follows:

- Pick $u_{Fake} \in_R \mathbb{Z}_q^*$.
- Compute $D_{Fake} = \left((PK_{T,1})^{H_4(PK_{T,2})}PK_{T,2} \right)^{u_{Fake}}$.
- Pick $r_{Rand} \in_R \mathbb{Z}_q^*$. Here it should be noted that r_{Rand} does not follow the actual algorithm, instead it is picked at random from $\mathbb{Z}_q^*$.
- Pick $\omega_{Fake} \in_R \{0,1\}^{l_1}$ and compute $r_{Fake} = H_1(m, \omega_{Fake})$. Note that in the **Encrypt** algorithm, r_{Fake} is the output of H_1 oracle on giving a message and a random string (ω_{Fake}) of size $\{0,1\}^{l_1}$ as input.
- Compute $E_{Fake} = \left((PK_{T,1})^{H_4(PK_{T,2})}PK_{T,2} \right)^{r_{Rand}}$.

- Choose $F_{Rand} \in_R \{0, 1\}^{l_0+l_1}$. In the **Encrypt** algorithm, F is the encryption of the message along with a random string ω_{Fake} of length $\{0, 1\}^{l_1}$. But in the construction of σ_{Fake}, we note that F_{Rand} is chosen at random.
- Compute $s_{Fake} = u_{Fake} + r_{Rand}H_3(D_{Fake}, E_{Fake}, F_{Rand}) \mod q$.
- Output the ciphertext $\sigma_{Fake} = (D_{Fake}, E_{Fake}, F_{Rand}, s_{Fake})$. Note that, σ_{Fake} passes the ciphertext validation test of Eq. (1).

The important properties possessed by σ_{Real} and σ_{Fake} are:

1. Output of **Decrypt**$(\sigma_{Real}, PK_T, SK_T, PARAMS)$ is m and the output of $ODec(PK_T, \sigma_{Real})$ is m. This is because σ_{Real} is a legitimate ciphertext of m produced by *Encrypt* algorithm.
2. Output of **Decrypt**$(\sigma_{Fake}, PK_T, SK_T, PARAMS)$ and $ODec(PK_T, \sigma_{Fake})$ is $\bot$ as σ_{Fake} fails to satisfy the validity check for the obtained message.
3. σ_{Real} is a valid ciphertext and σ_{Fake} is an invalid ciphertext with respect to both **Decrypt** algorithm and $ODec$ oracle. Therefore, the simulation of the decryption algorithm is perfect.
4. Both σ_{Real} and σ_{Fake} are valid ciphertexts corresponding to the **ReEncrypt** algorithm. This is because σ_{Real} is a legitimate ciphertext of m produced by the *Encrypt* algorithm. Again, σ_{Fake} passes the ciphertext verification test of Eq. (1) and the algorithm computes the re-encrypted ciphertext $\hat{\sigma}_{Fake} = (\hat{E}, F, V, W)$ as per the protocol where $\hat{E}_{Fake} = g^{r_{Fake}h}$.

Next, we re-encrypt both σ_{Real} and σ_{Fake} under the public key PK_j of a corrupt user. Let us consider the following notations.

- $\hat{\sigma}_{Real}^{(Scheme)} \leftarrow$ **ReEncrypt**$(\sigma_{Real}, PK_T, PK_j, RK_{T \to j}, PARAMS)$.
- $\hat{\sigma}_{Real}^{(Oracle)} \leftarrow OReE(PK_T, PK_j, \sigma_{Real})$.
- $\hat{\sigma}_{Fake}^{(Scheme)} \leftarrow$ **ReEncrypt**$(\sigma_{Fake}, PK_T, PK_j, RK_{T \to j}, PARAMS)$.
- $\hat{\sigma}_{Fake}^{(Oracle)} \leftarrow OReE(PK_T, PK_j, \sigma_{Fake})$.

Observations on $\hat{\sigma}_{Real}^{(Scheme)}$ **and** $\hat{\sigma}_{Real}^{(Oracle)}$:

1. $\hat{\sigma}_{Real}^{(Scheme)} = \hat{\sigma}_{Real}^{(Oracle)}$.
2. $\hat{\sigma}_{Fake}^{(Scheme)} \neq \hat{\sigma}_{Fake}^{(Oracle)}$.

The reason for observation 1 follows directly from the fact that σ_{Real} is a valid ciphertext. The reason for the violation in observation 2 is that the **ReEncrypt** algorithm is only a function of the re-encryption key but $OReE$ oracle makes use of the knowledge of r_{Fake} to generate $\hat{\sigma}_{Fake}^{(Oracle)}$. However, in the construction of σ_{Fake}, r_{Rand} is used in the generation of $\hat{\sigma}_{Fake}^{(Oracle)}$. The question here is, how will the adversary find this difference, that is $\hat{\sigma}_{Fake}^{(Scheme)} \neq \hat{\sigma}_{Fake}^{(Oracle)}$. Let us now demonstrate how the adversary captures this difference shown by the $OReE$ oracle simulation and the **ReEncrypt** algorithm.

Distinguishing the Oracle from the Real Algorithm:

1. $\mathcal{C}$ provides the system parameters $PARAMS$ to $\mathcal{A}$.
2. After getting training in **Phase-I**, $\mathcal{A}$ provides two messages m_0 and m_1 of equal length and a target public key PK_T to $\mathcal{C}$.
3. $\mathcal{C}$ generates the challenge ciphertext σ_T and gives as challenge to $\mathcal{A}$.
4. $\mathcal{A}$ now does the following:
   (a) Generate σ_{Fake} = $\mathbf{Encrypt}_{Fake}(m_0, PK_T)$ = $(D_{Fake}, E_{Fake}, F_{Rand}, s_{Fake})$. Here $\mathcal{A}$ knows r_{Rand} and u_{Fake}.
   (b) $\mathcal{A}$ queries $OReE(\sigma_{Fake}, PK_T, PK_j, RK_{T \to j}, PARAMS)$. It should noted that v, V, h, π, W are fixed for $T \to j$ delegation.
   (c) **Test:** If $\bot \leftarrow OReE((\sigma_{Fake}, PK_T, PK_j, RK_{T \to j}, PARAMS))$, then $ReEncrypt \neq OReE$ and $\mathcal{A}$ knows that it is not the real system and will abort. Else, $\mathcal{A}$ learns no clue about the simulation.

2.3   Fixing the Flaw

Note that modifying the re-encryption algorithm to fix the flaw is not possible since re-encryption of a valid ciphertext σ_T will always require the knowledge of $r = H_1(m, \omega)$ as no other trapdoor exists to obtain a re-encrypted ciphertext $\hat{\sigma}_j$. Again, the knowledge of the private key of the delegator is necessary to generate the re-encryption keys and re-encrypted ciphertexts. Consequently, we cannot provide a trivial fix to the scheme in order to address the problem. As a solution, we propose a new collusion-resistant unidirectional proxy re-encryption scheme without any pairing operation. We have incorporated additional information to the existing *Encrypt* algorithm along with ciphertext validity checks in both the *Re-Encrypt* and the *Decrypt* algorithm.

3   A Unidirectional Proxy Re-encryption Scheme

- **Setup**(λ): Choose two primes p and q such that $q|p-1$ and the bit-length of q is the security parameter λ. Let $\mathbb{G}$ be a subgroup of $\mathbb{Z}_p^*$ with order q. g is a generator of the group $\mathbb{G}$. Choose five hash functions $H_1 : \{0,1\}^{l_0} \times \{0,1\}^{l_1} \to \mathbb{Z}_q^*, H_2 : \mathbb{G} \to \{0,1\}^{l_0+l_1}, H_3 : \{0,1\}^* \to \mathbb{Z}_q^*, H_4 : \mathbb{G} \to \mathbb{Z}_q^*, H_5 : \mathbb{G}^4 \times \{0,1\}^{l_0+l_1} \to \mathbb{G}$. The hash functions are modelled as random oracles in the security proof reduction. Here l_0 and l_1 are security parameters determined by λ, and the message space $\mathcal{M}$ is $\{0,1\}^{l_0}$. Return the public parameters $PARAM = (q, \mathbb{G}, g, H_1, H_2, H_3, H_4, H_5, l_0, l_1)$.
- **KeyGen**($U_i, PARAMS$): To generate the private key (SK_i) and the corresponding public key (PK_i) of user U_i:
  - Pick $x_{i,1}, x_{i,2} \in_R \mathbb{Z}_q^*$ and set $SK_i = (x_{i,1}, x_{i,2})$.
  - Compute $PK_i = (PK_{i,1}, PK_{i,2}) = (g^{x_{i,1}}, g^{x_{i,2}})$.
- **ReKeyGen**($SK_i, PK_i, PK_j, PARAMS$): On input of a user i's private key $SK_i = (x_{i,1}, x_{i,2})$ and public key $PK_i = (PK_{i,1}, PK_{i,2})$ and user j's public key $PK_j = (PK_{j,1}, PK_{j,2})$, generate the re-encryption key $RK_{i \to j}$ as shown:

- Pick $h \in_R \{0,1\}^{l_0}$, $\pi \in_R \{0,1\}^{l_1}$.
- Compute $v = H_1(h, \pi)$, $V = PK_{j,2}^v$ and $W = H_2(g^v) \oplus (h||\pi)$.
- Define $RK_{i \to j}^{\langle 1 \rangle} = \frac{h}{x_{i,1}H_4(PK_{i,2})+x_{i,2}}$.
- Return $RK_{i \to j} = (RK_{i \to j}^{\langle 1 \rangle}, V, W)$.
- **Encrypt**$(m, PK_i, PARAMS)$: To encrypt a message $m \in \mathcal{M}$:
  - Pick $u \in_R \mathbb{Z}_q^*$, $\omega \in_R \{0,1\}^{l_1}$.
  - Compute $D = \left(PK_{i,1}^{H_4(PK_{i,2})} PK_{i,2}\right)^u$.
  - Compute $\bar{D} = H_5(PK_{i,1}, PK_{i,2}, D, E, F)^u$.
  - Compute $r = H_1(m, \omega)$.
  - Compute $E = \left(PK_{i,1}^{H_4(PK_{i,2})} PK_{i,2}\right)^r$.
  - Compute $\bar{E} = H_5(PK_{i,1}, PK_{i,2}, D, E, F)^r$.
  - Compute $F = H_2(g^r) \oplus (m||\omega)$.
  - Compute $s = u + r \cdot H_3(D, \bar{E}, F) \mod q$.
  - Output the ciphertext $\sigma_i = (D, \bar{E}, F, s)$.
- **ReEncrypt**$(\sigma_i, PK_i, PK_j, RK_{i \to j}, PARAMS)$: On input of an original ciphertext $\sigma_i = (E, \bar{E}, F, s)$ encrypted under public key $PK_i = (PK_{i,1}, PK_{i,2})$, the public keys PK_i and PK_j, a re-encryption key $RK_{i \to j} = (RK_{i \to j}^{\langle 1 \rangle}, V, W)$, re-encrypt σ_i into a ciphertext $\hat{\sigma}_j$ under the public key $PK_j = (PK_{j,1}, PK_{j,2})$ as follows:
  - Compute D and $\bar{D}$ as follows:

$$D = \left(PK_{i,1}^{H_4(PK_{i,2})} PK_{i,2}\right)^s \cdot (E^{H_3(E,\bar{E},F)})^{-1}$$
$$= \left(PK_{i,1}^{H_4(PK_{i,2})} PK_{i,2}\right)^u.$$
$$\bar{D} = H_5(PK_{i,1}, PK_{i,2}, D, E, F)^s \cdot (\bar{E}^{(E,\bar{E},F)})^{-1}$$
$$= H_5(PK_{i,1}, PK_{i,2}, D, E, F)^u.$$

  - Check the well-formedness of the ciphertext by verifying:

$$\left(PK_{i,1}^{H_4(PK_{i,2})} PK_{i,2}\right)^s \overset{?}{=} D \cdot E^{H_3(E,\bar{E},F)} \tag{2}$$

$$\left(H_5(PK_{i,1}, PK_{i,2}, D, E, F)\right)^s \overset{?}{=} \bar{D} \cdot \bar{E}^{H_3(E,\bar{E},F)} \tag{3}$$

  - If the above checks fail, return $\bot$. Else, compute $\bar{E} = E^{RK_{i \to j}^{\langle 1 \rangle}} = g^{rh}$.
  - Output $\hat{\sigma}_j = (\bar{E}, F, V, W)$.
- **Encrypt$_1$**$(m, PK_i, PARAMS)$: To generate a non-transformable ciphertext under public key PK_i of a message $m \in \mathcal{M}$:
  - Pick $h \in_R \{0,1\}^{l_0}$ and $\pi \in_R \{0,1\}^{l_1}$.
  - Compute $v = H_1(h, \pi)$, $V = PK_{j,2}^v$ and $W = H_2(g^v) \oplus (h||\pi)$.
  - Pick $\omega \in_R \{0,1\}^{l_1}$ and compute $r = H_1(m, \omega)$.
  - Compute $\hat{E} = (g^r)^h$ and $F = H_2(g^r) \oplus (m||\omega)$.
  - Output the non-transformable ciphertext $\hat{\sigma}_j = (\hat{E}, F, V, W)$.

- **Decrypt**($\sigma_i, PK_i, SK_i, PARAMS$): On input a ciphertext σ_i, public key PK_i and private key $SK_i = (x_{i,1}, x_{i,2})$, decrypt according to two cases:
  - Original ciphertext of the form $\sigma_i = (E, \bar{E}, F, s)$:
    * Check if the ciphertext is well-formed by computing the values of D and $\bar{D}$ and checking if Eqs. (2) and (3) holds.
    * If the conditions hold, extract $(m||\omega) = F \oplus H_2(E^{\overline{x_{i,1}H_4(PK_{i,2})+x_{i,2}}})$, else return $\perp$.
    * Return m if the following checks hold, else return $\perp$.

$$E \overset{?}{=} \left(PK_{i,1}^{H_4(PK_{i,2})} PK_{i,2}\right)^{H_1(m,\omega)}$$

$$\bar{E} \overset{?}{=} H_5(PK_{i,1}, PK_{i,2}, D, E, F)^{H_1(m,\omega)}$$

  - Transformed or non-transformable ciphertext of the form $\sigma_i = (\hat{E}, F, V, W)$:
    * Compute $(h||\pi) = W \oplus H_2(V^{1/SK_{i,2}})$, extract $(m||\omega) = F \oplus H_2(\hat{E}^{1/h})$.
    * Return m if $V \overset{?}{=} PK_{i,2}^{H_1(h,\pi)}$, $\hat{E} \overset{?}{=} g^{H_1(m,\omega)\cdot h}$ holds; else return $\perp$.

3.1 Correctness

Due to space constraints, the correctness of our scheme is given in the full version of the paper [9].

3.2 Security Proof

Original Ciphertext Security:

Theorem 1. *The proposed scheme is CCA-secure for the original ciphertext under the DCDH assumption and the $EUF-CMA$ security of Schnorr signature scheme [8]. If a $(t, \epsilon)IND\text{-}PRE\text{-}CCA$ $\mathcal{A}$ with an advantage ϵ breaks the IND-PRE-CCA security of the given scheme in time t, $\mathcal{C}$ can solve the DCDH problem with advantage ϵ' within time t' where:*

$$\epsilon' \geq \frac{1}{q_{H_2}}\left(\frac{\epsilon}{e(q_{RK}+1)} - \frac{q_{H_1}}{2^{l_1}} - \frac{q_{H_3}+q_{H_5}}{2^{l_0+l_1}} - q_d\left(\frac{q_{H_1}+q_{H_2}}{2^{l_0+l_1}} + \frac{2}{q}\right) - \epsilon_1 - \epsilon_2\right),$$

$$t' \leq t + (q_{H_1} + q_{H_2} + q_{H_3} + q_{H_4} + q_{H_5} + n_h + n_c + q_{RK} + q_{RE} + q_d)O(1)$$
$$+ (2n_h + 2n_c + 2q_{RK} + 5q_{RE} + 2q_d + q_{H_1}q_{RE} + (2q_{H_2} + 2q_{H_1})q_d)t_e,$$

We note that e is the base of natural logarithm, ϵ_1 denotes the advantage in breaking the CCA security of the hashed Elgamal encryption scheme and ϵ_2 denotes the advantage in breaking the EUF-CMA security of the Schnorr Signature scheme and t_e denotes the time taken for exponentiation in group $\mathbb{G}$.

Proof. Due to space constraints, the proof of the theorem is shown in the full version of the paper [9].

Transformed Ciphertext Security:

Theorem 2. *The proposed scheme is CCA-secure for the transformed ciphertext under the CDH assumption and the $EUF-CMA$ security of Schnorr signature scheme [8]. If a $(t,\epsilon)IND\text{-}PRE\text{-}CCA$ $\mathcal{A}$ with an advantage ϵ breaks the IND-CPRE-CCA security of the given scheme, $\mathcal{C}$ can solve the DCDH problem with advantage ϵ' within time t' where:*

$$\epsilon' \geq \frac{1}{q_{H_2}}\left(\frac{2\epsilon}{e(2+q_{RK})^2} - \frac{q_{H_1}}{2^{l_1}} - q_d\left(\frac{q_{H_1}+q_{H_2}}{2^{l_0+l_1}} + \frac{2}{q}\right) - \epsilon_2\right),$$

$$t' \leq t + (q_{H_1} + q_{H_2} + q_{H_3} + q_{H_4} + q_{H_5} + n_h + n_c + q_{RK} + q_{RE} + q_d)O(1)$$
$$+ (2n_h + 2n_c + 2q_{RK} + 3q_{RE} + 2q_d + (2q_{H_2} + 2q_{H_1})q_d)t_e,$$

Proof. The proof of the theorem is shown in the full version of the paper [9]. □

Non-transformable Ciphertext Security:

Theorem 3. *The proposed scheme is CCA-secure for the non-transformable ciphertext under the CDH assumption. If a $(t, \epsilon - \epsilon_2)IND\text{-}PRE\text{-}CCA$ $\mathcal{A}$ with an advantage $\epsilon - \epsilon_2$ breaks the IND-PRE-CCA security of the given scheme, $\mathcal{C}$ can solve the CDH problem with advantage ϵ' within time t' where:*

$$\epsilon' \geq \frac{1}{q_{H_2}}\left(\epsilon - \epsilon_2 - \frac{q_{H_1}}{2^{l_1}} - q_d\left(\frac{q_{H_1}+q_{H_2}}{2^{l_0+l_1}} + \frac{2}{q}\right)\right),$$

$$t' \leq t + (q_{H_1} + q_{H_2} + q_{H_3} + q_{H_4} + q_{H_5} + n_h + n_c + q_{RK} + q_d)O(1)$$
$$+ (2n_h + 2n_c + 2q_{RK} + 2q_d + (2q_{H_2} + 2q_{H_1})q_d)t_e,$$

Proof. The proof of the theorem is shown in the full version of the paper [9].

Delegator Secret Security:

Theorem 4. *The proposed scheme is DSK-secure under the DL assumption. If a $(t, \epsilon)DSK$ $\mathcal{A}$ with an advantage ϵ breaks the DSK security of the given scheme in time t, $\mathcal{C}$ can solve the DL problem with advantage ϵ within time t' where:*

$$t' \leq t + O(2q_{RK} + 2n_h + 2n_c)t_e,$$

Proof. The proof of the theorem is shown in the full version of the paper [9]. □

4   Efficiency Comparison

We give a comparison of our scheme with the modified scheme of Chow *et al.* by Canard *et al.* [2] in Table 1. We show the computational efficiency of our PRE scheme, and use t_e to denote the time for exponentiation operation. Note that $l = O(\log \lambda)$ denotes the number of commitments generated by the signer in the NIZK proof in the encryption protocol in [2]. The comparison shows that our proposed design is more efficient than the existing fix to the pairing-free unidirectional PRE scheme of Chow *et al.* constructed by Canard *et al.* [2].

Table 1. Comparative analysis of the modified pairing-free PRE scheme due to Canard *et al.* and our scheme. Note that $l = O(\log \lambda)$.

Algorithm	[2]	Our scheme
KeyGen	$2t_e$	$2t_e$
ReKeyGen	$2t_e$	$2t_e$
Encrypt$_1$	$7t_e$	$4t_e$
Encrypt	$(3 + l)t_e$	$5t_e$
ReEncrypt	$4t_e$	$6t_e$
Decrypt (original)	$7t_e$	$4t_e$
Decrypt (transformed)	$3t_e$	$8t_e$

5   Conclusion

Although pairing is an expensive operation, only one scheme due to Chow *et al.* [4] reported the pairing-free unidirectional property with collusion-resistance. In this paper, we point out that the security proof in the scheme is flawed, where the adversary is able to determine that the simulation provided by challenger is not consistent with real system. Also, we present a construction of unidirectional proxy re-encryption scheme without bilinear pairing that provides collusion-resistance, and show that our scheme is more efficient than the modified scheme of Chow *et al.* constructed by Canard *et al.* Our scheme is proven CCA-secure under a variant of the computational Diffie-Hellman assumption in the random oracle model.

References

1. Ateniese, G., Fu, K., Green, M., Hohenberger, S.: Improved proxy re-encryption schemes with applications to secure distributed storage. In: Proceedings of the Network and Distributed System Security Symposium, NDSS 2005, San Diego, California, USA (2005)
2. Canard, S., Devigne, J., Laguillaumie, F.: Improving the security of an efficient unidirectional proxy re-encryption scheme. J. Internet Serv. Inf. Secur. **1**(2/3), 140–160 (2011)
3. Canetti, R., Hohenberger, S.: Chosen-ciphertext secure proxy re-encryption. In: Proceedings of the 2007 ACM Conference on Computer and Communications Security, CCS 2007, Alexandria, Virginia, USA, 28–31 October 2007, pp. 185–194 (2007)
4. Chow, S.S.M., Weng, J., Yang, Y., Deng, R.H.: Efficient unidirectional proxy re-encryption. In: Bernstein, D.J., Lange, T. (eds.) AFRICACRYPT 2010. LNCS, vol. 6055, pp. 316–332. Springer, Heidelberg (2010). https://doi.org/10.1007/978-3-642-12678-9_19
5. Deng, R.H., Weng, J., Liu, S., Chen, K.: Chosen-ciphertext secure proxy re-encryption without pairings. In: Franklin, M.K., Hui, L.C.K., Wong, D.S. (eds.) CANS 2008. LNCS, vol. 5339, pp. 1–17. Springer, Heidelberg (2008). https://doi.org/10.1007/978-3-540-89641-8_1

6. Libert, B., Vergnaud, D.: Unidirectional chosen-ciphertext secure proxy re-encryption. IEEE Trans. Inf. Theory **57**(3), 1786–1802 (2011)
7. Mambo, M., Okamoto, E.: Proxy cryptosystems: Delegation of the power to decrypt ciphertexts. IEICE Trans. Fundam. Electron. Commun. Comput. Sci. **80**(1), 54–63 (1997)
8. Schnorr, C.-P.: Efficient signature generation by smart cards. J. Cryptol. **4**(3), 161–174 (1991)
9. Sharmila Deva Selvi, S., Paul, A., Rangan, C.P.: A provably-secure unidirectional proxy re-encryption scheme without pairing in the random oracle model (full version). Cryptology ePrint Archive, October 2017
10. Weng, J., Deng, R.H., Liu, S., Chen, K.: Chosen-ciphertext secure bidirectional proxy re-encryption schemes without pairings. Inf. Sci. **180**(24), 5077–5089 (2010)

Computational Aspects of Ideal (t, n)-Threshold Scheme of Chen, Laing, and Martin

Mayur Punekar[1](✉), Qutaibah Malluhi[1], Yvo Desmedt[2], and Yongee Wang[3]

[1] Department of Computer Science and Engineering, Qatar University, Doha, Qatar
mayur.punekar@ieee.org, qmalluhi@qu.edu.qa
[2] The University of Texas at Dallas, Richardson, TX, USA
yvo.desmedt@utdallas.edu
[3] Department of SIS, UNC Charlotte, Charlotte, NC, USA
yongge.wang@uncc.edu

Abstract. In CANS 2016, Chen, Laing, and Martin proposed an ideal (t, n)-threshold secret sharing scheme (the *CLM scheme*) based on random linear code. However, in this paper we show that this scheme is essentially same as the one proposed by Karnin, Greene, and Hellman in 1983 (the *KGH scheme*) from privacy perspective. Further, the authors did not analyzed memory or XOR operations required to either store or calculate an inverse matrix needed for recovering the secret. In this paper, we analyze computational aspects of the CLM scheme and discuss various methods through which the inverse matrix required during the secret recovery can be obtained. Our analysis shows that for $n \leq 30$ all the required inverse matrices can be stored in memory whereas for $30 \leq n < 9000$ calculating the inverse as and when required is more appropriate. However, the CLM scheme becomes impractical for $n > 9000$. Another method which we discuss to recover the secret in KGH scheme is to obtain only the first column of the inverse matrix using Lagrange's interpolation however, as we show, this method can not be used with the CLM scheme. Some potential application of the secret sharing schemes are also discussed. From our analysis we conclude that the CLM scheme is neither novel nor as practical as has been suggested by Chen *et al.* whereas the KGH scheme is better suited for practical applications with large n.

1 Introduction

Secret sharing refers to procedures in which a secret is distributed among a group of participants or players such that individual shares gives no information about the secret. In order to reconstruct the secret a sufficient number of participates must combine their shares together. A type of secret sharing scheme, known as (t, n)-threshold secret sharing scheme can be used to distribute secret k to n participants in such a way that any t participants can uniquely recover the secret and at the same time any set of $t - 1$ participants get no information about the

© Springer Nature Switzerland AG 2018
S. Capkun and S. S. M. Chow (Eds.): CANS 2017, LNCS 11261, pp. 470–481, 2018.
https://doi.org/10.1007/978-3-030-02641-7_22

secret. A secret sharing scheme for which the ratio of the size of the secret to the size of the largest share is 1 is called *ideal*. Shamir and Blakley independently introduced (t, n)-threshold schemes in 1979 [1,2]. Blakley's method is based on linear projective geometry where each share specifies a hyperplane and the secret k is the unique point of intersection of the n hyperplanes. Shamir's scheme relies on a polynomial of degree $t - 1$ to generate n shares and use Lagrange's interpolation method to recover the secret from t participants. However, Lagrange's interpolation is computationally intensive due to which efforts have been made to obtain schemes that have same properties as Shamir's scheme but can be implemented using XOR operations only. First such XOR based (t, n)-threshold scheme was proposed by Kurihara *et al.* in [3] which was a generalization of their earlier work on $(3, n)$-threshold scheme [4]. Some other XOR based scheme were also proposed by Lv *et al.* [5,6], Wang *et al.* [7], and Chen *et al.* [8].

In CANS 2016, Chen, Laing, and Martin [9] proposed a (t, n)-threshold secret sharing scheme which is based on a patent application by HP [8]. The scheme is defined over $GF(2^\lambda)$ which can be generalized to any Galois field. It uses random linear code and requires only XOR and shift operations for distribution and recovery of the secret. The proof for security of the scheme is also given and it has been proven that the scheme is ideal. However, we observe that the scheme by Chen *et al.* (the *CLM scheme*) is same as the one proposed earlier by Karnin, Grenne, and Hellman [10] in 1983 (the *KGH scheme*) from privacy perspective. Further, the CLM scheme relies on decoding a random linear code to recover the secret. The decoding method proposed by the authors inverts a $t \times t$ matrix G' which is derived by selecting t columns of the $t \times n$ generator matrix G used during distribution. Hence, the decoder either needs to calculate this inverse on the fly or has to store all possible such inverse matrices in memory. However, this important computational aspect of the CLM scheme has not been discussed by the authors in [9]. Their complexity analysis and comparison with other secret sharing schemes in [9] neither discuss the memory required to store all inverse matrices nor the number of XOR operations needed to obtain such an inverse.

In this paper, we analyze the CLM scheme and focus on the computational aspects of it. First, four issues related to the CLM scheme are discussed, namely, similarity of the CLM with the KGH scheme, computational aspects related to the inverse of G', XOR operations required for vector and matrix multiplication and inability of the CLM scheme to detect and correct erroneous shares due to lack of efficient decoding algorithm. Further, we suggest and discuss three ways in which a inverse matrix can be obtained which is required in the CLM scheme: (1) store all possible inverse matrices, (2) calculate the inverse on the fly, (3) calculate only the first column of the inverse matrix. We also give estimate of the memory required to store all inverse matrices and the runtime required to invert a matrix for different n.

The rest of the paper is structured as follows. In Sect. 2 we discuss Shamir's, the CLM and the KGH schemes. The CLM scheme is analyzed in detail in Sect. 3. Different approaches for inverse of G' are proposed and discussed in Sect. 4.

Section 5 discuss some potential applications of the CLM and KGH schemes. We conclude the paper in Sect. 6.

2   Background

2.1   Shamir's Scheme

In Sharmir's (t, n)-threshold secret sharing scheme the secrets and shares are the elements of finite field $GF(q)$ for some prime-power $q > n$. n distinct non-zero elements $\alpha_1, \alpha_2, \ldots, \alpha_n \in GF(q)$ are selected which are known to all parties. Suppose a secret $k \in GF(q)$ needs to be shared among n parties. Then, $t - 1$ random elements $a_1, \ldots, a_{t-1} \in GF(q)$ are chosen independently with uniform distribution. The secret along with the random numbers define a polynomial $P(x) = k + \sum_{i=1}^{t} a_i x^i$. The share of party $v_j = P(\alpha_j)$ where $P(x)$ is evaluated using the arithmetic of $GF(q)$.

The secret k can be recovered from t shares $v_{i_1}, \ldots, v_{i_t}$ using the following

$$Q(x) = \sum_{l=1}^{t} v_{i_l} \prod_{1 \le j \le t, j \le l} \frac{\alpha_{i_j} - x}{\alpha_{i_j} - \alpha_{i_l}} \tag{1}$$

where the secret is given by $k = P(0) = Q(0)$.

2.2   Karnin, Greene, and Hellman Scheme

Karnin, Greene, and Hellman proposed ideal secret sharing scheme (*KGH scheme*) in [10]. The KGH scheme uses a vector u of length t whose first element is same as the secret $k \in GF(q^\lambda)$ and rest of the $t - 1$ entries are generated randomly from $GF(q^\lambda)$ where q is a prime. A $t \times n + 1$ systematic generator matrix G for some linear code whose entries are from $GF(q^\lambda)$ is used for encoding the vector u. The selected linear code must be MDS, i.e., any $t \times t$ submatrix derived from G by selecting any t columns must be invertible. The following relationship exit between the vector u and shares v [10],

$$v = uG \tag{2}$$

such that $v_0 = u_1 = k$ which is secret to be protected and $v_1, \ldots, v_n$ are the n shares to be distributed. It has been shown by the authors that the Shamir's scheme [1] is a special case of the proposed method. The secret can be recovered when t shares are available for which the t linear equation over $GF(2^\lambda)$ with t unknown have to be solved.

2.3   Chen, Laing, and Martin Scheme

We now describe ideal (t, n)-threshold the CLM scheme proposed in [9]. This scheme uses two algorithms referred to as *Share* and *Recover*.

The $Share(k)$ algorithm [9] takes a secret $k \in \{0, 1\}^{\lambda}$ and parse it into t words where each word consists of $\lceil \frac{\lambda}{t} \rceil$ bits. If λ is not divisible by t, then k is padded with $(-\lambda) \mod t$ elements so that each word is an element in $F = GF(2^{\lceil \frac{\lambda}{t} \rceil})$ and $k \in F^t$. Then, $r_1, \ldots, r_{t-1} \in GF(2^{\lambda})$ dummy keys are randomly generated and again parsed into t words of $\lceil \frac{\lambda}{t} \rceil$ bits. These dummy keys are XORed with secret k to produce k'. All dummy keys $r_1, \ldots, r_{t-1}$ and modified secret k' are dispersed using $Share^{\mathrm{IDA}}$ algorithm [9, Sect. 2.3]. This algorithm multiplies the input vector with a systematic generator matrix G of an MDS linear code. The vector $\boldsymbol{K}'$ is obtained by applying $Share^{\mathrm{IDA}}$ to K'. Similarly, $\boldsymbol{R}_1, \ldots, \boldsymbol{R}_{t-1}$ are obtained from $r_1, \ldots, r_{t-1}$ using $Share^{\mathrm{IDA}}$. A new $t \times n$ matrix M is created by using vectors K' and $\boldsymbol{R}_1, \ldots, \boldsymbol{R}_{t-1}$ as its rows. Then the elements of row $i, 0 \leq i \leq t$ are shifted to the left by i places to obtain a new matrix M'. The elements in column i are then concatenated and are used as shares.

The $Recover$ algorithm [9] is used to recover the secret when at least t out of n shares are available. The algorithm retrieves the secret k using $Recover^{\mathrm{IDA}}$ algorithm which is essentially an algorithm to decode the MDS linear code. However, before $Recover^{\mathrm{IDA}}$ is used, the i-th share is shifted to the right by i places. $Recover^{\mathrm{IDA}}$ algorithm creates a t-vector C' by using t pooled shares. Then, a $t \times t$ matrix G' is formed which consists of the t rows of generator matrix G corresponding to the t shares pooled. The matrix G' is then inverted and multiplied by the vector C' to obtain k' and $r_1, \ldots, r_{k-1}$ from which the secret k can be obtained using $k = k' \oplus r_1 \oplus \cdots \oplus r_{t-1}$.

3   Analysis of CLM Scheme

In this section we explain four issues which we observed with CLM scheme.

First, we would like to point out that the CLM scheme is same as the KGH scheme explained in Sect. 2.2 from privacy perspective. The only difference between these two schemes lies in the fact that, Chen $et\ al.$ [9] add randomly generated words $r_1, \ldots r_{k-1}$ to the secret k before encoding it with $Share^{\mathrm{IDA}}$ whereas this step is not used by Karnin $et\ al.$ [10]. Also, it is not clear as to what advantage this additional step provides in terms of security or efficiency.

The second problem we observed is related to the decoder used in $Recover^{\mathrm{IDA}}$ algorithm. The decoder needs to either compute the inverse of G' on the fly or store all possible $t \times t$ G' matrices in memory. In the first case, the XOR operations required to compute the inverse needs to be considered. For the second case, the memory required to store all possible matrices needs to be analyzed. The comparison of the CLM scheme to other schemes as given in [9] neither includes memory required to store all inverse matrices nor the number of XOR operations required to invert a matrix on the fly.

The third issue is with vector and matrix multiplication used in $Recover^{\mathrm{IDA}}$ algorithm. The number of XOR required for this multiplication has not been included in the comparison of CLM scheme with other schemes in [9].

The fourth problem is related to the connections between the CLM and Shamir's scheme. It has been shown by Karnin $et\ al.$ in [10] that if generator matrix G is chosen appropriately then KGH scheme is equivalent to the

Shamir's scheme. However, in the CLM scheme the secret k is XORed with random words $r_1, \ldots, r_{k-1}$ to generate first word k' before encoding. This process though reduce the number of random words required in the CLM scheme has a downside that regardless of which generator matrix is chosen, the scheme does not have any relation to Shamir's scheme anymore. As shown by McEliece *et al.* in [12], Shamir's scheme corresponds to Reed-Solomon codes and hence Shamir's scheme has error detection and correction capability when more than t shares are available and Reed-Solomon decoding algorithm [13] is used to retrieve the secret. If one or more shares are modified by adversary or due to storage error then such error detection and correction capability can be used to retrieve the secret using the KGH scheme. In particular, for the (t, n)-threshold KGH scheme, more than $\lfloor (n-t)/2 \rfloor$ shares must be tampered with or in error so that the legitimate user is unable to retrieve the secret [12]. On the other hand, CLM scheme does not have any relation to Shamir's scheme and consequently Reed-Solomon decoding can not be used to detect and correct erroneous shares.

The CLM scheme like KGH scheme also uses a linear code and due to that some decoding algorithm can be used to detect and correct erroneous shares to retrieve the secret. However, it has been shown by Berlekamp *et al.* in [14] that, the decoding of an arbitrary code, e.g., codes with a random generator matrix, is NP-complete. The CLM scheme uses random generator matrix and the authors in [9] also did not present any efficient decoding algorithm for the linear code. Hence, even if more than t shares are available in the CLM scheme, it is impossible to detect and correct erroneous shares to recover the secret.

4   Computational Approaches for Inverse of G' in CLM Scheme

We now discuss various approaches that can be used to obtain inverse of G' matrix which is required by $Recover^{\text{IDA}}$ algorithm. There are mainly three methods: first, the algorithm can store precomputed inverse of all possible G' matrices. Second, inverse of G' can be calculated on the fly and third, only the first column of the inverse of G' can be calculated using a algorithm which do not invert the whole matrix. These approaches are discussed in detail in the following.

4.1   Precompute Inverse

The CLM (t, n)-threshold scheme can recover secret k from any t pooled shares. The decoder used in $Recover^{\text{IDA}}$ algorithm needs a $t \times t$ matrix G' which consists of the t columns of generator matrix G corresponding to the available t shares. In this approach, inverse of all possible G' are precomputed and stored so that $Recover^{\text{IDA}}$ algorithm can use them as and when needed. For example, if $n = 6, t = 3$ then the decoder needs to store inverse of $\binom{6}{3} = 20$ 3×3 G' matrices. Since all the required inverse of matrices are available to the algorithm, this approach is the fastest among the three discussed here. However, the number of

G' matrices increases exponentially as the maximum value of $\binom{n}{t}$ grows exponentially with n. The worst case occurs for $t = n/2$ when $\binom{n}{t}$ reach its maximum value. Even for moderate values of n, e.g., for $n = 30, t = 15$ the decoder requires inverse of $\binom{n}{t} = 155,117,520$ G' matrices ! Clearly, it would be very difficult to store all matrices in this case. Hence, this scheme becomes impractical even for moderate values of k and t. On a positive side, for $n = 6, t = 3$ the algorithm needs to store 20 matrices. Since each such matrix is of size 3×3, it has 9 entries in total. Let us assume that each entry from a matrix is stored with 4 Bytes. Then, storage of an inverse would require only 36 Bytes and in total 720 Bytes of RAM is required to store 20 matrices. This memory requirement is quite low and hence this approach is preferred when n is small enough. The memory required to store inverse of all possible G' for (t,n) threshold scheme where $t = \lfloor n/2 \rfloor$. is shown in Fig. 1. As can be observed, for $n = 26, t = 13$ the memory requirement is already 13 GB which makes this approach impractical for $n \geq 26, t = \lfloor n/2 \rfloor$.

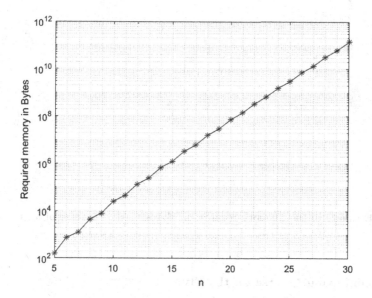

Fig. 1. Memory required in Bytes to store inverse of all possible G' for different n, here $t = \lfloor n/2 \rfloor$.

Other possibility is to store inverse matrices in hard disk drive (HDD) and read them as and when necessary from HDD. HDDs are in general much slower than RAM and hence the read time would increase. However, since large HDDs are relative cheap, e.g., 1TB HDD cost around \$ 40 US, it would be possible to store all inverse matrices for $n > 26$ using HDD. To estimate the memory required to store inverse matrices for higher n value, we use following approximation of $\binom{n}{t}$[15],

$$\binom{n}{\frac{n}{2}} \sim \sqrt{\frac{2}{m}} \frac{1}{\sqrt{n}} 2^n, \tag{3}$$

where $f(n) \sim g(n)$ if and only if $\lim_{n \to \infty} \frac{f(n)}{g(n)} = 1$. The results of our calculation using (3) is shown in Fig. 2. As can be observed, the memory requirement continues to grow exponentially for higher n values. For $n = 36$, our estimate shows that the required memory is around 48 Terabyte. Hence, even when HDD is used to store all inverse matrices, it is difficult to build a practical CLM scheme for $n > 36$.

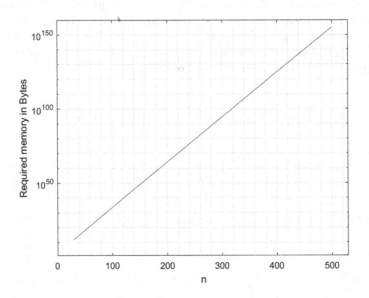

Fig. 2. Estimate of the Memory required in Bytes to store inverse of all possible G' for different n, here $t = \lfloor n/2 \rfloor$.

4.2   Computing Inverse on the Fly

The other possibility to obtain inverse of G' is to calculate it on the fly in $Recover^{\mathrm{IDA}}$ algorithm. Matrix inversion methods, e.g., Gaussian elimination, can be used for this purpose. Gaussian elimination is known to have complexity of $\mathcal{O}(n^3)$. Other possibility is to use LU decomposition to obtain lower and upper triangular matrices through which the inverse can be obtained. The complexity of LU decomposition is given by $\mathcal{O}(M(n))$ [16] where $M(n)$ is the time required to multiply two matrices of order n and $M(n) \geq n^a$ for some $a > 2$. Hence, if a faster matrix multiplication algorithm is used then the complexity of the LU decomposition can be reduced. For example, when a matrix multiplication is performed using the Coppersmith-Winograd algorithm [17] then the complexity of LU decomposition is given by $\mathcal{O}(n^{2.376})$. However, we remark that Coppersmith-Winograd algorithm achieves this improvements asymptotically and hence useful for theoretical analysis only. A more practical algorithm for matrix inversion is from Strassen [18] which has complexity of $\mathcal{O}(n^{2.81})$.

To get an idea of time required to compute the inverse of G', we computed the inverse of $t \times t$ matrices over $GF(2^4)$ for different t values using NTL library [19] with a C++ program. We carry out calculations on a PC with Intel Xeon E5-2640 CPU clocked at 2.5 GHz and 8 GB of RAM. As per the private correspondence with the author of the NTL library, the matrix inversion functions for matrices over $GF(2^E)$ in NTL use a variant of Gaussian elimination method. The time required to compute the inverse for different values of t is given in Fig. 3. It can be observed that as the value of t increases, the runtime required to invert the $t \times t$ matrix increases rapidly. E.g, for $t = 6000$ the runtime required to invert matrix is already close to 4 hours and should be more than 24 hours for $t = 9000$. Though much higher values of n can be achieved through this method compared to storing of all inverse matrices, due to its high runtime requirement this method is also impractical for larger n. Further, the inversion algorithm also requires significant memory, e.g., for $n = 6000$ the NTL required about 2 GB of RAM, which would also restrict the use of this method in practice.

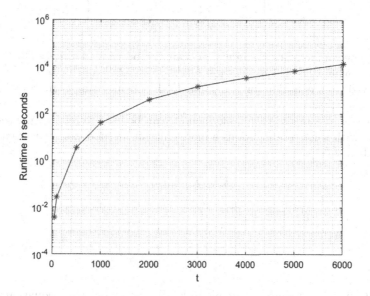

Fig. 3. Runtime in seconds to calculate inverse of $t \times t$ G' over $GF(2^4)$ for different t.

4.3 Computing only the First Column of the Inverse

As mentioned before, the KGH scheme can be converted to Shamir's scheme by using appropriate generator matrix. Shamir's scheme has an advantage that the values of $\alpha_{i_j} - \alpha_{i_l}$ required in (1) can be precomputed and stored to accelerate the recovery process. We show in the following that, the same values can be used to calculate the first column of the inverse of the submatrix G'. We remark that a method to find elements of the inverse matrix using Monte Carlo method was proposed by Forsythe and Leibler [20]. However, it does not exploit the special structure of the matrix used in the KGH scheme and hence not as efficient as the one discussed below.

Let us assume that a secret k has been protected using KGH scheme. When only t shares, i.e., $v_{i_1}, \ldots, v_{i_t}$ are available then we get following from (2), $u = (v_{i_1}, \ldots, v_{i_t}) G'^{-1}$. As can be observed from this equation the secret $k = u_1$ can be obtained from $u_1 = (v_{i_1}, \ldots, v_{i_t}) {G'_{.,1}}^{-1}$ where ${G'_{.,1}}^{-1}$ is the first column of G'^{-1}. Hence, in order to retrieve the secret $k = v_0 = u_1$ in KGH scheme, all the columns of the G'^{-1} are not required and instead only first column of the of G'^{-1} is enough. As discussed in Sect. 3, if G is chosen properly then KGH scheme is equivalent to the Sharmir's secret sharing scheme. The structure of G has to be selected as follows,

$$G = \begin{bmatrix} 1 & 1 & 1 & \cdots & 1 \\ 0 & \alpha & \alpha^2 & \cdots & \alpha^{q^\lambda - 1} \\ 0 & \alpha^2 & \alpha^4 & \cdots & \alpha^{(q^\lambda - 1)^2} \\ \vdots & \vdots & \vdots & \vdots & \vdots \\ 0 & \alpha^{k-1} & \alpha^{2k-1} & \cdots & \alpha^{(q^\lambda - 1)(k-1)} \end{bmatrix} \tag{4}$$

Then the components of v in $v = uG$ can also be evalued as $v_i = D(\alpha^i)$, $i = 1, 2, \ldots, n$, where $D(x) = u_1 + u_2 x + u_3 x^2 + \cdots + u_x^{k-1}$ and $v_0 = u_1 = k$. It can be observed from equation for $D(x)$ that KGH scheme using G from (4) is same as Shamir's scheme.

If t out of n shares are available then (1) can be used to retrieve the secret k. However, with some modifications a more efficient method can be derived which is given below [11].

$$k = \sum_{l=1}^{t} v_{i_l} \cdot \beta_l \tag{5}$$

where $v_{i_1}, \ldots, v_{i_t}$ are the t shares and

$$\beta_l = \prod_{1 \le j \le t, j \ne l} \frac{\alpha^{i_j}}{\alpha^{i_j} - \alpha^{i_l}} \tag{6}$$

Similarly, when t shares available in KGH scheme following equation can be used to retrieve the secret, $u = v' \cdot \tilde{G}$ where $v' = (v_{i_1}, \ldots, v_{i_t})$ is the vector derived from available t shares and $\tilde{G} = G'^{-1}$ is derived as per Sect. 2.3. However, we are interested in u_0 only and hence it is sufficient to use following equation instead,

$$u_0 = v' \cdot \tilde{G}_{.,1} \tag{7}$$

where $\tilde{G}_{.,1}$ is the first column of $\tilde{G}$. The similarities between (5) and (7) are easy to observe. The vector $\beta = (\beta_1, \ldots, \beta_t)$ is same as the column $\tilde{G}_{.,1}$. Hence, the first column of $\tilde{G}$ can be calculated using (6). Further, the computations in (6) can be accelerated by precomputing and storing $\frac{\alpha^{i_j}}{\alpha^{i_j} - \alpha^{i_l}}$ in memory. With such improvements, computation of the secret k can be accelerated substantially.

However, we remark that since the above mentioned improvement is based on Shamir's scheme, this improved method can be used with KGH scheme only.

5   Applications

Since Snowden revelations, secret sharing is now used in storage environments, such as backup. The typical operating system is such that people authorized to make a backup have full read access to all data. If 2-out-of-2 secret sharing is used, then two persons are responsible for the backup and one person alone is unable to understand the data being stored. A disadvantage of the use of 2-out-of-2 secret sharing is that it is vulnerable to destruction. When one of the 2 shares is destroyed the data is lost forever. For this reason 2-out-of-3 or in general t-out-of-n (where $t < n$) is recommended.

The paper of Shamir on secret sharing has more than 11,000 citations (Google Scholar). So, there are many other applications for secret sharing than just for backup. A major topic of research is to make secure multiparty computation practical. A major approach to achieve this, is the use of secret sharing. Another application of secret sharing is threshold cryptography. In this setting, no single party is able to use the secret key of a cryptosystem. Another application is the use of secret sharing to make communications robust against both eavesdropping and against an adversary on the network that can modify the data. This topic is usually called Private and Secure Message Transmission, although the word Secure might be better replaced by Reliable to reflect its full power. More recently, secret sharing has been proposed as a technology to achieve e-voting when computers used by voters could be hacked.

Finally note that secret sharing predates the internet. The idea of the use of combinatorics to require that no single party can open a mechanical saves was already mentioned by Liu in his 1968 book (pages 8 and 9) [21].

One important aspect of the secret sharing schemes is the practical values of n and t that may be required for the real world applications. Cramer *et al.* in [22] discussed threshold RSA, which is a type of threshold cryptosystem[1] and similar to threshold secret sharing scheme, with large n. They defined "reasonable values of n" as these "for n up to 4096." While in 2005, this bound on n may have been reasonable, nowadays in the age of social networks, we find *Facebook* groups of close to 6 million users [23] (roughly 2^{23}). Obviously, with *Facebook* having over 1 billion users, the size of the largest group will likely continue to grow. Note that since many real world systems use the same platform (operating system, hardware, etc.), a large value of t is also reasonable. Hence some applications, such as the one mentioned above, may have n in millions and similarly a very

---

[1] A cryptosystem is called a "threshold cryptosystem", if in order to decrypt an encrypted message, several parties (more than some threshold number) must cooperate in the decryption protocol [24].

large value of t. However, the CLM scheme can not be used for such state of the art applications as it can not support $n > 9000$.

6 Conclusion

In this paper, we analyzed the (t, n)-threshold secret sharing scheme proposed by Chen et al. in CANS 2016. First, we showed that this scheme is same as the one proposed earlier by Karnin et al. in 1983 from privacy perspective. Then, we made three more observations: (1) the authors in [9] did not consider the memory and XOR operations needed for obtaining an inverse matrix required during recovery of the secret, (2) XOR operations needed to compute matrix and vector multiplication for the secret recovery are also not considered in their analysis, (3) since the CLM scheme lack efficient decoding algorithm, it can not detect or correct erroneous shares whereas the KGH scheme can be designed for the same. From these observations we conclude that the authors did not provide detailed computational analysis of the CLM scheme in [9] and it is not as efficient in practice as has been suggested by the authors. Further, we proposed and discussed three methods to obtain inverse matrix required during secret recovery. We conclude that the CLM scheme is practical for $n \leq 30$ when all possible inverse matrices are stored in memory whereas up to $n = 9000$ can be obtained if the inverse matrix is calculated on the fly. The third method of obtaining only the first column of the inverse matrix through Lagrange's interpolation can be used only with the KGH scheme. The CLM scheme becomes impractical if $n > 9000$.

Acknowledgment. This publication was made possible by the NPRP award NPRP8-2158-1-423 from the Qatar National Research Fund (a member of The Qatar Foundation). The statements made herein are solely the responsibility of the authors.

References

1. Shamir, A.: How to share a secret. Commun. ACM **22**(11), 612–613 (1979)
2. Blakely, G.: Safeguarding cryptographic keys. In: Proceedings of the National Computer Conference, vol. 48, pp. 313–317 (1979)
3. Kurihara, J., Kiyomoto, S., Fukushima, K., Tanaka, T.: A new (k, n)-threshold secret sharing scheme and its extension. In: Wu, T.-C., Lei, C.-L., Rijmen, V., Lee, D.-T. (eds.) ISC 2008. LNCS, vol. 5222, pp. 455–470. Springer, Heidelberg (2008). https://doi.org/10.1007/978-3-540-85886-7_31
4. Kurihara, J., Kiyomoto, S., Fukushima, K., Tanaka, T.: A fast $(3, n)$-threshold secret sharing scheme using exclusive-or operations. IEICE Trans. Fundam. Electron. Commun. Comput. Sci. **91**(1), 127–138 (2008)
5. Lv, C., Jia, X., Tian, L., Jing, J., Sun, M.: Efficient ideal threshold secret sharing schemes based on exclusive-or operations. In: Proceedings of 4th International Conference on Network and System Security (NSS), pp. 136–143 (2010)

6. Lv, C., Jia, X., Lin, J., Jing, J., Tian, L., Sun, M.: Efficient secret sharing schemes. In: Park, J.J., Lopez, J., Yeo, S.-S., Shon, T., Taniar, D. (eds.) STA 2011. CCIS, vol. 186, pp. 114–121. Springer, Heidelberg (2011). https://doi.org/10.1007/978-3-642-22339-6_14

7. Wang, Y., Desmedt, Y.: Efficient secret sharing schemes achieving optimal information rate. In: Proceedings of IEEE Information Theory Workshop (ITW) 2014, Tasmania, Australia, pp. 516–520, November 2014

8. Chen, L., Camble, P.T., Watkins, M.R., Henry, I.J.: Utilizing error correction (ECC) for secure secret sharing. Hewlett Packard Enterprise Development LP, World Intellectual Property Organisation. Patent Number WO2016048297 (2016). https://www.google.com/patents/WO2016048297A1?cl=en

9. Chen, L., Laing, T.M., Martin, K.M.: Efficient, XOR-based, ideal (t, n)- threshold schemes. In: Foresti, S., Persiano, G. (eds.) CANS 2016. LNCS, vol. 10052, pp. 467–483. Springer, Cham (2016). https://doi.org/10.1007/978-3-319-48965-0_28

10. Karnin, E., Greene, J., Hellman, M.: On secret sharing systems. IEEE Trans. Inf. Theory **29**(1), 35–41 (1983)

11. Beimel, A.: Secret-sharing schemes: a survey. In: Chee, Y.M., et al. (eds.) IWCC 2011. LNCS, vol. 6639, pp. 11–46. Springer, Heidelberg (2011). https://doi.org/10.1007/978-3-642-20901-7_2

12. McEliece, R.J., Sarwate, D.V.: On sharing secrets and Reed-Solomon codes. Commun. ACM **24**(9), 583–584 (1981)

13. Berlekamp, E.R.: Algebraic Coding Theory, Revised edn. Aegean Park Press, Laguna Hills (1984). Previous publisher. McGraw-Hill, New York [1968]. ISBN 0-89412-063-8

14. Berlekamp, E., McEliece, R., van Tilborg, H.: On the inherent intractability of certain coding problems. IEEE Trans. Inf. Theory **24**(3), 384–386 (1978)

15. Worsch, T.: Lower and Upper Bounds for (Sums of) Binomial Coefficients. http://citeseerx.ist.psu.edu/viewdoc/summary?doi=10.1.1.44.9677

16. Bunch, J.R., Hopcroft, J.E.: Triangular factorization and inversion by fast matrix multiplication. Math. Comput. **28**(125), 231–236 (1974)

17. Coppersmith, D., Winograd, S.: Matrix multiplication via arithmetic progressions. J. Symb. Comput. **9**(3), 251–280 (1990)

18. Strassen, V.: Gaussian elimination is not optimal. Numerische Mathematik **13**(4), 354–356 (1969)

19. NTL: A Library for doing Number Theory. http://www.shoup.net/ntl/

20. Forsythe, G.E., Leibler, R.A.: Matrix inversion by a Monte Carlo method. Math. Tables Other Aids Comput. **4**(31), 127–129 (1950)

21. Liu, C.L.: Introduction to Combinatorial Mathematics. McGraw-Hill, New York (1968)

22. Cramer, R., Fehr, S., Stam, M.: Black-box secret sharing from primitive sets in algebraic number fields. In: Shoup, V. (ed.) CRYPTO 2005. LNCS, vol. 3621, pp. 344–360. Springer, Heidelberg (2005). https://doi.org/10.1007/11535218_21

23. http://www.adweek.com/digital/the-25-facebok-groups-with-over-1-million-members/

24. https://en.wikipedia.org/wiki/Threshold_cryptosystem

(Finite) Field Work: Choosing the Best Encoding of Numbers for FHE Computation

Angela Jäschke$^{(\boxtimes)}$ and Frederik Armknecht

University of Mannheim, Mannheim, Germany
{jaeschke,armknecht}@uni-mannheim.de

Abstract. Fully Homomorphic Encryption (FHE) schemes operate over finite fields while many use cases call for real numbers, requiring appropriate encoding of the data into the scheme's plaintext space. However, the choice of encoding can tremendously impact the computational effort on the encrypted data. In this work, we investigate this question for applications that operate over integers and rational numbers using p-adic encoding and the extensions p's Complement and Sign-Magnitude, based on three natural metrics: the number of finite field additions, multiplications, and multiplicative depth. Our results are partly constructive and partly negative: For the first two metrics, an optimal choice exists and we state it explicitly. However, for multiplicative depth the optimum does not exist globally, but we do show how to choose this best encoding depending on the use-case.

Keywords: Fully Homomorphic Encryption · Encoding · Efficiency

1 Introduction

Fully Homomorphic Encryption (FHE) schemes allow arbitrary computations on encrypted data. Though many works have focused on improving the efficiency of FHE schemes themselves, an often overlooked aspect that strongly impacts performance is *how* an FHE scheme is applied. For instance, most FHE schemes operate over finite fields $GF(p^k)$ for an arbitrary prime p and $k \geq 1$, while many use cases call for natural, integer, rational or even real numbers, requiring appropriate encoding of the data into the scheme's plaintext space.

Thus, a naturally arising question is *how to best encode the plaintext data so that later, operations on the encrypted data incur an overhead as small as possible.* In this work, we analyze the effort for FHE computation subject to different p-adic encoding choices like the size of p and the embedding into $\mathbb{Q}$. We base our analysis on the following three natural cost metrics that arise when embedding the plaintext data into the $GF(p^k)$ structure:

Multiplicative Depth: All current FHE schemes are noise-based with each multiplication doubling the amount of noise, and when a noise threshold is

© Springer Nature Switzerland AG 2018
S. Capkun and S. S. M. Chow (Eds.): CANS 2017, LNCS 11261, pp. 482–492, 2018.
https://doi.org/10.1007/978-3-030-02641-7_23

passed, ciphertexts cannot be decrypted correctly anymore. *Bootstrapping* can remove some of the noise before it exceeds this threshold, but this is a very costly operation. Thus, the goal is often to minimize the number of bootstrappings that are necessary by minimizing the number of consecutive multiplications, also referred to as *multiplicative depth*. For this reason, multiplicative depth has been the standard cost metric and is naturally part of our analysis.

Number of Field Multiplications: FHE multiplications are much more expensive than additions for all current schemes, so keeping track of this number is an obvious choice. In addition, the multiplicative depth in p-adic encoding quickly becomes so large that bootstrapping is unavoidable, so that minimizing the total number of multiplications can speed up performance significantly.

Number of Field Additions: For all schemes today, field additions cost almost nothing compared to field multiplications. However, there is no theoretical reason why this must be the case, so we include this metric because it might be valuable in the future for a different kind of scheme.

Our contributions are:

- We derive a generic formula that allows to express each digit when adding two numbers in p^k-adic encoding[1].
- Based on the generic formula, we analyze the costs for adding two encrypted natural numbers in p^k-adic encoding with the following results:
  1. For the required number of field additions or field multiplications when $k = 1$, the efforts for additions and multiplications of encrypted integers strictly increase with p, making $p = 2$ by far the best choice.
  2. For the depth metric when $k = 1$, the optimal p depends heavily on the use case and our formulas show how to compute it.
  3. For p^k-adic encoding with $k > 1$, we show that performance is always worse compared to p-adic encoding, so setting $k = 1$ is the best choice.
- We then extended our analysis to negative and rational numbers.

2   Related Work

This section presents only the most relevant publications, a more comprehensive version can be found in the extended paper [10]. The recent increase in papers regarding encoding for FHE shows its importance: [5] encodes rational numbers through continued fractions (only positive rationals and evaluating linear multivariate polynomials), whereas [6] focuses on efficiently embedding the computation into a single large plaintext space. A work that explores similar ideas as [6] and offers an implementation is [8]. [1] allows floating point numbers, and [3] gives a high-level overview of arithmetic methods for FHE, but restricted to positive numbers. In [9], arithmetic operations and different binary encodings for

---

[1] The term p^k-adic encoding denotes the natural extension of p-adic encoding to the field $GF(p^k)$ for $k \geq 1$ and is explained in Sect. 6.

rational numbers are examined and compared in their effort. [2] explores a non-integral base encoding, and [13] presents different arithmetic algorithms including a costly division, though apparently limited to positive numbers. Lastly, [4] allows approximate operations by utilizing noise from the encryption itself. To our knowledge, there are no papers concerned with the costs of encoding in a base other than $p = 2$ except [11], which exclusively analyzes [12] and uses different cost metrics. The latter also presents a formula for the carry of a half adder, but merely considers $GF(p^k)$ for $k = 1$ in the context of homomorphically computing the decryption step (needed for bootstrapping) of their variation of [7], and does not include an effort analysis.

3   Formula for Computing Carry Values over $\mathbb{Z}_p$

In this section and the following Sect. 4, we lay the foundation for the effort analysis starting from Sect. 5. We derive in this section the formulas for the digits of the sum of two numbers in p-adic encoding. We will see that the carry is particularly important, so we investigate it more closely in Sect. 4.

Let $A = a_n a_{n-1} \ldots a_1 a_0$ and $B = b_n b_{n-1} \ldots b_1 b_0$ be two p-adically encoded natural numbers. If we wish to add these numbers in this encoding, we can write

$$
\begin{array}{r}
a_n\ a_{n-1}\ \ldots\ a_2\ a_1\ a_0 \\
+\ \ \ \ \ \ b_n\ b_{n-1}\ \ldots\ b_2\ b_1\ b_0 \\
\hline
=\ c_{n+1}\ c_n\ c_{n-1}\ \ldots\ c_2\ c_1\ c_0
\end{array}
\tag{1}
$$

To homomorphically evaluate a function on encrypted data, we need to express the result as a polynomial in the inputs – we need to be able to write

$$
c_i = c_i(a_n, b_n, a_{n-1}, b_{n-1}, ..., a_1, b_1, a_0, b_0)
\tag{2}
$$

for any i, where $c_i(\ldots)$ refers to some polynomial. Clearly, it holds that $c_i = a_i + b_i + r_i$, where $r_0 = 0$, and for $i > 0$, r_i is the *carry* from position $i - 1$. Our goal in this section is to express $r_i(a_{i-1}, b_{i-1}, r_{i-1})$ as a polynomial, which will constitute Theorem 1. Addition is defined mod p, and we will often write r_i instead of $r_i(a_{i-1}, b_{i-1}, r_{i-1})$ for simplicity.

Theorem 1. *The formula for computing the carry* $r_i(a_{i-1}, b_{i-1}, r_{i-1})$ *is*

$$
r_i(a, b, r) = \sum_{k=1}^{p-1} \left(l_k(b) \cdot \sum_{j=1}^{k} l_{p-j}(a) \right) + r \cdot (p-1) \cdot l_{p-1}(a+b) := f_1(a,b) + r \cdot f_2(a,b)
$$

where $l_i(x) = \prod\limits_{j=0, j \neq i}^{p-1} (x - j)$. *This polynomial is unique in that there is no other polynomial of smaller or equal degree which also takes on the correct values for* r_i *at all points* $(a_{i-1}, b_{i-1}, r_{i-1})$ *with* $a_{i-1}, b_{i-1} \in \{0, \ldots, p-1\}, r_{i-1} \in \{0, 1\}$.

The proof is given in the extended version of this paper [10].

4   The Effort of Computing the Carry

In this section, we present the effort required to compute each digit c_i when adding two natural numbers encoded p-adically. Due to space constraints, the detailed derivations of these numbers can be found in the extended version of this paper [10]. Recall that $c_i = a_i + b_i + r_i$, and r_i can be computed as $r_i = f_1(a_{i-1}, b_{i-1}) + r_{i-1} \cdot f_2(a_{i-1}, b_{i-1})$. Note that there cannot be any cancellation between the terms of $f_1(a_{i-1}, b_{i-1})$ and $r_{i-1} \cdot f_2(a_{i-1}, b_{i-1})$ due to the variable r_{i-1}. Thus, to compute c_i, we must compute f_1 and f_2, and additionally perform 3 field additions and 1 multiplication. Regarding depth, note that $r_0 = 0$ and $r_1 = f_1(a_0, b_0)$ (thus having the depth of f_1), and subsequent r_i have a depth of $\max\{D(f_1), \max\{D(r_{i-1}), D(f_2)\} + 1\}$. Using this formula for r_2, we get

$$D(r_2) = \max\{\lceil \log_2(p) \rceil, \lceil \log_2(p-1) \rceil\} + 1 = \lceil \log_2(p) \rceil + 1$$

From here on, it is clear that $D(r_i) > D(f_1) \geq D(f_2)$, so the depth will increase by 1 with each i, leaving us with a total depth of $D(r_i) = \lceil \log_2(p) \rceil + i - 1$. The effort required to compute each digit c_i can be found in Table 1.

Table 1. Effort for computing c_i.

Effort	Field additions	Field multiplications	Depth
f_1	$4p - 6$	$2p \cdot \log_2(p) + p - 2 \cdot \log_2(p) - 3$	$\lceil \log_2(p) \rceil$
f_2	$p - 1$	$p - 2$	$\lceil \log_2(p-1) \rceil$
Additional	3	1	$+1$
Total c_i	$\mathbf{5p - 4}$	$\mathbf{2p \log_2(p) + 2p - 2 \log_2(p) - 4}$	$\lceil \log_2(p) \rceil + i - 1$

Special Cases: The effort for $c_0 = a_0 + b_0$ is only 1 field addition, and that for $c_1 = a_1 + b_1 + r_1 = a_1 + b_1 + f_1(a_0), b_0)$ is $4p + 4$ additions, $2p \cdot \log_2(p) + p - 2 \cdot \log_2(p) + 1$ multiplications, and $\lceil \log_2(p) \rceil$ depth. Another special case is $c_{n+1} = r_{n+1}$, which has 2 field additions less than the other $c_i, i > 1$.

5   Cost Analysis for Encrypted Natural Numbers

5.1   The Cost of Adding Two Natural Numbers

Suppose we have some natural number $x > p$ (in decimal representation). Thus, x will be encoded with $2 \leq \ell := \lfloor \log_p(x) \rfloor + 1$ digits p-adically and the result will have $\ell + 1$ digits. We have $c_0 = a_0 + b_0$ with an effort of 1 addition. Next, we have $c_1 = a_1 + b_1 + r_1 = a_1 + b_1 + f_1(a_0, b_0)$ with an effort of $4p - 4$ additions, $2p \cdot \log_2(p) + p - 2 \cdot \log_2(p) - 3$ multiplications and a depth of $\lceil \log_2(p) \rceil$. The last digit $c_\ell = r_\ell$ has a cost of $5p - 6$ additions, $2p \cdot \log_2(p) + 2p - 2 \cdot \log_2(p) - 4$ multiplications, and a depth of $\lceil \log_2(p) \rceil + 1$. The remaining $\ell - 2$ middle digits c_i have the normal effort of $5p - 4$ additions, $2p \cdot \log_2(p) + 2p - 2 \cdot \log_2(p) - 4$ multiplications and a depth of $\lceil \log_2(p) \rceil + i - 1$.
In total, the cost of adding two ℓ-digit numbers, $\ell > 2$, is:

- $9p - 9 + (\ell - 2) \cdot (5p - 4) = (5\ell - 1) \cdot p - (3\ell + 2)$ **field additions**
- $4p \cdot \log_2(p) + 3p - 4 \cdot \log_2(p) - 7 + (\ell - 2) \cdot (2p \cdot \log_2(p) + 2p - 2 \cdot \log_2(p) - 4)$
  $= 2\ell \cdot p \cdot \log_2(p) + (2\ell - 1) \cdot p - 2\ell \cdot \log_2(p) - 4 \cdot \ell + 1$ **field multiplications**
- **A multiplicative depth of** $\lceil \log_2(p) \rceil + \ell - 1$.

Note that for $0 \le x \le p - 1$, we only require one digit, which incurs a lower effort: **In total, the cost of adding two 1-digit numbers is $4p-5$ additions,** $2p \cdot \log_2(p) + p - 2 \cdot \log_2(p) - 3$ **multiplications and a depth of** $\lceil \log_2(p) \rceil$.

Recalling $\ell := \lfloor \log_p(x) \rfloor + 1$, we can now state the main result of this paper:

Theorem 2. *Using total number of additions or multiplications (or a balance between total number of multiplications and depth) as the cost metric, $p = 2$ is the most efficient encoding for adding two natural numbers in p-adic encoding.*

Proof. We can see that while the required encoding length $\ell = \lfloor \log_p(x) \rfloor + 1 = \lfloor \frac{\log_2(x)}{\log_2(p)} \rfloor + 1$ only decreases logarithmically, the effort grows with p as $\mathcal{O}(\ell \cdot p) = \mathcal{O}((\lfloor \frac{\log_2(x)}{\log_2(p)} \rfloor + 1) \cdot p) \approx \mathcal{O}(p + \frac{p}{\log_2(p)})$ (for additions) and as $\mathcal{O}(\ell \cdot p \cdot \log_2(p)) = \mathcal{O}((\lfloor \frac{\log_2(x)}{\log_2(p)} \rfloor + 1) \cdot p \cdot \log_2(p)) \approx \mathcal{O}(p \cdot \log_2(p) + p)$ (for multiplications). The depth $\lceil \log_2(p) \rceil + \ell - 1 = \lceil \log_2(p) \rceil + \lfloor \frac{\log_2(x)}{\log_2(p)} \rfloor$ also increases logarithmically.

We again point out that if the function being evaluated is known beforehand, choosing p so large that computations do not wrap around $\mod p$ is likely to be faster – however, this is not *Fully* Homomorphic Encryption but rather *Somewhat* Homomorphic Encryption. Theorem 2 holds for p-adic encoding with true FHE.

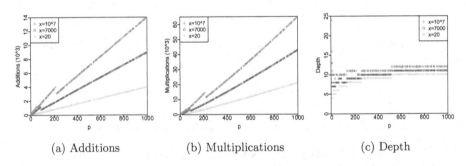

(a) Additions   (b) Multiplications   (c) Depth

Fig. 1. Number of field additions, multiplications and depth for adding $x = 20/7000/10^7$ to a number of same size. The x-axis is the encoding base p, the y-axis is number of operations/depth, and the plots correspond to the three numbers.

We illustrate this Theorem through Fig. 1, which shows the effort as p grows for selected values of x. We see that the number of additions, multiplications and the depth increase significantly as the encoding base p increases. Note that the jags in the first two diagrams occur when the base prime becomes so large that one digit less is required for encoding than under the previous prime, so the effort drops briefly before increasing again. The diagram for depth shows us an interesting phenomenon that is hidden in the asymptotic analysis: For low primes, the required number of digits dominates the total depth cost. This problem becomes more pronounced the larger the encoded number is, and vanishes after the first few primes as the expected asymptotic cost takes over. This means that if depth is the only cost metric that is being considered, choosing a slightly larger prime than 2 yields better results at the cost of significantly increased multiplications. Also, the optimal choice of p depends heavily on the numbers that are being encoded. For example, in Fig. 1c, the depth-optimal choices for adding x would be $p = 3$ for $x = 20$, $p = 7$ for $x = 7000$, and $p = 29$ for $x = 10^7$.

5.2 The Cost of Multiplying Two Natural Numbers

We now analyze the cost of multiplying two natural numbers in p-adic encoding with the standard multiplication algorithm. In performing this multiplication, there are two main steps: First, we perform a one digit multiplication of each b_i with all of $a_{\ell-1}a_{\ell-2}\ldots a_1 a_0$, shifting one space to the left with each increasing i. In the second step, we add the rows we obtained using the addition from the previous subsection as a building block. For the first step, except in the case of $p = 2$ (where $b_i \in \{0, 1\}$, so the rows are $(a_{\ell-1} \cdot b_i)\ldots(a_1 \cdot b_i)(a_0 \cdot b_i)$), this actually requires some computational effort because we have a carry r_i into the next digit. Similarly to Theorem 1, we can obtain the formula for this carry digit through a 3-fold Lagrange approximation over the variables a_i, b_i and r_i.

The second step consists of adding all the rows that we computed in the first step. We apply the improvement from [9] where we copy the digits of the upper row over the blank spaces on the right to the result, and apply a depth-optimal ordering in adding the rows. The exact formula can be found in the extended version [10], we instead illustrate our results here through Fig. 2.

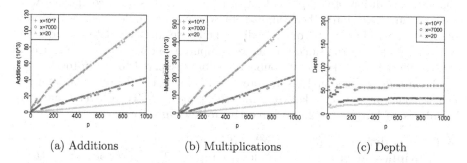

(a) Additions	(b) Multiplications	(c) Depth

Fig. 2. Number of field additions, field multiplications and multiplicative depth for multiplying $x = 20/7000/10^7$ to a number of same size.

We can see that for additions and multiplications, the effort is lowest at $p = 2$ and increases with p, though there are again some sharp drops when the required number of digits decreases. As expectes, the depth issue has propagated from addition, which we used as a building block in multiplication: The best depth for $x = 20$ would be $p = 23$, the best depth for $x = 7000$ would be $p = 89$, and the best depth for $x = 10^7$ would be $p = 59$. We would like to point out that these values are not the same values that were optimal for addition (e.g., $p = 89$ is far from optimal for adding $x = 7000$) - thus, if one were to use depth as the sole metric, the optimal choice of p not only depends on the size of the numbers one is working with, but also on the number of additions vs. multiplications one wants to perform on these inputs. In the context of outsourced information, it is also important to note that optimizing the choice of p in this way could leak unwanted information, depending on the specific outsourcing scenario.

6   Using $GF(p^k)$ as Encoding Base

We now generalize our analysis to arbitrary finite fields as encoding bases. Much in the same way as in Sects. 4 and 5, we have also analyzed the effort incurred when using $GF(p^k)$ for a prime p and a $k > 1$ as an encoding base. First, recall that $GF(p^k) \cong \mathbb{Z}_p[X]/(f(x))$ with $f(x)$ irreducible of degree k. We embed a decimal number between 0 and $p^k - 1$ into $GF(p^k)$, whose elements are polynomials over $\mathbb{Z}_p$, through the insertion homomorphism: The element $a = \sum_{i=0}^{k-1} \alpha_i X^i \in GF(p^k)$ (with $\alpha_i \in \mathbb{Z}_p$) encodes the number $\tilde{a} = \sum_{i=0}^{k-1} \alpha_i p^i \in \mathbb{N}$. Generalizing this to numbers larger than $p^k - 1$ is straightforward: We will represent a number $\tilde{a} = \sum_{j=0}^{n} \tilde{a}_j (p^k)^j \in \mathbb{N}$ as $a_n a_{n-1} \cdots a_1 a_0$ where $a_j \in GF(p^k)$.

We now analyze the effort of adding two natural numbers in this encoding. Intuitively, we do not expect this to perform better than the encoding through $\mathbb{Z}_p$: The carry bit should roughly have the same effort as for $\mathbb{Z}_{p'}$ with p' of size comparable to p^k, but the addition is now more complicated. Concretely, the native addition of $GF(p^k)$ is that of $(\mathbb{Z}_p)^k$, i.e., it is done component-wise with no carry-over into other components, whereas we would need the addition of $\mathbb{Z}_{p^k}$ to natively support our encoding. Thus, we must emulate the addition $c_i = a_i + b_i + r_i$ in the same way as we compute the carry bit, so we expect a similar effort here and at least double the effort compared to $\mathbb{Z}_{p'}$ in total.

The results of this analysis are presented below – the detailed computation can be found in the extended version of this paper [10]. **To add two natural numbers of lenght $\ell \geq 2$, we have the following effort:**

- **Field additions:** $= (3\ell - 1) \cdot p^{2k} + (6\ell - 4) \cdot p^k - 6\ell$
- **Multiplications:** $= (6\ell - 2) \cdot p^k \cdot \log_2(p^k) + (4\ell - 2) \cdot p^k - 2\ell \cdot \log_2(p^k) - 11\ell + 3$
- **Constant multiplications:** $= (2\ell - 1) \cdot p^{2k} + \ell - 1$
- **Multiplicative depth:** $\lceil \log_2(p^k - 1) \rceil + \ell$

The case where the inputs are have only one digit is again slightly less expensive and has been omitted for brevity. We now compare the calculated effort to:

1. Encoding the number in base p instead of p^k and performing the addition.
2. Encoding the number in base p' with p' close to p^k.

Figure 3 shows the effort of adding two numbers of same size ($x = 20/7000/10^7$) in p^k-adic encoding for p^k up to 1000. Blue crosses are $\mathbb{Z}_p$, pink circles p^2, yellow triangles p^3, and the black square groups all bases p^k with $k \geq 4$, since the primes p with $p^k \leq 1000$ for increasing k become very few. We see that the p^k-encoding performs poorly regarding all metrics, and using $\mathbb{Z}_p$ as an encoding base is the better choice. Recall from Sect. 5.1 that the smaller the encoding base p for a plaintext space of $\mathbb{Z}_p$, the smaller the cost in terms of ciphertext additions and multiplications, and that the optimal base in terms of multiplicative depth varies. However, the factor that induces this variation is the required encoding length, and since we can choose a prime p' close to p^k (thus requiring roughly the same length) which requires much less effort as shown in Fig. 3, there is no case where choosing p^k as an encoding base with $k > 1$ brings any benefit, so we do not continue with its analysis.

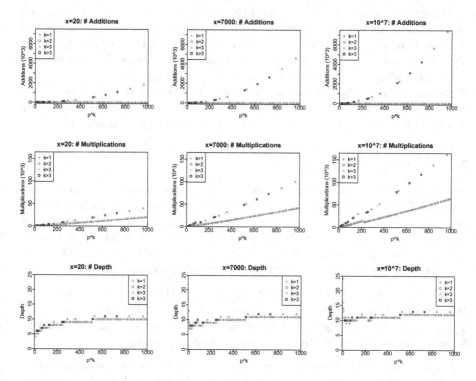

Fig. 3. Number of field additions, field multiplications and multiplicative depth for multiplying $x = 20$ (first column)/$x = 7000$ (second column)/$x = 10^7$ (right column) to a number of same size for encoding base $GF(p^k)$.

7 Rational Numbers and Integers

7.1 Representing Rational Numbers by Scaling

Let the encoding base p be an arbitrary prime. Given a rational number that we wish to encode (and assuming for the moment that negative numbers are no problem), we need to transform this rational into an integer, which we can then encode p-adically in the next step. The most straightforward approach is to introduce a scaling factor by picking a precision (i.e., there are σ p-adic digits after the point) σ with which we want to work with in the following and multiplying the rational with p^σ, rounding to obtain an integer to encode.

The importance of choosing the scaling factor as a power of p rather than any other number is as follows: Suppose that we have two rational numbers A and B which we scale and round to $\tilde{A} = A \cdot p^\sigma$ and $\tilde{B} = B \cdot p^\sigma$. After encoding and encrypting the individual digits, multiplying yields $\tilde{A} \cdot \tilde{B} = A \cdot B \cdot p^{2\sigma}$. Thus, after decrypting, the data owner needs to know what number to divide the result by, leaking unwanted information about the function that was applied, which might be the computing party's secret. Also the required number of digits increases to accommodate the extra precision digits, making all future computations less efficient. However, if we have a number in p-adic encoding, deleting the last σ digits corresponds to dividing by p^σ and truncating the result. This way, using p^σ as the scaling factor, the computing party can delete the σ least significant digits after each multiplication and thus keep the precision at a constant σ bits, increasing efficiency (by using less digits) and privacy (because the data owner now divides the result by p^σ regardless of the function that was applied).

7.2 Encoding Integers

Having seen how to transform rationals into integers, we need to incorporate negative numbers into our p-adic encoding. Generalizing from $p = 2$, for which these encodings are well known, we investigate two main approaches: p's Complement and Sign-Magnitude. Note that this question has been extensively studied in [9] but for the case $p = 2$ only. We state shortly how the results extend to the case $p > 2$, with a more detailed analysis in the extended version [10].

p's Complement: In this encoding, elements have the form $a_n \dots a_0$ with $a_i \in \{0, 1, \dots, p-1\}$ for $i = 0, \dots, n-1$, and $a_n \in \{0, 1\}$, where $x = -a_n \cdot p^n + \sum_{i=0}^{n-1} a_i \cdot p^i$. This means that the first digit encodes either 0 or $-p^n$ and the following digits correspond to the "normal" p-adic encoding.

Addition: The effort of adding two numbers in p's Complement encoding is comparable to twice the effort for adding two natural numbers derived in Sect. 5.1, except that the depth is twice as large. For $p = 2$, it is almost exactly the same effort as adding two natural numbers in binary encoding.

Multiplication: Multiplication p's Complement roughly requires the same effort as a natural number with twice as many digits, i.e., multiplying a natural number

x with ℓ digits in p's Complement encoding requires roughly the same effort as multiplying x^2 (which has 2ℓ digits) in "regular" p-adic encoding.

Sign-Magnitude: For the second encoding that we consider, Sign-Magnitude, the absolute value of the number is encoded p-adically as a natural number, and there is an extra digit (the MSB) which determines the sign. Concretely, elements in this encoding have the form $a_n a_{n-1} \ldots a_1 a_0$ with $a_i \in \{0, 1, \ldots, p-1\}$ for $i = 0, \ldots, n-1$, and $a_n \in \{0, 1\}$, where $x = (-1)^{a_n} \cdot \sum_{i=0}^{n-1} a_i \cdot p^i$. It is easy to see that for positive numbers, this encoding is the same as p's complement. This encoding suffers from having two representations of 0: $00 \ldots 0$ and $100 \ldots 0$.

Addition: Adding two numbers in Sign-Magnitude encoding costs roughly one p-adic addition, 2 comparisons (each costing about three additions) and 4 subtractions (about the same cost as addition). This yields 11 additions in "regular" p-adic encoding and is significantly more costly than $p's$ Complement encoding.

Multiplication: The effort of multiplying two numbers in Sign-Magnitude encoding is roughly the same as multiplying them with "regular" p-adic encoding.

Hybrid Encoding: We see that the choice of encoding can make a big difference in performance. As p's Complement addition is more efficient than Sign-Magnitude, but the latter is more efficient for multiplication, a hybrid approach like in [9] would be the best choice: One does all additions in p's Complement, and for multiplication switches the encoding to Sign-Magnitude. Using this, one can have roughly the same operation cost as for natural numbers in p-adic encoding (slightly more for additions), plus the cost of switching between encodings, which is roughly that of one p-adic addition. Of course, since this already holds true for natural numbers in p-adic encoding, the choice $p = 2$ by far incurs the least amount of field additions and multiplications in these two encodings and the hybrid encoding also, while the optimal depth choice remains variable.

References

1. Arita, S., Nakasato, S.: Fully homomorphic encryption for point numbers. IACR Cryptology ePrint Archive 2016/402 (2016)
2. Bonte, C., Bootland, C., Bos, J.W., Castryck, W., Iliashenko, I., Vercauteren, F.: Faster homomorphic function evaluation using non-integral base encoding. IACR Cryptology ePrint Archive 2017/333 (2017)
3. Chen, Y., Gong, G.: Integer arithmetic over ciphertext and homomorphic data aggregation. In: CNS (2015)
4. Cheon, J.H., Kim, A., Kim, M., Song, Y.: Homomorphic encryption for arithmetic of approximate numbers. IACR Cryptology ePrint Archive 2016/421 (2016)
5. Chung, H., Kim, M.: Encoding rational numbers for FHE-based applications. IACR Cryptology ePrint Archive 2016/344(2016)
6. Costache, A., Smart, N.P., Vivek, S., Waller, A.: Fixed point arithmetic in SHE scheme. IACR Cryptology ePrint Archive 2016/250 (2016)
7. van Dijk, M., Gentry, C., Halevi, S., Vaikuntanathan, V.: Fully homomorphic encryption over the integers. In: Gilbert, H. (ed.) EUROCRYPT 2010. LNCS, vol. 6110, pp. 24–43. Springer, Heidelberg (2010). https://doi.org/10.1007/978-3-642-13190-5_2

8. Dowlin, N., Gilad-Bachrach, R., Laine, K., Lauter, K., Naehrig, M., Wernsing, J.: Manual for using homomorphic encryption for bioinformatics. Technical report. MSR-TR-2015-87, Microsoft Research (2015)

9. Jäschke, A., Armknecht, F.: Accelerating homomorphic computations on rational numbers. In: Manulis, M., Sadeghi, A.-R., Schneider, S. (eds.) ACNS 2016. LNCS, vol. 9696, pp. 405–423. Springer, Cham (2016). https://doi.org/10.1007/978-3-319-39555-5_22

10. Jäschke, A., Armknecht, F.: (Finite) field work: choosing the best encoding of numbers for FHE Computation. IACR Cryptology ePrint Archive 2017/582 (2017)

11. Kim, E., Tibouchi, M.: FHE over the integers and modular arithmetic circuits. In: CANS, pp. 435–450 (2016)

12. Nuida, K., Kurosawa, K.: (Batch) fully homomorphic encryption over integers for non-binary message spaces. In: Oswald, E., Fischlin, M. (eds.) EUROCRYPT 2015. LNCS, vol. 9056, pp. 537–555. Springer, Heidelberg (2015). https://doi.org/10.1007/978-3-662-46800-5_21

13. Xu, C., Chen, J., Wu, W., Feng, Y.: Homomorphically encrypted arithmetic operations over the integer ring. In: Bao, F., Chen, L., Deng, R.H., Wang, G. (eds.) ISPEC 2016. LNCS, vol. 10060, pp. 167–181. Springer, Cham (2016). https://doi.org/10.1007/978-3-319-49151-6_12

An Efficient Attribute-Based Authenticated Key Exchange Protocol

Suvradip Chakraborty[1($\boxtimes$)], Y. Sreenivasa Rao[2],
and Chandrasekaran Pandu Rangan[1]

[1] Department of Computer Science and Engineering,
Indian Institute of Technology Madras, Chennai, India
{suvradip,rangan}@cse.iitm.ac.in
[2] Indian Institute of Information Technology Design and Manufacturing Kurnool,
Kurnool, India
sreenivasarao@iiitk.ac.in

Abstract. In this paper, we present a new and efficient construction of an Attribute-Based Authenticated Key Exchange (ABAKE) protocol, providing fine-grained access control over data. The state-of-the-art constructions of ABAKE protocols rely on extensive pairing and exponentiation operations (both polynomial in the size of the access policies) over appropriate groups equipped with bilinear maps. Our new construction of ABAKE protocol reduces the number of pairing operations to be *constant* (to be precise only 7) and the number of exponentiations to be linear in the number of clauses in the disjunctive normal form representing the general access policies. The main workhorse of our ABAKE construction is an Attribute-Based Signcryption (ABSC) scheme with constant number of pairings (only 7), which we construct. This also gives the first construction of ABSC schemes with constant number of pairings for general purpose access policies in the standard model. Our ABAKE protocol is also *round-optimal*, i.e., it is a single round protocol consisting of only a single message flow among the parties involved, and is *asynchronous* in nature, i.e., the message sent by one party does not depend on the incoming message from the other party. The security of our ABAKE protocol is proved under a variant of the Bilinear Diffie-Hellman Exponent assumption, in the Attribute-Based extended Canetti-Krawzyck (ABeCK) model, which is an extension of the extended Canetti-Krawzyck (eCK) model for attribute-based framework.

Keywords: Attribute-based signcryption
Authenticated key exchange · Bilinear pairing
Attribute-Based extended Canetti-Krawzyck (ABeCK) model

1 Introduction

Attribute-Based Encryption (ABE), introduced by Sahai and Waters [12] provides fine grained access control over encrypted data. Existing ABE constructions falls under two broad categories: (i) key-policy attribute-based systems

© Springer Nature Switzerland AG 2018
S. Capkun and S. S. M. Chow (Eds.): CANS 2017, LNCS 11261, pp. 493–503, 2018.
https://doi.org/10.1007/978-3-030-02641-7_24

[5,9], in which users' secret keys are associated with access policy (a.k.a. access structure) over an universe of attributes and the ciphertexts are associated with sets of attributes and (ii) ciphertext-policy attribute-based systems [1,11,14], in which users' private keys are associated with the attributes and the ciphertexts are associated with access policies. In this work, we consider ciphertext-policy attribute-based systems.

Attribute-Based Authenticated Key Exchange (ABAKE) is a new variant of the AKE protocols that allows users to establish a shared key and achieve mutual authentication over an insecure channel by using their attributes (unlike in the PKI settings where the users authenticate each other using their public keys), thereby providing fine-grained access control over transmitted data. Attribute-based key exchange finds its application in distributed collaborative systems where it is more convenient for users to communicate with other users using their roles or responsibilities which can be described by attributes. It is also applicable in scenarios like interactive chat rooms, online forums where a user can have read/write access to threads only if they have desired attributes. Hence an authenticated key exchange protocol that critically uses attributes can be employed in these settings. Despite its tremendous potential real-world applicability, the existing solutions for ABAKE protocols do not cater to its need. Most of the ABAKE protocols directly use an ABE as their underlying building block, and hence are computationally very expensive. In particular, for all the existing ABAKE solutions the pairing and exponentiation operations are polynomial in the size of the share generating matrix representing the access policy of users. This presents a major block for real-world deployment of these protocols. Hence, our initial quest was to construct an ABAKE protocol where the number of pairings is constant or even sub-linear in the size of the access policy.

To this end, we show how to construct an ABAKE protocol where the number of pairing operations is *independent* of the size of the access policy. In particular, our ABAKE construction uses only a *constant* number of pairings, and the number of exponentiations is *linear* in the number of clauses of the Disjunctive Normal Form (DNF) access formula representing the access policies of users.

1.1 Our Contribution

The main contributions of our paper are outlined as below:

1. We propose a new ABAKE protocol that employs only *constant* number of pairings (to be precise only 7 per user). The exponentiation operations are linear in $(m + n)$, where m denotes the number of clauses of the DNF access formula representing the access policies of users, and n denotes the size of the signing attribute set. The proposed ABAKE protocol is also *round-optimal*, i.e., it only needs *two-pass* interaction in a session. Our ABAKE protocol is proven secure in the ABeCK model [15], which is a natural extension of the eCK model [6] for attribute-based settings, assuming the intractability of the Gap modified Bilinear Diffie-Hellman Exponent (GmBDHE) problem. A comparison of our protocol with the existing attribute-based two-party and group

key exchange protocols are shown in Table 1. We leave open the problem of constructing an ABAKE protocol in the ABeCK model in the standard model achieving the same or comparable level of efficiency/ computational complexity as our construction. The related works relevant to ABAKE protocols is presented in the full version of our paper.

2. We propose a new construction of a ciphertext-policy ABSC (CP-ABSC) scheme with constant number of pairings. To the best of our knowledge, our CP-ABSC construction is the *first* such construction employing only a constant number of pairings. We use the ABSC scheme for our ABAKE construction, since one of the crucial requirement of an ABAKE protocol is mutual authentication of parties involved in a session. In Table 2, we compare our proposed CP-ABSC scheme with the state-of-the-art CP-ABSC schemes. The related works regarding CP-ABSC schemes is given in the full version of our paper.

Table 1. Comparison with existing ABAKE protocols

Scheme	Type of ABAKE	No. of rounds	No. of pairings (each party)	Basic building blocks	Security model	Assumptions
[4]	ABGKE	1	$2(\ell_M \times n_M) + 3$	IND-CCA secure ABE [1]	BR	GGM, RO
[13]	ABGKE	2	$\ell_M + 5$	ABSC	ABCK	CDH, RO
[15]	2-party ABAKE	1	$\ell_M^2 \times n_M$	Waters ABE [14]	ABeCK	GBDH, RO
[16]	2-party ABAKE	1	$4\ell_M + 2$	Waters ABE [14], PRF, Ext., OTS	ABCK	DPBDHE, Std
Proposed ABAKE	2-party ABAKE	1	7	proposed ABSC	ABeCK	GmBDHE, RO

BR: Bellare Rogaway, GGM: Generic Group Model, RO: Random Oracle, Std: Standard model, PRF: Pseudo-random Function, Ext.: Randomness extractor, GBDH: Gap Bilinear Diffie-Hellman Assumption, DPBDHE: Decisional Parallel Bilinear Diffie-Hellman Exponent Assumption, GmBDHE: Gap Modified Bilinear Diffie-Hellman Exponent Assumption, OTS: One-time signature, ABGKE: Attribute-based Group Key Exchange, $\ell_M(n_M)$: maximum number of rows (columns) of M.

Table 2. Comparison with existing CP-ABSC schemes

Scheme	Ciphertext size	Signcryption cost	Unsigncryption cost	
		Exp	Exp	Pairings
[2]	$(u_e + \ell_s + v_k + 2)B_G + s_{ot} + v_k + msg$	$u_e + 4\ell_s + v_k + 3$	0	$u_e + \ell_s + v_k + 3$
[7]	$(\ell_s + w_s + \ell_e + 3)B_G + msg$	$2\ell_s w_s + 3\ell_s + 2\ell_e + 4$	$2\ell_s w_s + w_s + \phi_e$	$\ell_s w_s + w_s + 2\phi_e + 4$
[11]	$(\ell_s + \ell_e + 4)B_G + B_{tt} + msg$	$4\ell_s + 2\ell_e + 7$	$\ell_s + 2\phi_e + 2$	$\ell_s + 5$
[10]	$(2\ell_s + 2\ell_e + 4)B_G + msg$	$4\ell_s + 3\ell_e + 12$	$\ell_s + \phi_e + 11$	$2\ell_s + 2\phi_e + 6$
Proposed ABSC	$(m + 3)B_G + B_Z + msg$	$m + \ell_s + 6$	$\ell_s + 4$	7

$\ell_s(\ell_e)$: number of attributes in a signing (encryption) access policy, u_e : size of encryption attribute universe, v_k : bit length of verification key of one-time signature scheme, ϕ_e : number of encryption attributes required in unsigncryption process, w_s : number of columns in signing linear secret-sharing matrix, $B_G(B_Z)$: bit length of an element of the group $\mathbb{G}$ ($\mathbb{Z}_p$), B_{tt}: bit length of time stamp, msg: bit length of a message or plaintext, s_{ot}: bit length of the signature generated by one-time signature scheme, m: number of clauses in the encryption DNF access policy.

2   Background

Notation: Throughout this work, we denote the security parameter by κ. We denote by $x \xleftarrow{R} X$ that x is randomly chosen from the finite set X according to uniform distribution. We use $[k]$ to denote the set $\{1, 2, \ldots, k\}$. We assume familiarity with the definitions of lagrange interpolation, bilinear pairings over elliptic curve groups. We refer the readers to the full version of our paper for these definitions.

Access Policy: Let U be the universe of attributes. Let 2^U be the collection of all non-empty subsets of U.

- Each nonempty subset of 2^U is called an access policy.
- An access policy $\Gamma \subset 2^U$ is called a monotone access policy if it satisfies the following property: for any $A \in 2^U$, $A \supseteq B$ for some $B \in \Gamma$ implies $A \in \Gamma$. That is, $A \in \Gamma \iff A \supseteq B$ for some $B \in \Gamma$.
  Note: From now on, by an access policy, we mean monotone access policy.
- For an attribute set $A \subset U$, if $A \in \Gamma$, we say that A satisfies the access policy Γ which is denoted by $\Gamma(A) = true$. (If $\Gamma(A) = false$, we say that A does not satisfy Γ.) Hence, $\Gamma(A) = true \iff A \in \Gamma \iff A \supseteq B$ for some $B \in \Gamma$.

Every access policy can be represented in *disjunctive normal form* (DNF) as follows. If $\Gamma := \{B_1, B_2, \ldots, B_k\}$, one can equivalently represent Γ as $\Gamma := B_1 \vee B_2 \vee \cdots \vee B_m$. In this case, $\Gamma(A) = true \iff A \supseteq B_k$ for some $k \in \{1, 2, \ldots, m\}$.

Complexity Assumption: The complexity assumption required for our paper is the Gap Modified q-BDHE (Gmq-BDHE) assumption, which is the gap version of the Decision Modified q-BDHE Assumption introduced in [8].

3   Attribute-Based Signcryption

An Attribute-Based Signcryption (ABSC) scheme comprises the following five algorithms **Setup, DecKeyGen, SignKeyGen, Signcrypt, Unsigncrypt**. The essential security notions for an ABSC scheme are *message confidentiality* and *ciphertext unforgeability*. We refer to the full version of our paper for the detailed definition of the above algorithms and the security models.

3.1   Proposed ABSC Scheme

This section describes our new ABSC scheme which utilizes only a constant number of pairings. In this construction, we logically combine the CP-ABE [8] and the (n, n)-threshold ABS [3] schemes to realize decryption efficient signcryption scheme in the attribute-based setting. The construction uses general monotone access policies represented in DNF to encrypt a message and (n, n)-threshold policies to sign a message. As in the underlying CP-ABE [8], we assume all the attributes in encryption policy are distinct. In security analysis, the challenge access policy submitted by the adversary is converted to Linear Secret-Sharing Scheme (LSSS) matrix in order to properly answer the adversary's queries. In our construction, each attribute is allowed both as an encryption attribute and a signature attribute. We generate two public keys $h_i^{(e)}$ and $h_i^{(s)}$ for every attribute i, where $\{h_i^{(e)}\}$ are used to encrypt messages whereas $\{h_i^{(s)}\}$ are used to sign a message. The proposed ABSC scheme is detailed in the following algorithms.

Setup$(1^\kappa, U)$ This algorithm takes as input the security parameter κ and an attribute universe description U, and creates the system public and secret parameters as follows.

- Select secure bilinear pairing parameters $\Sigma := [p, \mathbb{G}, \mathbb{G}_T, e]$ such that $\mathbb{G} = \langle g \rangle$, where g is a generator of the group $\mathbb{G}$.
- Sample $\theta, b \xleftarrow{R} \mathbb{Z}_p^*$ and set $g_1 := g^\theta, g_2 := g^b$ and $Y := e(g_1, g_2)$.
- Choose $z_i, \breve{\delta}_i \xleftarrow{R} \mathbb{Z}_p^*$ and set $h_i^{(e)} := g^{z_i}, h_i^{(s)} := g^{\breve{\delta}_i}$ for each $i \in U$.
- Pick $h_0, \omega_1, \omega_2, \omega_3, u_0, u_1, \ldots, u_\ell \xleftarrow{R} \mathbb{G}$.
- Choose $H_1 : \{0,1\}^* \to \{0,1\}^\ell$ and $H_2 : \{0,1\}^* \to \mathbb{Z}_p^*$ from appropriate families of collision-resistant hash functions.
- Define a function $\mathcal{H} : \{0,1\}^\ell \to \mathbb{G}$ by $\mathcal{H}(x) := u_0 \prod_{i=1}^\ell u_i^{x_i}$, where $x := (x_1, \ldots, x_\ell) \in \{0,1\}^\ell$.
- Let $\Pi_{\mathsf{SE}} := (\mathsf{SE\text{-}Enc}, \mathsf{SE\text{-}Dec})$ be a one-time symmetric-key encryption scheme with key space $\mathcal{K} := \{0,1\}^\tau$ and message space $\mathcal{M} := \{0,1\}^*$. Let KDF be the key derivation function with the output length τ.

  The system public parameters $\mathcal{PP}$ are set as $\mathcal{PP} := [\Sigma, g, g_2, Y, \{h_i^{(e)}, h_i^{(s)}\}_{i \in U}, h_0, \omega_1, \omega_2, \omega_3, u_0, \{u_i\}_{i=1}^\ell, H_1, H_2, \mathcal{H}, \Pi_{\mathsf{SE}}, \mathsf{KDF}, \mathcal{M}, U]$ The system master secret key $\mathcal{MK}$ is set as $\mathcal{MK} := g_1 = g^\theta$.

DecKeyGen$(\mathcal{PP}, \mathcal{MK}, A_d)$ This algorithm takes as input $\mathcal{PP}, \mathcal{MK}$ and a set $A_d \subset U$ of attributes, and computes the decryption key $\mathcal{DK}_{A_d} := [D, D', D_i, i \in A_d]$ as follows: Pick $r \xleftarrow{R} \mathbb{Z}_p^*$ and set $D := g_2^\theta h_0^r, D' := g^r, D_i := (h_i^{(e)})^r, i \in A_d$. Note that we compute g_2^θ without the knowledge of θ as $g_2^\theta = g_1^b$.

SignKeyGen$(\mathcal{PP}, \mathcal{MK}, A_s)$ This algorithm takes as input $\mathcal{PP}, \mathcal{MK}$ and a set $A_s \subset U$ of attributes, and computes the signing key $\mathcal{SK}_{A_s} := [S_i, i \in A_s]$ as follows: Let f be a random polynomial of degree $n-1$ such that $f(0) = \theta$. Compute $S_i := g_2^{f(i)+\breve{\delta}_i} h_0^{-\breve{\delta}_i}$ for each $i \in A_s$. Note that we treat every attribute used in signature primitive as an element of $\mathbb{Z}_p^*$.

Signcrypt$(\mathcal{PP}, \Gamma_e, \mathcal{SK}_{A_s}, W_s, msg)$ This algorithm takes as input $\mathcal{PP}$, an encryption access policy $\Gamma_e := B_1 \vee B_2 \vee \cdots \vee B_m$, a signing attribute set $W_s \subset A_s$ with $|W_s| = n$, the signing key $\mathcal{SK}_{A_s}$ and a message $msg \in \mathcal{M}$. Then it caries out the following steps.

- Sample $\alpha, \gamma \xleftarrow{R} \mathbb{Z}_p^*$ and set
  - $c := \mathsf{SE\text{-}Enc}(\mathsf{KDF}(Y^\alpha || \Gamma_e || W_s), msg)$
  - $C := g^\alpha, C_1 := \left(h_0 \prod_{i \in B_1} h_i^{(e)}\right)^\alpha, C_2 := \left(h_0 \prod_{i \in B_2} h_i^{(e)}\right)^\alpha, \ldots, C_m := \left(h_0 \prod_{i \in B_m} h_i^{(e)}\right)^\alpha$

- $ct := [c, C, C_1, C_2, \ldots, C_m]$
- $\sigma := \mathcal{H}(x)^\alpha \prod_{i \in W_s} S_i^{\Delta_{i,W_s}(0)}$, where $x = H_1(c||\Gamma_e||W_s)$
- $C_e := (\omega_1^\beta \omega_2^\gamma \omega_3)^\alpha$, where $\beta = H_2(ct||\sigma||\Gamma_e||W_s)$

The ciphertext is $\mathcal{CT} := [\Gamma_e, ct, \sigma, C_e, \gamma]$.

Unsigncrypt($\mathcal{PP}, \mathcal{CT}, W_v, \mathcal{DK}_{A_d}$) Given $\mathcal{PP}, \mathcal{CT}$, a verification attribute set W_v of size n and $\mathcal{DK}_{A_d}$, this algorithm proceeds as follows.

- Compute $\beta = H_2(ct||\sigma||\Gamma_e||W_v)$, $x = H_1(c||\Gamma_e||W_v)$ and check whether $e(g, C_e) \stackrel{??}{=} e(C, \omega_1^\beta \omega_2^\gamma \omega_3)$ and $e(g, \sigma) \cdot e(h_0 g^{-1}, \prod_{i \in W_v}(h_i^{(s)})^{\Delta_{i,W_v}(0)}) \stackrel{??}{=} Y \cdot e(\mathcal{H}(x), C)$. If any one of these two equations does not hold, output $\perp$. Otherwise, execute the subsequent steps. Now, $W_v = W_s$.
- If $\Gamma_e(A_d) = false$, output $\perp$.
- Otherwise, find B_k from $\Gamma_e := B_1 \vee B_2 \vee \cdots \vee B_m$ such that $B_k \subseteq A_d$ and then compute $e(C, D \prod_{i \in B_k} D_i) \cdot e(D', C_k)^{-1} = Y^\alpha$.
- Recover the correct message $msg = \mathsf{SE\text{-}Dec}(\mathsf{KDF}(Y^\alpha||\Gamma_e||W_v), c)$.

It can be verified the correctness of the ABSC scheme in a manner similar to the underlying CP-ABE and ABS primitives [3,8].

The security analysis of the proposed ABSC scheme is given in the full version of our paper.

4   Attribute-Based Authenticated Key Exchange

4.1   Security Model for Attribute-Based Authenticated Key Exchange

In this section, we present a strong security model for Attribute-Based Authenticated Key Exchange (in short, ABAKE) protocols, namely the Attribute-Based extended Canetti-Krawzyck (ABeCK) model. The ABeCK model can be seen as a natural extension of the extended Canetti-Krawzyck (eCK) model for the conventional public-key setting. The ABeCK is different from the eCK model in the following ways: (i) In the ABeCK model, there are no user public-keys (as in eCK model); instead each party P has a set of attributes $\mathbb{S}_P$, which also defines the session identifiers, (ii) The freshness condition for revealing the long-term/static secret keys of parties are different. Before giving the ABeCK security model, we explain the syntax of a ABAKE protocol following [15].

An ABAKE protocol comprises of three PPT algorithms – Setup, KeyGen and KeyExchange. These algorithms are discussed below.

1. Setup(1^κ): The setup algorithm takes as input the implicit security parameter κ and the attribute universe U and outputs the master public key MPK and master secret key MSK.
2. KeyGen($MSK, MPK, \mathbb{S}_P$): The key generation algorithm takes in the master secret key MSK, the master public key MPK, and a set of attributes $\mathbb{S}_P$ of a party P, and outputs a static secret key $SK_{\mathbb{S}_P}$ corresponding to $\mathbb{S}_P$.

3. KeyExchange: This algorithm is run between two or more users or parties in the system (in our case the number of users is two as it is two-party setting). Let us assume the two parties party A and party B share a session key by performing the following n-pass protocol. Party A (resp. B) selects a policy $\mathbb{A}_A$ (resp. $\mathbb{A}_B$) as an access structure.

For $i = 1, 2, \ldots, n$, upon receiving the $(i-1)^{th}$ message m_{i-1}, the party P (P is either A or B) computes the ith message by algorithm Message. Message takes as input MPK, the set of attributes $\mathbb{S}_P$ of party P, static secret key $SK_{\mathbb{S}_P}$, the access policy $\mathbb{A}_P$, all the sent and received messages so far, i.e., $m_1, \cdots, m_{i-1}$, and outputs the ith message m_i. The party P then sends m_i to the other party $\overline{P}$. Upon receiving the final message m_n, party P computes the session key by algorithm SessionKey. SessionKey takes as input MPK, the set of attributes $\mathbb{S}_P$, the static secret key $SK_{\mathbb{S}_P}$, the policy $\mathbb{A}_P$, the transcript $m_1, \cdots, m_n$, and outputs a session key K. Both parties P and $\overline{P}$ compute the same session key K if and only if $\mathbb{S}_A \in \mathbb{A}_B$ and $\mathbb{S}_B \in \mathbb{A}_A$. Note that by $\mathbb{S}_A \in \mathbb{A}_B$ and $\mathbb{S}_B \in \mathbb{A}_A$, we mean that $\mathbb{A}_B(\mathbb{S}_A) = true$ and $\mathbb{A}_A(\mathbb{S}_B) = true$, respectively.

The ABeCK model can be seen as an attribute-based variant of the original eCK model [6]. It allows the adversary to completely control the communication channel, apart from passive eavesdropping. It allows him to register arbitrary parties (set of attributes) into the system, obtain the session keys of (completed) sessions, secret keys corresponding to attribute of parties, the master secret key of the system, and also the session-specific or ephemeral randomness of parties. A notion of freshness of a session is also defined, that disallows the adversary to trivially compute the session key corresponding to that session. The goal of the adversary is then to distinguish the session key of a fresh session from a random key. We refer the reader to the full version of our paper for the detailed ABeCK model.

4.2   Proposed ABAKE Protocol

We now give the details of our ABAKE protocol.

Design Rationale: In our construction of the ABAKE protocol, we use the ABSC scheme proposed in Sect. 3.1 as the main building block. We construct our ABAKE protocol by suitably combining the ABSC scheme and the NAXOS technique [6]. In the ABeCK model the adversary can get the attribute secret key of a user involved in the test session or the ephemeral secret key of the test session, but not both. Hence, we use the NAXOS technique to bind together the static secret key and the ephemeral secret key of parties using a hash function modeled as a random oracle. Since, the adversary will not know any one of these secret values, the extracted value is uniformly random. Specifically, we convert the ephemeral secret keys $\breve{\alpha}_A$ and $\breve{\alpha}_B$ of parties A and B respectively to the pseudo-ephemeral values α_A and α_B respectively, using the hash function H_3.

In the KeyExchange algorithm party A (resp. B) chooses an access policy $\mathbb{A}_A$ (resp. $\mathbb{A}_B$) and a signing attribute set W_A (resp. W_B) and run the underlying **Signcrypt** algorithm to produce ephemeral public keys EPK_A (resp. EPK_B). If the attributes of party B (resp. A) satisfy the access policy $\mathbb{A}_A$ contained in EPK_A (resp. $\mathbb{A}_B$ contained in EPK_B) and the signatures σ_A and σ_B are valid, then party B (resp. A) can extract the encapsulated key Y^{α_A} (resp. Y^{α_B}) by running the decryption algorithm of the underlying ABSC scheme. Our ABSC scheme satisfies the additional property of *public verifiability* of ciphertext, which guarantees the ciphertext integrity of both the parties A and B. The authenticity of the sender is ensured by the signatures σ_A and σ_B respectively. In our ABAKE protocol, we use $\Upsilon_1 = Y^{\alpha_A}$, $\Upsilon_2 = Y^{\alpha_B}$, where α_A and α_B are derived from the ephemeral secret keys of A and B respectively. However, only Υ_1 and Υ_2 are not enough to achieve the security in the ABeCK model. The ABeCK model allows the adversary also to reveal the master secret key of the system. We cannot prove the security in such a case because the simulator cannot embed the BDHE instance to the master secret key and cannot extract information to obtain the answer of the Gmq-BDHE problem only from Y^{α_A} and Y^{α_B}. Thus, we add the seed $\Upsilon_3 = g^{\alpha_A \alpha_B}$ to the seed of the session key in order to simulate such a case.

Construction: We now give the detailed description of our *one-round* (two-pass) ABAKE protocol. Let U be the universe of attributes used in the system, and let (Setup, DecKeyGen, SignKeyGen, Signcrypt, Unsigncrypt) be the ABSC scheme described in Sect. 3.1. The Setup, KeyGen and KeyExchange algorithms of our ABAKE protocol are detailed below. For the KeyExchange algorithm please refer to Table 3.

1. Setup(1^κ): Run Setup($1^\kappa, U$) of the underlying ABSC scheme to generate the system public and secret parameters $\mathcal{PP}$ and $\mathcal{MK}$, respectively. Here, $\mathcal{PP} := [\Sigma, g, g_2, Y, \{h_i^{(e)}, h_i^{(s)}\}_{i\in U}, h_0, \omega_1, \omega_2, \omega_3, u_0, \{u_i\}_{i=1}^\ell, H_1, H_2, \mathcal{H}, \Pi_{\mathsf{SE}}, \mathsf{KDF}, \mathcal{M}, U]$ and $\mathcal{MK} := g^\theta$. Choose two collision-resistant hash functions $H_3 : \{0,1\}^* \to \mathbb{Z}_p^*$ and $H_4 : \{0,1\}^* \to \{0,1\}^\kappa$. Note that H_2 and H_3 are independent. The system public parameters MPK are set as $MPK := [\Sigma = (p, \mathbb{G}, \mathbb{G}_T, e), g, g_2, Y, \{h_i^{(e)}, h_i^{(s)}\}_{i\in U}, h_0, \omega_1, \omega_2, \omega_3, u_0, \{u_i\}_{i=1}^\ell, H_1, H_2, H_3, H_4, \mathcal{H}, U]$, and the master secret key MSK is set as $MSK = \mathcal{MK} := g^\theta$.

2. KeyGen($MSK, MPK, \mathbb{S}_P$): Here $P \in \{A, B\}$. Run DecKeyGen($MPK, MSK, \mathbb{S}_P$) and SignKeyGen($MPK, MSK, \mathbb{S}_P$), where $\mathbb{S}_P$ is the set of attributes of party P. Note that for each party P, $|\mathbb{S}_P| \geq n$. The secret key $SK_{\mathbb{S}_P}$ of party P is set as $SK_{\mathbb{S}_P} = \{\mathcal{DK}_{\mathbb{S}_P}, \mathcal{SK}_{\mathbb{S}_P}\} := \{[D_P, D_P', \{D_{P_i}\}_{i\in\mathbb{S}_P}], \{S_{P_i}\}_{i\in\mathbb{S}_P}\}$. Here, $D_P = g_2^\theta h_0^r$, $D_P' = g^r$, $D_{P_i} = (h_i^{(e)})^r$, $S_{P_i} = g_2^{f(i)+\check{\delta}_i} h_0^{-\delta_i}$, where $r \xleftarrow{R} \mathbb{Z}_p^*$ and f is a random polynomial of degree $n - 1$ such that $f(0) = \theta$.

   Similarly, party B also decides an access structure $\mathbb{A}_B$ and he hopes that the set of attributes $\mathbb{S}_A$ of party A satisfies $\mathbb{A}_B$. It also chooses a signing attribute set $W_B \subset \mathbb{S}_B$. The shared session key obtained by both the parties A and

Table 3. Proposed ABAKE protocol

Party A	Party B
Access policy: $\mathbb{A}_A = (A_1 \vee A_2 \vee \cdots \vee A_m)$;	Access policy: $\mathbb{A}_B = (B_1 \vee B_2 \vee \cdots \vee B_{m'})$;
Signing attribute set: $W_A \subset \mathbb{S}_A$ with $|W_A| = n$	Signing attribute set: $W_B \subset \mathbb{S}_B$ with $|W_B| = n$
1. Sample $\check{\alpha}_A, \gamma_A \xleftarrow{R} \mathbb{Z}_p^*$, and calculate	1. Sample $\check{\alpha}_B, \gamma_B \xleftarrow{R} \mathbb{Z}_p^*$, and compute
$\alpha_A := H_3(\check{\alpha}_A \| SK_{\mathbb{S}_A})$	$\alpha_B := H_3(\check{\alpha}_B \| SK_{\mathbb{S}_B})$
2. Compute $C_A := g^{\alpha_A}, C_{A_1} := \left(h_0 \prod_{i\in A_1} h_i^{(e)}\right)^{\alpha_A}$,	2. Compute $C_B := g^{\alpha_B}, C_{B_1} := \left(h_0 \prod_{i\in B_1} h_i^{(e)}\right)^{\alpha_B}$,
$C_{A_2} := \left(h_0 \prod_{i\in A_2} h_i^{(e)}\right)^{\alpha_A}, \ldots,$	$C_{B_2} := \left(h_0 \prod_{i\in B_2} h_i^{(e)}\right)^{\alpha_B}, \ldots,$
$C_{A_m} := \left(h_0 \prod_{i\in A_m} h_i^{(e)}\right)^{\alpha_A}$	$C_{B_{m'}} := \left(h_0 \prod_{i\in B_{m'}} h_i^{(e)}\right)^{\alpha_B}$
Let $ct_A := [C_A, C_{A_1}, C_{A_2}, \ldots, C_{A_m}]$.	Let $ct_B := [C_B, C_{B_1}, C_{B_2}, \ldots, C_{B_{m'}}]$.
3. Compute $\sigma_A := \mathcal{H}(x_A)^{\alpha_A} \prod_{i\in W_A} S_{A_i}^{\Delta_i, W_A(0)}$,	3. Compute $\sigma_B := \mathcal{H}(x_B)^{\alpha_B} \prod_{i\in W_B} S_{B_i}^{\Delta_i, W_B(0)}$,
where $x_A = H_1(C_A \| \mathbb{A}_A \| W_A)$.	where $x_B = H_1(C_B \| \mathbb{A}_B \| W_B)$.
4. Compute $C_{A_e} := (\omega_1^{\beta_A} \omega_2^{\gamma_A} \omega_3)^{\alpha_A}$,	4. Compute $C_{B_e} := (\omega_1^{\beta_B} \omega_2^{\gamma_B} \omega_3)^{\alpha_B}$,
where $\beta_A = H_2(ct_A \| \sigma_A \| \mathbb{A}_A \| W_A)$.	where $\beta_B = H_2(ct_B \| \sigma_B \| \mathbb{A}_B \| W_B)$.
Erase α_A.	Erase α_B.
$\xrightarrow{EPK_A := [\mathbb{A}_A, ct_A, \sigma_A, C_{A_e}, \gamma_A, W_A]}$	
$\xleftarrow{EPK_B := [\mathbb{A}_B, ct_B, \sigma_B, C_{B_e}, \gamma_B, W_B]}$	
5. Compute $\beta_B' = H_2(ct_B \| \sigma_B \| \mathbb{A}_B \| W_B)$, and	5. Compute $\beta_A' = H_2(ct_A \| \sigma_A \| \mathbb{A}_A \| W_A)$, and
$x_B' = H_1(C_B \| \mathbb{A}_B \| W_B)$.	$x_A' = H_1(C_A \| \mathbb{A}_A \| W_A)$.
6. Check whether:	6. Check whether:
(a). $e(g, C_{B_e}) \overset{??}{=} e(C_B, \omega_1^{\beta_B'} \omega_2^{\gamma_B} \omega_3)$.	(a). $e(g, C_{A_e}) \overset{??}{=} e(C_A, \omega_1^{\beta_A'} \omega_2^{\gamma_A} \omega_3)$
(b). $e(g, \sigma_B) \cdot e(h_0 g^{-1}, \prod_{i\in W_A}(h_i^{(s)})^{\Delta_i, W_A(0)}) \overset{??}{=}$ $Y \cdot e(\mathcal{H}(x_B'), C_B)$	(b). $e(g, \sigma_A) \cdot e(h_0 g^{-1}, \prod_{i\in W_B}(h_i^{(s)})^{\Delta_i, W_B(0)}) \overset{??}{=}$ $Y \cdot e(\mathcal{H}(x_A'), C_A)$
7. If $\neg(a)$ or $\neg(b)$, output $\bot$,	7. If $\neg(a)$ or $\neg(b)$, output $\bot$,
8. Else if $\mathbb{A}_B(\mathbb{S}_A) = false$, output $\bot$.	8. Else if $\mathbb{A}_A(\mathbb{S}_B) = false$, output $\bot$.
9. Otherwise, $\mathbb{S}_A \in \mathbb{A}_B$, thereby find B_k from $\mathbb{A}_B := (B_1 \vee \cdots \vee B_{m'})$ such that $B_k \subseteq \mathbb{S}_A$.	9. Otherwise, $\mathbb{S}_B \in \mathbb{A}_A$, thereby find A_k from $\mathbb{A}_A := (A_1 \vee \cdots \vee A_m)$ such that $A_k \subseteq \mathbb{S}_B$.
10. Calculate	10. Calculate
$\Upsilon_1 := Y^{H_3(\check{\alpha}_A \| SK_{\mathbb{S}_A})}$,	$\Upsilon_1 := e(C_A, D_B \prod_{i\in A_k} D_{B_i}) \cdot e(D_B', C_{A_k})^{-1} = Y^{\alpha_A}$,
$\Upsilon_2 := e(C_B, D_A \prod_{i\in B_k} D_{A_i}) \cdot e(D_A', C_{B_k})^{-1} = Y^{\alpha_B}$,	$\Upsilon_2 := Y^{H_3(\check{\alpha}_B \| SK_{\mathbb{S}_B})}$,
$\Upsilon_3 := C_B^{H_3(\check{\alpha}_A \| SK_{\mathbb{S}_A})}$.	$\Upsilon_3 := C_A^{H_3(\check{\alpha}_B \| SK_{\mathbb{S}_B})}$.
11. Compute $K := H_4(\Upsilon_1 \| \Upsilon_2 \| \Upsilon_3 \| EPK_A \| EPK_B)$	11. Compute $K := H_4(\Upsilon_1 \| \Upsilon_2 \| \Upsilon_3 \| EPK_A \| EPK_B)$.

B after successful completion of the protocol is denoted by K. The shared secrets that both parties compute are

$$\Upsilon_2 = e\left(g^{\alpha_B}, g_2^{\theta} h_0^r \prod_{i\in B_k}(h_i^{(e)})^r\right) \cdot e\left(g^r, \left(h_0 \prod_{i\in B_k}(h_i^{(e)})\right)^{\alpha_B}\right)^{-1}$$

$$= e(g^{\theta}, g_2)^{\alpha_B} \cdot e\left(g^r, \left(h_0 \prod_{i\in B_k}(h_i^{(e)})\right)^{\alpha_B}\right) \cdot e\left(g^r, \left(h_0 \prod_{i\in B_k}(h_i^{(e)})\right)^{\alpha_B}\right)^{-1}$$

$$= Y^{\alpha_B} = Y^{H_3(\check{\alpha}_B \| SK_{\mathbb{S}_B})}$$

Similarly, $\Upsilon_1 = e(C_A, D_B \prod_{i\in A_k} D_{B_i}) \cdot e(D_B', C_{A_k})^{-1} = Y^{\alpha_A} = Y^{H_3(\check{\alpha}_A \| SK_{\mathbb{S}_A})}$, $\Upsilon_3 = C_B^{H_3(\check{\alpha}_A \| SK_{\mathbb{S}_A})} = (g^{\alpha_B})^{\alpha_A} = g^{\alpha_A \alpha_B} = (g^{\alpha_A})^{\alpha_B} = C_A^{H_3(\check{\alpha}_B \| SK_{\mathbb{S}_B})}$.

Hence they can compute the *same* session key $K = H_4(\Upsilon_1||\Upsilon_2||\Upsilon_3||$ $EPK_A||EPK_B)$.

4.3 Security Analysis of Proposed ABAKE Protocol

We now show that our ABAKE scheme (see Sect. 4.2) is secure in ABeCK model.

Theorem 1. *Assume H_1 and H_2 are collision-resistant hash functions and the hash functions H_3 and H_4 are random oracles. Suppose the Gmq-BDHE assumption holds. Assume the number of rows and columns in the challenge LSSS matrix are at most q. Then our ABAKE scheme is selectively secure in the ABeCK model.*

Due to space constraints, we refer to the full version of our paper for the detailed proof.

5 Conclusion

In this paper, we proposed a one-round Attribute-Based Authenticated Key Exchange (ABAKE) protocol. Our ABAKE protocol employs only constant number of pairing operations and the exponentiation operations are linear in the number of clauses in the DNF access formula. This is in contrast to the other ABAKE protocols, where both of these operations are polynomial in the size of the access policy. To this end, we also presented the first construction of an Attribute-Based Signcryption (ABSC) scheme with constant number of pairings in the ciphertext-policy setting for general access control policies. Our ABAKE protocol is proven secure in the ABeCK model under the random oracle assumption. We leave open the problem of constructing an ABAKE protocol in the ABeCK model in the standard model achieving the same or comparable level of efficiency/ computational complexity as our construction.

References

1. Bethencourt, J., Sahai, A., Waters, B.: Ciphertext-policy attribute-based encryption. In: 2007 IEEE Symposium on Security and Privacy. SP 2007, pp. 321–334. IEEE (2007)
2. Emura, K., Miyaji, A., Rahman, M.S.: Dynamic attribute-based signcryption without random oracles. Int. J. Appl. Cryptol. 2(3), 199–211 (2012)
3. Gagné, M., Narayan, S., Safavi-Naini, R.: Short pairing-efficient threshold-attribute-based signature. In: Abdalla, M., Lange, T. (eds.) Pairing 2012. LNCS, vol. 7708, pp. 295–313. Springer, Heidelberg (2013). https://doi.org/10.1007/978-3-642-36334-4_19
4. Gorantla, M.C., Boyd, C., González Nieto, J.M.: Attribute-based authenticated key exchange. In: Steinfeld, R., Hawkes, P. (eds.) ACISP 2010. LNCS, vol. 6168, pp. 300–317. Springer, Heidelberg (2010). https://doi.org/10.1007/978-3-642-14081-5_19

5. Goyal, V., Pandey, O., Sahai, A., Waters, B.: Attribute-based encryption for fine-grained access control of encrypted data. In: Proceedings of the 13th ACM Conference on Computer and Communications Security, pp. 89–98. ACM (2006)
6. LaMacchia, B., Lauter, K., Mityagin, A.: Stronger security of authenticated key exchange. In: Susilo, W., Liu, J.K., Mu, Y. (eds.) ProvSec 2007. LNCS, vol. 4784, pp. 1–16. Springer, Heidelberg (2007). https://doi.org/10.1007/978-3-540-75670-5_1
7. Liu, J., Huang, X., Liu, J.K.: Secure sharing of personal health records in cloud computing: ciphertext-policy attribute-based signcryption. Futur. Gener. Comput. Syst. **52**, 67–76 (2015). Special Section: Cloud Computing: Security, Privacy and Practice
8. Malluhi, Q.M., Shikfa, A., Trinh, V.C.: A ciphertext-policy attribute-based encryption scheme with optimized ciphertext size and fast decryption. In: Proceedings of the 2017 ACM on Asia Conference on Computer and Communications Security, pp. 230–240. ACM (2017)
9. Ostrovsky, R., Sahai, A., Waters, B.: Attribute-based encryption with non-monotonic access structures. In: Proceedings of the 14th ACM Conference on Computer and Communications Security, pp. 195–203. ACM (2007)
10. Pandit, T., Pandey, S.K., Barua, R.: Attribute-based signcryption : signer privacy, strong unforgeability and IND-CCA2 security in adaptive-predicates attack. In: Chow, S.S.M., Liu, J.K., Hui, L.C.K., Yiu, S.M. (eds.) ProvSec 2014. LNCS, vol. 8782, pp. 274–290. Springer, Cham (2014). https://doi.org/10.1007/978-3-319-12475-9_19
11. Rao, Y.S.: A secure and efficient ciphertext-policy attribute-based signcryption for personal health records sharing in cloud computing. Futur. Gener. Comput. Syst. **67**, 133–151 (2017)
12. Sahai, A., Waters, B.: Fuzzy identity-based encryption. In: Cramer, R. (ed.) EUROCRYPT 2005. LNCS, vol. 3494, pp. 457–473. Springer, Heidelberg (2005). https://doi.org/10.1007/11426639_27
13. Steinwandt, R., Corona, A.S.: Attribute-based group key establishment. IACR Cryptology ePrint Archive, vol. 2010, p. 235 (2010)
14. Waters, B.: Ciphertext-policy attribute-based encryption: an expressive, efficient, and provably secure realization. In: Catalano, D., Fazio, N., Gennaro, R., Nicolosi, A. (eds.) PKC 2011. LNCS, vol. 6571, pp. 53–70. Springer, Heidelberg (2011). https://doi.org/10.1007/978-3-642-19379-8_4
15. Yoneyama, K.: Strongly secure two-pass attribute-based authenticated key exchange. In: Joye, M., Miyaji, A., Otsuka, A. (eds.) Pairing 2010. LNCS, vol. 6487, pp. 147–166. Springer, Heidelberg (2010). https://doi.org/10.1007/978-3-642-17455-1_10
16. Yoneyama, K.: Two-party round-optimal session-policy attribute-based authenticated key exchange without random oracles. In: Kim, H. (ed.) ICISC 2011. LNCS, vol. 7259, pp. 467–489. Springer, Heidelberg (2012). https://doi.org/10.1007/978-3-642-31912-9_31

Server-Aided Revocable Attribute-Based Encryption Resilient to Decryption Key Exposure

Baodong Qin[1,2], Qinglan Zhao[3], Dong Zheng[1,4(✉)], and Hui Cui[5]

[1] National Engineering Laboratory for Wireless Security,
Xi'an University of Posts and Telecommunications,
Xi'an 710121, People's Republic of China
qinbaodong@foxmail.com, zhengdong@xupt.edu.cn
[2] State Key Laboratory of Cryptology,
P.O. Box 5159, Beijing 100878, People's Republic of China
[3] Shanghai Jiao Tong University,
Shanghai 200240, People's Republic of China
zhaoqinglan@foxmail.com
[4] Westone Cryptologic Research Center, Beijing 100070, People's Republic of China
[5] School of Science, RMIT University, Melbourne, Australia
hui.cui@rmit.edu.au

Abstract. Attribute-based encryption (ABE) is a promising approach that enables scalable access control on encrypted data. However, one of the main efficiency drawbacks of ABE is the lack of practical user revocation mechanisms. In CCS 2008, Boldyreva, Goyal and Kumar put forward an efficient way to revoke users. But, it requires each data user storing a (non-constant) number of long-term private keys and periodically communicating with the key generation center to update his/her decryption keys. In ESORICS 2016, Cui et al. proposed the first server-aided revocable ABE scheme to address the above two issues. It involves an untrusted server to transform any non-revoked user's ABE ciphertexts into short ciphertexts using user's short-term transformation keys. The data user can fully decrypt the transformed ciphertexts using his/her local decryption keys. Cui et al. also introduced the decryption key exposure (DKE) attacks on transformation keys. However, if the untrusted server colludes with an adversary, the scheme may be insecure against DKE attacks on user's *local decryption keys*. In this paper, we first revisit Cui et al. security model, and enhance it by capturing the DKE attacks on user's local decryption keys and allowing the adversary to fully corrupt the server simultaneously. We then construct a server-aided revocable ABE based on Rouselakis-Waters ciphertext-policy ABE (CCS 2013). We show that our scheme is secure against local decryption key exposure attacks, and maintains the outstanding properties of efficient user revocation, short local ciphertext size and fast local decryption.

Keywords: Attribute-based encryption · Revocation
Decryption key exposure · Server-aided

S. Capkun and S. S. M. Chow (Eds.): CANS 2017, LNCS 11261, pp. 504–514, 2018.
https://doi.org/10.1007/978-3-030-02641-7_25

1   Introduction

Attribute-based encryption (ABE) [15] is a promising solution to enable scalable access control on encrypted data. In an ABE, a trusted key generation center (KGC) issues a secret key for each user according to that user's attributes. A data owner can encrypt data and embed an access policy into the ciphertext. The data owner then uploads the ciphertext into the cloud. A data user can decrypt a ciphertext if his/her attributes satisfy the embedded access policy. Usually, such an ABE system is called ciphertext-policy attribute-based encryption (CP-ABE) system. The opposite of CP-ABE is the key-policy attribute-based encryption (KP-ABE) [8]. In this paper, we focus on ciphertext-policy ABE.

Today, ABE has become a powerful and promising tool for secure cloud data sharing. However, there exist two main drawbacks in current ABE schemes. One is the efficiency issue as the size of ciphertext and the decryption time grows with the complexity of the access structure. The other is the user revocation issue. In USENIX 2011, Green et al. [9] suggested to address efficiency problem by involving a cloud. In particular, a data owner can store ABE ciphertexts in the cloud and a data user can provide the cloud with a single transformation key for translating any ABE ciphertext satisfied by that user's attributes into a short ciphertext, without leaking any part of the user's message to the cloud.

Revocation is very important for ABE as well as PKI-based cryptosystems. Although there are numerous studies on efficient revocation mechanisms in the traditional PKI setting [1,6,7,11], there are only a few studies on ABE setting. In ACM CCS 2008, Boldyreva, Goyal and Kumar [3] put forward an efficient way to revoke users by combining the fuzzy identity-based encryption (IBE) [15] and the binary-tree data structure (shorted as BGK scheme). In this method, the KGC first issues a long-term secret key for each user and then publishes the common key update information during each time period. Consequently, only non-revoked users can generate decryption keys from their long-term secret keys and the key updates. Compared with the naive revocation way proposed by Boneh and Franklin [4], the size of the key update information in [3] is significantly reduced from $O(N - R)$ to $O(R \log N/R)$ on average. Here, N is the total number of system users and R is the number of revoked users. Nevertheless, it also introduces other issues: (1) The BGK scheme as well as the follow-up work [10] is vulnerable to decryption key exposure (DKE); and (2) The size of a user's long-term secret key increases from $O(1)$ to $O(\log N)$, which depends on the number of system users. In PKC 2013, Seo and Emura [16] proposed a simple revocable IBE scheme with DKE resistance. Later, Qin et al. [12] proposed the notion of server-aided revocable IBE (SR-IBE). Like in [9], it involves an untrusted server to partially decrypt a user's ciphertext using a time-based short-term transformation key. But, the short-term transformation keys are generated by the server in a similar way as that of user's decryption keys in [16], i.e., the server in advance stores some long-term transformation keys, and then periodically receives key update information from the KGC to generate the corresponding short-term transformation keys of non-revoked users. A local user just holds a long-term secret key, and can use it to generate time-based short-term decryption keys by himself/herself. If short-term decryption keys for some time periods are leaked, they do not affect the security of other periods decryption keys. Fortunately, the SR-IBE scheme can delegate almost all workloads on data users to the untrusted server and withstand DKE attacks on user's local short-term decryption keys. Moreover, the data user keeps just one long-term secret key of constant-size. Recently, Cui et al. [5] extended SR-IBE to the notion of server-aided revocable ABE (SR-ABE) for supporting both decryption outsourcing and efficient user revocation. Interestingly, their SR-ABE scheme does not require any secure channels for key distribution, as data users generated their secret keys by themselves.

It should be noted that Cui et al. have already considered the decryption key exposure attacks in their security model. But the DKE attack is in fact defined for user's short-term transformation keys, as the decryption key generated by a data user in a normal ABE scheme is now created by the server and renamed as transformation key in their SR-ABE scheme. However, the untrusted server could be operated by anyone, including the adversary, and hence all short-term transformation keys could be exposed to the adversary. In addition, a user's local short-term decryption keys in [5] are always the same as his/her long-term secret key. So, their scheme cannot allow a user's local decryption keys being exposed if the user is not revoked. Seo and Emura [16] showed that decryption key exposure is a very realistic threat on many revocable cryptosystems. Therefore, it is necessary to design an server-aided revocable ABE scheme with DKE resistance on user's local decryption keys.

1.1 Our Contribution

Our contribution consists of three parts. First, we revisit Cui et al. system framework for SR-ABE and make a modification to their decryption key generation algorithm for capturing DKE attacks on user's local decryption keys. The new system framework is given in Fig. 1. As in [5], a user's ABE ciphertext should be first transformed by an untrusted server using a corresponding short-term transformation key. If the user is revoked, the server can not assist him/her in transforming ciphertexts any more. But, we separate a user's local decryption capacity into two parts: one is a long-term secret key that can decrypt any time period transformed ciphertexts, and the other is a time-based short-term decryption key that can just decrypt a single time period ciphertexts, transformed by the untrusted server. A user can use his/her long-term secret key to generate a short-term decryption key for any specified time-period. We present a formal security model for such an SR-ABE framework in Subsect. 2.2. Clearly, an SR-ABE scheme with DKE resistance allows a data user to delegate his/her decryption capacity to others, such as a laptop, for a specified time period.

Second, we construct a server-aided revocable CP-ABE scheme based on (non-revocable) Rouselakis-Waters CP-ABE scheme [13]. To achieve server-aided revocation, it is possible to adopt the idea of Qin et al. SR-IBE scheme [12]. In Qin et al. SR-IBE, a master secret key is actually split into two keys: one serves as a master key of a normal revocable IBE between the KGC and the server, and

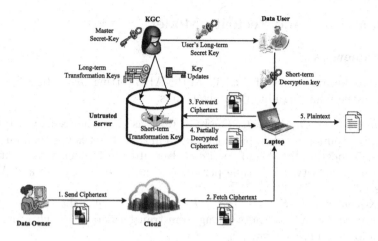

Fig. 1. System framework of our SR-ABE

the other serves as a master key of a two-level hierarchical IBE scheme between the KGC and the data users. This method can shrink the size of user's long-term secret key, but it does not achieve the advantage of decryption outsourcing. To conquer this challenge, we introduce the random splitting technique. Specifically, a master ABE key is also split into two parts. But, in contrast to [12], this decomposition is random and independent for different users. The first key serves as a master key of a revocable ABE between the KGC and that user, and will be used to generate user's short-term transformation keys. Due to such a random splitting technique, the second key just serves as a master key of an IBE scheme. The IBE master key will be used to generate that user's short-term decryption keys. Applying a transformation key to an ABE ciphertext, the result is actually an IBE ciphertext under that user's master (public) key and a time as identity. Thus, that user's local decryption only requires one IBE decryption, which only contains two pairing computations in our concrete construction. The key splitting approach is similar to that used in Cui et al.'s construction. But, the second key is used in a different way.

Third, we show that the security of our SR-ABE scheme can be efficiently reduced to the original Rouselakis-Waters CP-ABE scheme in a one-user setting. The details are given in Subsect. 3.2.

Notes. Due to space limitation, some related work is introduced in the full version of this paper. Some basic cryptographic notions used in this paper are also omitted, including bilinear groups, access structures, linear secret-sharing schemes (LSSS), binary tree and the node selection algorithm KUNodes. They can be found in [5] or the full version of this paper.

2 Framework and Security Model

2.1 Framework

We extend the framework of SR-IBE to the ABE setting. An SR-ABE scheme consists of the following nine PPT algorithms.

- Setup(1^{κ}) → (pp, msk, rl, st): The KGC takes as input the security parameter 1^{κ} and outputs the public parameter pp, a master secret key msk, an initially empty revocation list rl and a state st. Hereafter, S is a set of attributes, id is a user's identity, t is a time period, $\mathbb{A}$ is an access structure and M is a message.
- UserKG(pp, msk, id, S, st) → (pk_{id}, sk_{id}): The KGC runs the user key generation algorithm and outputs a long-term transformation key pk_{id} for server and a long-term secret key sk_{id} for user.
- TKeyUp(pp, msk, t, rl, st) → (ku_t, st): The KGC runs the transformation key update algorithm and outputs a key update ku_t for server and an updated state st.
- TranKG(pp, id, pk_{id}, ku_t) → $tk_{id,t}$: The server runs the transformation key generation algorithm and outputs a short-term transformation key $tk_{id,t}$.
- DecKG(pp, id, sk_{id}, t) → $dk_{id,t}$: The data user runs the decryption key generation algorithm and outputs a short-term decryption key $dk_{id,t}$.
- Encrypt($pp, \mathbb{A}, t, M$) → CT: The data owner runs the encryption algorithm and outputs a ciphertext CT.
- Transform($pp, id, S, tk_{id,t}, CT$) → $CT'/\perp$: The server runs the ciphertext transformation algorithm and outputs either a partially decrypted ciphertext CT' for data user or $\perp$ indicating the failure of the transformation.
- Decrypt($pp, id, dk_{id,t}, CT'$) → $M/\perp$: The data user runs the decryption algorithm and outputs the message M or a failure symbol $\perp$.
- Revoke(id, t, rl, st) → rl: The KGC runs the revocation algorithm and outputs an updated revocation list rl.

2.2 Security Model

We describe an IND-CPA security model for SR-ABE that captures both user revocation and local decryption key exposure. The following game is played between an adversary $\mathcal{A}$ and the challenger $\mathcal{C}$.

- **Setup.** The challenger runs the setup algorithm and gives pp to the adversary. The challenger keeps msk, rl and st.
- **Phase 1.** $\mathcal{A}$ can adaptively issues any of the following queries to $\mathcal{C}$.
  - Create(id, S): $\mathcal{C}$ runs UserKG on (id, S) to obtain the pair (pk, sk), and stores in table T the entry (id, S, pk, sk). It returns pk to $\mathcal{A}$.
  - Corrupt(id): If there exists an entry indexed by id in table T, then $\mathcal{C}$ obtains the entry (id, S, pk, sk) and sets $D = D \cup \{(id, S)\}$. It returns sk to $\mathcal{A}$. If no such entry exists, then it returns $\perp$.
  - TKeyUp(t): $\mathcal{C}$ runs TKeyUp(pp, msk, t, rl, st) and gives ku_t to $\mathcal{A}$.

- DecKG(id, t): If $\mathcal{A}$ issues a decryption key query on (id, t) and there exists an entry indexed by id in table T, then $\mathcal{C}$ obtains the entry (id, S, pk, sk). It then runs DecKG(pp, id, sk, t) and gives the decryption key $dk_{id,t}$ to $\mathcal{A}$. If no such entry exists, then it returns $\perp$.
  - Revocation(id, t): When $\mathcal{A}$ issues a revocation query on (id, t), $\mathcal{C}$ runs Revoke(id, t, rl, st) and outputs an updated revocation list rl.
- **Challenge.** The adversary submits two equal length message M_0, M_1, an access structure $\mathbb{A}^*$ and a time period t^* satisfying the following constraints (Suppose that the attribute set associated with id^* is S^*).
  1. If there exists a tuple $(id^*, S^*) \in D$ and $S^* \in \mathbb{A}^*$, then the adversary must query the revocation oracle on (id^*, t) at or before time period t^*.
  2. If there exists a tuple $(id^*, S^*, pk^*, sk^*) \in T$, $S^* \in \mathbb{A}^*$ and id^* is not revoked at or before time period t^*, then the adversary cannot query the decryption oracle on (id^*, t^*).

  The challenger picks a random bit $b \in \{0, 1\}$ and sends the challenge ciphertext $CT^* \leftarrow$ Encrypt($pp, \mathbb{A}^*, t^*, M_b$) to the adversary.
- **Phase 2.** The adversary continues issuing queries to the challenger as in Phase 1, with the same restrictions defined in the Challenge phase.
- **Guess.** The adversary makes a guess b' for b and wins the game if $b' = b$. The advantage of $\mathcal{A}$ in this game is defined as $|\Pr[b' = b] - 1/2|$.

Definition 1 (IND-CPA Secure SR-ABE). *An SR-ABE scheme is IND-CPA secure if for all polynomial-time adversaries have at most a negligible advantage in the IND-CPA game defined above.*

Selective Security. We say an SR-ABE scheme is selectively IND-CPA secure if we add an Init stage before the Setup phase where the adversary $\mathcal{A}$ commits to the challenge access structure $\mathbb{A}^*$ and the challenge time period t^*.

3 Server-Aided Revocable CP-ABE with DKE Resistance

3.1 The Construction

Our scheme consists of the following nine algorithms:

- Setup(1^κ) $\rightarrow (pp, msk, rl, st)$: It first generates the descriptions of groups and the bilinear map $\mathcal{D} = (p, G, G_T, e)$, where p is the prime order of groups G and G_T. Both the attribute space and the time space are Z_p.
  Then, it randomly chooses $g, u, h, u_0, h_0, w, v \in G$ and $\alpha \in Z_p$. Let rl be an empty list storing revoked users and BT be a binary tree with at least N leaf nodes. N is the maximal number of system users. Define two functions $F_1(y) = u^y h$ and $F_2(y) = u_0^y h_0$ to map any element y in Z_p to an element in G. Finally, it outputs $pp = (\mathcal{D}, g, w, v, u, h, u_0, h_0, e(g, g)^\alpha)$ and $msk = \alpha$, along with the revocation list rl and the initial state $st = $ BT.

- UserKG$(pp, msk, id, S = \{A_1, A_2, \ldots, A_k\}, \mathsf{BT}) \to (pk_{id}, sk_{id})$: Initially, it picks a random exponent $\beta_{id} \in Z_p$ and sets $sk_{id} = g^{\beta_{id}}$. Then, it chooses an undefined leaf node θ from the binary tree BT, and stores id in this node. Next, for each node $x \in \mathsf{Path}(\theta)$, it runs as follows.
  1. It fetches g_x from the node x. If x has not been defined, it randomly chooses $g_x \in G$, computes $g'_x = g^{\alpha - \beta_{id}}/g_x$, and store g_x in the node x.
  2. It picks $k + 1$ random exponents $r_x, r_{x,1}, r_{x,2}, \ldots, r_{x,k} \in Z_p$. Then it computes $P_{x,0} = g'_x w^{r_x}$, $P_{x,1} = g^{r_x}$, and for every $i \in [k]$, $P^i_{x,2} = g^{r_{x,i}}$ and $P^i_{x,3} = F_1(A_i)^{r_{x,i}} v^{-r_x}$.
  3. It outputs the following long-term transformation key and secret key $pk_{id} = \{x, P_{x,0}, P_{x,1}, P^i_{x,2}, P^i_{x,3}\}_{x \in \mathsf{Path}(\theta), i \in [k]}$ and $sk_{id} = g^{\beta_{id}}$.
- TKeyUp$(pp, msk, t, rl, st) \to (ku_t, st)$: Firstly, it fetches g_x from each node $x \in \mathsf{KUNodes}(\mathsf{BT}, rl, t)$. Then, it picks a random exponent $s_x \in Z_p$ and computes $Q_{x,0} = g_x F_2(t)^{s_x}$, $Q_{x,1} = g^{s_x}$. The transformation key update information is $ku_t = \{x, Q_{x,0}, Q_{x,1}\}_{x \in \mathsf{KUNodes}(\mathsf{BT}, rl, t)}$.
- TranKG$(pp, id, pk_{id}, ku_t) \to tk_{id,t}$: Suppose that θ is the leaf node storing the identity id. Let $I = \{x : x \in \mathsf{Path}(\theta)\}$ and $J = \{x : x \in \mathsf{KUNodes}(\mathsf{BT}, rl, t)\}$. If $I \cap J = \varnothing$, it returns $\bot$. Otherwise, there must be exactly one node $x \in I \cap J$. Then, it computes $tk_0 = P_{x,0} \cdot Q_{x,0}$, $tk_1 = P_{x,1}$, $tk_{2,i} = P^i_{x,2}$, $tk_{3,i} = P^i_{x,3}$, and $tk_4 = Q_{x,1}$. The algorithm returns $tk_{id,t} = (tk_0, tk_1, \{tk_{2,i}, tk_{3,i}\}_{i \in [k]}, tk_4)$.
- DecKG$(pp, id, sk_{id}, t) \to dk_{id,t}$: It picks a random exponent r_t and computes the short-term decryption key $dk_{id,t} = (D_0, D_1) = (g^{\beta_{id}} F_2(t)^{r_t}, g^{r_t})$.
- Encrypt$(pp, (\mathbb{M}, \rho), t, M) \to CT$: Given an LSSS access structure $(\mathbb{M}, \rho) \in (Z_p^{\ell \times n}, \mathcal{F}([\ell] \to Z_p))$, it randomly chooses a vector $\boldsymbol{v} = (\gamma, y_2, \ldots, y_n)^{\perp} \in Z_p^n$ and computes the shares $\boldsymbol{\lambda} = (\lambda_1, \ldots, \lambda_\ell)^{\perp} = \mathbb{M} \cdot \boldsymbol{v}$. It picks ℓ random exponents $\mu_1, \ldots, \mu_\ell \in Z_p$ and computes $CT = (C, C_0, \{C_{i,1}, C_{i,2}, C_{i,3}\}_{i \in [\ell]}, C_4)$, where $C = M \cdot e(g, g)^{\alpha\gamma}$, $C_0 = g^{\gamma}$, $C_4 = F_2(t)^{\gamma}$ and for $i \in [\ell]$, $C_{i,1} = w^{\lambda_i} \cdot v^{\mu_i}$, $C_{i,2} = F_1(\rho(i))^{-\mu_i}$ and $C_{i,3} = g^{\mu_i}$.
- Transform$(pp, id, S, tk_{id,t}, CT) \to CT'/\bot$: If the user is revoked at time period t or the attribute set S does not satisfy the access policy $(\mathbb{M}, \rho)$ associated with the ciphertext CT, the algorithm returns $\bot$. Otherwise, it computes the set $I = \{i : \rho(i) \in S\}$ and the constants $\{\omega_i \in Z_p\}_{i \in I}$ such that $\sum_{i \in I} \omega_i \mathbb{M}_i = (1, 0, \ldots, 0)$, where $\mathbb{M}_i$ is the i-th row of $\mathbb{M}$. Next, the algorithm computes $B = \frac{e(C_0, tk_0) \cdot e(C_4, tk_4)^{-1}}{\prod_{i \in I} (e(C_{i,1}, tk_1) \cdot e(C_{i,2}, tk_{2,\tau}) \cdot e(C_{i,3}, tk_{3,\tau}))^{\omega_i}}$, where τ is the index of attribute $\rho(i)$ in S, i.e., $A_\tau = \rho(i)$. Finally, the algorithm returns the transformed ciphertext $CT' = (C', C'_0, C'_4)$, where $C' = C/B$, $C'_0 = C_0$ and $C'_4 = C_4$.
- Decrypt$(pp, id, dk_{id,t}, CT') \to M/\bot$: Given a transformed ciphertext $CT' = (C', C'_0, C'_4)$ and a decryption key $dk_{id,t} = (D_0, D_1)$, the algorithm computes and returns $M = C' \cdot e(C'_4, D_1)/e(C'_0, D_0)$.
- Revoke$(id, t, rl, st) \to rl$: If a user id is revoked at time period t, the algorithm adds (x, t) to rl for all nodes x associated with identity id.

3.2   Security Analysis

We prove the security in the so-called one-user setting. In such setting, only one "target user" has the capacity to access to the challenge ciphertext and the adversary can corrupt either his/her long-term secret key or his/her short-term decryption keys. So, the adversary falls into the following two distinct classes.

- Type-1: The target user is revoked at or before the challenge time period t^* and thus $\mathcal{A}$ is allowed to corrupt the target user's long-term secret key.
- Type-2: The target user is not revoked, so $\mathcal{A}$ is not allowed to corrupt the target user's long-term secret key. But $\mathcal{A}$ may obtain the target user's short-term decryption keys for time periods except the challenge time period t^*.

Clearly, the above two cases do not cross with each other in the one-user setting. So, we can randomly guess which type of attacks the target user is suffered from with (non-negligible) probability $1/2$. More importantly, we can efficiently simulate the security game for these two kinds of adversaries separately. The proof can be extended to the setting with many target users if these two types of adversaries do not cross with each other, i.e., the adversary either corrupts all target users' long-term secret keys or corrupts all target users' short-term decryption keys. Generally, there two different adversaries may exist simultaneously even for two target users. It is not clear how to prove the security of our scheme in the general multi-user setting. We leave it as our future work. Alternatively, we can design a new SR-ABE scheme secure in the multi-user setting.

Theorem 1 (Selectively IND-CPA Security). *If the Rouselakis-Waters CP-ABE scheme is selectively IND-CPA secure, then our server-aided revocable CP-ABE scheme is also selectively IND-CPA secure in one-user setting.*

The proof of Theorem 1 is given in the full version of the paper.

3.3   Comparison

In recent years, numerous work [2,3,5,14,17] develop new cryptographic techniques to construct revocable ABE schemes in prime-order bilinear groups. Table 1 summarizes the properties of these revocable ABE schemes and ours. Firstly, with the exception of [17], all ABE schemes support indirect revocation. This removes the requirement of holding an authenticated revocation list by the data owner or the server. Secondly, among those ABE schemes that involve a third server during revocation, ours and [5] do not need a trusted server while [17] needs. That is, in our scheme, any user can collude with the third server, while in [17] revoked users cannot. Thirdly, with the exception of ours, all these ABE schemes are vulnerable to local decryption key exposure. Though our security is only proved in one-user setting, it does not imply that the other ciphertexts cannot be shared by multiple users. In addition, if all target users suffer from other Type-1 attacks or Type-2 attacks separately, our scheme is also secure in the multi-user setting.

To show that our scheme indeed achieves outsourcing functionality, we compare it with the original (non-revocable) Rouselakis-Waters CP-ABE scheme in terms of basic performance such as ciphertext size in Table 2. We also make a

Table 1. Property comparison among existing revocable ABE schemes. Let N and R denote the number of all system users and the revoked users, respectively. Let ℓ denote the number of attributes presented in an access structure, k denote the size of an attribute set associated with a user and "-" denote not-applicable. "One/Multi-User Setting" means a challenge ciphertext allows a single user or many users to access to.

Properties	BGK08 [3]	AI09 [2]	SSW12 [14]	YDL+13 [17]	CDL+16 [5]	Section 3
Revocation mode	Indirect	Indirect & Direct	Indirect	Direct	Indirect	Indirect
Type of ABE	KP-ABE	KP-ABE	KP/CP-ABE	CP-ABE	CP-ABE	CP-ABE
Third server	-	-	-	Semi-trust	Untrust	Untrust
Security	Selective	Selective	Selective	Selective	Selective	Selective
Key update size	$O(R \log \frac{N}{R})$	$O(R \log \frac{N}{R})$	$O(R \log \frac{N}{R})$	-	$O(R \log \frac{N}{R})$	$O(R \log \frac{N}{R})$
Secret key size	$O(\ell \log N)$	$O(\ell \log N)$	$O(l \log N)$ & $O(k \log N)$	$O(1)$	$O(1)$	$O(1)$
Secure channel	Yes	Yes	Yes	Yes	No	Yes/no
DKE resistance	No	No	No	No	No	Yes
User setting	Multi-user	Multi-user	Multi-user	Multi-user	Multi-user	One-user

Table 2. Efficiency Comparison Among RW-Type ABE Schemes. Only the dominant operations of exponentations and pairings are listed. The parameters ℓ, k and m denote the number of rows of the matrix policy, the size of the attribute set, and the rows utilized during decryption or transformation, respectively. Let "-" denote not-applicable. Let $\#Z_p$, $\#G$, $\#G_T$ denote the number of elements in groups Z_p, G and G_T respectively. Let $\#E_G$ (resp. $\#E_{G_T}$) denote the number of operations of exponentations in group G (resp. G_T), and let $\#P$ denote the number of operations of pairings.

	RW13 [13]	CDL+16 [5]	Section 3
ABE ciphertext size $[\#G, \#G_T]$	$[3\ell + 1, 1]$	$[3\ell + 2, 1]$	$[3\ell + 2, 1]$
Transformed ciphertext size $[\#G, \#G_T]$	-	$[1, 1]$	$[2, 1]$
User's secret-key size $[\#Z_p, \#G]$	$[0, 2k + 2]$	$[1, 0]$	$[0, 1]$
Encryption $[\#E_G, \#E_{G_T}]$	$[5\ell + 1, 1]$	$[5\ell + 3, 1]$	$[5\ell + 3, 1]$
ABE transform $[\#E_{G_T}, \#P]$	-	$[m, 3m + 2]$	$[m, 3m + 2]$
Final ABE decryption $[\#E_{G_T}, \#P]$	$[m, 3m + 1]$	$[1, 0]$	$[0, 2]$

comparison with Cui et al. SR-CP-ABE scheme, as their scheme is also realized from the Rouselakis-Waters CP-ABE scheme. It is showed that both of the server-aided ABE schemes outsource most ABE ciphertext and user's secret key into the cloud or the server, leaving only constant-size ciphertext and secret key to the local user. The heavy operations during decryption are also outsourced to the server, and the local user needs just two pairing operations. To achieve local DKE resistance, our scheme increases the partially-decrypted ciphertext size and the operation time of final ABE decryption in comparison with Cui et al. SR-ABE scheme. Nevertheless, the increments are only constant and very few, e.g., the transformed ciphertext increases only one group element.

4 Conclusion

This paper proposed a new security model for server-aided revocable ABE scheme to capture a realistic threat, namely local decryption key exposure, introduced by Seo and Emura in PKC 2013. Based on a standard (non-revocable) Rouselakis-Waters CP-ABE scheme, we presented an SR-CP-ABE scheme supporting both decryption outsourcing and local DKE resistance. As our scheme is proved to be selective IND-CPA secure in just one-user setting, a nature question is how to construct an SR-ABE scheme with DKE resistance in a multi-user setting.

Acknowledgments. This work was supported by the National Natural Science Foundation of China (Grant No. 61502400 and Grant No. 61602378), and the Science Foundation of Sichuan Educational Committee (Grant No. 16ZB0140).

References

1. Aiello, W., Lodha, S., Ostrovsky, R.: Fast digital identity revocation. In: Krawczyk, H. (ed.) CRYPTO 1998. LNCS, vol. 1462, pp. 137–152. Springer, Heidelberg (1998). https://doi.org/10.1007/BFb0055725
2. Attrapadung, N., Imai, H.: Attribute-based encryption supporting direct/indirect revocation modes. In: Parker, M.G. (ed.) IMACC 2009. LNCS, vol. 5921, pp. 278–300. Springer, Heidelberg (2009). https://doi.org/10.1007/978-3-642-10868-6_17
3. Boldyreva, A., Goyal, V., Kumar, V.: Identity-based encryption with efficient revocation. In: Ning, P., Syverson, P.F., Jha, S. (eds.) CCS 2008, pp. 417–426. ACM, New York (2008)
4. Boneh, D., Franklin, M.: Identity-based encryption from the weil pairing. In: Kilian, J. (ed.) CRYPTO 2001. LNCS, vol. 2139, pp. 213–229. Springer, Heidelberg (2001). https://doi.org/10.1007/3-540-44647-8_13
5. Cui, H., Deng, R.H., Li, Y., Qin, B.: Server-aided revocable attribute-based encryption. In: Askoxylakis, I., Ioannidis, S., Katsikas, S., Meadows, C. (eds.) ESORICS 2016. LNCS, vol. 9879, pp. 570–587. Springer, Cham (2016). https://doi.org/10.1007/978-3-319-45741-3_29
6. Gentry, C.: Certificate-based encryption and the certificate revocation problem. In: Biham, E. (ed.) EUROCRYPT 2003. LNCS, vol. 2656, pp. 272–293. Springer, Heidelberg (2003). https://doi.org/10.1007/3-540-39200-9_17
7. Goyal, V.: Certificate revocation using fine grained certificate space partitioning. In: Dietrich, S., Dhamija, R. (eds.) FC 2007. LNCS, vol. 4886, pp. 247–259. Springer, Heidelberg (2007). https://doi.org/10.1007/978-3-540-77366-5_24
8. Goyal, V., Pandey, O., Sahai, A., Waters, B.: Attribute-based encryption for fine-grained access control of encrypted data. In: Juels, A., Wright, R.N., di Vimercati, S.D.C. (eds.) CCS 2006, pp. 89–98. ACM, New York (2006)
9. Green, M., Hohenberger, S., Waters, B.: Outsourcing the decryption of ABE ciphertexts. In: 2011 20th USENIX Security Symposium. USENIX Association (2011)
10. Libert, B., Vergnaud, D.: Adaptive-ID secure revocable identity-based encryption. In: Fischlin, M. (ed.) CT-RSA 2009. LNCS, vol. 5473, pp. 1–15. Springer, Heidelberg (2009). https://doi.org/10.1007/978-3-642-00862-7_1
11. Micali, S.: Efficient certificate revocation. Technical report, Cambridge, MA, USA (1996)

12. Qin, B., Deng, R.H., Li, Y., Liu, S.: Server-aided revocable identity-based encryption. In: Pernul, G., Ryan, P.Y.A., Weippl, E. (eds.) ESORICS 2015. LNCS, vol. 9326, pp. 286–304. Springer, Cham (2015). https://doi.org/10.1007/978-3-319-24174-6_15
13. Rouselakis, Y., Waters, B.: Practical constructions and new proof methods for large universe attribute-based encryption. In: Sadeghi, A., Gligor, V.D., Yung, M. (eds.) CCS 2013, pp. 463–474. ACM, New York (2013)
14. Sahai, A., Seyalioglu, H., Waters, B.: Dynamic credentials and ciphertext delegation for attribute-based encryption. In: Safavi-Naini, R., Canetti, R. (eds.) CRYPTO 2012. LNCS, vol. 7417, pp. 199–217. Springer, Heidelberg (2012). https://doi.org/10.1007/978-3-642-32009-5_13
15. Sahai, A., Waters, B.: Fuzzy identity-based encryption. In: Cramer, R. (ed.) EUROCRYPT 2005. LNCS, vol. 3494, pp. 457–473. Springer, Heidelberg (2005). https://doi.org/10.1007/11426639_27
16. Seo, J.H., Emura, K.: Revocable identity-based encryption revisited: security model and construction. In: Kurosawa, K., Hanaoka, G. (eds.) PKC 2013. LNCS, vol. 7778, pp. 216–234. Springer, Heidelberg (2013). https://doi.org/10.1007/978-3-642-36362-7_14
17. Yang, Y., Ding, X., Lu, H., Wan, Z., Zhou, J.: Achieving revocable fine-grained cryptographic access control over cloud data. In: Desmedt, Y. (ed.) ISC 2013. LNCS, vol. 7807, pp. 293–308. Springer, Cham (2015). https://doi.org/10.1007/978-3-319-27659-5_21

A New Direction for Research on Data Origin Authentication in Group Communication

Robert Annessi$^{(\boxtimes)}$ (iD), Tanja Zseby$^{(\boxtimes)}$ (iD), and Joachim Fabini$^{(\boxtimes)}$

Institute of Telecommunications, TU Wien, Vienna, Austria
{robert.annessi,tanja.zseby,joachim.fabini}@tuwien.ac.at

Abstract. Group communication facilitates efficient data transmission to numerous receivers by reducing data replication efforts both at the sender and in the network. Group communication is used in today's communication networks in many ways, such as broadcasting in cellular networks, IP multicast on the network layer, or as application layer multicast. Despite many efforts in providing data origin authentication for specific application areas in group communication, no efficient and secure all-purpose solution has been proposed so far.

In this paper, we analyze data origin authentication schemes from 25 years of research. We distinguish three general approaches to address the challenge and assign six conceptually different classes to these three approaches. We show that each class comprises trade-offs from a specific point of view that prevent the class from being generally applicable to group communication. We then propose to add a new class of schemes based on recent high-performance digital signatures. We argue that the high-speed signing approach is secure, resource efficient, and can be applied with acceptable communication overhead. This new class therefore provides a solution that is generally applicable and should be the foundation of future research on data origin authentication for group communication.

1 Introduction

Group communication is ubiquitous in today's communication networks. It facilitates transparent and efficient data transmission to numerous receivers by minimizing data replication efforts both at the sender and in the network. In this paper, we use the term group communication for any one-to-many communication such as multicast, broadcast, or point-to-multipoint communication. Group communication is a generic concept and can be implemented on different layers: data link (Ethernet, Asynchronous Transfer Mode (ATM), or Infiniband), network (IPv4, IPv6) and application layer using overlay networks). It is used in content broadcasting, video conferencing, information distribution (stock-market, software updates, etc.) and Massively Multiplayer Online Games (MMOGs). It is applied in Content Delivery Networks (CDNs), Peer to Peer (P2P), cellular and wireless sensor networks. In this way, the high-speed signing class proposed

© Springer Nature Switzerland AG 2018
S. Capkun and S. S. M. Chow (Eds.): CANS 2017, LNCS 11261, pp. 515–525, 2018.
https://doi.org/10.1007/978-3-030-02641-7_26

in this paper is applicable to a huge number of use cases as it is not tied to specific communication networks technologies, network topologies, or applications. While this work focuses on one-to-many communication, the high-speed signing class may be also applied to many-to-many communication settings as long as each sender uses its own signature or the signing key is shared among senders.

Group communication comprises various challenges, many of which stem from its unidirectional nature and dynamic group membership. Some challenges can be solved easier on higher layers, such as guaranteeing reliable delivery of packets. Other issues tend to reoccur, such as efficient and secure authentication of the sender, no matter on which layer group communication functionality is implemented. This reoccurring, fundamental problem in group communication – the authentication of the sender – is called data origin authentication[1]. Despite more than 25 years of research on data origin authentication for group communication, during which various ideas have been proposed, no sufficiently efficient and secure scheme as yet exists that could be deployed generally on a large scale. For this reason, application-specific solutions are developed that may employ sub-optimal or even insecure data origin authentication schemes.

In this position paper, we argue that the most promising research path for securing group communication is to design faster authentication schemes. Furthermore, we show that high-performance signature schemes significantly elevate the solution space and provide a general solution to the authentication challenge in group communication. We are especially concerned that unsuitable data origin authentication schemes may be deployed in future protocols, e.g., in time synchronization protocols [2] or Smart Grid communication [3,4] and hope that this paper provides valuable details to future protocol designers to guide their decisions and encourages them to use high-speed signatures to solve the authentication challenge in their group communication scenarios.

2   Background

Since most communication networks lack strict access control and network nodes may be compromised, cryptographic methods are needed to assure receivers that packets have been indeed sent by a legitimate sender and data has not been modified by unauthorized entities. For group communication, two types of authentication have to be distinguished [5]: group authentication and data origin authentication.

Group Authentication assures that data originates from a legitimate but unidentified group member and has not been modified by entities outside the group. Message Authentication Codes (MACs) with a key shared by all group members are a well understood and efficient method for achieving group authentication. However, receivers cannot distinguish between the individual group members because they are all sharing the same key and can therefore generate valid MACs. This is of particular importance in group communication since there are usually many receivers involved, and a single dishonest or compromised

---

[1] Sometimes still referred to as *source authentication*, a term considered deprecated [1].

receiver is sufficient to impersonate the sender. Besides this security issue, MACs are also rather inefficient in group communication as the shared key needs to be renewed and redistributed every time a receiver leaves or joins the group.

Data origin authentication allows receivers to verify that data was indeed sent by a specific sender (non-repudiation). This can be achieved by digital signatures using asymmetric cryptographic, because only the sender is in possession of the secret key required to generate signatures. Table 1 summarizes the security properties provided by group authentication and data origin authentication. The main downside of today's digital signature schemes such as RSA [6], DSA [7], and ECDSA [8] is, however, that they come at high computational cost and therefore introduce substantial penalty in terms of delay, both in the sender and in the receiver. Consequently, it is widely believed that digital signatures are roughly 2 to 3 magnitudes slower than MACs [9] so that signing each packet is not a practical solution. We will revise this assumption as we show the potential of recently proposed high-performance digital signature schemes as foundation for data origin authentication in group communication.

Table 1. Group authentication vs. data origin authentication

Security property	Group authentication	Data origin authentication
Integrity	✓	✓
Non-repudiation	✗	✓
Authenticity	Group	Sender

3 Data Origin Authentication Schemes

Data origin authentication schemes for group communication have matured over more than 25 years, and many ideas were proposed to solve this challenging problem. None of the proposed schemes, however, satisfies all constraints and requirements of applications so that naming a single superior scheme seems non-trivial [10]. Challal, Bettahar, and Bouabdallah identified six distinct classes [11] in the sheer number of data origin authentication schemes: deferred signing[2], signature propagation, signature dispersal, secret-information asymmetry, time-based asymmetry, and hybrid asymmetry. In addition to these six classes, we wish to suggest a new class – high-speed signing – where recently proposed high-speed signature schemes are employed.

We identify three conceptional distinct approaches among the proposed data origin authentication schemes. The first approach aims to extend symmetric

---

[2] Challal, Bettahar, and Bouabdallah originally used the term *"differed signing"* but we think that they actually meant *"deferred signing"* as it makes more sense in this context.

schemes to data origin authentication. The other two approaches aim to overcome the computational intensive nature of public-key based authentication schemes: reducing the cost of conventional signatures schemes and designing fast authentication schemes. Table 2 shows the six previously proposed classes as well as our new high-speed signing class assigned to the three approaches we identified.

In this section, we briefly introduce all six classes of data origin authentication schemes and show that each of them comprises a trade-off from a specific point of view. We then argue in Sect. 4 that our high-speed signing class does not require any of those trade-offs.

Table 2. Approaches to Data Origin Authentication and Classes of Schemes

Approach	Class
Extend symmetric schemes to data origin authentication	Secret-information asymmetry
Reduce the cost of conventional signature schemes	Deferred Signing
	Signature propagation
	Signature dispersal
Design fast authentication schemes	Time-based asymmetry
	Hybrid asymmetry
	High-speed signing

3.1   Extending Symmetric Schemes for Data Origin Authentication

Secret-Information Asymmetry. With secret-information asymmetry schemes, such as k-MAC [12], the sender shares a set of keys with receivers (instead of only one key). The sender knows the entire set of keys and therefore can generate valid authentication information but each receiver's partial view allows just to verify (but not to generate) authentication information. The k-MAC scheme uses distinct keys to calculate receiver-specific MACs. Then, all MACs are appended to a packet. Upon reception of this packet, each receiver can verify one MAC it has the key for but cannot create valid authentication information on behalf of the sender as all the other keys are unknown.

The class of secret-information asymmetry schemes entails information-theoretically secure schemes, which means that they do not provide enough information to enable attacks. In this way, these schemes protect against adversaries with potentially unlimited computational power. However, secret-information asymmetry schemes are prone to collusion of receivers, where fraudulent receivers collaborate in order to reconstruct the sender's entire set of keys [11]. Furthermore, secret-information asymmetry schemes require substantial computational resources for signing (and for verification) and also need to distribute new keys individually to each receiver frequently.

3.2   Reducing the Cost of Conventional Signature Schemes

Deferred Signing. With deferred signing, such as offline/online signing [13], the signing process is split into two steps: a slow offline and a fast online step. In the online step, each packet is signed using a one-time signature scheme, which is computationally very efficient. The one-time keys need to be certified to ensure that they originate from the claimed sender. For this purpose, a conventional signature scheme with a certified public key is used in the offline step to sign each one-time key. The generation and signing of the one-time keys is independent of the actual packet to be signed and, therefore, can be conducted offline in advance.

High performance in the online signing part can be achieved because packets are signed with a computationally very efficient one-time signature scheme. The computationally expensive part, precomputing the one-time keys and signing each of them with a conventional signature scheme, is conducted offline. The computational effort required in the offline part, however, is substantial and the communication overhead is large because of the size of the one-time signatures.

Signature Propagation. Another approach to reduce the cost of conventional signatures is followed by signature propagation schemes, such as Receiver driven Layered Hash-chaining (RLH) [14]. Instead of signing each packet individually, a signature from a conventional signature scheme is appended to one packet only, the signature packet. Hashes of non-signature packets are included in preceding packets such that a chain of packets is built in which each packet carries the hash of the subsequent packet. In this way, the digital signature propagates through all packets so that the computational cost of its generation is amortized as hash operations are computationally inexpensive. Signature propagation schemes, however, require packets to be buffered at the sender or at the receiver before they can be signed and their signature be verified, respectively. Such buffering introduces additional delay that may be intolerable to specific applications, such as real-time applications. Receiver-side buffering additionally increases the risk for Denial of Service (DoS) attacks as buffers may be filled with bogus packets by an attacker with access to the network. Furthermore, signature propagation schemes rely on the successful reception of signature packets and are, therefore, hardly resistant to packet loss.

Signature Dispersal. The basic idea behind signature dispersal schemes, such as [15], is that packets are divided into fixed-size blocks, and each block is signed independently with a digital signature. The signature of a block is split, and each part of the signature is appended to one packet (from the same block). Also, additional information is appended to each packet, which helps receivers to reconstruct the signature even if some packets were lost. In this way, signature dispersal schemes improve packet loss resistance compared to signature propagation schemes that entirely rely on the reception of signature packets. Computational efficiency is reduced, however, and receivers need to wait for the whole block before they can verify its authenticity.

3.3   Designing Faster Authentication Schemes

Compared to reducing the computational cost of digital signature schemes, a conceptional distinct approach is designing fast authentication schemes. We distinguish three different classes that follow the approach of designing faster authentication schemes: time-based asymmetry, hybrid asymmetry, and the high-speed signing class we wish to suggest.

Time-Based Asymmetry. In time-based asymmetry schemes, such as Timed Efficient Stream Loss-tolerant Authentication (TESLA) [16,17], key asymmetry is achieved through a common notion of time. In TESLA, the secret and the public key are identical - they are only separated through time. While the key is secret, it is used to sign messages. Meanwhile, clients buffer messages and can verify their authenticity after the (secret) key has been disclosed and therefore becomes public. Once the key is disclosed, the sender has to switch to another key to sign new messages. A common notion of time guarantees that a key is known by clients only after it is not used anymore for signing messages. The keys are associated by a one-way chain such that only the initial key needs to be signed with a (certified) key from a conventional signature scheme.

Computational efficiency is achieved by basically employing symmetric keys (and introducing asymmetry through time). Also, packet loss resistance is provided as packets are signed independently from each other and receivers can recover from having lost keys due to the one-way chain. However, at some point the last key from the chain is used, and new keys need to be generated and distributed securely. Such secure out-of-band channel for key distribution may not be available to all applications. Furthermore, the clocks of the sender and of receivers are assumed to be strictly synchronized such that the accuracy of time synchronization becomes a security requirement. In case the assumed time synchronization accuracy does not hold, the security of the authentication scheme breaks entirely, which is a severe drawback for those applications that cannot guarantee accurately synchronized clocks.

Hybrid Asymmetry. The aim of schemes in the hybrid asymmetry class, such as Time Valid Hash to Obtain Random Subsets (TV-HORS) [18], is to combine the strengths of secret-information asymmetry schemes (immediate signing and verification) and time-based asymmetry schemes (computational efficiency) while mitigating their limitations (no resistance to collusion attacks and strict dependency on time synchronization). Hybrid asymmetry schemes are computationally efficient, but they introduce additional communication overhead and still depend on loose time synchronization between sender and receivers. Like in time-based asymmetry schemes, the keys used in hybrid asymmetry schemes can only sign a fixed number of packets. Once this limit is reached, a new key has to be generated and distributed securely in order to sign more packets. Again, such secure out-of-band channel may not be available to all applications.

4   High-Speed Signing

Two classes of data origin authentication schemes, time-based asymmetry and hybrid asymmetry, already go into – what we consider to be – the right direction as they do not aim to reduce the computational cost of conventional signature schemes but aim to design fast authentication schemes in the first place. An implicit assumption from schemes in those classes is that digital signature schemes are computationally too expensive by nature. This assumption, however, only holds for conventional but not for novel high-performance signature schemes. For this reason, we argue to sign every single packet independently despite the common assumption that such approach is impractical due to the computationally expensive nature of (conventional) signature schemes. Employing high-performance signature schemes can mitigate the negative performance impact of conventional schemes.

For this purpose, signature schemes that offer previously unrivaled performance are needed such as Ed25519 [19], an elliptic-curve signature scheme *"carefully engineered at several levels of design and implementation to achieve very high speed without compromising security"* [19], or MQQ-SIG [20] a signature scheme based on multivariate-quadratic (MQ) quasigroups. Both schemes are designed to provide extremely fast signing and verification operations. Since many MQ signature schemes have been broken (including MQQ-SIG [21]) and some of them have been fixed and broken again, it is safe to say that MQ schemes involve serious security challenges. For this reason, we do not recommend to use MQQ-SIG specifically in practice. Nevertheless, we include MQQ-SIG in our evaluation since MQ schemes have very attractive properties (specifically post-quantum security and high-performance), and MQQ-SIG is one of the fastest of MQ signature schemes. Furthermore, we hope that highlighting group communication use-cases spurs future research on MQ schemes even more.

Performance Measurement. In a small experiment, we measured the speed of these high-performance signature schemes on Commercial Off-The-Shelf (COTS) hardware, an Intel Celeron CPU clocked at 2.26 GHz running Debian Linux 8 32-bit. We disabled Intel's Hyper-threading and Turbo Boost, CPU-frequency scaling, and CPU-sleep states to not interfere with the measurement. Ed25519 signed and verified about 13k packets per second and has a communication overhead of 64 B per packet. MQQ-SIG signed and verified over 36k packets per second with a communication overhead of 32 B per packet. In this way, high-speed signing outperforms[3] TV-HORS from the hybrid asymmetry class, which can sign and verify only 5k packets per second with a communication overhead of 106 B per packet according to [4]. Table 3 summarizes the measurement results.

Because of this high computational efficiency and low communication overhead, there is no need to trade-off other properties like in all the other classes of data origin authentication schemes. High-speed signing provides immediate signing and verification as neither the sender nor the receivers need to buffer

---

[3] Admittedly, the measurements were not conducted under the exact same conditions.

Table 3. Measurement Results

Scheme	Signing and verification	Overhead
Ed25519	13k packets/s	64 B/packet
MQQ-SIG	36k packets/s	32 B/packet
TV-HORS [4]	5k packets/s	106 B/packet

packets. It provides collusion resistance since every receiver has identical information, the sender's public key. Authentication schemes can obviously not be completely independent of time synchronization since the validity of the sender's public key needs to be verified. However, while the other classes that follow the same approach (of designing fast authentication schemes) depend on the time synchronization's accuracy in the order of seconds to minutes, the high-speed signing class' dependency is in the order of months to years and therefore practically as independent as possible. Furthermore, high-speed signing provides resistance to packet loss as each packet carries independent authentication information. Table 4 provides a summary of the classes of data origin authentication schemes.

Table 4. Summary of data origin authentication classes

Class	Computational efficiency	Low communication overhead	Immediate signing and verification	Collusion resistance	Resistance against packet loss	Independence of time synchronization	Only initial key distribution	Information-theoretical security
Secret-Information Asymmetry	✗	✗	✓	✗	✓	✓	✗	✓
Deferred Signing	~	✗	✓	✓	✓	✓	✓	✗
Signature Propagation	✓	~	✗	✓	✗	✓	✓	✗
Signature Dispersal	~	~	✗	✓	~	✓	✓	✗
Time-Based Asymmetry	✓	✓	~	✓	✓	✗	✗	✗
Hybrid Asymmetry	✓	~	✓	✓	✓	~	✗	✗
High-Speed Signing	✓	✓	✓	✓	✓	✓	✓	✗

Property is either satisfied (✓), somewhat satisfied (~), or unsatisfied (✗).

5 Discussion

As highlighted in this paper, each previously existing class of data origin authentication schemes comprises a trade-off from a specific point of view. Secret-asymmetry schemes trade-off information-theoretical security against collusion resistance, which means that they protect against adversaries with unlimited computational resources but are prone to fraudulent receivers who collaborate in order to impersonate the sender. Deferred signing schemes trade-off online computational resources against communication overhead and offline computational resources. Signature propagation schemes trade-off computational efficiency and communication overhead against packet loss resistance as they rely on the successful reception of signature packets - from the moment a signature packet is missing the receiver cannot authenticate any more packets. Signature dispersal schemes trade-off packet loss resistance against computational efficiency and immediate signing and verification such that the sender and the receivers need to wait before they can sign and verify packets, respectively, which is a drawback to applications with real-time requirements. Time-based asymmetry schemes trade-off computational efficiency and communication overhead against independency of time synchronization and require a secure out-of-band channel for key distribution. Hybrid asymmetry schemes trade-off computational efficiency against a secure out-of-band channel for key distribution as well (and, by a smaller degree, also against independency of time synchronization). The high-speed signing class, on the other hand, provides all desired properties (except information-theoretical security) without having to trade-off one against the other.

6 Conclusion

In this position paper, we tackled a fundamental challenge in secure group communication – data origin authentication. We identified three basic approaches to data origin authentication: extending symmetric schemes to data origin authentication, reducing the cost of conventional digital signature schemes, and designing fast authentication schemes. For every approach, we investigated the associated classes of data origin authentication schemes and showed that schemes from each class comprise a trade-off from a specific point of view.

We introduced a new class of data origin authentication schemes, high-speed signing, that follows the approach of designing fast authentication schemes. This high-speed signing class employs a simple yet new approach to data origin authentication for group communication – signing every packet independently with a high-performance digital signature scheme. Signing every packet is commonly assumed to be impractical due to the high computational cost of conventional digital signature schemes. We revised this assumption, however, as we showed that recently proposed high-performance digital signature schemes are perfectly suitable as foundation to data origin authentication as they achieve computational efficiency, low communication overhead, as well as all other desired properties (besides information-theoretical security).

We hope that this position paper helps to avoid employing unsuitable data origin authentication schemes in various fields in the future such as in time synchronization [2], where a time-based asymmetry scheme is currently proposed in standardization, or in Smart Grids [3,4], where a hybrid-asymmetry scheme and reducing the computational cost of a conventional digital signature scheme have been proposed just recently. Concluding, we argue that designing fast authentication schemes for group communication is generally the right direction but research should focus on high-speed digital signature schemes instead of other classes in order to solve the problem of data origin authentication for secure group communication.

References

1. Shirey, R.: Internet Security Glossary, Version 2. RFC 4949 (Informational). Internet Engineering Task Force, August 2007. http://www.ietf.org/rfc/rfc4949.txt
2. Sibold, D., Roettger, S., Teichel, K.: Network Time Security. Internet-Draft draft-IETF-NTP-network-time-security-15. IETF Secretariat, September 2016. https://tools.ietf.org/html/draft-ietf-ntp-network-time-security-15. Accessed 08 Mar 2017
3. Law, Y.W., et al.: Comparative study of multicast authentication schemes with application to wide-area measurement system. In: ACM SIGSAC Symposium on Information, Computer and Communications Security, ASIACCS 2013, pp. 287–298. ACM, NY (2013). https://doi.org/10.1145/2484313.2484349, ISBN 978-1-4503-1767-2
4. Tesfay, T., Le Boudec, J.-Y.: Experimental comparison of multicast authentication for wide area monitoring systems. IEEE Trans. Smart Grid 9, 4394–4404 (2017). https://doi.org/10.1109/TSG.2017.2656067. ISSN 1949–3053, 1949–3061
5. Hardjono, T., Tsudik, G.: IP multicast security: issues and directions. Annales des télécommunications 55(7–8), 324–340 (2000)
6. Rivest, R.L., Shamir, A., Adleman, L.: A method for obtaining digital signatures and public-key cryptosystems. Commun. ACM 21(2), 120–126 (1978)
7. ElGamal, T.: A public key cryptosystem and a signature scheme based on discrete logarithms. In: Blakley, G.R., Chaum, D. (eds.) CRYPTO 1984. LNCS, vol. 196, pp. 10–18. Springer, Heidelberg (1985). https://doi.org/10.1007/3-540-39568-7_2
8. Johnson, D., Menezes, A., Vanstone, S.: The elliptic curve digital signature algorithm (ECDSA). Int. J. Inf. Secur. 1(1), 36–63 (2001)
9. Katz, J.: Digital Signatures. Springer, Boston (2010). https://doi.org/10.1007/978-0-387-27712-7. ISBN 978-0-387-27711-0, 978-0-387-27712-7
10. Steinwandt, R., Villányi, V.I.: A one-time signature using run-length encoding. Inf. Process. Lett. 108(4), 179–185 (2008). https://doi.org/10.1016/j.ipl.2008.05.004. ISSN 0020–0190
11. Challal, Y., Bettahar, H., Bouabdallah, A.: A taxonomy of multicast data origin authentication: issues and solutions. IEEE Commun. Surv. Tutor. 6(3), 34–57 (2004). https://doi.org/10.1109/COMST.2004.5342292. ISSN 1553–877X
12. Canetti, R., et al.: Multicast security: a taxonomy and some efficient constructions. In: Eighteenth Annual Joint Conference of the IEEE Computer and Communications Societies, INFOCOM 1999, Vol. 2, pp. 708–716, March 1999. https://doi.org/10.1109/INFCOM.1999.751457
13. Even, S., Goldreich, O., Micali, S.: On-line/off-line digital signatures. J. Cryptol. 9(1), 35–67 (1996)

14. Challal, Y., Bouabdallah, A., Hinard, Y.: RLH: receiver driven layered hash-chaining for multicast data origin authentication. Comput. Commun. **28**(7), 726–740 (2005)
15. Tartary, C., Wang, H., Ling, S.: Authentication of digital streams. IEEE Trans. Inf. Theory **57**(9), 6285–6303 (2011). https://doi.org/10.1109/TIT.2011.2161960. ISSN 0018-9448
16. Perrig, A., et al.: Efficient authentication and signing of multicast streams over lossy channels. In: IEEE Symposium on Security and Privacy (S&P), pp. 56–73 (2000)
17. Perrig, A., et al.: Timed Efficient Stream Loss-Tolerant Authentication (TESLA): Multicast Source Authentication Transform Introduction. RFC 4082 (Informational). Internet Engineering Task Force, June 2005. http://www.ietf.org/rfc/rfc4082.txt
18. Wang, Q., et al.: Time valid one-time signature for time-critical multicast data authentication. In: IEEE INFOCOM 2009, pp. 1233–1241, April 2009. https://doi.org/10.1109/INFCOM.2009.5062037
19. Bernstein, D.J., et al.: High-speed high-security signatures. J. Cryptogr. Eng. **2**(2), 77–89 (2012)
20. Gligoroski, D., et al.: MQQ-SIG. In: Chen, L., Yung, M., Zhu, L. (eds.) INTRUST 2011. LNCS, vol. 7222, pp. 184–203. Springer, Heidelberg (2012). https://doi.org/10.1007/978-3-642-32298-3_13
21. Faugère, J.-C., Gligoroski, D., Perret, L., Samardjiska, S., Thomae, E.: A polynomial-time key-recovery attack on MQQ cryptosystems. In: Katz, J. (ed.) PKC 2015. LNCS, vol. 9020, pp. 150–174. Springer, Heidelberg (2015). https://doi.org/10.1007/978-3-662-46447-2_7

Modelling Traffic Analysis in Home Automation Systems

Frederik Möllers[1(✉)], Stephanie Vogelgesang[2], Jochen Krüger[1], Isao Echizen[3], and Christoph Sorge[1]

[1] CISPA, Saarland Informatics Campus, Saarbrücken, Germany
frederik.moellers@uni-saarland.de
[2] Ministry of Justice, Saarbrücken, Saarland, Germany
[3] National Institute of Informatics, Tokyo, Japan

Abstract. The threat of attacks on Home Automation Systems (HASs) is increasing. Research has shown that passive adversaries can detect user habits and interactions. Despite encryption and other measures becoming a standard, traffic analysis remains an unsolved problem. In this paper, we show that existing solutions from different research areas cannot be applied to this scenario. We establish a model for traffic analysis in Home Automation Systems which allows the analysis and comparison of attacks and countermeasures. We also take a look at legal aspects, highlighting problem areas and recent developments.

1 Introduction

HASs are an emerging trend in consumer electronics. The benefits are promising: an increased comfort of living; savings on energy and resource consumption; increased safety and security. However, HASs have been developed with a focus on usability, energy efficiency and low cost. In addition to active attacks, adversaries can passively intercept communication using cheap, available hardware. Smart homes can thus actually facilitate privacy breaches.

Encryption and other methods do not completely solve this problem. Traffic analysis attacks disclose habits as well as presence or absence of users. In order to counter this, dummy traffic can be generated by the system. However, generating too much dummy traffic negatively affects the lifetime of battery-powered devices and can exceed regulatory thresholds.

Our contributions in this paper are as follows:

- We establish system and attacker models for traffic analysis attacks in Home Automation Systems. They allow modelling realistic attack scenarios such as those shown in previous works and are extensible to account for new findings.
- We formulate privacy goals for Home Automation Systems using ideas from the field of Private Information Retrieval. By building on established definitions, we can leverage research that has been conducted in related fields.
- We illustrate the application of our definitions of privacy goals by examining two approaches to dummy traffic.

S. Capkun and S. S. M. Chow (Eds.): CANS 2017, LNCS 11261, pp. 526–536, 2018.
https://doi.org/10.1007/978-3-030-02641-7_27

– We sketch issues and current developments in the interaction between technology and legal frameworks (especially criminal law). We briefly describe how technology and international law are intertwined and can work together as a comprehensive concept for data protection.

2 Related Work

Research so far has tackled related problems in different scenarios. For HASs, problems have been identified but no solutions have been proposed so far.

2.1 Wireless Sensor Networks

In both Wireless Sensor Networks (WSNs) and HASs, devices have little computational power and communication is costly in terms of power consumption. However, research on privacy in WSNs mainly focuses on location privacy [1–3] instead of hiding the existence of communication.

Yang et al. have proposed a scheme employing constant-rate dummy traffic generation which could be applied to systems not using multi-hop routing [4]. We examine such a scheme in Sect. 5.2. The authors also propose a scheme for random dummy traffic generation requiring much less overall traffic [5], but this introduces delays which are to be avoided in HASs. Furthermore, the behaviour of the inhabitants might not conform to the distribution assumed in their work.

2.2 MIX Networks

Early approaches of dummy traffic generation [6] use constant-rate traffic, which we examine in Sect. 5.2. More recently, there has been significant research on traffic analysis in low-latency MIX networks [7]. Shmatikov and Wang [8] have developed an approach which aims to find a balance between the amount of dummy traffic and the success rate of traffic analysis attacks. Their approach may be applicable to HAS networks with modifications, though their evaluation is based on HTTP traffic samples. It is unclear how much of the approach can be applied to HASs and whether the same goals can be achieved.

2.3 Differential Privacy

Definitions of Differential Privacy are used to develop techniques which provide unobservability of events or user data. However, ε-Differential Privacy is a property of a specific function [9]. In our scenario we do not know the informational value of specific data points and we do not know which computations an attacker will perform on captured traffic.

Dwork et al. have developed an approach to continuously monitor an event source and count its events with the counter guaranteeing ε-Differential Privacy even if its internal state is visible to the attacker at some point in time [10], However, the guarantees do not necessarily apply to other functions.

2.4 Steganography and Covert Channels

Steganography can hide communication without generating any dummy traffic. However, since there is no cover data available in HAS communication, many steganography approaches cannot be applied to this setting. Hiding the traffic among noise of a wireless channel has been investigated by Bash et al. [11]. It is unclear whether approaches on higher layers of the network stack can either exceed some fundamental limitations or achieve the same results at lower (manufacturing or communication) costs. Furthermore, the approach was developed for wireless networks and might not be applicable to wired systems.

2.5 Home Automation Systems

Möllers et al. have shown that unencrypted, wireless HAS communication leaks automation rules and user habits to passive observers [12]. Mundt et al. have shown that wired systems are also susceptible to the same eavesdropping attacks [13]. Encryption and padding do not protect against statistical disclosure attacks, as was shown by Möllers et al. [14]. Other authors have focused on other aspects of security, for example confidentiality of information and access controls [15].

3 System and Attacker Model

In this section, we first list our assumptions and then describe the model.

3.1 Assumptions

System: Network Topology and Routing. We assume that the network graph is a clique with respect to intended communication, i.e. no routing is necessary. The reason behind this assumption is that routing introduces a set of problems, but also opportunities which are already well understood.

System: Encryption. We assume that the HAS uses padding and encryption for both message payloads and addressing information. The attacker cannot break the encryption in reasonable time and is unable to learn information about the sender, receiver or contents of a message. If state-of-the-art approaches are applied correctly, this assumption is reasonable. Even though we do not know of any system actually using both full packet encryption and padding, we consider this to be an engineering problem rather than a research one.

Attacker: Mode of Operation. The attacker in our scenario is passive. No active attacks such as traffic injection, node compromise or DoS are launched.

Traffic injection does not lead to a reaction by the system if encryption and authentication are correctly implemented. Node compromise requires either the presence of vulnerabilities in the nodes' software or physical intervention by the attacker. Denial of service attacks offer no benefit for the attacker (messages are resent after the attack ends) and are detectable.

Attacker: Reception. The attacker has perfect reception as well as an accurate clock. They are able to capture all traffic, do not suffer from reception errors and can save the time at which a message was captured. Both experiments on real-world installations of HASs [12,13] have proven this to be realistic.

Attacker: Limits. The attacker does not launch triangulation or device fingerprinting attacks [16]. These attacks require a considerable amount of effort from the attacker and countermeasures are a separate area of research.

Attacker: Awareness and Knowledge. The attacker is aware of the underlying algorithms of countermeasures, but does not know runtime information (e.g. the internal state of PRNGs). We model privacy goals with respect to a given set of tasks that the user might perform. These may differ vastly depending on the setting, so any assumption reduces the utility of the model.

3.2 System Model

Communication packets from the HAS are observed by the adversary. According to the assumptions, the only information available to the attacker is a *set of message timestamps* (or fingerprints) F.

$$F = R \cup E \cup D \tag{1}$$

where

- R is a set of *regular messages*. These can be from automation rules or reactions to environmental events (e.g. temperature changes) and are of no particular interest to the attacker.
- E is a set of *interesting events* such as direct user interaction (e.g. pressing a ⁀light switch) or anything that is of particular interest to the attacker.
- D is a set of *dummy messages* carrying no information.

Messages from other use cases can be put into either group, depending on the scenario. Events from remote user interaction (using an internet gateway) can e.g. be put into R as they leak no information about user presence.

Any of the subsets can be empty. If the HAS consists of a single actuator with a remote control (i.e. no automation rules), then $R = \emptyset$. If it consists of a single temperature sensor periodically sending data to a base station, then $E = \emptyset$.

We assume that $R \cap E \cap D = \emptyset$. If the system supports multiple concurrent channels and the channel does not leak information about a message's contents, the timestamps can be annotated with the channel ID.

3.3 Attacker Model

For a given capture interval $[x, y]$ (x and y are timestamps), the attacker observes the HAS's output $f^{x,y} \subseteq F$. If $t(m)$ denotes the timestamp of message m,

$$f^{x,y} = \{m \in F \mid t(m) \geq x \land t(m) \leq y\} \tag{2}$$

We define all possible subsets $f \subseteq F$ satisfying this condition as the *Subsequence Set* $\mathbb{S}(F)$. Note that when modelling HASs, message timestamps follow a random distribution, so $\mathbb{S}(F)$ is a set of random distributions (one for every possible capture interval) rather than a set of sequences (or sets of concrete timestamps).

4   Privacy Goals

Toledo et al. have presented a model for information leakage in Private Information Retrieval settings [17]. We formulate our definitions similar to theirs in order to facilitate the development and analysis of countermeasures.

The majority of [17] is not applicable to our setting. In particular, we do not have equivalents for (corrupted) databases/servers. We instead focus on the user (U_t in [17]) and the adversary (A). Furthermore, our Subsequence Set $\mathbb{S}(F)$ relates to the adversarial observation space (Ω): For a given time frame $[x, y]$, the observation space $\Omega_{x,y}$ is the distribution $f^{x,y} \in \mathbb{S}(F)$. An observation ($O$) is a sample from this random distribution. Corresponding to the queries (Q_i, Q_j), the attacker in the HAS scenario provides the user with two *tasks* T_i, T_j (e.g. "Interact with the system during the time $[x, y]$.") of which the user randomly chooses one to execute. The attacker then captures the HAS's traffic (obtaining a sample from the random distribution $f^{x,y}$) and tries to identify the task.

We can thus formulate a notion of privacy in Home Automation Systems.

Definition 1. *A Home Automation System provides (ε-δ)-private communication if there are constants $\varepsilon \geq 0$ and $0 \leq \delta < 1$, such that for any possible adversary-provided tasks T_i, T_j and for all possible adversarial observations O (being a particular random sample of the distribution $f^{x,y} \in \mathbb{S}(F)$) we have that*

$$Pr(O|T_i) \leq e^{\varepsilon} \cdot Pr(O|T_j) + \delta$$

We assume that timestamps are discrete. If they are continuous and $f^{x,y} \in \mathbb{S}(F)$ is a density function, the same definition can be used by substituting the probability for the value of the density function.

As in [17], if $\delta = 0$ we call the stronger property ε-private communication. Note that we require $\delta < 1$. This only affects some cases where a particular observation is certain for one task and impossible for another and prevents the definition from being overly broad.

4.1   Indistinguishability and Unobservability

In practice, if the tasks can be arbitrary, the attacker may choose them to be e.g. "Press the light switch for 10 times in 2 s." and "Do not interact with the system for 10 min", producing distinguishable patterns. In order to account for this, we define a slightly weaker property.

Definition 2. *A Home Automation System provides (ε-δ)-indistinguishability for a set of tasks $\mathbb{T}$ if there are constants $\varepsilon \geq 0$ and $0 \leq \delta < 1$, such that* *for all tasks $T_i, T_j \in \mathbb{T}$ and for all possible adversarial observations O (being a particular random sample of the distribution $f^{x,y} \in \mathbb{S}(F)$) we have that*

$$Pr(O|T_i) \leq e^{\varepsilon} \cdot Pr(O|T_j) + \delta$$

Probability density functions can also be used here for continuous timestamps. If $\delta = 0$, we call the stronger property ε-indistinguishability.

(ε-δ)-indistinguishability is only defined for a limited set of tasks $\mathbb{T}$. Some tasks are theoretically possible, but unlikely to be encountered in practice. For example, a system might be able to provide unobservability of the user pressing a light switch twice within 10 min by making sure that there are always at least two messages in every 10-min interval. While this does not fulfil the goal of (ε-δ)-private communication, it covers much of the everyday activity and might be more energy efficient than a system offering full (ε-δ)-private communication.

When considering real-world attack scenarios like the detection of user presence [14], the tasks provided by the attacker follow a particular pattern. Instead of choosing two unrelated tasks, the attacker wants to extract a certain piece of binary information from the captured data (such as "Did the user interact with the system?"). In this case, the tasks T_i and T_j from the definition are complementary: $T_j = \bar{T}_i$ (i.e. if T_i is "Interact with the system.", then $T_j = \bar{T}_i$ is "Do not interact with the system."). Due to this being an important special case of (ε-δ)-indistinguishability, we define a separate property:

Definition 3. *A HAS provides (ε-δ)-unobservability of a set of tasks $\mathbb{T}$ if*

$$\forall T \in \mathbb{T} : \bar{T} \in \mathbb{T}$$

and the system provides (ε-δ)-indistinguishability for $\mathbb{T}$.

If $\delta = 0$ we call the stronger property ε-unobservability.

These definitions capture our models as well as real attacks [12–14]. Using existing models of user behaviour, one can prove privacy guarantees of a dummy traffic generation scheme.

5   Examples

For trivial approaches it is easy to see whether or not they fulfil the privacy goals.

5.1   No Dummy Traffic

Möllers et al. have analysed a system which does not produce dummy traffic at all [14]. They have shown that the system does not offer ε-unobservability for the tasks "Interact with the system during a one-hour period." and "Do not interact with the system for one hour." if the attacker knows certain thresholds.

In their experiment, the attacker was able to determine conditions which, if met by the adversarial observation O, would reliably indicate user activity or inactivity. If the predicates $P(O)$ and $A(O)$ denote these conditions, then

$$\forall O : P(O) \Rightarrow$$
$$Pr(O|\text{"Interact with the system"}) > 0 \ \wedge \ Pr(O|\text{"Do not interact"}) = 0$$
$$\forall O : A(O) \Rightarrow$$
$$Pr(O|\text{"Interact with the system"}) = 0 \ \wedge \ Pr(O|\text{"Do not interact"}) > 0$$

As $\nexists \varepsilon : e^{\varepsilon} \cdot 0 > 0$, the system does not offer ε-unobservability, ε-indistinguishability or ε-private communication in general.

In their experiment, the probability of obtaining an adversarial observation meeting the condition if the user performed the given task was less than 1. Thus, the system *may* offer (ε-δ)-unobservability.

5.2   Constant-Rate (Dummy) Traffic

Next, we analyse the concept of *Constant Rate (Dummy) Traffic*. We assume that the system is generating dummy traffic if (and only if) there are no genuine messages to send. Time is divided into slots of fixed length. At the end of every timeslot, either one genuine or one dummy message is transmitted.

Formally, if M is the set containing the ending time of each timeslot and messages in R and E are delayed so that they only occur at the end of a timeslot ($R \subseteq M$, $E \subseteq M$), then dummy traffic is generated by the system so that $D = M \setminus (R \cup E)$.

By construction, the output of the system $F = R \cup E \cup D = M$ is exactly the same, no matter how the timestamps of genuine messages in R and E are distributed. Thus, for any interval $[x, y]$ the adversarial observation will be $O = M \cap [x, y]$, which is stochastically independent from the distributions of R and E. Consequently, for any task T to be executed by the user, it holds that $Pr(O|T) = Pr(O) = 1$. In conclusion, a system using constant-rate traffic provides (ε-δ)-private communication with $\varepsilon = \delta = 0$ (or (0-0)-private communication).

In practice, using Constant Rate Traffic poses a problem. In order to keep the delay for user interaction reasonably low, the overall traffic rate must be very high (i.e. ≥ 1 message per second). However, this can lead to the system violating regulatory thresholds or draining the battery of connected devices. In wired systems, this is generally not an issue. In a wireless setting, other approaches which minimise the generated amount of traffic have to be evaluated. The development of such a system is left for future work.

6   The Legal Framework

As with most topics in the field of security and privacy, technical protective measures and the legal framework for prosecution of offenders are intertwined. On the one hand, technical countermeasures make attacks more difficult. On the other hand, an effective legal framework can even act as a deterrent to potential offenders. In this context, we highlight problematic areas in legal frameworks using the German Criminal Code as an example. We then present an international effort aiming to improve the legal situation in over 50 countries.

6.1 Legal Challenges

The interception of traffic from HASs (or any other private network, for that matter) can be considered "data theft". However, the definition of theft in the German Criminal Code (Sect. 242) only applies to *chattels* and not to incorporeal data [18]. Even if the adversary actively intrudes and asserts control of the HAS, the attack does not qualify as trespassing according to Sect. 123 of the German Criminal Code. The law requires physical entry into a spatially delimited area [19, Sect. 123, Recital 15].

These two cases show a fundamental problem with many legal frameworks: Criminal law has been developed in times of limited technical prevalence.

6.2 Legal Reforms

In order to update the law and to keep up with new technical developments, legislators have passed reforms. For the German Criminal Code, these are e.g. Sect. 202a (Data espionage) and 202b (Data Interception). In our scenario, Sect. 202b is to be applied. It punishes the illegal interception of data from a non-public data processing facility or from the electromagnetic broadcast of a data processing facility. While it can be argued that air is a broadcast medium and that wireless transmissions are public by nature, the German Criminal Code bases the definition on the intention of the sender.

As criminal law only applies to the respective country, a detailed examination of the Section's contents is outside the scope of this paper. However, Sects. 202a and 202b are mere examples of a global development: By changing Sect. 202a and introducing Sect. 202b in 2007, the German legislator has implemented the Convention of Cybercrime of the Council of Europe. This guideline, also known as the *Budapest Convention*, has been opened for signature in 2001 and has since been ratified by over 50—European and non-European—countries.

The Budapest Convention is of dogmatic importance for cybercrime and has ramifications on a global scale, especially in the following two areas.

Prosecution. Attacks involving computers ("Cyber Attacks") often reach across borders. National solo efforts to combat these are rarely promising. Instead, a coordinated concept supported by as many countries as possible is necessary.

Technical and Legal Terms. One part of this coordinated concept is a collection of common terms. In this paper, we assume that the attacker can only access traffic (meta-)data. The distinction between content and traffic (or meta-) data is important for the legal framework and can be explicitly found in the Budapest Convention. In Article 2 d), the convention states that

> "traffic data" means any computer data relating to a communication by means of a computer system, generated by a computer system that formed a part in the chain of communication, indicating the communication's origin, destination, route, time, date, size, duration, or type of underlying service.

Establishing common terms and ensuring consistent usage is not a trivial task. For example, the German Telecommunications Act considers location data for mobile devices to be traffic data, but this classification cannot be found in the Budapest Convention.

It remains an open question whether the term *data* as used in Article 3 of the Budapest Convention refers to both traffic and content data or only to content data. Answering this question as well as problems resulting from this are beyond the scope of this paper.[1]

6.3   Summary of the Legal Situation

The Budapest Convention contains regulations about criminal law. Certain actions such as the illegal interception of data are to be penalised. This reveals a central idea: Criminal law can be considered part of a comprehensive data protection concept which aims to optimise both technical and legal aspects.

This understanding does not only hold for the German Criminal Code we used as an example here. Instead, it holds for all countries which have signed or ratified the treaty—implementing it in national law. The list of signatures[2] does not only include major European countries such as France, the UK and Italy. It also features many others such as Australia, Canada, Japan or the USA. The questions sketched here regarding the relationship of technology (in this context, especially Home Automation Technology) and criminal law are therefore of international importance.

7   Conclusion

We have established a model for traffic analysis attacks in Home Automation Systems, keeping assumptions general and adapting existing definitions. The model is suitable for developing dummy traffic generation schemes not only for Home Automation Systems, but for networks with similar properties as well. The definitions ensure that privacy guarantees can be mathematically proven.

We have also shown how technology and attacks using it have forced legal reforms and new laws that surpass national borders. The Budapest Convention

---

[1] This especially holds for questions regarding data retention.
[2] https://www.coe.int/en/web/conventions/full-list/-/conventions/treaty/185/signatures, last accessed 10 July 2017.

serves as an important step towards an internationally agreed understanding of terms and necessary actions. While discrepancies between technology and the laws governing it are unavoidable, we have shown that the two can work together towards the goal of protecting people's privacy.

References

1. Chan, H., Perrig, A.: Security and privacy in sensor networks. Computer **36**(10), 103–105 (2003)
2. Conti, M., Willemsen, J., Crispo, B.: Providing source location privacy in wireless sensor networks: a survey. IEEE Commun. Surv. Tutorials **15**(3), 1238–1280 (2013)
3. Matos, A., Aguiar, R.L., Girao, J., Armknecht, F.: Toward dependable networking: secure location and privacy at the link layer. IEEE Wirel. Commun. **15**(5), 30–36 (2008)
4. Yang, Y., Shao, M., Zhu, S., Urgaonkar, B., Cao, G.: Towards event source unobservability with minimum network traffic in sensor networks. In: Proceedings of WiSec 2008, pp. 77–88. ACM (2008)
5. Shao, M., Yang, Y., Zhu, S., Cao, G.: Towards statistically strong source anonymity for sensor networks. ACM TOSN **9**(3), 34:1–34:23 (2008)
6. Pfitzmann, A., Pfitzmann, B., Waidner, M.: ISDN-mixes: untraceable communication with very small bandwidth overhead. In: Effelsberg, W., Meuer, H.W., Müller, G. (eds.) Kommunikation in verteilten Systemen. Informatik-Fachberichte, vol. 267, pp. 451–463. Springer, Heidelberg (1991). https://doi.org/10.1007/978-3-642-76462-2_32
7. Levine, B.N., Reiter, M.K., Wang, C., Wright, M.: Timing attacks in low-latency mix systems. In: Juels, A. (ed.) FC 2004. LNCS, vol. 3110, pp. 251–265. Springer, Heidelberg (2004). https://doi.org/10.1007/978-3-540-27809-2_25
8. Shmatikov, V., Wang, M.-H.: Timing analysis in low-latency mix networks: attacks and defenses. In: Gollmann, D., Meier, J., Sabelfeld, A. (eds.) ESORICS 2006. LNCS, vol. 4189, pp. 18–33. Springer, Heidelberg (2006). https://doi.org/10.1007/11863908_2
9. Dwork, C.: Differential Privacy. In: Bugliesi, M., Preneel, B., Sassone, V., Wegener, I. (eds.) ICALP 2006. LNCS, vol. 4052, pp. 1–12. Springer, Heidelberg (2006). https://doi.org/10.1007/11787006_1
10. Dwork, C., Naor, M., Pitassi, T., Rothblum, G.N.: Differential privacy under continual observation. In: Proceedings of ACM STOC 2010, pp. 715–724. ACM (2010)
11. Bash, B.A., Goeckel, D., Guha, S., Towsley, D.: Hiding information in noise: fundamental limits of covert wireless communication. IEEE Commun. Mag. **53**(12), 26–31 (2015)
12. Möllers, F., Seitz, S., Hellmann, A., Sorge, C.: Extrapolation and prediction of user behaviour from wireless home automation communication. In: Proceedings of WiSec 2014, pp. 195–200. ACM (2014)
13. Mundt, T., Dähn, A., Glock, H.W.: Forensic analysis of home automation systems. In: HotPETs (2014)
14. Möllers, F., Sorge, C.: Deducing user presence from inter-message intervals in home automation systems. In: Hoepman, J.-H., Katzenbeisser, S. (eds.) SEC 2016. IAICT, vol. 471, pp. 369–383. Springer, Cham (2016). https://doi.org/10.1007/978-3-319-33630-5_25

15. Bergstrom, P., Driscoll, K., Kimball, J.: Making home automation communications secure. Computer **34**(10), 50–56 (2001)
16. Bratus, S., Cornelius, C., Kotz, D., Peebles, D.: Active behavioral fingerprinting of wireless devices. In: Proceedings of WiSec 2008, pp. 56–61. ACM (2008)
17. Toledo, R.R., Danezis, G., Goldberg, I.: Lower-cost ε-private information retrieval. Proc. Priv. Enhancing Technol. **4**, 184–201 (2016)
18. Vogelgesang, S.: Datenspeicherung in modernen Fahrzeugen - wem "gehören" die im Fahrzeug gespeicherten Daten? juris - Die Monatszeitschrift **3**(1), 2–8 (2016)
19. Fischer, T.: Strafgesetzbuch: StGB. 64 edn. C.H.BECK (2017)

VisAuth: Authentication over a Visual Channel Using an Embedded Image

Jack Sturgess$^{(\boxtimes)}$ and Ivan Martinovic

Department of Computer Science, University of Oxford, Oxford, UK
{jack.sturgess,ivan.martinovic}@cs.ox.ac.uk

Abstract. Mobile payment systems are pervasive; their design is driven by convenience and security. In this paper, we identify five common problems in existing systems: (i) specialist hardware requirements, (ii) no reader-to-user authentication, (iii) use of invisible channels, (iv) dependence on a client-server connection, and (v) no inherent fraud detection. We then propose a novel system which overcomes these problems, so as to mutually authenticate a user, a point-of-sale reader, and a verifier over a visual channel, using an embedded image token to transport information, while providing inherent unauthorised usage detection. We show our system to be resilient against replay and tampering attacks.

1 Introduction

The popularity of cashless payments has risen sharply in recent years, surpassing cash payments in some places [1]. Consumers moved from cash to cashless payment cards primarily for convenience, not security—the first generation of magnetic strip payment cards were authenticated with an easily-forged, handwritten signature. These systems were widely replaced with Europay, MasterCard, and Visa (EMV)[1] payment card systems, protected by a chip and secret personal identification number (PIN). The payment card is inserted into a specialist reader and the PIN is entered and verified by the chip to authenticate the user and authorise the payment; more recent cards support contactless payments using near field communication (NFC) between card and reader.

Attacks (*e.g.*, [2,3]) on payment card systems and incidences of fraud continue to occur, so consumers move to new cashless systems with the promise of greater security and convenience. Furthermore, carrying a dedicated payment card is increasingly regarded as inconvenient [4], so newer systems integrate directly with a device which the user would already be carrying, such as a smartphone. Strong reasons for the widespread adoption of tap-and-pay systems include usability and security [5]. Tap-and-pay systems require the user to install an app on a device and provision a payment card to a virtual wallet. In Apple Pay[2], a token is created for each card and stored in the device's secure

---

[1] www.emvco.com/about_emvco.aspx (last accessed: June 2017).
[2] www.apple.com/business/docs/iOS_Security_Guide.pdf (last accessed: June 2017).

S. Capkun and S. S. M. Chow (Eds.): CANS 2017, LNCS 11261, pp. 537–546, 2018.
https://doi.org/10.1007/978-3-030-02641-7_28

element; payments are made between the client and a compatible reader over NFC, with the user authenticating to the app using TouchID (fingerprint) or a passcode to authorise the payment. Android Pay[3] and Samsung Pay[4] are similar, but all card data processing and tokenisation is handled on a cloud server (due to the greater range of supported devices and their differing levels of security), so the client must connect to the server regularly to acquire new tokens; this has a negative impact on usability in areas where WiFi availability is poor.

It is difficult to prevent eavesdropping over invisible channels (*e.g.*, [6,7]) and limiting the range to reduce the risk is not reliable (*e.g.*, [8,9]). Some systems communicate over a visual channel between the client and reader, giving the user more control over the broadcast, potentially making it difficult for an adversary to intercept without being noticed. In Yoyo Wallet[5], a virtual wallet is hosted on a cloud server; the user authenticates to the app using a PIN, then a QR token is passed between the client and a specialist reader to authorise a payment. Each QR token may be used up to three times before the client needs to reconnect to the server, meaning it is more connection-dependent than tap-and-pay systems. Two similar systems, WeChat[6] and AliPay[7], both currently very popular in China, support QR codes and barcodes to transfer information.

There is little compatibility between newer payment systems, and exclusionary business models often mandate the use of specialist or dedicated point-of-sale readers. Merchants struggle to accept them all, so brand loyalty and local trends may factor into consumer decisions, detracting focus from security. Furthermore, the physical presence of a specialist reader may give a false perception of trust to a user that the reader is legitimate (a rogue reader could easily be dressed to look genuine). Anti-phishing systems exist in other forms [10,11] and some banking interfaces (*e.g.*, DoubleSafe[8]) use a personalised greeting message to authenticate to the user before requesting a PIN or password, but we are yet to see this feature in point-of-sale readers—instead, we see measures such as payment limits, which mitigate damage at the expense of usability. Purnomo *et al.* [12] present a system where mutual authentication is achieved via a trusted third party, however it requires a connection throughout.

None of the systems offer an inherent mechanism whereby unauthorised usage is easily detected by the user, aside from manually checking the account balance. Yoyo Wallet comes close: if an adversary were to expend the three uses of a stolen QR token, it would become useless and so indicate a problem when the user next tries to use it (unless the user is online, in which case the token will automatically refresh before use).

---

[3] support.google.com/androidpay (last accessed: June 2017).
[4] www.samsung.com/us/support/answer/ANS00043790 (last accessed: June 2017).
[5] www.yoyowallet.com/support.html (last accessed: June 2017).
[6] pay.weixin.qq.com/index.php/public/wechatpay (last accessed: June 2017).
[7] global.alipay.com/products/spot (last accessed: June 2017).
[8] www.tangerine.ca/en/security (last accessed: Oct. 2017).

In this paper, due to space limitations, we focus on popular, real world systems. We identify five common drawbacks in these systems and propose a new mobile payment scheme with the goal of overcoming those drawbacks.

2   Objectives and Assumptions

Design Objectives. The purpose of our system is to authenticate a user to a verifier using a client and via a point-of-sale reader to authorise a payment; the system should also provide the following features to overcome the common drawbacks identified in existing systems:

- *No specialist hardware requirement.*
- *Mutual authentication:* the system should authenticate the user to the verifier via the reader; it should also authenticate the verifier and the reader to the user *before* he authenticates to it, so as not to reveal secrets to a rogue reader.
- *Visual channel:* the system should operate over a visual channel between the client and the reader.
- *No client-to-verifier connection requirement.*
- *Unauthorised usage detection:* the user should be told if an unauthorised user has impersonated him, in a way that the intruder cannot avoid or erase.

System Model. The system consists of four components: a *user* (prover); a *client*—i.e., a user device, such as a smartphone, with a camera, a screen, and our app installed on it; a *verifier*, such as an authentication server, which maintains a database of users' cryptographic materials (see Sect. 3); and a point-of-sale *reader* with a camera, a screen, and a means to enter a PIN (such as a touchscreen).

During enrollment, we assume that there is a secure channel between the client and the verifier over which they exchange cryptographic materials. The user will choose a PIN and a personalised message; we assume the former can be reset only by connecting with the verifier, whereas the latter can be changed for freshness on the client at any time without connecting to the verifier.

During authentication, we assume that there is a visual channel between the client and the reader, and a secure channel between the reader and the verifier. The client captures the payment amount and embeds data into an image to transfer it over the visual channel. For any given user, we assume that only one authentication attempt may be active at a time; simultaneous attempts should be rejected by the verifier. A visualisation of the system is shown in Fig. 1.

The system verifies three factors to authenticate the user: possession of the client (something he has), verified by the data it embeds into the image, and knowledge of the image and PIN (something he knows). It is recommended, but not assumed, that the client require the user to authenticate to it using a biometric, such as a fingerprint, to add a fourth factor (something he is).

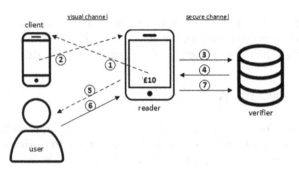

Fig. 1. The system model. The client reads the payment amount (1), embeds the amount, a message, and some authentication data into an image, and displays the image to the reader (2), which sends it to the verifier (3). The verifier returns the message (4), which authenticates it to the user (5), who then enters his PIN (6), which authenticates him to the verifier (7).

Threat Model. We assume that the adversary can watch and modify communications between the user and the verifier, such as by deploying a rogue reader. We assume that he knows everything about the user and may have access to any of the images used—*e.g.*, the user may have shared them on social media, with or without embedded data.

The goal of the adversary is to impersonate a legitimate user and either authorise or modify a payment without that user's knowledge. In the first case, the adversary is a rogue user who may attempt to perform a *replay attack*, by re-using a previous image token to authorise a new payment. In the second case, the adversary is a rogue merchant who may attempt to perform a *tampering attack*, by using a rogue reader to authorise a payment for an amount different to what is displayed by modifying the image token.

In this paper, we will not consider attacks on the client, such as physical theft, cloning, or malware—these are all to be covered in future work. We also do not consider attacks that take place during the enrollment phase, attacks on the verifier, or denial-of-service attacks.

3   System Architecture

Cryptographic Materials. During enrollment, the verifier exchanges some cryptographic materials with the client. It shares its public key, I_e, and its public symmetric transposition key, M, used for embedding. It generates the user a unique identifier, ID, and two secret block cipher keys[9], K, L: K is used by the client to encrypt data while embedding it into the cover-image; L is not shared with the client. It also generates the user two blocks of secret binary data, p, u: p is used to authenticate the client to the verifier and its size should be sufficiently

---

[9] An authenticated encryption algorithm should be chosen, such as AES-EAX.

large to authenticate with confidence (*e.g.*, 1,000 bits); u is used to update K and p after each authentication and its size should be large enough to cover both.

The user chooses a secret PIN of memorable size (*e.g.*, 4 digits); PIN is stored on the verifier and $\{PIN\}_L$ is returned and stored on the client. The user also chooses a personalised greeting message, m, used to authenticate the reader and the verifier to the user; it need not be remembered, only recognised, but its size should be bounded to fit on a reader's screen (*e.g.*, up to 40 characters).

A summary of the materials used in the system is shown in Table 1.

Table 1. A summary of the materials used in the system.

	Stored on verifier	Stored on client	Purpose
ID	✓	✓	Identifies user
I_e	✓	✓	Verifier's public key
I_d	✓	✗	Verifier's private key
M	✓	✓	Verifier's public transposition key
K	✓	✓	Secret key; used to encrypt a, m, p
L	✓	✗	Secret key; used to encrypt PIN
p	✓	✓	Authenticates client to verifier
u	✓	✓	Updates K, p
H	✓	✓	Hash function; modifies u
PIN	✓	✗	Authenticates user to verifier
m	✗	✓	Authenticates verifier and reader to user
a	✗	✗	payment amount

Embedding Data. In this paper, we will restrict our attention to embedding data in the spatial domain with a simple LSB-embedding algorithm. To embed some data d into an image using a transposition cipher M, we apply M to the image to rearrange its pixels, then we embed d into them sequentially; we denote this by $[d]_M$. After embedding, we recreate the image by inverting the rearrangement. To extract d from the embedded image, we reapply the rearrangement.

To protect the confidentiality of the data as it passes over the visual channel, we encrypt it before embedding it using some key k; we denote this by $[\{d\}_k]_M$.

To protect the integrity of the embedded data as it passes over the visual channel, we strengthen the algorithm with repeat embedding. To do so, we choose an odd number $n > 1$ and then embed $\{d\}_k$ into the image n times sequentially; we denote this by $[\{d\}_k^n]_M$. When extracting, we treat any discrepant bit as having whichever value was extracted for it most frequently.

Image and Storage. To use the client, the user must authenticate to it by selecting the correct cover-image from a set containing decoy images (benefi-cially, this is more human-usable than recalling a password [13]). To prevent an adversary from identifying the cover-image by metadata examination, we embed junk data into the decoys whenever the cover-image is updated.

We will store ID and $\{PIN\}_L$ embedded in the cover-image, such that an adversary with access to the client would need to identify it to find them. For this embedding, we can either use M or define a local key. Other cryptographic materials are stored securely within the client, using a secure element if available.

Unauthorised Usage Detection. After authentication, we update K and p for freshness; this provides resistance to replay attacks by making each embedded image good for only one use. To do so, we modify u using a hash function, H, which (i) preserves the size of u and (ii) ensures that its future values are not predictable; we use a to achieve the latter, since a is known to both the client and the verifier at the time of hashing,

$$u = H(u\|a).$$

We then apply our new u as a stream cipher to update K and p,

$$\{K,p\} = u \oplus \{K,p\}.$$

We do this on both the client and the verifier to keep the values synchronised.

By updating K and p after each authentication, we achieve inherent unauthorised usage detection. If an adversary were to successfully impersonate the user, the K and p values on his client and the verifier would update, unavoidably de-synchronising the user's client's K and p values from the verifier's. The system can be reset by exchanging new cryptographic materials with the verifier over a secure channel. In the case of unauthorised usage, the verifier can identify the last legitimate transaction by using the client's value of u and recreating the verifier's current value of u, since the latter is the result of deterministic hashes of the former with ordered payment values known to the verifier.

Enrollment Protocol. A secure channel is established between the client and the verifier to register an account and exchange cryptographic materials, such that the user can later authenticate. The protocol is shown in Fig. 2.

Steps 1–5. When the user registers an account, the verifier generates him a new ID, K, L, p, and u and stores them. The verifier knows M, I_e, and I_d and has values set for n and H. The verifier sends I_e, M, ID, K, p, and u to the client, which stores them.

Steps 6–12. The client prompts the user to choose PIN and m; it stores m, sends PIN to the verifier, and gets $\{PIN\}_L$ back, which it stores.

Authentication Protocol. A visual channel between the client and the reader and a secure channel between the reader and the verifier are required for the system to achieve mutual authentication. For any given user, only one authentication attempt may be active at a time. The protocol is shown in Fig. 3.

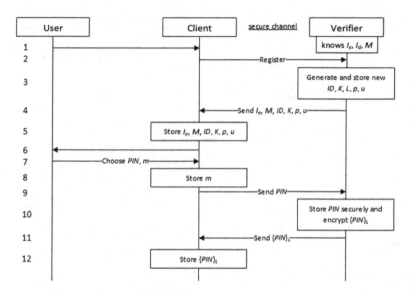

Fig. 2. The enrollment protocol.

Steps 1–6: User Authenticates to Client. The user authenticates to the client by selecting the cover-image; the client extracts ID and $\{PIN\}_L$. The reader sends a to the verifier to initiate the transaction and displays a. The client captures a using its camera and embeds $[\{ID\}_{I_e}^n, \{PIN\}_L^n, \{a, m, p\}_K^n]_M$ into the image to create the image token. The user may change m before embedding the data, since the required keys are stored on the client.

Steps 7–11: Client Authenticates to Verifier. In a commitment scheme [14], the user displays the image token to the reader, which sends it to the verifier. Firstly, the verifier knows M and so extracts $\{ID\}_{I_e}$, $\{PIN\}_L$, and $\{a, m, p\}_K$. Secondly, it uses ID to look up K to decrypt a, m, and p. Thirdly, it verifies that a and p match its expectations; this authenticates the client to the verifier.

Steps 12–14: Reader and Verifier Authenticate to User. The reader and verifier authenticate to the user (and confirm a) by displaying m on the reader.

Steps 15–18: User Authenticates to Verifier. The user enters PIN to the reader, which sends it to the verifier. The verifier compares the entered PIN, the $\{PIN\}_L$ extracted from the image token, and its own stored version to ensure that all three of them match; this authenticates the user to the verifier.

Steps 19–20: Client and Verifier Modify u and Update K and p. The verifier modifies u using a hash function $H(u\|a)$ and uses u to update its K and p values. The client does likewise such that their respective values remain the same.

4   Discussion

The system provides real-time authentication and achieves its design objectives. Firstly, it requires no specialist hardware: the reader needs only a camera, a

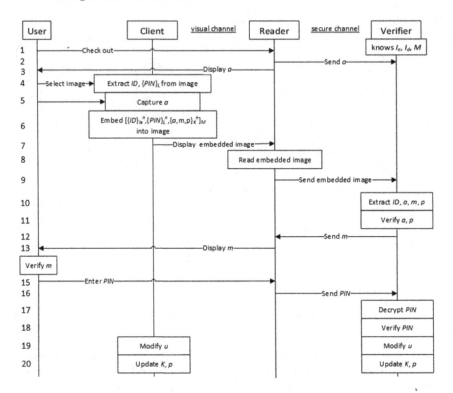

Fig. 3. The authentication protocol.

screen, and a touchscreen, which can be satisfied by any modern smartphone or tablet, making it easily deployable. Secondly, the authentication protocol provides resistance to phishing attacks in the form of mutual authentication by ensuring that the user authenticates to the client, the client to the verifier, the reader and verifier to the user (before PIN needs to be revealed), and the user to the verifier. The system authenticates the user to the verifier by transferring information, embedded in an image token, over a visual channel, meaning that it does not require a client-to-verifier connection nor the use of invisible channels, such as NFC, which risk interception. The user is informed of unauthorised usage as an inherent part of the protocol. Furthermore, since the image token contains the payment amount, the system does not only authenticate the user (which might be misused), but also binds the amount to be authorised.

The system relies on the user in two ways. Firstly, it is conceivable that a poor choice of cover-image weakens user-to-client authentication [15]; this reliance is alleviated if a biometric is used to authenticate the user to the client. Secondly, one benefit of using a visual channel over an invisible channel is that the user, with reasonable care, has greater control over who or what can see the image token; however, such care cannot be reasonably expected of all users.

For the *replay attack*, the adversary attempts to impersonate a legitimate user to authorise a payment by replaying a captured image token that was previously sent to the verifier. At the end of the transaction from which it was captured, the data embedded in the image token became out-of-date. In order to update it, the adversary would need to know K, p, u, and a. While the adversary may know a, and then be able to determine K and p by brute force, u is not stored in the image at all. To glean any useful knowledge of u by observing its effects over time would require an impractical number of transactions. Therefore, the system is resistant to replay attacks.

For the *tampering attack*, the adversary attempts to have a legitimate user authorise a payment for a different amount $a' \neq a$ by using a rogue reader to modify the embedded data in the image token before sending it to the verifier. At step 2 of the authentication protocol, the adversary sends a' to the verifier while displaying a to the user at step 3. At step 8, the adversary can extract $\{a, m, p\}_K$ from the image since M is public; however, he would need to compromise the secret key, K, to change a or the attack will fail at step 11 when the verifier decrypts and compares it with a' from step 2. Assuming K is a strong, authenticated encryption key, then it is unlikely that the adversary will be able to compromise it in the short time available before the user becomes suspicious. Therefore, the system is resistant to tampering attacks.

5 Conclusion and Future Work

Our system meets its design objectives by providing a novel means to mutually authenticate a user, reader, and verifier over a visual channel without any specialist hardware nor a connection between client and verifier; it is resistant to replay and tampering attacks and offers inherent and unavoidable unauthorised usage detection. It makes use of three factors in a convenient manner: a device which would be carried anyway, a chosen image familiar to the user, and a PIN.

In future work, we intend to investigate attacks on the client, including physical theft, cloning, and malware. We note that PIN is not stored on the client, meaning that the system provides some resistance to indiscriminate malware infections; the adversary would need to observe the user entering PIN separately in each case, increasing the work required in such an attack. We also plan to run user studies to better understand user requirements, such as acceptable transaction durations and system intuitiveness, and to study in greater detail the technical constraints of using an embedded image, such as allowable noise and the use of gridlines to handle rotation.

References

1. British Retail Consortium: Debit Cards Overtake Cash to Become Number One Payment Method in the UK (2017)
2. Bond, M., Choudary, O., Murdoch, S.J., Skorobogatov, S., Anderson, R.: Chip and skim: cloning EMV cards with the pre-play attack. In: IEEE Symposium on Security and Privacy (SP) (2014)

3. Emms, M., Arief, B., Freitas, L., Hannon, J., van Moorsel, A.: Harvesting high value foreign currency transactions from EMV contactless credit cards without the PIN. In: ACM Conference on Computer and Communications Security (CCS) (2014)
4. Jupiter Research. Integrated Handsets: Balancing Device Functionality with Consumer Desires (2005)
5. Huh, J.H., Verma, S., Rayala, S.S.V., Bobba, R.B., Beznosov, K., Kim, H.: I Don't Use Apple Pay because it's less secure...: perception of security and usability in mobile tap-and-pay. In: Proceedings of the Workshop on Usable Security (USEC) (2017)
6. Murdoch, S.J., Drimer, S., Anderson, R., Bond, M.: Chip and PIN is broken. In: IEEE Symposium on Security and Privacy (SP) (2010)
7. Francis, L., Hancke, G., Mayes, K., Markantonakis, K.: On the security issues of NFC enabled mobile phones. Int. J. Internet Technol. Secur. Trans. **2**, 336–356 (2010)
8. Kortvedt, H., Mjolsnes, S.: Eavesdropping near field communication. In: The Norwegian Information Security Conference (NISK) (2009)
9. Diakos, T.P., Briffa, J.A., Brown, T.W.C., Wesemeyer, S.: Eavesdropping near-field contactless payments: a quantitative analysis. J. Eng. **2013**, 48–54 (2013)
10. Schechter, S.E., Dhamija, R., Ozment, A., Fischer, I.: The Emperor's new security indicators. In: IEEE Symposium on Security and Privacy (2007)
11. Marforio, C., Masti, R.J, Soriente, C., Kostiainen, K., Čapkun, S.: Evaluation of personalized security indicators as an anti-phishing mechanism for smartphone applications. In: CHI Conference on Human Factors in Computing Systems, pp. 540–551 (2016)
12. Purnomo, A.T., Gondokaryono, Y.S., Kim, C.-S.: Mutual authentication in securing mobile payment system using encrypted QR code based on public key infrastructure. In: IEEE 6th International Conference on System Engineering and Technology (ICSET) (2016)
13. Biddle, R., Chiasson, S., Oorschot, P.C.: Graphical passwords: learning from the first twelve years. ACM Comput. Surv. (CSULR) **44**, 19 (2012)
14. Brassard, G., Chaum, D., Crepeau, C.: Minimum disclosure proofs of knowledge. J. Comput. Syst. Sci. **37**, 156–189 (1988)
15. Davis, D., Monrose, F., Reiter, M.K.: On user choice in graphical password schemes. In: USENIX Security Symposium 13, p. 11 (2004)

Author Index

Printed in the United States
By Bookmasters